TABLE F.1 Formulas for Unit Conversions*

Name, Symbol, Dimensions			Conversion Formula
Length	L	L	**1 m** = 3.281 ft = 1.094 yd = 39.37 in = km/1000 = $10^6\,\mu$m **1 ft** = 0.3048 m = 12 in = mile/5280 = km/3281 **1 mm** = m/1000 = in/25.4 = 39.37 mil = 1000 μm = 10^7Å
Speed	V	L/T	**1 m/s** = 3.600 km/hr = 3.281 ft/s = 2.237 mph = 1.944 knots **1 ft/s** = 0.3048 m/s = 0.6818 mph = 1.097 km/hr = 0.5925 knots
Mass	m	M	**1 kg** = 2.205 lbm = 1000 g = slug/14.59 = (metric ton or tonne or Mg)/1000 **1 lbm** = lbf $\cdot$ s²/(32.17 ft) = kg/2.205 = slug/32.17 = 453.6 g = 16 oz = 7000 grains = short ton/2000 = metric ton (tonne)/2205
Density	ρ	M/L^3	**1000 kg/m³** = 62.43 lbm/ft³ = 1.940 slug/ft³ = 8.345 lbm/gal (US)
Force	F	ML/T^2	**1 lbf** = 4.448 N = 32.17 lbm $\cdot$ ft/s² **1 N** = kg $\cdot$ m/s² = 0.2248 lbf = 10^5 dyne

Pressure, Shear Stress	p, τ	M/LT^2	**1 Pa** = N/m² = kg/m $\cdot$ s² = 10^{-5} bar = 1.450×10^{-4} lbf/in² = inch H$_2$O/249.1
			     = 0.007501 torr = 10.00 dyne/cm²
			 1 atm = 101.3 kPa = 2116 psf = 1.013 bar = 14.70 lbf/in² = 33.90 ft of water
			     = 29.92 in of mercury = 10.33 m of water = 760 mm of mercury = 760 torr
			 1 psi = atm/14.70 = 6.895 kPa = 27.68 in H$_2$O = 51.71 torr

Volume	V	L^3	**1 m³** = 35.31 ft³ = 1000 L = 264.2 U.S. gal **1 ft³** = 0.02832 m³ = 28.32 L = 7.481 U.S. gal = acre-ft/43,560 **1 U.S. gal** = 231 in³ = barrel (petroleum)/42 = 4 U.S. quarts = 8 U.S. pints = 3.785 L = 0.003785 m³
Volume Flow Rate (Discharge)	Q	L^3/T	**1 m³/s** = 35.31 ft³/s = 2119 cfm = 264.2 gal (US)/s = 15850 gal (US)/m **1 cfs** = 1 ft³/s = 28.32 L/s = 7.481 gal (US)/s = 448.8 gal (US)/m
Mass Flow Rate	$\dot{m}$	M/T	**1 kg/s** = 2.205 lbm/s = 0.06852 slug/s
Energy and Work	E, W	ML^2/T^2	**1 J** = kg $\cdot$ m²/s² = N $\cdot$ m = W $\cdot$ s = volt $\cdot$ coulomb = 0.7376 ft $\cdot$ lbf = 9.478×10^{-4} Btu = 0.2388 cal = 0.0002388 Cal = 10^7 erg = kWh/3.600×10^6
Power	$P, \dot{E}, \dot{W}$	ML^2/T^3	**1 W** = J/s = N $\cdot$ m/s = kg $\cdot$ m²/s³ = 1.341×10^{-3} hp = 0.7376 ft $\cdot$ lbf/s = 1.0 volt-ampere = 0.2388 cal/s = 9.478×10^{-4} Btu/s **1 hp** = 0.7457 kW = 550 ft $\cdot$ lbf/s = 33,000 ft $\cdot$ lbf/min = 2544 Btu/h
Angular Speed	ω	T^{-1}	**1.0 rad/s** = 9.549 rpm = 0.1591 rev/s
Viscosity	μ	M/LT	**1 Pa $\cdot$ s** = kg/m $\cdot$ s = N $\cdot$ s/m² = 10 poise = 0.02089 lbf $\cdot$ s/ft² = 0.6720 lbm/ft $\cdot$ s
Kinematic Viscosity	ν	L^2/T	**1 m²/s** = 10.76 ft²/s = 10^6 cSt
Temperature	T	Θ	**K** = °C + 273.15 = °R/1.8 **°C** = (°F − 32)/1.8 **°R** = °F + 459.67 = 1.8 K **°F** = 1.8°C + 32

*A useful online reference is www.onlineconversion.com

TABLE F.2 Commonly Used Equations

Ideal gas law

$p = \rho RT$ (Eq. 1.10, p. 14)

Specific weight

$\gamma = \rho g$ (Eq. 2.3, p. 31)

Specific gravity

$S = \dfrac{\rho}{\rho_{H_2O \text{ at } 4°C}} = \dfrac{\gamma}{\gamma_{H_2O \text{ at } 4°C}}$ (Eq. 2.5, p. 32)

Kinematic viscosity

$\nu = \mu/\rho$ (Eq. 2.15, p. 38)

Definition of viscosity

$\tau = \mu \dfrac{dV}{dy}$ (Eq. 2.16, p. 39)

Pressure equations

$p_{gage} = p_{abs} - p_{atm}$ (Eq. 3.3a; p. 62)
$p_{vacuum} = p_{atm} - p_{abs}$ (Eq. 3.3b, p. 62)

Hydrostatic equation

$\dfrac{p_1}{\gamma} + z_1 = \dfrac{p_2}{\gamma} + z_2 = \text{constant}$ (Eq. 3.10a, p. 66)

$p_z = p_1 + \gamma z_1 = p_2 + \gamma z_2 = \text{constant}$ (Eq. 3.10b, p. 66)

$\Delta p = -\gamma \Delta z$ (Eq. 3.10c, p. 66)

Manometer equations

$p_2 = p_1 + \displaystyle\sum_{down} \gamma_i h_i - \sum_{up} \gamma_i h_i$ (Eq. 3.21, p. 74)

$h_1 - h_2 = \Delta h(\gamma_B/\gamma_A - 1)$ (Eq. 3.22, p. 75)

Hydrostatic force equations (flat panels)

$F_P = \bar{p}A$ (Eq. 3.28, p. 80)

$y_{cp} - \bar{y} = \dfrac{\bar{I}}{\bar{y}A}$ (Eq. 3.33, p. 81)

Buoyant force (Archimedes equation)

$F_B = \gamma V_D$ (Eq. 3.41a, p. 87)

The Bernoulli equation

$\left(\dfrac{p_1}{\gamma} + \dfrac{V_1^2}{2g} + z_1\right) = \left(\dfrac{p_2}{\gamma} + \dfrac{V_2^2}{2g} + z_2\right)$ (Eq. 4.21b, p. 133)

$\left(p_1 + \dfrac{\rho V_1^2}{2} + \rho g z_1\right) = \left(p_2 + \dfrac{\rho V_2^2}{2} + \rho g z_2\right)$ (Eq. 4.21a, p. 133)

Coefficient of pressure

$C_p = \dfrac{p_z - p_{zo}}{\rho V_o^2/2} = \dfrac{h - h_o}{V_o^2/(2g)}$ (Eq. 4.47, p. 147)

Volume flow rate equation

$Q = \bar{V}A = \dfrac{\dot{m}}{\rho} = \displaystyle\int_A V dA = \int_A \mathbf{V} \cdot \mathbf{dA}$ (Eq. 5.10, p. 174)

Mass flow rate equation

$\dot{m} = \rho A \bar{V} = \rho Q = \displaystyle\int_A \rho V dA = \int_A \rho \mathbf{V} \cdot \mathbf{dA}$ (Eq. 5.11, p. 174)

Continuity equation

$\dfrac{d}{dt}\displaystyle\int_{cv} \rho dV + \int_{cs} \rho \mathbf{V} \cdot \mathbf{dA} = 0$ (Eq. 5.28, p. 183)

$\dfrac{d}{dt}M_{cv} + \displaystyle\sum_{cs} \dot{m}_o - \sum_{cs} \dot{m}_i = 0$ (Eq. 5.29, p. 183)

$\rho_2 A_2 V_2 = \rho_1 A_1 V_1$ (Eq. 5.33, p. 189)

Momentum equation

$\displaystyle\sum \mathbf{F} = \dfrac{d}{dt}\int_{cv} \mathbf{v}\rho \, dV + \int_{cs} \mathbf{v}\rho \mathbf{V} \cdot \mathbf{dA}$ (Eq. 6.7, p. 213)

$\displaystyle\sum \mathbf{F} = \dfrac{d(m_{cv}\mathbf{v}_{cv})}{dt} + \sum_{cs} \dot{m}_o\mathbf{v}_o - \sum_{cs} \dot{m}_i\mathbf{v}_i$ (Eq. 6.10, p. 213)

Energy equation

$\left(\dfrac{p_1}{\gamma} + \alpha_1\dfrac{\bar{V}_1^2}{2g} + z_1\right) + h_p = \left(\dfrac{p_2}{\gamma} + \alpha_2\dfrac{\bar{V}_2^2}{2g} + z_2\right) + h_t + h_L$

(Eq. 7.29; p. 262)

The power equation

$P = FV = T\omega$ (Eq. 7.3, p. 255)
$P = \dot{m}gh = \gamma Qh$ (Eq. 7.31; p. 265)

Efficiency of a machine

$\eta = \dfrac{P_{output}}{P_{input}}$ (Eq. 7.32; p. 267)

Reynolds number (pipe)

$Re_D = \dfrac{VD}{\nu} = \dfrac{\rho VD}{\mu} = \dfrac{4Q}{\pi D\nu} = \dfrac{4\dot{m}}{\pi D\mu}$ (Eq. 10.1, p. 361)

Combined head loss equation

$h_L = \displaystyle\sum_{pipes} f\dfrac{L}{D}\dfrac{V^2}{2g} + \sum_{components} K\dfrac{V^2}{2g}$ (Eq. 10.45, p. 382)

Friction factor f (Resistance coefficient)

$f = \dfrac{64}{Re_D}$ $Re \le 2000$ (Eq. 10.34, p. 370)

$f = \dfrac{0.25}{\left[\log_{10}\left(\dfrac{k_s}{3.7D} + \dfrac{5.74}{Re_D^{0.9}}\right)\right]^2}$ (Re $\ge$ 3000) (Eq. 10.39, p. 375)

Drag force equation

$F_D = C_D A\left(\dfrac{\rho V_0^2}{2}\right)$ (Eq. 11.5, p. 409)

Lift force equation

$F_L = C_L A\left(\dfrac{\rho V_0^2}{2}\right)$ (Eq. 11.17, p. 424)

TABLE F.3 Useful Constants

Name of Constant	Value
Acceleration of gravity	$g = 9.81 \text{ m/s}^2 = 32.2 \text{ ft/s}^2$
Universal gas constant	$R_u = 8.314 \text{ kJ/kmol} \cdot \text{K} = 1545 \text{ ft} \cdot \text{lbf/lbmol} \cdot {}^\circ\text{R}$
Standard atmospheric pressure	$P_{atm} = 1.0 \text{ atm} = 101.3 \text{ kPa} = 14.70 \text{ psi} = 2116 \text{ psf} = 33.90 \text{ ft of water}$ $P_{atm} = 10.33 \text{ m of water} = 760 \text{ mm of Hg} = 29.92 \text{ in of Hg} = 760 \text{ torr} = 1.013 \text{ bar}$

TABLE F.4 Properties of Air $[T = 20^\circ\text{C} (68^\circ\text{F}), p = 1 \text{ atm}]$

Property	SI Units	Traditional Units
Specific gas constant	$R_{air} = 287.0 \text{ J/kg} \cdot \text{K}$	$R_{air} = 1716 \text{ ft} \cdot \text{lbf/slug} \cdot {}^\circ\text{R}$
Density	$\rho = 1.20 \text{ kg/m}^3$	$\rho = 0.0752 \text{ lbm/ft}^3 = 0.00234 \text{ slug/ft}^3$
Specific weight	$\gamma = 11.8 \text{ N/m}^3$	$\gamma = 0.0752 \text{ lbf/ft}^3$
Viscosity	$\mu = 1.81 \times 10^{-5} \text{ N} \cdot \text{s/m}^2$	$\mu = 3.81 \times 10^{-7} \text{ lbf} \cdot \text{s/ft}^2$
Kinematic viscosity	$\nu = 1.51 \times 10^{-5} \text{ m}^2/\text{s}$	$\nu = 1.63 \times 10^{-4} \text{ ft}^2/\text{s}$
Specific heat ratio	$k = c_p/c_v = 1.40$	$k = c_p/c_v = 1.40$
Specific heat	$c_p = 1004 \text{ J/kg} \cdot \text{K}$	$c_p = 0.241 \text{ Btu/lbm} \cdot {}^\circ\text{R}$
Speed of sound	$c = 343 \text{ m/s}$	$c = 1130 \text{ ft/s}$

TABLE F.5 Properties of Water $[T = 15^\circ\text{C} (59^\circ\text{F}), p = 1 \text{ atm}]$

Property	SI Units	Traditional Units
Density	$\rho = 999 \text{ kg/m}^3$	$\rho = 62.4 \text{ lbm/ft}^3 = 1.94 \text{ slug/ft}^3$
Specific weight	$\gamma = 9800 \text{ N/m}^3$	$\gamma = 62.4 \text{ lbf/ft}^3$
Viscosity	$\mu = 1.14 \times 10^{-3} \text{ N} \cdot \text{s/m}^2$	$\mu = 2.38 \times 10^{-5} \text{ lbf} \cdot \text{s/ft}^2$
Kinematic viscosity	$\nu = 1.14 \times 10^{-6} \text{ m}^2/\text{s}$	$\nu = 1.23 \times 10^{-5} \text{ ft}^2/\text{s}$
Surface tension (water–air)	$\sigma = 0.073 \text{ N/m}$	$\sigma = 0.0050 \text{ lbf/ft}$
Bulk modulus of elasticity	$E_v = 2.14 \times 10^9 \text{ Pa}$	$E_v = 3.10 \times 10^5 \text{ psi}$

TABLE F.6 Properties of Water $[T = 4^\circ\text{C} (39^\circ\text{F}), p = 1 \text{ atm}]$

Property	SI Units	Traditional Units
Density	$\rho = 1000 \text{ kg/m}^3$	$\rho = 62.4 \text{ lbm/ft}^3 = 1.94 \text{ slug/ft}^3$
Specific weight	$\gamma = 9810 \text{ N/m}^3$	$\gamma = 62.4 \text{ lbf/ft}^3$

www.wileyplus.com

WileyPLUS is a research-based online environment for effective teaching and learning.

WileyPLUS builds students' confidence because it takes the guesswork out of studying by providing students with a clear roadmap:

- **what to do**
- **how to do it**
- **if they did it right**

It offers interactive resources along with a complete digital textbook that help students learn more. With *WileyPLUS*, students take more initiative so you'll have greater impact on their achievement in the classroom and beyond.

For more information, visit www.wileyplus.com

ENGINEERING FLUID MECHANICS

SI VERSION

TENTH EDITION

Donald F. Elger
University of Idaho, Moscow

Barbara C. Williams
University of Idaho, Moscow

Clayton T. Crowe
Washington State University, Pullman

John A. Roberson
Washington State University, Pullman

WILEY

This 10th Edition is dedicated to our friend
and colleague Clayton T. Crowe (1933–2012)

CONTENTS

PREFACE   xi

CHAPTER 1   Building a Solid Foundation   1
1.1 Defining Engineering Fluid Mechanics   2
1.2 Describing Liquids and Gases   3
1.3 Idealizing Matter   5
1.4 Dimensions and Units   6
1.5 Carrying and Canceling Units   9
1.6 Applying the Ideal Gas Law (IGL)   13
1.7 The Wales-Woods Model   15
1.8 Checking for Dimensional
     Homogeneity (DH)   19
1.9 Summarizing Key Knowledge   22

CHAPTER 2   Fluid Properties   28
2.1 Defining the System   28
2.2 Characterizing Mass and Weight   30
2.3 Modeling Fluids as Constant Density   32
2.4 Finding Fluid Properties   34
2.5 Describing Viscous Effects   35
2.6 Applying the Viscosity Equation   39
2.7 Characterizing Viscosity   42
2.8 Characterizing Surface Tension   45
2.9 Predicting Boiling Using Vapor
     Pressure   50
2.10 Characterizing Thermal Energy in
     Flowing Gases   51
2.11 Summarizing Key Knowledge   52

CHAPTER 3   Fluid Statics   60
3.1 Describing Pressure   61
3.2 Calculating Pressure Changes Associated
     with Elevation Changes   65
3.3 Measuring Pressure   72
3.4 Predicting Forces on Plane Surfaces
     (Panels)   77
3.5 Calculating Forces on Curved
     Surfaces   83
3.6 Calculating Buoyant Forces   85
3.7 Predicting Stability of Immersed and
     Floating Bodies   88
3.8 Summarizing Key Knowledge   92

CHAPTER 4   The Bernoulli Equation and Pressure Variation   111
4.1 Describing Streamlines, Streaklines,
     and Pathlines   112
4.2 Characterizing Velocity of a Flowing Fluid   114
4.3 Describing Flow   117
4.4 Acceleration   123
4.5 Applying Euler's Equation to Understand
     Pressure Variation   127
4.6 Applying the Bernoulli Equation along
     a Streamline   132
4.7 Measuring Velocity and Pressure   139
4.8 Characterizing Rotational Motion
     of a Flowing Fluid   142
4.9 The Bernoulli Equation for Irrotational Flow   146
4.10 Describing the Pressure Field for Flow
     over a Circular Cylinder   147
4.11 Calculating the Pressure Field for
     a Rotating Flow   149
4.12 Summarizing Key Knowledge   152

CHAPTER 5   Control Volume Approach and Continuity Equation   169
5.1 Characterizing the Rate of Flow   170
5.2 The Control Volume Approach   176
5.3 Continuity Equation (Theory)   182
5.4 Continuity Equation (Application)   184
5.5 Predicting Caviation   191
5.6 Summarizing Key Knowledge   194

CHAPTER 6   Momentum Equation   208
6.1 Understanding Newton's Second
     Law of Motion   209
6.2 The Linear Momentum Equation: Theory   213
6.3 Linear Momentum Equation: Application   216
6.4 The Linear Momentum Equation for
     a Stationary Control Volume   218
6.5 Examples of the Linear Momentum
     Equation (Moving Objects)   228
6.6 The Angular Momentum Equation   233
6.7 Summarizing Key Knowledge   236

CHAPTER 7   The Energy Equation   252
7.1 Energy Concepts   253
7.2 Conservation of Energy   255
7.3 The Energy Equation   257
7.4 The Power Equation   265
7.5 Mechanical Efficiency   267
7.6 Contrasting the Bernoulli Equation
     and the Energy Equation   270
7.7 Transitions   270
7.8 Hydraulic and Energy Grade Lines   273
7.9 Summarizing Key Knowledge   277

CHAPTER 8   Dimensional Analysis
             and Similitude   292
8.1 Need for Dimensional Analysis   292
8.2 Buckingham Π Theorem   294
8.3 Dimensional Analysis   295
8.4 Common π-Groups   299
8.5 Similitude   302
8.6 Model Studies for Flows without
     Free-Surface Effects   305
8.7 Model-Prototype Performance   308
8.8 Approximate Similitude at High
     Reynolds Numbers   309
8.9 Free-Surface Model Studies   312
8.10 Summarizing Key Knowledge   315

CHAPTER 9   Predicting Shear Force   324
9.1 Uniform Laminar Flow   325
9.2 Qualitative Description of the Boundary Layer 330
9.3 Laminar Boundary Layer   331
9.4 Boundary Layer Transition   335
9.5 Turbulent Boundary Layer   336
9.6 Pressure Gradient Effects of
     Boundary Layers   347
9.7 Summarizing Key Knowledge   349

CHAPTER 10   Flow in Conduits   359
10.1 Classifying Flow   360
10.2 Specifying Pipe Sizes   363
10.3 Pipe Head Loss   363
10.4 Stress Distributions in Pipe Flow   366
10.5 Laminar Flow in a Round Tube   367
10.6 Turbulent Flow and the Moody Diagram   371
10.7 Strategy for Solving Problems   375
10.8 Combined Head Loss   379
10.9 Nonround Conduits   384
10.10 Pumps and Systems of Pipes   385
10.11 Key Knowledge   391

CHAPTER 11   Drag and Lift   406
11.1 Relating Lift and Drag to Stress
      Distributions   407
11.2 Calculating Drag Force   408
11.3 Drag of Axisymmetric and 3-D Bodies   413
11.4 Terminal Velocity   418
11.5 Vortex Shedding   419
11.6 Reducing Drag by Streamlining   420
11.7 Drag in Compressible Flow   421
11.8 Theory of Lift   422
11.9 Lift and Drag on Airfoils   426
11.10 Lift and Drag on Road Vehicles   432
11.11 Summarizing Key Knowledge   435

CHAPTER 12   Compressible Flow   445
12.1 Wave Propagation in Compressible
      Fluids   445
12.2 Mach Number Relationships   451
12.3 Normal Shock Waves   455
12.4 Isentropic Compressible Flow Through
      a Duct with Varying Area   460
12.5 Summarizing Key Knowledge   471

CHAPTER 13   Flow Measurements   478
13.1 Measuring Velocity and Pressure   478
13.2 Measuring Flow Rate (Discharge)   486
13.3 Measurement in Compressible Flow   501
13.4 Accuracy of Measurements   505
13.5 Summarizing Key Knowledge   506

CHAPTER 14   Turbomachinery   517
14.1 Propellers   518
14.2 Axial-Flow Pumps   523
14.3 Radial-Flow Machines   527
14.4 Specific Speed   531
14.5 Suction Limitations of Pumps   532
14.6 Viscous Effects   534
14.7 Centrifugal Compressors   535
14.8 Turbines   538
14.9 Summarizing Key Knowledge   547

CHAPTER 15   Flow in Open Channels   554
15.1 Description of Open-Channel Flow   555
15.2 Energy Equation for Steady
      Open-Channel Flow   557
15.3 Steady Uniform Flow   558
15.4 Steady Nonuniform Flow   567
15.5 Rapidly Varied Flow   567
15.6 Hydraulic Jump   577

15.7 Gradually Varied Flow   582
15.8 Summarizing Key Knowledge   590

CHAPTER 16   **Modeling of Fluid Dynamics Problems**   **598**
16.1 Models in Fluid Mechanics   599
16.2 Foundations for Learning Partial Differential Equations (PDEs)   603
16.3 The Continuity Equation   612
16.4 The Navier-Stokes Equation   619

16.5 Computational Fluid Dynamics (CFD)   623
16.6 Examples of CFD   628
16.7 A Path for Moving Forward   631
16.8 Summarizing Key Knowledge   632

Appendix   639

Answers   651

Index   661

PREFACE

Audience

This book is written for engineering students of all majors who are taking a first or second course in fluid mechanics. Students should have background knowledge in physics (mechanics), chemistry, statics, and calculus.

Why We Wrote This Book

As students and as teachers, we love fluid mechanics. So, we wrote a book to share our passion for the discipline. As educators our motivation is to present knowledge so that students can learn in depth.

Approach

Knowledge. The chapters are organized to lay out the knowledge for students. Each chapter begins with statements of what is important to learn. These learning outcomes are formulated in terms of what *Students will be able to do*. Then, the sections present the knowledge. Last, the knowledge is summarized at the end of each chapter.

Practice with Feedback. Learning anything takes a lot of practice. Thus, this text has more than 1,000 end-of-chapter problems. In addition, we are building online problems that provide practice with feedback. The online resources, presented in *WileyPLUS*, are organized so that the grading and record keeping are done by the computer, thereby freeing up the teacher to focus on more important tasks such as mentoring students. *WileyPLUS* is an online learning system; see www.wiley.com/college/elger for more information.

Features of this Book

Learning Outcomes. Each chapter begins with learning outcomes so students can identify what knowledge they should gain by studying the chapter. These learning outcomes are formulated in terms of what *Students will be able to do*.

Rationale. Each section describes what content is presented and why this content is relevant so students can connect their learning to what is important to them.

Visual Approach. The text uses sketches and photographs to help students learn more effectively by connecting images to words and equations.

Foundational Concepts. This text presents major concepts in a clear and concise format. These concepts form building blocks for higher levels of learning.

Seminal Equations. This text emphasizes technical derivations so that students can learn to do the derivations on their own, increasing their levels of knowledge. Features include the following:

- Derivations of each main equation are presented in a step-by-step fashion.
- The assumptions associated with each main equation are stated during the derivation and after the equation is developed.
- The holistic meaning of main equations is explained using words.
- Main equations are named and listed in Table F.2.

- Main equations are summarized in tables in the chapters.
- A process for applying each main equation is presented in the relevant chapter.

Chapter Summaries. Each chapter concludes with a summary so students can review the key knowledge in the chapter.

Online Problems. We have created many online problems that provide immediate feedback to students while also ensuring that students complete the assigned work on time. These problems are available in *WileyPLUS* at instructor's discretion.

Process Approach. A process is a method for getting results. A process approach involves figuring out how experts do things and adapting this same approach. This textbook presents multiple processes.

Wales-Woods Model. The Wales-Woods Model represents how experts solve problem. This model is presented in Chapter 1 and used in example problems throughout the text.

Grid Method. This text presents a systematic process, called the grid method, for carrying and canceling units. Unit practice is emphasized because it helps engineers spot and fix mistakes and because it helps engineers put meaning on concepts and equations.

Example Problems. Each chapter has approximately 10 example problems, each worked out in detail. The purpose of these examples is to show how the knowledge is used in context and to present essential details needed for application.

Solutions Manual. The text includes a detailed solutions manual for instructors. Many solutions are presented with the Wales-Woods model.

Image Gallery. The figures from the text are available in PowerPoint format, for easy inclusion in lecture presentations. This resource is available only to instructors. To request access to this password-protected resource, visit the Instructor Companion Site portion of the Web site located at www.wiley.com/college/elger, and register for a password.

Interdisciplinary Approach. Historically, this text was written for the civil engineer. We are retaining this approach while adding material so that the text is also appropriate for other engineering disciplines. For example, the text presents the Bernoulli equation using both head terms (civil engineering approach) and terms with units of pressure (the approach used by chemical and mechanical engineers). We include problems that are relevant to product development as practiced by mechanical and electrical engineers. Some problems feature other disciplines, such as exercise physiology. The reason for this interdisciplinary approach is that the world of today's engineer is becoming more and more interdisciplinary.

What is New in the 10th Edition

- Chapters 1, 2, 4, 5, 6, and 7 were rewritten to present knowledge in a simpler, more-accessible way.
- *Checkpoint Problems* were added to the text so students can assess their understanding as they are reading.
- Learning objectives were rewritten to communicate to students what they should be able to do after they have learned the material.
- Learning objectives were rewritten so that they align with sections.
- Summary sections for each chapter were rewritten so that they align with the learning objectives.
- Chapter 16 was added. This chapter describes modeling, the partial differential equation approach, and computational fluid dynamics.
- The Wales-Woods model was formally introduced and incorporated into the text.
- Approximately 30% of the end-of-chapter problems are new or modified.
- Many additional online problems with feedback have been created and are available in WileyPLUS at instructor's discretion.

Author Team

The book was originally written by Professor John Roberson, with Professor Clayton Crowe adding the material on compressible flow. Professor Roberson retired from active authorship after the 6th edition, Professor Donald Elger joined on the 7th edition, and Professor Barbara Williams joined on the 9th edition. Professor Crowe retired from active authorship after the 9th edition. This 10th edition was developed by Professors Elger and Williams with Professor Crowe reviewing chapters for technical accuracy. Professor Crowe passed away on February 5, 2012. He was active on the book until a few weeks before his death and will be missed by his co-authors.

Donald Elger, Barbara Williams, and Clayton Crowe (Photo by Archer Photography: www.archerstudio.com)

Acknowledgments

We acknowledge our colleagues and mentors. Donald Elger acknowledges his Ph.D. mentor, Ronald Adams, who always asked why and how. He also acknowledges Ralph Budwig, fluid mechanics researcher and colleague, who has provided many hours of delightful inquiry about fluid mechanics. Barbara Williams acknowledges Wilfried Brutsuert at Cornell University and George Bloomsburg at the University of Idaho, who inspired her passion for fluid mechanics.

Contact Us

We welcome feedback and ideas for interesting end-of-chapter problems. Please contact us at the e-mail addresses given below.

Donald F. Elger (delger@uidaho.edu)

Barbara C. Williams (barbwill@uidaho.edu)

BUILDING A SOLID FOUNDATION 1

FIGURE 1.1

As engineers, we get to design cool systems like this glider. This is exciting! (© Ben Blankenburg/Corbis RF/Age Fotostock America, Inc.)

Chapter Road Map

The purpose of this chapter is to help students build a foundation for learning fluid mechanics. The chapter has three main objectives: to define engineering, to describe fluids, and to introduce skills that are useful for solving engineering problems.

Learning Objectives

STUDENTS WILL BE ABLE TO

- Define engineering, fluid mechanics, and learning. (§1.1)
- Define fluid, liquid, and gas. (§1.2)
- Describe the characteristics of liquids and gases. (§1.2)
- Explain macroscopic and microscopic descriptions. (§1.3)
- Explain the continuum assumption. (§1.3)
- Define a fluid particle. (§1.3)
- Describe units and dimensions. (§1.4)
- Determine if a set of units are consistent. (§1.4)
- Apply the grid method to carry and cancel units. (§1.5)
- Apply the ideal gas law, or IGL. (§1.6)
- Describe the Wales-Woods model, or WWM. (§1.7)
- Check for dimensional homogeneity, or DH. (§1.8)
- Define a π-group. Define the derivative and the integral. (§1.8)

"Begin difficult things while they are easy. Do great things when they are small. The difficult things of the world must have once been easy. The great things must have once been small. A thousand-mile journey begins with one step."
-Lao-tzu (Chinese philosopher who founded Taoism in about 600 BC)

In this chapter, we invite you to take the first steps of your journey in learning fluid mechanics. We have been on this journey most of our lives, and we love to share our passion and our knowledge with you as you *walk your path*.

1.1 Defining Engineering Fluid Mechanics

As engineers we ought to be able to explain to a layperson what our discipline is about. Thus, this section defines engineering fluid mechanics and defines learning.

Engineering

Engineers design systems that benefit people. For example, Fig. 1.1 shows a glider and Fig. 1.2a shows wind turbines being used to generate electrical power for a community. Fig. 1.2b shows people working on slow-sand filter technology. This technology is used to produce safe drinking water for families. The person in the center of the photo is a mechanical engineering student who worked on this project during his senior year. We also design hydroelectric power systems such as Hoover Dam. We design oil pipelines, artificial hearts, jet engines, and cooling systems for buildings. *Engineers design the technology of the world.*

FIGURE 1.2

(a) Commercial wind turbines in Oregon.
(b) Engineering slow sand filter technology near Nairobi Kenya. (Photos by Donald Elger)

(a)   (b)

The National Research Council (1) states that "*engineering is the process of designing the human made world.*" They assert that science involves study of the natural world, whereas engineering involves modifying the world to meet human needs. Of course, science, math, and engineering are interwoven. Thus, the central purpose of engineering education is to teach engineering students how to design the human-made world in ways that integrate and capitalize on math and science while considering foremost the needs of people.

Regarding **math**, this can be defined as the abstract and logical study of numbers, quantities, and space. **Science** is the systematic study of the physical world through observation and experiment. Science differs from math in that math is about abstraction and symbols, whereas science is about understanding the physical world. Science is the music and math is means of writing the music down. **Technology** is the collection of machinery, equipment, and tools developed from scientific knowledge. By applying existing technology, engineers leverage the progress of those who have come before.

In addition to math, science, and technology, engineers apply knowledge from other fields such as economics, sociology, and psychology. Although these fields are applied to a lesser degree than math and science, they are still important. Thus, we say that engineers apply the knowledge (i.e., collective wisdom) of humankind.

When Cegnar (2), a practicing engineer, saw Katehi et al.'s description (1), he suggested that engineering requires more that just math, science, and technology. Cegnar stated that solutions also involve creativity and innovation. Solutions involve persistence and struggle in the face of challenges. Solutions involve constraints such as time and money. Solutions involve the ability to simplify and idealize that which is complex. The skills that professionals use to be creative, to handle adversity, to manage constraints, and to idealize are called the *art* of engineering.

Fig. 1.3 summarizes ideas about engineering. The upper row summarizes what engineering is. The lower row summarizes how engineering is done and why engineering is done. The term **process** means a systematic and effective method for getting results.

FIGURE 1.3

A summary of ideas about engineering.

Definition of Fluid Mechanics

Mechanics is the field of science focused on the motion of material bodies. Mechanics involves force, energy, motion, deformation, and material properties. When mechanics applies to material bodies in the solid phase, the discipline is called **solid mechanics**. When the material body is in the gas or liquid phase, the discipline is called fluid mechanics.

In summary, **fluid mechanics** is the science of energy, motion, deformation, and properties when the material is in the gas or liquid phase.

Definition of Learning

Researchers at the Harvard Graduate School of Education (3, 4) define **understanding** as the ability to carry out performances that show one's grasp of a subject and advance it at the same time. Understanding is about being able to apply knowledge in new ways. Based on these ideas, we define **learning** as the process of developing (or improving) one's abilities to do something useful while also advancing one's ability to learn in the future.

Summary To learn engineering fluid mechanics means to develop the ability to design systems that involve fluids while also advancing one's abilities to learn in the future.

1.2 Describing Liquids and Gases

Designers need to understand the nature of the materials they work with. Thus, this section describes fluids. A wonderful starting point is the atomic hypothesis as stated by the Nobel-prize-winning physicist Richard Feynman (5):

If, in some cataclysm, all of scientific knowledge were to be destroyed, and only one sentence passed on to the next generation of creatures, what statement would contain the most information in the fewest words? I believe it is the atomic hypothesis (or atomic fact, or whatever you wish to call it) that all things are made of atoms—little particles that move around in perpetual motion, attracting each other when they are a little distance apart, but repelling upon being squeezed into one another. In that one sentence, you will see, there is an enormous amount of information about the world, if just a little imagination and thinking are applied.

A **fluid** is a substance whose molecules move freely past each other. More specifically, a fluid is a substance that will continuously deform (i.e., flow) under the action of a shear stress. Alternatively, a solid will deform under the action of a shear stress but will not flow like a fluid. Both liquids and gases are classified as fluids.

Because of differences in the forces between molecules, liquids and gases behave differently. As shown in the first row of Table 1.1, a liquid will take the shape of a container, whereas a gas will expand to fill a closed container. The behavior of the liquid is produced by strong attractive force between the molecules. This strong attractive force also explains why the density of a liquid is much higher than the density of gas (see the fourth row). The attributes in Table 1.1 can be generalized by defining a gas and liquid based on the differences in the attractive forces between molecules. A **gas** is a phase of material in which molecules are widely spaced, molecules move about freely, and forces between molecules are minuscule, except during collisions. Alternatively, a **liquid** is a phase of material in which molecules are closely spaced, molecules move about, and there are strong attractive forces between molecules.

TABLE 1.1 Comparison of Solids, Liquids, and Gases

Attribute	Solid	Liquid	Gas
Typical Visualization			
Description	Solids hold their shape; no need for a container	Liquids take the shape of the container and will stay in open container	Gases expand to fill a closed container
Mobility of Molecules	Molecules have low mobility because they are bound in a structure by strong intermolecular forces	Molecules move around freely even though there are strong intermolecular forces between molecules	Molecules move around freely with little interaction except during collisions; this is why gases expand to fill their container
Typical Density	Often high; e.g., density of steel is 7700 kg/m^3	Medium; e.g., density of water is 1000 kg/m^3	Small; e.g., density of air at sea level is 1.2 kg/m^3
Molecular Spacing	Small—molecules are close together	Small—molecules are held close together by intermolecular forces	Large—on average, molecules are far apart
Effect of Shear Stress	Produces deformation	Produces flow	Produces flow
Effect of Normal Stress	Produces deformation that may associate with volume change; can cause failure	Produces deformation associated with volume change	Produces deformation associated with volume change
Viscosity	NA	High; decreases as temperature increases	Low; increases as temperature increases
Compressibility	Difficult to compress; bulk modulus of steel is 160 × 10^9 Pa	Difficult to compress; bulk modulus of liquid water is 2.2 × 10^9 Pa	Easy to compress; bulk modulus of a gas at room conditions is about 1.0 × 10^5 Pa

1.3 Idealizing Matter

Engineers apply idealized* models to characterize material behavior. Thus, this section presents ideas for understanding materials and their behaviors.

The Microscopic and Macroscopic Viewpoints

A **microscopic viewpoint** describes material behavior by characterizing the behavior of atoms and molecules, often using statistical methods to characterize average molecular behavior. Alternatively, a **macroscopic viewpoint** describes material behavior without resulting to models at the atomic level. The macroscopic viewpoint is simpler, so it is used more often.

Matter can be studied from a macroscopic viewpoint or a microscopic viewpoint. Most engineering models are based on a macroscopic viewpoint. However, in selected cases such as the kinetic theory of gases, the microscopic viewpoint is useful. In addition, the microscopic model is useful for understanding phenomena such as surface tension and viscosity.

The Continuum Assumption

Because a body of fluid is comprised of molecules, properties are due to average molecular behavior. That is, a fluid usually behaves as if it were comprised of continuous matter that is infinitely divisible into smaller and smaller parts. This idea is called the **continuum assumption**.

When the continuum assumption is valid, engineers can apply limit concepts from differential calculus. A limit concept typically involves letting a length, an area, or a volume approach zero. Because of the continuum assumption, fluid properties such as density and velocity can be considered continuous functions of position with a value at each point in space.

To gain insight into the validity of the continuum assumption, consider a hypothetical experiment to find density. Fig. 1.4a shows a container of gas in which a volume ΔV has been identified. The idea is to find the mass of the molecules Δm inside the volume and then to calculate density by

$$\rho = \frac{\Delta m}{\Delta V}$$

The calculated density is plotted in Fig. 1.4b. When the measuring volume ΔV is very small (approaching zero), the number of molecules in the volume will vary with time because of the random nature of molecular motion. Thus, the density will vary as shown by the wiggles in the blue line. As volume increases, the variations in calculated density will decrease until the calculated density is independent of the measuring volume. This condition corresponds to the

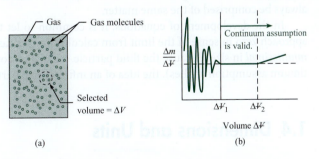

(a)   (b)

FIGURE 1.4

When a measuring volume ΔV is large enough for random molecular effects to average out, the continuum assumption is valid.

---

*Engineers idealize because this makes things easier and faster. To *idealize* means to simplify an entity (an idea, a physical system, a mathematical model, etc.) by removing extraneous details that have little impact on utility.

vertical line at ΔV_1. If the volume is too large, as shown by ΔV_2, then the value of density may change due to spatial variations.

In most applications, the continuum assumption is valid as shown by the next example.

EXAMPLE. Probability theory shows that including 10^6 molecules in a volume will allow the determination of density to within 1%. Thus, a cube that contains 10^6 molecules should be large enough to accurately estimate macroscopic properties such as density and velocity. Find the length of a cube that contains 10^6 molecules. Assume room conditions. Do calculations for (a) water, and (b) air.

Solution. (a) The number of moles of water is $10^6/6.02 \times 10^{23} = 1.66 \times 10^{-18}$ mol. The mass of the water is $(1.66 \times 10^{-18} \text{ mol})(0.0180 \text{ kg/mol}) = 2.99 \times 10^{-20}$ kg. The volume of the cube is $(2.99 \times 10^{-20} \text{ kg})/(999 \text{ kg/m}^3) = 2.99 \times 10^{-23} \text{ m}^3$. Thus, the length of the side of a cube is 3.1×10^{-8} m. (b) Repeating this calculation with air gives a length of 3.5×10^{-7} m.

Review. For the continuum assumption to apply, the object being analyzed would need to be larger than the lengths calculated in the solution. If we select 100 times larger as our criteria, then the *continuum assumption applies* to objects with:

- Length $(L) > 3.1 \times 10^{-6}$ m (for liquid water at room conditions)
- Length $(L) > 3.5 \times 10^{-5}$ m (for air at room conditions)

Given the two length scales just calculated, it is apparent that the *continuum assumption applies to most problems of engineering importance.* However, there are a few situations where the problem length scales are too small.

EXAMPLE. When air is in motion at a very low density, such as when a spacecraft enters the earth's atmosphere, then the spacing between molecules is significant in comparison to the size of the spacecraft.

EXAMPLE. When a fluid flows through the tiny passages in nanotechnology devices, then the spacing between molecules is significant compared to the size of these passageways.

The Fluid Particle

When developing equations or visualizing the flow of a fluid, it is useful to visualize a small unit of fluid that is part of a larger body. A **fluid particle** is defined as a small quantity of fluid with fixed identity. *Small* means that the lengths of the particle are much smaller that the characteristic length(s) of the problem under study. The words *fixed identity* mean that the particle is always comprised of the same matter. Typically, a fluid particle in a flow will change shape (i.e., deform) and change orientation in response to forces. However, the fluid particle will always be comprised of the same matter.

In the development of equations, it is common to let the dimensions of a fluid particle approach zero in sense of the limit from calculus. In this case, we say that the fluid particle is *infinitesimal* in size. Because the fluid particle is a macroscopic concept (i.e., assume the continuum assumption applies), the idea of an infinitesimal particle is valid.

1.4 Dimensions and Units

As engineers we record data; we measure things. The foundation of measurement is the *dimension* and the *unit*. Thus, this section introduces these topics.

Dimensions

A **dimension** is a category for measurement. For example, engineers measure power, so power is a dimension. Dimensions can be identified by asking the question: *what are we interested in measuring?* Answers to this question can include force, length, volume, work, and viscosity. Thus, these variables are dimensions.

Dimensions can be related by using equations. For example, Newton's second law, $F = ma$, relates the dimensions of force, mass, and acceleration. Because dimensions can be related, engineers and scientists can express dimensions using a limited set that are called *primary dimensions*. Table 1.2 lists one set of primary dimensions.

TABLE 1.2 Primary Dimensions

Dimension	Symbol	Unit (SI)
Length	L	meter (m)
Mass	M	kilogram (kg)
Time	T	second (s)
Temperature	θ	kelvin (K)
Electric current	i	ampere (A)
Amount of light	C	candela (cd)
Amount of matter	N	mole (mol)

A *secondary dimension* is any dimension that can be expressed using primary dimensions. For example, the secondary dimension "force" is expressed in primary dimensions by using $F = ma$. The primary dimensions of acceleration are L/T^2, so

$$[F] = [ma] = M\frac{L}{T^2} = \frac{ML}{T^2} \tag{1.1}$$

In Eq. (1.1), the square brackets means "dimensions of." Thus $[F]$ means "the dimension of force. Similarly, $[ma]$ means the dimensions of mass times acceleration. This equation reads "the primary dimensions of force are mass times length divided by time squared." Notice that primary dimensions are not enclosed in brackets. For example, ML/T^2 is not enclosed in brackets.

One can find primary dimensions by applying a known equation.

EXAMPLE. Suppose the goal is to find the primary dimensions of work.

Step 1: Find an equation.

$$(\text{work}) = (\text{force})(\text{distance})$$
$$W = Fd$$

Step 2: Use the equation to relate the secondary dimensions:
$$[W] = [Fd] = [F][d]$$

Step 3: Insert primary dimensions and do algebra.

$$[W] = [F][d] = \frac{ML}{T^2} \times L = \frac{ML^2}{T^2}$$

One can also find primary dimensions by looking them up. For example, Table F.1 (front of book) shows that, the primary dimensions of viscosity are *M/LT*. Similarly, Table A.6 (back of book) lists primary dimensions for symbols used in this text.

Units

A **unit** is a standard for measurement so that size or magnitude can be characterized. Units allow quantification. For example, to quantify how much volume (a dimension), one selects from a variety of units: liters, cubic meters, etc. For example, one might state that a tank has a volume of 42 liters. The dimension describes what (i.e., the volume) and the unit describes how much (42 liters). Similarly, measurement of energy (a dimension) can be expressed using units of joules or units of calories. The relationship between units and dimensions is illustrated in Fig. 1.5. As shown, a dimension can be visualized as a number line, and a unit is a way to increment a dimension so that magnitude can be measured.

FIGURE 1.5

The relationship between units and a dimension.

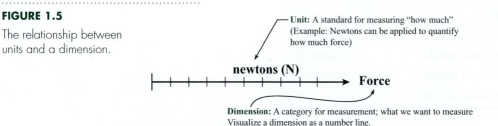

Unit: A standard for measuring "how much" (Example: Newtons can be applied to quantify how much force)

newtons (N)

Force

Dimension: A category for measurement; what we want to measure
Visualize a dimension as a number line.

✔ CHECKPOINT PROBLEM 1.1

Weightwatchers, Inc. has developed "Points"™, which are used to track food intake. Points are calculated as a function of calories, grams of fat, and grams of fiber. You're only supposed to eat a certain number of Points™ in a day. Is the Point™ a dimension or a unit?

Unit Systems

This text uses the International System of Units (abbreviated SI from the French "Le Système International d'Unités"), which is based on the meter, kilogram, and second. The SI system is the international standard for measurement.

Consistent Units

Consistent units are defined as a set of units for which the conversion factors only contain the number 1.0. This means, for example, that:

$$(1 \text{ unit of force}) = (1 \text{ unit of mass})(1 \text{ unit of acceleration})$$

$$(1 \text{ unit of power}) = (1 \text{ unit of work})/(1 \text{ unit of time})$$ **(1.2)**

$$(1 \text{ unit of speed}) = (1 \text{ unit of distance})/(1 \text{ unit of time})$$

Table 1.3 lists consistent units in the SI system and in the traditional system.

TABLE 1.3 Consistent Units

Dimension	SI system
length	meter (m)
mass	kilogram (kg)
time	second (s)
force	newton (N)
pressure	pascal (Pa)
density	kilogram per meter cubed (kg/m³)
volume	cubic meters (m³)
power	watt (W)

Regarding unit practice, three recommendations are

• Use consistent units because this eliminates extraneous unit conversions.

• Use the SI system whenever possible because this system is the international standard, and this system is simpler and leads to more accurate work for most people.

• Become proficient with traditional units because these units are still commonly used.

Organizing Units and Dimensions

Table F.1 (front of book) shows how units and dimensions fit together in fluid mechanics. Four primary dimensions (M, L, T, θ) are used to build approximately 12 secondary dimensions (flow rate, pressure, power, etc.). Each of these dimensions can be quantified with many different units.

1.5 Carrying and Canceling Units

Carrying and canceling units in engineering is beneficial, if not essential. Thus, this section introduces a method called the grid method, developed by Wales and Stager (6). Although other methods are available, the grid method is presented here because it is simple and clear.

Example of the Grid Method

The grid method is illustrated in Fig. 1.6. As shown, this calculation is an estimate of the power P required to ride a bicycle at a speed of $V = 20$ mph. The engineer estimated that the required force to move against wind drag is $F = 4.0$ lbf and applied the equation $P = FV$. As shown, the calculation reveals that the power is 159 watts.

FIGURE 1.6
Grid method.

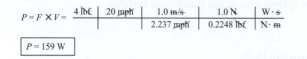

$$P = F \times V = \frac{4 \text{ lbf}}{} \; \bigg| \; \frac{20 \text{ mph}}{} \; \bigg| \; \frac{1.0 \text{ m/s}}{2.237 \text{ mph}} \; \bigg| \; \frac{1.0 \text{ N}}{0.2248 \text{ lbf}} \; \bigg| \; \frac{W \cdot s}{N \cdot m}$$

$$P = 159 \text{ W}$$

The idea of the grid method is to keep multiplying the right side of the equation by the number 1.0 until the units are the desired units. For example in Fig. 1.6, the engineer multiplied the right side of the equation by 1.0 three times.

$$1.0 = \frac{1 \text{ m/s}}{2.237 \text{ mph}} \quad \text{(first time)}$$

$$1.0 = \frac{1.0 \text{ N}}{0.2249 \text{ lbf}} \quad \text{(second time)}$$

$$1.0 = \frac{W \cdot s}{N \cdot m} \quad \text{(third time)}$$

Finding Unity Conversion Ratios

Each equation listed above is called a *unity conversion ratio* (**conversion ratio** for short) because the pure number 1.0 without units appears on the left side. There are three methods for finding unity conversion ratios. The first method is to derive a formula.

Step 1. Start with a definition:

$$\text{power} = \frac{\text{work}}{\text{time}}$$

Step 2. List the units of each variable.

$$1.0 \text{ W} = 1.0 \text{ watt} = \frac{1.0 \text{ joule}}{1.0 \text{ second}} = \frac{1.0 \text{ newton-meter}}{\text{second}} = \frac{1.0 \text{ N-m}}{s}$$

Step 3. Do algebra.

$$1.0 = \frac{W \cdot s}{N \cdot m}$$

The second method is look up a formula in front of this book.

> **EXAMPLE.** Find the row labeled "speed" in Table F.1 and note that 1.0 m/s = 2.237 mph. This formula can be rearranged to give
>
> $$1.0 = \frac{1 \text{ m/s}}{2.237 \text{ mph}}$$

The third method is use a memorized fact. For example, if one can remember that 1.00 inch is equal to 2.54 centimeters, one can write

$$1.0 = \frac{1 \text{ inch}}{2.54 \text{ cm}}$$

Examples of the Grid Method

The steps of the grid method are listed in the first column of Table 1.4. Examples showing how to apply the steps are presented in the second and third columns.

TABLE 1.4 Applying the Grid Method (Two Examples)

Step	Example 1	Example 2
Problem Statement =>	**Situation:** Convert a pressure of 2.00 psi to pascals.	**Situation:** Find the force in newtons that is needed to accelerate a mass of 10 g at a rate of 15 ft/s².
Step 1. Write the equation down	not applicable	$F = ma$
Step 2. Insert numbers and units	$p = 2.00$ psi	$F = ma = (0.01 \text{ kg})(15 \text{ ft/s}^2)$
Step 3. Look up conversion ratios (see Table F.1)	$1.0 = \dfrac{1 \text{ Pa}}{1.45 \times 10^{-4} \text{ psi}}$	$1.0 = \dfrac{1.0 \text{ m}}{3.281 \text{ ft}} \qquad 1.0 = \dfrac{\text{N} \cdot \text{s}^2}{\text{kg} \cdot \text{m}}$
Step 4. Multiply terms and cancel units.	$p = \left[2.00 \text{ psi} \right] \left[\dfrac{1 \text{ Pa}}{1.45 \times 10^{-4} \text{ psi}} \right]$	$F = \left[0.01 \text{ kg} \right] \left[\dfrac{15 \text{ ft}}{s^2} \right] \left[\dfrac{1.0 \text{ m}}{3.281 \text{ ft}} \right] \left[\dfrac{\text{N} \cdot s^2}{\text{kg} \cdot \text{m}} \right]$
Step 5. Do calculations.	$p = 13.8$ kPa	$F = 0.0457$ N

Using Pounds-Mass and Slugs

Engineers often use pounds-mass (lbm) and slugs in calculations. Thus, this subsection shows how to use these units.

Table F.1 shows how mass units are related. One kilogram of mass is equivalent to 2.2 pounds mass (1 kg = 2.2 lbm). One pound of mass is equivalent to 454 grams (1.0 lbm = 453.6 g). One slug of mass is equivalent to 32.2 pounds mass or 14.6 kilograms (1.0 slug = 32.17 lbm = 14.59 kg).

Mass units can be related to force units by application of $F = ma$. In the SI unit system, a force of 1.0 N is defined as the magnitude of force that will accelerate a mass of 1.0 kg at a rate of 1.0 m/s². Thus,

$$(1.0 \text{ N}) \equiv (1.0 \text{ kg})(1.0 \text{ m/s}^2)$$

Rewriting this expression gives a conversion ratio

$$1.0 = \frac{\text{kg} \cdot \text{m}}{\text{N} \cdot \text{s}^2} \tag{1.3}$$

When the mass unit is the slug, a force of 1.0 pound-force (lbf) is defined as the force that will accelerate a mass of 1.0 slugs at a rate of 1 ft/s². Thus,

$$(1.0 \text{ lbf}) \equiv (1.0 \text{ slug})(1.0 \text{ ft/s}^2)$$

Rewriting this expression gives the conversion factor

$$1.0 = \frac{\text{slug} \cdot \text{ft}}{\text{lbf} \cdot \text{s}^2} \tag{1.4}$$

When the mass unit is lbm, a force of 1.0 lbf is defined as the magnitude of force that will accelerate a mass of 1.0 lbm at a rate of 32.2 ft/s². So,

$$(1.0 \text{ lbf}) \equiv (1.0 \text{ lbm})(32.2 \text{ ft/s}^2)$$

Thus, the conversion ratio relating force and mass units becomes

$$1.0 = \frac{32.2 \text{ lbm} \cdot \text{ft}}{\text{lbf} \cdot \text{s}^2} \tag{1.5}$$

Example 1.1 shows how to use the grid method.

EXAMPLE 1.1

Grid Method Applied to Calculating Thrust from a Rocket

Problem Statement

A water rocket is fabricated by attaching fins to a 1-liter plastic bottle. The rocket is partially filled with water, and the air space above the water is pressurized, causing water to jet out of the rocket and propel the rocket upward. The thrust force T from the water jet is given by $T = \dot{m}V$, where $\dot{m}$ is the rate at which the water flows out of the rocket in units of mass per time and V is the speed of the water jet. Estimate the thrust force in newtons for a jet velocity of $V = 30$ m/s where the mass flow rate is $\dot{m} = 9$ kg/s. Apply the grid method during your calculations.

Define the Situation

A rocket is propelled by a water jet.

Thrust Force $= T = \dot{m}V$

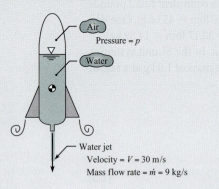

Water jet
Velocity = V = 30 m/s
Mass flow rate = $\dot{m}$ = 9 kg/s

State the goal

$T(\text{N})$, $T(\text{lbf})$ ◀ thrust force in newtons and pounds-force

Generate Ideas and Make a Plan

Apply the process given in Table 1.4.

Take Action (Execute the Plan)

1. Thrust force (SI units)
$$T = \dot{m}V$$

- Insert numbers and units:
$$T\,(\text{N}) = \dot{m}V = (9 \text{ kg/s})(30 \text{ m/s})$$

- Insert conversion ratios and cancel units:
$$T\,(\text{N}) = \left[\frac{9 \text{ kg}}{s}\right]\left[\frac{30 \text{ m}}{s}\right]\left[\frac{\text{N} \cdot s^2}{\text{kg} \cdot \text{m}}\right]$$

$$T = 270 \text{ N}$$

1.6 Applying the Ideal Gas Law (IGL)

The design of systems that involve gases (e.g., airbags, shock absorbers, combustion systems, aircraft) often involve application of the IGL. Thus, this section presents this topic.

Theoretical Development of the IGL

Brown et al. (7) states that the IGL was developed empirically. An *empirical* equation is one that was developed by the logical process called induction. *Induction* is the process of making many experimental observations and then concluding that something is always true because every experiment indicates this truth. For example, if a person concludes that the sun will rise tomorrow because it has risen every day in the past, this is an example of inductive reasoning.

The IGL was developed by combining three empirical equations that had been discovered previously. The first of these equation, called Boyle's law, states that when temperature T is held constant, the pressure p and volume $\forall$ of a fixed quantity of gas are related by:

$$p\forall = \text{constant} \qquad \text{(Boyle's law)}$$

The second equation, Charles's law, states that when pressure is held constant, the temperature and volume V of a fixed quantity of gas are related by:

$$\frac{V}{T} = \text{constant} \quad \text{(Charles's law)}$$

The third equation was derived by a hypothesis formulated by Avogadro: *Equal volumes of gases at the same temperature and pressure contain equal number of molecules.* When Boyle's law, Charles's law, and Avogadro's law are combined, the result is the ideal gas equation in this form:

$$pV = nR_u T \qquad (1.6)$$

where n is the amount of gas measured in units of moles. A **mole** is defined as the amount of matter that contains as many particles as there are atoms in 12 g of carbon-12. This means that a mole of gas will contain 6.02214×10^{23} particles. In Eq. (1.6), R_u is a constant called the universal gas constant; some useful values are

$$R_u = 8.314 \frac{kJ}{kmol \cdot K}$$

To make the ideal gas law more useful, it can be rearranged to use mass units instead of mole units. To relate moles and mass, let

$$n(\text{moles}) \times \mathcal{M} \frac{(\text{grams})}{(\text{mole})} = m(\text{grams}) \qquad (1.7)$$

where $\mathcal{M}$ is the molar mass of the gas and m is mass of the gas. To develop the mass form of the ideal gas equation, substitute Eq. (1.7) into Eq. (1.6).

$$pV = \frac{m}{\mathcal{M}} R_u T = m \left(\frac{R_u}{\mathcal{M}} \right) T = mRT \qquad (1.8)$$

$$pV = mRT$$

where the specific gas constant R is given by

$$R = (\text{specific gas constant}) = \frac{R_u}{\mathcal{M}} = \frac{\text{ideal gas constant}}{\text{molar mass}} \qquad (1.9)$$

To introduce density into the IGL, rewrite Eq. (1.8) and then introduce the definition of density ρ:

$$p = \left(\frac{m}{V} \right) RT = \rho RT \qquad (1.10)$$

Validity of the IGL

An equation is *valid* when calculated values closely match (say within 5%) values that would be measured if an experiment was done. Regarding the validity of the IGL, some useful tips are presented here.

- For gases near atmospheric conditions, the IGL is a good approximation.
- When both the liquid phase and the gas phase are present (e.g., propane in a tank used for a barbecue), one can consult thermodynamic tables (8) to find the density of the gas phase.

- When a gas is very hot such as the exhaust stream of a rocket, then the gas can ionize or disassociate. Both of these effects can invalidate the ideal gas law.
- To determine if a gas can be characterized with the IGL, one can calculate the compressibility factor, which is commonly given the symbol Z and presented in thermodynamics texts (8).

Working Equations

An equation that is used for applications is called a *working equation*. Working equations for fluid mechanics are presented in Table F.2 in the front of the book. In addition, many of these working equations are described in more detail; see, for example, Table 1.5 for the IGL.

Table 1.5 lists the most useful forms of the IGL and lists the variables. Notice the tips in the last column of the table. Tips are identified by parenthesis.

TABLE 1.5 Summary of the Ideal Gas Law Equations

Description	Equation		Variables
Density form of the IGL	$p = \rho R T$	(1.10)	p = pressure (Pa) (use absolute pressure, not gage or vacuum pressure) ρ = density (kg/m³) R = specific gas constant (J/(kg · K)) (look up R in Table A.2) T = temperature (K) (use absolute temperature)
Mass form of the IGL	$p V = m R T$	(1.8)	V = volume (m³) m = mass (kg)
Mole form of the IGL	$p V = n R_u T$	(1.6)	n = number of moles R_u = universal gas constant (R_u = 8.314 J/(mol · K))
This equation is used to relate gas constants	$R = \dfrac{R_u}{\mathcal{M}}$	(1.9)	$\mathcal{M}$ = molar mass (kg/mol)

1.7 The Wales-Woods Model

Engineers use calculations to figure things out. Thus, this section presents a model, called the Wales-Woods model, that reveals how professionals do calculations.

Rationale for the Wales-Woods Model (WWM)

An expert is a person who does things well with minimal effort. For example, an expert golf player hits a golf ball far with little effort. It is human nature to desire the ability to create *great results with minimal effort*.

Learning to do something well is facilitated by *deliberate practice* according to Dr. Anders Ericsson and his colleagues (9). Dr. Ericsson is the Conradi Eminent Scholar of Psychology at Florida State University and an international authority on the development of expertise. He asserts that it is deliberate practice, not innate talent, that leads to expertise. Deliberate practice involves understanding how experts do things and then practicing these fundamentals over a long period of time. Thus, the rationale for the WWM is to reveal how experts solve technical problems so that *students can practice these skills and develop themselves over time into professionals who solve difficult problems with minimal effort.*

The WWM is based on the research of Professors Charles Wales, Anni Nardi, Robert Stager, and Donald Woods (6, 10-17). These researchers studied how experts solved problems, and then they figured out how to teach these patterns to students.

The WWM is effective. Based on 5 years of data, Wales (11) reports that when students were taught problem solving as freshman, the graduation rate increased by 32% and the average grade point average increased by 25%, as compared to the control group, who were not taught these skills. Based on 20 years of data, Woods (17) reports that students taught problem-solving skills, as compared to control groups, showed significant gains in confidence, problem-solving ability, attitude toward lifetime learning, self-assessment, and recruiter response.

Introduction to the WWM

Experts have a method or process that they apply to solve problems. Thus, this subsection introduces this process in the context of solving a textbook problem.

Example 1.2 shows the WWM applied to a textbook problem. The left column shows the problem and the solution. The right column explains how to apply the WWM and lists skills (i.e., actions) that are used in the model.

EXAMPLE 1.2

Applying the IGL to Predict Weight

PROBLEM AND THE SOLUTION

Problem Statement

Find the total weight of a 0.481 m³ tank of nitrogen if the nitrogen is pressurized to 3.45 MPa abs., the tank itself weighs 222 N, and the temperature is 20°C.

Define the Situation

A tank contains nitrogen. $W_{tank} = 222$ N

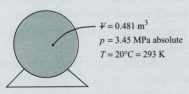

$\Psi = 0.481\ m^3$
$p = 3.45$ MPa absolute
$T = 20°C = 293$ K

Assumption: The IGL applies.

Nitrogen: (Table A.2) $R_{N_2} = 297$ (J/kg · K).

EXPLANATION OF THE WWM

To the left is a typical problem statement from a textbook.

Experts read and interpret the problem statement. Experts present their own interpretation of the problem.

To **define the situation** is to summarize the problem in a way that shows *how you are idealizing the problem.* Actions:

- Visualize the problem as if it exists in the real world. A useful question to ask is, *what am I looking at?*

- Identify scientific concepts that may be useful. A useful question to ask is, *what are the physics?*

- Summarize the physical situation (write down 1 to 2 sentences).

- List known values of variables.

- Sketch the situation; this sketch is called a *situation diagram.* Use engineering conventions on this diagram.

- Convert units to *consistent units.*

- State main assumptions.

- List fluid properties (see Section 2.4)

State the Goal

$W_T(N)$ ⬅ Weight total (nitrogen + tank)

To **state the goal** is to summarize the results you intend to create. Actions:

- List the variable(s) to be solved for.
- List the units on these variables.
- Describe each variable(s) with a short statement.

Generate Ideas and Make a Plan

Because weight is the goal, let

$$W_T = W_{tank} + W_{N_2} \qquad (a)$$

In Eq. (a), W is known and W_{N_2} is unknown, so it becomes the new goal. Select Newton's law of gravitation because this equation has the new goal in it.

$$W_{N_2} = m_{N_2} g \qquad (b)$$

In Eq. (b), identify that m_{N_2} is unknown. This parameter can be found by applying the ideal gas law.

$$p V = mRT \qquad (c)$$

In Eq. (c), all new variables are known. Thus, the problem is cracked. There are three equations (a, b, and c) and three unknown variables (weight of nitrogen, mass of nitrogen, and total weight of the tank). The step-by-step plan is

1. Calculate mass of nitrogen using Eq. (c).
2. Calculate weight of nitrogen using Eq. (b).
3. Calculate the total weight using Eq. (a).

To **generate ideas** is to consider alternative approaches to reach your goal(s) and to select the best ideas.

The actions that work on most problems are listed here. These steps from Wales et al. (6) can be remembered with the acronym GENI.

- Step 1. Start with **G**oal
- Step 2. Identify an **E**quation that contains the goal
- Step 3. In this equation, identify the unknowns (**N**eeds)
- Step 4. In this equation, identify the knowns (**I**nformation)
- Step 5. Repeat steps 1 to 4 until the number of equations is equal to the number of unknowns. At this point the problem is figured out (we say *the problem is cracked*)

To **make a plan** is to figure out the steps to reach your goals.

- Identify the easiest and fastest way to get to your goal.
- List the steps.

Note: Most of the time, the steps of the plan are in reverse order of the steps of the reasoning process.

Take Action (Execute the Plan)

1. Ideal gas law (mass form)

$$m_{N_2} = \frac{p V}{RT}$$

$$= \left(\frac{3.45 \times 10^6 \text{ N}}{\text{m}^2} \right) \left(0.481 \text{ m}^3 \right) \left(\frac{\text{kg} \cdot \text{K}}{297 \text{ N} \cdot \text{m}} \right) \left(\frac{1}{293 \text{ K}} \right)$$

$$= 19.1 \text{ kg}$$

2. Newton's law of gravity

$$W_{N_2} = mg = (19.1 \text{ kg})(9.81 \text{ m/s}^2) = 187 \text{ N}$$

3. Total weight

$$W_T = W_{tank} + W_{N_2} = (222 \text{ N}) + (187 \text{ N}) = \boxed{409 \text{ N}}$$

To **take action** is to execute the steps of the plan. Actions:

- On each step, list the name of the main equation or give another descriptive label.
- Carry and cancel units with the grid method. (Note: Unit cancellations are not shown in the text or solution manual because we have not yet found a simple way to do this.)
- Box the final answer(s).

Review the Solution and the Process

1. *Knowledge.* Use the mass form of the IGL when mass is the goal.
2. *Knowledge.* $W = mg$ can be derived from Newton's law of gravity. Thus, this equation is a special case of this law.
3. *Validate.* To check the IGL assumption, we calculated the compressibility factor and found that the IGL was accurate to within about 98%.
4. *Implications.* For this problem, the weight of the gas is significant as compared to the weight of the tank.
5. *Skill.* To save time, add problem information to the situation diagram.

To **review the solution and the process** is to think critically and then to write one to three useful or insightful thoughts. An effective approach is to ask questions. Examples:

- *Validate.* How can I check (validate) my solution? Does my solution make sense? Why?
- *Implications.* What did I learn? What might my result mean in the real world?
- *Skill(s).* What skills helped me solve this problem? What skills will help me solve problems in the future?
- *Knowledge.* What knowledge was useful for solving this problem? What new ideas did I gain?
- *Discussion.* What aspects of the solution are worthwhile to point out?

Structure of the Wales-Woods Model (WWM)

As shown by Example 1.2, the WWM is comprised of six thinking operations. A *thinking operation* (Table 1.6) is a collection of skills for achieving a certain outcome. Notice that each thinking operation has an outcome and a rationale.

TABLE 1.6 Structure of the WWM (Thinking Operations)

Thinking Operation	Outcome of This Thinking Operation	Why Do This Thinking Operation? (Rationale)
Define the situation	The model (idealization) used to solve the problem is clear, specific, and organized.	So you know how you are idealizing the problem.
State the goal	The goal is specific and actionable (not vague).	So you know where you are at (i.e., the situation) and where you need to go (i.e., the goal).
Generate ideas	The ideas for solving the problem are clear and specific. In addition, there is a logical process that shows how the problem solver was able to find a path to the solution.	Because the reasoning process reveals how the problem can be solved. This gives one the ability to solve unfamiliar problems and reduces or eliminates the need to memorize solutions. In addition, this gives one the satisfaction that *I cracked the problem!*
Make a plan	There is a list of steps for reaching the goal.	To find a simple and effective solution method. To create an organized plan of attack.
Take action (Execution)	The steps for reaching the goal have been executed, and the goal has been attained.	To reach the goals and enjoy the satisfaction of completing the problem.
Review the solution and the process	One to three insightful statements are written down.	To grow. This growth can take multiple forms. Examples: to become better at problem solving, to increase knowledge, to increase abilities to validate, to increase abilities to think critically, and to increase self-awareness of problem solving.

Applying the WWM to a Design Problem

The WWM can be applied to a design problem, for example, redesigning a bike pump (see Fig. 1.7). Suppose that a conventional bike pump take too many strokes to inflate a tire, and a designer wishes to redesign the pump to solve this problem. Example 1.3 illustrates how to apply the WWM to this task.

FIGURE 1.7

A bike pump being used to inflate a mountain bike tire. (Photo by Donald Elger)

EXAMPLE 1.3

The Wales-Woods Model Applied to a Design Problem

Problem Statement

Size a bike pump that will inflate a typical mountain bike tire in 20 strokes.

Define the Situation

Redesign a bike pump to inflate a bike tire in 20 strokes. Idealize the bike tire as a volumetric region.

Air: (Table A.2) $R_{air} = 287$ (J/kg · K)

Assumptions:

- Idealize the tire as a cylinder of length $L = 1.94$ m and diameter $D = 0.045$ m.
- Assume that $p_{inflate} = 50$ psig ≈ 450 kPa absolute.
- Isothermal compression: $T = 20\ °C = 293$ K

State the Goal

$V_{pump}(L) \Longleftarrow$ Volume of pump cylinder in liters. (Note: Using this volume, a designer can select a pump diameter and then calculate a stroke length.)

Generate Ideas and Make a Plan

Because the goal is V, apply the IGL to the pump.

$$V_{pump} = \frac{m_{pump} R_{air} T}{p_{pump}} \qquad (a)$$

In Eq. (a), all parameters are known except for the mass of air inside the pump = (m_{pump}). To find this variable, apply conservation of mass:

$$(\text{mass of air in tire}) = \left(\frac{\text{mass of air}}{\text{stroke}}\right)(\text{number of strokes}) \qquad (b)$$

$$m_{tire} = m_{pump} N$$

In Eq. (b), the unknown, (m_{tire}), can be found using the IGL. Thus, the problem is cracked! The steps for doing calculations are

1. Calculate the mass of air inside the tire using the IGL.
2. Relate masses using: $m_{tire} = (m_{pump})(20\ \text{strokes})$.
3. Calculate the volume of the pump using the IGL.

Take Action (Execute the Plan)

1. IGL (apply to tire)

$$V_{tire} = \left(\frac{\pi D^2}{4}\right)L = \frac{\pi(0.45\ \text{m})^2}{4}(1.94\ \text{m}) = 3.085 \times 10^{-3}\ \text{m}^3$$

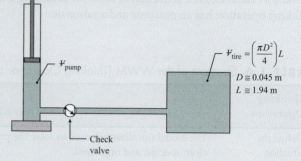

$$m_{tire} = \frac{p_{tire} V_{tire}}{R_{air} T}$$

$$= \left(\frac{450 \times 10^3\ \text{N}}{\text{m}^2}\right)\left(0.003085\ \text{m}^3\right)\left(\frac{\text{kg} \cdot \text{K}}{287\ \text{N} \cdot \text{m}}\right)\left(\frac{1}{293\ \text{K}}\right)$$

$$= 0.0165\ \text{kg}$$

2. Conservation of mass (Eq. b)

$$m_{pump} = \frac{m_{tire}}{N} = \frac{0.0165\ \text{kg}}{20} = 0.000825\ \text{kg}$$

3. IGL (apply to pump cylinder):

$$V_{pump} = \frac{m_{pump} R_{air} T}{p_{pump}}$$

$$= \frac{(0.000825\ \text{kg})}{(101 \times 10^3\ \text{Pa})}\left(\frac{287\ \text{J}}{\text{kg} \cdot \text{K}}\right)\left(\frac{293\ \text{K}}{1}\right)$$

$$= 0.687\ \text{L}$$

Review the Solution and the Process

1. *Skills.* Notice how the system was idealized: a piston/cylinder, a check valve, and a volume to hold air.

2. *Discussion.* The calculated volume is slightly less than the volume of a typical wine bottle (750 mL).

3. *Knowledge.* The specific gas constant R was found in Table A.2. Note that R is different than the universal gas constant R_u.

4. *Discussion.* To estimate the size of bike pump, assume the typical user can comfortably apply a downward force of about 125 N. Thus, the area of the piston (using gage pressure) is about

$$A = F/p \approx (125\ \text{N})/(350 \times 10^3\ \text{Pa}) = 0.00036\ \text{m}^2.$$

The corresponding length of the pump is

$$L_{pump} = V/A$$

$$= (0.000687\ \text{m}^3)/(0.00036\ \text{m}^2) = 1.92\ \text{m}$$

A pump that is nearly 2 meters tall is not practical, so we would not recommend this solution.

Learning the Wales-Woods Model

Learning the WWM is straightforward. Practice the six thinking operations and the embedded skills. Get feedback from teachers or coaches. Recognize that learning the WWM requires a lot of time and patience. It is much like learning the golf swing. Understanding the golf swing is easy, but learning to swing the golf club consistently requires practice over a long period of time.

1.8 Checking for Dimensional Homogeneity (DH)

Checking for DH is a simple and effective approach for checking an equation. Because engineers frequently check for validity, this topic is presented next.

Dimensional Homogeneity (DH)

When the primary dimensions of each term of an equation are the same, the equation is **Dimensionally Homogeneous**, or DH for short. Example 1.4 shows how to check an equation for dimensional homogeneity.

EXAMPLE 1.4

Applying Dimensional Homogeneity to the Ideal Gas Law

Problem Statement

Show that the ideal gas law (density form) is dimensionally homogeneous.

Define the Situation

The ideal gas law (density form) is $p = \rho RT$.

State the Goal

Prove that the ideal gas law is DH.

Generate Ideas and Make a Plan

To check for DH, show that the primary dimensions of each term are the same. The steps are

1. Find the primary dimensions of the first term.
2. Find the primary dimensions of the second term.
3. Prove dimensional homogeneity by comparing the terms.

Take Action (Execute the Plan)

1. Primary dimensions (first term)
   - From Table A.6, the primary dimensions are

$$[p] = \frac{M}{LT^2}$$

2. Primary dimensions (second term)
   - From Table A.6, the primary dimensions are

$$[\rho] = M/L^3$$
$$[R] = L^2/\theta T^2$$
$$[T] = \theta$$

   - Thus

$$[\rho RT] = \left(\frac{M}{L^3}\right)\left(\frac{L^2}{\theta T^2}\right)(\theta) = \frac{M}{LT^2}$$

3. Conclusion: The ideal gas law is dimensionally homogeneous because the primary dimensions of each term are the same.

Dimensionless Groups

Engineers often arrange variables so that primary dimensions cancel out. For example, consider a pipe with an inside diameter D and length L. These variables can be grouped to form a new variable L/D, which is an example of a dimensionless group. A *dimensionless group* is any arrangement of variables in which the primary dimensions cancel.

EXAMPLE. The Mach number M, which relates fluid speed V to the speed of sound c, is a common dimensionless group.

$$M = \frac{V}{c}$$

EXAMPLE. Another common dimensionless group is named the Reynolds number and given the symbol Re. The Reynolds number involves density, velocity, length, and viscosity μ:

$$Re = \frac{\rho V L}{\mu} \tag{1.11}$$

The convention in this text is to use the symbol [-] to indicate that the primary dimensions of a dimensionless group cancel out. For example,

$$[Re] = \left[\frac{\rho V L}{\mu}\right] = [-] \tag{1.12}$$

Dimensionless groups are also called *π–groups*.

Primary Dimensions of Derivative and Integral Terms

Because many equations in fluid mechanics involve derivatives or integrals, this subsection shows how to analyze these terms and introduces the definition of the derivative and integral.

Let's start with the derivative. In calculus, the *derivative* is defined as a ratio:

$$\frac{df}{dy} \equiv \lim_{\Delta y \to 0} \frac{\Delta f}{\Delta y}$$

where Δf is an amount or change in a dependent variable and Δy is an amount or change in a independent variable. Thus, the primary dimensions of a first-order derivative can be found by using a ratio:

$$\left[\frac{df}{dy}\right] = \left[\frac{f}{y}\right] = \frac{[f]}{[y]} \tag{1.13}$$

The primary dimensions for a higher-order derivative can also be found by using the basic definition of the derivative. The resulting formula for a second-order derivative is

$$\left[\frac{d^2 f}{dy^2}\right] = \lim_{\Delta y \to 0} \frac{\Delta(df/dy)}{\Delta y} = \left[\frac{f}{y^2}\right] = \frac{[f]}{[y^2]} \tag{1.14}$$

For example, applying Eq. (1.14) to acceleration shows that

$$\left[\frac{d^2 y}{dt^2}\right] = \left[\frac{y}{t^2}\right] = \frac{L}{T^2}$$

To find primary dimensions of an integral, recall from calculus that an *integral* is defined as an infinite sum of terms that are very small (i.e., *infinitesimal*).

$$\int f \, dy \equiv \lim_{N \to \infty} \sum_{i=1}^{N} f \Delta y_i$$

Thus,

$$\left[\iint f\, dy\right] = [f][y] \qquad \text{(1.15)}$$

For example, position is given by the integral of velocity with respect to time. Checking primary dimensions for this integral gives

$$\left[\int V\, dt\right] = [V][t] = \frac{L}{T} \cdot T = L$$

Summary One can find primary dimensions on derivative and integral terms by applying fundamental definitions from calculus. This process is illustrated by Example 1.5.

EXAMPLE 1.5

Finding the Primary Dimensions for a Derivative and an Integral

Problem Statement

Find the primary dimensions of $\mu \dfrac{d^2 u}{dy^2}$, where μ is viscosity, u is fluid velocity, and y is distance. Repeat for $\dfrac{d}{dt}\displaystyle\int_{\mathcal{V}} \rho\, d\mathcal{V}$ where t is time, $\mathcal{V}$ is volume, and ρ is density.

Generate Ideas and Make a Plan

1. Because a second-order derivative is involved in term 1, apply Eq. (1.14).

2. Because a first-order derivative and an integral is involved in term 2, apply Eqs. (1.13) and (1.15).

Take Action (Execute the Plan)

1. Primary dimensions of $\mu \dfrac{d^2 u}{dy^2}$

   - From Table A.6:
   $$[\mu] = M/LT$$
   $$[u] = L/T$$
   $$[y] = L$$

   - Apply Eq. (1.14):
   $$\left[\frac{d^2 u}{dy^2}\right] = \left[\frac{u}{y^2}\right] = \frac{L/T}{L^2}$$

Define the Situation

Term 1 is $\mu \dfrac{d^2 u}{dy^2}$. Term 2 is $\dfrac{d}{dt}\displaystyle\int_{\mathcal{V}} \rho\, d\mathcal{V}$.

State the Goal

Find the primary dimensions on term 1 and term 2.

- Combine the previous two steps:
$$\left[\mu \frac{d^2 u}{dy^2}\right] = [\mu]\left[\frac{d^2 u}{dy^2}\right] = \left(\frac{M}{LT}\right)\left(\frac{L/T}{L^2}\right) = \boxed{\frac{M}{L^2 T^2}}$$

2. Primary dimensions of $\dfrac{d}{dt}\displaystyle\int_{\mathcal{V}} \rho\, d\mathcal{V}$

   - Find primary dimensions from Table A.6:
   $$[t] = T$$
   $$[\rho] = M/L^3$$
   $$[\mathcal{V}] = L^3$$

   - Apply Eqs. (1.13) and (1.15) together:
   $$\left[\frac{d}{dt}\int_{\mathcal{V}} \rho\, d\mathcal{V}\right] = \left[\frac{\rho \mathcal{V}}{t}\right] = \left(\frac{M}{L^3}\right)\left(\frac{L^3}{T}\right) = \boxed{\frac{M}{T}}$$

Primary Dimensions of a Constant

Some equations have constants, so this subsection shows how to analyze these terms. The method is illustrated by the next two examples.

EXAMPLE. The hydrostatic equation (below) relates pressure p, density ρ, the gravitational constant g, and elevation z. Find the primary dimensions on the constant C.

$$p + \rho g z = \text{constant} = C$$

Solution. For DH, the constant C needs to have the same primary dimensions as p and $\rho g z$. Thus the dimensions of C are $[C] = M/LT^2$.

EXAMPLE. Suppose velocity V is given as a function of distance y using two constants a and b (below). Find the primary dimensions of the constants.

$$V(y) = ay(b - y)$$

Solution. For dimensional homogeneity both sides of this equation need to have primary dimensions of velocity: $[L/T]$. By inspection, one can conclude that $[b] = L$ and $[a] = L^{-1}T^{-1}$. To validate this solution, check the primary dimensions on the right side of the given equation.

$$[ay(b - y)] = [a][y][b - y] = \left(\frac{1}{L \cdot T}\right)(L)(L) = \frac{L}{T}$$

Because these dimensions match the dimensions on velocity, the equation is DH.

1.9 Summarizing Key Knowledge

Definition of Engineering Fluid Mechanics and Learning

- *Engineering* is an art and a process for applying math, science, and technology to design products that benefit humankind.
- *Fluid Mechanics* is the branch of physics that is concerned with forces, motion, and energy as these ideas apply to materials that are in the liquid or gas phases.
- *Learning* is the process of (a) developing (or improving) one's abilities to do something useful, while also (b) increasing one's capacity for future learning.

Fluids, Liquids, and Gases

- Both liquids and gasses are classified as fluids. A *fluid* is defined as a material that deforms continuously under the action of a shear stress.
- A significant difference between gases and liquids is that the molecules in liquids experience strong intermolecular forces, whereas the molecules in gases move about freely with little or no interactions except during collisions.
- Liquids and gases differ in many important respects. Gases expand to fill their containers, whereas liquids will occupy a fixed volume. Gases have much smaller values of density than liquids. For other differences, see Table 1.1 (p. 4).

Ideas for Idealizing Material Behavior

- A *microscopic viewpoint* involves understanding material behavior by understanding what the molecules are doing.
- A *macroscopic viewpoint* involves understanding material behavior without the need to consider what the molecules are doing.
- In the *continuum assumption*, matter is idealized as consisting of continuous material that can be broken into smaller and smaller parts.

- The continuum assumptions applies to most fluid flows.
- A *fluid particle* is a small quantity of fluid with fixed identity and with length dimensions that are very small (e.g., 1/100th) as compared to problem dimensions.

Units and Dimensions

- Dimensions and units are the basis for measurement.
- A *dimension* is a category for measurement. Examples include mass, force, and energy.
- *Units* are the divisions by which a dimension is measured.
- Each dimension can be quantified using a variety of different units. For example, energy can be quantified using joules and N-m.
- All dimensions can be expressed using a limited set of *primary dimensions*. Dimensions that are not primary dimensions are called secondary dimensions.
- Fluid mechanics uses four primary dimensions: mass (M), length (L), time (T), and temperature (θ).

The Grid Method

- The *grid method* is a systematic way to carry and cancel units.
- The main idea of the grid method is to multiply terms in equations by the pure number 1.0 (called a conversion ratio).
- A *conversion ratio* is an equality relationship between units such that the pure number 1.0 appears on one side of an equation. Examples of conversion ratios are 1.0 = (1.0 kg)/(2.2 lbm) and 1.0 = (1.0 lbf)/(4.45 N).

The Ideal Gas Law (IGL)

- Many real gases can be idealized as an ideal gas.
- In the IGL, temperature must be in absolute temperature units (Kelvin or Rankine).
- In the IGL, pressure must be absolute pressure, not gage or vacuum pressure.
- Three useful and equivalent ways to express the IGL are given here.

$$p = \rho RT \quad \text{(density form)}$$
$$p = mRT \quad \text{(mass form)}$$
$$p\forall = nR_u T \quad \text{(mole form)}$$

- The universal gas constant R_u and the specific gas constant R are related by $R = R_u/\mathcal{M}$ where $\mathcal{M}$ is the molar mass (kg/mol) of the gas.
- For a summary of the equations of the IGL, see Table 1.5 (p. 14).

The Wales-Woods Model (WWM)

- The WWM is an idealization of what experts do when they solve problems.
- The WWM is comprised of six *thinking operations*: define the situation, state the goal, generate ideas, make a plan, take action, and review the process and the results. Table 1.6 on page 17 summarizes the thinking operations.
- Each thinking operation can be broken down into specific actions (skills); see Example 1.2 (p. 16) for a listing of relevant skills.

Dimensional Homogeneity (DH)

- *Dimension homogeneity* means that each term in an equation has the same primary dimensions. This means that each term will also have the same units.

- To check to see if an equation is DH, calculate the primary dimensions on each term.

- A *dimensionless group* (also known as a π-*group*) is a group of variables arranged so that the primary dimensions cancel out.

- From calculus

  ▸ The derivative is defined as a ratio:

$$\frac{df}{dy} = \lim_{\Delta y \to 0} \frac{\Delta f}{\Delta y}$$

  ▸ The integral is defined as a infinite sum of small terms:

$$\int f \, dy = \lim_{N \to \infty} \sum_{i=1}^{N} f \, \Delta y_i$$

- To find the primary dimensions on a derivative or integral term, apply the definitions of these operations.

REFERENCES

1. Katehi, L., Pearson, G., and Feder, M., Eds., *Understanding the Status and Improving the Prospects: Committee on K–12 Engineering Education*, National Academies Press, Washington D.C., 2010, 27.

2. Personal communication with Erik Cegnar, practicing engineering at IVUS Innovations Technology, 2010.

3. Blythe, T. *The Teaching for Understanding Guide*. Jossey-Bass, San Francisco, 1998, 13.

4. Wiske, M.S., Ed. *Teaching for Understanding: Linking Research with Practice*. San Francisco: Jossey-Bass, 1998.

5. Feynman, R.P., Leighton, R.B., and Sands, M. *The Feynman Lectures on Physics*, Vol. 1, Reading, MA: Addison-Wesley, p. 1–2.

6. Wales, C.E., and Stager, R.A. *Thinking with Equations,* Center for Guided Design, West Virginia University, Morgantown, WV., 1990.

7. Brown, T.L., LeMay, H.E., and Bursten, B.E. *Chemistry: The Central Science*. Englewood Cliffs, NJ: Prentice-Hall, 1991.

8. Moran, M.J., Shapiro, H.N., Boettner, D.D., and Bailey M.B. *Fundamentals of Engineering Thermodynamics*. Hoboken, NJ: Wiley, 2011, p. 122.

9. Ericsson, K.A., Prietula, M.J., and Cokely, E.T. "The Making of an Expert." *Harvard Business Review* (July–August, 2007).

10. Wales, C.E. "Guided Design: Why & How You Should Use It," *Engineering Education*, vol. 62, no. 8 (1972).

11. Wales, C.E. "Does How You Teach Make a Difference?" *Engineering Education*, vol. 69, no. 5, 81–85 (1979).

12. Wales, C.E., Nardi, A.H., and Stager, R.A. *Thinking Skills: Making a Choice*. Center for Guided Design, West Virginia University, Morgantown, WV., 1987.

13. Wales, C.E., Nardi, A.H., and Stager, R.A. *Professional-Decision-Making*, Center for Guided Design, West Virginia University, Morgantown, WV., 1986.

14. Wales, C.E., and Stager, R.A., (1972b). "The Design of an Educational System." *Engineering Education*, vol. 62, no. 5 (1972).

15. Wales, C.E., and Stager, R.A. *Thinking with Equations*. Center for Guided Design, West Virginia University, Morgantown, WV., 1990.

16. Woods, D.R. "How I Might Teach Problem-Solving?" in J.E. Stice, (ed.) *Developing Critical Thinking and Problem-Solving Abilities*. New Directions for Teaching and Learning, no. 30, San Francisco: Jossey-Bass, 1987.

17. Woods, D.R. "An Evidence-Based Strategy for Problem Solving." *Engineering Education*, vol. 89. no. 4, 443–459 (2000).

PROBLEMS

Defining Engineering Fluid Mechanics (§1.1)

1.1 Read the definition of engineering in §1.1. How does this compare with your ideas of what engineering is? What is similar? What is different?

1.2 Given the definition of engineering in §1.1, what do you think that you should be learning? How do you know if you have learned it?

1.3 Should the definition of engineering in §1.1 include the idea that engineers also need to be very good with humanities and social sciences? What do you believe? Why?

1.4 Select an engineered design (e.g., hydroelectric power as in a dam, an artificial heart) that involves fluid mechanics and is also highly motivating to you. Write a one-page essay that addresses the following questions. Why is this application motivating to you? How does the system you selected work? What role did engineers play in the design and development of this system?

1.5 Many engineering students believe that fixing a washing machine is an example of engineering because it involves solving a problem. Write a brief essay in which you address the following questions: Is fixing a washing machine an example of engineering? Why or why not? How do your ideas align or misalign with the definition of engineering given in §1.1?

Describing Liquids and Gases (§1.2)

1.6 Propose three new rows for Table 1.1, (p. 4, §1.2) and fill them in.

1.7 Based on molecular mechanisms, explain why aluminum melts at 660°C, whereas ice will melt at 0°C.

Idealizing Matter (§1.3)

1.8 **PLUS** The continuum assumption (select all that apply)

   a. applies in a vacuum such as in outer space

   b. assumes that fluids are infinitely divisible into smaller and smaller parts

   c. is a bad assumption when the length scale of the problem or design is similar to the spacing of the molecules

   d. means that density can idealized as a continuous function of position

   e. only applies to gases

1.9 **PLUS** A fluid particle

   a. is defined as one molecule

   b. is small given the scale of the problem being considered

   c. is so small that the continuum assumption does not apply

Dimensions and Units (§1.4)

1.10 **PLUS** For each variable given, list three common units.

   a. Volume flow rate (Q), mass flow rate ($\dot{m}$), and pressure (p)

   b. Force, energy, power

   c. Viscosity

1.11 **PLUS** In Table F.2 (front of book), find the hydrostatic equation. For each form of the equation that appears, list the name, symbol, and primary dimensions of each variable.

1.12 In the context of measurement, a dimension is:

   a. a category for measurement

   b. a standard of measurement for size or magnitude

   c. an increment for measuring "how much"

Carrying/Canceling Units: Grid Method (§1.5)

1.13 **PLUS** In your own words, describe what actions need to be taken in each step of the grid method.

1.14 **PLUS** Apply the grid method to calculate the density of an ideal gas using the formula $\rho = p/RT$. Express your answer in kg/m³. Use the following data: absolute pressure is $p = 400$ kPa, the gas constant is $R = 287$ J/kg · K, and the temperature is $T = 82°C$.

1.15 **PLUS** The pressure rise Δp associated with wind hitting a window of a building can be estimated using the formula $\Delta p = \rho(V^2/2)$, where ρ is density of air and V is the speed of the wind. Apply the grid method to calculate pressure rise for $\rho = 1.2$ kg/m³ and $V = 100$ km/h. Express your answer in pascals.

1.16 Apply the grid method to calculate force using $F = ma$. Find force in newtons for $m = 10$ kg and $a = 10$ m/s².

1.17 **PLUS** When a bicycle rider is traveling at a speed of $V = 40$ km/h, the power P she needs to supply is given by $P = FV$, where $F = 22$ N is the force necessary to overcome aerodynamic drag. Apply the grid method to calculate:

   a. power in watts.

   b. energy in food calories to ride for 1 hour.

1.18 **GO** Apply the grid method to calculate the cost in U.S. dollars to operate a pump for one year. The pump power is 1500 kW. The pump operates for 20 hr/day, and electricity costs $0.10 per kWh.

Ideal Gas Law (IGL) (§1.6)

1.19 Start with the ideal gas law and prove that

   a. Boyle's law is true.

   b. Charles's law is true.

1.20 Calculate the number of molecules in

    a. one cubic centimeter of liquid water at room conditions

    b. one cubic centimeter of air at room conditions

1.21 Start with the mole-form of the ideal gas law and show the steps to prove that the mass form is correct.

1.22 Start with the universal gas constant and show that $R_{N_2} = 297$ J/(kg · K).

1.23 PLUS A spherical tank holds CO_2 at a pressure of 303.9 kPa and a temperature of 20°C. During a fire, the temperature is increased by a factor of 4 to 80°C. Does the pressure also increase by a factor of 4? Justify your answer using equations.

1.24 An engineer living at an elevation of 762 m is conducting experiments to verify predictions of glider performance. To process data, density of ambient air is needed. The engineer measures temperature (23.5°C) and atmospheric pressure (92.2 kPa). Calculate density in units of kg/m³. Compare the calculated value with data from Table A.2 and make a recommendation about the effects of elevation on density; that is, are the effects of elevation significant?

1.25 GO Calculate the density and specific weight of carbon dioxide at a pressure of 300 kN/m² absolute and 60°C.

1.26 Determine the density of methane gas at a pressure of 300 kN/m² absolute and 60°C.

1.27 GO A spherical tank is being designed to hold 10 moles of methane gas at a pressure of 2 kPa and a temperature of 21°C. What diameter spherical tank should be used?

1.28 GO Natural gas is stored in a spherical tank at a temperature of 10°C. At a given initial time, the pressure in the tank is 100 kPa gage, and the atmospheric pressure is 100 kPa absolute. Some time later, after considerably more gas is pumped into the tank, the pressure in the tank is 200 kPa gage, and the temperature is still 10°C. What will be the ratio of the mass of natural gas in the tank when $p = 200$ kPa gage to that when the pressure was 100 kPa gage?

1.29 PLUS At a temperature of 100°C and an absolute pressure of 500 kPa, what is the ratio of the density of water to the density of air, ρ_w/ρ_a?

1.30 GO Find the total weight of a 0.17 m³ tank of oxygen if the oxygen is pressurized to 2758 kPa abs, the tank itself weighs 400 N, and the temperature is 21°C.

1.31 GO A 4 m³ oxygen tank is at 20°C and 700 kPa. The valve is opened, and some oxygen is released until the pressure in the tank drops to 500 kPa. Calculate the mass of oxygen that has been released from the tank if the temperature in the tank does not change during the process.

1.32 PLUS What is the (a) specific weight, and (b) density of air at an absolute pressure of 600 kPa and a temperature of 50°C?

1.33 PLUS Meteorologists often refer to air masses in forecasting the weather. Estimate the mass of 4.17 km³ of air in kilograms. Make your own reasonable assumptions with respect to the conditions of the atmosphere.

1.34 A bicycle rider has several reasons to be interested in the effects of temperature on air density. The aerodynamic drag force decreases linearly with density. Also, a change in temperature will affect the tire pressure.

    a. To visualize the effects of temperature on air density, write a computer program that calculates the air density at atmospheric pressure for temperatures from −10°C to 50°C.

    b. Also assume that a bicycle tire was inflated to an absolute pressure of 450 kPa at 20°C. Assume the volume of the tire does not change with temperature. Write a program to show how the tire pressure changes with temperature in the same temperature range, −10°C to 50°C.

Prepare a table or graph of your results for both problems. What engineering insights do you gain from these calculations?

1.35 A design team is developing a prototype CO_2 cartridge for a manufacturer of rubber rafts. This cartridge will allow a user to quickly inflate a raft. A typical raft is shown in the sketch.

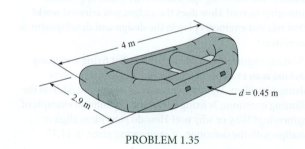

PROBLEM 1.35

Assume a raft inflation pressure of 21 kPa (this means that the absolute pressure is 21 kPa greater than local atmospheric pressure). Estimate the volume of the raft and the mass of CO_2 in grams in the prototype cartridge.

1.36 A team is designing a helium-filled balloon that will fly to an altitude of 25,000 m. As the balloon ascends, the upward force (buoyant force) will need to exceed the total weight. Thus, weight is critical. Estimate the weight (in newtons) of the helium inside the balloon. The balloon is inflated at a site where the atmospheric pressure is 89 kPa and the temperature is 22°C. When inflated prior to launch, the balloon is spherical (radius 1.3 m) and the inflation pressure equals the local atmospheric pressure.

Engineering Calculations and the WWM (§1.7)

1.37 In Example 1.2 (p. 16, §1.7), what are the three steps that an engineer takes to "State the Goal"?

Dimensional Homogeneity (DH) (§1.8)

1.38 The hydrostatic equation is $p/\gamma + z = C$, where p is pressure, γ is specific weight, z is elevation, and C is a constant. Prove that the hydrostatic equation is dimensionally homogeneous.

1.39 PLUS Find the primary dimensions of each of the following terms.

 a. $(\rho V^2)/2$ (kinetic pressure), where ρ is fluid density and V is velocity

 b. T (torque)

 c. P (power)

 d. $(\rho V^2 L)/\sigma$ (Weber number), where ρ is fluid density, V is velocity, L is length, and σ is surface tension

1.40 The power provided by a centrifugal pump is given by $P = \dot{m}gh$, where $\dot{m}$ is mass flow rate, g is the gravitational constant, and h is pump head. Prove that this equation is dimensionally homogeneous.

1.41 PLUS Find the primary dimensions of each of the following terms.

 a. $\int_A \rho V^2 \, dA$, where ρ is fluid density, V is velocity, and A is area.

 b. $\dfrac{d}{dt} \int_V \rho V \, dV$, where $\dfrac{d}{dt}$ is the derivative with respect to time, ρ is density, and V is volume.

2 FLUID PROPERTIES

FIGURE 2.1

This photo shows engineers observing a flume. A *flume* is an artificial channel for conveying water. The flume shown is situated in Boise, Idaho, and is used to study sediment transport in rivers. (Photo courtesy of Professor Ralph Budwig of the Center for Ecohydraulics Research, University of Idaho.)

Chapter Road Map

This chapter introduces ideas for idealizing real-world problems, introduces fluid properties, and presents the viscosity equation.

Learning Objectives

STUDENTS WILL BE ABLE TO

- Define system, boundary, surroundings, state, process, and property. (§2.1)
- Define density, specific gravity, and specific weight. Relate these properties using calculations. (§2.2)
- Explain the meaning of a constant density flow and discuss the relevant issues. (§2.3)
- Look up fluid properties; document the results. (§2.4)
- Define viscosity, shear stress, shear force, velocity gradient, velocity profile, the no-slip condition, and kinematic viscosity. (§2.4)
- Apply the shear stress equation to problem solving. (§2.6)
- Describe a Newtonian and non-Newtonian fluid. (§2.7)
- Describe surface tension; solve relevant problems. (§2.8)
- Describe vapor pressure; look up data for water. (§2.9)

2.1 Defining the System

To solve real-world problems, engineers idealize the physical world. One aspect of the engineering process is to create a precise definition of what is being analyzed. A **system** is whatever is being studied or analyzed by the engineer. A system can be a collection of matter, or it can be a region in space. Anything that is not part of the system is considered to be part of the **surroundings**. The **boundary** is the imaginary surface that separates the system from its surroundings.

EXAMPLE. For the flume shown in Fig. 2.1, the water that is situated inside the flume could be defined as the system. For this system, the surroundings would be the flume walls, the air above the flume, etc. Notice that *engineers are specific about what the system is, what the surroundings are, and what boundary is.*

EXAMPLE. Suppose an engineer is analyzing the air flow from a tank being used by a SCUBA diver. As shown in Fig. 2.2, the engineer might select a system comprised of the tank and the regulator. For this system, everything that is external to the tank and regulator is the surroundings. Notice that *the system is defined with a sketch* because this is good professional practice.

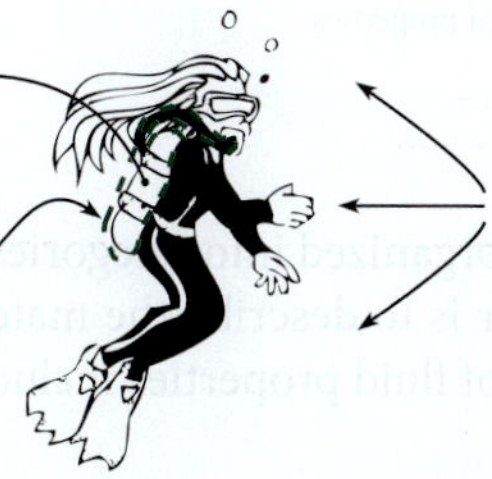

System: What the engineer selects for study (tank plus regulator in this example)

Surroundings: Everything that is not part of the system (in this example, the air bubbles, water, diver, etc.)

Boundary: The surface separating the system and the surroundings (shown by dotted blue line in this example)

FIGURE 2.2

Example of a system, its surroundings, and the boundary.

Engineers select systems in ways that make problem solving the easiest and most correct. Although the choice of system must fit the problem at hand, there are often multiple possibilities for which system to select. This topic will be revisited throughout the text as various kinds of systems are introduced and applied.

Systems are described by specifying numbers that characterize the system. The numbers are called properties. A **property** is a measurable characteristic of a system that depends only on the present conditions within the system.

EXAMPLE. In Fig. 2.2, some examples of properties (i.e., measurable characteristics) are

- The pressure of the air inside the tank
- The density of air inside the tank
- The weight of the system (tank plus air plus regulator)

Some parameters in engineering are measurable, yet they are not properties. For example, work is not a property because the quantity of work depends on how a system interacts with its surroundings. Similarly, neither force nor torque are properties because these parameters depend on the interaction between a system and its surroundings. Heat transfer is not a property. Mass flow rate is not a property.

The **state** of a system means the condition of the system as defined by specifying its properties.

EXAMPLE. Fig. 2.3 shows air being compressed by a piston in a cylinder. The air inside the cylinder is defined as the system. At state 1, the conditions of the system are defined by specifying properties such as pressure, temperature, and density. Similarly, state 2 is defined by specifying these same properties.

The change of a system from one state to another state is called a **process**.

EXAMPLE. When air is compressed (Fig. 2.3), this is a process because the air (i.e., the system) has changed from one set of conditions (state 1) to another set of conditions (state 2). Engineers label processes that commonly occur. For example, an *isothermal process* is one in which the temperature of the system is held constant. For example, an *adiabatic process* is one in which there is no heat transfer between the system and the surroundings.

FIGURE 2.3

Air in a cylinder being compressed by a piston. State 1 is a label for the conditions of the system prior to compression. State 2 is a label for the conditions of the system after compression.

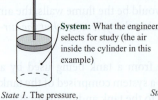

System: What the engineer selects for study (the air inside the cylinder in this example)

State 1. The pressure, temperature, volume, etc. of the air before compression.

State 2. The pressure, temperature, volume, etc. of the air after compression.

State: The condition of a system as specified by giving values of properties

Properties can be organized into categories. One category is called *material properties*. The purpose of this chapter is to describe the material properties of fluids, which are called **fluid properties**. Examples of fluid properties include density, viscosity, and surface tension.

2.2 Characterizing Mass and Weight

Engineers characterize the weight and mass of a fluid with three properties: density, specific weight, and specific gravity.

Mass Density, ρ

Mass Density, ρ (rho), gives the ratio of mass to volume at a point. In particular, select a point (x, y, z) in space and select a small volume ΔV surrounding the point. The mass of the matter within the volume is Δm, and the density is

$$\rho = \left(\frac{\text{mass}}{\text{volume}}\right)_{\text{point}} = \lim_{\Delta V \to 0} \frac{\Delta m}{\Delta V} \tag{2.1}$$

For simplicity, the label *mass density* is shortened to *density* for the remainder of this book.

The reason that density is defined at a point is that density can vary with location.

> **EXAMPLE.** In a lake, the temperature of the water varies with depth. Therefore, the density will also vary with depth because density changes with temperature.

Based on Eq. (2.1), the dimensions of density are

$$[\rho] = \frac{[\text{mass}]}{[\text{volume}]} = \frac{M}{L^3}$$

The SI unit of density is the kilogram per cubic meter (kg/m³). The traditional units of density are slugs per cubic foot (slug/ft³) or pounds-mass per cubic foot (lbm/ft³).

In general, density varies with both temperature and pressure. However for liquids, the density is changed very little by changes in pressure, so engineers assume that density depends on temperature only. Fig. 2.4 compares density values for common liquids.

In Fig. 2.4, notice the following:

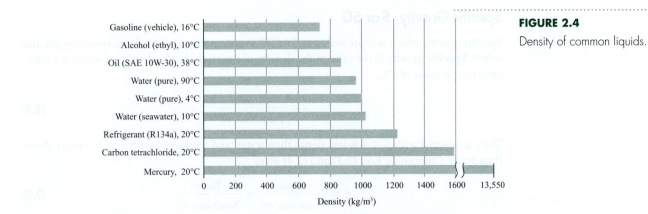

FIGURE 2.4
Density of common liquids.

- When water is heated, the density goes down slightly; in particular, density drops 3.5% for a temperature change of 90°C.
- Most, but not all, liquids have a density within 30% of the density of water.
- Seawater is slightly heavier than freshwater.
- Some liquids, such as oil and gasoline, are lighter than water. When such liquids are immiscible with water, they will float on water.

Because water is common in application, some useful values to memorize are

$$\rho_{\text{water, 4°C}} = 1000 \text{ kg/m}^3 = 1 \text{ kilogram/liter} = 1 \text{ gram/milliliter}$$

For easy reference, these properties along with similar data for air are presented in front of the book in Tables F.3, F.4, and F.5. Additional property data are presented in the appendices in Tables A.2 to A.5.

Specific Weight, γ

Specific weight is represented by the Greek symbol γ (gamma). **Specific weight** is the ratio of weight to volume at a point. In particular, select a point (x, y, z) in space and image a small volume ΔV surrounding the point. The weight of the matter within the volume is ΔW, and the specific weight is:

$$\gamma = \left(\frac{\text{weight}}{\text{volume}} \right)_{\text{point}} = \lim_{\Delta V \to 0} \frac{\Delta W}{\Delta V} \tag{2.2}$$

To relate γ and ρ, recall that weight and mass are related by $W = mg$. Divide this equation by volume to give

$$\gamma = \rho g \tag{2.3}$$

Because water is common in application, a useful value to memorize is

$$\gamma_{\text{water, 15°C}} = 9800 \text{ N/m}^3$$

Other values of γ are presented in Tables F.4 to F.6 (front of book) and in Tables A.2 to A.5 (appendices in back of book).

Specific Gravity, *S* or SG

Specific gravity, which is represented as *S* or SG, is commonly used to characterize liquids and solids. **Specific gravity** is the ratio of the density of a material to the density of water at a reference temperature of 4°C.

$$S = \frac{\rho_{material}}{\rho_{liquid\ water,\ 4°C}} \tag{2.4}$$

Thus, a material with $S < 1$ is less dense than water, and a material with $S > 1$ is more dense than water. Combining Eqs. (2.3) and (2.4) gives

$$S = \frac{\rho_{liquid}}{\rho_{liquid\ water,\ 4°C}} = \frac{\gamma_{liquid}}{\gamma_{liquid\ water,\ 4°C}} \tag{2.5}$$

The properties ρ, γ, and SG are related; if one of these properties is known, the other two can be calculated.

EXAMPLE. Specific weight for mercury is $\gamma_{mercury} = 133$ kN/m³. Calculate the density and specific gravity. Use SI units.

Solution. Applying Eq. (2.3) gives density:

$$\rho_{mercury} = \frac{\gamma_{mercury}}{g} = \frac{(133,000\ \text{N/m}^3)}{(9.81\ \text{m/s}^2)} = 13,600\ \text{kg/m}^3$$

Applying Eq. (2.5) and the reference value for γ_{H_2O} from Table F.6 gives

$$S_{mercury} = \frac{\gamma_{mercury}}{\gamma_{liquid\ water,\ 4°C}} = \frac{(133,000\ \text{N/m}^3)}{(9810\ \text{N/m}^3)} = 13.6$$

Review. To validate the calculated values of ρ and S, one can consult Table A.4. Note that S has no units because it is a ratio.

2.3 Modeling Fluids as Constant Density

Engineers decide if they will idealize a fluid as *constant density* or as *variable density*. This section introduces concepts that are useful for making informed decisions.

The Bulk Modulus of Elasticity

All fluids are compressible. To characterize compressibility, engineers use **bulk modulus of elasticity**, E_v(kPa)

$$E_v = -\frac{dp}{dV/V} = -\frac{\text{change in pressure}}{\text{fractional change in volume}} \tag{2.6}$$

where dp is the differential pressure change, dV is the differential volume change, and V is the volume of fluid. Because dV/V is negative for a positive dp, a negative sign is used in the definition to yield a positive E_v.

The bulk modulus of elasticity of water is approximately 2.2 GN/m², which corresponds to a volume change of about 1/20 of 1% for pressure change of about 1013.25 kPa. Thus, water and most liquids are assumed to be incompressible.

Two useful formulas for the bulk modulus of an ideal gas are

$$E_v = p \quad \text{(isothermal process)}$$

$$E_v = kp \quad \text{(adiabatic process)}$$

(2.7)

where p is pressure and $k = c_p/c_v$ is the specific heat ratio.

EXAMPLE. Compare the compressibility of air and water at room conditions.

Solution. Assume the air is at 100 kPa and that the air is being compressed isothermally. With these assumptions, the bulk modulus of air is $E_v(\text{air}) = 1 \times 10^5$ Pa. The bulk modulus of water is $E_v(\text{water}) = 2.2 \times 10^9$ Pa. Thus,

$$\frac{\text{compressibility (air)}}{\text{compressibility (water)}} = \frac{(1/E_v)_{\text{air}}}{(1/E_v)_{\text{water}}} = \frac{2.2 \times 10^9 \text{ Pa}}{1.0 \times 10^5 \text{ Pa}} = 22{,}000$$

Review. This value (22,000) means that the volume change of air will be 22,000 that of water for the same applied pressure change.

The Constant Density Assumption

Constant density means that the density of a flowing fluid can be assumed to be constant spatially and temporally without causing significant changes (say 5%) in numbers that are calculated.

Because liquids have a high value of bulk modulus, they are commonly assumed to be incompressible. **Incompressible** means that the density of each fluid particle is independent of pressure.

A fluid that is incompressible can still have a **variable density**, meaning that density differs at various points in space or time.

EXAMPLE. When saline and freshwater are mixing as in estuaries, density variations occur even though the water can assumed to be incompressible.

Regarding gases, it is common to assume that a flowing gas has a constant density. The reason this assumption works is that pressure variations within the flow are not large enough to cause significant density variations.

High-speed flows of gases, such as the flow around a jet airplane, need to be modeled as compressible flows (see Chapter 12). To distinguish *constant density gas flow* from *variable density gas flow*, engineers use the Mach number M. The Mach number is the ratio of the speed of the flowing fluid V to the speed at which sound travels in the fluid c:

$$\text{Mach number} = \text{M} \equiv \frac{V}{c}$$

A *criterion for idealizing a gas as constant density* is:

$$(\text{M} < 0.3)$$

(2.8)

When flow is steady and Eq. (2.8) is satisfied, the density variation is less than 5% (2).

EXAMPLE. To gain a feel what Eq. (2.8) means, consider air at 20°C. The speed of sound is $c \approx 340$ m/s. Thus, a flow of air can be assumed to be incompressible for $V < \approx 100$ m/s (220 mph). Because this is quite fast, the majority of gas flows in industrial applications can be idealized as constant density.

Both liquids and gases can have significant density variations when the fluid is being heated or cooled. Common engineering practice is to assume constant density and then look up property values at an appropriate average temperature.

EXAMPLE. When liquid water enters a heat exchanger at 20°C and exits at 80°C, common practice is to assume constant density and look up a value at 50°C. This value, from Table A.5, is $\rho = 988$ kg/m^3.

In summary, it is common to assume constant density when solving fluid mechanics problems. Most problems and methods in this text are based on this assumption. An important exception is the high-speed flow of gases, a topic presented in Chapter 12.

2.4 Finding Fluid Properties

One can look up fluid properties in engineering handbooks, textbooks, or off the Internet. A fluid property often depends on the temperature and pressure of the fluid. Thus, it is good engineering practice to document as shown in Fig. 2.5. Six aspects of good practice are

1. List the name of the fluid.
2. List the temperature and pressure at which the property was reported by the source.
3. Cite the source of the fluid property.
4. List relevant assumptions.
5. List the value and units of the fluid property.
6. Be concise; write down the minimum information required to get the job done.

FIGURE 2.5

Recommended practices for documenting fluid properties.

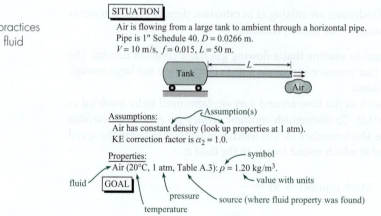

SITUATION

Air is flowing from a large tank to ambient through a horizontal pipe.
Pipe is 1" Schedule 40. $D = 0.0266$ m.
$V = 10$ m/s, $f = 0.015$, $L = 50$ m.

Tank — L — Air

Assumptions: Assumption(s)
Air has constant density (look up properties at 1 atm).
KE correction factor is $\alpha_2 = 1.0$.

Properties: symbol
Air (20°C, 1 atm, Table A.3): $\rho = 1.20$ kg/m^3.
 value with units

fluid GOAL
pressure source (where fluid property was found)
temperature

While looking up fluid properties, many details are important. The details are summarized in Table 2.2 (page 53). As shown in the next example, Table 2.2 will be used throughout this chapter.

EXAMPLE. Where are values of specific gravity (SG) tabulated? Does specific gravity depend on temperature, on pressure, or on both?

Solution. Table 2.2 shows that values of SG are tabulated in Table A.4. Table 2.2 indicates that SG goes down as temperature increases and that SG is constant with pressure.

2.5 Describing Viscous Effects

Viscous effects influence energy loss, drag force, flow separation, and other parameters of interest. Thus, this section introduces concepts that are useful for describing and characterizing viscous effects.

Viscosity

Viscosity, μ (mu), is the fluid property that characterizes resistance to flow.

EXAMPLE. Fluids resist being forced to flow through pipes, so pumps are added to drive the flow through the pipe. For the same flow rate, a fluid with high viscosity (e.g., molasses) will require more power from a pump than a fluid with low viscosity (e.g., water).

EXAMPLE. Fluids resist the motion of immersed objects through them. A small force will easily push a spoon through a bowl of water. This same force will barely move a spoon through a bowl of honey because the viscosity of honey is much higher than the viscosity of water.

Viscosity is also referred to as *absolute viscosity* and *dynamic viscosity*. Viscosity is defined mathematically as the ratio of shear stress to the rate of shear strain at a point.

$$\text{viscosity}(\mu) \equiv \frac{\text{shear stress}}{\text{rate of shear strain}} = \frac{\tau}{\left(\dfrac{dV}{dy}\right)} \tag{2.9}$$

The symbols on the right side of Eq. (2.9) are described in the next subsections. Eq. (2.9) is called *The Viscosity Equation* in this textbook. Other engineering references call this equation "*Newton's Law of Viscosity*."

Shear Force and Shear Stress

Viscosity leads to forces that are analogous to frictional forces. For example, when fluid flows past a flat plate as shown in Fig. 2.6 the flowing fluid causes a drag force that is called the shear force.

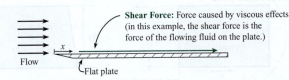

Shear Force: Force caused by viscous effects (in this example, the shear force is the force of the flowing fluid on the plate.)

Flow

Flat plate

FIGURE 2.6

Viscosity causes a force that is called the shear force.

The shear force is a *distributed force* meaning that the force is spread out over an area. Because the force per unit area is not the same at each x location, a concept called shear stress is used. **Shear stress** is the ratio of tangential force to area at a point on a surface:

$$\text{shear stress}(\tau) \equiv \left(\frac{\text{tangential force}}{\text{surface area}}\right)_{\text{point on a surface}} = \lim_{\Delta A \to 0} \frac{\Delta F_{\text{tangential}}}{\Delta A} \tag{2.10}$$

The terms in Eq. (2.10) are illustrated in Fig. 2.7.

Now that τ has been introduced, we can define **shear force** F_s as the net force on a body due to shear stress acting over the body.

FIGURE 2.7

This sketch shows terms that appear in the definition of shear stress.

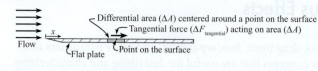

Rate of Shear Strain

Viscosity causes a fluid particle to continuously change shape or deform. This deformation is illustrated in Fig. 2.8. As shown, a fluid particle that is rectangular at time t will deform so that it is nonrectangular at time $t + \Delta t$. The deformation occurs because the fluid at the top of the particle is moving faster than the fluid at the bottom of the particle. In particular the fluid at the top is moving with speed $V + \Delta V$, and the fluid at the bottom is moving with speed V. This change in velocity over distance (called velocity gradient) is linked to deformation of the fluid particle.

FIGURE 2.8

Depiction of strain caused by a shear stress (force per area) in a fluid. The rate of strain is the rate of change of the interior angle of the original rectangle.

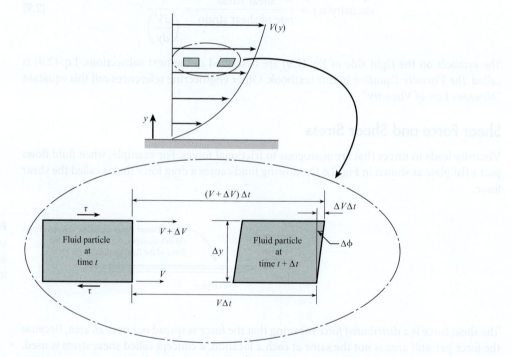

The **rate of shear strain** describes the change in an angle of a particle as a function of time. The mathematical definition is

$$\text{rate of shear strain} \equiv \lim_{\Delta t \to 0} \frac{\Delta \phi}{\Delta t} \qquad (2.11)$$

where the angle $\Delta\phi$ is defined in the lower sketch of Fig. 2.8. To evaluate Eq. (2.11), use the sketch in Fig. 2.8 to write

$$\tan(\Delta\phi) = \frac{\Delta V \Delta t}{\Delta y} \tag{2.12}$$

In the limit as $\Delta t \to 0$, apply the small angle approximation to write

$$\tan(\Delta\phi) \approx \Delta\phi \tag{2.13}$$

Combine Eqs. (2.11) to (2.13) to give

$$\text{rate of shear strain} \equiv \lim_{\Delta t \to 0} \frac{\Delta\phi}{\Delta t} = \lim_{\Delta t \to 0} \frac{\Delta V}{\Delta y} = \frac{dV}{dy} \tag{2.14}$$

Eq. (2.14) shows that the rate of shear strain of a fluid particle is equal to the derivative of the velocity with respect to distance. The derivative on the right side of Eq. (2.14) is called the *velocity gradient*, which is the next topic.

The Velocity Profile

Viscous effects cause the velocity of a flowing fluid to vary with distance y as shown in Fig. 2.9. Notice that y measures distance from the wall. The variation of velocity with distance is called a **velocity profile**. The change in velocity is often called a **velocity gradient** because the gradient in mathematics describes a change in a variable with respect to distance.

··

FIGURE 2.9

Viscosity causes fluid near a wall to slow down, thereby creating the no-slip condition and a velocity profile. The velocity profile causes a fluid particle to experience shear stresses as shown.

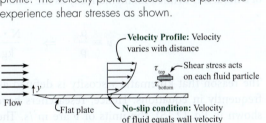

The presence of a velocity gradient indicates that shear stresses act on fluid particles.

EXAMPLE. Fig. 2.9 shows a fluid particle (blue shading) situated in a velocity profile. The fluid above this particle is moving fast relative to the particle, and this causes the particle to experience a force that acts to the right. Similarly, the fluid below this particle is moving slower, and this causes the particle to experience a force acting to the left. These forces can be represented as shear stresses.

The No-Slip Condition

In addition to causing shear stress, viscosity causes the **no-slip condition**, which is labeled in Fig. 2.9. The no-slip condition, originally deduced from experiments, tells us that the *velocity of fluid in contact with a solid will equal the velocity of the solid.* Therefore, in Fig. 2.9 the velocity of the fluid at the surface of the plate will equal zero: $V(y = 0) = 0$. If a body is moving, for example a wing moving through the air, the velocity of the air at a point on the wing surface will equal the velocity of the wing at this same point.

Finding Values of Viscosity

Table 2.2 (page 53) summarizes information for looking up viscosity values. The following example illustrates how to use this table.

EXAMPLE. Find the *dynamic viscosity* of air at a pressure of 506.625 kPa and a temperature of 120°C. Assume that air at these conditions is an ideal gas.

Ideas/Plan. Table 2.2 (page 53) indicates that (a) viscosity can be looked up in Table A.3, and (b) viscosity will vary with temperature, but not pressure. Thus look up viscosity at $T = 120$°C and $p = 1$ atm.

Action. From Table A.3, the viscosity of air at $T = 120$°C and $p = 1$ atm is $\mu = 2.26 \times 10^{-5}$ N · s/m².

✔ **CHECKPOINT PROBLEM 2.1**

An inventor is considering two lubricants: glycerin and SAE 10W-30 oil.
 (a) Which lubricant has a higher viscosity at 338.71 K?
 (b) Which lubricant has a higher viscosity at 383.15 K?

Kinematic Viscosity, v

Kinematic viscosity, v (nu), is a property that combines the viscous and the mass characteristics of a fluid. It is defined mathematically as the ratio of viscosity to density:

$$v \equiv \frac{\mu}{\rho} \quad \Rightarrow \quad \frac{N \cdot s/m^2}{kg/m^3} = m^2/s \qquad (2.15)$$

The reason that kinematic viscosity is defined as a property is that the ratio μ/ρ occurs frequently in equations. Hence, researchers have identified it as a distinct property. As shown in Eq. (2.15), the units of v are m²/s. The units can be helpful for distinguishing kinematic viscosity from viscosity. Be careful not to mix up μ and v; they are different properties!

EXAMPLE. Find the *kinematic viscosity* of water at a pressure of 506.625 kPa and a temperature of 80°C.

Ideas/Plan. From Table 2.2 (page 53), the kinematic viscosity of water can be found in Table A.5. Also, the kinematic viscosity of a liquid is independent of pressure.

Action. From Table A.5, the kinematic viscosity of water at $T = 80$°C and $p = 101.3$ kPa is $v = 3.64 \times 10^{-7}$ m²/s.

✔ **CHECKPOINT PROBLEM 2.2**

What is the kinematic viscosity of nitrogen at 709.275 kPa of pressure (absolute) and a temperature of 15°C?

2.6 Applying the Viscosity Equation

This section shows how to solve problems when shear stress is a parameter. The working equation is the viscosity equation, Eq. (2.9), which is usually written as

$$\tau = \mu \frac{dV}{dy} \tag{2.16}$$

The equation tells us that shear stress, τ, in a flowing fluid is linearly related to the velocity gradient (dV/dy). The constant of proportionality is the viscosity (μ). Terms in Eq. (2.16) are summarized in Table 2.1.

TABLE 2.1 Summary of the Viscosity Equation

Equation		Terms
$\tau = \mu \dfrac{dV}{dy}$	(2.16)	τ = shear stress (N/m²)
		μ = viscosity (Pa · s) (also called dynamic viscosity or absolute viscosity)
		$\dfrac{dV}{dy}$ = velocity gradient (s⁻¹) (also called the rate of shear strain)

One type of problem involves specifying two of the three variables in the viscosity equation and asking for the third variable. This case is illustrated in Example 2.1.

EXAMPLE 2.1

Applying the Viscosity Equation to Calculate Shear Stress in a Poiseuille Flow

Problem Statement

A famous solution in fluid mechanics, called Poiseuille flow, involves laminar flow in a round pipe (See Chapter 10 for details). Consider Poiseuille flow with a velocity profile in the pipe given by

$$V(r) = V_o(1 - (r/r_o)^2)$$

where r is radial position as measured from the centerline, V_o is the velocity at the center of the pipe, and r_o is the pipe radius. Find the shear stress at the center of the pipe, at the wall, and where $r = 1$ cm. The fluid is water (15°C), the pipe diameter is 4 cm, and $V_o = 1$ m/s.

Define the Situation

Water flows in a round pipe (Poiseuille flow).

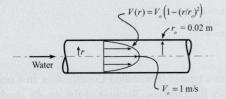

Water (15°C, 1 atm, Table A.5): $\mu = 1.14 \times 10^{-3}$ N · s/m².

State the Goal

Calculate the shear stress at three points:

$\tau(r = 0.00$ m) (N/m²) ⬅ pipe centerline

$\tau(r = 0.01$ m) (N/m²) ⬅ middle of the pipe

$\tau(r = 0.02$ m) (N/m²) ⬅ the wall

Generate Ideas and Make a Plan

Because the goal is τ, select the *Viscosity Equation* (Eq. 2.16).

Let the position variable be r instead of y.

$$\tau = -\mu \frac{dV}{dr} \tag{a}$$

Regarding the minus sign in Eq. (a), the y in the viscosity equation is measured from the wall (see Fig. 2.9) The coordinate r is in the opposite direction. The sign change occurs when the variable is changed from y to r.

To find μ, use Table 2.2 on page 53 to identify that

• Viscosity of water at 15°C can be found in Table F.5.

• Viscosity of a liquid is independent of pressure.

To find the velocity gradient in Eq. (a), differentiate the given velocity profile.

$$\frac{dV(r)}{dr} = \frac{d}{dr}(V_o(1 - (r/r_o)^2)) = \frac{-2V_o r}{r_o^2} \tag{b}$$

Now, the goal can be found. **Plan.** Apply Eq. (b) to find the velocity gradient. Then, substitute into Eq. (a).

Take Action (Execute the Plan)

1. Viscosity Equation ($r = 0$ m)

$$\left.\frac{dV(r)}{dr}\right|_{r=0\,m} = \frac{-2V_o(0\,m)}{r_o^2} = \frac{-2(1\,m/s)(0\,m)}{(0.02\,m)^2} = 0.0\,s^{-1}$$

$$\tau(r = 0\,m) = -\mu\left.\frac{dV(r)}{dr}\right|_{r=0\,m}$$

$$= (1.14 \times 10^{-3}\,N \cdot s/m^2)(0.0\,s^{-1})$$

$$= \boxed{0.0\,N/m^2}$$

2. Viscosity Equation ($r = 0.01$ m)

$$\left.\frac{dV(r)}{dr}\right|_{r=0.01\,m} = \frac{-2V_o(0.01\,m)}{r_o^2}$$

$$\frac{-2(1\,m/s)(0.01\,m)}{(0.02\,m)^2} = -50\,s^{-1}$$

Next, calculate shear stress.

$$\tau(r = 0.01\,m) = -\mu\left.\frac{dV(r)}{dr}\right|_{r=0.01\,m}$$

$$= (1.14 \times 10^{-3}\,N \cdot s/m^2)(50\,s^{-1})$$

$$= \boxed{0.0570\,N/m^2}$$

3. Viscosity Equation ($r = 0.02$ m)

$$\left.\frac{dV(r)}{dr}\right|_{r=0.02\,m} = \frac{-2V_o(0.02\,m)}{r_o^2}$$

$$= \frac{-2(1\,m/s)(0.02\,m)}{(0.02\,m)^2} = -100\,s^{-1}$$

Next, calculate shear stress.

$$\tau(r = 0.02\,m) = -\mu\left.\frac{dV(r)}{dr}\right|_{r=0.02\,m}$$

$$= (1.14 \times 10^{-3}\,N \cdot s/m^2)(100\,s^{-1})$$

$$= \boxed{0.114\,N/m^2}$$

Review the Solution and the Process

1. *Tip.* On most problems, including this example, carrying and canceling units is useful, if not critical.

2. *Notice.* Shear stress varies with location. For this example, τ is zero on the centerline of the flow and nonzero everywhere else. The maximum value of shear stress occurs at the wall of the pipe.

3. *Notice.* For flow in a round pipe, the viscosity equation has a minus sign and uses the position coordinate r.

$$\tau = -\mu\frac{dV}{dr}$$

Example 2.1 showed that the *magnitude of shear stress is proportional the velocity gradient.* This idea is illustrated in Fig. 2.10. Notice that the figure is drawn vertically so that the variable V is upward and the variable r is horizontal. This was done so that a slope of zero is a horizontal line and an infinite slope is a vertical line. Another way to show the relationship between slope and shear stress is an equation.

$$\tau\uparrow = \mu\frac{dV}{dr}\uparrow \qquad (2.17)$$

Eq. (2.17) means that if dV/dy increases, then shear stress τ will increase. Or, that if dV/dy decreases then τ will decrease.

FIGURE 2.10

The velocity profile from Example 2.1. This figure shows that one can make qualitative predictions of the shear stress by examining the slope of the velocity profile.

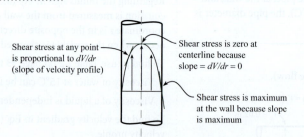

Shear stress at any point is proportional to dV/dr (slope of velocity profile)

Shear stress is zero at centerline because slope = dV/dr = 0

Shear stress is maximum at the wall because slope is maximum

A second category of problems involves a type of flow, called Couette flow. In Couette flow, as shown in Fig. 2.11, a moving plate causes fluid to flow. Because of the *no-slip condition*, the velocity of the fluid at the top is equal to the velocity of the moving plate. Similarly, the velocity of the fluid at the bottom is zero because the bottom plate is stationary. In the region between the plates, the velocity profile is linear. Additional details about Couette flow are presented in Chapter 9.

When the viscosity equation (Eq. 2.16) is applied to the Couette flow that is shown in Fig. 2.11, the derivative can be replaced with a ratio because the velocity gradient is linear.

$$\tau = \mu \frac{dV}{dy} = \mu \frac{\Delta V}{\Delta y} \qquad (2.18)$$

The terms on the right side of Eq. (2.18) can be analyzed as follows

$$\tau = \mu \frac{\Delta V}{\Delta y} = \mu \frac{V_o - 0}{H - 0} = \mu \frac{V_o}{H}$$

Rewrite the equation to give a form of the *Viscosity Equation* that is useful for Couette flow.

$$\tau \big|_{\text{Couette Flow}} = \text{constant} = \mu \frac{V_o}{H} \qquad (2.19)$$

Eq. (2.19) reveals that the shear stress at all points in a Couette flow is constant with a magnitude of $\mu V_o/H$. Example 2.2 presents a typical problem that involves Couette flow.

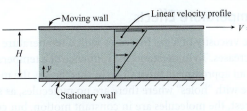

FIGURE 2.11

Couette flow is a flow that is driven by a moving wall. The velocity profile in the fluid is linear.

EXAMPLE 2.2

Applying the Viscosity Equation to Couette Flow

Problem Statement

A board 1 m by 1 m that weighs 25 N slides down an inclined ramp (slope = 20°) with a constant velocity of 2.0 cm/s. The board is separated from the ramp by a thin film of oil with a viscosity of 0.05 N · s/m². Assuming that the oil can be modeled as a Couette flow, calculate the space between the board and the ramp.

Define the Situation

A board slides down an oil film on a inclined plane.

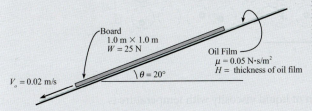

Assumptions. (1) Couette flow. (2) Board has constant speed.

State the Goal

H(mm) ◀ Thickness of the film of oil

Generate Ideas and Make a Plan

Because the goal is H, apply the *Viscosity Equation* (Eq. 2.19):

$$H = \mu \frac{V_o}{\tau} \qquad (a)$$

To find the shear stress τ in Eq. (a), draw a *Free Body Diagram* (FBD) of the board. In the FBD, W is the weight, N is the normal force, and F_{shear} is shear force. Because shear stress is constant with x, the shear force can be expressed as $F_{shear} = \tau A$.

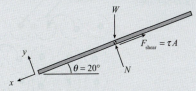

Because the board moves at constant speed, the forces are in balance. Thus, apply *force equilibrium*.

$$\sum F_x = 0 = W \sin\theta - \tau A \qquad (b)$$

Rewrite Eq. (b) as

$$\tau = (W \sin\theta)/A \qquad (c)$$

Eq. (c) can be solved for τ. The plan is

1. Calculate τ using Force Equilibrium (Eq. c).
2. Calculate H using the Shear Stress Equation (Eq. a).

Take Action (Execute the Plan)

1. Force equilibrium

$$\tau = (W\sin\theta)/A = (25 \text{ N})(\sin 20°)/(1.0 \text{ m}^2) = 8.55 \text{ N/m}^2$$

2. Shear stress equation

$$H = \mu\frac{V_o}{\tau} = (0.05 \text{ N} \cdot \text{s/m}^2)\frac{(0.02 \text{ m/s})}{(8.55 \text{ N/m}^2)} = \boxed{0.117 \text{ mm}}$$

Review the Solution and the Process

1. *H* is about 12% of a millimeter; this is quite small.

2. *Tip.* Solving this problem involved drawing an FBD. The FBD is useful for most problems involving Couette flow.

2.7 Characterizing Viscosity

This section presents ideas about developing equations for viscosity as a function of temperature. This section also introduces the non-Newtonian fluid.

Temperature Effects

The viscosity of a gas increases with a temperature rise. In comparison, the viscosity of liquid decreases. To understand the influence of a temperature change on a liquid, it is helpful to rely on an approximate theory (3). In this theory, the molecules in a liquid form a latticelike structure with "holes" where there are no molecules, as shown in Fig. 2.12. Even when the liquid is at rest, the molecules are in constant motion, but confined to cells, or "cages." The cage or lattice structure is caused by attractive forces between the molecules. The cages may be thought of as energy barriers. When the liquid is flowing, there is a shear stress, τ, imposed by one layer on another in the fluid. This shear stress assists a molecule in overcoming the energy barrier so that it can more easily move into the next hole. At a higher temperature the size of the energy barrier is smaller, and it is easier for molecules to make the jump, so that the net effect is that the fluid has a a lower viscosity.

FIGURE 2.12

Visualization of molecules in a liquid.

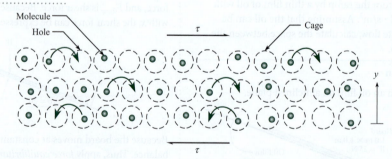

An equation for the variation of liquid viscosity with temperature is

$$\mu = Ce^{b/T} \tag{2.20}$$

where *C* and *b* are empirical constants that require viscosity data at two temperatures for evaluation. This equation should be used primarily for data interpolation.

The variation of viscosity (dynamic and kinematic) for other fluids is given in Figs. A.2 and A.3. Example 2.3 shows how to find the constants that appear in Eq. (2.20).

EXAMPLE 2.3

Developing an Algebraic Equation for Viscosity of a Liquid as a Function of Temperature

Problem Statement

The dynamic viscosity of water at 293 K is 1.00×10^{-3} N · s/m^2, and the viscosity at 313 K is 6.53×10^{-4} N · s/m^2. Using Eq. (2.20), estimate the viscosity at 303 K.

Define the Situation

An equation for the viscosity of water can be found by using Eq. (2.20) plus the following two data points:

- Water (293 K, 1 atm): $\mu = 1.00 \times 10^{-3}$ N · s/m^2.
- Water (313 K, 1 atm): $\mu = 6.53 \times 10^{-4}$ N · s/m^2.

State the Goal

1. Find an equation for $\mu(T)$ for water.
2. Calculate the viscosity of water at $T = 303$ K = 30°C.

Generate Ideas and Make a Plan

Because Eq. (2.20) has two unknown constants (C and b), and there are two known values for viscosity, the two constants can be found. The plan is

1. Linearize Eq. (2.20) by taking the logarithm.
2. Plug values of μ and T into the linearized equation.
3. Solve for C and b using the linear equations from step 2.
4. Find the equation for $\mu(T)$ and then find $\mu(T = 303$ K).

Take Action (Execute the Plan)

1. Take logarithms of both sides of Eq. (2.20)

$$\ln \mu = \ln C + b/T \tag{a}$$

2. Plug data points into Eq. (a)

$$-6.908 = \ln C + 0.00341\, b$$
$$-7.334 = \ln C + 0.00319\, b \tag{b}$$

3. Solve equations in step 2 for C and b

$$\ln C = -13.51 \qquad b = 1936 \text{ K}$$
$$C = e^{-13.51} = 1.357 \times 10^{-6} \text{ (N · s/m}^2)$$

4. Substitute C and b into Eq. (2.20)

$$\boxed{\mu(T) = (1.357 \times 10^{-6} \text{ N · s/m}^2)e^{(1936 \text{ K})/T}} \tag{c}$$

Solve for viscosity at 303 K.

$$\mu = \boxed{8.08 \times 10^{-4} \text{ N · s/m}^2}$$

Review the Solution and the Process

1. *Validation.* The calculated value can be checked by comparing to data in Table A.5. The result differs by 1% from the table value.

2. *Tip.* This solution required absolute temperature units of Kelvin (K). Some problems cannot be solved correctly if one uses temperature units in Celsius (°C).

3. *Tip.* Notice how units were applied in this solution. Using units is good engineering practice.

As compared to liquids, gases do not have zones or cages to which molecules are confined by intermolecular bonding. Gas molecules are always undergoing random motion. If this random motion of molecules is superimposed on two layers of gas, where the top layer is moving faster than the bottom layer, periodically a gas molecule will randomly move from one layer to the other. This behavior of a molecule in a low-density gas is analogous to people jumping back and forth between two conveyor belts moving at different speeds, as shown in Fig. 2.13. When people jump from the high-speed belt to the low-speed belt, a restraining (or braking) force has to be applied to slow the person down (analogous to viscosity). If the people are heavier, or are moving faster, a greater braking force must be applied. This analogy also applies for gas molecules translating between fluid layers where a shear force is needed to maintain the layer speeds. As the gas temperature increases, more of the molecules will be making random jumps. Just as the jumping person causes a braking action on the belt, highly mobile gas molecules have momentum, which must be resisted by the layer to which the molecules jump. Therefore, as the temperature increases, the viscosity, or resistance to shear, also increases.

FIGURE 2.13

Analogy of people moving between conveyor belts and gas molecules translating between fluid layers.

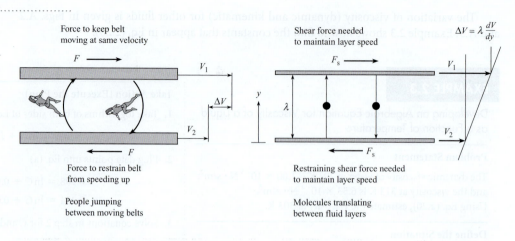

An estimate for the variation of gas viscosity with temperature is **Sutherland's equation**,

$$\frac{\mu}{\mu_0} = \left(\frac{T}{T_0}\right)^{3/2} \frac{T_0 + S}{T + S} \tag{2.21}$$

where μ_0 is the viscosity at temperature T_0, and S is Sutherland's constant. All temperatures are absolute. Sutherland's constant for air is 111 K; values for other gases are given in Table A.2. Using Sutherland's equation for air yields viscosities with an accuracy of $\pm 2\%$ for temperatures between 170 K and 1900 K. In general, the effect of pressure on the viscosity of common gases is minimal for pressures less than 1013.25 kPa.

Newtonian versus non-Newtonian Fluids

Fluids for which the shear stress is directly proportional to the rate of strain are called **Newtonian fluids**. Because shear stress is directly proportional to the shear strain, dV/dy, a plot relating these variables (see Fig. 2.14) results in a straight line passing through the origin. The slope of this line is the value of the dynamic (absolute) viscosity. For some fluids the

FIGURE 2.14

Shear stress relations for different types of fluids.

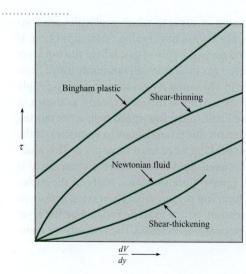

shear stress may not be directly proportional to the rate of strain; these are called **non-Newtonian fluids**. One class of non-Newtonian fluids, shear-thinning fluids, has the interesting property that the ratio of shear stress to shear strain decreases as the shear strain increases (see Fig. 2.14). Some common shear-thinning fluids are toothpaste, catsup, paints, and printer's ink. Fluids for which the viscosity increases with shear rate are shear-thickening fluids. Some examples of these fluids are mixtures of glass particles in water and gypsum-water mixtures. Another type of non-Newtonian fluid, called a Bingham plastic, acts like a solid for small values of shear stress and then behaves as a fluid at higher shear stress. The shear stress versus shear strain rate for a Bingham plastic is also shown in Fig. 2.14.

In general, non-Newtonian fluids have molecules that are more complex than Newtonian fluids. Thus, if you are working with a fluid that may be non-Newtonian, consider doing some research. The reason is that many of the equations and math models presented in textbooks (including this one) only apply to Newtonian fluids.

To learn more about non-Newtonian fluids, watch the film entitled *Rheological Behavior of Fluids* (4). For more information on the theory of flow of non-Newtonian fluids, see references (5) and (6).

2.8 Characterizing Surface Tension*

Engineers need to be able to predict and characterize surface tension effects because they affect many industrial problems. Some examples of surface tension effects:

- *Wicking.* Water will wick into a paper towel. Ink will wick into paper. Polypropylene, an excellent fiber for cold-weather aerobic activity, wicks perspiration away from the body.
- *Capillary Rise.* A liquid will rise in a small-diameter tube. Water will rise in soil.
- *Capillary Instability.* A liquid jet will break up into drops.
- *Drop and Bubble Formation.* Water on a leaf beads up. A leaky faucet drips. Soap bubbles form.
- *Excess Pressure*: The pressure inside a water drop is higher than ambient pressure. The pressure inside a vapor bubble during boiling is higher than ambient pressure.
- *Walking on Water.* The water strider, an insect, can walk on water. Similarly, a metal paper clip or a metal needle can be positioned to float (through the action of surface tension) on the surface of water.
- *Detergents.* Soaps and detergents improve the cleaning of clothes because they lower the surface tension of water so that the water can more easily wick into the pores of the fabric.

Many experiments have shown that the surface of liquid behaves like a stretched membrane. The material property that characterizes this behavior is **surface tension**, σ (sigma). Surface tension can be expressed in terms of force:

$$\text{Surface Tension}(\sigma) = \frac{\text{Force along an interface}}{\text{Length of the interface}} \qquad \textbf{(2.22)}$$

---

*The authors acknowledge and thank Dr. Eric Aston for his feedback and inputs on this section. Dr. Aston is a Chemical Engineering Professor at the University of Idaho.

FIGURE 2.17

Water wets glass because adhesion is greater than cohesion. Wetting is associated with a contact angle less than 90°.

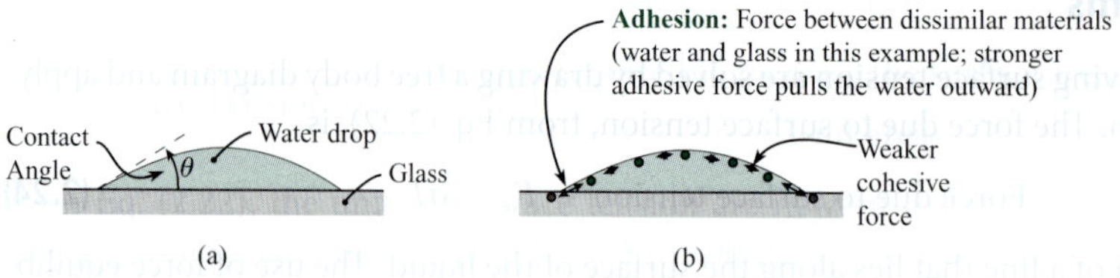

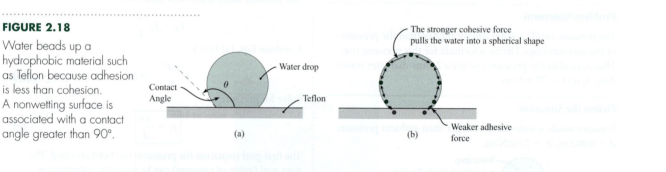

(a)   (b)

On some surfaces such as Teflon and wax paper, a drop of water will bead up (Fig 2.18) because *adhesion between the water and teflon is less than cohesion of the water*. A surface in which water beads up is called *hydrophobic* (water hating). Surface such as glass in which water drops spread out are called *hydrophillic* (water loving).

FIGURE 2.18

Water beads up a hydrophobic material such as Teflon because adhesion is less than cohesion. A nonwetting surface is associated with a contact angle greater than 90°.

(a)   (b)

Capillary action describes the tendency of a liquid to rise in narrow tubes or to be drawn into small openings. Capillary action is responsible for water being drawn into the narrow openings in soil or into the narrow openings between the fibers of a dry paper towel.

When a capillary tube is placed into a container of water, the water rises up the tube (Fig 2.19) because the adhesive force between the water and the glass pulls the water up the tube. This is called capillary rise. Notice how the contact angle for the water is the same in Figs. 2.17 and 2.19. Alternatively, when a fluid is nonwetting such as mercury on glass, then the liquid will display capillary repulsion.

FIGURE 2.19

Water will rise up a glass tube (capillary rise), whereas mercury will move downward (capillary repulsion).

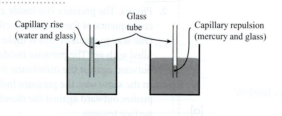

To derive an equation for capillary rise (see Fig. 2.20), define a system comprised of the water inside the capillary tube. Then, draw a free body diagram (FBD). As shown, the pull of surface tension is lifting the column of water. Applying force equilibrium gives

$$\text{Weight} = \text{Surface Tension Force}$$

$$\gamma \left(\frac{\pi d^2}{4} \right) \Delta h = \sigma \pi d \cos \theta \qquad \textbf{(2.25)}$$

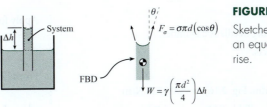

FIGURE 2.20

Sketches used for deriving an equation for capillary rise.

Assume the contact angle is nearly zero so cos θ ≈ 1.0. Note that this is a good assumption for a water/glass interface. Eq. (2.25) simplifies to

$$\Delta h = \frac{4\sigma}{\gamma d} \qquad (2.26)$$

EXAMPLE. Calculate the capillary rise for water (20°C) in a glass tube of diameter $d = 1.6$ mm.

Solution. From Table A.5, $\gamma = 9790$ N/m³. From Fig. 2.16, $\sigma = 0.0728$. Now, calculate capillary rise using Eq. (2.26):

$$\Delta h = \frac{4(0.0728 \text{ N/m})}{(9790 \text{ N/m}^3)(1.6 \times 10^{-3} \text{ m})} = \boxed{18.6 \text{ mm}}$$

✔ CHECKPOINT PROBLEM 2.3

Two capillary tubes are placed in a liquid. The diameter of tube A is twice the diameter of tube B. Which statement is true?

    a. Capillary rise in both tubes is the same.

    b. Capillary rise in tube A is twice that of tube B.

    c. Capillary rise in tube B is twice that of tube A.

    d. None of the above.

Example 2.5 shows a case involving a non-wetting surface.

EXAMPLE 2.5

Applying Force Equilibrium to Determine the Size of a Sewing Needle That Can Be Supported by Surface Tension

Problem Statement

The Internet shows examples of sewing needles that appear to be "floating" on top of water. This effect is due to surface tension supporting the needle. Determine the largest diameter of sewing needle that can be supported by water. Assume that the needle material is stainless steel with $SG_{ss} = 7.7$.

Define the Situation

A sewing needle is supported by the surface tension of a water surface.

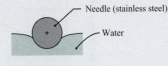

Assumptions
- Assume the sewing needle is a cylinder.
- Neglect end effects.

Properties
- Water (20°C, 1 atm, Fig. 2.16): $\sigma = 0.0728$ N/m
- Water (4°C, 1 atm, Table F.6): $\gamma_{H2O} = 9810$ N/m³
- SS: $\gamma_{ss} = (7.7)(9810$ N/m³$) = 75.5$ kN/m³

State the Goal

d(mm) ◀ Diameter of the largest needle that can be supported by the water.

Generate Ideas and Make a Plan

Because the weight of the needle is supported by the surface tension force, draw a *Free Body Diagram* (FBD). Select a system comprised of the needle plus the surface layer of the water. The FBD is

Apply force equilibrium.

Force due to surface tension = Weight of needle
$$F_\sigma = W \tag{a}$$
From Eq. (2.24)
$$F_\sigma = \sigma 2L \cos\theta \tag{b}$$
where L is the length of the needle. The weight of the needle is
$$W = \left(\frac{\text{weight}}{\text{volume}}\right)[\text{volume}] = \gamma_{ss}\left[\left(\frac{\pi d^2}{4}\right)L\right] \tag{c}$$

Combine Eqs. (a), (b), and (c). Also, assume the angle θ is zero because this gives the maximum possible diameter:
$$\sigma 2L = \gamma_{ss}\left(\frac{\pi d^2}{4}\right)L \tag{d}$$

Plan. Solve Eq. (d) for d and then plug numbers in.

Take Action (Execute the Plan)

$$d = \sqrt{\frac{8\sigma}{\pi\gamma_{ss}}} = \sqrt{\frac{8(0.0728 \text{ N/m})}{\pi(75.5 \times 10^3 \text{ N/m}^3)}} = \boxed{1.57 \text{ mm}}$$

Review the Solution and the Process

Notice. When applying specific gravity, look up water properties at the reference temperature of 4°C.

2.9 Predicting Boiling Using Vapor Pressure

A liquid, even at a low temperature, can boil as it flows through a system. This boiling can reduce performance and damage equipment. Thus, engineers need to be able to predict when boiling will occur. This prediction is based on the vapor pressure.

    Vapor pressure, p_v(kPa), is the pressure at which the liquid phase and the vapor phase of a material will be in thermal equilibrium. Vapor pressure is also called *saturation pressure,* and the corresponding temperature is called *saturation temperature.*

    Vapor pressure can be visualized on a *phase diagram.* A phase diagram for water is shown in Fig. 2.21. As shown, water will exist in the liquid phase for any combination of temperature and pressure that lies above the blue line. Similarly, the water will exist in the vapor phase for points below the blue line. Along the blue line, the liquid and vapor phases are in thermal equilibrium. When boiling occurs, the pressure and temperature of the water will be given by one of the points on the blue line. In addition to Fig. 2.21, data for vapor pressure of water are tabulated in Table A.5.

FIGURE 2.21

A phase diagram for water.

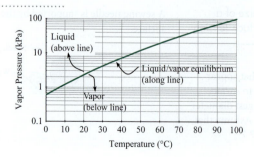

EXAMPLE. Water at 20°C flows through a venturi nozzle and boils. Explain why. Also, give the value of pressure in the nozzle.

Solution. The water is boiling because the pressure has dropped to the vapor pressure. Table 2.2 (page 53) indicates that p_v can be looked up in Table A.5. Thus, the vapor pressure at 20°C (Table A.5) is $p_v = 2.34$ kPa absolute. This value can be validated by using Fig. 2.21.

Review. Vapor pressure is commonly expressed used *absolute pressure*. Absolute pressure is the value of pressure as measured relative to a pressure of absolute zero.

2.10 Characterizing Thermal Energy in Flowing Gases

Engineers characterize thermal energy changes using properties introduced in this section. Thermal energy is the energy associated with molecules in motion. This means that thermal energy is associated with temperature change (sensible energy change) and phase change (latent energy change). For most fluids problems, thermal properties are not important. However, thermal properties are used for compressible flow of gases (Chapter 12).

Specific Heat, c

Specific heat characterizes the amount of thermal energy that must be transferred to a unit mass of substance to raise its temperature by one degree. The dimensions of specific heat are energy per unit mass per degree temperature change, and the corresponding units are J/kg · K.

The magnitude of c depends on the process. If a gas is heated at *constant volume*, less energy is required than if the gas is heated at *constant pressure*. This is because a gas that is heated at constant pressure must do work as it expands against its surroundings.

The *constant volume specific heat*, c_v, applies to a process carried out at constant volume. The *constant pressure specific heat*, c_p, applies to process carried out at constant pressure. The ratio c_p/c_v is called the **specific heat ratio** and given the symbol k. Values for c_p and k for various gases are given in Table A.2.

Internal Energy

Internal energy includes all the energy in matter except for the kinetic energy and potential energy. Thus, internal energy includes multiple forms of energy such as chemical energy, electrical energy, and thermal energy. Specific internal energy, u, has dimensions of energy per unit mass. The units are J/kg.

Enthalpy

When a material is heated at constant pressure, the energy balance is

$$\text{(energy added)} = \left(\begin{array}{c}\text{energy to increase}\\\text{thermal energy}\end{array}\right) + \left(\begin{array}{c}\text{energy to do work}\\\text{as the material expands}\end{array}\right)$$

The work term is needed because the material is exerting a force over a distance as it pushes its surroundings away during the process of thermal expansion.

Enthalpy is a property that characterizes the amount of energy associated with a constant temperature heating or cooling process. Enthalpy per unit mass is defined mathematically by

$$\text{(enthalpy)} = \text{(internal energy)} + \text{(pressure over density)}$$
$$h = u + p/\rho$$

Ideal Gas Behavior

For an ideal gas, the properties h, u, c_p, and c_v depend only on temperature, not on pressure.

2.11 Summarizing Key Knowledge

Systems and Associated Concepts

- The *system* is the matter that the engineer selects for study.
- The *surroundings* are everything else that is not part of the system.
- The *boundary* is the surface that separates the system from its surroundings.
- The *state* of a system is the condition of the system as specified by values of the properties of the system.
- A *process* is a change of a system from one state to another.
- A *property* is a measurable characteristic of a system that depends only on the present state.

Constant Density Assumption

- All fluids, including liquids, are compressible.
- Modeling a fluid as *constant density* means that one assumes the density is constant with position and time. *Variable density* means the density can change with position or time.
- Modeling a fluid as incompressible means one assumes that the density of each fluid particle is constant.
- Most fluid problems are idealized as constant density problems. A notable exception is the high speed flow of gases.
- A gas should be modeled as compressible when the Mach number is greater than 0.3.

Viscosity Concepts

- *Viscosity* μ is also called dynamic viscosity or absolute viscosity.
- Viscosity is related to *kinematic viscosity* by $\nu = \mu/\rho$. Viscosity and kinematic are different properties.
- The *velocity profile* is a plot or equation that shows how velocity varies with position.
- The *no-slip condition* means that the velocity of fluid in contact with a solid surface will equal the velocity of the surface.
- The *shear force*, F_s is the net force due to shear stress.
- *Shear stress* τ is the tangential force per area at a point.
- A *Newtonian liquid* is one in which a plot of τ versus dV/dy is a straight line.
- A *non-Newtonian liquid* has a stress-strain relationship that is nonlinear. In general non-Newtonian liquids have more complex molecular structures than Newtonian fluids; examples of non-Newtonian liquids include paint, toothpaste, and molten plastics.
- Equations developed for Newtonian fluids (i.e., many textbook equations) do not apply to non-Newtonian fluids.

The Viscosity Equation

The viscosity equation (Eq. 2.16) relates shear stress to velocity gradient. The equation is

$$\tau = \mu \frac{dV}{dy}$$

Terms in the viscosity equation are summarized in Table 2.1 (page 39). Problems that are solvable with the viscosity equation can be classified into two categories.

- *Direct Calculations Problems.* Problems in this category specify two of the three variables in the equation and ask for the third variable. See Example 2.1 on page 39.
- *Couette Flow Problems.* Problems in this category involve a linear velocity profile in a small gap. See Example 2.2 on page 41.

Miscellaneous Topics

- A liquid flowing in a system can boil when the pressure drops to the vapor pressure. This boiling is typically detrimental to a design.
- To document a fluid property, list the source, temperature, pressure, and main assumptions.
- Surface tension problems are usually solved by drawing an FBD and summing forces.
- The formula for capillary rise of water in a glass tube is $\Delta h = (4\sigma)/(\gamma d)$.

Fluid Properties

Table 2.2 summarizes the most useful fluid properties. Columns 1 and 2 describe the property. Columns 3 and 4 describe how the property varies with temperature and pressure. Blue shading is used to distinguish between gases and liquids. For example, look in the row for viscosity. The viscosity of gases increases with a temperature rise, whereas the viscosity of liquids decreases with a temperature rise. The *Notes* column gives tips and lists the locations in this text where fluid properties can be found.

TABLE 2.2 Summary of Fluid Properties

Property	Units (SI)	Temperature Effects	Pressure Effects (common trends)	Notes
Density (ρ): Ratio of mass to volume at a point	$\dfrac{kg}{m^3}$	$\rho\downarrow$ as $T\uparrow$ if gas is free to expand	$\rho\uparrow$ as $p\uparrow$ if gas is compressed.	• *Air.* Find ρ in Table F.4 or Table A.3. • *Other Gases.* Find ρ in Table A.2. • *Caution!* Tables for gases are for $p = 101.3$ kPa. For other pressures, find ρ using the ideal gas law.
		$\rho\downarrow$ as $T\uparrow$ for liquids	ρ of liquids are constant with pressure	• *Water.* Find ρ in Table F.5 or Table A.5. • *Note.* For water, $\rho\uparrow$ as $T\uparrow$ for temperatures from 0 to about 4°C. Maximum density of water is at $T \approx 4°C$. • *Other Liquids.* Find ρ in Table A.4.
Specific Weight (γ): Ratio of weight to volume at a point	$\dfrac{N}{m^3}$	$\gamma\downarrow$ as $T\uparrow$ if fluid is free to expand	same trends as density	• Use same tables as for density. • ρ and γ can be related using $\gamma = \rho g$. • *Caution!* Tables for gases are for $p = 101.3$ kPa. For other pressures, find γ using the ideal gas law and $\gamma = \rho g$.
Specific Gravity (S or SG): Ratio of (density of a liquid) to (density of water at 4°C)	none	$SG\downarrow$ as $T\uparrow$	SG of liquids are constant with pressure	• Find SG data in Table A.4. • SG is used for liquids, not commonly used for gases. • Density of water (at 4°C) is listed in Table F.6.

(continued)

TABLE 2.2 Summary of Fluid Properties (*Continued*)

Property	Units (SI)	Temperature Effects	Pressure Effects (common trends)	Notes
Viscosity (μ): A property that characterizes resistance to shear stress and fluid friction	$\dfrac{N \cdot s}{m^2}$	$\mu\uparrow$ as $T\uparrow$ for gases.	μ of gases is independent of pressure	• *Air:* Find μ in Table F.4, Table A.3, Fig. A.2. • *Other gases:* Find properties in Table A.2, Fig. A.2. • *Hint:* Viscosity is also known as dynamic viscosity and absolute viscosity. • *Caution!* Avoid confusing viscosity and kinematic viscosity; these are different properties.
		$\mu\downarrow$ as $T\uparrow$ for liquids.	μ of liquids is independent of pressure	• *Water:* Find μ in Table F.5, Table A.5, Fig. A.2. • *Other Liquids.* Find μ in Table A.4, Fig. A.2.
Kinematic Viscosity (ν): A property that characterizes the mass and viscous properties of a fluid	$\dfrac{m^2}{s}$	$\nu\uparrow$ as $T\uparrow$ for gases	$\nu\uparrow$ as $p\uparrow$ for gases	• *Air:* Find μ in Table F.4, Table A.3, Fig. A.3. • *Other gases:* Find properties in Table A.2, Fig. A.3. • *Caution!* Avoid confusing viscosity and kinematic viscosity; these are different properties. • *Caution!* Gas tables are for $p = 101.3$ kPa. For other pressures, look up $\mu = \mu(T)$, then find ρ using the ideal gas law, and calculate ν using $\nu = \mu/\rho$.
		$\nu\downarrow$ as $T\uparrow$ for liquids.	ν of liquids is independent of pressure	• *Water:* Find ν in Table F.5, Table A.5, Fig. A.3. • *Other liquids:* Find ν in Table A.4, Fig. A.3.
Surface Tension (σ): A property that characterizes the tendency of a liquid surface to behave as a stretched membrane	$\dfrac{N}{m}, \dfrac{J}{m^2}$	$\sigma\downarrow$ as $T\uparrow$ for liquids.	σ of liquids is independent of pressure	• *Water:* Find σ in Fig. 2.16 (page 46). • *Other liquids:* Find σ in Table A.4. • Surface tension is a property of liquids (not gases). • Surface tension is greatly reduced by contaminates or impurities.
Vapor Pressure p_v: The pressure at which a liquid will boil	Pa	$p_v\uparrow$ as $T\uparrow$ for liquids	not applicable	• *Water:* Find p_v in Fig. 2.21 (page 50) and Table A-5.
Bulk Modulus of Elasticity E_v: A property that characterizes the compressibility of a fluid	Pa	not presented here	not presented here	• *Ideal gas (isothermal process):* $E_v = p = $ pressure. • *Ideal gas (adiabatic process):* $E_v = kp$; $k = c_p/c_v$. • *Water:* $E_v \approx 2.2 \times 10^9$ Pa.

REFERENCES

1. Center for Echohydraulics. Downloaded on 4/4/12 from http://www.uidaho.edu/engr/ecohydraulics/about

2. Pritchard, P. J. *Fox and McDonald's Introduction to Fluid Mechanics,* 8e, Wiley, New York, 2011, p. 42.

3. Bird, R. B., Stewart, W. E., and Lightfoot, E. N. *Transport Phenomena.* New York: John Wiley & Sons, 1960.

4. Fluid Mechanics Films, downloaded 7/31/11 from http://web.mit.edu/hml/ncfmf.html

5. Harris, J. *Rheology and non-Newtonian Flow.* New York: Longman, 1977.

6. Schowalter, W. R. *Mechanics of Non-Newtonian Fluids.* New York: Pergamon Press, 1978.

7. White, F. M. *Fluid Mechanics,* 7th ed. New York: McGraw-Hill, 2011, p. 828.

8. Fluid Mechanics Films, downloaded 7/31/11 from http://web.mit.edu/hml/ncfmf.html

9. Shaw, D. J. *Introduction to Colloid and Surface Chemistry,* 4e, Maryland Heights, MO: Butterworth-Heinemann, 1992.

PROBLEMS

Defining the System (§2.1)

2.1 A system is separated from its surrounding by a

    a. border

    b. boundary

    c. dashed line

    d. dividing surface

Characterizing Weight and Mass (§2.2)

2.2 How are density and specific weight related?

2.3 PLUS Density is (select all that apply)

    a. weight/volume

    b. mass/volume

    c. volume/mass

    d. mass/weight

2.4 PLUS Which of these are units of density? (select all that apply)

    a. kg/m^3

    b. mg/cm^3

2.5 PLUS Specific gravity (select all that apply)

    a. can have units of N/m^3

    b. is dimensionless

    c. increases with temperature

    d. decreases with temperature

2.6 If a liquid has a specific gravity of 1.7, what is the density in kg/m^3? What is the specific weight in N/m^3?

2.7 What are SG, γ, and ρ for mercury? State your answers in SI units.

2.8 PLUS If a gas has $\gamma = 15$ N/m^3, what is its density? State your answers in SI units.

Bulk Modulus of Elasticity (§2.3)

2.9 PLUS If you have a bulk modulus of elasticity that is a very large number, then a small change in pressure would cause

    a. a very large change in volume

    b. a very small change in volume

2.10 PLUS Dimensions of the bulk modulus of elasticity are

    a. the same as the dimensions of pressure/density

    b. the same as the dimensions of pressure/volume

    c. the same as the dimensions of pressure

2.11 The bulk modulus of elasticity of ethyl alcohol is 1.06×10^9 Pa. For water, it is 2.15×10^9 Pa. Which of these liquids is easier to compress?

    a. ethyl alcohol

    b. water

2.12 PLUS A pressure of 2×10^6 N/m^2 is applied to a mass of water that initially filled a 2000 cm^3 volume. Estimate its volume after the pressure is applied.

2.13 PLUS Calculate the pressure increase that must be applied to water to reduce its volume by 2%.

2.14 PLUS An open vat in a food processing plant contains 400 L of water at 20°C and atmospheric pressure. If the water is heated to 80°C, what will be the percentage change in its volume? If the vat has a diameter of 3 m, how much will the water level rise due to this temperature increase?

Finding Fluid Properties (§2.4)

2.15 Where in this text can you find:

    a. density data for such liquids as oil and mercury?

    b. specific weight data for air (at standard atmospheric pressure) at different temperatures?

    c. specific gravity data for sea water and kerosene?

2.16 PLUS Regarding water and seawater:

    a. Which is more dense, seawater or freshwater?

    b. Find the density of seawater (10°C, 3.3% salinity).

    c. What pressure is specified for the values in (b)?

2.17 PLUS If the density, ρ, of air (in an open system at atmospheric pressure) increases by a factor of 1.4x due to a temperature change,

    a. specific weight increases by 1.4x

    b. specific weight increases by 13.7x

    c. specific weight remains the same

Describing Viscous Effects (§2.5)

2.18 The following questions relate to viscosity.

    a. What are the primary dimensions of viscosity? What are five common units?

    b. What is the viscosity of SAE 10W-30 motor oil at 319.26 K?

2.19 PLUS Shear stress has dimensions of

    a. force/area

    b. dimensionless

2.20 PLUS The term dV/dy, the velocity gradient

    a. has dimensions of L/T and represents shear strain

    b. has dimensions of T^{-1} and represents the rate of shear strain

2.21 PLUS For the velocity gradient dV/dy

a. the change in velocity dV is in the direction of flow

b. the change in velocity dV is perpendicular to flow

2.22 PLUS The no-slip condition

a. only applies to ideal flow

b. only applies to rough surfaces

c. means velocity, V, is zero at the wall

d. means velocity, V, is the velocity of the wall

2.23 PLUS Kinematic viscosity (select all that apply)

a. is another name for absolute viscosity

b. is viscosity/density

c. is dimensionless because forces are canceled out

d. has dimensions of L^2/T

e. is only used with compressible fluids

2.24 What is the change in the viscosity and density of water between 10°C and 70°C? What is the change in the viscosity and density of air between 10°C and 70°C? Assume standard atmospheric pressure ($p = 101$ kN/m² absolute).

2.25 PLUS Determine the change in the kinematic viscosity of air that is heated from 10°C to 70°C. Assume standard atmospheric pressure.

2.26 PLUS Find the dynamic and kinematic viscosities of kerosene, SAE 10W-30 motor oil, and water at a temperature of 38°C.

2.27 What is the ratio of the dynamic viscosity of air to that of water at standard pressure and a temperature of 20°C? What is the ratio of the kinematic viscosity of air to that of water for the same conditions?

Applying the Viscosity Equation (§2.6)

2.28 PLUS At a point in a flowing fluid, the shear stress is 0.689 Pa, and the velocity gradient is 1 s⁻¹. Is this fluid more, or less, viscous than water?

2.29 PLUS SAE 10W-30 oil with viscosity 4.79×10^{-3} N · s/m² is used as a lubricant between two parts of a machine that slide past one another with a velocity difference of 1.83 m/s. What spacing, in cm, is required if you don't want a shear stress of more than 95.8 Pa?

2.30 The velocity distribution for water (20°C) near a wall is given by $u = a(y/b)^{1/6}$, where $a = 10$ m/s, $b = 2$ mm, and y is the distance from the wall in mm. Determine the shear stress in the water at $y = 1$ mm.

2.31 The velocity distribution for the flow of crude oil at 37.8°C ($\mu = 3.83 \times 10^{-3}$ Pa · s) between two walls is shown and is given by $u = 100y(0.1 - y)$ m/s, where y is measured in meters and the space between the walls is 3 cm. Plot the velocity distribution and determine the shear stress at the walls.

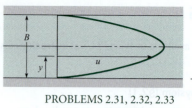

PROBLEMS 2.31, 2.32, 2.33

2.32 PLUS **(part a only)** A liquid flows between parallel boundaries as shown above. The velocity distribution near the lower wall is given in the following table:

y in mm	V in m/S
0.0	0.00
1.0	1.00
2.0	1.99
3.0	2.98

a. If the viscosity of the liquid is 10^{-3} N · s/m², what is the maximum shear stress in the liquid?

b. Where will the minimum shear stress occur?

2.33 GO Suppose that glycerin is flowing ($T = 20$°C) and that the pressure gradient dp/dx is -1.6 kN/m³. What are the velocity and shear stress at a distance of 12 mm from the wall if the space B between the walls is 5.0 cm? What are the shear stress and velocity at the wall? The velocity distribution for viscous flow between stationary plates is

$$u = -\frac{1}{2\mu}\frac{dp}{dx}(By - y^2)$$

2.34 PLUS Two plates are separated by a 3.175 cm space. The lower plate is stationary; the upper plate moves at a velocity of 7.62 m/s. Oil (SAE 10W-30, 338.7 K), which fills the space between the plates, has the same velocity as the plates at the surface of contact. The variation in velocity of the oil is linear. What is the shear stress in the oil?

2.35 PLUS The sliding plate viscometer shown below is used to measure the viscosity of a fluid. The top plate is moving to the right with a constant velocity of 10 m/s in response to a force of 3 N. The bottom plate is stationary. What is the viscosity of the fluid? Assume a linear velocity distribution.

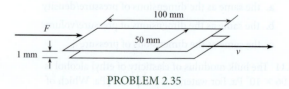

PROBLEM 2.35

2.36 A laminar flow occurs between two horizontal parallel plates under a pressure gradient dp/ds (p decreases in the positive s direction). The upper plate moves left (negative) at velocity u_t. The expression for local velocity u is given as

$$u = -\frac{1}{2\mu}\frac{dp}{ds}(Hy - y^2) + u_t\frac{y}{H}$$

a. Is the magnitude of the shear stress greater at the moving plate ($y = H$) or at the stationary plate ($y = 0$)?

b. Derive an expression for the y position of zero shear stress.

c. Derive an expression for the plate speed u_t required to make the shear stress zero at $y = 0$.

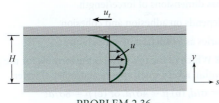

PROBLEM 2.36

2.37 This problem involves a cylinder falling inside a pipe that is filled with oil, as depicted in the figure. The small space between the cylinder and the pipe is lubricated with an oil film that has viscosity μ. Derive a formula for the steady rate of descent of a cylinder with weight W, diameter d, and length ℓ sliding inside a vertical smooth pipe that has inside diameter D. Assume that the cylinder is concentric with the pipe as it falls. Use the general formula to find the rate of descent of a cylinder 100 mm in diameter that slides inside a 100.5 mm pipe. The cylinder is 200 mm long and weighs 15 N. The lubricant is SAE 20W oil at 10°C.

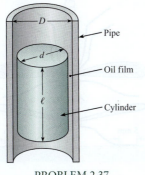

PROBLEM 2.37

2.38 WILEY GO The device shown consists of a disk that is rotated by a shaft. The disk is positioned very close to a solid boundary. Between the disk and the boundary is viscous oil.

a. If the disk is rotated at a rate of 1 rad/s, what will be the ratio of the shear stress in the oil at $r = 2$ cm to the shear stress at $r = 3$ cm?

b. If the rate of rotation is 2 rad/s, what is the speed of the oil in contact with the disk at $r = 3$ cm?

c. If the oil viscosity is 0.01 N · s/m² and the spacing y is 2 mm, what is the shear stress for the conditions noted in part (b)?

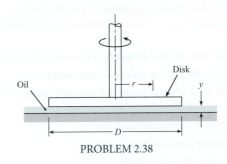

PROBLEM 2.38

2.39 Some instruments having angular motion are damped by means of a disk connected to the shaft. The disk, in turn, is immersed in a container of oil, as shown. Derive a formula for the damping torque as a function of the disk diameter D, spacing S, rate of rotation ω, and oil viscosity μ.

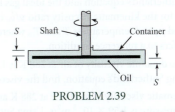

PROBLEM 2.39

2.40 One type of viscometer involves the use of a rotating cylinder inside a fixed cylinder The gap between the cylinders must be very small to achieve a linear velocity distribution in the liquid. (Assume the maximum spacing for proper operation is 1.27 mm). Design a viscometer that will be used to measure the viscosity of motor oil from 283.15 K to 366.48 K.

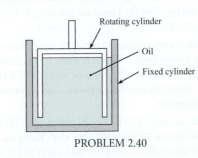

PROBLEM 2.40

Characterizing Viscosity (§2.7)

2.41 If temperature increases, does the viscosity of water increase or decrease? Why? If temperature increases, does the viscosity of air increase or decrease? Why?

2.42 Sutherland's equation (select all that apply):

    a. relates temperature and viscosity

    b. must be calculated using Kelvin

    c. requires use of a single universal constant for all gases

    d. requires use of a different constant for each gas

2.43 PLUS When looking up values for density, absolute viscosity, and kinematic viscosity, which statement is true for both liquids and gases?

    a. all three of these properties vary with temperature

    b. all three of these properties vary with pressure

    c. all three of these properties vary with temperature and pressure

2.44 Common Newtonian fluids are

    a. toothpaste, catsup, and paint

    b. water, oil, and mercury

    c. all of the above

2.45 Which of these flows (deforms) with even a small shear stress applied?

    a. a Bingham plastic

    b. a Newtonian fluid

2.46 Using Sutherland's equation and the ideal gas law, develop an expression for the kinematic viscosity ratio ν/ν_0 in terms of pressures p and p_0 and temperatures T and T_0, where the subscript 0 refers to a reference condition.

2.47 PLUS The dynamic viscosity of air at 288 K is 1.78×10^{-5} N · s/m^2. Using Sutherland's equation, find the viscosity at 373 K.

2.48 The kinematic viscosity of methane at 288 K and atmospheric pressure is 1.59×10^{-5} m^2/s. Using Sutherland's equation and the ideal gas law, find the kinematic viscosity at 474 K and 202.65 kPa.

2.49 The dynamic viscosity of nitrogen at 288 K is 1.72×10^{-5} N · s/m^2. Using Sutherland's equation, find the dynamic viscosity at 366.48 K.

2.50 PLUS The kinematic viscosity of helium at 288 K and 101.325 kPa is 1.13×10^{-4} m^2/s. Using Sutherland's equation and the ideal gas law, find the kinematic viscosity at 272 K and a pressure of 152 kPa.

2.51 Ammonia is very volatile, so it may be either a gas or a liquid at room temperature. When it is a gas, its absolute viscosity at 293 K is 9.91×10^{-6} N · s/m^2 and at 473 K is 1.66×10^{-5} N · s/m^2. Using these two data points, find Sutherland's constant for ammonia.

2.52 PLUS The viscosity of SAE 10W-30 motor oil at 311 K is 0.067 N · s/m^2 and at 372 K is 0.011 N · s/m^2. Using Eq. (2.20) (p. 42, §2.7) for interpolation, find the viscosity at 333 K. Compare this value with that obtained by linear interpolation.

2.53 The viscosity of grade 100 aviation oil at 310.9 K is 0.212 N · s/m^2 and at 372 K is 1.87×10^{-2} N · s/m^2. Using Eq. (2.20) (p. 42, §2.7), find the viscosity 338.7 K.

2.54 Find the kinematic and dynamic viscosities of air and water at a temperature of 40°C and an absolute pressure of 170 kPa.

2.55 PLUS Consider the ratio μ_{100}/μ_{50}, where μ is the viscosity of oxygen and the subscripts 100 and 50 are the temperatures of the oxygen in degrees Celsius. Does this ratio have a value (a) less than 1, (b) equal to 1, or (c) greater than 1?

Characterizing Surface Tension (§2.8)

2.56 PLUS Surface tension: (select all that apply)

    a. only occurs at an interface, or surface

    b. has dimensions of energy/area

    c. has dimensions of force/area

    d. has dimensions of force/length

    e. depends on adhesion and cohesion

    f. varies as a function of temperature

2.57 PLUS Which of the following is the formula for the gage pressure within a very small spherical droplet of water:

    (a) $p = \sigma/d$, (b) $p = 4\sigma/d$, or (c) $p = 8\sigma/d$?

2.58 A spherical soap bubble has an inside radius R, a film thickness t, and a surface tension σ. Derive a formula for the pressure within the bubble relative to the outside atmospheric pressure. What is the pressure difference for a bubble with a 4 mm radius? Assume σ is the same as for pure water.

2.59 PLUS A water bug is suspended on the surface of a pond by surface tension (water does not wet the legs). The bug has six legs, and each leg is in contact with the water over a length of 5 mm. What is the maximum mass (in grams) of the bug if it is to avoid sinking?

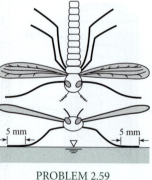

PROBLEM 2.59

2.60 A water column in a glass tube is used to measure the pressure in a pipe. The tube is 6.35 mm (1/4 in.) in diameter. How much of the water column is due to surface-tension effects? What would be the surface-tension effects if the tube were 3.2 mm or 0.8 mm in diameter?

2.61 Calculate the maximum capillary rise of water between two vertical glass plates spaced 1 mm apart.

PROBLEM 2.61

2.62 What is the pressure within a 1 mm spherical droplet of water relative to the atmospheric pressure outside?

2.63 By measuring the capillary rise in a tube, one can calculate the surface tension. The surface tension of water varies linearly with temperature from 0.0756 N/m at 0°C to 0.0589 N/m at 100°C. Size a tube (specify diameter and length) that uses capillary rise of water to measure temperature in the range from 0°C to 100°C. Is this design for a thermometer a good idea?

2.64 GO Capillary rise can be used to describe how far water will rise above a water table because the interconnected pores in the soil act like capillary tubes. This means that deep-rooted plants in the desert need only grow to the top of the "capillary fringe" in order to get water; they do not have to extend all the way down to the water table.

    a. Assuming that interconnected pores can be represented as a continuous capillary tube, how high is the capillary rise in a soil consisting of a silty soil, with pore diameter of 10 μm?

    b. Is the capillary rise higher in a soil with fine sand (pore diam. approx. 0.1 mm), or in fine gravel (pore diam. approx. 3 mm)?

    c. Root cells extract water from soil using capillarity. For root cells to extract water from the capillary zone, do the pores in a root need to be smaller than, or greater than, the pores in the soil? Ignore osmotic effects.

2.65 Consider a soap bubble 2 mm in diameter and a droplet of water, also 2 mm in diameter, that are falling in air. If the value of the surface tension for the film of the soap bubble is assumed to be the same as that for water, which has the greater pressure inside it? (a) the bubble, (b) the droplet, (c) neither—the pressure is the same for both.

2.66 A drop of water at 20°C is forming under a solid surface. The configuration just before separating and falling as a drop is shown in the figure. Assume the forming drop has the volume of a hemisphere. What is the diameter of the hemisphere just before separating?

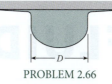

PROBLEM 2.66

2.67 PLUS The surface tension of a liquid is being measured with a ring as shown. The ring has an outside diameter of 10 cm and an inside diameter of 9.5 cm. The mass of the ring is 10 g. The force required to pull the ring from the liquid is the weight corresponding to a mass of 16 g. What is the surface tension of the liquid (in N/m)?

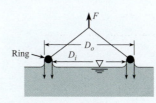

PROBLEM 2.67

Vapor Pressure (§2.9)

2.68 If liquid water at 30°C is flowing in a pipe and the pressure drops to the vapor pressure, what happens in the water?

    a. the water begins condensing on the walls of the pipe

    b. the water boils

    c. the water flashes to vapor

2.69 PLUS How does vapor pressure change with increasing temperature?

    a. it increases

    b. it decreases

    c. it stays the same

2.70 PLUS Water is at 20°C, and the pressure is lowered until vapor bubbles are noticed to be forming. What must the magnitude of the pressure be?

2.71 A student in the laboratory plans to exert a vacuum in the head space above a surface of water in a closed tank. She plans for the absolute pressure in the tank to be 10,400 Pa. The temperature in the lab is 20°C. Will water bubble into the vapor phase under these circumstances?

2.72 The vapor pressure of water at 100°C is 101 kPa because water boils under these conditions. The vapor pressure of water decreases approximately linearly with decreasing temperature at a rate of 3.1 kPa/°C. Calculate the boiling temperature of water at an altitude of 3000 m, where the atmospheric pressure is 69 kPa absolute.

3 FLUID STATICS

FIGURE 3.1
The first man-made structure to exceed the masonry mass of the Great Pyramid of Giza was the Hoover Dam. Design of dams involves calculations of hydrostatic forces. (Photo courtesy of U.S. Bureau of Reclamation, Lower Colorado Region)

Chapter Road Map

This chapter introduces concepts related to pressure and describes how to calculate forces associated with distributions of pressure. The emphasis is on fluids in hydrostatic equilibrium.

Learning Objectives

STUDENTS WILL BE ABLE TO

- Define hydrostatic equilibrium. Define pressure. (§3.1)
- Convert between gage, absolute, and vacuum pressure. (§3.1)
- Convert pressure units. (§3.1)
- List the steps to derive the hydrostatic differential equation. (§3.2)
- Describe the physics of the hydrostatics equation and the meaning of the variables that appear in the equation. Apply the hydrostatic equation. (§3.2)
- Explain how these instruments work: mercury barometer, piezometer, manometer, and Bourdon tube gage. (§3.3)
- Apply the manometer equations. (§3.3)
- Explain center-of-pressure and hydrostatically equivalent force. Describe how pressure is related to pressure force. (§3.4)
- Apply the panel equations to predict forces and moments. (§3.4)
- Solve problems that involve curved surfaces. (§3.5)
- Describe the physics of the buoyancy equation and the meaning of the variables that appear in the equation. Apply the buoyancy equation. (§3.6)
- Determine if floating objects are stable or unstable. (§3.7)

As shown in Fig. 3.2, the hydrostatic condition involves equilibrium of a fluid particle. **Hydrostatic equilibrium** means that each fluid particle is in force equilibrium with the net force due to pressure balancing the weight of the fluid particle. Equations in this chapter are based on an assumption of hydrostatic equilibrium.

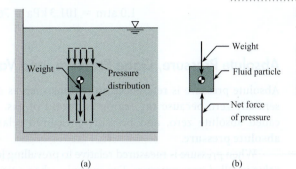

(a)   (b)

FIGURE 3.2

The hydrostatic condition.
(a) A fluid particle in a body of fluid.
(b) Forces acting on the fluid particle.

3.1 Describing Pressure

Because engineers use pressure in the solution of nearly all fluid mechanics problems, this section introduces fundamental ideas about pressure.

Pressure

Pressure is the ratio of normal force to area at a point.

$$p = \frac{\text{magnitude of normal force}}{\text{unit area}} \bigg|_{\substack{\text{at a point} \\ \text{due to a fluid}}} = \lim_{\Delta A \to 0} \frac{|\Delta \vec{F}_{\text{normal}}|}{\Delta A} \qquad (3.1)$$

Pressure is defined at a point because pressure typically varies with each (x, y, z) location in a flowing fluid.

Pressure is a scalar that produces a resultant force by its action on an area. The resultant force is normal to the area and acts in a direction toward the surface (compressive).

Pressure is caused by the molecules of the fluid interacting with the surface. For example, when a soccer ball is inflated, the internal pressure on the skin of the ball is caused by air molecules striking the wall.

Units of pressure can be organized into three categories:

- *Force per area*. The SI unit is the newtons per square meter or pascals (Pa). The traditional units include psi, which is pounds-force per square inch, and psf, which is pounds-force per square foot.

- *Liquid column height*. Sometimes pressure units give an equivalent height of a column of liquid. For example, pressure in a balloon will push a water column upward about 0.2 meter as shown in Fig. 3.3. Engineers state that the pressure in the balloon is 0.2 meter of water: $p = 0.2$ m-H_2O. When pressure is given in units of "height of a fluid column," the pressure value can be directly converted to other units using Table F.1. For example, the pressure in the balloon is

$$p = (0.2 \text{ m-}H_2O) \times (101.3 \text{ kPa}/10.33 \text{ m-}H_2O) = 1.96 \text{ kPa}$$

FIGURE 3.3

Pressure in a balloon causing a column of water to rise 0.2 meter.

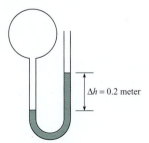

$\Delta h = 0.2$ meter

- *Atmospheres*. Sometimes pressure units are stated in terms of atmopheres where 1.0 atm is the air pressure at sea level at standard conditions. Another common unit is the bar, which is very nearly equal to 1.0 atm. (1.0 bar = 10^5 kPa)

Standard atmospheric pressure in various units is

$$1.0 \text{ atm} = 101.3 \text{ kPa} = 760 \text{ mm-Hg}$$

Absolute Pressure, Gage Pressure, and Vacuum Pressure

Absolute pressure is referenced to regions such as outer space, where the pressure is essentially zero because the region is devoid of gas. The pressure in a perfect vacuum is called absolute zero, and pressure measured relative to this zero pressure is termed **absolute pressure**.

When pressure is measured relative to prevailing local atmospheric pressure, the pressure value is called **gage pressure**. For example, when a tire pressure gage gives a value of 300 kPa this means that the absolute pressure in the tire is 300 kPa greater than local atmospheric pressure. To convert gage pressure to absolute pressure, add the local atmospheric pressure. For example, a gage pressure of 50 kPa recorded in a location where the atmospheric pressure is 100 kPa is expressed as either

$$p = 50 \text{ kPa gage} \quad \text{or} \quad p = 150 \text{ kPa abs} \qquad \textbf{(3.2)}$$

In SI units, gage and absolute pressures are identified after the unit as shown in Eq. (3.2).

When pressure is less than atmospheric, the pressure can be described using vacuum pressure. **Vacuum pressure** is defined as the difference between atmospheric pressure and actual pressure. Vacuum pressure is a positive number and equals the absolute value of gage pressure (which will be negative). For example, if a gage connected to a tank indicates a vacuum pressure of 31.0 kPa, this can also be stated as 70.0 kPa absolute, or −31.0 kPa gage.

Figure 3.4 provides a visual description of the three pressure scales. Notice that $p_B = 51$ kPa abs is equivalent to −50 kPa gage and 50 kPa vacuum. Notice that $p_A =$ of 301 kPa abs

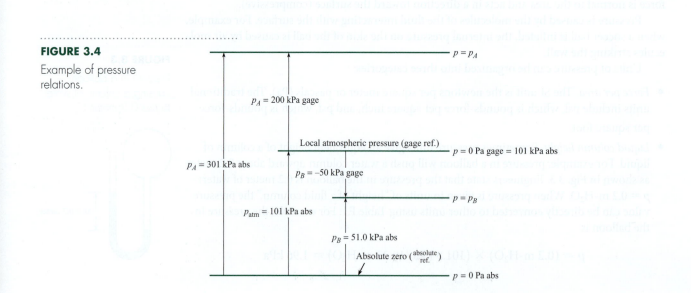

FIGURE 3.4

Example of pressure relations.

is equivalent to 200 kPa gage. Gage, absolute, and vacuum pressure can be related using equations labeled as the "pressure equations."

$$p_{\text{gage}} = p_{\text{abs}} - p_{\text{atm}} \qquad (3.3a)$$

$$p_{\text{vacuum}} = p_{\text{atm}} - p_{\text{abs}} \qquad (3.3b)$$

$$p_{\text{vacuum}} = -p_{\text{gage}} \qquad (3.3c)$$

EXAMPLE. Suppose the pressure in a car tire is specified as 3 bar. Find the absolute pressure in units of kPa.

Solution. Recognize that tire pressure is commonly specified in gage pressure. Thus, convert the gage pressure to absolute pressure.

$$p_{\text{abs}} = p_{\text{atm}} + p_{\text{gage}} = (101.3 \text{ kPa}) + (3 \text{ bar})\frac{(101.3 \text{ kPa})}{(1.013 \text{ bar})} = 401 \text{ kPa absolute}$$

Hydraulic Machines

A **hydraulic machine** uses a fluid to transmit forces or energy to assist in the performance of a human task. An example of a hydraulic machine is a hydraulic car jack in which a user can supply a small force to a handle and lift an automobile. Other examples of hydraulic machines include braking systems in cars, forklift trucks, power steering systems in cars, and airplane control systems (3).

The hydraulic machine provides a mechanical advantage (Fig. 3.5). **Mechanical advantage** is defined as the ratio of output force to input force:

$$(\text{mechanical advantage}) \equiv \frac{(\text{output force})}{(\text{input force})} \qquad (3.4)$$

Mechanical advantage of a lever (Fig. 3.5) is found by summing moments about the fulcrum to give $F_1 L_1 = F_2 L_2$, where L denotes the length of the lever arm.

$$(\text{mechanical advantage; lever}) \equiv \frac{(\text{output force})}{(\text{input force})} = \frac{F_2}{F_1} = \frac{L_1}{L_2} \qquad (3.5)$$

To find mechanical advantage of the hydraulic machine, apply force equilibrium to each piston (Fig. 3.5) to give $F_1 = p_1 A_1$ and $F_2 = p_2 A_2$, where p is pressure in the cylinder and A is face area of the piston. Next, let $p_1 = p_2$ and solve for the mechanical advantage

$$(\text{mechanical advantage; hydraulic machine}) \equiv \frac{(\text{output force})}{(\text{input force})} = \frac{F_2}{F_1} = \frac{A_2}{A_1} = \frac{D_2^2}{D_1^2} \qquad (3.6)$$

The hydraulic machine is often used to illustrate Pascal's principle. This principle states that when there is an increase in pressure at any point in a confined fluid, there is an equal increase at every other point in the container. This principle is evident when a balloon is inflated because the balloon expands evenly in all directions. The principle is also evident in the hydraulic machine (Fig. 3.6).

FIGURE 3.5
Both the lever and hydraulic machine provide a mechanical advantage.

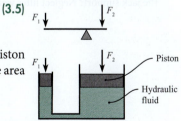

FIGURE 3.6

The figures show how the hydraulic machine can be used to illustrate Pascal's principle.

Pascal's principle. An applied force creates a pressure change that is transmitted to every point in the fluid and to the walls of the container

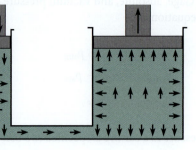

✔ **CHECKPOINT PROBLEM 3.1**

What is the mechanical advantage of this hydraulic machine? (neglect pressure changes due to elevation changes)

$W = 2$ tons, $S = 0.9$

$h = 3$ cm, $D_2 = 6$ cm, $D_1 = 1$ cm

  a. 2:1

  b. 4:1

  c. 6:1

  d. 16:1

  e. 36:1

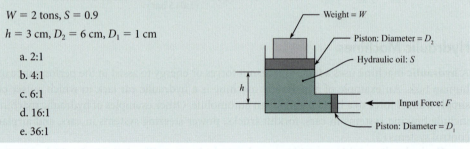

Weight = W

Piston: Diameter = D_2

Hydraulic oil: S

Input Force: F

Piston: Diameter = D_1

EXAMPLE 3.1

Applying Force Equilibrium to a Hydraulic Jack

Problem Statement

A hydraulic jack has the dimensions shown. If one exerts a force F of 100 N on the handle of the jack, what load, F_2, can the jack support? Neglect lifter weight.

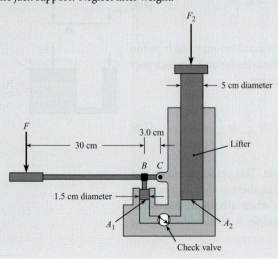

Define the Situation

A force of $F = 100$ N is applied to the handle of a jack.

Assumption: Weight of the lifter (see sketch) is negligible.

State the Goal

$F_2(\text{N}) \Longleftarrow$ Load that the jack can lift

Generate Ideas and Make a Plan

Because the goal is F_2, apply force equilibrium to the lifter. Then, analyze the small piston and the handle. The plan is

1. Calculate force acting on the small piston by applying moment equilibrium.

2. Calculate pressure p_1 in the hydraulic fluid by applying force equilibrium.

3. Calculate the load F_2 by applying force equilibrium.

Take Action (Execute the Plan)

1. Moment equilibrium (handle)

$$\sum M_c = 0$$

$$(0.33 \text{ m}) \times (100 \text{ N}) - (0.03 \text{ m})F_1 = 0$$

$$F_1 = \frac{0.33 \text{ m} \times 100 \text{ N}}{0.03 \text{ m}} = 1100 \text{ N}$$

2. Force equilibrium (small piston)

$$\sum F_{\text{small piston}} = p_1 A_1 - F_1 = 0$$

$$p_1 A_1 = F_1 = 1100 \text{ N}$$

Thus

$$p_1 = \frac{F_1}{A_1} = \frac{1100 \text{ N}}{\pi d^2/4} = 6.22 \times 10^6 \text{ N/m}^2$$

3. Force equilibrium (lifter)

Note that $p_1 = p_2$ because they are at the same elevation (this fact will be established in the next section).

$$\sum F_{\text{lifter}} = F_2 - p_1 A_2 = 0$$

$$F_2 = p_1 A_2 = \left(6.22 \times 10^6 \frac{\text{N}}{\text{m}^2}\right)\left(\frac{\pi}{4} \times (0.05 \text{ m})^2\right) = \boxed{12.2 \text{ kN}}$$

Review the Results and the Process

1. *Discussion.* The jack in this example, which combines a lever and a hydraulic machine, provides an output force of 12,200 N from an input force of 100 N. Thus, this jack provides a mechanical advantage of 122 to 1.

2. *Knowledge.* Hydraulic machines are analyzed by applying force and moment equilibrium. The force of pressure is typical given by $F = pA$.

3.2 Calculating Pressure Changes Associated with Elevation Changes

Pressure changes when elevation changes. For example, as a submarine dives to deeper depth, water pressure increases. Conversely, as an airplane gains elevation, air pressure decreases. Because engineers predict pressure changes associated with elevation change, this section introduces the relevant equations.

Theory: The Hydrostatic Differential Equation

All equations in fluid statics are based on the hydrostatic differential equation, which is derived in this subsection. To begin the derivation, visualize any region of static fluid (e.g., water behind a dam), isolate a cylindrical body, and then sketch a free-body diagram (FBD) as shown in Fig. 3.7. Notice that the cylindrical body is oriented so that its longitudinal axis is parallel to an arbitrary ℓ direction. The body is $\Delta\ell$ long, ΔA in cross-sectional area, and inclined at an angle α with the horizontal. Apply force equilibrium in the ℓ direction:

$$\sum F_\ell = 0$$

$$F_{\text{Pressure}} - F_{\text{Weight}} = 0$$

$$p\Delta A - (p + \Delta p)\Delta A - \gamma\Delta A\Delta\ell\sin\alpha = 0$$

Simplify and divide by the volume of the body $\Delta\ell\Delta A$ to give

$$\frac{\Delta p}{\Delta\ell} = -\gamma\sin\alpha$$

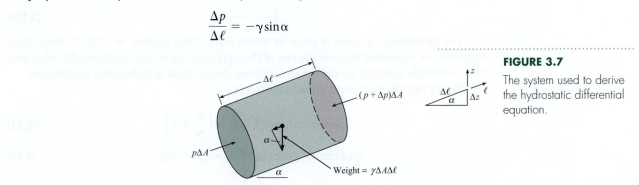

FIGURE 3.7

The system used to derive the hydrostatic differential equation.

From Fig. 3.7, the sine of the angle is given by

$$\sin \alpha = \frac{\Delta z}{\Delta \ell}$$

Combining the previous two equations and letting Δz approach zero gives

$$\lim_{\Delta z \to 0} \frac{\Delta p}{\Delta z} = -\gamma$$

The final result is

$$\frac{dp}{dz} = -\gamma \quad \text{(hydrostatic differential equation)} \qquad \textbf{(3.7)}$$

Equation (3.7) is valid in a body of fluid when the force balance shown in Fig. 3.2 is satisfied.

Equation (3.7) means that changes in pressure correspond to changes in elevation. If one travels upward in the fluid (positive z direction), the pressure decreases; if one goes downward (negative z), the pressure increases; if one moves along a horizontal plane, the pressure remains constant. Of course, these pressure variations are exactly what a diver experiences when ascending or descending in a lake or pool.

Derivation of the Hydrostatic Equation

This subsection shows how to derive the hydrostatic equation, which is used to calculate pressure variations in a fluid with constant density. To begin, assume that specific weight γ is constant and integrate Eq. (3.7) to give

$$p + \gamma z = p_z = \text{constant} \qquad \textbf{(3.8)}$$

where the term z is the elevation (vertical distance) above a fixed horizontal reference plane called a datum, and p_z is **piezometric pressure**. Dividing Eq. (3.8) by γ gives

$$\frac{p_z}{\gamma} = \left(\frac{p}{\gamma} + z\right) = h = \text{constant} \qquad \textbf{(3.9)}$$

where h is the **piezometric head**. Because h is constant Eq. (3.9) can be written as:

$$\frac{p_1}{\gamma} + z_1 = \frac{p_2}{\gamma} + z_2 \qquad \textbf{(3.10a)}$$

where the subscripts 1 and 2 identify any two points in a static fluid of constant density. Multiplying Eq. (3.10a) by γ gives

$$p_1 + \gamma z_1 = p_2 + \gamma z_2 \qquad \textbf{(3.10b)}$$

In Eq. (3.10b), letting $\Delta p = p_2 - p_1$ and letting $\Delta z = z_2 - z_1$ gives

$$\Delta p = -\gamma \Delta z \qquad \textbf{(3.10c)}$$

The hydrostatic equation is given by either Eq. (3.10a), (3.10b), or (3.10c). These three equations are equivalent because any one of the equations can be used to derive the other two. The hydrostatic equation is valid for any constant density fluid in hydrostatic equilibrium.

Notice that the hydrostatic equation involves

$$\text{piezometric head} = h \equiv \left(\frac{p}{\gamma} + z\right) \qquad \textbf{(3.11)}$$

$$\text{piezometric pressure} = p_z \equiv (p + \gamma z) \qquad \textbf{(3.12)}$$

To calculate piezometric head or piezometric pressure, an engineer identifies a specific location in a body of fluid and then uses the value of pressure and elevation at that location. Piezometric pressure and head are related by

$$p_z = h\gamma \tag{3.13}$$

FIGURE 3.8
Oil floating on water.

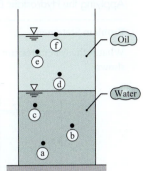

Piezometric head, h, a property that is widely used in fluid mechanics, characterizes hydrostatic equilibrium. When hydrostatic equilibrium prevails in a body of fluid of constant density, then h will be constant at all locations. For example, Fig. 3.8 shows a container with oil floating on water. Because piezometric head is constant in the water, $h_a = h_b = h_c$. Similarly the piezometric head is constant in the oil: $h_d = h_e = h_f$. Notice that piezometric head is not constant when density changes. For example, $h_c \neq h_d$ because points c and d are in different fluids with different values of density.

✔ **CHECKPOINT PROBLEM 3.2**

In the glass of water shown, which location has the highest value of piezeometric head? Which location has the highest value of the piezometric pressure?

   a. A

   b. B

   c. C

   d. None of the above

Hydrostatic Equation: Working Equations and Examples

The hydrostatic equation is summarized in Table 3.1.

TABLE 3.1 Summary of the Hydrostatic Equation

Name and Description	Equation	Terms
Head Form: Physics: (pressure head + elevation head at point 1) = (pressure head + elevation head at point 2). Another way to state the physics: The piezometric head in a static fluid with uniform density is constant at every point.	$\dfrac{p_1}{\gamma} + z_1 = \dfrac{p_2}{\gamma} + z_2$ (3.10)	p = pressure (Pa) (use absolute or gage pressure; not vacuum pressure) (p/γ is also called pressure head) z = elevation (m) (sketch a datum and measure z from this datum) (z is also called elevation head) γ = specific weight (N/m³) $p/\gamma + z$ = piezometric head (m)
Pressure Change (Δp) Form: Physics: For an elevation change of Δz, the pressure in a static fluid with uniform density will change by $\gamma \Delta z$.	$\Delta p = -\gamma \Delta z = -\rho g \Delta z$ (3.10)	Δp = change in pressure between points 1 & 2 (Pa) Δz = change in elevation between points 1 & 2 (m) ρ = density (kg/m³) g = gravitational constant (9.81 m/s²)

Example 3.2 shows the process for applying the hydrostatic equation.

EXAMPLE 3.2

Applying the Hydrostatic Equation to Find Pressure in a Tank

Problem Statement

What is the water pressure at a depth of 10.67 m in the tank shown?

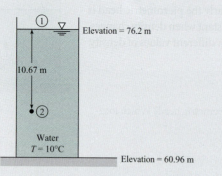

① ▽ Elevation = 76.2 m

10.67 m

② ●

Water
$T = 10°C$

Elevation = 60.96 m

Define the Situation

Water is contained in a tank that is 15.24 m deep.

Properties. Water (10 °C, 1 atm, Table A.5): $\gamma = 9810 \text{ N/m}^3$.

State the Goal

p_2 (kPa gage) ← Water pressure at point 2.

Generate Ideas and Make a Plan

Apply the idea that piezometric head is constant. Steps:

1. Equate piezometric head at elevation 1 with piezometric head at elevation 2 (i.e., apply Eq. 3.10a).
2. Analyze each term in Eq. (3.10a).
3. Solve for the pressure at elevation 2.

Take Action (Execute the Plan)

1. Hydrostatic equation (Eq. 3.10a)

$$\frac{p_1}{\gamma} + z_1 = \frac{p_2}{\gamma} + z_2$$

2. Term-by-term analysis of Eq. (3.10a) yields:
   - $p_1 = p_{atm} = 0$ kPa gage
   - $z_1 = 76.2$ m
   - $z_2 = 65.5$ m

3. Combine steps 1 and 2; solve for p_2

$$\frac{p_1}{\gamma} + z_1 = \frac{p_2}{\gamma} + z_2$$

$$0 + 76.2 \text{ m} = \frac{p_2}{9810 \text{ N/m}^3} + 65.5 \text{ m}$$

$$p_2 = 105 \text{ kPa gage}$$

Review the Solution and the Process

1. *Validation.* The calculated pressure change (105 kPa gage) is slightly greater than 1 atm (101.3 kPa). Because one atmosphere corresponds to a water column of 10.33 m and this problem involves 10.67 m of water column, the solution appears correct.

2. *Skill.* This example shows how to write down a governing equation and then analyze each term. This skill is called *term-by-term analysis.*

3. *Knowledge.* The gage pressure at the free surface of a liquid in contact with the atmosphere is zero ($p_1 = 0$ in this example).

4. *Skill.* Label a pressure as absolute or gage or vacuum. For this example, the pressure unit (kPa gage) denotes a gage pressure.

5. *Knowledge.* The hydrostatic equation is valid when density is constant. This condition is met on this problem.

Example 3.3 shows how to find pressure by applying the idea of "constant piezometric head" to a problem involving several fluids. Notice the continuity of pressure across a planar interface.

EXAMPLE 3.3

Applying the Hydrostatic Equation to Oil and Water in a Tank

Problem Statement

Oil with a specific gravity of 0.80 forms a layer 0.90 m deep in an open tank that is otherwise filled with water (10°C). The total depth of water and oil is 3 m. What is the gage pressure at the bottom of the tank?

Problem Definition

Oil and water are contained in a tank.

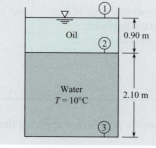

▽ ①
Oil
② 0.90 m

Water
$T = 10°C$ 2.10 m

③

Water (10°C, 1 atm, Table A.5) $\gamma_{\text{water}} = 9810 \text{ N/m}^3$.

Oil. $\gamma_{\text{oil}} = S\gamma_{\text{water, 4°C}} = 0.8(9810 \text{ N/m}^3) = 7850 \text{ N/m}^3$.

State the Goal

p_3 (kPa gage) ← pressure at bottom of the tank

Generate Ideas and Make a Plan

Because the goal is p_3, apply the hydrostatic equation to the water. Then, analyze the oil. The plan steps are

1. Find p_2 by applying the hydrostatic equation (3.10a).
2. Equate pressures across the oil–water interface.
3. Find p_3 by applying the hydrostatic equation given in Eq. (3.10a).

Solution

1. Hydrostatic equation (oil)

$$\frac{p_1}{\gamma_{\text{oil}}} + z_1 = \frac{p_2}{\gamma_{\text{oil}}} + z_2$$

$$\frac{0 \text{ Pa}}{\gamma_{\text{oil}}} + 3 \text{ m} = \frac{p_2}{0.8 \times 9810 \text{ N/m}^3} + 2.1 \text{ m}$$

$$p_2 = 7.063 \text{ kPa}$$

2. Oil–water interface

$$p_2\big|_{\text{oil}} = p_2\big|_{\text{water}} = 7.063 \text{ kPa}$$

3. Hydrostatic equation (water)

$$\frac{p_2}{\gamma_{\text{water}}} + z_2 = \frac{p_3}{\gamma_{\text{water}}} + z_3$$

$$\frac{7.063 \times 10^3 \text{ Pa}}{9810 \text{ N/m}^3} + 2.1 \text{ m} = \frac{p_3}{9810 \text{ N/m}^3} + 0 \text{ m}$$

$$\boxed{p_3 = 27.7 \text{ kPa gage}}$$

Review

Validation: Because oil is less dense than water, the answer should be slightly smaller than the pressure corresponding to a water column of 3 m. From Table F.1, a water column of 10 m ≈ 1 atm. Thus, a 3 m water column should produce a pressure of about 0.3 atm = 30 kPa. The calculated value appears correct.

Pressure Variation in the Atmosphere

This subsection describes how to calculate pressure, density and temperature in the atmosphere for applications such as modeling of atmospheric dynamics and the design of gliders, airplanes, balloons, and rockets.

Equations for pressure variation in the earth's atmosphere are derived by integrating the hydrostatic differential equation (3.7). To begin the derivation, write the ideal gas law (2.5):

$$\rho = \frac{p}{RT} \qquad \text{(3.14)}$$

Multiply by g:

$$\gamma = \frac{pg}{RT} \qquad \text{(3.15)}$$

Equation (3.15) requires temperature-versus-elevation data for the atmosphere. It is common practice to use the U.S. Standard Atmosphere (1). The **U.S. Standard Atmosphere** defines values for atmospheric temperature, density, and pressure over a wide range of altitudes. The first model was published in 1958; this was updated in 1962, 1966, and 1976. The U.S. Standard Atmosphere gives average conditions over the United States at 45° N latitude in July.

The U.S. Standard Atmosphere also gives average conditions at sea level. The sea level temperature is 15°C, the pressure is 101.33 kPa abs, and the density is 1.225 kg/m³.

Temperature data for the U.S. Standard Atmosphere are given in Fig. 3.9 for the lower 30 km of the atmosphere. The atmosphere is about 1000 km thick and is divided into five layers, so Fig. 3.9 only gives data near the earth's surface. In the **troposphere**, defined as the

FIGURE 3.9

Temperature variation with altitude for the U.S. standard atmosphere in July (1).

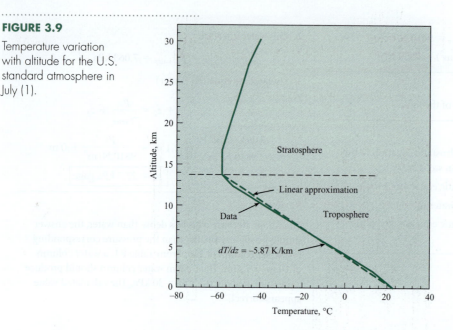

layer between sea level and 13.7 km, the temperature decreases nearly linearly with increasing elevation at a lapse rate of 5.87 K/km. The **stratosphere** is the layer that begins at the top of the troposphere and extends up to about 50 km. In the lower regions of the stratosphere, the temperature is constant at $-57.5°C$, to an altitude of 16.8 km, and then the temperature increases monotonically to $-38.5°C$ at 30.5 km.

Pressure Variation in the Troposphere

Let the temperature T be given by

$$T = T_0 - \alpha(z - z_0) \tag{3.16}$$

In this equation T_0 is the temperature at a reference level where the pressure is known, and α is the lapse rate. Combine Eq. (3.15) with the hydrostatic differential equation (3.7) to give

$$\frac{dp}{dz} = -\frac{pg}{RT} \tag{3.17}$$

Substituting Eq. (3.16) into Eq. (3.17) gives

$$\frac{dp}{dz} = -\frac{pg}{R[T_0 - \alpha(z - z_0)]}$$

Separate the variables and integrate to obtain

$$\frac{p}{p_0} = \left[\frac{T_0 - \alpha(z - z_0)}{T_0}\right]^{g/\alpha R}$$

Thus, the atmospheric pressure variation in the troposphere is

$$p = p_0 \left[\frac{T_0 - \alpha(z - z_0)}{T_0}\right]^{g/\alpha R} \tag{3.18}$$

Example 3.7 shows how to apply Eq. (3.18) to find pressure at a specified elevation in the troposphere.

Pressure Variation in the Lower Stratosphere

In the lower part of the stratosphere (13.7 to 16.8 km above the earth's surface as shown in Fig. 3.9), the temperature is approximately constant. In this region, integration of Eq. (3.17) gives

$$\ln p = \frac{zg}{RT} + C$$

At $z = z_0$, $p = p_0$, so the preceding equation reduces to

$$\frac{p}{p_0} = e^{-(z-z_0)g/RT}$$

so the atmospheric pressure variation in the stratosphere takes the form

$$p = p_0 e^{-(z-z_0)g/RT} \tag{3.19}$$

where p_0 is pressure at the interface between the troposphere and stratosphere, z_0 is the elevation of the interface, and T is the temperature of the stratosphere. Example 3.5 shows how to apply Eq. (3.19) to find pressure at a specified elevation in the troposphere.

EXAMPLE 3.4

Predicting Pressure in the Troposphere

Problem Statement

If the sea level pressure and temperature are 101.3 kPa and 23°C, what is the pressure at an elevation of 2000 m, assuming that standard atmospheric conditions prevail?

Situation

Standard atmospheric conditions prevail at an elevation of 2000 m.

Goal

p(kPa absolute) ◀ atmospheric pressure at $z = 2000$ m

Plan

Calculate pressure using Eq. (3.18).

Action

$$p = p_0 \left[\frac{T_0 - \alpha(z - z_0)}{T_0} \right]^{g/\alpha R}$$

where $p_0 = 101{,}300$ N/m², $T_0 = 273 + 23 = 296$ K, $\alpha = 5.87 \times 10^{-3}$ K/m, $z - z_0 = 2000$ m, and $g/\alpha R = 5.823$. Then

$$p = 101.3 \left(\frac{296 - 5.87 \times 10^{-3} \times 2000}{296} \right)^{5.823}$$

$$= \boxed{80.0 \text{ kPa absolute}}$$

EXAMPLE 3.5

Calculating Pressure in the Lower Stratosphere

Problem Statement

If the pressure and temperature are 15.9 kPa absolute and −57.5°C at an elevation of 13.72 km, what is the pressure at 16.77 km, assuming isothermal conditions over this range of elevation?

Situation

Standard atmospheric conditions prevail at an elevation of 16.77 km.

Goal

p ◀ Atmospheric pressure (kPa absolute) at an elevation of 16.77 km

Plan

Calculate pressure using Eq. (3.19).

Action

For isothermal conditions,

$$T = -57.5°C = 215.65 \text{ K}$$

$$p = p_0 e^{-(z-z_0)g/RT} = 15.9 e^{-(3.05)(9.81)/(287 \times 215.5)}$$

$$= 15.9 e^{-0.483}$$

Therefore the pressure at 16.77 km is

$$\boxed{p = 9.81 \text{ kPa abs}}$$

3.3 Measuring Pressure

When engineers design and conduct experiments, pressure nearly always needs to be measured. Thus, this section describes five scientific instruments for measuring pressure.

Barometer

FIGURE 3.10

A mercury barometer.

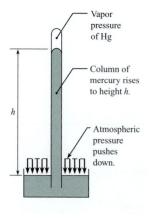

An instrument that is used to measure atmospheric pressure is called a **barometer**. The most common types are the mercury barometer and the aneroid barometer. A mercury barometer is made by inverting a mercury-filled tube in a container of mercury as shown in Fig. 3.10. The pressure at the top of the mercury barometer will be the vapor pressure of mercury, which is very small: $p_v = 2.43 \times 10^{-4}$ kPa at 20°C. Thus, atmospheric pressure will push the mercury up the tube to a height h. The mercury barometer is analyzed by applying the hydrostatic equation:

$$p_{atm} = \gamma_{Hg} h + p_v \approx \gamma_{Hg} h \qquad (3.20)$$

Thus, by measuring h, local atmospheric pressure can be determined using Eq. (3.20).

An aneroid barometer works mechanically. An aneroid is an elastic bellows that has been tightly sealed after some air was removed. When atmospheric pressure changes, this causes the aneroid to change size, and this mechanical change can be used to deflect a needle to indicate local atmospheric pressure on a scale. An aneroid barometer has some advantages over a mercury barometer because it is smaller and allows data recording over time.

Bourdon-Tube Gage

A **Bourdon-tube** gage, Fig. 3.11, measures pressure by sensing the deflection of a coiled tube. The tube has an elliptical cross section and is bent into a circular arc, as shown in Fig. 3.11b. When atmospheric pressure (zero gage pressure) prevails, the tube is undeflected, and for this

FIGURE 3.11

Bourdon-tube gage. (a) View of typical gage. (Photo by Donald Elger) (b) Internal mechanism (schematic).

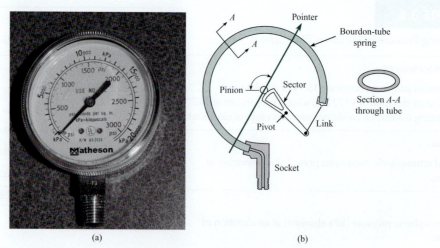

(a)   (b)

condition the gage pointer is calibrated to read zero pressure. When pressure is applied to the gage, the curved tube tends to straighten (much like blowing into a party favor to straighten it out), thereby actuating the pointer to read a positive gage pressure. The Bourdon-tube gage is common because it is low cost, reliable, easy to install, and available in many different pressure ranges. There are disadvantages: dynamic pressures are difficult to read accurately; accuracy of the gage can be lower than other instruments; and the gage can be damaged by excessive pressure pulsations.

Piezometer

A **piezometer** is a vertical tube, usually transparent, in which a liquid rises in response to a positive gage pressure. For example, Fig. 3.12 shows a piezometer attached to a pipe. Pressure in the pipe pushes the water column to a height h, and the gage pressure at the center of the pipe is $p = \gamma h$, which follows directly from the hydrostatic equation (3.10c). The piezometer has several advantages: simplicity, direct measurement (no need for calibration), and accuracy. However, a piezometer cannot easily be used for measuring pressure in a gas, and a piezometer is limited to low pressures because the column height becomes too large at high pressures.

FIGURE 3.12

Piezometer attached to a pipe.

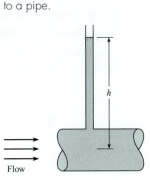

Flow

h

Manometer

A **manometer**, often shaped like the letter "U," is a device for measuring pressure by raising or lowering a column of liquid. For example, Fig. 3.13 shows a U-tube manometer that is being used to measure pressure in a flowing fluid. In the case shown, positive gage pressure in the pipe pushes the manometer liquid up a height Δh. To use a manometer, engineers relate the height of the liquid in the manometer to pressure as illustrated in Example 3.6.

FIGURE 3.13

U-tube manometer.

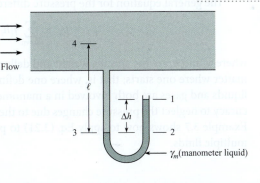

Flow

ℓ

Δh

γ_m(manometer liquid)

EXAMPLE 3.6

Pressure Measurement (U-Tube Manometer)

Problem Statement

Water at 10°C is the fluid in the pipe of Fig. 3.13, and mercury is the manometer fluid. If the deflection Δh is 60 cm and ℓ is 180 cm, what is the gage pressure at the center of the pipe?

Define the Situation

Pressure in a pipe is being measured using a U-tube manometer.

Properties:

Water (10°C), Table A.5, $\gamma = 9810$ N/m³.

Mercury, Table A.4: $\gamma = 133,000$ N/m³.

State the Goal

Calculate gage pressure (kPa) in the center of the pipe.

Generate Ideas and Make a Plan

Start at point 1 and work to point 4 using ideas from Eq. (3.10c). When fluid depth increases, add a pressure change. When fluid depth decreases, subtract a pressure change.

Take Action (Execute the Plan)

1. Calculate the pressure at point 2 using the hydrostatic equation (3.10c).

$$p_2 = p_1 + \text{pressure increase between 1 and 2} = 0 + \gamma_m \Delta h_{12}$$

$$= \gamma_m(0.6 \text{ m}) = (133{,}000 \text{ N/m}^3)(0.6 \text{ m})$$

$$= 79.8 \text{ kPa}$$

2. Find the pressure at point 3.

- The hydrostatic equation with $z_3 = z_2$ gives

$$p_3\big|_{\text{water}} = p_2\big|_{\text{water}} = 79.8 \text{ kPa}$$

- When a fluid-fluid interface is flat, pressure is constant across the interface. Thus, at the oil–water interface

$$p_3\big|_{\text{mercury}} = p_3\big|_{\text{water}} = 79.8 \text{ kPa}$$

3. Find the pressure at point 4 using the hydrostatic equation given in Eq. (3.10c).

$$p_4 = p_3 - \text{pressure decrease between 3 and 4} = p_3 - \gamma_w \ell$$

$$= 79{,}800 \text{ Pa} - (9810 \text{ N/m}^3)(1.8 \text{ m})$$

$$= 62.1 \text{ kPa gage}$$

Once one is familiar with the basic principle of manometry, it is straightforward to write a single equation rather than separate equations as was done in Example 3.6. The single equation for evaluation of the pressure in the pipe of Fig 3.13 is

$$0 + \gamma_m \Delta h - \gamma \ell = p_4$$

One can read the equation in this way: Zero pressure at the open end, plus the change in pressure from point 1 to 2, minus the change in pressure from point 3 to 4, equals the pressure in the pipe. The main concept is that pressure increases as depth increases and decreases as depth decreases.

The general equation for the pressure difference measured by the manometer is:

$$p_2 = p_1 + \sum_{\text{down}} \gamma_i h_i - \sum_{\text{up}} \gamma_i h_i \tag{3.21}$$

where γ_i and h_i are the specific weight and deflection in each leg of the manometer. It does not matter where one starts; that is, where one defines the initial point 1 and final point 2. When liquids and gases are both involved in a manometer problem, it is well within engineering accuracy to neglect the pressure changes due to the columns of gas. This is because $\gamma_{\text{liquid}} \gg \gamma_{\text{gas}}$. Example 3.7 shows how to apply Eq. (3.21) to perform an analysis of a manometer that uses multiple fluids.

EXAMPLE 3.7

Manometer Analysis

Problem Statement

What is the pressure of the air in the tank if $\ell_1 = 40$ cm, $\ell_2 = 100$ cm, and $\ell_3 = 80$ cm?

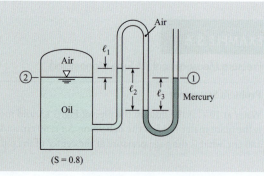

Define the Situation

A tank is pressurized with air.

Assumptions: Neglect the pressure change in the air column.

Properties:

- Oil: $\gamma_{oil} = S\gamma_{water} = 0.8 \times 9810 \text{ N/m}^3 = 7850 \text{ N/m}^3$.
- Mercury, Table A.4: $\gamma = 133,000 \text{ N/m}^3$.

State the Goal

Find the pressure (kPa gage) in the air.

Generate Ideas and Make a Plan

Apply the manometer equation (3.21) from location 1 to location 2.

Take Action (Execute the Plan)

Manometer equation

$$p_1 + \sum_{down} \gamma_i h_i - \sum_{up} \gamma_i h_i = p_2$$

$$p_1 + \gamma_{mercury}\ell_3 - \gamma_{air}\ell_2 + \gamma_{oil}\ell_1 = p_2$$

$$0 + (133,000 \text{ N/m}^3)(0.8 \text{ m}) - 0 + (7850 \text{ N/m}^3)(0.4 \text{ m}) = p_2$$

$$\boxed{p_2 = p_{air} = 110 \text{ kPa gage}}$$

Because the manometer configuration shown in Fig. 3.14 is common, it is useful to derive an equation specific to this application. To begin, apply the manometer equation (3.21) between points 1 and 2:

$$p_1 + \sum_{down} \gamma_i h_i - \sum_{up} \gamma_i h_i = p_2$$

$$p_1 + \gamma_A(\Delta y - \Delta h) - \gamma_B\Delta h - \gamma_A(\Delta y + z_2 - z_1) = p_2$$

Simplifying gives

$$(p_1 + \gamma_A z_1) - (p_2 + \gamma_A z_2) = \Delta h(\gamma_B - \gamma_A)$$

Dividing through by γ_A gives

$$\left(\frac{p_1}{\gamma_A} + z_1\right) - \left(\frac{p_2}{\gamma_A} + z_2\right) = \Delta h\left(\frac{\gamma_B}{\gamma_A} - 1\right)$$

Recognize that the terms on the left side of the equation are piezometric head and rewrite to give the final result:

$$h_1 - h_2 = \Delta h\left(\frac{\gamma_B}{\gamma_A} - 1\right) \tag{3.22}$$

Equation (3.22) is valid when a manometer is used as shown in Fig. 3.14. Example 3.8 shows how this equation is used.

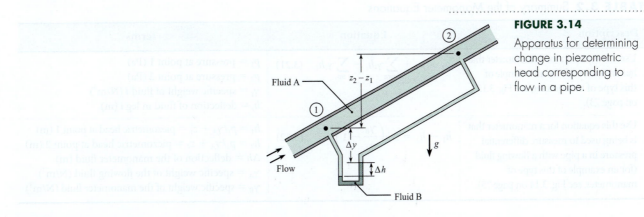

FIGURE 3.14

Apparatus for determining change in piezometric head corresponding to flow in a pipe.

EXAMPLE 3.8

Change in Piezometric Head for Pipe Flow

Problem Statement

A differential mercury manometer is connected to two pressure taps in an inclined pipe as shown in Fig. 3.14. Water at 10°C is flowing through the pipe. The deflection of mercury in the manometer is 2.5 cm. Find the change in piezometric pressure and piezometric head between points 1 and 2.

Define the Situation

Water is flowing in a pipe.

Properties:
1. Water (10°C), Table A.5, $\gamma_{water} = 9810$ N/m^3.
2. Mercury, Table A.4, $\gamma_{Hg} = 133{,}000$ N/m^3.

State the Goal

Find the
- Change in piezometric head (m) between points 1 and 2.
- Change in piezometric pressure (kPa) between 1 and 2.

Generate Ideas and Make a Plan

1. Find difference in the piezometric head using Eq. (3.22).
2. Relate piezometric head to piezometric pressure using Eq. (3.13).

Take Action (Execute the Plan)

1. Difference in piezometric head

$$h_1 - h_2 = \Delta h\left(\frac{\gamma_{Hg}}{\gamma_{water}} - 1\right) = 0.025 \text{ m}\left(\frac{133{,}000 \text{ N/m}^3}{9810 \text{ N/m}^3} - 1\right)$$

$$= \boxed{0.31 \text{ m}}$$

2. Piezometric pressure

$$p_z = h\gamma_{water}$$
$$= 0.31 \text{ m} \times 9810 \text{ N/m}^3 = \boxed{3.04 \text{ kPa}}$$

Summary of the Manometer Equations

These manometer equations are summarized in Table 3.2. Because the equations were derived from the hydrostatic equation, they have the same assumptions: constant fluid density and hydrostatic conditions.

The process for applying the manometer equations is

Step 1. For measurement of pressure at a point, select Eq. (3.21). For measurement of pressure or head change between two points in a pipe, select Eq. (3.22).
Step 2. Select points 1 and 2 where you know information or where you want to find information.
Step 3. Write the general form of the manometer equation.
Step 4. Perform a "term-by-term analysis."

TABLE 3.2 Summary of the Manometer Equations

Description	Equation	Terms
Use this equation for a manometer that has an open end (for an example of this type of manometer, see Fig. 3.13 on page 73).	$p_2 = p_1 + \sum_{down}\gamma_i h_i - \sum_{up}\gamma_i h_i$ (3.21)	p_1 = pressure at point 1 (Pa) p_2 = pressure at point 2 (Pa) γ_i = specific weight of fluid i (N/m^3) h_i = deflection of fluid in leg i (m)
Use this equation for a manometer that is being used to measure differential pressure in a pipe with a flowing fluid (for an example of this type of manometer, see Fig. 3.14 on page 75).	$h_1 - h_2 = \Delta h\left(\frac{\gamma_B}{\gamma_A} - 1\right)$ (3.22)	$h_1 = p_1/\gamma_A + z_1$ = piezometric head at point 1 (m) $h_2 = p_2/\gamma_A + z_2$ = piezometric head at point 2 (m) Δh = deflection of the manometer fluid (m) γ_A = specific weight of the flowing fluid (N/m^3) γ_B = specific weight of the manometer fluid (N/m^3)

Pressure Transducers

A **pressure transducer** is a device that converts pressure to an electrical signal. Modern factories and systems that involve flow processes are controlled automatically, and much of their operation involves sensing of pressure at critical points of the system. Therefore, pressure-sensing devices, such as pressure transducers, are designed to produce electronic signals that can be transmitted to oscillographs or digital devices for record-keeping or to control other devices for process operation. Basically, most transducers are tapped into the system with one side of a small diaphragm exposed to the active pressure of the system. When the pressure changes, the diaphragm flexes, and a sensing element connected to the other side of the diaphragm produces a signal that is usually linear with the change in pressure in the system. There are many types of sensing elements; one common type is the resistance-wire strain gage attached to a flexible diaphragm as shown in Fig. 3.15. As the diaphragm flexes, the wires of the strain gage change length, thereby changing the resistance of the wire. This change in resistance is converted into a voltage change that can then be used in various ways.

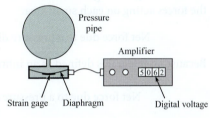

FIGURE 3.15

Schematic diagram of strain-gage pressure transducer.

Another type of pressure transducer used for measuring rapidly changing high pressures, such as the pressure in the cylinder head of an internal combustion engine, is the piezoelectric transducer (2). These transducers operate with a quartz crystal that generates a charge when subjected to a pressure. Sensitive electronic circuitry is required to convert the charge to a measurable voltage signal.

Computer data acquisition systems are used widely with pressure transducers. The analog signal from the transducer is converted (through an A/D converter) to a digital signal that can be processed by a computer. This expedites the data acquisition process and facilitates storing data.

3.4 Predicting Forces on Plane Surfaces (Panels)

Engineers predict hydrostatic forces on large structures such as dams. Thus, this section explains how to relate pressure to force. Next, this section describes how to calculate hydrostatic forces on panels, where a panel is a flat surface.

The Pressure Distribution

A **pressure distribution** (Fig. 3.16) is a visual or mathematical description that shows how pressure varies from point to point along a surface. For example, in the figure the pressure will be high in the front of the cylinder and low in the back of the cylinder. Notice that the pressure distribution is *always compressive* and that pressure is *always normal to the surface.*

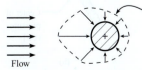

FIGURE 3.16

The pressure distribution caused by a fluid flowing over a circular cylinder.

Relating Pressure to Force

To relate pressure to force, select a small area dA (Fig. 3.17) on a surface. Then, define a normal vector $\mathbf{n}$ that is positive in a direction outward from the surface. The magnitude of the force is $dF = pdA$, and the direction of the force is inward toward the surface. Thus, the force $d\mathbf{F}$ is

$$d\mathbf{F} = (-p)\,\mathbf{n}dA$$

FIGURE 3.17

Terms used to define the pressure force.

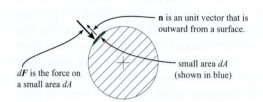

n is an unit vector that is outward from a surface.

small area dA (shown in blue)

$d\mathbf{F}$ is the force on a small area dA

where the negative sign is used because the force acts inward. To obtain the total force, add up the forces acting on each small area:

$$\text{Net force due to a pressure distribution} = \mathbf{F}_p = \sum d\mathbf{F} = \sum(-p)\mathbf{n}dA$$

Because an integral is defined as an infinite sum, this equation can be written as

$$\text{Net force due to a pressure distribution} \equiv \mathbf{F}_p = \int_{\text{Area}}(-p)\mathbf{n}dA \qquad \textbf{(3.23)}$$

In summary, the net force due to pressure can be found by integrating pressure over area while using a normal vector to keep track of the direction of incremental force on each unit of area.

Force of a Uniform Pressure Distribution

When pressure is the same at every point, as shown in Fig. 3.18a, the pressure distribution is called a **uniform pressure distribution**. For a uniform pressure distribution, Eq. (3.23) reduces to

$$F_p = \int_A pdA = pA$$

The resultant force of pressure F_p passes through a point called the **center of pressure (CP)**. Notice that the CP is represented using a circle with a "plus symbol" inside. For a uniform pressure distribution on a panel, the CP is located at the centroid of area.

FIGURE 3.18

(a) Uniform pressure distribution, and (b) equivalent force.

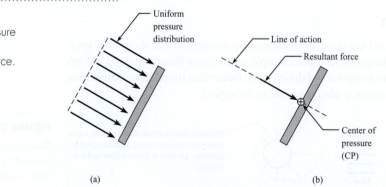

Uniform pressure distribution

Line of action

Resultant force

Center of pressure (CP)

(a)   (b)

Hydrostatic Pressure Distribution

When a pressure distribution is produced by a fluid in hydrostatic equilibrium (Fig. 3.19a), then the pressure distribution is called a **hydrostatic pressure distribution**. Notice that a hydrostatic pressure distribution is linear with depth. In Fig. 3.19b, the pressure distribution is represented by a resultant force that acts at the CP. Notice that the CP is located below the centroid of area.

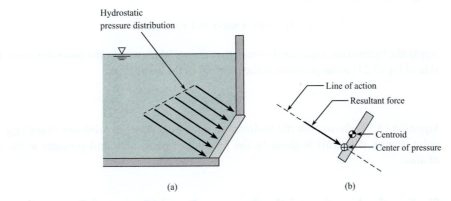

(a)   (b)

FIGURE 3.19

(a) Hydrostatic pressure distribution, and
(b) Resultant force **F** acting at the center of pressure.

Force on a Panel (Magnitude)

Next, we will show how to find the force on one face of a panel (e.g., a gate, a wall, a dam) that is acted on by a hydrostatic pressure distribution. To begin, sketch a panel of arbitrary shape submerged in a liquid (Fig. 3.20). Line AB is the edge view of a panel. The plane of the panel intersects the horizontal liquid surface at axis 0-0 with an angle α. The distance from the axis 0-0 to the horizontal axis through the centroid of the area is given by $\bar{y}$. The distance from 0-0 to the differential area dA is y.

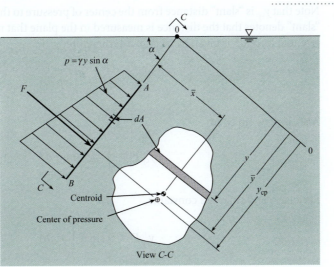

View *C-C*

FIGURE 3.20

Distribution of hydrostatic pressure on a plane surface.

The force due to pressure is given by Eq. (3.23), which reduces to

$$F_p = \int_A p\,dA \tag{3.24}$$

In Eq. (3.24), the pressure can be found with the hydrostatic equation:

$$p = \gamma \Delta z = \gamma y \sin \alpha \tag{3.25}$$

Combine Eqs (3.24) and (3.25) to give

$$F_p = \int_A p\,dA = \int_A \gamma y \sin \alpha\,dA = \gamma \sin \alpha \int_A y\,dA \qquad \text{(3.26)}$$

Because the integral on the right side of Eq. (3.24) is the first moment of the area, replace the integral by its equivalent, $\bar{y}A$. Therefore

$$F_p = \gamma \bar{y} A \sin \alpha = (\gamma \bar{y} \sin \alpha)A \qquad \text{(3.27)}$$

Apply the hydrostatic equation to show that the variables within the parentheses on the right side of Eq. (3.27) is the pressure at the centroid of the area. Thus,

$$F_p = \bar{p}A \qquad \text{(3.28)}$$

Equation (3.28) shows that the hydrostatic force on a panel of arbitrary shape (e.g., rectangular, round, elliptical) is given by the product of panel area and pressure at the centroid of area.

Finding the Location of the Force on Panel (Center of Pressure)

This subsection shows how to derive an equation for the vertical location of the center of pressure (CP). For the panel shown in Fig. 3.20 to be in moment equilibrium, the torque due to the resultant force F_p must balance the torque due to each differential force.

$$y_{cp}F_p = \int y\,dF$$

Note that y_{cp} is "slant" distance from the center of pressure to the surface of the liquid. The label "slant" denotes that the distance is measured in the plane that runs through the panel. The differential force dF is given by $dF = p\,dA$; therefore,

$$y_{cp}F = \int_A yp\,dA$$

Also, $p = \gamma y \sin \alpha$, so

$$y_{cp}F = \int_A \gamma y^2 \sin \alpha\,dA \qquad \text{(3.29)}$$

Because γ and $\sin \alpha$ are constants,

$$y_{cp}F = \gamma \sin \alpha \int_A y^2\,dA \qquad \text{(3.30)}$$

The integral on the right-hand side of Eq. (3.30) is the second moment of the area (often called the area moment of inertia). This shall be identified as I_0. However, for engineering applications it is convenient to express the second moment with respect to the horizontal centroidal axis of the area. Hence by the parallel-axis theorem,

$$I_0 = \bar{I} + \bar{y}^2 A \qquad \text{(3.31)}$$

Substitute Eq. (3.31) into Eq. (3.30) to give

$$y_{cp}F = \gamma \sin \alpha (\bar{I} + \bar{y}^2 A)$$

However, from Eq. (3.25), $F = \gamma \bar{y} \sin \alpha A$. Therefore,

$$y_{cp}(\gamma \bar{y} \sin \alpha\, A) = \gamma \sin \alpha (\bar{I} + \bar{y}^2 A) \tag{3.32}$$

$$y_{cp} = \bar{y} + \frac{\bar{I}}{\bar{y}A}$$

$$y_{cp} - \bar{y} = \frac{\bar{I}}{\bar{y}A} \tag{3.33}$$

In Eq. (3.33), the area moment of inertia $\bar{I}$ is taken about a horizontal axis that passes through the centroid of area. Formulas for $\bar{I}$ are presented in Fig. A.1. The slant distance $\bar{y}$ measures the length from the surface of the liquid to the centroid of the panel along an axis that is aligned with the "slant of the panel" as shown in Fig. 3.20.

Equation (3.33) shows that the Center of Pressure (CP) will be situated below the centroid. The distance between the CP and the centroid depends on the depth of submersion, which is characterized by $\bar{y}$, and on the panel geometry, which is characterized by $\bar{I}/A$.

Due to assumptions in the derivations, Eqs. (3.28) and (3.33) have several limitations. First, they only apply to a single fluid of constant density. Second, the pressure at the liquid surface needs to be $p = 0$ gage to correctly locate the CP. Third, Eq. (3.33) gives only the vertical location of the CP, not the lateral location.

Summary of the Panel Equations

The panel equations (Table 3.3) are used to calculate the force on a flat plate that is subjected to a hydrostatic pressure distribution.

TABLE 3.3 Summary of the Panel Equations

Description	Equation		Terms
Apply this equation to predict the magnitude of the hydrostatic force.	$F_p = \bar{p}A$	(3.28)	F_p = resultant force due to pressure distribution (N) $\bar{p}$ = pressure at the depth of the centroid (Pa) A = area of the surface of the plate (m²)
Apply this equation to locate the center of pressure (CP).	$y_{cp} - \bar{y} = \dfrac{\bar{I}}{\bar{y}A}$	(3.33)	$(y_{cp} - \bar{y})$ = slant distance from the centroid to the center of pressure (m) $\bar{I}$ = area moment of inertia of panel about centroidal axis (m⁴) (for formulas, see Fig. A.1 on page A-1) $\bar{y}$ = slant distance from centroid to liquid surface (m)

This figure defines terms.

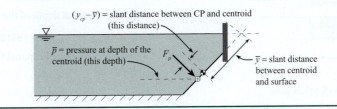

EXAMPLE 3.9

Hydrostatic Force Due to Concrete

Problem Statement

Determine the force acting on one side of a concrete form 2.44 m high and 1.22 m wide that is used for pouring a basement wall. The specific weight of concrete is 23.6 kN/m³.

Define the Situation

Concrete in a liquid state acts on a vertical surface.

The vertical wall is 2.44 m high and 1.22 m wide

Assumptions: Freshly poured concrete can be represented as a liquid.

Properties: Concrete: $\gamma = 23.6$ kN/m³.

State the Goal

Find the resultant force (kN) acting on the wall.

Plan

Apply the panel equation (3.28).

Solution

1. Panel equation

$$F = \bar{p}A$$

2. Term-by-term analysis

   • $\bar{p}$ = pressure at depth of the centroid

   $\bar{p} = (\gamma_{\text{concrete}})(z_{\text{centroid}}) = (23.6 \text{ kN/m}^3)(2.44/2 \text{ m})$

   $= 28.79$ kPa

   • A = area of panel

   $$A = (2.44 \text{ m})(1.22 \text{ m}) = 2.977 \text{ m}^2$$

3. Resultant force

   $$F = \bar{p}A = (28.79 \text{ kPa})(2.977 \text{ m}^2) = \boxed{85.7 \text{ kN}}$$

EXAMPLE 3.10

Force to Open an Elliptical Gate

Problem Statement

An elliptical gate covers the end of a pipe 4 m in diameter. If the gate is hinged at the top, what normal force F is required to open the gate when water is 8 m deep above the top of the pipe and the pipe is open to the atmosphere on the other side? Neglect the weight of the gate.

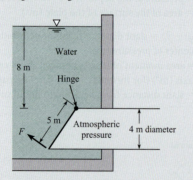

Define the Situation

Water pressure is acting on an elliptical gate.

Properties: Water (10°C), Table A.5: $\gamma = 9810$ N/m³.

Assumptions:

1. Neglect the weight of the gate.
2. Neglect friction between the bottom on the gate and the pipe wall.

State the Goal

F(N) ◀ Force needed to open gate.

Generate Ideas and Make a Plan

1. Calculate resultant hydrostatic force using $F = \bar{p}A$.
2. Find the location of the center of pressure using Eq. (3.33).
3. Draw an FBD of the gate.
4. Apply moment equilibrium about the hinge.

Take Action (Execute the Plan)

1. Hydrostatic (resultant) force

   • $\bar{p}$ = pressure at depth of the centroid

   $$\bar{p} = (\gamma_{\text{water}})(z_{\text{centroid}}) = (9810 \text{ N/m}^3)(10 \text{ m}) = 98.1 \text{ kPa}$$

   • A = area of elliptical panel (using Fig. A.1 to find formula)

   $$A = \pi ab$$
   $$= \pi(2.5 \text{ m})(2 \text{ m}) = 15.71 \text{ m}^2$$

- Calculate resultant force

$$F_p = \bar{p}A = (98.1 \text{ kPa})(15.71 \text{ m}^2) = \boxed{1.54 \text{ MN}}$$

2. Center of pressure
- $\bar{y} = 12.5$ m, where $\bar{y}$ is the slant distance from the water surface to the centroid.
- Area moment of inertia $\bar{I}$ of an elliptical panel using a formula from Fig. A.1

$$\bar{I} = \frac{\pi a^3 b}{4} = \frac{\pi (2.5 \text{ m})^3 (2 \text{ m})}{4} = 24.54 \text{ m}^4$$

- Finding center of pressure

$$y_{cp} - \bar{y} = \frac{\bar{I}}{\bar{y}A} = \frac{25.54 \text{ m}^4}{(12.5 \text{ m})(15.71 \text{ m}^2)} = 0.125 \text{ m}$$

3. FBD of the gate:

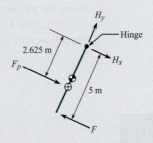

4. Moment equilibrium

$$\sum M_{\text{hinge}} = 0$$

$$1.541 \times 10^6 \text{ N} \times 2.625 \text{ m} - F \times 5 \text{ m} = 0$$

$$F = \boxed{809 \text{ kN}}$$

3.5 Calculating Forces on Curved Surfaces

As engineers, we calculate forces on curved surfaces when we are designing components such as tanks, pipes, and curved gates. Thus, this topic is described in this section.

Consider the curved surface AB in Fig. 3.21a. The goal is to represent the pressure distribution with a resultant force that passes through the center of pressure. One approach is to integrate the pressure force along the curved surface and find the equivalent force. However, it is easier to sum forces for the free body shown in the upper part of Fig. 3.21b. The lower sketch in Fig. 3.21b shows how the force acting on the curved surface relates to the force F acting on the free body. Using the FBD and summing forces in the horizontal direction shows that

$$F_x = F_{AC} \tag{3.34}$$

The line of action for the force F_{AC} is through the center of pressure for side AC.

The vertical component of the equivalent force is

$$F_y = W + F_{CB} \tag{3.35}$$

where W is the weight of the fluid in the free body and F_{CB} is the force on the side CB.

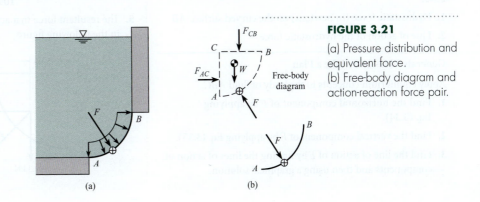

(a)

(b)

FIGURE 3.21

(a) Pressure distribution and equivalent force.

(b) Free-body diagram and action-reaction force pair.

The force F_{CB} acts through the centroid of surface CB, and the weight acts through the center of gravity of the free body. The line of action for the vertical force may be found by summing the moments about any convenient axis.

Example 3.11 illustrates how curved surface problems can be solved by applying equilibrium concepts together with the panel force equations.

EXAMPLE 3.11

Hydrostatic Force on a Curved Surface

Problem Statement

Surface AB is a circular arc with a radius of 2 m and a width of 1 m into the paper. The distance EB is 4 m. The fluid above surface AB is water, and atmospheric pressure prevails on the free surface of the water and on the bottom side of surface AB. Find the magnitude and line of action of the hydrostatic force acting on surface AB.

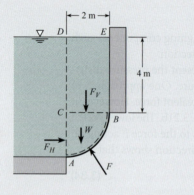

Define the Situation

Situation: A body of water is contained by a curved surface.

Properties: Water (10°C), Table A.5: $\gamma = 9810$ N/m³.

State the Goal

Find:

1. Hydrostatic force (in newtons) on the curved surface AB.
2. Line of action of the hydrostatic force.

Generate Ideas and Make a Plan

Apply equilibrium concepts to the body of fluid ABC.

1. Find the horizontal component of F by applying Eq. (3.34).
2. Find the vertical component of F by applying Eq. (3.35).
3. Find the line of action of F by finding the lines of action of components and then using a graphical solution.

Take Action (Execute the Plan)

1. Force in the horizontal direction

$$F_x = F_H = \bar{p}A = (5 \text{ m})(9810 \text{ N/m}^3)(2 \times 1 \text{ m}^2)$$
$$= 98.1 \text{ kN}$$

2. Force in the vertical direction
   - Vertical force on side CB

$$F_V = \bar{p}_0 A = 9.81 \text{ kN/m}^3 \times 4 \text{ m} \times 2 \text{ m} \times 1 \text{ m} = 78.5 \text{ kN}$$

   - Weight of the water in volume ABC

$$W = \gamma V_{ABC} = (\gamma)(\tfrac{1}{4}\pi r^2)(w)$$
$$= (9.81 \text{ kN/m}^3) \times (0.25 \times \pi \times 4 \text{ m}^2)(1 \text{ m}) = 30.8 \text{ kN}$$

   - Summing forces

$$F_y = W + F_V = 109.3 \text{ kN}$$

3. Line of action (horizontal force)

$$y_{cp} = \bar{y} + \frac{\bar{I}}{\bar{y}A} = (5 \text{ m}) + \left(\frac{1 \times 2^3/12}{5 \times 2 \times 1} \text{ m}\right)$$
$$y_{cp} = 5.067 \text{ m}$$

4. The line of action (x_{cp}) for the vertical force is found by summing moments about point C:

$$x_{cp}F_y = F_V \times 1 \text{ m} + W \times \bar{x}_w$$

The horizontal distance from point C to the centroid of the area ABC is found using Fig. A.1: $\bar{x}_w = 4r/3\pi = 0.849$ m. Thus,

$$x_{cp} = \frac{78.5 \text{ kN} \times 1 \text{ m} + 30.8 \text{ kN} \times 0.849 \text{ m}}{109.3 \text{ kN}} = 0.957 \text{ m}$$

5. The resultant force that acts on the curved surface is shown in the following figure.

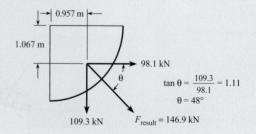

The central idea of this section is that *forces on curved surfaces may be found by applying equilibrium concepts to systems comprised of the fluid in contact with the curved surface.* Notice how equilibrium concepts are used in each of the following situations.

Consider a sphere holding a gas pressurized to a gage pressure p_i as shown in Fig. 3.22. The indicated forces act on the fluid in volume ABC. Applying equilibrium in the vertical direction gives

$$F = p_i A_{AC} + W$$

Because the specific weight for a gas is quite small, engineers usually neglect the weight of the gas:

$$F = p_i A_{AC} \tag{3.36}$$

Another example is finding the force on a curved surface submerged in a reservoir of liquid as shown in Fig. 3.23a. If atmospheric pressure prevails above the free surface and on the outside of surface AB, then force caused by atmospheric pressure cancels out, and equilibrium gives

$$F = \gamma V_{ABCD} = W\downarrow \tag{3.37}$$

Hence the force on surface AB equals the weight of liquid above the surface, and the arrow indicates that the force acts downward.

Now consider the situation where the pressure distribution on a thin curved surface comes from the liquid underneath, as shown in Fig. 3.23b. If the region above the surface, volume $abcd$, were filled with the same liquid, the pressure acting at each point on the upper surface of ab would equal the pressure acting at each point on the lower surface. In other words, there would be no net force on the surface. Thus, the equivalent force on surface ab is given by

$$F = \gamma V_{abcd} = W\downarrow \tag{3.38}$$

where W is the weight of liquid needed to fill a volume that extends from the curved surface to the free surface of the liquid.

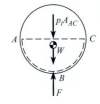

FIGURE 3.22

Pressurized spherical tank showing forces that act on the fluid inside the marked region.

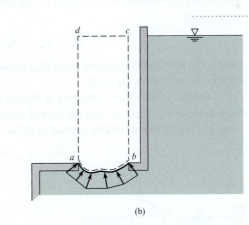

(a) (b)

FIGURE 3.23

Curved surface with (a) liquid above and (b) liquid below. In (a), arrows represent forces acting on the liquid. In (b), arrows represent the pressure distribution on surface ab.

3.6 Calculating Buoyant Forces

Engineers calculate buoyant forces for applications such as the design of ships, sediment transport in rivers, and fish migration. Buoyant forces are sometimes significant in problems involving gases, for example, a weather balloon. Thus, this section describes how to calculate the buoyant force on an object.

A **buoyant force** is defined as an upward force (with respect to gravity) on a body that is totally or partially submerged in a fluid, either a liquid or gas. Buoyant forces are caused by the hydrostatic pressure distribution.

The Buoyant Force Equation

To derive an equation, consider a body $ABCD$ submerged in a liquid of specific weight γ (Fig. 3.24). The sketch on the left shows the pressure distribution acting on the body. As shown by Eq. (3.38), pressures acting on the lower portion of the body create an upward force equal to the weight of liquid needed to fill the volume above surface ADC. The upward force is

$$F_{up} = \gamma(V_b + V_a)$$

FIGURE 3.24

Two views of a body immersed in a liquid.

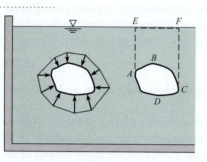

where V_b is the volume of the body (i.e., volume $ABCD$) and V_a is the volume of liquid above the body (i.e., volume $ABCFE$). As shown by Eq. (3.37), pressures acting on the top surface of the body create a downward force equal to the weight of the liquid above the body:

$$F_{down} = \gamma V_a$$

Subtracting the downward force from the upward force gives the net or buoyant force F_B acting on the body:

$$F_B = F_{up} - F_{down} = \gamma V_b \tag{3.39}$$

Hence, the net force or buoyant force (F_B) equals the weight of liquid that would be needed to occupy the volume of the body.

Consider a body that is floating as shown in Fig. 3.25. The marked portion of the object has a volume V_D. Pressure acts on curved surface ADC causing an upward force equal to the weight of liquid that would be needed to fill volume V_D. The buoyant force is given by

$$F_B = F_{up} = \gamma V_D \tag{3.40}$$

FIGURE 3.25

A body partially submerged in a liquid.

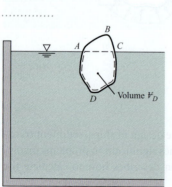

Hence, the buoyant force equals the weight of liquid that would be needed to occupy the volume $\mathcal{V}_D$. This volume is called the displaced volume. Comparison of Eqs. (3.39) and (3.40) shows that one can write a single equation for the buoyant force:

$$F_B = \gamma \mathcal{V}_D \qquad (3.41a)$$

In Eq. (3.41a), $\mathcal{V}_D$ is the volume that is displaced by the body. If the body is totally submerged, the displaced volume is the volume of the body. If a body is partially submerged, the displaced volume is the portion of the volume that is submerged.

Eq. (3.41b) is only valid for a single fluid of uniform density. The general principle of buoyancy is called **Archimedes' principle**:

$$(\text{buoyant force}) = F_B = (\text{weight of the displaced fluid}) \qquad (3.41b)$$

The buoyant force acts at a point called the center of buoyancy, which is located at the center of gravity of the displaced fluid.

✔ **CHECKPOINT PROBLEM 3.3**

Consider a balloon filled with helium (case A) and a balloon filled with air (case B). Which statement is correct?

  a. Buoyant force (case A) > Buoyant force (case B)

  b. Buoyant force (case A) < Buoyant force (case B)

  c. Buoyant force (case A) = Buoyant force (case B)

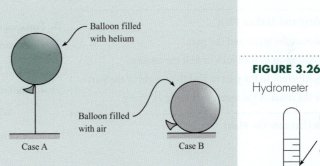

Balloon filled with helium

Balloon filled with air

Case A   Case B

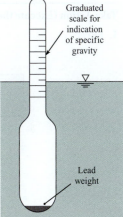

FIGURE 3.26

Hydrometer

Graduated scale for indication of specific gravity

Lead weight

The Hydrometer

A **hydrometer** (Fig. 3.26) is an instrument for measuring the specific gravity of liquids. It is typically made of a glass bulb that is weighted on one end so the hydrometer floats in an upright position. A stem of constant diameter is marked with a scale, and the specific weight of the liquid is determined by the depth at which the hydrometer floats. The operating principle of the hydrometer is buoyancy. In a heavy liquid (i.e., high γ), the hydrometer will float shallower because a lesser volume of the liquid must be displaced to balance the weight of the hydrometer. In a light liquid, the hydrometer will float deeper.

EXAMPLE 3.12

Buoyant Force on a Metal Part

Problem Statement

A metal part (object 2) is hanging by a thin cord from a floating wood block (object 1). The wood block has a specific gravity $S_1 = 0.3$ and dimensions of $50 \times 50 \times 10$ mm. The metal part has a volume of 6600 mm^3. Find the mass m_2 of the metal part and the tension T in the cord.

Define the Situation

A metal part is suspended from a floating block of wood.

Properties:

Water (15°C), Table A.5: $\gamma = 9800$ N/m^3.

Wood: $S_1 = 0.3$.

State the Goal

- Find the mass (in grams) of the metal part.
- Calculate the tension (in newtons) in the cord.

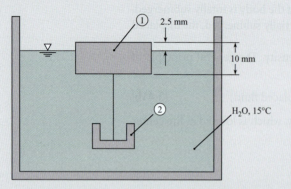

Generate Ideas and Make a Plan

1. Draw FBDs of the block and the part.
2. Apply equilibrium to the block to find the tension.
3. Apply equilibrium to the part to find the weight of the part.
4. Calculate the mass of the metal part using $W = mg$.

Take Action (Execute the Plan)

1. FBDs

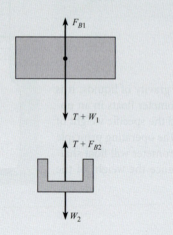

2. Force equilibrium (vertical direction) applied to block

$$T = F_{B1} - W_1$$

- Buoyant force $F_{B1} = \gamma V_{D1}$, where V_{D1} is the submerged volume

$$
\begin{aligned}
F_{B1} &= \gamma V_{D1} \\
&= (9800 \text{ N/m}^3)(50 \times 50 \times 7.5 \text{ mm}^3)(10^{-9} \text{ m}^3/\text{mm}^3) \\
&= 0.184 \text{ N}
\end{aligned}
$$

- Weight of the block

$$
\begin{aligned}
W_1 &= \gamma S_1 V_1 \\
&= (9800 \text{ N/m}^3)(0.3)(50 \times 50 \times 10 \text{ mm}^3)(10^{-9} \text{ m}^3/\text{mm}^3) \\
&= 0.0735 \text{ N}
\end{aligned}
$$

- Tension in the cord

$$T = (0.184 - 0.0735) = \boxed{0.110 \text{ N}}$$

3. Force equilibrium (vertical direction) applied to metal part

- Buoyant force

$$F_{B2} = \gamma V_2 = (9800 \text{ N/m}^3)(6600 \text{ mm}^3)(10^{-9}) = 0.0647 \text{ N}$$

- Equilibrium equation

$$W_2 = T + F_{B2} = (0.110 \text{ N}) + (0.0647 \text{ N})$$

4. Mass of metal part

$$m_2 = W_2/g = \boxed{17.8 \text{ g}}$$

Review the Solution and the Process

Discussion. Notice that tension in the cord (0.11 N) is less than the weight of the metal part (0.18 N). This result is consistent with the common observation that an object will "weigh less in water than in air."

Tip. When solving problems that involve buoyancy, draw an FBD.

3.7 Predicting Stability of Immersed and Floating Bodies

Engineers calcuate whether an object will tip over or remain in an upright position when placed in a liquid, for example for the design of ships and buoys. Thus, stability is presented in this section.

Immersed Bodies

When a body is completely immersed in a liquid, its stability depends on the relative positions of the center of gravity of the body and the centroid of the displaced volume of fluid, which is called the **center of buoyancy**. If the center of buoyancy is above the center of gravity (see Fig. 3.27a) any tipping of the body produces a righting couple, and consequently, the body is stable. Alternatively, if the center of gravity is above the center of buoyancy, any tipping produces an overturning moment, thus causing the body to rotate through 180° (see Fig. 3.27c). If the center of buoyancy and center of gravity are coincident, the body is neutrally stable—that is, it lacks a tendency for righting itself or for overturning (see Fig. 3.27b).

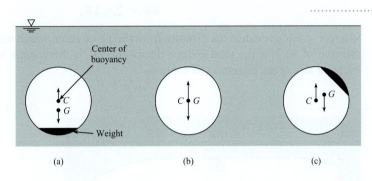

(a)   (b)   (c)

FIGURE 3.27

Conditions of stability for immersed bodies.
(a) Stable. (b) Neutral.
(c) Unstable.

Floating Bodies

The question of stability is more involved for floating bodies than for immersed bodies because the center of buoyancy may take different positions with respect to the center of gravity, depending on the shape of the body and the position in which it is floating. For example, consider the cross section of a ship shown in Fig. 3.28a. Here the center of gravity G is above the center of buoyancy C. Therefore, at first glance it would appear that the ship is unstable and could flip over. However, notice the position of C and G after the ship has taken a small angle of heel. As shown in Fig. 3.28b, the center of gravity is in the same position, but the center of buoyancy has moved outward of the center of gravity, thus producing a righting moment. A ship having such characteristics is stable.

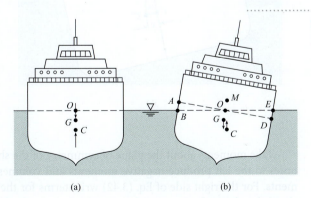

(a)   (b)

FIGURE 3.28

Ship stability relations.

The reason for the change in the center of buoyancy for the ship is that part of the original buoyant volume, as shown by the wedge shape *AOB*, is transferred to a new buoyant volume *EOD*. Because the buoyant center is at the centroid of the displaced volume, it follows that for this case the buoyant center must move laterally to the right. The point of intersection of the lines of action of the buoyant force before and after heel is called the *metacenter M,* and the distance *GM* is called the *metacentric height.* If *GM* is positive—that is, if *M* is above *G*—the ship is stable; however, if *GM* is negative, the ship is unstable. Quantitative relations involving these basic principles of stability are presented in the next paragraph.

Consider the ship shown in Fig. 3.29, which has taken a small angle of heel α. First evaluate the lateral displacement of the center of buoyancy, *CC'*; then it will be easy by simple trigonometry to solve for the metacentric height *GM* or to evaluate the righting moment. Recall that the center of buoyancy is at the centroid of the displaced volume. Therefore, resort to the fundamentals of centroids to evaluate the displacement *CC'*. From the definition of the centroid of a volume,

$$\bar{x}V = \Sigma x_i \Delta V_i \qquad (3.42)$$

where $\bar{x} = CC'$, which is the distance from the plane about which moments are taken to the centroid of V; V is the total volume displaced; ΔV_i is the volume increment; and x_i is the moment arm of the increment of volume.

FIGURE 3.29

(a) Plan view of ship at waterline.
(b) Section A-A of ship.

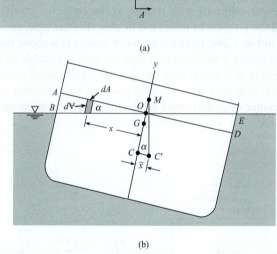

(a)

(b)

Take moments about the plane of symmetry of the ship. Recall from mechanics that volumes to the left produce negative moments and volumes to the right produce positive moments. For the right side of Eq. (3.42) write terms for the moment of the submerged volume about the plane of symmetry. A convenient way to do this is to consider the moment of the volume before heel, subtract the moment of the volume represented by the wedge *AOB*, and add the moment represented by the wedge *EOD*. In a general way this is given by the following equation:

$$\bar{x}V = \text{moment of } V \text{ before heel} - \text{moment of } V_{AOB} + \text{moment of } V_{EOD} \qquad (3.43)$$

Because the original buoyant volume is symmetrical with y-y, the moment for the first term on the right is zero. Also, the sign of the moment of V_{AOB} is negative; therefore, when this negative moment is subtracted from the right-hand side of Eq. (3.43), the result is

$$\bar{x}V = \sum x_i \Delta V_{iAOB} + \sum x_i \Delta V_{iEOD} \tag{3.44}$$

Now, express Eq. (3.44) in integral form:

$$\bar{x}V = \int_{AOB} x \, dV + \int_{EOD} x \, dV \tag{3.45}$$

But it may be seen from Fig. 3.29b that dV can be given as the product of the length of the differential volume, $x \tan \alpha$, and the differential area, dA. Consequently, Eq. (3.45) can be written as

$$\bar{x}V = \int_{AOB} x^2 \tan \alpha \, dA + \int_{EOD} x^2 \tan \alpha \, dA$$

Here $\tan \alpha$ is a constant with respect to the integration. Also, because the two terms on the right-hand side are identical except for the area over which integration is to be performed, combine them as follows:

$$\bar{x}V = \tan \alpha \int_{A_{waterline}} x^2 \, dA \tag{3.46}$$

The second moment, or moment of inertia of the area defined by the waterline, is given the symbol I_{00}, and the following is obtained:

$$\bar{x}V = I_{00} \tan \alpha$$

Next, replace $\bar{x}$ by CC' and solve for CC':

$$CC' = \frac{I_{00} \tan \alpha}{V}$$

From Fig. 3.29b,

$$CC' = CM \tan \alpha$$

Thus eliminating CC' and $\tan \alpha$ yields

$$CM = \frac{I_{00}}{V}$$

However,

$$GM = CM - CG$$

Therefore the *metacentric height* is

$$GM = \frac{I_{00}}{V} - CG \tag{3.47}$$

Equation (3.47) is used to determine the stability of floating bodies. As already noted, if GM is positive, the body is stable; if GM is negative, it is unstable.

Note that for small angles of heel α, the righting moment or overturning moment is given as follows:

$$RM = \gamma V GM \alpha \tag{3.48}$$

However, for large angles of heel, direct methods of calculation based on these same principles would have to be employed to evaluate the righting or overturning moment.

EXAMPLE 3.13

Stability of a Floating Block

Problem Statement

A block of wood 30 cm square in cross section and 60 cm long weighs 318 N. Will the block float with sides vertical as shown?

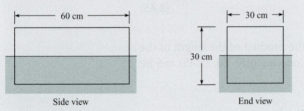

Side view   End view

Define the Situation

A block of wood is floating in water.

State the Goal

Determine the stable configuration of the block of wood.

Generate Ideas and Make a Plan

1. Apply force equilibrium to find the depth of submergence.
2. Determine if block is stable about the long axis by applying Eq. (3.47).
3. If block is not stable, repeat steps 1 and 2.

Take Action (Execute the Plan)

1. Equilibrium (vertical direction)

$$\sum F_y = 0$$

$$-\text{weight} + \text{buoyant force} = 0$$

$$-318 \text{ N} + 9810 \text{ N/m}^3 \times 0.30 \text{ m} \times 0.60 \text{ m} \times d = 0$$

$$d = 0.18 \text{ m} = 18 \text{ cm}$$

2. Stability (longitudinal axis)

$$GM = \frac{I_{00}}{\mathcal{V}} - CG = \frac{\frac{1}{12} \times 60 \times 30^3}{18 \times 60 \times 30} - (15 - 9)$$

$$= 4.167 - 6 = -1.833 \text{ cm}$$

Because the metacentric height is negative, the block is not stable about the longitudinal axis. Thus a slight disturbance will make it tip to the orientation shown below.

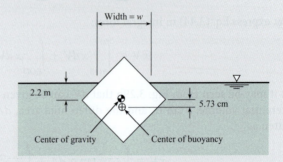

3. Equilibrium (vertical direction—see preceding figure)

$$-\text{weight} + \text{buoyant force} = 0$$

$$-(318 \text{ N}) + (9810 \text{ N/m}^3)(\mathcal{V}_D) = 0$$

$$\mathcal{V}_D = 0.0324 \text{ m}^3$$

4. Find the dimension w.

(Displaced volume)
= (Block volume) − (Volume above the waterline).

$$\mathcal{V}_D = 0.0324 \text{ m}^3 = (0.3^2)(0.6) \text{ m}^3 - \frac{w^2}{4}(0.6 \text{ m})$$

$$w = 0.379 \text{ m}$$

5. Moment of inertia at the waterline

$$I_{00} = \frac{bh^3}{12} = \frac{(0.6 \text{ m})(0.379 \text{ m})^3}{12} = 0.00273 \text{ m}^4$$

6. Metacentric height

$$GM = \frac{I_{00}}{\mathcal{V}} - CG = \frac{0.00273 \text{ m}^4}{0.0324 \text{ m}^3} - 0.0573 \text{ m} = 0.027 \text{ m}$$

Because the metacentric height is positive, the block will be stable in this position.

3.8 Summarizing Key Knowledge

Pressure and Hydrostatic Equilibrium

- A *hydrostatic condition* means that the weight of each fluid particle is balanced by the net pressure force.
- *Pressure p* is ratio of (magnitude of normal force due to a fluid) to (area) at a point.

▸ Pressure always acts to compress the material that is in contact with the fluid exerting the pressure.

▸ Pressure is a scalar quantity; not a vector.

• Engineers express pressure with gage pressure, absolute pressure, and vacuum pressure.

▸ Absolute pressure is measured relative to absolute zero.

▸ Gage pressure gives the magnitude of pressure relative to atmospheric pressure.

$$p_{\text{abs}} = p_{\text{atm}} + p_{\text{gage}}$$

▸ Vacuum pressure gives the magitude of the pressure below atmospheric pressure.

$$p_{\text{vacuum}} = p_{\text{atm}} - p_{\text{abs}}$$

Describing Pressure and Hydrostatic Equilibrium

• The weight of a fluid causes pressure to increase with increasing depth, giving the *hydrostatic differential equation*. The equations that are used in hydrostatics are derived from this equation. The hydrostatic differential equation is

$$\frac{dp}{dz} = -\gamma = -\rho g$$

• If density is constant, the hydrostatic differential equation can be integrated to give the hydrostatic equation. The meaning (i.e., physics) of the hydrostatic equation is that pizeometric head (or piezometric pressure) is everywhere constant in a static body of fluid.

$$\frac{p}{\gamma} + z = \text{constant}$$

Pressure Distributions and Forces Due to Pressure

• A fluid in contact with a surface produces a *pressure distribution*, which is a mathematical or visual description of how the pressure varies along the surface.

• To find the force due to a pressure distribution, integrate the pressure distribution over area using a normal vector to track the direction of the force acting on dA.

$$\text{Net force due to a pressure distribution} = \mathbf{F}_p = \int_A (-p)\mathbf{n}\,dA$$

• A pressure distribution is often represented as a statically equivalent force $\mathbf{F}_p$ acting at the *center of pressure* (CP)

• A *uniform pressure distribution* means that the pressure is the same at every point on a suface. Pressure distributions due to gases are typically idealized as uniform pressure distributions.

• A *hydrostatic pressure distibution* means that the pressure varies according to $dp/dz = -\gamma$

Force on a Flat Surface (Hydrostatic Pressure Distribution)

• For a panel subjected to a hydrostatic pressure distribution, the hydrostatic force is

$$F_p = \bar{p}A$$

- This hydrostatic force
  ▸ Acts *at* the centroid of area for a uniform pressure distribution
  ▸ Acts *below* the centroid of area for a hydrostatic pressure distibution. The slant distance between the center of pressure and the centroid of area is given by

$$y_{cp} - \bar{y} = \frac{\bar{I}}{\bar{y}A}$$

Hydrostatic Forces on a Curved Surface

- When a surface is curved, one can find the pressure force by applying force equilibrium to a free body comprised of the fluid in contact with the surface.

The Buoyant Force

- The *buoyant force* is the pressure force on a body that is partially or totally submerged in a fluid.
- The magnitude of the buoyant force is given by

$$\text{Buoyant force} = F_B = \text{Weight of the displaced fluid}$$

- The center of buoyancy is located at the center of gravity of the displaced fluid. The direction of the buoyant force is opposite the gravity vector.
- When the buoyant force is due to a single fluid with constant density, the magnitude of the buoyant force is:

$$F_B = \gamma \mathcal{V}_D$$

Hydrodynamic Stability

- Hydrodynamic stability means that if an object is displaced from equilibrium then there is a moment that causes the object to return to equilibrium.
- The criteria for stability are
  ▸ *Immersed object.* The body is stable if the center of gravity is below the center of buoyancy.
  ▸ *Floating object.* The body is stable if the metacentric height is positive.

REFERENCES

1. U.S. Standard Atmosphere Washington, DC: U.S. Government Printing Office, 1976.

2. Holman, J. P., and W. J. Gajda, Jr. *Experimental Methods for Engineers.* New York: McGraw-Hill, 1984.

3. Wikipedia contributors "Hydraulic machinery," Wikipedia, The Free Encyclopedia, http://en.wikipedia.org/w/index. php?title=Hydraulic_machinery&oldid=161288040 (accessed October 4, 2007).

PROBLEMS

PLUS Problem available in *WileyPLUS* at instructor's discretion.

GO Guided Online (GO) Problem, available in *WileyPLUS* at instructor's discretion.

Describing Pressure (§3.1)

3.1 PLUS A 100 mm diameter sphere contains an ideal gas at 20°C. Apply the grid method (§1.5 in Ch. 1) to calculate the density in units of kg/m^3.

   a. Gas is helium. Gage pressure is 50.8 cm H$_2$O.

   b. Gas is methane. Vacuum pressure is 20.7 kPa.

3.2 PLUS For the questions below, assume standard atmospheric pressure.

   a. For a vacuum pressure of 30 kPa, what is the absolute pressure? Gage pressure?

   b. For a pressure of 200 kPa gage, what is the absolute pressure in kPa?

3.3 PLUS The local atmospheric pressure is 99.0 kPa. A gage on an oxygen tank reads a pressure of 300 kPa gage. What is the pressure in the tank in kPa abs?

3.4 Using §3.1 and other resources, answer the following questions. Strive for depth, clarity, and accuracy while also combining sketches, words, and equations in ways that enhance the effectiveness of your communication.

   a. What are five important facts that engineers need to know about pressure?

   b. What are five common instances in which people use gage pressure?

   c. What are the most common units for pressure?

   d. Why is pressure defined using a derivative?

   e. How is pressure similar to shear stress? How does pressure differ from shear stress?

3.5 GO The Crosby gage tester shown in the figure is used to calibrate or to test pressure gages. When the weights and the piston together weigh 140 N, the gage being tested indicates 200 kPa. If the piston diameter is 30 mm, what percentage of error exists in the gage?

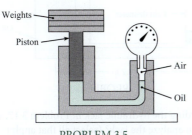

Weights

Piston

Air

Oil

PROBLEM 3.5

3.6 PLUS As shown, a mouse can use the mechanical advantage provided by a hydraulic machine to lift up an elephant.

   a. Derive an algebraic equation that gives the mechanical advantge of the hydraulic machine shown. Assume the pistons are frictionless and massless.

   b. A mouse can have a mass of 25 g and an elephant a mass of 7500 kg. Determine a value of D_1 and D_2 so that the mouse can support the elephant.

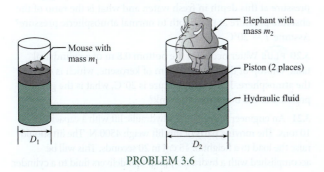

Mouse with mass m_1

Elephant with mass m_2

Piston (2 places)

Hydraulic fluid

D_1 D_2

PROBLEM 3.6

3.7 Find a parked automobile for which you have information on tire pressure and weight. Measure the area of tire contact with the pavement. Next, using the weight information and tire pressure, use engineering principles to calculate the contact area. Compare your measurement with your calculation and discuss.

Deriving and Applying the Hydrostatic Equation (§3.2)

3.8 PLUS To derive the hydrostatic equation, which of the following must be assumed? (Select all that are correct.)

   a. the specific weight is constant

   b. the fluid has no charged particles

   c. the fluid is at equilibrium

3.9 Imagine two tanks. Tank A is filled to depth h with water. Tank B is filled to depth h with oil. Which tank has the largest pressure? Why? Where in the tank does the largest pressure occur?

3.10 Consider Figure 3.8 on p. 67 of §3.2.

   a. Which fluid has the larger density?

   b. If you graphed pressure as a function of z in these two layered liquids, in which fluid does the pressure change more with each incremental change in z?

3.11 PLUS Apply the grid method (§1.5 in Ch. 1) with the hydrostatic equation ($\Delta p = \gamma \Delta z$) to each of the following cases.

   a. Predict the pressure change Δp in kPa for an elevation change Δz of 305 cm in a fluid with a density of 1.5 g/cm³.

   b. Predict the pressure change in kPa for a fluid with S = 0.8 and an elevation change of 22 m.

   c. Predict pressure change in meter of water for a fluid with a density of 1.2 kg/m³ and an elevation change of 305 m.

   d. Predict the elevation change in millimeters for a fluid with S = 13 that corresponds to a change in pressure of 17 kPa.

3.12 PLUS Using §3.2 and other resources, answer the following questions. Strive for depth, clarity, and accuracy while also combining sketches, words, and equations in ways that enhance the effectiveness of your communication.

   a. What does hydrostatic mean? How do engineers identify whether a fluid is hydrostatic?

   b. What are the common forms on the hydrostatic equation? Are the forms equivalent or are they different?

   c. What is a datum? How do engineers establish a datum?

   d. What are the main ideas of Eq. (3.10) on p. 66 of §3.2? That is, what is the meaning of this equation?

   e. What assumptions need to be satisfied to apply the hydrostatic equation?

3.13 GO Apply the grid method to each situation.

   a. What is the change in air pressure in pascals between the floor and the ceiling of a room with walls that are 305 cm tall.

   b. A diver in the ocean (S = 1.03) records a pressure of 253 kPa on her depth gage. How deep is she?

   c. A hiker starts a hike at an elevation where the air pressure is 94 kPa, and he ascends 366 m to a mountain summit.

Assuming the density of air is constant, what is the pressure in kPa at the summit?

d. Lake Pend Oreille, in northern Idaho, is one of the deepest lakes in the world, with a depth of 350 m in some locations. This lake is used as a test facility for submarines. What is the maximum pressure that a submarine could experience in this lake?

e. A 70 m tall standpipe (a standpipe is vertical pipe that is filled with water and open to the atmosphere) is used to supply water for fire fighting. What is the maximum pressure in the standpipe?

3.14 PLUS As shown, an air space above a long tube is pressurized to 50 kPa vacuum. Water (20°C) from a reservoir fills the tube to a height h. If the pressure in the air space is changed to 25 kPa vacuum, will h increase or descrease and by how much? Assume atmospheric pressure is 100 kPa.

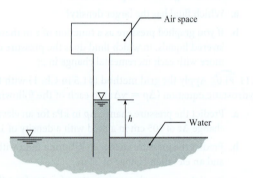

PROBLEM 3.14

3.15 PLUS For the closed tank with Bourdon-tube gages tapped into it, what is the specific gravity of the oil and the pressure reading on gage C?

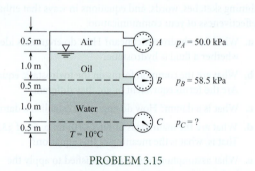

PROBLEM 3.15

3.16 This manometer contains water at room temperature. The glass tube on the left has an inside diameter of 1 mm ($d = 1.0$ mm). The glass tube on the right is three times as large. For these conditions, the water surface level in the left tube will be (a) higher than the water surface level in the right tube, (b) equal to the water surface level in the right tube, or (c) less than the water surface level in the right tube. State your main reason or assumption for making your choice.

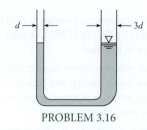

PROBLEM 3.16

3.17 PLUS If a 200 N force F_1 is applied to the piston with the 4 cm diameter, what is the magnitude of the force F_2 that can be resisted by the piston with the 10 cm diameter? Neglect the weights of the pistons.

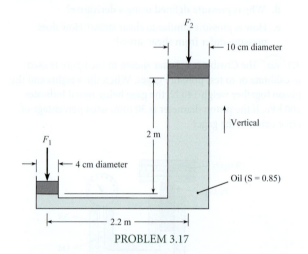

PROBLEM 3.17

3.18 Regarding the hydraulic jack in Problem 3.17, which ideas were used to analyze the jack? (select all that apply)

a. pressure = (force)(area)

b. pressure increases linearly with depth in a hydrostatic fluid

c. the pressure at the very bottom of the 4-cm chamber is larger than the pressure at the very bottom of the 10-cm chamber

d. when a body is stationary, the sum of forces on the object is zero

e. when a body is stationary, the sum of moments on the object is zero

f. pressure = (weight/volume)(change in elevation)

3.19 Some skin divers go as deep as 50 m. What is the gage pressure at this depth in fresh water, and what is the ratio of the absolute pressure at this depth to normal atmospheric pressure? Assume $T = 20°C$.

3.20 PLUS Water occupies the bottom 0.8 m of a cylindrical tank. On top of the water is 0.3 m of kerosene, which is open to the atmosphere. If the temperature is 20°C, what is the gage pressure at the bottom of the tank?

3.21 An engineer is designing a hydraulic lift with a capacity of 10 tons. The moving parts of this lift weigh 4500 N. The lift should raise the load to a height of 15 cm in 20 seconds. This will be accomplished with a hydraulic pump that delivers fluid to a cylinder.

Hydraulic cylinders with a stroke of 200 cm are available with bore sizes from 5 to 20 cm. Hydraulic piston pumps with an operating pressure range from 1380 to 20,700 kPa gage are available with pumping capacities of 20, 38, and 56 liters per minute. Select a hydraulic pump size and a hydraulic cylinder size that can be used for this application.

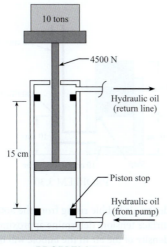

PROBLEM 3.21

3.22 (GO) A tank with an attached manometer contains water at 20°C. The atmospheric pressure is 100 kPa. There is a stopcock located 1 m from the surface of the water in the manometer. The stopcock is closed, trapping the air in the manometer, and water is added to the tank to the level of the stopcock. Find the increase in elevation of the water in the manometer assuming the air in the manometer is compressed isothermally.

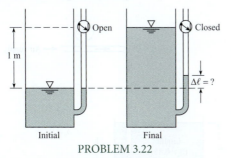

PROBLEM 3.22

3.23 (PLUS) A tank is fitted with a manometer on the side, as shown. The liquid in the bottom of the tank and in the manometer has a specific gravity (S) of 3.0. The depth of this bottom liquid is 20 cm. A 15 cm layer of water lies on top of the bottom liquid. Find the position of the liquid surface in the manometer.

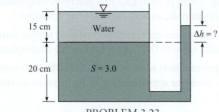

PROBLEM 3.23

3.24 (PLUS) As shown, a load acts on a piston of diameter D_1. The piston rides on a reservoir of oil of depth h_1 and specific gravity S. The reservoir is connected to a round tube of diameter D_2 and oil rises in the tube to height h_2. The oil in the tube is open to atmosphere. Derive an equation for the height h_2 in terms of the weight W of the load and other relevant variables. Neglect the weight of the piston.

3.25 As shown, a load of mass 5 kg is situated on a piston of diameter $D_1 = 120$ mm. The piston rides on a reservoir of oil of depth $h_1 = 42$ mm and specific gravity S = 0.8. The reservoir is connected to a round tube of diameter $D_2 = 5$ mm and oil rises in the tube to height h_2. Find h_2. Assume the oil in the tube is open to atmosphere and neglect the weight of the piston.

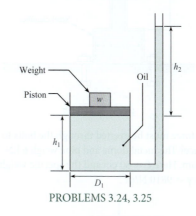

PROBLEMS 3.24, 3.25

3.26 (GO) What is the maximum gage pressure in the odd tank shown in the figure? Where will the maximum pressure occur? What is the hydrostatic force acting on the top (CD) of the last chamber on the right-hand side of the tank? Assume $T = 10°C$.

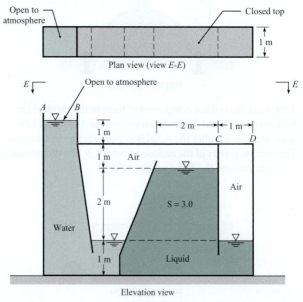

PROBLEM 3.26

3.27 PLUS The steel pipe and steel chamber shown in the figure together weigh 2700 N. What force will have to be exerted on the chamber by all the bolts to hold it in place? The dimension ℓ is equal to 75 cm. *Note:* There is no bottom on the chamber—only a flange bolted to the floor.

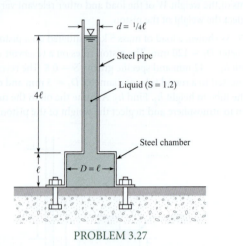

PROBLEM 3.27

3.28 What force must be exerted through the bolts to hold the dome in place? The metal dome and pipe weigh 6 kN. The dome has no bottom. Here $\ell = 80$ cm and the specific weight of the water is $\gamma = 9810$ N/m³.

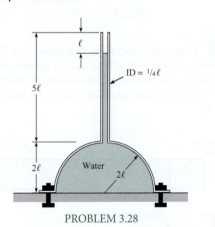

PROBLEM 3.28

3.29 Find the vertical component of force in the metal at the base of the spherical dome shown when gage A reads 35 kPa gage. Indicate whether the metal is in compression or tension. The specific gravity of the enclosed fluid is 1.5. The dimension L is 60 cm. Assume the dome weighs 4450 N.

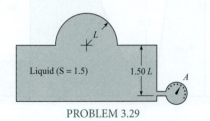

PROBLEM 3.29

3.30 GO The piston shown weighs 45 N. In its initial position, the piston is restrained from moving to the bottom of the cylinder by means of the metal stop. Assuming there is neither friction nor leakage between piston and cylinder, what volume of oil (S = 0.85) would have to be added to the 2.5 cm tube to cause the piston to rise 2.5 cm from its initial position?

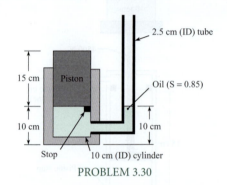

PROBLEM 3.30

3.31 Consider an air bubble rising from the bottom of a lake. Neglecting surface tension, determine approximately what the ratio of the density of the air in the bubble will be at a depth of 10 m to its density at a depth of 2.5 m.

3.32 One means of determining the surface level of liquid in a tank is by discharging a small amount of air through a small tube, the end of which is submerged in the tank, and reading the pressure on the gage that is tapped into the tube. Then the level of the liquid surface in the tank can be calculated. If the pressure on the gage is 15 kPa, what is the depth d of liquid in the tank?

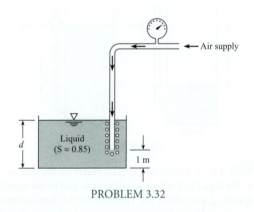

PROBLEM 3.32

Calculating Pressure in the Atmosphere (§3.2)

3.33 For Fig. 3.9 on p. 70 of §3.2 that describes temperature variation with altitude, answer the following questions.

  a. Does the linear approximation relating temperature to altitude apply in the troposphere or the stratosphere?

  b. At approximately what altitude in the earth's atmosphere does the linear approximation for temperature variation fail?

3.34 The boiling point of water decreases with elevation because of the pressure change. What is the boiling point of water at an elevation of 2000 m and at an elevation of 4000 m for standard atmospheric conditions?

3.35 From a depth of 10 m in a lake to an elevation of 4000 m in the atmosphere, plot the variation of absolute pressure. Assume that the lake water surface elevation is at mean sea level and assume standard atmospheric conditions.

3.36 PLUS Assume that a woman must breathe a constant mass rate of air to maintain her metabolic processes. If she inhales and exhales 16 times per minute at sea level, where the temperature is 15°C and the pressure is 101 kPa, what would you expect her rate of breathing at 5486 m to be? Use standard atmospheric conditions.

3.37 A pressure gage in an airplane indicates a pressure of 95 kPa at takeoff, where the airport elevation is 1 km and the temperature is 10°C. If the standard lapse rate of 5.87°C/km is assumed, at what elevation is the plane when a pressure of 75 kPa is read? What is the temperature for that condition?

3.38 Denver, Colorado, is called the "mile-high" city. What are the pressure, temperature, and density of the air when standard atmospheric conditions prevail?

3.39 PLUS An airplane is flying at 10 km altitude in a U.S. standard atmosphere. If the internal pressure of the aircraft interior is 100 kPa, what is the outward force on a window? The window is flat and has an elliptical shape with lengths of 300 mm along the major axis and 200 mm along the minor axis.

3.40 The mean atmospheric pressure on the surface of Mars is 0.7 kPa, and the mean surface temperature is -63°C. The atmosphere consists primarily of CO_2 (95.3%) with small amounts of nitrogen and argon. The acceleration due to gravity on the surface is 3.72 m/s². Data from probes entering the Martian atmosphere show that the temperature variation with altitude can be approximated as constant at -63°C from the Martian surface to 14 km, and then a linear decrease with a lapse rate of 1.5°C/km up to 34 km. Find the pressure at 8 km and 30 km altitude. Assume the atmosphere is pure carbon dioxide. Note that the temperature distribution in the atmosphere of Mars differs from that of Earth because the region of constant temperature is adjacent to the surface and the region of decreasing temperature starts at an altitude of 14 km.

3.41 Design a computer program that calculates the pressure and density for the U.S. standard atmosphere from 0 to 30 km altitude. Assume the temperature profiles are linear and are approximated by the following ranges, where z is the altitude in kilometers:

0–13.72 km	$T = 23.1 - 5.87z$ (°C)
13.7–16.8 km	$T = -57.5$°C
16.8–30 km	$T = -57.5 + 1.387(z - 16.8)$°C

Measuring Pressure (§3.3)

3.42 Match the following pressure-measuring devices with the correct name. The device names are: barometer, Bourdon gage, piezometer, manometer, and pressure transducer.

a. A vertical or U-shaped tube where changes in pressure are documented by changes in relative elevation of a liquid that is usually denser than the fluid in the system measured; can be used to measure vacuum.

b. Typically contains a diaphragm, a sensing element, and conversion to an electric signal.

c. A round face with a scale to measure needle deflection, where the needle is deflected by changes in extension of a coiled hollow tube.

d. A vertical tube where a liquid rises in response to a positive gage pressure.

e. An instrument used to measure atmospheric pressure; of various designs.

Applying the Manometer Equations (§3.3)

3.43 PLUS Which is the more correct way to describe the two summation (Σ) terms of the manometer equation, Eq (3.21) on p. 74 of §3.3?

a. Add the downs and subtract the ups.

b. Subtract the downs and add the ups.

3.44 PLUS Using the Internet and other resources, answer the following questions:

a. What are three common types of manometers? For each type, make a sketch and give a brief description.

b. How would you build a manometer from materials that are commonly available? Sketch your design concept.

3.45 PLUS Is the gage pressure at the center of the pipe (a) negative, (b) zero, or (c) positive? Neglect surface tension effects and state your rationale.

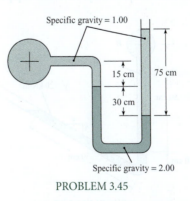

PROBLEM 3.45

3.46 Determine the gage pressure at the center of the pipe (point A) in pascal when the temperature is 21°C with $h_1 = 40$ cm and $h_2 = 5$ cm.

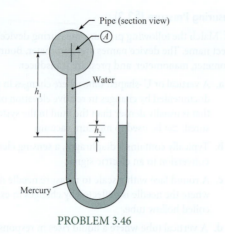

PROBLEM 3.46

3.47 WILEY PLUS Considering the effects of surface tension, estimate the gage pressure at the center of pipe A for $h = 120$ mm and $T = 20°C$.

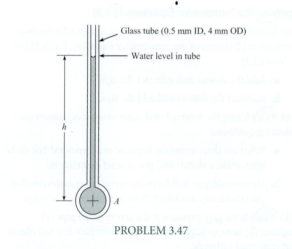

PROBLEM 3.47

3.48 WILEY PLUS What is the pressure at the center of pipe B?

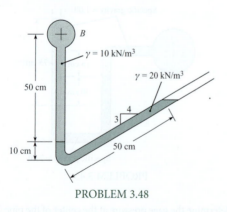

PROBLEM 3.48

3.49 The ratio of container diameter to tube diameter is 8. When air in the container is at atmospheric pressure, the free surface in

the tube is at position 1. When the container is pressurized, the liquid in the tube moves 40 cm up the tube from position 1 to position 2. What is the container pressure that causes this deflection? The liquid density is 1200 kg/m³.

3.50 The ratio of container diameter to tube diameter is 10. When air in the container is at atmospheric pressure, the free surface in the tube is at position 1. When the container is pressurized, the liquid in the tube moves 1 m up the tube from position 1 to position 2. What is the container pressure that causes this deflection? The specific weight of the liquid is 7850 N/m³.

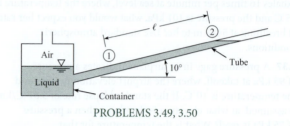

PROBLEMS 3.49, 3.50

3.51 WILEY PLUS Determine the gage pressure at the center of pipe A in kilopascals.

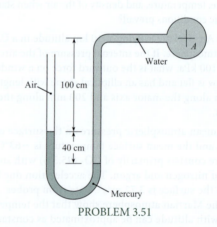

PROBLEM 3.51

3.52 A device for measuring the specific weight of a liquid consists of a U-tube manometer as shown. The manometer tube has an internal diameter of 0.5 cm and originally has water in it. Exactly 2 cm³ of unknown liquid is then poured into one leg of the manometer, and a displacement of 5 cm is measured between the surfaces as shown. What is the specific weight of the unknown liquid?

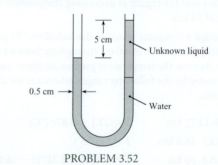

PROBLEM 3.52

3.53 Mercury is poured into the tube in the figure until the mercury occupies 375 mm of the tube's length. An equal volume of water is then poured into the left leg. Locate the water and mercury surfaces. Also determine the maximum pressure in the tube.

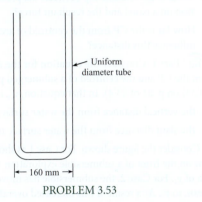

PROBLEM 3.53

3.54 ^{WILEY} PLUS Find the pressure at the center of pipe A. $T = 10°C$.

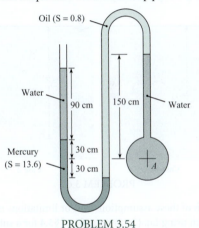

PROBLEM 3.54

3.55 Determine (a) the difference in pressure and (b) the difference in piezometric head between points A and B. The elevations z_A and z_B are 10 m and 11 m, respectively, $\ell_1 = 1$ m, and the manometer deflection ℓ_2 is 50 cm.

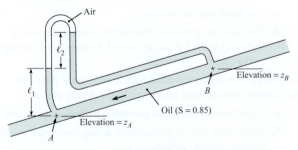

PROBLEM 3.55

3.56 The deflection on the manometer is h meters when the pressure in the tank is 150 kPa absolute. If the absolute pressure in the tank is doubled, what will the deflection on the manometer be?

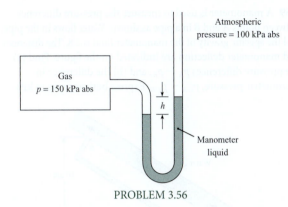

PROBLEM 3.56

3.57 ^{WILEY} PLUS A vertical conduit is carrying oil (S = 0.95). A differential mercury manometer is tapped into the conduit at points A and B. Determine the difference in pressure between A and B when $h = 7.5$ cm. What is the difference in piezometric head between A and B?

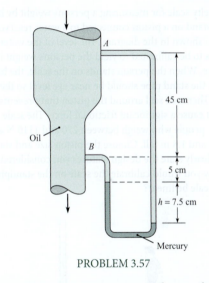

PROBLEM 3.57

3.58 Two water manometers are connected to a tank of air. One leg of the manometer is open to 100 kPa pressure (absolute) while the other leg is subjected to 90 kPa. Find the difference in deflection between both manometers, $\Delta h_a - \Delta h_b$.

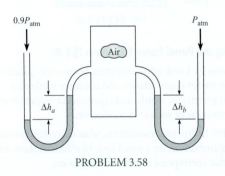

PROBLEM 3.58

3.59 A manometer is used to measure the pressure difference between points A and B in a pipe as shown. Water flows in the pipe, and the specific gravity of the manometer fluid is 2.8. The distances and manometer deflection are indicated on the figure. Find (a) the pressure differences $p_A - p_B$, and (b) the difference in piezometric pressure, $p_{z,A} - p_{z,B}$. Express both answers in kPa.

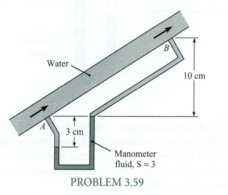

PROBLEM 3.59

3.60 A novelty scale for measuring a person's weight by having the person stand on a piston connected to a water reservoir and stand pipe is shown in the diagram. The level of the water in the stand pipe is to be calibrated to yield the person's weight in pounds force. When the person stands on the scale, the height of the water in the stand pipe should be near eye level so the person can read it. There is a seal around the piston that prevents leaks but does not cause a significant frictional force. The scale should function for people who weigh between 270 and 1110 N and are between 1.2 and 1.8 m tall. Choose the piston size and standpipe diameter. Clearly state the design features you considered. Indicate how you would calibrate the scale on the standpipe. Would the scale be linear?

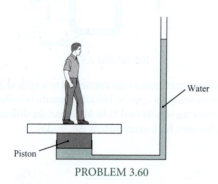

PROBLEM 3.60

Applying the Panel Force Equations (§3.4)

3.61 Using §3.4 and other resources, answer the questions below. Strive for depth, clarity, and accuracy while also combining sketches, words, and equations in ways that enhance the effectiveness of your communication.

    a. For hydrostatic conditions, what do typical pressure distributions on a panel look like? Sketch three examples that correspond to different situations.

    b. What is a center of pressure (CP)? What is a centroid of area?

    c. In Eq. (3.28) on p. 80 of §3.4, what does $\bar{p}$ mean? What factors influence the value of $\bar{p}$?

    d. What is the relationship between the pressure distribution on a panel and the resultant force?

    e. How far is the CP from the centroid of area? What factors influence this distance?

3.62 GO Part 1. Consider the equation for the distance between the CP and the centroid of a submerged panel (Eq. (3.33) on p. 81 of §3.4). In that equation, y_{cp} is

    a. the vertical distance from the water surface to the CP.

    b. the slant distance from the water surface to the CP.

Part 2. Consider the figure shown. For Case 1 as shown, the viewing window on the front of a submersible exploration vehicle is at a depth of y_1. For Case 2, the submersible has moved deeper in the ocean, to y_2. As a result of this increased overall depth of the submersible and its window, does the spacing between the CP and centroid (a) get larger, (b) stay the same, or (c) get smaller?

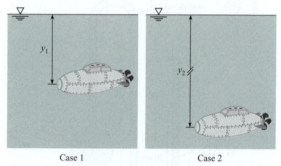

Case 1   Case 2

PROBLEM 3.62

3.63 Which of these assumptions and/or limitations must be known when using Eq. (3.33) on p. 81 of §3.4 for a submerged surface or panel to calculate the distance between the centroid of the panel and the center of pressure of the hydrostatic force (select all that apply):

    a. The equation only applies to a single fluid of constant density

    b. The pressure at the surface must be p = 0 gage

    c. The panel must be vertical

    d. The equation gives only the vertical location (as a slant distance) to the CP, not the lateral distance from the edge of the body

3.64 PLUS Two cylindrical tanks have bottom areas A and $4A$ respectively, and are filled with water to the depths shown.

    a. Which tank has the higher pressure at the bottom of the tank?

    b. Which tank has the greater force acting downward on the bottom circular surface?

3.65 PLUS What is the force acting on the gate of an irrigation ditch if the ditch and gate are 1.2 m wide, 1.2 m deep, and the ditch is competely full of water? There is no water on the other side of the gate. The weather has been hot for weeks, so the water is 21°C.

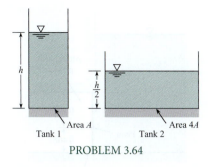

PROBLEM 3.64

3.66 WILEY PLUS Consider the two rectangular gates shown in the figure. They are both the same size, but gate A is held in place by a horizontal shaft through its midpoint and gate B is cantilevered to a shaft at its top. Now consider the torque T required to hold the gates in place as H is increased. Choose the valid statement(s): (a) T_A increases with H. (b) T_B increases with H. (c) T_A does not change with H. (d) T_B does not change with H.

3.67 WILEY PLUS For gate A, choose the statements that are valid: (a) The hydrostatic force acting on the gate increases as H increases. (b) The distance between the CP on the gate and the centroid of the gate decreases as H increases. (c) The distance between the CP on the gate and the centroid of the gate remains constant as H increases. (d) The torque applied to the shaft to prevent the gate from turning must be increased as H increases. (e) The torque applied to the shaft to prevent the gate from turning remains constant as H increases.

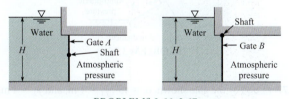

PROBLEMS 3.66, 3.67

3.68 WILEY PLUS As shown, water (15°C) is in contact with a square panel; $d = 1$ m and $h = 2$ m.

a. Calculate the depth of the centroid

b. Calculate the resultant force on the panel

c. Calculate the distance from the centroid to the CP.

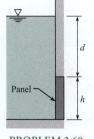

PROBLEM 3.68

3.69 WILEY GO As shown, a round viewing window of diameter $D = 0.5$ m is situated in a large tank of seawater ($S = 1.03$). The top of the window is 1.5 m below the water surface, and the window is angled at 60° with respect to the horizontal. Find the hydrostatic force acting on the window and locate the corresponding CP.

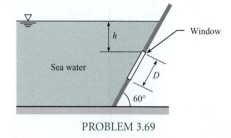

PROBLEM 3.69

3.70 WILEY PLUS Find the force of the gate on the block. See sketch.

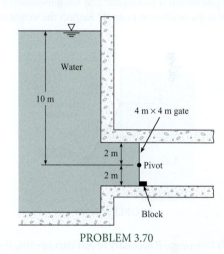

PROBLEM 3.70

3.71 Assume that wet concrete ($\gamma = 23{,}600$ N/m³) behaves as a liquid. Determine the force per meter of length exerted on the forms. If the forms are held in place as shown, with ties between vertical braces spaced every 60 cm, what force is exerted on the bottom tie?

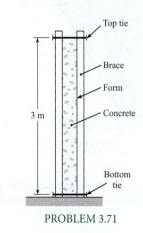

PROBLEM 3.71

3.72 PLUS A rectangular gate is hinged at the water line, as shown. The gate is 1.2 m high and 1.8 m wide. The specific weight of water is 9810 N/m³. Find the necessary force applied at the bottom of the gate to keep it closed.

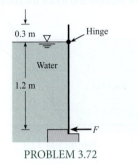

PROBLEM 3.72

3.73 The gate shown is rectangular and has dimensions 6 m by 4 m. What is the reaction at point A? Neglect the weight of the gate.

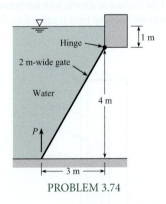

PROBLEM 3.73

3.74 PLUS Determine P necessary to just start opening the 2 m–wide gate.

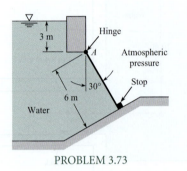

PROBLEM 3.74

3.75 PLUS The square gate shown is eccentrically pivoted so that it automatically opens at a certain value of h. What is that value in terms of ℓ?

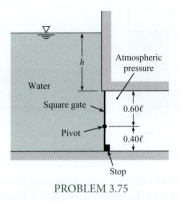

PROBLEM 3.75

3.76 GO This 3 m–diameter butterfly valve is used to control the flow in a 3 m–diameter outlet pipe in a dam. In the position shown, it is closed. The valve is supported by a horizontal shaft through its center. What torque would have to be applied to the shaft to hold the valve in the position shown?

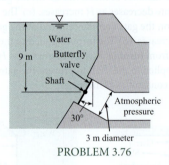

PROBLEM 3.76

3.77 PLUS For the gate shown, $\alpha = 45°$, $y_1 = 1$ m, and $y_2 = 4$ m. Will the gate fall or stay in position under the action of the hydrostatic and gravity forces if the gate itself weighs 150 kN and is 1.0 m wide? Assume $T = 10°C$. Use calculations to justify your answer.

3.78 PLUS For this gate, $\alpha = 45°$, $y_1 = 1$ m, and $y_2 = 2$ m. Will the gate fall or stay in position under the action of the hydrostatic and gravity forces if the gate itself weighs 80,000 N and is 1 m wide? Assume $T = 10°C$. Use calculations to justify your answer.

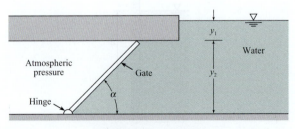

PROBLEMS 3.77, 3.78

3.79 Determine the hydrostatic force F on the triangular gate, which is hinged at the bottom edge and held by the reaction R_T at the upper corner. Express F in terms of γ, h, and W. Also determine the ratio R_T/F. Neglect the weight of the gate.

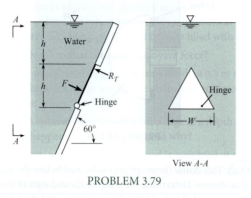

View A-A

PROBLEM 3.79

3.80 PLUS In constructing dams, the concrete is poured in lifts of approximately 1.5 m ($y_1 = 1.5$ m). The forms for the face of the dam are reused from one lift to the next. The figure shows one such form, which is bolted to the already cured concrete. For the new pour, what moment will occur at the base of the form per meter of length (normal to the page)? Assume that concrete acts as a liquid when it is first poured and has a specific weight of 24 kN/m³.

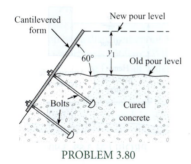

PROBLEM 3.80

3.81 The plane rectangular gate can pivot about the support at B. For the conditions given, is it stable or unstable? Neglect the weight of the gate. Justify your answer with calculations.

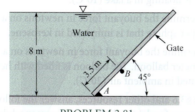

PROBLEM 3.81

Calculating Pressure on Curved Surfaces (§3.5)

3.82 PLUS Two hemispheric shells are perfectly sealed together, and the internal pressure is reduced to 25% of atmospheric pressure. The inner radius is 10.5 cm, and the outer radius is 10.75 cm. The seal is located halfway between the inner and

outer radius. If the atmospheric pressure is 101.3 kPa, what force is required to pull the shells apart?

3.83 If exactly 20 bolts of 2.5 cm diameter are needed to hold the air chamber together at A-A as a result of the high pressure within, how many bolts will be needed at B-B? Here $D = 40$ cm and $d = 20$ cm.

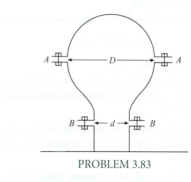

PROBLEM 3.83

3.84 For the plane rectangular gate ($\ell \times w$ in size), figure (a), what is the magnitude of the reaction at A in terms of γ_w and the dimensions ℓ and w? For the cylindrical gate, figure (b), will the magnitude of the reaction at A be greater than, less than, or the same as that for the plane gate? Neglect the weight of the gates.

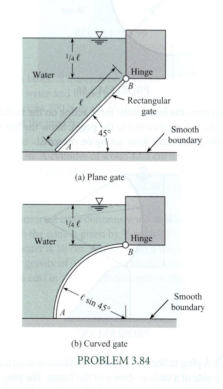

(a) Plane gate

(b) Curved gate

PROBLEM 3.84

3.85 Water is held back by this radial gate. Does the resultant of the pressure forces acting on the gate pass above the pin, through the pin, or below the pin?

3.103 PLUS The floating platform shown is supported at each corner by a hollow sealed cylinder 1 m in diameter. The platform itself weighs 30 kN in air, and each cylinder weighs 1.0 kN per meter of length. What total cylinder length L is required for the platform to float 1 m above the water surface? Assume that the specific weight of the water (brackish) is 10,000 N/m³. The platform is square in plan view.

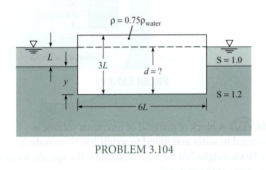

PROBLEM 3.103

3.104 To what depth d will this rectangular block (with density 0.75 times that of water) float in the two-liquid reservoir?

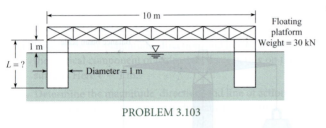

PROBLEM 3.104

3.105 PLUS Determine the minimum volume of concrete ($\gamma = 23.6$ kN/m³) needed to keep the gate (1 m wide) in a closed position, with $\ell = 2$ m. Note the hinge at the bottom of the gate.

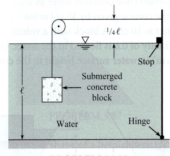

PROBLEM 3.105

3.106 A cylindrical container 1.2 m high and 0.6 m in diameter holds water to a depth of 0.6 m. How much does the level of the water in the tank change when a 22 N block of ice is placed in the container? Is there any change in the water level in the tank when the block of ice melts? Does it depend on the specific gravity of the ice? Explain all the processes.

3.107 PLUS The partially submerged wood pole is attached to the wall by a hinge as shown. The pole is in equilibrium under the action of the weight and buoyant forces. Determine the density of the wood.

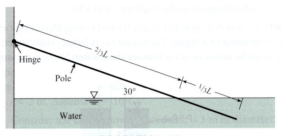

PROBLEM 3.107

3.108 A gate with a circular cross section is held closed by a lever 1 m long attached to a buoyant cylinder. The cylinder is 25 cm in diameter and weighs 200 N. The gate is attached to a horizontal shaft so it can pivot about its center. The liquid is water. The chain and lever attached to the gate have negligible weight. Find the length of the chain such that the gate is just on the verge of opening when the water depth above the gate hinge is 10 m.

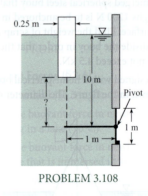

PROBLEM 3.108

3.109 A balloon is to be used to carry meteorological instruments to an elevation of 4570 m where the air pressure is 56 kPa abs. The balloon is to be filled with helium, and the material from which it is to be fabricated weighs 0.5 N/m². If the instruments weigh 35 N, what diameter should the spherical balloon have?

3.110 A weather balloon is constructed of a flexible material such that the internal pressure of the balloon is always 10 kPa higher than the local atmospheric pressure. At sea level the diameter of the balloon is 1 m, and it is filled with helium. The balloon material, structure, and instruments have a mass of 100 g. This does not include the mass of the helium. As the balloon rises, it will expand. The temperature of the helium is always equal to the local atmospheric temperature, so it decreases as the balloon gains altitude. Calculate the maximum altitude of the balloon in a standard atmosphere.

Measuring ρ, γ, and S with Hydrometers (§3.6)

3.111 PLUS The hydrometer shown weighs 0.015 N. If the stem sinks 6.0 cm in oil ($z = 6.0$ cm), what is the specific gravity of the oil?

3.112 PLUS The hydrometer shown sinks 5.3 cm ($z = 5.3$ cm) in water (15°C). The bulb displaces 1.0 cm³, and the stem area is 0.1 cm². Find the weight of the hydrometer.

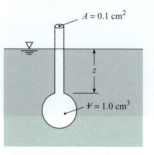

$A = 0.1 \text{ cm}^2$

z

$V = 1.0 \text{ cm}^3$

PROBLEMS 3.111, 3.112

3.113 GO A common commercial hydrometer for measuring the amount of antifreeze in the coolant system of an automobile engine consists of a chamber with differently colored balls. The system is calibrated to give the range of specific gravity by distinguishing between the balls that sink and those that float. The specific gravity of an ethylene glycol-water mixture varies from 1.012 to 1.065 for 10% to 50% by weight of ethylene glycol. Assume there are six balls, 1 cm in diameter each, in the chamber. What should the weight of each ball be to provide a range of specific gravities between 1.01 and 1.06 with 0.01 intervals?

3.114 PLUS A hydrometer with the configuration shown has a bulb diameter of 2 cm, a bulb length of 8 cm, a stem diameter of 1 cm, a length of 8 cm, and a mass of 40 g. What is the range of specific gravities that can be measured with this hydrometer?

(*Hint:* Liquid levels range between bottom and top of stem.)

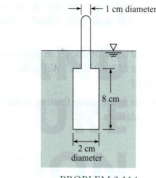

1 cm diameter

8 cm

2 cm diameter

PROBLEM 3.114

Predicting Stability (§3.7)

3.115 A barge 6 m wide and 12 m long is loaded with rocks as shown. Assume that the center of gravity of the rocks and barge is located along the centerline at the top surface of the barge. If the rocks and the barge weigh 1780 kN, will the barge float upright or tip over?

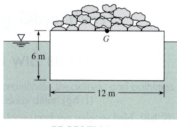

G

6 m

12 m

PROBLEM 3.115

3.116 A floating body has a square cross section with side w as shown in the figure. The center of gravity is at the centroid of the cross section. Find the location of the water line, ℓ/w, where the body would be neutrally stable ($GM = 0$). If the body is floating in water, what would be the specific gravity of the body material?

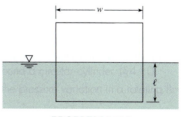

w

ℓ

PROBLEM 3.116

4.1 Describing Streamlines, Streaklines, and Pathlines

To visualize and describe flowing fluids, engineers use the streamline, streakline, and pathline. Hence, these topics are introduced in this section.

Pathlines and Streaklines

The **pathline** is the path of a fluid particle as it moves through a flow field. For example, when the wind blows a leaf, this provides an idea about what the flow is doing. If we imagine that the leaf is tiny and attached to a particle of air as this particle moves, then the motion of the leaf will reveal the motion of the particle. Another way to think of a pathline is to imagine attaching a light to a fluid particle. A time exposure photograph taken of the moving light would be the pathline. One way to reveal pathlines in a flow of water is to add tiny beads that are neutrally buoyant so that bead motion is the same as motion of fluid particles. Observing these beads as they move through the flow reveals the pathline of each particle.

The **streakline** is the line generated by a tracer fluid, such as a dye, continuously injected into a flow field at a starting point. For example, if smoke is introduced into a flow of air, the resulting lines are streaklines. Streaklines are shown in Fig. 4.1. These streaklines were produced by vaporizing mineral oil on a vertical wire that was heated by passing an electrical current through the wire.

Streamlines

The **streamline** is defined as a line that is everywhere tangent to the local velocity vector.

> **EXAMPLE.** The flow pattern for water draining through an opening in a tank (Fig. 4.2a) can be visualized by examining streamlines. Notice that velocity vectors at points a, b, and c are tangent to the streamlines. Also, the streamlines adjacent to the wall follow the contour of the wall because the fluid velocity is parallel to the wall. The generation of a flow pattern is an effective way of illustrating the flow field.

Streamlines for flow around an airfoil (Fig. 4.2b) reveal that part of the flow goes over the airfoil and part goes under. The flow is separated by the **dividing streamline**. At the location

FIGURE 4.2

(a) Flow through an opening in a tank.
(b) Flow over an airfoil section.

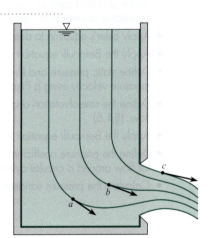

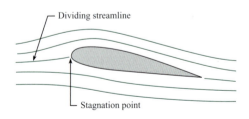

Dividing streamline

Stagnation point

(a)

(b)

where the dividing streamline intersects the body, the velocity will be zero with respect to the body. This is called the stagnation **point**.

Streamlines for flow over an Volvo ECC prototype (Fig. 4.3) allow engineers to assess aerodynamic features of the flow and possibly change the shape to achieve better performance, such as reduced drag.

FIGURE 4.3

Predicted streamline pattern over the Volvo ECC prototype. (Courtesy of Analytical Methods, VSAERO software, Volvo Concept Center.)

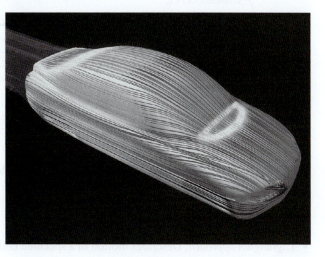

Comparing Streamlines, Streaklines, and Pathlines

When flow is steady, the pathline, streakline, and streamline look the same so long as they all pass through the same point. Thus, the streakline, which can be revealed by experimental means, will show what the streamline looks like. Similarly, a particle in the flow will follow a line traced out of a streakline.

When flow is unsteady, then the streamline, streaklines, and pathlines look different. A captivating film entitled *Flow Visualization* (1) shows how and why the streamline, streakline, and pathline differ in unsteady flow.

EXAMPLE. To show how pathlines, streaklines, and streamlines differ in unsteady flow, consider a two-dimensional flow that initially has horizontal streamlines (Fig. 4.4). At a given time, t_0, the flow instantly changes direction, and the flow moves upward to the right at 45° with no further change. A fluid particle is tracked from the starting point, and up to time t_0, the pathline is the horizontal line segment shown on Fig. 4.4a. After time t_0, the particle continues to follow the streamline and moves up the right as shown in Fig. 4.4b. Both line segments constitute the pathline. Notice in Fig. 4.4b that the pathline (black dotted line) differs from an streamline for ($t < t_0$) and any streamline for ($t > t_0$). Thus, the pathline and the streamline are not the same.

Next consider the streakline by introducing black tracer fluid as shown in Figures 4.4c and d. As shown, the streakline in Fig. 4.4d differs from the pathline and from any streamline.

FIGURE 4.4

Streamlines, pathlines, and streakline for an unsteady flow field.

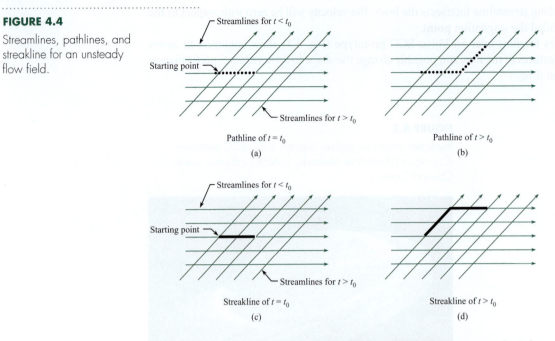

FIGURE 4.4

Streamlines, pathlines, and streakline for an unsteady flow field.

4.2 Characterizing Velocity of a Flowing Fluid

This section introduces *velocity* and the *velocity field*. Then, these ideas are used to introduce two alternative methods for describing motion.

- *Lagrangian approach*: Describes motion of a specified collection of matter.
- *Eulerian approach*. Describes motion at locations in space.

Describing Velocity

Velocity, a property of a fluid particle, gives the speed and direction of travel of the particle at an instant in time. The mathematical definition of velocity is:

$$\mathbf{V}_A = \frac{d\mathbf{r}_A}{dt} \tag{4.1}$$

where $\mathbf{V}_A$ is the velocity of particle A, and $\mathbf{r}_A$ is the position of particle A at time t.

> **EXAMPLE.** When water drains from a tank (Fig. 4.5a), $\mathbf{V}_A$ gives the speed and direction of travel of the particle at point A. The velocity $\mathbf{V}_A$ is the time rate of change of the vector $\mathbf{r}_A$.

FIGURE 4.5

Water draining out of a tank. (a) The velocity of Particle A is the time derivative of the position. (b) The velocity field represents the velocity of each fluid particle throughout the region of flow.

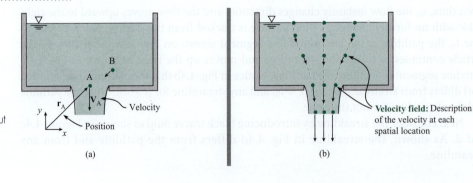

Velocity Field

A description of the velocity of each fluid particle in a flow is called a **velocity field**. In general each fluid particle in a flow has a different velocity. For example, particles A and B in Fig 4.5a have different velocities. Thus, the velocity field describes how the velocity varies with position (see Fig. 4.5b).

A velocity field can be described visually (Fig. 4.5b) or mathematically as shown by the following example.

EXAMPLE. A steady, two-dimensional velocity field in a corner is given by

$$\mathbf{V} = (2x \text{ s}^{-1})\mathbf{i} - (2y \text{ s}^{-1})\mathbf{j} \tag{4.2}$$

where x and y are position coordinates measured in meters, and $\mathbf{i}$ and $\mathbf{j}$ are unit vectors in the x and y directions, respectively.

When a velocity field is given by an equation, a plot can help one visualize the flow. For example, select the location $(x, y) = (1, 1)$ and then substitute $x = 1.0$ meter and $y = 1.0$ meter into Eq. (4.2) to give the velocity as

$$\mathbf{V} = (2 \text{ m/s})\mathbf{i} - (2 \text{ m/s})\mathbf{j} \tag{4.3}$$

Plot this point and repeat this process at other points to create Fig. 4.6a. Last, one can use definition of the streamline (line that is everywhere tangent to the velocity vector) to create a streamline pattern (Fig. 4.6b).

FIGURE 4.6

The velocity field specified by Eq. (4.2): (a) velocity vectors, and (b) the streamline pattern.

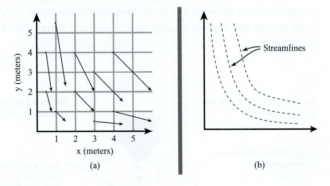

Summary The velocity field describes the velocity of each fluid particle in a spatial region. The velocity field can be shown visually as in Figs. 4.5 and 4.6 or described mathematically as in Eq. 4.2.

The concept of a field can be generalized. A **field** is a mathematical or visual description of a variable as a function of position and time.

EXAMPLES. A pressure field describes the distribution of pressure at various points in space and time. A temperature field describes the distribution of temperature at various points in space and time.

A field can be scalar valued (e.g., temperature field, pressure field) or a field can be vector valued (e.g., velocity field, acceleration field).

✔ CHECKPOINT PROBLEM 4.1

A velocity field is given as $\mathbf{V} = (ax + by)\mathbf{i}$ where $a = b = 2\text{ s}^{-1}$ and (x, y) is the position in the field in meters. A particle moving in this field

    a. Moves in the x-direction only

    b. Moves in the y-direction only

    c. Moves in both the x- and y-directions.

The Eulerian and Lagrangian Approaches

In solid mechanics, it is straightforward to describe the motion of a particle or a rigid body. In contrast, the particles in a flowing fluid move in more complicated ways and it is not practical to track the motion of each particle. Thus, researchers invented a second way to describe motion.

The first way to describe motion (called the **Lagrangian approach**) involves selecting a body and then describing the motion of this body. The second way (called the **Eulerian approach**) involves selecting a region in space and then describing the motion that is occurring at points in space. In addition, the Eulerian approach allows properties to be evaluated at spatial locations as a function of time. This is because the Eulerian approach uses fields.

> **EXAMPLE.** Consider falling particles (Fig. 4.7). The Lagrangian approach uses equations that describe an individual particle. The Eulerian approach uses an equation for the *velocity field*. Although the equations of the two approaches are different, they predict the same values of velocity. Note that the equation $v = \sqrt{2g|z|}$ in Fig. 4.7 was derived by letting the kinetic energy of the particle equal the change in gravitational potential energy.

FIGURE 4.7

This figure shows small particles released from rest and falling under the action of gravity. Equations on the left side of the image show how motion is described using a Lagrangian approach. Equations on the right side show an Eulerian approach.

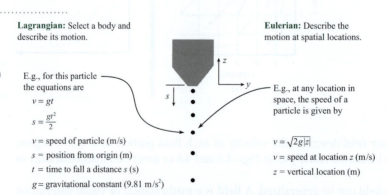

Lagrangian: Select a body and describe its motion.

E.g., for this particle the equations are

$v = gt$

$s = \dfrac{gt^2}{2}$

v = speed of particle (m/s)

s = position from origin (m)

t = time to fall a distance s (s)

g = gravitational constant (9.81 m/s²)

Eulerian: Describe the motion at spatial locations.

E.g., at any location in space, the speed of a particle is given by

$v = \sqrt{2g|z|}$

v = speed at location z (m/s)

z = vertical location (m)

When the ideas in Fig. 4.7 are generalized, the independent variables of the Lagrangian approach are initial position and time. The independent variable of the Eulerian approach are position in the field and time. Table 4.1 compares the Lagrangian and the Eulerian approaches.

TABLE 4.1 Comparison of the Lagrangian and the Eulerian Approaches

Feature	Lagrangian Approach	Eulerian Approach
Basic idea	Observe or describe the motion of a body of matter of fixed identity.	Observe or describe the motion of matter at spatial locations.
Solid mechanics (application)	Used in dynamics.	Used in elasticity. Can be used to model the flow of materials.
Fluid mechanics (application)	Fluid mechanics uses Eulerian ideas (e.g., fluid particle, streakline, acceleration of a fluid particle). Equations in fluid mechanics are often derived from an Lagrangian viewpoint.	Nearly all mathematical equations in fluid mechanics are written using the Eulerian approach.
Independent variables	Initial position (x_0, y_0, z_0) and time (t).	Spatial location (x, y, z) and time (t).
Mathematical complexity	Simpler.	More complex; e.g., partial derivatives and nonlinear terms appear.
Field concept	Not used in the Lagrangian approach.	The field is an Eulerian concepts. When fields are used, the mathematics often includes the divergence, gradient, and curl.
Types of systems used	Closed systems, particles, rigid bodies, system-of-particles.	Control volumes.

Representing Velocity Using Components

When the velocity field is represented in Cartesian components the mathematical form is

$$\mathbf{V} = u(x, y, z, t)\mathbf{i} + u(x, y, z, t)\mathbf{j} + u(x, y, z, t)\mathbf{k} \qquad (4.4)$$

where $u = u(x, y, z, t)$ is the x-component of the velocity vector in and $\mathbf{i}$ is a unit vector in the x direction. The coordinates (x, y, z) give the spatial location in the field and t is time. Similarly, the components v and w give the y- and z-components of the velocity vector.

Another way to represent a velocity is to use *normal and tangential components*. In this approach (Fig. 4.8), unit vectors are attached to the particle and move with the particle. The tangential unit vector $\mathbf{u}_t$ is tangent to the path of the particle and the normal unit vector $\mathbf{u}_n$ is normal to path and directed inward toward the center of curvature. The position coordinate s measures distance traveled along the path. The velocity of a fluid particle is represented as $\mathbf{V} = V(s, t)\mathbf{u}_t$ where V is the speed of the particle and t is time.

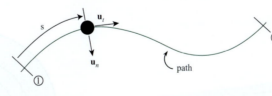

FIGURE 4.8

Describing motion of a fluid particle using normal and tangential components.

4.3 Describing Flow

Engineers use many words to describe flowing fluids. Speaking and understanding this language is seminal to professional practice. Thus, this section introduces concepts for describing flowing fluids. Because there are many ideas, a summary table is presented (see Table 4.4 on page 153).

Uniform and Nonuniform Flow

To introduce uniform flow, consider a velocity field of the form

$$\mathbf{V} = \mathbf{V}(s, t)$$

where s is distance traveled by a fluid particle along a path, and t is time (Fig. 4.9). This mathematical representation is called *normal and tangential components*. This approach is useful when the path of a particle is known.

In a **uniform flow**, the velocity is constant in magnitude and direction along a streamline at each instant in time. In uniform flow the streamlines must be rectilinear, which means straight and parallel (see Fig. 4.10). Uniform flow can be described by an equation.

$$\left(\frac{\partial \mathbf{V}}{\partial s}\right)_t = \frac{\partial \mathbf{V}}{\partial s} = 0 \qquad \text{(uniform flow)} \tag{4.5}$$

Regarding notation in this text, we omit the variables that are held constant when writing partial derivatives. For example, in Eq. (4.5), the leftmost terms show the formal way to write a partial derivative, and the middle term shows a simpler notation. The rationale for the simpler notation is that variables that are held constant can be inferred from the context.

In **nonuniform flow**, the velocity changes along a streamline either in magnitude, direction, or both. It follows that any flow with streamline curvature is nonuniform. Also, any flow in which the speed of the flow is changing spatially is also nonuniform.

$$\frac{\partial \mathbf{V}}{\partial s} \neq 0 \qquad \text{(nonuniform flow)}$$

EXAMPLES. Nonuniform flow occurs in the converging duct in Fig. 4.11a because the speed increases as the duct converges. Nonuniform flow occurs for the vortex in Fig. 4.11b because the streamlines are curved.

FIGURE 4.9

Fluid particle moving along a pathline.

s

V

Fluid particle

Particle path

Initial point
($s = 0, t = 0$)

FIGURE 4.10

Uniform flow in a pipe.

FIGURE 4.11

Flow patterns for nonuniform flow.
(a) Converging flow.
(b) Vortex flow.

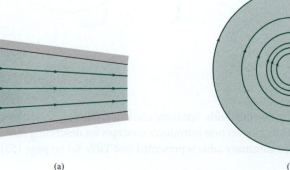

(a)

(b)

Steady and Unsteady Flow

In general, a velocity field **V** depends of position **r** and time t: $\mathbf{V} = \mathbf{V}(\mathbf{r}, t)$. However, in many situations, the velocity is constant with time, so $\mathbf{V} = \mathbf{V}(\mathbf{r})$. This is called steady flow. **Steady flow** means that velocity at each location in space is constant with time. This idea can be written mathematically as:

$$\left.\frac{\partial \mathbf{V}}{\partial t}\right|_{\text{all points in velocity field}} = \mathbf{0}$$

In an **unsteady flow** the velocity is changing, at least at some points, in the velocity field. This idea can be represented with an equation.

$$\frac{\partial \mathbf{V}}{\partial t} \neq 0$$

EXAMPLE. If the flow in a pipe changed with time due to a valve opening or closing, the flow would be unsteady; that is, the velocity at locations in the velocity field would be increasing or decreasing with time.

✔ **CHECKPOINT PROBLEM 4.2**

As shown, water drains out of a small opening in a container. Which statement is true?

  a. The flow in the container is steady.

  b. The flow in the container is unsteady.

Flow

Laminar and Turbulent Flow

In a famous experiment, Osborne Reynolds showed that there are two different kinds of flow that can occur in a pipe.* The first type, called **laminar flow**, is a well-ordered state of flow in which adjacent fluid layers move smoothly with respect to each other. The flow occurs in layers or laminae. An example of laminar flow is the flow of thick syrup (Fig. 4.12a).

FIGURE 4.12

Examples of laminar and turbulent flow (a) the flow of maple syrup is laminar (Lauri Patterson/The Agency Collection/Getty Images) (b) the flow of steam out of a smokestack is turbulent (Photo by Donald Elger)

(a)

(b)

---

*Reynolds experiment is described in Chapter 10.

The second type of flow identified by Reynolds is called **turbulent flow**, which is an unsteady flow characterized by eddies of various sizes and intense cross-stream mixing. Turbulent flow can be observed in the wake of a ship. Also, turbulent flow can be observed for a smokestack (Fig. 4.12b). Notice that the mixing of the turbulent flow is apparent because the plume widens and disperses.

Laminar flow in a pipe (Fig. 4.13a) has a smooth parabolic velocity distribution. Turbulent flow (Fig. 4.13b) has a plug-shaped velocity distribution because eddies mix the flow, which tends to keep the distribution uniform. In both laminar and turbulent flow, the no-slip condition applies.

FIGURE 4.13

Laminar and turbulent flow in a straight pipe.
(a) Laminar flow.
(b) Turbulent flow.
Both sketches assume fully developed flow.

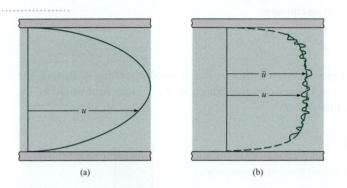

(a)   (b)

Time-Averaged Velocity

Turbulent flow is unsteady, so the standard approach is to represent the velocity as a time-averaged velocity $\bar{u}$ plus a fluctuating component u'. Thus, the velocity is expressed as $u = \bar{u} + u'$ (see Fig. 4.13b). Thus, the fluctuating component is defined as the difference between the local velocity and the time-averaged velocity. A turbulent flow is designated as "steady" if the time-averaged velocity is unchanging with time. For an interesting look at turbulent flows, see the film entitled *Turbulence* (3). Table 4.2 compares laminar and turbulent flows.

TABLE 4.2 Comparison of the Laminar and Turbulent Flow

Feature	Laminar Flow	Turbulent Flow
Basic description	Smooth flow in layers (laminae).	The flow has many eddies of various sizes. The flow appears random, chaotic, and unsteady.
Velocity profile in a pipe	Parabolic; ratio of mean velocity to centerline velocity is 0.5 for fully developed flow.	Pluglike; ratio of mean velocity to centerline velocity is between 0.8 and 0.9.
Mixing of materials added to the flow	Low levels of mixing. Difficult to get a material to mix with a fluid in laminar flow.	High levels of mixing. Easy to get a material to mix; e.g., visualize cream mixing with coffee.
Variation with time	Can be steady or unsteady.	Always unsteady.
Dimensionality of flow	Can be 1D, 2D, or 3D.	Always 3D.
Availability of mathematical solutions	In principle, any laminar flow can be solved with an analytical or computer solution. There are many existing analytical solutions. Solutions are very close to what would be measured with an experiment.	There is no complete theory of turbulent flow. There are a limited number of semiempirical solution approaches. Many turbulent flows cannot be accurately predicted with computer models or analytical solutions. Engineers often rely on experiments to characterize turbulent flow.

Feature	Laminar Flow	Turbulent Flow
Practical importance	Although many problems of practical problems involve laminar flow, these problems are not nearly as common as problems that involve turbulent flow.	The majority of practical problems involve turbulent flow. Typically, the flow of air and water in piping systems is turbulent. Most flows of water in open channels are turbulent.
Occurrence (Reynolds number)	Occurs at lower values of Reynolds numbers. (Reynolds number is introduced in Chapter 8.)	Occurs at higher values of Reynolds numbers.

One-Dimensional and Multidimensional Flows

The dimensionality of a flow field can be illustrated by example. Fig. 4.14a shows the velocity distribution for an axisymmetric flow in a circular duct. The flow is uniform, or fully developed, so the velocity does not change in the flow direction (z). The velocity depends on only one spatial dimension, namely the radius r, so the flow is one-dimensional or 1D. Fig. 4.14b shows the velocity distribution for uniform flow in a square duct. In this case the velocity depends on two dimensions, namely x and y, so the flow is two dimensional. Figure 4.14c also shows the velocity distribution for the flow in a square duct but the duct cross-sectional area is expanding in the flow direction so the velocity will be dependent on z as well as x and y. This flow is three-dimensional, or 3-D.

Another good example of three-dimensional flow is turbulence because the velocity components at any one time depend on the three coordinate directions. For example, the velocity component u at a given time depends on x, y, and z; that is, $u(x, y, z)$. Turbulent flow is unsteady, so the velocity components also depend on time.

Another definition frequently used in fluid mechanics is quasi-one-dimensional flow. By this definition it is assumed that there is only one component of velocity in the flow direction and that the velocity profiles are uniformly distributed; that is, constant velocity across the duct cross section.

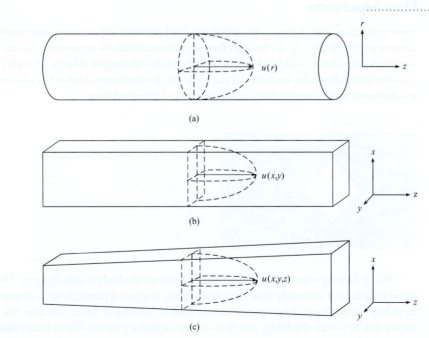

FIGURE 4.14

Flow dimensionality, (a) one-dimensional flow, (b) two-dimensional flow, and (c) three-dimensional flow.

Viscous and Inviscid Flow

In a **viscous flow** the forces associated with viscous shear stresses are large enough to effect the dynamic motion of the particles that comprise the flow. For example, when a fluid flows in a pipe as shown in Fig. 4.13, this is a viscous flow. Indeed, both laminar and turbulent flows are types of viscous flows.

In a **inviscid flow** the forces associated with viscous shear stresses are small enough so that they do not affect the dynamic motion of the particles that comprise the flow. Thus, in inviscid flow, the viscous stresses can be neglected in the equations for motion.

Boundary Layer, Wake, and Potential Flow Regions

To idealize many complex flows, engineers use ideas that can be illustrated by flow over a sphere (Fig. 4.15). As shown, the flow is divided into three regions: an inviscid flow region, a wake, and a boundary layer.

FIGURE 4.15

Flow pattern around a sphere when the Reynolds number is high. The sketch shows the regions of flow.

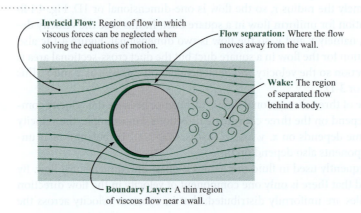

Inviscid Flow: Region of flow in which viscous forces can be neglected when solving the equations of motion.

Flow separation: Where the flow moves away from the wall.

Wake: The region of separated flow behind a body.

Boundary Layer: A thin region of viscous flow near a wall.

Flow Separation

Flow separation (Fig. 4.15) occurs when the fluid particles adjacent to a body deviate from the contours of the body. Fig. 4.16 shows flow separation behind a square rod. Notice that the flow separates from the shoulders of the rod and that the wake region is large. In both Figs. 4.15 and 4.16 the flow follows the contours of the body on the upstream sides of the objects. The region in which a flow follows the body contour is called **attached flow**.

FIGURE 4.16

Flow pattern past a square rod illustrating separation at the edges.

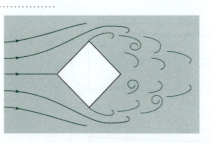

When flow separates (Fig. 4.16), the drag force on the body is usually large. Thus, designers strive to reduce or eliminate flow separation when designing products such as automobiles and airplanes. In addition, flow separation can lead to structural failure because the wake is unsteady due to vortex shedding, and this creates oscillatory forces. These forces cause structural vibrations, which can lead to failure when the structure's natural frequency is closely matched to the vortex shedding frequency. In a famous example, vortex shedding associated with flow separation caused the Tacoma Narrows Bridge near Seattle, Washington, to oscillate wildly and to fail catastrophically.

Fig. 4.17 shows flow separation for an airfoil (an airfoil is a body with the cross sectional shape of a wing). Flow separation occurs when the airfoil is rotated to an angle of attack that is too high. Flow separation in this context causes an airplane to stall, which means that the lifting force drops dramatically and the wings can no longer keep the airplane in level flight. Stall is to be avoided.

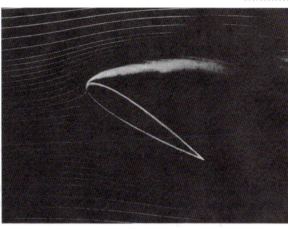

FIGURE 4.17

Smoke traces showing separation on an airfoil section at a large angle of attack. (Courtesy Education Development Center, Inc., Waltham, MA)

Flow separation can occurs inside pipes. For example, flow passing through an orifice in a pipe will separate (see Fig. 13.14 in Section 13.2). In this case, the zone of separated flow is usually called a recirculating zone. Separating flow within a pipe is usually undesirable because it causes energy losses, low pressure zones that can lead to cavitation and vibrations.

Summary *Attached flow* means that flow is moving parallel to walls of a body. *Flow separation*, which occurs in both internal and external flows, means the flow moves away from the wall. Flow separation is related to phenomenon of engineering interest such as drag, structural vibrations, and cavitation.

4.4 Acceleration

Predicting forces is important to the designer. Because forces are related to acceleration, this section describes what acceleration means in the context of a flowing fluid.

Definition of Acceleration

Acceleration is a property of a fluid particle that characterizes the change in speed of the particle and the change in the direction of travel at an instant in time. The mathematical definition of acceleration is:

$$\mathbf{a} = \frac{d\mathbf{V}}{dt} \tag{4.6}$$

where $\mathbf{V}$ is the velocity of the particle and t is time.

To begin, select a fluid particle (Fig. 4.22a) and orient the particle in an arbitrary direction ℓ and at an angle α with respect to the horizontal plane (Fig. 4.22b). Assume that viscous forces are zero. Assume the particle is in a flow and that the particle is accelerating. Now, apply Newton's second law in the ℓ-direction:

$$\sum F_\ell = ma_\ell$$

$$F_{pressure} + F_{gravity} = ma_\ell \tag{4.12}$$

The mass of the particle is

$$m = \rho \Delta A \Delta \ell$$

The net force due to pressure in the ℓ-direction is

$$F_{pressure} = p\Delta A - (p + \Delta p)\Delta A = -\Delta p \Delta A$$

The force due to gravity is

$$F_{gravity} = -\Delta W_\ell = -\Delta W \sin\alpha \tag{4.13}$$

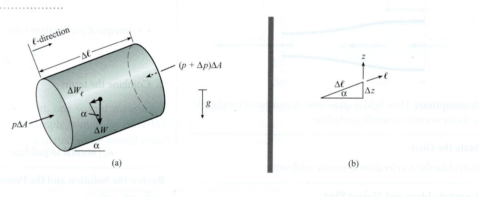

(a)   (b)

From Fig. 4.22b note that $\sin\alpha = \Delta z/\Delta \ell$, so Eq. (4.13) becomes

$$F_{gravity} = -\Delta W \frac{\Delta z}{\Delta \ell}$$

The weight of the particle is $\Delta W = \gamma \Delta \ell \Delta A$. Substituting the mass of the particle and the forces on the particle into Eq. (4.12) yields

$$-\Delta p \Delta A - \gamma \Delta \ell \Delta A \frac{\Delta z}{\Delta \ell} = \rho \Delta \ell \Delta A a_\ell$$

Dividing through by the volume of the particle $\Delta A \Delta \ell$ results in

$$-\frac{\Delta p}{\Delta \ell} - \gamma \frac{\Delta z}{\Delta \ell} = \rho a_\ell$$

Taking the limit as $\Delta \ell$ approaches zero (reduce the particle to an infinitesimal size) leads to

$$-\frac{\partial p}{\partial \ell} - \gamma \frac{\partial z}{\partial \ell} = \rho a_\ell \tag{4.14}$$

Assume a constant density flow, so γ is constant and Eq. (4.14) reduces to

$$-\frac{\partial}{\partial \ell}(p + \gamma z) = \rho a_\ell \tag{4.15}$$

Equation (4.15) is a scalar form of *Euler's equation*. Because this equation is true in any scalar direction, one can write this in an equivalent vector form:

$$-\nabla p_z = \rho \mathbf{a} \tag{4.16}$$

where ∇p_z is the gradient of the piezometric pressure, and $\mathbf{a}$ is the acceleration of the fluid particle.

Physical Interpretation of Euler's Equation

Euler's equation shows that the pressure gradient is colinear with the acceleration vector and opposite in direction.

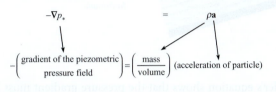

Thus, by using knowledge of acceleration, one can make inferences about the pressure variation. Three important cases are presented next. At this point, we recommend the film entitled *Pressure Fields and Fluid Acceleration* (4) because this film illustrates fundamental concepts using laboratory experiments.

Case 1: Pressure Variation Due to Changing Speed of a Particle

When a fluid particle is speeding up or slowing down as it moves along a streamline, then pressure will vary in a direction tangent to the streamline. For example Fig. 4.23 shows a fluid particle moving along a stagnation streamline. Because the particle is slowing down, the acceleration vector points to the left. Therefore the pressure gradient must point to the right. Thus, the pressure is increasing along the streamline, and the direction of increasing pressure is to the right. *Summary.* When a particle is changing speed, then pressure will vary in a direction that is tangent to the streamline.

FIGURE 4.23

This figure shows flow over a sphere. The blue object is a fluid particle moving along the stagnation streamline.

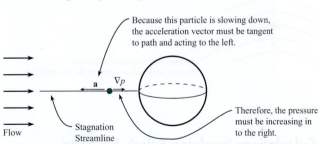

Case 2: Pressure Variation Normal to Rectilinear Streamlines

When streamlines are straight and parallel (Fig. 4.24), then *piezometric pressure will be constant along a line that is normal to the streamlines*. To prove this fact, draw a line that is normal to the streamlines (see Fig. 4.24). Then recognize that

$$a_n = \frac{V^2}{r} = \frac{V^2}{\infty} = 0$$

FIGURE 4.24

Flow with rectilinear streamlines. The numbered steps give the logic to show that pressure variation normal to rectilinear streamlines is hydrostatic.

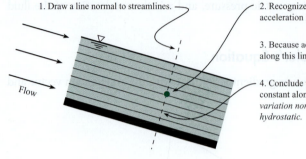

1. Draw a line normal to streamlines.

2. Recognize that the normal component of acceleration for this particle must be zero.

3. Because acceleration is zero, presssure gradient along this line must be zero.

4. Conclude that piezometric pressure must be constant along this line. Therefore, *pressure variation normal to rectilinear streamlines is hydrostatic.*

Flow

Because $a_n = 0$, Euler's equation shows that the pressure gradient must be zero: $\partial(p + \gamma z)/\partial n = 0$. Thus, conclude that piezometric pressure ($p + \gamma z$) is constant along any line that is normal to the streamlines. *Summary.* Pressure variation normal to rectilinear streamlines is hydrostatic.

Case 3: Pressure Variation Normal to Curved Streamlines

When streamlines are curved (Fig. 4.25), then *piezometric pressure will increase along a line that is normal to the streamlines.* The direction of increasing pressure will be outward from the center of curvature of the streamlines. Fig. 4.25 shows why pressure will vary. A fluid particle on a curved streamline must have a component of acceleration inward. Therefore, the gradient of the pressure will point outward. Because the gradient points in the direction of increasing pressure, we conclude that pressure will increase along the line drawn normal to the streamlines. *Summary.* When streamlines are curved, then pressure increases outward from the center of curvature* of the streamlines.

FIGURE 4.25

Flow with curved streamlines. Assume that the fluid particle has constant speed. Thus, the acceleration vector points inward towards the center of curvature.

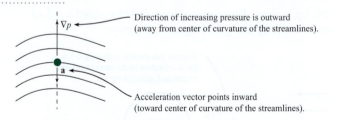

Direction of increasing pressure is outward (away from center of curvature of the streamlines).

∇p

a

Acceleration vector points inward (toward center of curvature of the streamlines).

Calculations Involving Euler's Equation

In most cases, calculations involving Euler's equation are beyond the scope of this book. However, when a fluid is accelerating as a rigid body, then Euler's equation can be applied in a simple way. Examples 4.2 and 4.3 show how to do this.

*Each streamline has a center of curvature at each point along the streamline. There is not a single center of curvature of a group of streamlines.

EXAMPLE 4.2

Applying Euler's equation to a Column of Fluid being Accelerated Upward

Problem Statement

A column water in a vertical tube is being accelerated by a piston in the vertical direction at 100 m/s². The depth of the water column is 10 cm. Find the gage pressure on the piston. The water density is 10^3 kg/m³.

Define the Situation

A column of water is being accelerated by a piston.

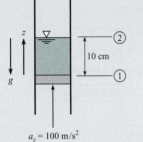

$a_z = 100$ m/s²

Assumptions:

- Acceleration is constant.
- Viscous effects are unimportant.
- Water is incompressible.

Properties: $\rho = 10^3$ kg/m³

State the Goal

Find: The gage pressure on the piston.

Generate Ideas and Make a Plan

1. Apply Euler's equation, Eq. (4.15), in the z-direction.
2. Integrate between locations 1 and 2.
3. Set pressure equal to zero (gage pressure) at section 2.
4. Calculate the pressure on the piston.

Take Action (Execute the Plan)

1. Because the acceleration is constant, there is no dependence on time, so the partial derivative in Euler's equation can be replaced by an ordinary derivative. Euler's equation becomes:

$$\frac{d}{dz}(p + \gamma z) = -\rho a_z$$

2. Integration between sections 1 and 2:

$$\int_1^2 d(p + \gamma z) = \int_1^2 (-\rho a_z)\, dz$$

$$(p_2 + \gamma z_2) - (p_1 + \gamma z_1) = -\rho a_z(z_2 - z_1)$$

3. Algebra:

$$p_1 = (\gamma + \rho a_z)\Delta z = \rho(g + a_z)\Delta z$$

4. Evaluation of pressure:

$$p_1 = 10^3 \text{ kg/m}^3 \times (9.81 + 100) \text{ m/s}^2 \times 0.1 \text{ m}$$

$$p_1 = \boxed{10.9 \times 10^3 \text{ Pa} = 10.9 \text{ kPa, gage}}$$

EXAMPLE 4.3

Applying Euler's Equation to Gasoline in a Decelerating Tanker

Problem Statement

The tank on a trailer truck is filled completely with gasoline, which has a specific weight of 6.60 kN/m³. The truck is decelerating at a rate of 3.05 m/s².

a. If the tank on the trailer is 6.1 m long and if the pressure at the top rear end of the tank is atmospheric, what is the pressure at the top front?

b. If the tank is 1.83 m high, what is the maximum pressure in the tank?

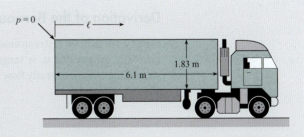

Define the Situation

Situation: Decelerating tank of gasoline with pressure equal to zero gage at top rear end.

Assumptions:

1. Deceleration is constant.

2. Gasoline is incompressible.

Properties: $\gamma = 6.60 \text{ kN/m}^3$

State the Goal

Find:

1. Pressure (kPa, gage) at top front of tank.

2. Maximum pressure (kPa, gage) in tank.

Make a Plan

1. Apply Euler's equation, Eq. (4.15), along top of tank. Elevation, z, is constant.

2. Evaluate pressure at top front.

3. Maximum pressure will be at front bottom. Apply Euler's equation from top to bottom at front of tank.

4. Using result from step 2, evaluate pressure at front bottom.

Take Action (Execute the Plan)

1. Euler's equation along the top of the tank

$$\frac{dp}{d\ell} = -\rho a_\ell$$

Integration from back (1) to front (2)

$$p_2 - p_1 = -\rho a_\ell \Delta\ell = -\frac{\gamma}{g} a_\ell \Delta\ell$$

2. Evaluation of p_2 with $p_1 = 0$

$$p_2 = -\left(\frac{6.60 \text{ kN/m}^3}{9.81 \text{ m/s}^2}\right) \times (-3.05 \text{ m/s}^2) \times 6.1 \text{ m}$$

$$= \boxed{12.5 \text{ (kPa gage)}}$$

3. Euler's equation in vertical direction

$$\frac{d}{dz}(p + \gamma z) = -\rho a_z$$

4. For vertical direction, $a_z = 0$. Integration from top of tank (2) to bottom (3):

$$p_2 + \gamma z_2 = p_3 + \gamma z_3$$

$$p_3 = p_2 + \gamma(z_2 - z_3)$$

$$p_3 = 12.5 \text{ kN/m}^2 + 6.6 \text{ kN/m}^3 \times 1.83 \text{ m}$$

$$p_3 = \boxed{24.6 \text{ kPa gage}}$$

4.6 Applying the Bernoulli Equation along a Streamline

Because the Bernoulli equation is used frequently in fluid mechanics, this section introduces this topic.

Derivation of the Bernoulli Equation

Select a particle on a streamline (Fig. 4.26). The position coordinate s gives the particle's position. The unit vector $\mathbf{u}_t$ is tangent to the streamline, and the unit vector $\mathbf{u}_n$ is normal to the streamline. Assume steady flow so the velocity of the particle depends on position only. That is, $V = V(s)$.

FIGURE 4.26

Sketch used for the derivation of the Bernoulli equation.

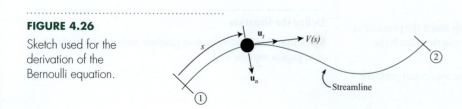

Assume that viscous forces on the particle can be neglected. Then, apply Euler's equation (Eq. 4.15) to the particle in the u_t direction.

$$-\frac{\partial}{\partial s}(p + \gamma z) = \rho a_t \qquad (4.17)$$

Acceleration is given by Eq. (4.11). Because the flow is steady, $\partial V / \partial t = 0$, and Eq. (4.11) gives

$$a_t = V\frac{\partial V}{\partial s} + \frac{\partial V}{\partial t} = V\frac{\partial V}{\partial s} \qquad (4.18)$$

Because p, z, and V in Eqs. (4.17) and (4.18) depend only on position s, the partial derivatives become ordinary derivatives (i.e., functions only of a single variable). Thus, write the these derivatives as ordinary derivatives and combine Eqs. (4.17) and (4.18) to give

$$-\frac{d}{ds}(p + \gamma z) = \rho V\frac{dV}{ds} = \rho\frac{d}{ds}\left(\frac{V^2}{2}\right) \qquad (4.19)$$

Move all the terms to one side:

$$\frac{d}{ds}\left(p + \gamma z + \rho\frac{V^2}{2}\right) = 0 \qquad (4.20)$$

When the derivative of an expression is zero, the expression is equal to a constant. Thus, rewrite Eq. (4.20) as:

$$p + \gamma z + \rho\frac{V^2}{2} = C \qquad (4.21a)$$

where C is a constant. Eq. (4.21a) is the *pressure form of the Bernoulli equation*. This is called the pressure form because all terms have units of pressure. Dividing Eq. (4.21a) by the specific weight yields the *head form of the Bernoulli equation,* which is given as Eq. (4.21b). In the head form, all terms have units of length.

$$\frac{p}{\gamma} + z + \frac{V^2}{2g} = C \qquad (4.21b)$$

Physical Interpretation #1 (Energy Is Conserved)

One way to interpret the Bernoulli equation leads to the idea that *when the Bernoulli equation applies, the total head of the flowing fluid is a constant along a streamline.* To develop this interpretation, recall that piezometric head, introduced in Chapter 3, is defined as

$$\text{piezometric head} = h \equiv \frac{p}{\gamma} + z \qquad (4.22)$$

Introduce Eq. (4.22) into Eq. (4.21b)

$$h + \frac{V^2}{2g} = \text{Constant} \qquad (4.23)$$

Now, velocity head is defined by

$$\text{velocity head} \equiv \frac{V^2}{2g} \qquad (4.24)$$

Combine Eqs. (4.22) to (4.24) to give

$$\left(\begin{array}{c}\text{Piezometric}\\\text{head}\end{array}\right) + \left(\begin{array}{c}\text{Velocity}\\\text{head}\end{array}\right) = \left(\begin{array}{c}\text{Constant along}\\\text{streamline}\end{array}\right) \tag{4.25}$$

Eq. (4.25) is shown visually in Fig. 4.27. Notice that piezometric head (blue lines) and the velocity head (gray lines) are changing, but the sum of the piezometric head plus velocity head is everywhere constant. Thus, the total head is constant for all points along a streamline when the Bernoulli equation applies.

FIGURE 4.27

Water flowing through a Venturi nozzle. The piezometers shows the piezometric head at locations 1, 2, and 3.

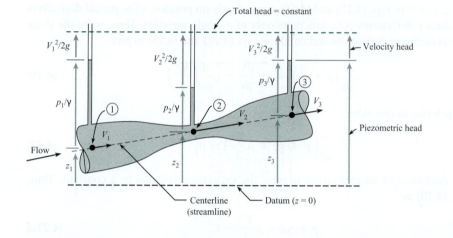

The previous discussion introduced head. **Head** *is a concept that is used to characterize the balance of work and energy in a flowing fluid.* As shown in Fig. 4.27, *head can be visualized as the height of a column of liquid.* Each type of head describes a work or energy term. Velocity head characterizes the kinetic energy in a flowing fluid, elevation head characterizes the gravitational potential energy of a fluid, and pressure head is related to work done by the pressure force. As shown in Fig. 4.27, the total head is constant. This means that when the Bernoulli equation applies, the fluid is not losing energy as it flows. The reason is that viscous effects are the cause of energy loses, and viscous effects are negligible when the Bernoulli equation applies.

Physical Interpretation #2 (Velocity and Pressure Vary Inversely)

A second way to interpret the Bernoulli equation leads to the idea that *when velocity increases, then pressure will decrease.* To develop this interpretation, recall that piezometric pressure, introduced in Chapter 3, is defined as

$$\text{piezometric pressure} = p_z \equiv p + \gamma z \tag{4.26}$$

Introduce Eq. (4.26) into Eq. (4.21a)

$$p_z + \frac{\rho V^2}{2} = \text{Constant} \tag{4.27}$$

For Eq. (4.27) to be true, piezometric pressure and velocity must vary inversely so that the sum of p_z and $(V^2/2g)$ is a constant. Thus, the pressure form of the Bernoulli equation shows that *piezometric pressure varies inversely with velocity.* In regions of high velocity, piezometric pressure will be low; in regions of low velocity, piezometric pressure will be high.

EXAMPLE. Fig. 4.28 shows a Vinturi™ red wine aerator, which is a product that is used to add air to wine. When wine flows through the Vinturi™, the shape of the device causes an increase in the wine's velocity and a corresponding decrease in its pressure. At the throat, the pressure is below atmospheric pressure so air flows inward through two inlet ports and mixes with the wine to create aerated wine, which tastes better to most people.

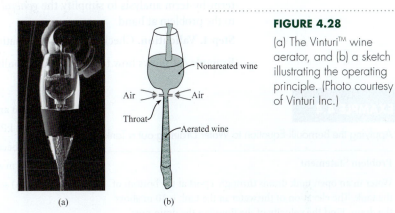

(a)   (b)

FIGURE 4.28

(a) The Vinturi™ wine aerator, and (b) a sketch illustrating the operating principle. (Photo courtesy of Vinturi Inc.)

Working Equations and Process

Table 4.3 summarizes the Bernoulli equation.

TABLE 4.3 Summary of the Bernoulli Equation

Description	Equation	Terms
Bernoulli equation (head form) Recommend form to use for liquids	$\left(\dfrac{p_1}{\gamma} + \dfrac{V_1^2}{2g} + z_1\right) = \left(\dfrac{p_2}{\gamma} + \dfrac{V_2^2}{2g} + z_2\right)$ Eq. (4.21b)	p = static pressure (Pa) (use gage pressure or abs pressure) (avoid vacuum pressure; will be wrong) γ = specific weight (N/m^3) V = speed (m/s) g = gravitational constant = 9.81 m/s^3 z = elevation or elevation head (m) $\dfrac{p}{\gamma}$ = pressure head (m)
Bernoulli equation (pressure form) Recommend form to use for gases	$\left(p_1 + \dfrac{\rho V_1^2}{2} + \rho g z_1\right) = \left(p_2 + \dfrac{\rho V_2^2}{2} + \rho g z_2\right)$ Eq. (4.21a)	$\dfrac{V^2}{2g}$ = velocity head (m) $\dfrac{p}{\gamma} + z$ = piezometric head (m) $p + \gamma z$ = piezometric pressure (Pa) $\dfrac{\rho V^2}{2}$ = kinetic pressure (Pa)

The process for applying the Bernoulli equation is

Step 1. Selection. Select the head form or the pressure form. Check that the assumptions are satisfied.

Step 2. Sketching. Select a streamline. Then, select points 1 and 2 where you know information or where you want to find information. Annotate your documentation to show the streamline and points.

Step 3. General Equation. Write the general form of the Bernoulli equation. Perform a term-by-term analysis to simplify the *general equation* to a *reduced equation* that applies to the problem at hand.

Step 4. Validation. Check the reduced equation to ensure that it makes physical sense.

Example 4.4 shows how to apply the Bernoulli equation to a draining tank of water.

EXAMPLE 4.4

Applying the Bernoulli Equation to Water Draining out a Tank

Problem Statement

Water in an open tank drains through a port at the bottom of the tank. The elevation of the water in the tank is 10 m above the drain. Find the velocity of the liquid in the drain port.

Define the Situation

Water flows out of a tank.

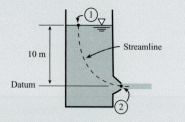

Assumptions:

- Steady flow.
- Viscous effects are negligible.

State the Goal

V_2 (m/s) ◀— Velocity at the exit port.

Generate Ideas and Make a Plan

Selection. Select the head form of the Bernoulli equation because the fluid is a liquid. Document assumptions (see above).

Sketching. Select point 1 where information is known and point 2 where information is desired. On the situation diagram (see above), sketch the streamline, label points 1 and 2, and label the datum.

General Equation.

$$\left(\frac{p_1}{\gamma} + \frac{V_1^2}{2g} + z_1 \right) = \left(\frac{p_2}{\gamma} + \frac{V_2^2}{2g} + z_2 \right) \quad \text{(a)}$$

Term-by-term analysis.

- $p_1 = p_2 = 0$ kPa gage
- Let $V_1 = 0$ because $V_1 \ll V_2$
- Let $z_1 = 10$ m and $z_2 = 0$ m

Reduce Eq. (a) so it applies to the problem at hand.

$$(0 + 0 + 10 \text{ m}) = \left(0 + \frac{V_2^2}{2g} + 0 \right) \quad \text{(b)}$$

Simplify Eq. (b):

$$V_2 = \sqrt{2g(10 \text{ m})} \quad \text{(c)}$$

Because Eq. (c) has only one unknown, the plan is to use this equation to solve for V_2.

Take Action (Execute the Plan)

$$V_2 = \sqrt{2g(10 \text{ m})}$$
$$V_2 = \sqrt{2(9.81 \text{ m/s}^2)(10 \text{ m})}$$
$$\boxed{V_2 = 14 \text{ m/s}}$$

Review the Solution and the Process

1. *Knowledge.* Notice that the same answer would be calculated for an object dropped from the same elevation as the water in the tank. This is because both problems involve equating gravitational potential energy at 1 with kinetic energy at 2.

2. *Validate.* The assumption of the small velocity at the liquid surface is generally valid. It can be shown (Chapter 5) that

$$\frac{V_1}{V_2} = \frac{D_2^2}{D_1^2}$$

For example, a diameter ratio of 10 to 1 ($D_2/D_1 = 0.1$) results in the velocity ratio of 100 to 1 ($V_1/V_2 = 1/100$).

When the Bernoulli equation is applied to a gas, it is common to neglect the elevation terms because these terms are negligibly small as compared to the pressure and velocity terms. An example of applying the Bernoulli equation to a flow of air is presented in Example 4.5.

EXAMPLE 4.5

Applying the Bernoulli Equation to Air Flowing around a Bicycle Helmet

Problem Statement

The problem is to estimate the pressure at locations A and B so these values can be used to estimate the ventilation in a bicycle helmet that is being designed. Assume an air density of $\rho = 1.2$ kg/m^3 and an air speed of 12 m/s relative to the helmet. Point A is a stagnation point, and the velocity of air at point B is 18 m/s.

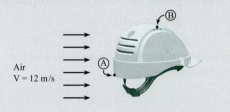

Air
$V = 12$ m/s

Define the Situation

Idealize flow around a bike helmet as flow around the upper half of a sphere. Assume steady flow. Assume that point B is outside the boundary layer. Relabel the points as shown in the situation diagram because this makes application of the Bernoulli equation easier.

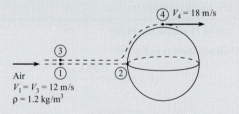

④ $V_4 = 18$ m/s

③
① ②

Air
$V_1 = V_3 = 12$ m/s
$\rho = 1.2$ kg/m^3

State the Goal

p_2(Pa gage) ⬅ Pressure at the forward stagnation point.
p_4(Pa gage) ⬅ Pressure at the shoulder.

Generate Ideas and Make a Plan

Selection. Select the pressure form of the Bernoulli equation because the flow is air. Then write the Bernoulli equation along the stagnation streamline (i.e., from point 1 to point 2).

$$\left(p_1 + \frac{\rho V_1^2}{2} + \rho g z_1\right) = \left(p_2 + \frac{\rho V_2^2}{2} + \rho g z_2\right) \quad \text{(a)}$$

Conduct a **term-by-term analysis**.

- $p_1 = 0$ kPa gage because the external flow is at atmospheric pressure.
- $V_1 = 12$ m/s
- let $z_1 = z_2 = 0$ because elevation terms are negligibly small for a gas flow such as a flow of air
- let $V_2 = 0$ because this is a stagnation point.

Now, simplify Eq. (a).

$$0 + \frac{\rho V_1^2}{2} + 0 = p_2 + 0 + 0 \quad \text{(b)}$$

Eq. (b) has only a single unknown (p_2).

Next, apply the Bernoulli equation to the streamline that connects points 3 and 4.

$$\left(p_3 + \frac{\rho V_3^2}{2} + \rho g z_3\right) = \left(p_4 + \frac{\rho V_4^2}{2} + \rho g z_4\right) \quad \text{(c)}$$

Do a term-by-term analysis to give:

$$\left(0 + \frac{\rho V_3^2}{2} + 0\right) = \left(p_4 + \frac{\rho V_4^2}{2} + 0\right) \quad \text{(d)}$$

Eq. (d) has only one unknown (p_4). The plan is
1. Calculate (p_2) using Eq. (b).
2. Calculate (p_4) using Eq. (d).

Take Action (Execute the Plan)

1. Bernoulli equation (point 1 to point 2)

$$p_2 = \frac{\rho V_1^2}{2} = \frac{(1.2 \text{ kg/m}^3)(12 \text{ m/s})^2}{2}$$

$$\boxed{p_2 = 86.4 \text{ Pa gage}}$$

2. Bernoulli equation (point 3 to point 4)

$$p_2 = \frac{\rho(V_3^2 - V_4^2)}{2} = \frac{(1.2 \text{ kg/m}^3)(12^2 - 18^2)(\text{m/s})^2}{2}$$

$$\boxed{p_2 = -108 \text{ Pa gage}}$$

Review the Solution and the Process

1. *Discussion.* Notice that where the velocity is high (i.e., point 4), the pressure is low (negative gage pressure).
2. *Knowledge.* Remember to specify pressure units in gage pressure or absolute pressure.
3. *Knowledge.* Theory shows that the velocity at the shoulder of a sphere is 3/2 the velocity in the free stream.

Example 4.6 involves a venturi. A **venturi** (also called a venturi nozzle) is a constricted section as shown in this example. As fluid flows through a venturi, the pressure is reduced in the narrow area, called the throat. This drop in pressure is called the venturi effect.

The venturi can be used to entrain liquid drops into a flow of gas as in a carburetor. The venturi can also be used to measure the flow rate. The venturi is commonly analyzed with the Bernoulli equation.

EXAMPLE 4.6

Applying the Bernoulli Equation to Flow through a Venturi Nozzle

Problem Statement

Piezometric tubes are tapped into a venturi section as shown in the figure. The liquid is incompressible. The upstream piezometric head is 1 m, and the piezometric head at the throat is 0.5 m. The velocity in the throat section is twice as large as in the approach section. Find the velocity in the throat section.

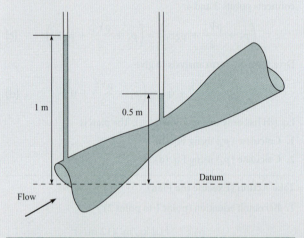

Define the Situation

A liquid flows through a venturi nozzle.

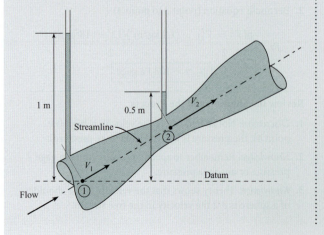

State the Goal

V_2(m/s) ◀━ Velocity at point 2.

Generate Ideas and Make a Plan

Select the Bernoulli equation because the problem involves flow through a nozzle. Select the head form because a liquid is involved. Select a streamline and points 1 and 2. **Sketch** these choices on the situation diagram.

Write the general form of the Bernoulli equation.

$$\frac{p_1}{\gamma} + z_1 + \frac{V_1^2}{2g} = \frac{p_2}{\gamma} + z_2 + \frac{V_2^2}{2g} \quad \text{(a)}$$

Introduce piezometric head because this is what the piezometer measures:

$$h_1 + \frac{V_1^2}{2g} = h_2 + \frac{V_2^2}{2g}$$

$$(1.0 \text{ m}) + \frac{V_1^2}{2g} = (0.5 \text{ m}) + \frac{V_2^2}{2g}$$

Let $V_1 = 0.5\, V_2$

$$(1.0 \text{ m}) + \frac{(0.5\, V_2)^2}{2g} = (0.5 \text{ m}) + \frac{V_2^2}{2g} \quad \text{(b)}$$

Plan. Use Eq. (b) to solve for V_2.

Take Action (Execute the Plan)

Bernoulli equation (i.e., Eq. b):

$$(0.5 \text{ m}) = \frac{0.75\, V_2^2}{2g}$$

Thus,

$$V_2 = \sqrt{\frac{2g(0.5 \text{ m})}{0.75}}$$

$$V_2 = \sqrt{\frac{2(9.81 \text{ m/s}^2)(0.5 \text{ m})}{0.75}}$$

$$V_2 = \boxed{3.62 \text{ m/s}}$$

Review the Solution and the Process

1. *Knowledge.* Notice how a piezometer is used to measure piezometric head in the nozzle.

2. *Knowledge.* A piezometer could not be used to measure the piezometric head if the pressure anywhere in the line were subatmospheric. In this case, pressure gages or manometers could be used.

4.7 Measuring Velocity and Pressure

The piezometer, stagnation tube, and Pitot-static tube have long been used to measure pressure and velocity. Indeed, many concepts in measurement are based on these instruments. Thus, this section describes these instruments.

Static Pressure

Static pressure is the pressure in a flowing fluid. A common way to measure static pressure is to drill a small hole in the wall of a pipe and then connect a piezometer or pressure gage to this port (see Fig. 4.29). This port is called a **pressure tap**. The reason that a pressure tap is useful is that it provides a way to measure static pressure that does not disturb the flow.

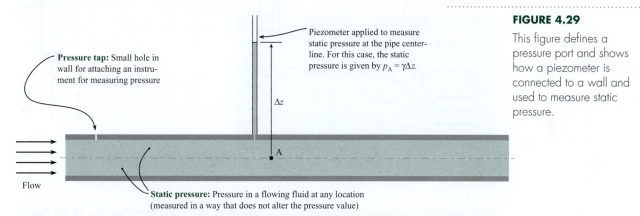

Pressure tap: Small hole in wall for attaching an instrument for measuring pressure

Piezometer applied to measure static pressure at the pipe centerline. For this case, the static pressure is given by $p_A = \gamma \Delta z$.

Δz

A

Flow

Static pressure: Pressure in a flowing fluid at any location (measured in a way that does not alter the pressure value)

FIGURE 4.29

This figure defines a pressure port and shows how a piezometer is connected to a wall and used to measure static pressure.

✔ CHECKPOINT PROBLEM 4.3

Restaurants often use large coffee dispensers (see sketch). The sight glass shows the level of coffee. If the valve is opened, what happens to the level of coffee that is visible in the sight glass? Will the level go up, go down, or stay the same? Why?

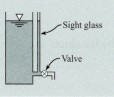

Sight glass

Valve

Stagnation Tube

A **stagnation tube** (also known as a total head tube) is an open-ended tube directed upstream in a flow (see Fig. 4.30). A stagnation tube measures the sum of static pressure and kinetic pressure.

Kinetic pressure is defined at an arbitrary point A as:

$$\left(\begin{matrix} \text{kinetic pressure} \\ \text{at point A} \end{matrix} \right) = \frac{\rho V_A^2}{2}$$

Next, we will derive an equation for velocity in an open channel flow. For the stagnation tube in Fig. 4.30, select points 0 and 1 on the streamline, and let $z_0 = z_1$. The Bernoulli equation reduces to

$$p_1 + \frac{\rho V_1^2}{2} = p_0 + \frac{\rho V_0^2}{2} \tag{4.28}$$

The velocity at point 1 is zero (stagnation point). Hence, Eq. (4.28) simplifies to

$$V_0^2 = \frac{2}{\rho}(p_1 - p_0) \tag{4.29}$$

FIGURE 4.30

Stagnation tube.

FIGURE 4.31

Pitot-static tube.

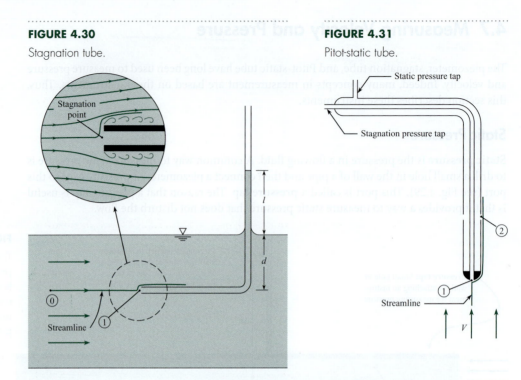

Next, apply the hydrostatic equation: $p_0 = \gamma d$ and $p_1 = \gamma(l + d)$. Therefore, Eq. (4.29) can be written as

$$V_0^2 = \frac{2}{\rho}(\gamma(l + d) - \gamma d)$$

which reduces to

$$V_0 = \sqrt{2gl} \qquad (4.30)$$

Pitot-Static Tube

The **Pitot-static tube,** named after the eighteenth-century French hydraulic engineer who invented it, is based on the same principle as the stagnation tube, but it is much more versatile than the stagnation tube. The Pitot-static tube, shown in Fig. 4.31, has a pressure tap at the upstream end of the tube for sensing the kinetic pressure. There are also ports located several tube diameters downstream of the front end of the tube for sensing the static pressure in the fluid where the velocity is essentially the same as the approach velocity. When the Bernoulli equation, Eq. (4.21a), is applied between points 1 and 2 along the streamline shown in Fig. 4.31, the result is

$$p_1 + \gamma z_1 + \frac{\rho V_1^2}{2} = p_2 + \gamma z_2 + \frac{\rho V_2^2}{2}$$

But $V_1 = 0$, so solving that equation for V_2 gives an equation for velocity.

$$V_2 = \left[\frac{2}{\rho}(p_{z,1} - p_{z,2})\right]^{1/2} \qquad (4.31)$$

Here $V_2 = V$, where V is the velocity of the stream and $p_{z,1}$ and $p_{z,2}$ are the piezometric pressures at points 1 and 2, respectively.

By connecting a pressure gage or manometer between the pressure taps shown in Fig. 4.31 that lead to points 1 and 2, one can easily measure the flow velocity with the Pitot-static tube. A major advantage of the Pitot-static tube is that it can be used to measure velocity in a pressurized pipe; a stagnation tube is not convenient to use in such a situation.

If a differential pressure gage is connected across the taps, the gage measures the difference in piezometric pressure directly. Therefore Eq. (4.31) simplifies to

$$V = \sqrt{2\Delta p/\rho} \qquad\qquad (4.32)$$

where Δp is the pressure difference measured by the gage.

More information on Pitot-static tubes and flow measurement is available in the *Flow Measurement Engineering Handbook* (5). Example 4.7 illustrates the application of the Pitot-static tube with a manometer. Then, Example 4.8 illustrates application with a pressure gage.

EXAMPLE 4.7

Applying a Pitot-Static Tube (pressure measured with a manometer).

Problem Statement

A mercury manometer is connected to the Pitot-static tube in a pipe transporting kerosene as shown. If the deflection on the manometer is 18 cm, what is the kerosene velocity in the pipe? Assume that the specific gravity of the kerosene is 0.81.

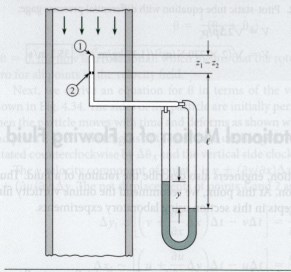

Define the Situation

A Pitot-static tube is mounted in a pipe and connected to a manometer.

Assumptions: Pitot-static tube equation is applicable.

Properties: $S_{kero} = 0.81$, from Table A.4, $S_{Hg} = 13.55$.

State the Goal

Find: Flow velocity (m/s).

Generate Ideas and Make a Plan

1. Find difference in piezometric pressure using the manometer equation.
2. Substitute in Pitot-static tube equation.
3. Evaluate velocity.

Take Action (Execute the Plan)

1. Manometer equation between points 1 and 2 on Pitot-static tube:

$$p_1 + (z_1 - z_2)\gamma_{kero} + \ell\gamma_{kero} - y\gamma_{Hg} - (\ell - y)\gamma_{kero} = p_2$$

or

$$p_1 + \gamma_{kero}z_1 - (p_2 + \gamma_{kero}z_2) = y(\gamma_{Hg} - \gamma_{kero})$$

$$p_{z,1} - p_{z,2} = y(\gamma_{Hg} - \gamma_{kero})$$

2. Substitution into the Pitot-static tube equation:

$$V = \left[\frac{2}{\rho_{kero}}y(\gamma_{Hg} - \gamma_{kero})\right]^{1/2}$$

$$= \left[2gy\left(\frac{\gamma_{Hg}}{\gamma_{kero}} - 1\right)\right]^{1/2}$$

FIGURE 4.34

Translation and deformation of a fluid particle.

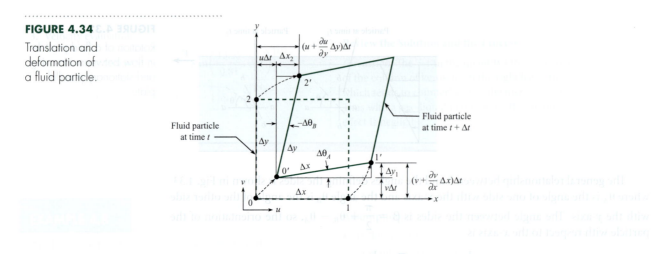

Referring to Fig. 4.34, the angles $\Delta\theta_A$ and $\Delta\theta_B$ are given by

$$\Delta\theta_A = \operatorname{asin}\left(\frac{\Delta y_1}{\Delta x}\right) \sim \frac{\Delta y_1}{\Delta x} \sim \frac{\partial v}{\partial x}\Delta t$$

$$-\Delta\theta_B = \operatorname{asin}\left(\frac{\Delta x_2}{\Delta y}\right) \sim \frac{\Delta x_2}{\Delta y} \sim \frac{\partial u}{\partial y}\Delta t$$

(4.35)

Dividing the angles by Δt and taking the limit as $\Delta t \to 0$,

$$\dot{\theta}_A = \lim_{\Delta t \to 0} \frac{\Delta\theta_A}{\Delta t} = \frac{\partial v}{\partial x}$$

$$\dot{\theta}_B = \lim_{\Delta t \to 0} \frac{\Delta\theta_B}{\Delta t} = -\frac{\partial u}{\partial y}$$

(4.36)

Substituting these results into Eq. (4.33) gives the rotational rate of the particle about the z-axis (normal to the page),

$$\dot{\theta} = \frac{1}{2}\left(\frac{\partial v}{\partial x} - \frac{\partial u}{\partial y}\right)$$

This component of rotational velocity is defined as Ω_z, so

$$\Omega_z = \frac{1}{2}\left(\frac{\partial v}{\partial x} - \frac{\partial u}{\partial y}\right)$$

(4.37a)

Likewise, the rotation rates about the other axes are

$$\Omega_x = \frac{1}{2}\left(\frac{\partial w}{\partial y} - \frac{\partial v}{\partial z}\right)$$

(4.37b)

$$\Omega_y = \frac{1}{2}\left(\frac{\partial u}{\partial z} - \frac{\partial w}{\partial x}\right)$$

(4.37c)

The rate-of-rotation vector is

$$\Omega = \Omega_x \mathbf{i} + \Omega_y \mathbf{j} + \Omega_z \mathbf{k}$$

(4.38)

An irrotational flow ($\Omega = 0$) requires that

$$\frac{\partial v}{\partial x} = \frac{\partial u}{\partial y} \tag{4.39a}$$

$$\frac{\partial w}{\partial y} = \frac{\partial v}{\partial z} \tag{4.39b}$$

$$\frac{\partial u}{\partial z} = \frac{\partial w}{\partial x} \tag{4.39c}$$

The most extensive application of these equations is in ideal flow theory. An ideal flow is the flow of an irrotational, incompressible fluid. Flow fields in which viscous effects are small can often be regarded as irrotational. In fact, if a flow of an incompressible, inviscid fluid is initially irrotational, it will remain irrotational.

Vorticity

The most common way to describe rotation is to use **vorticity**, which is a vector equal to twice the rate-of-rotation vector. The magnitude of the vorticity indicates the rotationality of a flow and is very important in flows where viscous effects dominate, such as boundary layer, separated, and wake flows. The vorticity equation is

$$\omega = 2\Omega$$

$$= \left(\frac{\partial w}{\partial y} - \frac{\partial v}{\partial z}\right)\mathbf{i} + \left(\frac{\partial u}{\partial z} - \frac{\partial w}{\partial x}\right)\mathbf{j} + \left(\frac{\partial v}{\partial x} - \frac{\partial u}{\partial y}\right)\mathbf{k} \tag{4.40}$$

$$= \nabla \times \mathbf{V}$$

where $\nabla \times \mathbf{V}$ is the curl of the velocity field.

An irrotational flow signifies that the vorticity vector is everywhere zero. Example 4.9 illustrates how to evaluate the rotationality of a flow field, and Example 4.10 evaluates the rotation of a fluid particle.

EXAMPLE 4.9

Evaluating Rotation

Problem Statement

The vector $\mathbf{V} = 10x\mathbf{i} - 10y\mathbf{j}$ represents a two-dimensional velocity field. Is the flow irrotational?

Define the Situation

Velocity field is given.

State the Goal

Determine if flow is irrotational.

Generate Ideas and Make a Plan

Because $w = 0$ and $\dfrac{\partial}{\partial z} = 0$, apply Eq. (4.39a) to evaluate rotationality.

Take Action (Execute the Plan)

Velocity components and derivatives

$$u = 10x \qquad \frac{\partial u}{\partial y} = 0$$

$$v = -10y \qquad \frac{\partial v}{\partial x} = 0$$

Thus, flow is irrotational.

EXAMPLE 4.10

Rotation of a Fluid Particle

Problem Definition

A fluid exists between stationary and moving parallel flat plates, and the velocity is linear as shown. The distance between the plates is 1 cm, and the upper plate moves at 2 cm/s. Find the amount of rotation that the fluid particle located at 0.5 cm will undergo after it has traveled a distance of 1 cm.

Sketch:

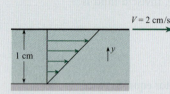

Define the Situation

This problem involves Couette flow.

Assumptions: Planar flow ($w = 0$ and $\dfrac{\partial}{\partial z} = 0$).

State the Goal

Find the rotation of a fluid particle (in radians) at the midpoint after traveling 1 cm.

Generate Ideas and Make a Plan

1. Use Eq. (4.37a) to evaluate rotational rate with $v = 0$.
2. Find time for particle to travel 1 cm.
3. Calculate amount of rotation.

Take Action (Execute the Plan)

1. Velocity distribution

$$u = 0.02 \text{ m/s} \times \frac{y}{0.01 \text{ m}} = 2y \ (1/\text{s})$$

Rotational rate

$$\Omega_z = \frac{1}{2}\left(\frac{\partial v}{\partial x} - \frac{\partial u}{\partial y}\right) = -1 \text{ rad/s}$$

2. Time to travel 1 cm:

$$u = 2 \ (1/\text{s}) \times 0.005 \text{ m} = 0.01 \text{ m/s}$$

$$\Delta t = \frac{\Delta x}{u} = \frac{0.01 \text{ m}}{0.01 \text{ m/s}} = 1 \text{ s}$$

3. Amount of rotation

$$\Delta\theta = \Omega_z \times \Delta t = -1 \times 1 = -1 \text{ rad}$$

Review the Solution and the Process

Discussion. Note that the rotation is negative (in clockwise direction).

4.9 The Bernoulli Equation for Irrotational Flow

When flow is irrotational, the Bernoulli equation can be applied between any two points in this flow. That is, the points do not need to be on the same streamline. This *irrotational form* of the Bernoulli equation is used extensively in applications such as classical hydrodynamics, the aerodynamics of lifting surfaces (wings), and atmospheric winds. Thus, this section describes how to derive the Bernoulli equation for an irrotational flow.

To begin the derivation, apply the Euler equation, Eq. (4.15), in the n direction (normal to the streamline)

$$-\frac{\mathrm{d}}{\mathrm{d}n}(p + \gamma z) = \rho a_n \tag{4.41}$$

where the partial derivative of n is replaced by the ordinary derivative because the flow is assumed steady (no time dependence). Two adjacent streamlines and the direction n is shown in Fig. 4.35. The local fluid speed is V, and the local radius of curvature of the streamline is r. The acceleration normal to the streamline is the centripetal acceleration, so

$$a_n = -\frac{V^2}{r} \tag{4.42}$$

FIGURE 4.35

Two adjacent streamlines showing direction n between lines.

where the negative sign occurs because the direction n is outward from the center of curvature and the centripetal acceleration is toward the center of curvature. Using the irrotationality condition, the acceleration can be written as

$$a_n = -\frac{V^2}{r} = -V\left(\frac{V}{r}\right) = V\frac{dV}{dr} = \frac{d}{dr}\left(\frac{V^2}{2}\right) \tag{4.43}$$

Also the derivative with respect to r can be expressed as a derivative with respect to n by

$$\frac{d}{dr}\left(\frac{V^2}{2}\right) = \frac{d}{dn}\left(\frac{V^2}{2}\right)\frac{dn}{dr} = \frac{d}{dn}\left(\frac{V^2}{2}\right)$$

because the direction of n is the same as r so $dn/dr = 1$. Eq. (4.43) can be rewritten as

$$a_n = \frac{d}{dn}\left(\frac{V^2}{2}\right) \tag{4.44}$$

Substituting the expression for acceleration into Euler's equation, Eq. (4.41), and assuming constant density results in

$$\frac{d}{dn}\left(p + \gamma z + \rho\frac{V^2}{2}\right) = 0 \tag{4.45}$$

or

$$p + \gamma z + \rho\frac{V^2}{2} = C \tag{4.46}$$

which is the Bernoulli equation, and C is constant in the n direction (across streamlines).

Summary For an irrotational flow, the constant C in the Bernoulli equation is the same across streamlines as well as along streamlines, so it is the same everywhere in the flow field. Thus, *when applying the Bernoulli equation for irrotational flow, one can select points 1 and 2 at any locations, not just along a streamline.*

4.10 Describing the Pressure Field for Flow over a Circular Cylinder

Flow over a circular cylinder is a paradigm (i.e., model) for external flow over many objects. Thus, this flow is described in this section.

The Pressure Coefficient

To describe the pressure field, engineers often use a dimensionless group called the **pressure coefficient:**

$$C_p = \frac{p_z - p_{zo}}{\rho V_o^2/2} = \frac{h - h_o}{V_o^2/(2g)} \tag{4.47}$$

FIGURE 4.36

Irrotational flow past a cylinder. (a) Streamline pattern. (b) Pressure distribution.

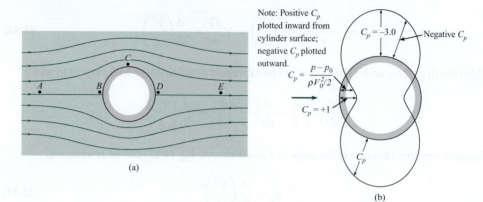

(a)

(b)

Pressure Distribution for an Ideal Fluid

An **ideal fluid** is defined as a fluid that is nonviscous and that has constant density. If we assume an irrotational flow of an ideal fluid, then calculations reveal the results shown in Fig. 4.36a. Features to notice in this figure are

- The pressure distribution is symmetric on the front and back of the cylinder.
- The pressure coefficient is sometimes negative (plotted outward), which corresponds to negative gage pressure.
- The pressure coefficient is sometimes positive (plotted inward), which corresponds to positive gage pressure.
- The maximum pressure ($C_p = +1.0$) occurs on the front and back of the cylinder at the stagnation points (points B and D).
- The minimum pressure ($C_p = -3.0$) occurs at the midsection, where the velocity is highest (point C).

Next, we introduce the concepts of a favorable and adverse pressure gradient. To begin, apply Euler's equation while neglecting gravitational effects:

$$\rho a_t = -\frac{\partial p}{\partial s}$$

One notes that $a_t > 0$ if $\partial p / \partial s < 0$; that is, the fluid particle accelerates if the pressure decreases with distance along a pathline. This is a **favorable pressure gradient**. On the other hand, $a_t < 0$ if $\partial p / \partial s > 0$, so the fluid particle decelerates if the pressure increases along a pathline. This is an **adverse pressure gradient**. The definitions of pressure gradient are summarized in the table.

| Favorable pressure gradient | $\partial p / \partial s < 0$ | $a_t > 0$ (acceleration) |
| Adverse pressure gradient | $\partial p / \partial s > 0$ | $a_t < 0$ (deceleration) |

Visualize the motion of a fluid particle in Fig. 4.36a as it travels around the cylinder from A to B to C to D and finally to E. Notice that it first decelerates from the free-stream velocity to zero velocity at the forward stagnation point as it travels in an adverse pressure gradient. Then as it passes from B to C, it is in a favorable pressure gradient, and it accelerated to its highest speed. From C to D the pressure increases again toward the rearward stagnation point, and the particle decelerates but has enough momentum to reach D. Finally, the pressure decreases from D to E, and this favorable pressure gradient accelerates the particle back to the free-stream velocity.

Pressure Distribution for a Viscous Flow

Consider the flow of a real (viscous) fluid past a cylinder as shown in Fig. 4.37. The flow pattern upstream of the midsection is very similar to the pattern for an ideal fluid. However, in a viscous fluid the velocity at the surface is zero (no-slip condition), whereas with the flow of an inviscid fluid the surface velocity need not be zero. Because of viscous effects, a boundary layer forms next to the surface. The velocity changes from zero at the surface to the free-stream velocity across the boundary layer. Over the forward section of the cylinder, where the pressure gradient is favorable, the boundary layer is quite thin.

FIGURE 4.37

Flow of a real fluid past a circular cylinder. (a) Flow pattern. (b) Pressure distribution.

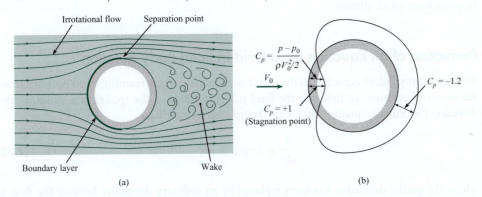

(a)   (b)

Downstream of the midsection, the pressure gradient is adverse, and the fluid particles in the boundary layer, slowed by viscous effects, can only go so far and then are forced to detour away from the surface. The particle is pushed off the wall by pressure force associated with the adverse pressure gradient. The point where the flow leaves the wall is called the separation point. A recirculatory flow called a wake develops behind the cylinder. The flow in the wake region is called separated flow. The pressure distribution on the cylinder surface in the wake region is nearly constant, as shown in Fig. 4.37b. The reduced pressure in the wake leads to increased drag.

4.11 Calculating the Pressure Field for a Rotating Flow

This section describes how to relate pressure and velocity for a *fluid in a solid body rotation*. To understand solid body rotation, consider a cylindrical container of water (Fig. 4.38a) which is stationary. Imagine that the container is placed into rotational motion about an axis (Fig. 4.38b)

FIGURE 4.38

Sketch used to define a
fluid in solid body rotation.

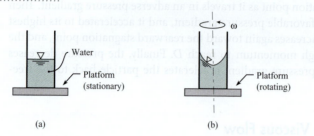

(a)   (b)

and allowed to reach steady state with an angular speed of ω. At steady state, the fluid particles
will be at rest with respect to each other. That is, the distance between any two fluid particles
will be constant. This condition also describes rotation of a rigid body; thus, this type of motion is defined as a **fluid in a solid body rotation**.

Situations in which a fluid rotates as a solid body are found in many engineering applications. One common application is the centrifugal separator. The centripetal accelerations resulting from rotating a fluid separate the heavier particles from the lighter particles as the
heavier particles move toward the outside and the lighter particles are displaced toward the
center. A milk separator operates in this fashion, as does a cyclone separator for removing par-
ticulates from an air stream.

Derivation of an Equation for a Fluid in Solid Body Rotation

To begin, apply Euler's equation in the direction normal to the streamlines and outward from
the center of rotation. In this case the fluid particles rotate as the spokes of a wheel, so the
direction ℓ in Euler's equation, Eq. (4.15), is replaced by r giving

$$-\frac{d}{dr}(p + \gamma z) = \rho a_r \tag{4.48}$$

where the partial derivative has been replaced by an ordinary derivative because the flow is
steady and a function only of the radius r. From Eq. (4.11), the acceleration in the radial direction (away from the center of curvature) is

$$a_r = -\frac{V^2}{r}$$

and Euler's equation becomes

$$-\frac{d}{dr}(p + \gamma z) = -\rho \frac{V^2}{r} \tag{4.49}$$

For solid body rotation about a fixed axis,

$$V = \omega r$$

Substituting this velocity distribution into Euler's equation results in

$$\frac{d}{dr}(p + \gamma z) = \rho r \omega^2 \tag{4.50}$$

Integrating Eq. (4.50) with respect to r gives

$$p + \gamma z = \frac{\rho r^2 \omega^2}{2} + \text{const} \qquad (4.51)$$

or

$$\frac{p}{\gamma} + z - \frac{\omega^2 r^2}{2g} = C \qquad (4.52a)$$

This equation can also be written as

$$p + \gamma z - \rho \frac{\omega^2 r^2}{2} = C \qquad (4.52b)$$

These equivalent equations describe the *pressure variation in rotating flow*. Example 4.11 shows how to apply the equation.

EXAMPLE 4.11

Calculating the Surface Profile of a Rotating Liquid

Problem Statement

A cylindrical tank of liquid shown in the figure is rotating as a solid body at a rate of 4 rad/s. The tank diameter is 0.5 m. The line AA depicts the liquid surface before rotation, and the line $A'A'$ shows the surface profile after rotation has been established. Find the elevation difference between the liquid at the center and the wall during rotation.

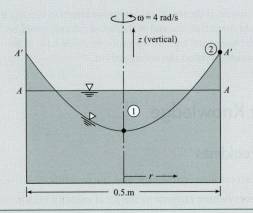

Define the Situation

A liquid is rotating in a cylindrical tank.

State the Goal

Calculate the elevation difference (in meters) between liquid at the center and at the wall.

Generate Ideas and Make a Plan

1. Apply Eq. (4.52a), between points 1 and 2.
2. Calculate the elevation difference.

Take Action (Execute the Plan)

1. Equation (4.52a).

$$\frac{p_1}{\gamma} + z_1 - \frac{\omega^2 r_1^2}{2g} = \frac{p_2}{\gamma} + z_2 - \frac{\omega^2 r_2^2}{2g}$$

The pressure at both points is atmospheric, so $p_1 = p_2$ and the pressure terms cancel out. At point 1, $r_1 = 0$, and at point 2, $r = r_2$. The equation reduces to

$$z_2 - \frac{\omega^2 r_2^2}{2g} = z_1$$

$$z_2 - z_1 = \frac{\omega^2 r_2^2}{2g}$$

2. Elevation difference:

$$z_2 - z_1 = \frac{(4\ \text{rad/s})^2 \times (0.25\ \text{m})^2}{2 \times 9.81\ \text{m/s}^2}$$

$$= \boxed{0.051\ \text{m or 5.1 cm}}$$

Review the Solution and the Process

Notice that the surface profile is parabolic.

Example 4.12 illustrates the analysis of a rotating flow in a manometer.

EXAMPLE 4.12

Evaluating a Rotating Manometer Tube

Problem Statement

When the U-tube is not rotated, the water stands in the tube as shown. If the tube is rotated about the eccentric axis at a rate of 8 rad/s, what are the new levels of water in the tube?

Define the Situation

A manometer tube is rotated around an eccentric axis.

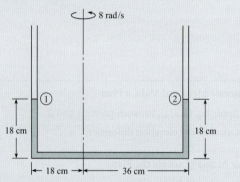

Assumptions: Liquid is incompressible.

State the Goal

Find the levels of water in each leg.

Generate Ideas and Make a Plan

The total length of the liquid in the manometer must be the same before and after rotation, namely 90 cm. Assume, to start with, that liquid remains in the bottom leg. The pressure at the top of the liquid in each leg is atmospheric.

1. Apply the equation for pressure variation in rotating flows, Eq. (4.52a), to evaluate difference in elevation in each leg.

2. Using constraint of total liquid length, find the level in each leg.

Take Action (Execute the Plan)

1. Application of Eq. (4.52a) between top of leg on left (1) and on right (2):

$$z_1 - \frac{r_1^2 \omega^2}{2g} = z_2 - \frac{r_2^2 \omega^2}{2g}$$

$$z_2 - z_1 = \frac{\omega^2}{2g}(r_2^2 - r_1^2)$$

$$= \frac{(8 \text{ rad/s})^2}{2 \times 9.81 \text{ m/s}^2}(0.36^2 \text{ m}^2 - 0.18^2 \text{ m}^2) = 0.317 \text{ m}$$

2. The sum of the heights in each leg is 36 cm.

$$z_2 + z_1 = 0.36 \text{ m}$$

Solution for the leg heights:

$$z_2 = 0.338 \text{ m}$$

$$z_1 = 0.022 \text{ m}$$

Review the Solution and the Process

Discussion. If the result was a negative height in one leg, it would mean that one end of the liquid column would be in the horizontal leg, and the problem would have to be reworked to reflect this configuration.

4.12 Summarizing Key Knowledge

Pathline, Streamlines, and Streaklines

- To visualize flow, engineers use the streamline, streakline, and the pathline.
  - ▸ The *streamline* is a curve that is everywhere tangent to the local velocity vector.
  - ▸ The *streamline* is a mathematical entity that cannot be observed in the physical world.
  - ▸ The configuration of streamlines in a flow field is called the *flow pattern*.
  - ▸ The *pathline* is the line (straight or curved) that a particle follows.
  - ▸ A *streakline* is the line produced by a dye or other marker fluid introduced at a point.
- In *steady flow*, pathlines, streaklines, and streamlines are coincident (i.e., on top of each other) if they share a common point.
- In *unsteady flow*, pathlines, streaklines, and streamlines are not coincident.

Velocity and Velocity Field

- In a flowing fluid, *velocity* is defined as the speed and direction of travel of a fluid particle.
- A *velocity field* is a mathematical or graphical description that shows the velocity at each point (i.e., spatial location) within a flow.

Eulerian and Lagrangian Descriptions

There are two ways to describe motion (Lagrangian and Eulerian).

- In the *Lagrangian approach*, the engineer identifies a specified collection of matter and describes its motion. For example, when a engineer is describing the motion of a fluid particle this is a Lagrangian-based description.
- In the *Eulerian approach*, the engineer identifies a region in space and describes the motion of matter that is passing by in terms of what is happening at various spatial locations. For example, the velocity field is an Eulerian-based concept.
  - ‣ The Eulerian approach uses fields. A *field* is a mathematical or graphical description that shows how a variable is distributed spatially. A field can be a *scalar field* or a *vector field*.
  - ‣ The Eulerian approach uses the divergence, gradient, and curl operators.
  - ‣ The Eulerian approach uses more complicated mathematics (e.g., partial derivatives) than the Lagrangian approach.

Describing Flow

Engineers describe flowing fluids using the ideas summarized in Table 4.4.

TABLE 4.4 How Engineers Describe Flowing Fluids

Description	Key Knowledge
Engineers classify flows as *uniform* or *nonuniform*.	• Uniform and nonuniform flow describe how velocity varies spatially. • *Uniform flow* means that the velocity at each point on a given streamline is the same. Uniform flow requires rectilinear streamlines (straight and parallel). • *Nonuniform flow* means that velocity at various points on a given streamline differs.
Engineers classify flows as *steady* or *unsteady*.	• *Steady flow* means the velocity is constant with respect to time at every point in space. • *Unsteady flow* means the velocity is changing with time at some or all points in space. • Engineers often idealize unsteady flows as steady flow. Example: A draining tank of water is commonly assumed to be a steady flow.
Engineers classify flows as *laminar* or *turbulent*.	• *Laminar flow* involves flow in smooth layers (laminae) with low levels of mixing between layers. • *Turbulent flow* involves flow that is dominated by eddies of various size. Flow is chaotic, unsteady, 3D. High levels of mixing. • Occasionally, engineers describe a flow as *transitional*. This means that the flow is changing from a laminar flow to a turbulent flow.

(Continued)

TABLE 4.4 How Engineers Describe Flowing Fluids (*Continued*)

Description	Key Knowledge
Engineers classify flows as *1D*, *2D*, or *3D*.	• *One-dimensional (1-D) flow* means the velocity depends on one spatial variable. E.g., velocity depends on radius r only. • *Three-dimensional (3-D) flow* means the velocity depends on three spatial variables. E.g., velocity depends on three position coordinates: $\mathbf{V} = \mathbf{V}(x, y, z)$.
Engineers classify flows as *viscous flow* or *inviscid flow*.	• In a *viscous flow*, the forces associated with viscous shear stresses are significant. Thus, viscous terms are included when solving the equations of motion. • In an *inviscid flow*, the forces associated with viscous shear stresses are insignificant. Thus, viscous terms are neglected when solving the equations of motion. The fluid behaves as if its viscosity were zero.
Engineers describe flows by describing an *inviscid flow region*, a *boundary layer*, and a *wake*.	• In the *inviscid flow region*, the streamlines are smooth and the flow can be analyzed with Euler's equation. • The *boundary layer* is a thin region of fluid next to wall. Viscous effects are significant in the boundary layer. • The *wake* is the region of separated flow behind a body.
Engineers describe flows as separated or attached.	• *Flow separation* is when fluid particles move away from the wall. • *Attached flow* is when fluid particles are moving along a wall or boundary. • The region of separated flow inside a pipe or duct is often called a *recirculation zone*.

Acceleration

- *Acceleration* is a property of a fluid particle that characterizes
  - ▸ The change in speed of the particle
  - ▸ The change in direction of travel of the particle
- *Acceleration* is defined mathematically as the derivative of the velocity vector.
- *Acceleration* of a fluid particle can be described qualitatively. Guidelines:
  - ▸ If a particle is traveling on a curved streamline, there will be a component of acceleration that is normal to the streamline and directed inwards toward the center of curvature.
  - ▸ If the particle is changing speed, there will be a component of acceleration that is tangent to the streamline.
- In an *Eulerian representation of acceleration*,
  - ▸ Terms that involve derivatives with respect to time are *local acceleration* terms.
  - ▸ All other terms are *convective acceleration* terms. Most of these terms involve derivatives with respect to position.

Euler's Equation

- *Euler's equation* is *Newton's second law of motion* applied to a fluid particle when the flow is inviscid and incompressible.

- Euler's equation can be written as a *vector equation*:

$$-\nabla p_z = \rho \mathbf{a}$$

- This vector form can be also be written as a *scalar equation* in an arbitrary ℓ direction.

$$-\frac{\partial}{\partial \ell}(p + \gamma z) = -\left(\frac{\partial p_z}{\partial \ell}\right) = \rho a_\ell$$

- *Physics of Euler's equation*: The gradient of piezometric pressure is colinear with acceleration and opposite in direction. This reveals how pressure varies:
  ▶ When streamlines are curved, pressure will increase outward from the center of curvature.
  ▶ When a streamline is rectilinear and a particle on the streamline is changing speed, then the pressure will change in a direction tangent to the streamline. The direction of increasing pressure is opposite of the acceleration vector.
  ▶ When streamlines are rectilinear, pressure variation normal to the streamlines is hydrostatic.

The Bernoulli Equation

- The *Bernoulli equation* is *conservation of energy* applied to a fluid particle. It is derived by integrating Euler's equation for steady, inviscid, and constant density flow.
- For the assumptions just stated, the Bernoulli equation is applied between any two points on the same streamline.
- The Bernoulli equations has two forms:
  ▶ *Head Form*: $p/\gamma + z + V^2/(2g) = $ constant
  ▶ *Pressure Form*: $p + \rho g z + (\rho V^2)/2 = $ constant
- There are two equivalent ways to describe the physics of the Bernoulli equation:
  ▶ When speed increases, then piezometric pressure decreases (along a streamline).
  ▶ The total head (velocity head plus piezometric head) is constant along a streamline. This means that energy is conserved as a fluid particle moves along a streamline.

Measuring Velocity and Pressure

- When pressure is measured at a *pressure tap* on the wall of a pipe, this provides a measurement of static pressure. This same measurement can also be used to determine pressure head or piezometric head.
- *Static pressure* is defined as the pressure in a flowing fluid. Static pressure must be measured in a way that does not change the value of the measured pressure.
- *Kinetic pressure* is $(\rho V^2)/2$.
- A *stagnation tube* provides a measurement of (static pressure) + (kinetic pressure):

$$p + (\rho V^2)/2$$

- The *Pitot-static* tube, provides a method to measure both static pressure and kinetic pressure at a point in a flowing fluid. Thus, this instrument provides a way to measure fluid velocity.

Fluid Rotation, Vorticity, and Irrotational Flow

- Rate of rotation Ω
  ▶ Is a property of a fluid particle that describes how fast the particle is rotating.
  ▶ Is defined by placing two perpendicular lines on a fluid particle and then averaging the rotational rate of these lines.
  ▶ Is a vector quantity with the direction of the vector given by the right-hand rule.

- A common way to describe rotation is to use the vorticity vector ω, which is twice the rotation vector: $\omega = 2\Omega$

- In Cartesian coordinates, the vorticity is given by

$$\omega = \left(\frac{\partial w}{\partial y} - \frac{\partial v}{\partial z}\right)\mathbf{i} + \left(\frac{\partial u}{\partial z} - \frac{\partial w}{\partial x}\right)\mathbf{j} + \left(\frac{\partial v}{\partial x} - \frac{\partial u}{\partial y}\right)\mathbf{k}$$

- An irrotational flow is one in which vorticity is everywhere zero.

- When applying the Bernoulli equation for irrotational flow, one can select points 1 and 2 at any locations, not just along a streamline.

Describing the Pressure Field

- The pressure field is often described using a π-group called the pressure coefficient.

- The pressure gradient near a body is related to flow separation.

  ▸ An adverse pressure gradient is associated with flow separation.

  ▸ A positive pressure gradient is associated with attached flow.

- The pressure field for flow over a circular cylinder is a paradigm for understanding external flows. The pressure along the front of the cylinder is high, and the pressure in the wake is low.

- When flow is rotating as a solid body, the pressure field p can be described using

$$p + \gamma z - \rho\frac{\omega^2 r^2}{2} = C$$

where ω is the rotational speed, and r is the distance from the axis of rotation to the point in the field.

Describing the Pressure Field (Summary)

Pressure variations in a flowing fluid are associated with three phenomenon:

- **Weight.** Due to the weight of a fluid, pressure increases with increasing depth (i.e., decreasing elevation). This topic is presented in Chapter 3 (Hydrostatics)

- **Acceleration.** When fluid particles are accelerating, there are usually pressure variations associated with the acceleration. In inviscid flow, the gradient of the pressure field is aligned in a direction opposite of the acceleration vector.

- **Viscous Effects.** When viscous effects are significant, there can be associated pressure changes. For example, there are pressure drops associated with flows in horizontal pipes and ducts. This topic is presented in Chapter 10 (Conduit Flow).

REFERENCES

1. *Flow Visualization*, Fluid Mechanics Films, downloaded 7/31/11 from http://web.mit.edu/hml/ncfmf.html

2. Hibbeler, R.C. *Dynamics*. Englewood Cliffs, NJ: Prentice Hall, 1995.

3. *Turbulence*, Fluid Mechanics Films, downloaded 7/31/11 from http://web.mit.edu/hml/ncfmf.html

4. *Pressure Fields and Fluid Acceleration*, Fluid Mechanics Films, downloaded 7/31/11 from http://web.mit.edu/hml/ncfmf.html

5. Miller, R.W. (ed) *Flow Measurement Engineering Handbook*, New York: McGraw-Hill, 1996.

6. *Vorticity, Part 1, Part 2*, Fluid Mechanics Films, downloaded 7/31/11 from http://web.mit.edu/hml/ncfmf.html

PROBLEMS

PLUS Problem available in *WileyPLUS* at instructor's discretion.

GO Guided Online (GO) Problem, available in *WileyPLUS* at instructor's discretion.

Streamlines, Streaklines, and Pathlines (§4.1)

4.1 If somehow you could attach a light to a fluid particle and take a time exposure photo, would the image you photographed be a pathline or streakline? Explain from definition of each.

4.2 Is the pattern produced by smoke rising from a chimney on a windy day analogous to a pathline or streakline? Explain from the definition of each.

4.3 PLUS A windsock is a sock-shaped device attached to a swivel on top of a pole. Windsocks at airports are used by pilots to see instantaneous shifts in the direction of the wind. If one drew a line co-linear with a windsock's orientation at any instant, the line would be best approximate a (a) pathline, (b) streakline, or (c) streamline.

4.4 PLUS For streamlines, streaklines, and streamlines to all be co-linear, the flow must be

    a. dividing

    b. stagnant

    c. steady

    d. a tracer

4.5 At time $t = 0$, dye was injected at point A in a flow field of a liquid. When the dye had been injected for 4 s, a pathline for a particle of dye that was emitted at the 4 s instant was started. The streakline at the end of 10 s is shown below. Assume that the speed (but not the velocity) of flow is the same throughout the 10 s period. Draw the pathline of the particle that was emitted at $t = 4$ s. Make your own assumptions for any missing information.

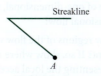

Streakline

A

PROBLEM 4.5

4.6 For a given hypothetical flow, the velocity from time $t = 0$ to $t = 5$ s was $u = 2$ m/s, $v = 0$. Then, from time $t = 5$ s to $t = 10$ s, the velocity was $u = +3$ m/s, $v = -4$ m/s. A dye streak was started at a point in the flow field at time $t = 0$, and the path of a particle in the fluid was also traced from that same point starting at the same time. Draw to scale the streakline, pathline of the particle, and streamlines at time $t = 10$ s.

4.7 At time $t = 0$, a dye streak was started at point A in a flow field of liquid. The speed of the flow is constant over a 10 s period, but the flow direction is not necessarily constant. At any particular instant the velocity in the entire field of flow is the same. The streakline produced by the dye is shown above. Draw (and label) a streamline for the flow field at $t = 8$ s.

    Draw (and label) a pathline that one would see at $t = 10$ s for a particle of dye that was emitted from point A at $t = 2$ s.

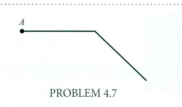

A

PROBLEM 4.7

Velocity and the Velocity Field (§4.2)

4.8 PLUS A velocity field is given mathematically as $\mathbf{V} = 2\mathbf{i} + 4y\mathbf{j}$. The velocity field is:

    a. 1D in x

    b. 1D in y

    c. 2D in x and y

The Eulerian and Lagrangian Approaches (§4.2)

4.9 PLUS There is a gasoline spill in a major river. The mayor of a large downstream city demands an estimate of how many hours it will take for the spill to get to the water supply plant intake. The emergency responders measure the speed of the leading edge of the spill, effectively focusing on one particle of fluid. Meanwhile, environmental engineers at the local university employ a computer model, which simulates the velocity field for any stage of the river, and for all locations (including steep narrow canyon sections with fast velocities, and an extremely wide reach with slow velocities). To compare these two mathematical approaches, which statement is most correct?

    a. The responders have an Eulerian approach, and the engineers have a Lagrangian one

    b. The responders have a Lagrangian approach, and the engineers have an Eulerian one.

Describing Flow (§4.3)

4.10 Identify five examples of an unsteady flow and explain what features classify them as an unsteady flow.

4.11 You are pouring a heavy syrup on your pancakes. As the syrup spreads over the pancake, would the thin film of syrup be a laminar or turbulent flow? Why?

4.12 PLUS A velocity field is given by $\mathbf{V} = 10xy\mathbf{i}$. It is

    a. 1-D and steady

    b. 1-D and unsteady

    c. 2-D and steady

    d. 2-D and unsteady

4.13 Which is the most correct way to characterize turbulent flow?

    a. 1D

    b. 2D

    c. 3D

4.14 In the system in the figure, the valve at C is gradually opened in such a way that a constant rate of increase in discharge is produced. How would you classify the flow at B while the valve is being opened? How would you classify the flow at A?

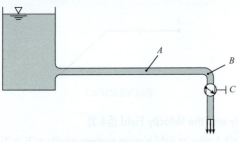

PROBLEM 4.14

4.15 Water flows in the passage shown. If the flow rate is decreasing with time, the flow is classified as (a) steady, (b) unsteady, (c) uniform, or (d) nonuniform.

PROBLEM 4.15

4.16 If a flow pattern has converging streamlines, how would you classify the flow?

4.17 Consider flow in a straight conduit. The conduit is circular in cross section. Part of the conduit has a constant diameter, and part has a diameter that changes with distance. Then, relative to flow in that conduit, correctly match the items in column A with those in column B.

A	B
Steady flow	$\partial V_s/\partial s = 0$
Unsteady flow	$\partial V_s/\partial s \neq 0$
Uniform flow	$\partial V_s/\partial t = 0$
Nonuniform flow	$\partial V_s/\partial t \neq 0$

4.18 Classify each of the following as a one-dimensional, two–dimensional, or three-dimensional flow.

a. Water flow over the crest of a long spillway of a dam.

b. Flow in a straight horizontal pipe.

c. Flow in a constant-diameter pipeline that follows the contour of the ground in hilly country.

d. Airflow from a slit in a plate at the end of a large rectangular duct.

e. Airflow past an automobile.

f. Airflow past a house.

g. Water flow past a pipe that is laid normal to the flow across the bottom of a wide rectangular channel.

Acceleration (§4.4)

4.19 Acceleration is the rate of change of velocity with time. Is the acceleration vector always aligned with the velocity vector? Explain.

4.20 For a rotating body, is the acceleration toward the center of rotation a centripetal or centrifugal acceleration? Look up word meanings and word roots.

4.21 PLUS In a flowing fluid, acceleration means that a fluid particle is

a. changing direction

b. changing speed

c. changing both speed and direction

d. any of the above

4.22 PLUS The flow passing through a nozzle is steady. The speed of the fluid increases between the entrance and the exit of the nozzle. The acceleration halfway between the entrance and the nozzle is

a. convective

b. local

c. both

4.23 PLUS Local acceleration

a. is close to the origin

b. is quasi nonuniform

c. occurs in unsteady flow

4.24 GO Figure 4.36 on p. 148 in §4.10 shows the flow pattern for flow past a circular cylinder. Assume that the approach velocity at A is constant (does not vary with time).

a. Is the flow past the cylinder steady or unsteady?

b. Is this a case of one-dimensional, two-dimensional, or three-dimensional flow?

c. Are there any regions of the flow where local acceleration is present? If so, show where they are and show vectors representing the local acceleration in the regions where it occurs.

d. Are there any regions of flow where convective acceleration is present? If so, show vectors representing the convective acceleration in the regions where it occurs.

4.25 PLUS The velocity along a pathline is given by $V \text{ (m/s)} = s^2 t^{1/2}$ where s is in meters and t is in seconds. The radius of curvature is 0.4 m. Evaluate the acceleration tangent and normal to the path at $s = 1.5$ m and $t = 0.5$ seconds.

4.26 Tests on a sphere are conducted in a wind tunnel at an air speed of U_0. The velocity of flow toward the sphere along the longitudinal axis is found to be $u = -U_0(1 - r_0^3/x^3)$, where r_0 is the radius of the sphere and x the distance from its center. Determine the acceleration of an air particle on the x-axis upstream of the sphere in terms of x, r_0, and U_0.

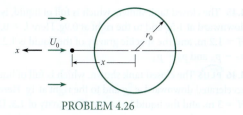

PROBLEM 4.26

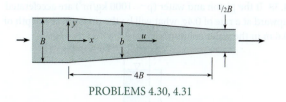

PROBLEMS 4.30, 4.31

4.27 GO In this flow passage the velocity is varying with time. The velocity varies with time at section A-A as

$$V = 5 \text{ m/s} - 2.25 \frac{t}{t_0} \text{ m/s}$$

At time $t = 0.50$ s, it is known that at section A-A the velocity gradient in the s direction is $+2$ m/s per meter. Given that t_0 is 0.5 s and assuming quasi-one-dimensional flow, answer the following questions for time $t = 0.5$ s.

   a. What is the local acceleration at A-A?

   b. What is the convective acceleration at A-A?

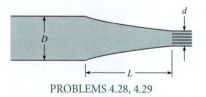

PROBLEM 4.27

4.28 PLUS The nozzle in the figure is shaped such that the velocity of flow varies linearly from the base of the nozzle to its tip. Assuming quasi-one-dimensional flow, what is the convective acceleration midway between the base and the tip if the velocity is 0.3 m/s at the base and 1.2 m/s at the tip? Nozzle length is 46 cm.

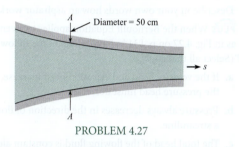

PROBLEMS 4.28, 4.29

4.29 PLUS In Prob. 4.28 the velocity varies linearly with time throughout the nozzle. The velocity at the base is $2t$ (m/s) and at the tip is $6t$ (m/s). What is the local acceleration midway along the nozzle when $t = 2$ s?

4.30 Liquid flows through this two-dimensional slot with a velocity of $V = 2(q_0/b)(t/t_0)$, where q_0 and t_0 are reference values. What will be the local acceleration at $x = 2B$ and $y = 0$ in terms of B, t, t_0, and q_0?

4.31 What will be the convective acceleration for the conditions of Prob. 4.30?

4.32 PLUS The velocity of water flow in the nozzle shown is given by the following expression:

$$V = 2t/(1 - 0.5x/L)^2,$$

where V = velocity in meters per second, t = time in seconds, x = distance along the nozzle, and L = length of nozzle = 1.2 m. When $x = 0.5L$ and $t = 3$ s, what is the local acceleration along the centerline? What is the convective acceleration? Assume quasi-one-dimensional flow prevails.

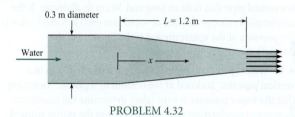

PROBLEM 4.32

Euler's Equation and Pressure Variation (§4.5)

4.33 State Newton's second law of motion. What are the limitations on the use of Newton's second law? Explain.

4.34 What is the differences between a force due to weight and a force due to pressure? Explain.

4.35 A pipe slopes upward in the direction of liquid flow at an angle of $30°$ with the horizontal. What is the pressure gradient in the flow direction along the pipe in terms of the specific weight of the liquid if the liquid is decelerating (accelerating opposite to flow direction) at a rate of $0.4\,g$?

4.36 PLUS What pressure gradient is required to accelerate kerosene (S = 0.81) vertically upward in a vertical pipe at a rate of $0.5\,g$?

4.37 The hypothetical liquid in the tube shown in the figure has zero viscosity and a specific weight of 10 kN/m^3. If $p_B - p_A$ is equal to 12 kPa, one can conclude that the liquid in the tube is being accelerated (a) upward, (b) downward, or (c) neither: acceleration = 0.

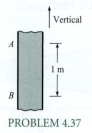

PROBLEM 4.37

4.38 If the piston and water ($\rho = 1000$ kg/m^3) are accelerated upward at a rate of $0.4g$, what will be the pressure at a depth of 0.6 m in the water column?

PROBLEMS 4.38, 4.39

4.39 *WILEY GO* Water ($\rho = 1000$ kg/m^3) stands at a depth of 3 m in a vertical pipe that is open at the top and closed at the bottom by a piston. What upward acceleration of the piston is necessary to create a pressure of 55 kPa gage immediately above the piston?

4.40 *WILEY PLUS* What pressure gradient is required to accelerate water ($\rho = 1000$ kg/m^3) in a horizontal pipe at a rate of 8 m/s^2?

4.41 Water ($\rho = 1000$ kg/m^3) is accelerated from rest in a horizontal pipe that is 80 m long and 30 cm in diameter. If the acceleration rate (toward the downstream end) is 5 m/s^2, what is the pressure at the upstream end if the pressure at the downstream end is 90 kPa gage?

4.42 Water ($\rho = 1000$ kg/m^3) stands at a depth of 3 m in a vertical pipe that is closed at the bottom by a piston. Assuming that the vapor pressure is zero (abs), determine the maximum downward acceleration that can be given to the piston without causing the water immediately above it to vaporize.

4.43 A liquid with a specific weight of 15,700 N/m^3 is in the conduit. This is a special kind of liquid that has zero viscosity. The pressures at points A and B are 8.1 kPa and 4.8 kPa, respectively. Which one (or more) of the following conclusions can one draw with certainty? (a) The velocity is in the positive ℓ direction. (b) The velocity is in the negative ℓ direction. (c) The acceleration is in the positive ℓ direction. (d) The acceleration is in the negative ℓ direction.

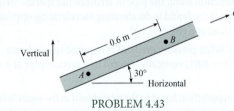

PROBLEM 4.43

4.44 If the velocity varies linearly with distance through this water nozzle, what is the pressure gradient, dp/dx, halfway through the nozzle? ($\rho = 1000$ kg/m^3).

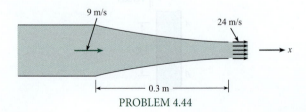

PROBLEM 4.44

4.45 The closed tank shown, which is full of liquid, is accelerated downward at $1.5g$ and to the right at $0.9g$. Here $L = 0.9$ m, $H = 1.2$ m, and the specific gravity of the liquid is 1.2. Determine $p_C - p_A$ and $p_B - p_A$.

4.46 *PLUS* The closed tank shown, which is full of liquid, is accelerated downward at $\frac{2}{3}g$ and to the right at $1g$. Here $L = 2.5$ m, $H = 3$ m, and the liquid has a specific gravity of 1.3. Determine $p_C - p_A$ and $p_B - p_A$.

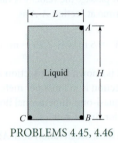

PROBLEMS 4.45, 4.46

Applying the Bernoulli Equation (§4.6)

4.47 Describe in your own words how an aspirator works.

4.48 *WILEY PLUS* When the Bernoulli Equation applies to a venturi, such as in Fig. 4.27 on p. 134 in §4.6, which of the following are true? (Select all that apply.)

   a. If the velocity head and elevation head increase, then the pressure head must decrease.

   b. Pressure always decreases in the direction of flow along a streamline.

   c. The total head of the flowing fluid is constant along a streamline.

4.49 *WILEY PLUS* A water jet issues vertically from a nozzle, as shown. The water velocity as it exits the nozzle is 18 m/s. Calculate how high h the jet will rise. (*Hint:* Apply the Bernoulli equation along the centerline.)

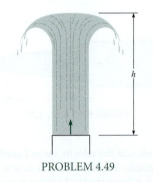

PROBLEM 4.49

4.50 A pressure of 10 kPa, gage, is applied to the surface of water in an enclosed tank. The distance from the water surface to the outlet is 0.5 m. The temperature of the water is 20°C. Find the velocity (m/s) of water at the outlet. The speed of the water surface is much less than the water speed at the outlet.

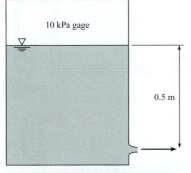

10 kPa gage

0.5 m

PROBLEM 4.50

4.51 Water flows through a vertical contraction (venturi) section. Piezometers are attached to the upstream pipe and minimum area section as shown. The velocity in the pipe is 3 m/s. The difference in elevation between the two water levels in the piezometers is 15 cm. The water temperature is 20°C. What is the velocity (m/s) at the minimum area?

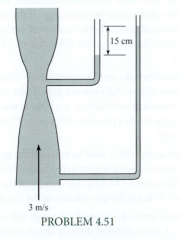

15 cm

3 m/s

PROBLEM 4.51

4.52 PLUS Kerosene at 20°C flows through a contraction section as shown. A pressure gage connected between the upstream pipe and throat section shows a pressure difference of 20 kPa. The gasoline velocity in the throat section is 8 m/s. What is the velocity (m/s) in the upstream pipe?

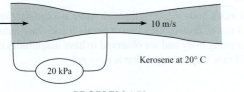

10 m/s

20 kPa

Kerosene at 20° C

PROBLEM 4.52

Stagnation Tubes and Pitot-Static Tubes (§4.7)

4.53 PLUS A stagnation tube placed in a river (select all that apply)

  a. can be used to determine air pressure

  b. can be used to determine fluid velocity

  c. measures kinetic pressure

4.54 PLUS A Pitot-static tube is mounted on an airplane to measure airspeed. At an altitude of 3048 m, where the temperature is −5°C and the pressure is 69 kPa abs, a pressure difference corresponding to 25 cm of water is measured. What is the airspeed?

4.55 PLUS A glass tube is inserted into a flowing stream of water with one opening directed upstream and the other end vertical. If the water velocity is 5 m/s, how high will the water rise in the vertical leg relative to the level of the water surface of the stream?

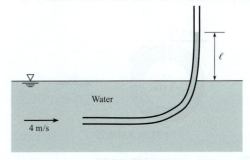

Water

ℓ

4 m/s

PROBLEM 4.55

4.56 A Bourdon-tube gage is taped into the center of a disk as shown. Then for a disk that is about 0.3 m in diameter and for an approach velocity of air (V_0) of 12 m/s, the gage would read a pressure intensity that is (a) less than $\rho V_0^2/2$, (b) equal to $\rho V_0^2/2$, or (c) greater than $\rho V_0^2/2$.

Bourdon-tube gage

PROBLEM 4.56

4.57 An air-water manometer is connected to a Pitot-static tube used to measure air velocity. If the manometer deflects 5 cm, what is the velocity? Assume $T = 15.5$°C and $p = 103$ kPa abs.

4.58 The flow-metering device shown consists of a stagnation probe at station 2 and a static pressure tap at station 1. The velocity at station 2 is 1.5 times that at station 1. Air with a density of 1.2 kg/m³ flows through the duct. A water manometer is connected between the stagnation probe and the pressure tap, and a deflection of 10 cm is measured. What is the velocity at station 2?

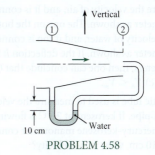

Vertical

① ②

10 cm   Water

PROBLEM 4.58

4.59 The "spherical" Pitot probe shown is used to measure the flow velocity in water ($\rho = 1000$ kg/m³). Pressure taps are located at the forward stagnation point and at 90° from the forward stagnation point. The speed of fluid next to the surface of the sphere varies as $1.5\, V_0 \sin \theta$, where V_0 is the free-stream velocity and θ is measured from the forward stagnation point. The pressure taps are at the same level; that is, they are in the same horizontal plane. The piezometric pressure difference between the two taps is 2 kPa. What is the free-stream velocity V_0?

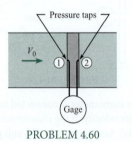

PROBLEM 4.59

4.60 ⒲ PLUS A device used to measure the velocity of fluid in a pipe consists of a cylinder, with a diameter much smaller than the pipe diameter, mounted in the pipe with pressure taps at the forward stagnation point and at the rearward side of the cylinder. Data show that the pressure coefficient at the rearward pressure tap is −0.3. Water with a density of 1000 kg/m³ flows in the pipe. A pressure gage connected by lines to the pressure taps shows a pressure difference of 500 Pa. What is the velocity in the pipe?

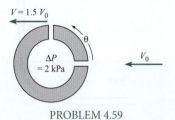

PROBLEM 4.60

4.61 Explain how you might design a spherical Pitot-static probe to provide the direction and velocity of a flowing stream. The Pitot-static probe will be mounted on a string that can be oriented in any direction.

4.62 ⒲ PLUS Two Pitot-static tubes are shown. The one on the top is used to measure the velocity of air, and it is connected to an air-water manometer as shown. The one on the bottom is used to measure the velocity of water, and it too is connected to an air-water manometer as shown. If the deflection h is the same for both manometers, then one can conclude that (a) $V_A = V_w$, (b) $V_A > V_w$, or (c) $V_A < V_w$.

4.63 A Pitot-static tube is used to measure the velocity at the center of a 30 cm pipe. If kerosene at 20°C is flowing and the deflection on a mercury-kerosene manometer connected to the Pitot tube is 10 cm, what is the velocity?

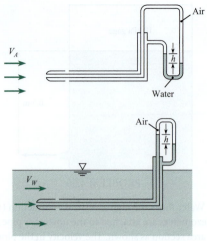

PROBLEM 4.62

4.64 ⒲ PLUS A Pitot-static tube used to measure air velocity is connected to a differential pressure gage. If the air temperature is 20°C at standard atmospheric pressure at sea level, and if the differential gage reads a pressure difference of 2 kPa, what is the air velocity?

4.65 A Pitot-static tube used to measure air velocity is connected to a differential pressure gage. If the air temperature is 15.5°C at standard atmospheric pressure at sea level, and if the differential gage reads a pressure difference of 718 Pa, what is the air velocity?

4.66 A Pitot-static tube is used to measure the gas velocity in a duct. A pressure transducer connected to the Pitot tube registers a pressure difference of 13.8 kPa. The density of the gas in the duct is 2.25 kg/m³. What is the gas velocity in the duct?

4.67 A sphere moves horizontally through still water at a speed of 3.35 m/s. A short distance directly ahead of the sphere (call it point A), the velocity, with respect to the earth, induced by the sphere is 0.3 m/s in the same direction as the motion of the sphere. If p_0 is the pressure in the undisturbed water at the same depth as the center of the sphere, then the value of the ratio p_A/p_0 will be (a) less than unity, (b) equal to unity, or (c) greater than unity.

4.68 ⒲ PLUS Body A travels through water at a constant speed of 13 m/s as shown. Velocities at points B and C are induced by the moving body and are observed to have magnitudes of 5 m/s and 3 m/s, respectively. What is $p_B - p_C$?

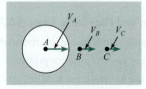

PROBLEM 4.68

4.69 Water in a flume is shown for two conditions. If the depth d is the same for each case, will gage A read greater or less than gage B? Explain.

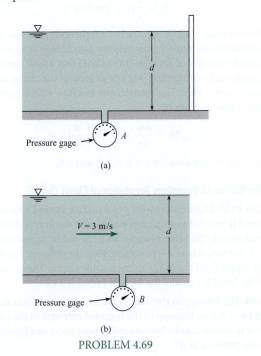

Pressure gage → A

(a)

$V = 3$ m/s

Pressure gage → B

(b)

PROBLEM 4.69

4.70 The apparatus shown in the figure is used to measure the velocity of air at the center of a duct having a 10 cm diameter. A tube mounted at the center of the duct has a 2 mm diameter and is attached to one leg of a slant-tube manometer. A pressure tap in the wall of the duct is connected to the other end of the slant-tube manometer. The well of the slant-tube manometer is sufficiently large that the elevation of the fluid in it does not change significantly when fluid moves up the leg of the manometer. The air in the duct is at a temperature of 20°C, and the pressure is 150 kPa. The manometer liquid has a specific gravity of 0.7, and the slope of the leg is 30°. When there is no flow in the duct, the liquid surface in the manometer lies at 2.3 cm on the slanted scale. When there is flow in the duct, the liquid moves up to 6.7 cm on the slanted scale. Find the velocity of the air in the duct. Assuming a uniform velocity profile in the duct, calculate the rate of flow of the air.

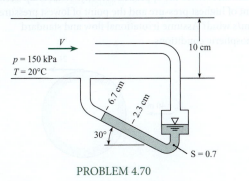

$p = 150$ kPa
$T = 20°C$

10 cm

6.7 cm

2.3 cm

30°

$S = 0.7$

PROBLEM 4.70

4.71 A rugged instrument used frequently for monitoring gas velocity in smokestacks consists of two open tubes oriented to the flow direction as shown and connected to a manometer. The pressure coefficient is 1.0 at A and -0.3 at B. Assume that water, at 20°C, is used in the manometer and that a 5 mm deflection is noted. The pressure and temperature of the stack gases are 101 kPa and 250°C. The gas constant of the stack gases is 200 J/kg K. Determine the velocity of the stack gases.

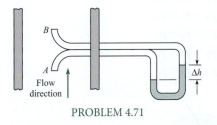

B

A
Flow
direction

Δh

PROBLEM 4.71

4.72 The pressure in the wake of a bluff body is approximately equal to the pressure at the point of separation. The velocity distribution for flow over a sphere is $V = 1.5\, V_0 \sin\theta$, where V_0 is the free-stream velocity and θ is the angle measured from the forward stagnation point. The flow separates at $\theta = 120°$. If the free-stream velocity is 100 m/s and the fluid is air ($\rho = 1.2$ kg/m^3), find the pressure coefficient in the separated region next to the sphere. Also, what is the gage pressure in this region if the free-stream pressure is atmospheric?

4.73 A Pitot-static tube is used to measure the airspeed of an airplane. The Pitot tube is connected to a pressure-sensing device calibrated to indicate the correct airspeed when the temperature is 17°C and the pressure is 101 kPa. The airplane flies at an altitude of 3000 m, where the pressure and temperature are 70 kPa and −6.3°C. The indicated airspeed is 70 m/s. What is the true airspeed?

4.74 An aircraft flying at 3048 m uses a Pitot-static tube to measure speed. The instrumentation on the aircraft provides the differential pressure as well as the local static pressure and the local temperature. The local static pressure is 67.6 kPa gage, and the air temperature is −3.9°C. The differential pressure is 3.5 kPa. Find the speed of the aircraft in km/s.

4.75 You need to measure air flow velocity. You order a commercially available Pitot-static tube, and the accompanying instructions state that the airflow velocity is given by

$$V(\text{m/s}) = 26314.7\sqrt{\frac{h_v}{d}}$$

where h_v is the "velocity pressure" in meters of water and d is the density in kilogram per cubic meter. The velocity pressure is the deflection measured on a water manometer attached to the static and total pressure ports. The instructions also state the density d can be calculated using

$$d\,(\text{kg/m}^3) = 4560.2\frac{p_a}{T}$$

where P_a is the barometric pressure in meters of mercury and T is the absolute temperature in Kelvin. Before you use the Pitot tube you want to confirm that the equations are correct. Determine if they are correct.

4.76 Consider the flow of water over the surfaces shown. For each case the depth of water at section D-D is the same (0.3 m), and the mean velocity is the same and equal to 3 m/s. Which of the following statements are valid?

a. $p_C > p_B > p_A$

b. $p_B > p_C > p_A$

c. $p_A = p_B = p_C$

d. $p_B < p_C < p_A$

e. $p_A < p_B < p_C$

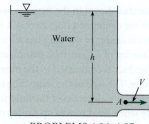

PROBLEM 4.76

Characterizing Rotational Motion of a Fluid (§4.8)

4.77 What is meant by rotation of a fluid particle? Use a sketch to explain.

4.78 Consider a spherical fluid particle in an inviscid fluid (no shear stresses). If pressure and gravitational forces are the only forces acting on the particle, can they cause the particle to rotate? Explain.

4.79 PLUS The vector $\mathbf{V} = 10x\mathbf{i} - 10y\mathbf{j}$ represents a two-dimensional velocity field. Is the flow irrotational?

4.80 The u and v velocity components of a flow field are given by $u = -\omega y$ and $v = \omega x$. Determine the vorticity and the rate of rotation of flow field.

4.81 The velocity components for a two-dimensional flow are

$$u = \frac{Cx}{(y^2 + x^2)} \qquad v = \frac{Cy}{(x^2 + y^2)}$$

where C is a constant. Is the flow irrotational?

4.82 PLUS A two-dimensional flow field is defined by $u = x^2 - y^2$ and $v = -2xy$. Is the flow rotational or irrotational?

4.83 Fluid flows between two parallel stationary plates. The distance between the plates is 1 cm. The velocity profile between the two plates is a parabola with a maximum velocity at the centerline of 2 cm/s. The velocity is given by

$$u = 2(1 - 4y^2)$$

where y is measured from the centerline. The cross-flow component of velocity, v, is zero. There is a reference line located 1 cm downstream. Find an expression, as a function of y, for the amount of rotation (in radian) a fluid particle will undergo when it travels a distance of 1 cm downstream.

4.84 A combination of a forced and a free vortex is represented by the velocity distribution

$$v_\theta = \frac{1}{r}[1 - \exp(-r^2)]$$

For $r \to 0$ the velocity approaches a rigid body rotation, and as r becomes large, a free-vortex velocity distribution is approached. Find the amount of rotation (in radians) that a fluid particle will experience in completing one circuit around the center as a function of r. *Hint:* The rotation rate in a flow with concentric streamlines is given by

$$2\dot\theta = \frac{dv_\theta}{dr} + \frac{v_\theta}{r} = \frac{1}{r}\frac{d}{dr}(v_\theta r)$$

Evaluate the rotation for $r = 0.5, 1.0,$ and 1.5.

The Bernoulli Equation (Irrotational Flow) (§4.9)

4.85 PLUS Liquid flows with a free surface around a bend. The liquid is inviscid and incompressible, and the flow is steady and irrotational. The velocity varies with the radius across the flow as $V = 1/r$ m/s, where r is in meters. Find the difference in depth of the liquid from the inside to the outside radius. The inside radius of the bend is 1 m and the outside radius is 3 m.

4.86 The velocity in the outlet pipe from this reservoir is 9 m/s and $h = 5.5$ m. Because of the rounded entrance to the pipe, the flow is assumed to be irrotational. Under these conditions, what is the pressure at A?

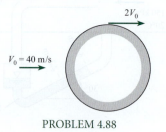

PROBLEMS 4.86, 4.87

4.87 PLUS The velocity in the outlet pipe from this reservoir is 8 m/s and $h = 19$ m. Because of the rounded entrance to the pipe, the flow is assumed to be irrotational. Under these conditions, what is the pressure at A?

4.88 The maximum velocity of the flow past a circular cylinder, as shown, is twice the approach velocity. What is Δp between the point of highest pressure and the point of lowest pressure in a 40 m/s wind? Assume irrotational flow and standard atmospheric conditions.

PROBLEM 4.88

4.89 The velocity and pressure are given at two points in the flow field. Assume that the two points lie in a horizontal plane and that the fluid density is uniform in the flow field and is equal to 1000 kg/m³. Assume steady flow. Then, given these data, determine which of the following statements is true. (a) The flow in the contraction is nonuniform and irrotational. (b) The flow in the contraction is uniform and irrotational. (c) The flow in the contraction is nonuniform and rotational. (d) The flow in the contraction is uniform and rotational.

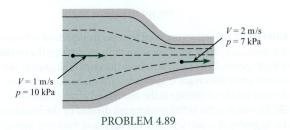

$V = 2$ m/s
$p = 7$ kPa

$V = 1$ m/s
$p = 10$ kPa

PROBLEM 4.89

4.90 Water ($\rho = 1000$ kg/m³) flows from the large orifice at the bottom of the tank as shown. Assume that the flow is irrotational. Point B is at zero elevation, and point A is at 0.3 m elevation. If $V_A = 1.2$ m/s at an angle of 45° with the horizontal and if $V_B = 3.6$ m/s vertically downward, what is the value of $p_A - p_B$?

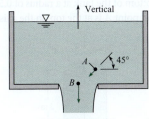

Vertical

A 45°

B

PROBLEM 4.90

4.91 ^{WILEY GO} Ideal flow theory will yield a flow pattern past an airfoil similar to that shown. If the approach air velocity V_0 is 80 m/s, what is the pressure difference between the bottom and the top of this airfoil at points where the velocities are $V_1 = 85$ m/s and $V_2 = 75$ m/s? Assume ρ_{air} is uniform at 1.2 kg/m³.

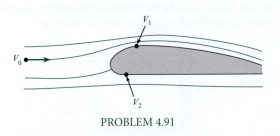

V_1

V_0

V_2

PROBLEM 4.91

4.92 Consider the flow of water between two parallel plates in which one plate is fixed as shown. The distance between the plates is h, and the speed of the moving plate is V. A person wishes to calculate the pressure difference between the plates and applies the Bernoulli equation between points 1 and 2,

$$z_1 + \frac{p_1}{\gamma} + \frac{V_1^2}{2g} = z_2 + \frac{p_2}{\gamma} + \frac{V_2^2}{2g}$$

and concludes that

$$p_1 - p_2 = \gamma(z_2 - z_1) + \rho\frac{V_2^2}{2}$$

$$= \gamma h + \rho\frac{V^2}{2}$$

Is this correct? Provide the reason for your answer.

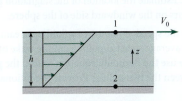

1   V_0

h

z

2

PROBLEM 4.92

4.93 Euler's equations for a planar (two-dimensional) flow in the xy-plane are

$$u\frac{\partial u}{\partial x} + v\frac{\partial u}{\partial y} = -g\frac{\partial h}{\partial x} \qquad x = \text{direction}$$

$$u\frac{\partial v}{\partial x} + v\frac{\partial v}{\partial y} = -g\frac{\partial h}{\partial y} \qquad y = \text{direction}$$

a. The slope of a streamline is given by

$$\frac{dy}{dx} = \frac{v}{u}$$

Using this relation in Euler's equation, show that

$$d\left(\frac{u^2 + v^2}{2g} + h\right) = 0$$

or

$$d\left(\frac{V^2}{2g} + h\right) = 0$$

which means that $V^2/2g + h$ is constant along a streamline.

b. For an irrotational flow,

$$\frac{\partial u}{\partial y} = \frac{\partial v}{\partial x}$$

Substituting this equation into Euler's equation, show that

$$\frac{\partial}{\partial x}\left(\frac{V^2}{2g} + h\right) = 0$$

$$\frac{\partial}{\partial y}\left(\frac{V^2}{2g} + h\right) = 0$$

which means that $V^2/2g + h$ is constant in all directions.

Pressure Field for a Circular Cylinder (§4.10)

4.94 PLUS A fluid is flowing around a cylinder as shown in Fig 4.37 on p. 149 in §4.10. A favorable pressure gradient can be found

  a. upstream of the stagnation point

  b. at the stagnation point

  c. between the stagnation point and separation point

4.95 The velocity distribution over the surface of a sphere upstream of the separation point is $u_\theta = 1.5\,U \sin\theta$, where U is the free stream velocity and θ is the angle measured from the forward stagnation point. A pressure of -6.35 cm-H_2O gage is measured at the point of separation on a sphere in a 30.5 m/s airflow with a density of 1.12 kg/m^3. The pressure far upstream of the sphere in atmospheric. Estimate the location of the stagnation point (θ). Separation occurs on the windward side of the sphere.

4.96 Knowing the speed at point 1 of a fluid upstream of a sphere and the average speed at point 5 cm the wake of in the sphere, can one use the Bernoulli equation to find the pressure difference between the two points? Provide the rationale for your decision.

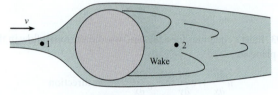

PROBLEM 4.96

Pressure Field for a Rotating Flow (§4.11)

4.97 Take a spoon and rapidly stir a cup of liquid. Report on the contour of the surface. Provide an explanation for the observed shape.

4.98 This closed tank, which is 1.2 m in diameter, is filled with water ($\rho = 1000$ kg/m^3) and is spun around its vertical centroidal axis at a rate of 10 rad/s. An open piezometer is connected to the tank as shown so that it is also rotating with the tank. For these conditions, what is the pressure at the center of the bottom of the tank?

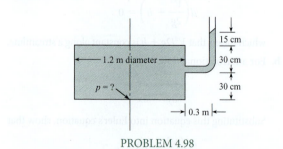

PROBLEM 4.98

4.99 A tank of liquid (S = 0.80) that is 0.3 m in diameter and 0.3 m high ($h = 0.3$ m) is rigidly fixed (as shown) to a rotating arm having a 0.6 m radius. The arm rotates such that the speed at point A is 6 m/s. If the pressure at A is 1.2 kPa, what is the pressure at B?

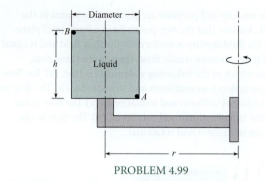

PROBLEM 4.99

4.100 PLUS Separators are used to separate liquids of different densities, such as cream from skim milk, by rotating the mixture at high speeds. In a cream separator the skim milk goes to the outside while the cream migrates toward the middle. A factor of merit for the centrifuge is the centrifugal acceleration force (RCF), which is the radial acceleration divided by the acceleration due to gravity. A cream separator can operate at 9000 rpm (rev/min). If the bowl of the separator is 20 cm in diameter, what is the centripetal acceleration if the liquid rotates as a solid body and what is the RCF?

4.101 A closed tank of liquid (S = 1.2) is rotated about a vertical axis (see the figure), and at the same time the entire tank is accelerated upward at 4 m/s^2. If the rate of rotation is 10 rad/s, what is the difference in pressure between points A and B ($p_B - p_A$)? Point B is at the bottom of the tank at a radius of 0.5 m from the axis of rotation, and point A is at the top on the axis of rotation.

PROBLEM 4.101

4.102 GO A U-tube is rotated about one leg, as shown. Before being rotated the liquid in the tube fills 0.25 m of each leg. The length of the base of the U-tube is 0.5 m, and each leg is 0.5 m long. What would be the maximum rotation rate (in rad/s) to ensure that no liquid is expelled from the outer leg?

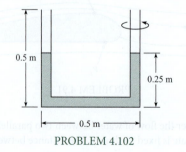

PROBLEM 4.102

4.103 An arm with a stagnation tube on the end is rotated at 100 rad/s in a horizontal plane 10 cm below a liquid surface as shown. The arm is 20 cm long, and the tube at the center of rotation extends above the liquid surface. The liquid in the tube is the same as that in the tank and has a specific weight of 10,000 N/m³. Find the location of the liquid surface in the central tube.

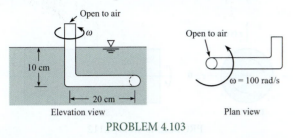

PROBLEM 4.103

4.104 A U-tube is rotated at 5.2 rad/s about one leg. The fluid at the bottom of the U-tube has a specific gravity of 3.0. The distance between the two legs of the U-tube is 0.3 m. A 15 cm height of another fluid is in the outer leg of the U-tube. Both legs are open to the atmosphere. Calculate the specific gravity of the other fluid.

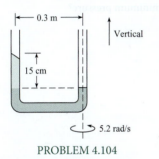

PROBLEM 4.104

4.105 PLUS A manometer is rotated around one leg, as shown. The difference in elevation between the liquid surfaces in the legs is 20 cm. The radius of the rotating arm is 10 cm. The liquid in the manometer is oil with a specific gravity of 0.8. Find the number of g's of acceleration in the leg with greatest amount of oil.

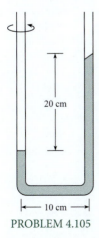

PROBLEM 4.105

4.106 A fuel tank for a rocket in space under a zero-g environment is rotated to keep the fuel in one end of the tank. The system is rotated at 3 rev/min. The end of the tank (point A) is 1.5 m from the axis of rotation, and the fuel level is 1 m from the rotation axis. The pressure in the nonliquid end of the tank is 0.1 kPa, and the density of the fuel is 800 kg/m³. What is the pressure at the exit (point A)?

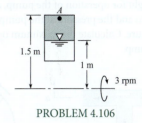

PROBLEM 4.106

4.107 Water stands in these tubes as shown when no rotation occurs. Derive a formula for the angular speed at which water will just begin to spill out of the small tube when the entire system is rotated about axis A-A.

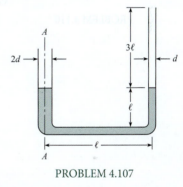

PROBLEM 4.107

4.108 PLUS Water ($\rho = 1000$ kg/m³) fills a slender tube 1 cm in diameter, 40 cm long, and closed at one end. When the tube is rotated in the horizontal plane about its open end at a constant speed of 50 rad/s, what force is exerted on the closed end?

4.109 Water ($\rho = 1000$ kg/m³) stands in the closed-end U-tube as shown when there is no rotation. If $\ell = 2$ cm and if the entire system is rotated about axis A-A, at what angular speed will

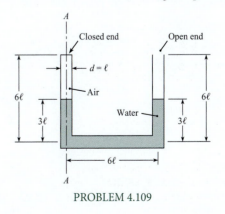

PROBLEM 4.109

water just begin to spill out of the open tube? Assume that the temperature for the system is the same before and after rotation and that the pressure in the closed end is initially atmospheric.

4.110 PLUS A simple centrifugal pump consists of a 10 cm disk with radial ports as shown. Water is pumped from a reservoir through a central tube on the axis. The wheel spins at 3000 rev/min, and the liquid discharges to atmospheric pressure. To establish the maximum height for operation of the pump, assume that the flow rate is zero and the pressure at the pump intake is atmospheric pressure. Calculate the maximum operational height z for the pump.

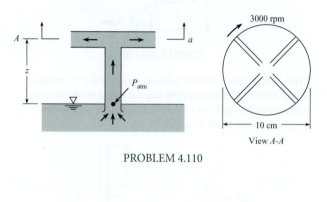

PROBLEM 4.110

4.111 A closed cylindrical tank of water ($\rho = 1000$ kg/m³) is rotated about its horizontal axis as shown. The water inside the tank rotates with the tank ($V = r\omega$). Derive an equation for dp/dz along a vertical-radial line through the center of rotation. What is dp/dz along this line for $z = -1$ m, $z = 0$, and $z = +1$ m when $\omega = 5$ rad/s? Here $z = 0$ at the axis.

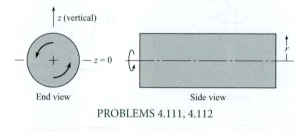

End view   Side view

PROBLEMS 4.111, 4.112

4.112 The tank shown is 1.2 m in diameter and 3.6 m long and is closed and filled with water ($\rho = 1000$ kg/m³). It is rotated about its horizontal-centroidal axis, and the water in the tank rotates with the tank ($V = r\omega$). The maximum velocity is 7.5 m/s. What is the maximum difference in pressure in the tank? Where is the point of minimum pressure?

CONTROL VOLUME APPROACH AND CONTINUITY EQUATION

5

FIGURE 5.1

The photo shows an evacuated-tube solar collector that is being tested to measure the efficiency. This project was done by undergraduate engineering students. The team applied the control volume concept, the continuity equation, the flow rate equations as well as knowledge from thermodynamics and heat transfer. (Photo by Donald Elger.)

Chapter Road Map

This chapter describes how conservation of mass can be applied to a flowing fluid. The resulting equation is called the *continuity equation*. The continuity equation is applied to a spatial region called a control volume, which is also introduced.

Learning Objectives

STUDENTS WILL BE ABLE TO

- Define mass flow rate and volume flow rate. (§5.1)
- Apply the flow rate equations. Describe how the flow rate equations are derived. (§5.1)
- Define and calculate the mean velocity. (§5.1)
- Describe the types of systems that engineers use for analysis. List the key differences between a CV and a closed system. (§5.2)
- Describe the purpose, application, and derivation of the Reynolds transport theorem. (§5.2)
- Describe and apply the continuity equation. Describe how the equation is derived. (§5.3, §5.4)
- Explain what cavitation means, describe why it is important, and list guidelines for designing to avoid cavitation. (§5.5)

5.1 Characterizing the Rate of Flow

Engineers characterize the rate of flow using the (a) mass flow rate, $\dot{m}$, and (b) the volume flow rate Q. Thus, these concepts and associated equations are introduced in this section.

Volume Flow Rate (Discharge)

Volume flow rate Q is the *ratio of volume to time at an instant in time*. In equation form,

$$Q = \left(\frac{\text{volume of fluid passing through a cross sectional area}}{\text{interval of time}}\right)_{\substack{\text{instant} \\ \text{in time}}} = \lim_{\Delta t \to 0} \frac{\Delta V}{\Delta t} \qquad (5.1)$$

EXAMPLE. To describe volume flow rate (Q) for a gas pump (Fig. 5.2a), select a cross-sectional area. Then, Q is the volume of gasoline that flowed across the specified section during a specified time interval (say one second) divided by the time interval. The units could be gallons per minute or liters per second.

EXAMPLE. To describe volume flow rate (Q) for a person inhaling while doing yoga (Fig. 5.2b), select a cross-sectional area as shown. Then, Q is the volume of air that flowed across the specified section during a specified time interval (say $\Delta t = 0.01$ s) divided by the time interval. Notice that the time interval should be short because the flow rate is continuously varying during breathing. The idea is to let $\Delta t \to 0$ so that the flow rate is characterized at an instant in time.

FIGURE 5.2

Sketches used to define volume flow rate
(a) gasoline flowing out of a valve at a filling station,
(b) air flowing inward to a person during inhalation.

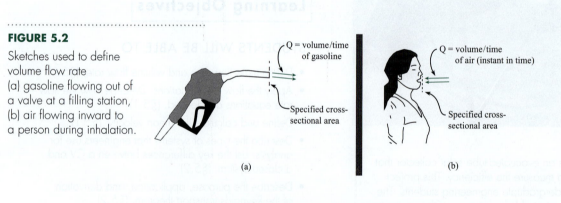

Q = volume/time of gasoline

Specified cross-sectional area

(a)

Q = volume/time of air (instant in time)

Specified cross-sectional area

(b)

Volume flow rate is often called *discharge*. Because these two terms are synonyms, this text uses both terms interchangeably.

The SI units of discharge are cubic meters of volume per second (m^3/s). In traditional units, the consistent unit is cubic feet of volume per second (ft^3/s). Often this unit is written as cfs, which stands for cubic feet per second.

Deriving Equations for Volume Flow Rate (Discharge)

This subsection shows how to derive useful equations for discharge Q in terms of fluid velocity and section area A.

To relate Q to velocity V, select a flow of fluid (Fig. 5.3) in which velocity is assumed to be constant across the pipe cross section. Suppose a marker is injected over the cross section at section A-A for a period of time Δt. The fluid that passes A-A in time Δt is represented by the marked volume. The length of the marked volume is $V\Delta t$ so the volume is $\Delta \mathcal{V} = AV\Delta t$. Apply the definition of Q:

$$Q = \lim_{\Delta t \to 0} \frac{\Delta \mathcal{V}}{\Delta t} = \lim_{\Delta t \to 0} \frac{AV\Delta t}{\Delta t} = VA \tag{5.2}$$

In Eq. (5.2), notice how the units work out:

$$Q = VA$$

$$\text{Flow Rate } (\text{m}^3/\text{s}) = \text{Velocity } (\text{m/s}) \times \text{Area } (\text{m}^2)$$

FIGURE 5.3

Volume of fluid in flow with uniform velocity distribution that passes section A-A in time Δt.

FIGURE 5.4

Volume of fluid that passes section A-A in time Δt.

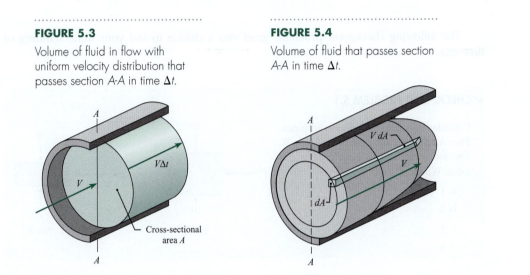

Because Eq. (5.2) is based on a uniform velocity distribution, consider a flow in which the velocity varies across the section (see Fig. 5.4). The blue shaded region shows the volume of fluid that passes across a differential area of the section. Using the idea of Eq. (5.2), let $dQ = V\,dA$. To obtain the total flow rate, add up the volume flow rate through each differential element and then apply the definition of the integral:

$$Q = \sum_{\text{section}} V_i\, dA_i = \int_A V\, dA \tag{5.3}$$

Eq. (5.3) means that *velocity integrated over section area gives discharge*. To develop another useful result, divide Eq. (5.3) by area A to give

$$\overline{V} = \frac{Q}{A} = \frac{1}{A}\int_A V\, dA \tag{5.4}$$

Eq. (5.4) provides a definition of $\overline{V}$, which is called the **mean velocity**. As shown, the mean velocity is an area-weighted average velocity. For this reason, mean velocity is sometimes called *area-averaged velocity*. This label is useful for distinguishing an area-averaged velocity from a *time-averaged velocity,* which is used for characterizing turbulent flow (see Section 4.3). Some useful values of mean velocity are summarized in Table 5.1.

TABLE 5.1 Values of Mean Velocity

Situation	Equation for Mean Velocity
Fully developed laminar flow in a round pipe. For more information, see Section 10.5.	$\overline{V}/V_{max} = 0.5$, where V_{max} is the value of the maximum velocity in the pipe. Note that V_{max} is the value of the velocity at the center of the pipe.
Fully developed laminar flow in a rectangular channel (channel has infinite width).	$\overline{V}/V_{max} = 2/3 = 0.667$
Fully developed turbulent flow in a round pipe. For more information, see Section 10.6.	$\overline{V}/V_{max} \approx 0.79$ to 0.86, where the ratio depends on Reynolds number.

The following checkpoint problems gives you a chance to test your understanding of flow rate.

✔ **CHECKPOINT PROBLEM 5.1**

Consider flow through two round pipes. Pipe A has twice the diameter of pipe B. The mean velocity in each pipe is the same. What is Q_A/Q_B?

a. 1
b. 2
c. 4
d. 8

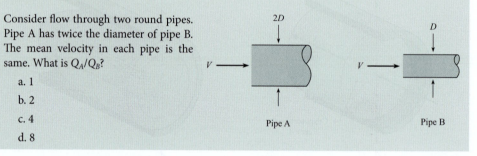

Pipe A

Pipe B

✔ **CHECKPOINT PROBLEM 5.2**

Consider flow through two round pipes. The maximum velocity in each pipe is the same. The only difference is the velocity distribution. Which pipe has the larger value of mean velocity? Why?

a. Pipe A
b. Pipe B
c. They both have the same mean velocity

Pipe A

Pipe B

Eq. (5.4) can be generalized by using the concept of the dot product. The dot product is useful when the velocity vector is aligned at an angle with respect to the section area (Fig. 5.5). The only component of velocity that contributes to the flow through the differential area dA is the component normal to the area, V_n. The differential discharge through area dA is

$$dQ = V_n\,dA$$

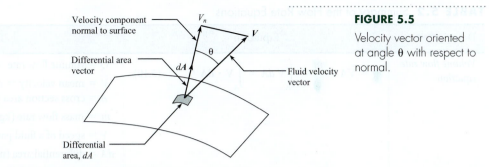

FIGURE 5.5

Velocity vector oriented at angle θ with respect to normal.

If the vector, **dA**, is defined with magnitude equal to the differential area, dA, and direction normal to the surface, then $V_n \, dA = |\mathbf{V}| \cos \theta \, dA = \mathbf{V} \cdot \mathbf{dA}$ where $\mathbf{V} \cdot \mathbf{dA}$ is the dot product of the two vectors. Thus a more general equation for the discharge or volume flow rate through a surface A is

$$Q = \int_A \mathbf{V} \cdot \mathbf{dA} \tag{5.5}$$

If the velocity is constant over the area and the area is a planar surface, then the discharge is

$$Q = \mathbf{V} \cdot \mathbf{A}$$

If, in addition, the velocity and area vectors are aligned, then

$$Q = VA$$

which reverts to the original equation developed for discharge, Eq. (5.2).

Mass Flow Rate

Mass flow rate $\dot{m}$ is the *ratio of mass to time at an instant in time.* In equation form,

$$\dot{m} = \left(\frac{\text{mass of fluid passing through a cross sectional area}}{\text{interval of time}} \right)_{\substack{\text{instant} \\ \text{in time}}} = \lim_{\Delta t \to 0} \frac{\Delta m}{\Delta t} \tag{5.6}$$

The common units for mass flow rate are kg/s, lbm/s, and slugs/s.

Using the same approach as for volume flow rate, the mass of the fluid in the marked volume in Fig. 5.3 is $\Delta m = \rho \Delta \mathcal{V}$, where ρ is the average density. Thus, one can derive several useful equations:

$$\dot{m} = \lim_{\Delta t \to 0} \frac{\Delta m}{\Delta t} = \rho \lim_{\Delta t \to 0} \frac{\Delta \mathcal{V}}{\Delta t} = \rho Q$$

$$= \rho A V \tag{5.7}$$

The generalized form of the mass flow equation corresponding to Eq. (5.5) is

$$\dot{m} = \int_A \rho \mathbf{V} \cdot \mathbf{dA} \tag{5.8}$$

where both the velocity and fluid density can vary over the cross-sectional area. If the density is constant, then Eq. (5.7) is recovered. Also if the velocity vector is aligned with the area vector, such as integrating over the cross-sectional area of a pipe, Eq. (5.8) reduces to

$$\dot{m} = \int_A \rho V \, dA \tag{5.9}$$

Working Equations

Table 5.2 summarizes the flow rate equations. Notice that multiplying Eq. (5.10) by density gives Eq. (5.11).

TABLE 5.2 Summary of the Flow Rate Equations

Description	Equation	Terms
Volume flow rate equation	$Q = \bar{V}A = \dfrac{\dot{m}}{\rho} = \displaystyle\int_A V\,dA = \int_A \mathbf{V} \cdot \mathbf{dA}$ (5.10)	Q = volume flow rate = discharge (m³/s) $\bar{V}$ = mean velocity = area averaged velocity (m/s) A = cross section area (m²) $\dot{m}$ = mass flow rate (kg/s) V = speed of a fluid particle (m/s) dA = differential area (m²) $\mathbf{V}$ = velocity of a fluid particle (m/s) $\mathbf{dA}$ = differential area vector (m²) (points outward from control surface)
Mass flow rate equation	$\dot{m} = \rho A \bar{V} = \rho Q = \displaystyle\int_A \rho V\,dA = \int_A \rho \mathbf{V} \cdot \mathbf{dA}$ (5.11)	$\dot{m}$ = mass flow rate (kg/s) ρ = mass density (kg/m³)

Example Problems

For most problems, application of the flow rate equation involves substituting numbers into the appropriate equation; see Example 5.1 for this case.

EXAMPLE 5.1

Applying the Flow Rate Equations to a Flow of Air in a Pipe

Problem Statement

Air that has a mass density of 1.24 kg/m³ flows in a pipe with a diameter of 30 cm at a mass rate of flow of 3 kg/s. What are the mean velocity and discharge in this pipe for both systems of units?

Define the Situation

Air flows in a pipe.

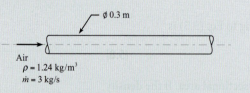

Air
$\rho = 1.24$ kg/m³
$\dot{m} = 3$ kg/s

$\varnothing\,0.3$ m

State the Goal

Q(m³/s) ← Volume flow rate (discharge)

$\bar{V}$(m/s) ← Mean velocity

Generate Ideas and Make a Plan

Because Q is the goal and $\dot{m}$ and ρ are known, apply the mass flow rate equation (Eq. 5.11):

$$\dot{m} = \rho Q \qquad \text{(a)}$$

To find the last goal ($\bar{V}$), apply the volume flow rate equation (Eq. 5.10):

$$Q = \bar{V}A \qquad \text{(b)}$$

The plan is

1. Calculate Q using Eq. (a).
2. Calculate $\bar{V}$ using Eq. (b).

Take Action (Execute the Plan)

1. Mass flow rate equation:

$$Q = \frac{\dot{m}}{\rho} = \frac{3 \text{ kg/s}}{1.24 \text{ kg/m}^3} = \boxed{2.42 \text{ m}^3/\text{s}}$$

2. Volume flow rate equation:

$$V = \frac{Q}{A} = \frac{2.42 \text{ m}^3/\text{s}}{\left(\frac{1}{4}\pi\right) \times (0.30 \text{ m})^2} = \boxed{34.2 \text{ m/s}}$$

When fluid passes across a control surface and the velocity vector is at an angle with respect to the surface normal vector, then one uses the dot product. This case is illustrated by Example 5.2.

EXAMPLE 5.2

Calculating the Volume Flow Rate by Applying the Dot Product

Problem Statement

Water flows in a channel that has a slope of 30°. If the velocity is assumed to be constant, 12 m/s, and if a depth of 60 cm is measured along a vertical line, what is the discharge per meter of width of the channel?

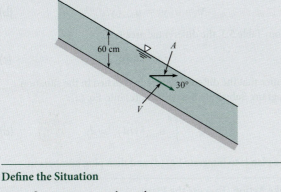

Define the Situation

Water flows in an open channel.

State the Goal

$Q(\text{m}^3/\text{s})$ ◀ discharge per meter of width of the channel

Generate Ideas and Make a Plan

Because V and A are not at right angles, apply

$Q = \mathbf{V} \cdot \mathbf{A} = VA \cos \theta$. Because all variables are known except Q, the plan is to substitute in values.

Take Action (Execute the Plan)

$$Q = \mathbf{V} \cdot \mathbf{A} = V(\cos 30°)A$$

$$= (12 \text{ m/s})(\cos 30°)(0.6 \text{ m})$$

$$= \boxed{6.24 \text{ m}^3/\text{s per meter}}$$

Review the Solution and the Process

1. *Knowledge.* This example involves a channel flow. A flow is a *channel flow* when a liquid (usually water) flows with open surface exposed to air under the action of gravity.

2. *Knowledge.* The discharge per unit width is usually designated as q.

Another important case is when velocity varies at different points on the control surface. In this case, one uses an integral to determine flow rate as specified by Eq. (5.10):

$$Q = \int_A V \, dA.$$

In this integral, the differential area dA depends on the physics of the problem. Two common cases are shown in Table 5.3. Analyzing a variable velocity is illustrated by Example 5.3.

TABLE 5.3 Differential Areas for Determining Flow Rate

Label	Sketch	Description
Channel Flow	$dA = wdy$	When velocity varies as $V = V(y)$ in a rectangular channel, then use a differential area dA given by $dA = wdy$ where w is the width of the channel and dy is a differential height.
Pipe Flow	$dA = 2\pi r dr$	When velocity varies as $V = V(r)$ in a round pipe, then use a differential area dA given by $dA = 2\pi r dr$ where r is the radius of the differential area and dr is a differential radius.

EXAMPLE 5.3

Determining Flow Rate by Integration

Problem Statement

The water velocity in the channel shown in the accompanying figure has a velocity distribution across the vertical section equal to $u/u_{max} = (y/d)^{1/2}$. What is the discharge in the channel if the water is 2 m deep ($d = 2$ m), the channel is 5 m wide, and the maximum velocity is 3 m/s?

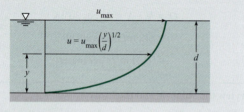

Define the Situation

Water flows in a channel.

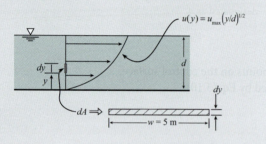

State the Goal

$Q(\text{m}^3/\text{s}) \longleftarrow$ Discharge (Volume Flow Rate)

Generate Ideas and Make a Plan

Because velocity is varying over the cross-sectional area, apply Eq. (5.10):

$$Q = \int_A V \, dA \tag{a}$$

Because Eq. (a) has two unknowns (V and dA), find equations for these unknowns. The velocity is given:

$$V = u(y) = u_{max}(y/d)^{1/2} \tag{b}$$

From Table 5.3, the differential area is

$$dA = w \, dy \tag{c}$$

Notice that the differential area is sketched in the situation diagram. Substitute Eqs. (b) and (c) into Eq. (a):

$$Q = \int_0^d u_{max}(y/d)^{1/2} w \, dy \tag{d}$$

The plan is to integrate Eq. (d) and then plug numbers in.

Take Action (Execute the Plan)

$$Q = \int_0^d u_{max}(y/d)^{1/2} w \, dy$$

$$= \frac{w u_{max}}{d^{1/2}} \int_0^d y^{1/2} \, dy$$

$$= \frac{w u_{max}}{d^{1/2}} \frac{2}{3} y^{3/2} \Big|_0^d = \frac{w u_{max}}{d^{1/2}} \frac{2}{3} d^{3/2}$$

$$= \frac{(5\text{ m})(3\text{ m/s})}{(2\text{ m})^{1/2}} \times \frac{2}{3} \times (2\text{ m})^{3/2} = \boxed{20\text{ m}^3/\text{s}}$$

5.2 The Control Volume Approach

Engineers solve problems in fluid mechanics using the *control volume approach*. Equations for this approach are derived using *Reynolds transport theorem*. These topics are presented in this section.

The Closed System and the Control Volume

As introduced in Section 2.1, a *system* is whatever the engineer selects for study. The *surroundings* are everything that is external to the system, and the *boundary* is the interface between the system and the surroundings. Systems can be classified into two categories: the closed system and the open system (also known as a control volume).

The **closed system** (also known as a *control mass*) is a fixed collection of matter that the engineer selects for analysis. By definition, mass cannot cross the boundary of a closed system.

The boundary of a closed system can move and deform.

EXAMPLE. Consider air inside a cylinder (see Fig. 5.6). If the goal is to calculate the pressure and temperature of the air during compression, engineers select a closed system comprised of the air inside the cylinder. The system boundaries would deform as the piston moves so that the closed system always contains the same matter. This is an example of a closed system because the mass within the system is always the same.

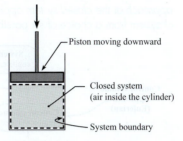

FIGURE 5.6
Example of a closed system.

Because the closed system involves selection and analysis of a specific collection of matter, the closed system is a Lagrangian concept.

The **control volume** (CV or cv; also known as an open system) is a specified volumetric region in space that the engineer selects for analysis. The matter inside a control volume is usually changing with time because mass is flowing across the boundaries. Because the control volume involves selection and analysis of a region in space, the CV is an Eulerian concept.

EXAMPLE. Suppose water is flowing through a tank (Fig. 5.7) and the goal is to calculate the depth of water h as a function of time. A key to solving this problem is to select a system, and the best choice of a system is a CV surrounding the tank. Note that the CV is always three dimensional because it is a volumetric region. However, CVs are usually drawn in two dimensions. The boundary surfaces of a CV are called the **control surface**. This is abbreviated as CS or cs.

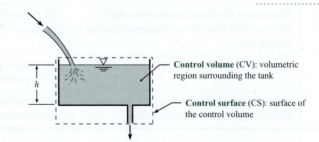

FIGURE 5.7
Water entering a tank through the top and exiting through the bottom.

A control volume can be defined so that it is deforming or fixed. When a **fixed CV** is defined, this means that the shape of the CV and its volume are constant with time. When a **deforming CV** is defined, the shape of the CV and its volume change with time, typically to mimic the volume of a region of fluid.

EXAMPLE. To model a rocket made from a balloon suspended on a string, one can define a deforming CV that surrounds the deflating balloon and follow the shape of the balloon during the process of deflation.

Summary When engineers analyze a problem, they select the type of system that is most useful (see Fig. 5.8). There are two approaches. Using the *control volume approach* the engineer selects a region in space and analyzes flow through this region. Using the *closed system approach*, the engineer selects a body of matter of fixed identity and analyzes this matter.

Table 5.4 compares the *Control Volume Approach* and *Closed System Approach*.

FIGURE 5.8

When engineers select a system, they choose either the *control volume approach* or the *closed system approach*. Then, they select the specific type of system from a choice of six possibilities.

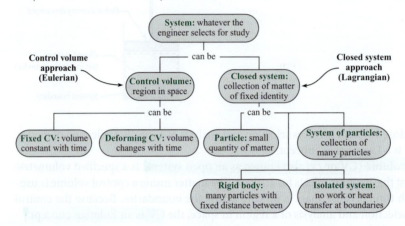

TABLE 5.4 Comparison of the Control Volume and the Closed System Approaches

Feature	Closed System Approach	Control Volume Approach
Basic idea	Analyze a body or fixed collection of matter.	Analyze a spatial region.
Lagrangian versus Eulerian	Lagrangian approach.	Eulerian approach.
Mass crossing the boundaries	Mass cannot cross the boundaries.	Mass is allowed to cross the boundaries.
Mass (quantity)	The mass of the closed system must stay constant with time; always the same number of kilograms.	The mass of the materials inside the CV can stay constant or can change with time.
Mass (identity)	Always contains the same matter.	Can contain the same matter at all times. Or the identity of the matter can vary with time.
Application	Solid mechanics, fluid mechanics, thermodynamics, and other thermal sciences.	Fluid mechanics, thermodynamics, and other thermal sciences.

Intensive and Extensive Properties

Properties, which are measurable characteristics of a system, can be classified into two categories. An **extensive property** is any property that depends on the amount of matter present. An **intensive property** is any property that is independent of the amount of matter present.

Examples (extensive). Mass, momentum, energy, and weight are extensive properties because each of these properties depends on the amount of matter present. **Examples (intensive).** Pressure, temperature, and density are intensive properties because each of these properties are independent on the amount of matter present.

Many intensive properties are obtained by taking the ratio of two extensive properties. For example, density is the ratio of mass to volume. Similarly, specific energy e is the ratio of energy to mass.

To develop a general equation to relate intensive and extensive properties, define a generic extensive property, B. Also, define a corresponding intensive property b.

$$b = \left(\frac{B}{\text{mass}} \right)_{\text{point in space}}$$

The amount of extensive property B contained in a control volume at a given instant is

$$B_{cv} = \int_{cv} b \, dm = \int_{cv} b \rho \, d\Psi \qquad \textbf{(5.12)}$$

where dm and $d\Psi$ are the differential mass and differential volume, respectively, and the integral is carried out over the control volume.

Property Transport across the Control Surface

Because flow transports mass, momentum, and energy across the control surface, the next step is to describe this transport. Consider flow through a duct (Fig. 5.9) and assume that the velocity is uniformly distributed across the control surface. Then, the mass flow rate through each section is given by

$$\dot{m}_1 = \rho_1 A_1 V_1 \qquad \dot{m}_2 = \rho_2 A_2 V_2$$

The rate of outflow minus the rate of inflow is

$$(\text{outflow minus inflow}) = (\text{net mass outflow rate}) = \dot{m}_2 - \dot{m}_1 = \rho_2 A_2 V_2 - \rho_1 A_1 V_1$$

Next, we'll introduce velocity. The same control volume is shown in Fig. 5.10 with each control surface area represented by a vector, $\mathbf{A}$, oriented outward from the control volume and with magnitude equal to the cross-sectional area. The velocity is represented by a vector, $\mathbf{V}$. Taking the dot product of the velocity and area vectors at both stations gives

$$\mathbf{V}_1 \cdot \mathbf{A}_1 = -V_1 A_1 \qquad \mathbf{V}_2 \cdot \mathbf{A}_2 = V_2 A_2$$

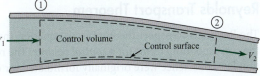

FIGURE 5.9

Flow through control volume in a duct.

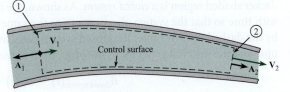

FIGURE 5.10

Control surfaces represented by area vectors and velocities by velocity vectors.

because at station 1 the velocity and area have the opposite directions while at station 2 the velocity and area vectors are in the same direction. Now the net mass outflow rate can be written as

$$\text{net mass outflow rate} = \rho_2 V_2 A_2 - \rho_1 V_1 A_1$$

$$= \rho_2 \mathbf{V}_2 \cdot \mathbf{A}_2 + \rho_1 \mathbf{V}_1 \cdot \mathbf{A}_1 \qquad (5.13)$$

$$= \sum_{cs} \rho \mathbf{V} \cdot \mathbf{A}$$

Equation (5.13) states that if the dot product $\rho \mathbf{V} \cdot \mathbf{A}$ is summed for all flows into and out of the control volume, the result is the net mass flow rate out of the control volume, or the net mass efflux (*efflux* means outflow). If the summation is positive, the net mass flow rate is out of the control volume. If it is negative, the net mass flow rate is into the control volume. If the inflow and outflow rates are equal, then

$$\sum_{cs} \rho \mathbf{V} \cdot \mathbf{A} = 0$$

To obtain the net rate of flow of an extensive property B across a section, write

$$\overbrace{\left(\frac{B}{\text{mass}}\right)}^{b} \overbrace{\left(\frac{\text{mass}}{\text{time}}\right)}^{\dot{m}} = \overbrace{\left(\frac{B}{\text{time}}\right)}^{\dot{B}}$$

Next, include all inlet and outlet ports:

$$\dot{B}_{\text{net}} = \sum_{cs} b \overbrace{\rho \mathbf{V} \cdot \mathbf{A}}^{\dot{m}} \qquad (5.14)$$

Equation (5.14) is applicable for all flows where the properties are uniformly distributed across the flow area. To account for property variation, replace the sum with an integral:

$$\dot{B}_{\text{net}} = \int_{cs} b \rho \mathbf{V} \cdot \mathbf{dA} \qquad (5.15)$$

Eq. (5.15) will be used in the derivation of the Reynolds transport theorem.

Reynolds Transport Theorem

The Reynolds transport theorem is an equation that relates a derivative for a *closed system* to the corresponding terms for a *control volume*. The reason for the theorem is that the conservation laws of science were originally formulated for closed systems. Over time, researchers figured out how to modify the equations so that they apply to a control volume. The result is the Reynolds transport theorem.

To derive the Reynolds transport theorem, consider a flowing fluid; see Fig. 5.11. The darker shaded region is a *closed system*. As shown, the boundaries of the closed system change with time so that the system always contains the same matter. Also, define a CV as identified by the dashed line. At time t the closed system consists of the material inside the control volume and the material going in, so the property B of the system at this time is

$$B_{\text{closed system}}(t) = B_{cv}(t) + \Delta B_{\text{in}} \qquad (5.16)$$

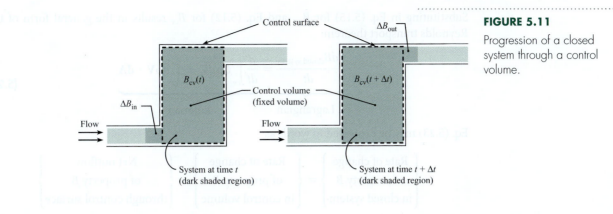

FIGURE 5.11

Progression of a closed system through a control volume.

At time $t + \Delta t$ the closed system has moved and now consists of the material in the control volume and the material passing out, so B of the system is

$$B_{\text{closed system}}(t + \Delta t) = B_{\text{cv}}(t + \Delta t) + \Delta B_{\text{out}} \qquad (5.17)$$

The rate of change of the property B is

$$\frac{dB_{\text{closed system}}}{dt} = \lim_{\Delta t \to 0} \left[\frac{B_{\text{closed system}}(t + \Delta t) - B_{\text{closed system}}(t)}{\Delta t} \right] \qquad (5.18)$$

Substituting in Eqs. (5.16) and (5.17) results in

$$\frac{dB_{\text{closed system}}}{dt} = \lim_{\Delta t \to 0} \left[\frac{B_{\text{cv}}(t + \Delta t) - B_{\text{cv}}(t) + \Delta B_{\text{out}} - \Delta B_{\text{in}}}{\Delta t} \right] \qquad (5.19)$$

Rearranging terms yields

$$\frac{dB_{\text{closed system}}}{dt} = \lim_{\Delta t \to 0} \left[\frac{B_{\text{cv}}(t + \Delta t) - B_{\text{cv}}(t)}{\Delta t} \right] + \lim_{\Delta t \to 0} \frac{\Delta B_{\text{out}}}{\Delta t} - \lim_{\Delta t \to 0} \frac{\Delta B_{\text{in}}}{\Delta t} \qquad (5.20)$$

The first term on the right side of Eq. (5.20) is the rate of change of the property B inside the control volume, or

$$\lim_{\Delta t \to 0} \left[\frac{B_{\text{cv}}(t + \Delta t) - B_{\text{cv}}(t)}{\Delta t} \right] = \frac{dB_{\text{cv}}}{dt} \qquad (5.21)$$

The remaining terms are

$$\lim_{\Delta t \to 0} \frac{\Delta B_{\text{out}}}{\Delta t} = \dot{B}_{\text{out}} \qquad \text{and} \qquad \lim_{\Delta t \to 0} \frac{\Delta B_{\text{in}}}{\Delta t} = \dot{B}_{\text{in}}$$

These two terms can be combined to give

$$\dot{B}_{\text{net}} = \dot{B}_{\text{out}} - \dot{B}_{\text{in}} \qquad (5.22)$$

or the net efflux, or net outflow rate, of the property B through the control surface. Equation (5.20) can now be written as

$$\frac{dB_{\text{closed system}}}{dt} = \frac{d}{dt} B_{\text{cv}} + \dot{B}_{\text{net}}$$

Substituting in Eq. (5.15) for $\dot{B}_{\text{net}}$ and Eq. (5.12) for B_{cv} results in the general form of the Reynolds transport theorem:

$$\underbrace{\frac{dB_{\text{closed system}}}{dt}}_{\text{Lagrangian}} = \underbrace{\frac{d}{dt}\int_{\text{cv}} b\rho\, d\mathcal{V} + \int_{\text{cs}} b\rho \mathbf{V} \cdot d\mathbf{A}}_{\text{Eulerian}} \tag{5.23}$$

Eq. (5.23) may be expressed in words as

$$\left\{\begin{array}{c}\text{Rate of change}\\ \text{of property } B\\ \text{in closed system}\end{array}\right\} = \left\{\begin{array}{c}\text{Rate of change}\\ \text{of property } B\\ \text{in control volume}\end{array}\right\} + \left\{\begin{array}{c}\text{Net outflow}\\ \text{of property } B\\ \text{through control surface}\end{array}\right\}$$

The left side of the equation is the Lagrangian form, that is, the rate of change of property B for the closed system. The right side is the Eulerian form, that is, the change of property B evaluated in the control volume and the flux measured at the control surface. This equation applies at the instant the system occupies the control volume and provides the connection between the Lagrangian and Eulerian descriptions of fluid flow. The velocity $\mathbf{V}$ is always measured with respect to the control surface because it relates to the mass flux across the surface.

A simplified form of the Reynolds transport theorem can be written if the mass crossing the control surface occurs through a number of inlet and outlet ports, and the velocity, density and intensive property b are uniformly distributed (constant) across each port. Then

$$\frac{dB_{\text{closed system}}}{dt} = \frac{d}{dt}\int_{\text{cv}} b\rho\, d\mathcal{V} + \sum_{\text{cs}} \rho b\mathbf{V} \cdot \mathbf{A} \tag{5.24}$$

where the summation is carried out for each port crossing the control surface.

An alternative form can be written in terms of the mass flow rates:

$$\frac{dB_{\text{closed system}}}{dt} = \int_{\text{cv}} \rho b\, d\mathcal{V} + \sum_{\text{cs}} \dot{m}_o b_o - \sum_{\text{cs}} \dot{m}_i b_i \tag{5.25}$$

where the subscripts i and o refer to the inlet and outlet ports, respectively, located on the control surface. This form of the equation does not require that the velocity and density be uniformly distributed across each inlet and outlet port, but the property b must be.

5.3 Continuity Equation (Theory)

The continuity equation is the law of *conservation of mass* applied to a control volume. Because this equation is commonly used by engineers, this section presents the relevant topics.

Derivation

The law of conservation of mass for a closed system can be written as

$$\frac{d(\text{mass of a closed system})}{dt} = \frac{dm_{\text{closed system}}}{dt} = 0 \tag{5.26}$$

To transform (Eq. 5.26) into an equation for a control volume, apply the Reynolds transport theorem, Eq. (5.23). In Eq. (5.23), the extensive property is mass, $B_{cv} = m_{\text{closed system}}$. The corresponding value intensive property is mass per unit mass, or simply, unity.

$$b = \frac{m_{\text{closed system}}}{m_{\text{closed system}}} = 1$$

Substituting for B_{cv} and b in Eq. (5.23) gives

$$\frac{dm_{\text{closed system}}}{dt} = \frac{d}{dt}\int_{cv} \rho \, d\mathcal{V} + \int_{cs} \rho \mathbf{V} \cdot \mathbf{dA} \qquad \textbf{(5.27)}$$

Combining Eq. (5.26) to Eq. (5.27) gives the *general form of the continuity equation.*

$$\frac{d}{dt}\int_{cv} \rho \, d\mathcal{V} + \int_{cs} \rho \mathbf{V} \cdot \mathbf{dA} = 0 \qquad \textbf{(5.28)}$$

If mass crosses the boundaries at a number of inlet and exit ports, then Eq. (5.28) reduces to give the *simplified form of the continuity equation:*

$$\frac{d}{dt}m_{cv} + \sum_{cs} \dot{m}_o - \sum_{cs} \dot{m}_i = 0 \qquad \textbf{(5.29)}$$

accumulation

Physical Interpretation of the Continuity Equation

Fig. 5.12 shows the meaning of the terms in the continuity equation. The top row gives the general form (Eq. 5.28), and the second row gives the simplified form (Eq. 5.29). The arrows shows which terms have the same conceptual meaning.

The **accumulation** term describes the changes in the quantity of mass inside the control volume (CV) with respect to time. Mass inside a CV can increase with time (accumulation is positive), decrease with time (accumulation is negative) or stay the same (accumulation is zero).

The **inflow and outflow** terms describe the rates at which mass is flowing across the surfaces of the control volume. Sometimes inflow and outflow are combined to give **efflux**, which is defined as the net positive rate at which is mass is flowing out of a CV. That is, (efflux) = (outflow) − (inflow). When efflux is positive, there is a net flow of mass out of the CV, and accumulation is negative. When efflux is negative, then accumulation is positive.

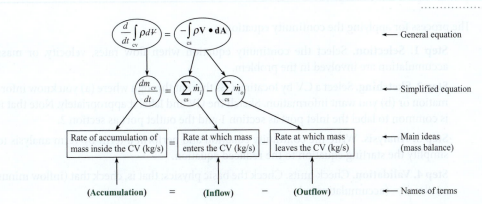

FIGURE 5.12

This figure shows the conceptual meaning of the continuity equation.

As shown in Fig. 5.12, the physics of the continuity equation can be summarized as:

$$\text{accumulation} = \text{inflow} - \text{outflow} \tag{5.30}$$

where all terms in Eq. (5.30) are rates (see Fig. 5.12)

Eq. (5.30) is called a balance equation because the ideas relate to our everyday experiences with how things balance. For example, the accumulation of cash in a bank account equals the inflows (deposits) minus the outflows (withdrawals). Because the continuity equation is a balance equation, it is sometimes called the *mass balance equation*.

The continuity equation is applied at an instant in time and the units are kg/s. Sometimes the continuity equation is integrated with respect to time and the units are kg. To recognize a problem that will involve integration, *look for a change in state during a time interval*.

5.4 Continuity Equation (Application)

This section describes how to apply the continuity equation and presents example problems.

Working Equations

Three useful forms of the continuity equations are summarized in Table 5.5.

TABLE 5.5 Summary of the Continuity Equation

Description	Equations	Terms
General form: valid for any problem.	$\dfrac{d}{dt}\displaystyle\int_{cv} \rho \, d\mathcal{V} + \int_{cs} \rho \mathbf{V} \cdot \mathbf{dA} = 0$ (Eq. 5.28)	t = time (s) ρ = density (kg/m^3)
Simplified form: useful when there are well defined inlet and exit ports.	$\dfrac{d}{dt} m_{cv} + \displaystyle\sum_{cs} \dot{m}_o - \sum_{cs} \dot{m}_i = 0$ (Eq. 5.29)	$d\mathcal{V}$ = differential volume (m^3) $\mathbf{V}$ = fluid velocity vector (m/s) (reference frame is the control surface)
Pipe flow form; valid for flow in a pipe. (*gases*: density can vary but the density must be uniform across sections 1 and 2). (*liquids*: the equation reduces to $A_2 V_2 = A_1 V_1$ for a constant density assumption).	$\rho_2 A_2 V_2 = \rho_1 A_1 V_1$ (Eq. 5.33)	$\mathbf{dA}$ = differential area vector (m^2) (positive direction of $\mathbf{dA}$ is outward from CS) m_{cv} = mass inside the control volume (kg) $\dot{m} = \rho A V$ = mass/time crossing CS (kg/s) A = area of flow (m^2) V = mean velocity (m/s)

The process for applying the continuity equation is

> **Step 1. Selection.** Select the continuity equation when flow rates, velocity, or mass accumulation are involved in the problem.

> **Step 2. Sketching.** Select a CV by locating CSs that cut through where (a) you know information or (b) you want information. Sketch the CV and label it appropriately. Note that it is common to label the inlet port as section 1 and the outlet port as section 2.

> **Step 3. Analysis.** Write the continuity equation and perform a term-by-term analysis to simplify the starting equation to the reduced equation.

> **Step 4. Validation.** Check units. Check the basic physics; that is, check that (inflow minus outflow) = (accumulation).

Example Problems

The first example problem (Example 5.4) shows how continuity is applied to a problem that involves accumulation of mass.

EXAMPLE 5.4

Applying the Continuity Equation to a Tank with an Inflow and an Outflow

---

Problem Statement

A stream of water flows into an open tank. The speed of the incoming water is $V = 7$ m/s, and the section area is $A = 0.0025$ m². Water also flows out of the tank at rate of $Q = 0.003$ m³/s. Water density is 1000 kg/m³. What is the rate at which water is being stored (or removed from) the tank?

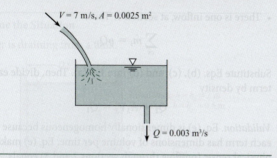

$V = 7$ m/s, $A = 0.0025$ m²

$Q = 0.003$ m³/s

Define the Situation

Water flows into a tank at the top and out at the bottom.

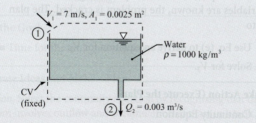

$V_1 = 7$ m/s, $A_1 = 0.0025$ m²

① ② Water $\rho = 1000$ kg/m³

CV (fixed)

② $Q_2 = 0.003$ m³/s

State the Goal

(dm_{cv}/dt) (kg/s) ⟵ rate of accumulation of water in tank

Generate Ideas and Make a Plan

Selection. Select the simplified form of the continuity equation (Eq. 5.29).

Sketching. Modify the situation diagram to show the CV and sections 1 and 2. Notice that the CV in the upper left corner is sketched so that it is at a right angle to the inlet flow.

Analysis. Write the continuity equation (simplified form)

$$\frac{d}{dt}m_{cv} + \sum_{cs} \dot{m}_o - \sum_{cs} \dot{m}_i = 0 \tag{a}$$

Analyze the outflow and inflow terms.

$$\sum_{cs} \dot{m}_o = \rho Q_2 \tag{b}$$

$$\sum_{cs} \dot{m}_i = \rho A_1 V_1 \tag{c}$$

Combine Eqs. (a), (b), and (c).

$$\frac{d}{dt}m_{cv} = \rho A_1 V_1 - \rho Q_2 \tag{d}$$

Validate. Each term has units of kilograms per second. Eq. (d) makes physical sense: (rate of accumulation of mass) = (rate of mass flow in) − (rate of mass flow out).

Because variables on the right side of Eq. (d) are known, the problem can be solved. The plan is:

1. Calculate the flow rates on the right side of Eq. (d).
2. Apply Eq. (d) to calculate the rate of accumulation.

Take Action (Execute the Plan)

1. Mass flow rates (inlet and outlet).

$$\rho A_1 V_1 = (1000 \text{ kg/m}^3)(0.0025 \text{ m}^2)(7 \text{ m/s}) = 17.5 \text{ kg/s}$$
$$\rho Q_2 = (1000 \text{ kg/m}^3)(0.003 \text{ m}^3/\text{s}) = 3 \text{ kg/s}$$

2. Accumulation

$$\frac{dm_{cv}}{dt} = 17.5 \text{ kg/s} - 3 \text{ kg/s}$$

$$= \boxed{14.5 \text{ kg/s}}$$

Review the Solution and the Process

1. *Discussion.* Because the accumulation is positive, the quantity of mass within the control volume is increasing with time.

2. *Discussion.* The rising level of water in the tank causes air to flow out of the CV. Because air has a density that is about 1/1000 of the density of water, this effect is negligible.

Example 5.7 shows another instance in which the continuity equation is integrated with respect to time.

EXAMPLE 5.7

Depressurization of Gas in Tank

Problem Definition

Methane escapes through a small (10^{-7} m^2) hole in a 10 m^3 tank. The methane escapes so slowly that the temperature in the tank remains constant at 23°C. The mass flow rate of methane through the hole is given by $\dot{m} = 0.66\, pA/\sqrt{RT}$, where p is the pressure in the tank, A is the area of the hole, R is the gas constant, and T is the temperature in the tank. Calculate the time required for the absolute pressure in the tank to decrease from 500 to 400 kPa.

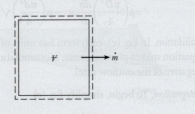

Define the Situation

Methane leaks through a 10^{-7} m^2 hole in 10 m^3 tank.

Assumptions.

1. Gas temperatures constant at 23°C during leakage.

2. Ideal gas law is applicable.

Properties: Table A.2, $R = 518$ J/kgK.

State the Goal

Find: Time (in seconds) for pressure to decrease from 500 kPa to 400 kPa.

Generate Ideas and Make a Plan

Select a CV that encloses whole tank.

1. Apply continuity equation, Eq. (5.29).

2. Analyze term by term.

3. Solve equation for elapsed time.

4. Calculate time.

Take Action (Execute the Plan)

1. Continuity equation

$$\frac{d}{dt} m_{cv} + \sum_{cs} \dot{m}_o - \sum_{cs} \dot{m}_i = 0$$

2. Term-by-term analysis.

- Rate of accumulation term. The mass in the control volume is the sum of the mass of the tank shell, M_{shell}, and the mass of methane in the tank,

$$m_{cv} = m_{shell} + \rho V$$

where V is the internal volume of the tank, which is constant. The mass of the tank shell is constant, so

$$\frac{dm_{cv}}{dt} = V\frac{d\rho}{dt}$$

- There is no mass inflow:

$$\sum_{cs} \dot{m}_i = 0$$

- Mass out flow rate is

$$\sum_{cs} \dot{m}_o = 0.66\frac{pA}{\sqrt{RT}}$$

Substituting terms into continuity equation

$$V\frac{d\rho}{dt} = -0.66\frac{pA}{\sqrt{RT}}$$

3. Equation for elapsed time:

- Use ideal gas law for ρ,

$$V\frac{d}{dt}\left(\frac{p}{RT}\right) = -0.66\frac{pA}{\sqrt{RT}}$$

- Because R and T are constant,

$$\frac{dp}{dt} = -0.66\frac{pA\sqrt{RT}}{V}$$

- Next, separate variables

$$\frac{dp}{p} = -0.66\frac{A\sqrt{RT}\,dt}{V}$$

- Integrating equation and substituting limits for initial and final pressure

$$t = \frac{1.52\,V}{A\sqrt{RT}}\ln\frac{p_0}{p_f}$$

4. Elapsed time

$$t = \frac{1.52\,(10\text{ m}^3)}{(10^{-7}\text{ m}^2)\left(518\dfrac{\text{J}}{\text{kg}\cdot\text{K}} \times 300\text{ K}\right)^{1/2}}\ln\frac{500}{400} = \boxed{8.6\times10^4\text{ s}}$$

Review the Solution and the Process

1. *Discussion.* The time corresponds to approximately one day.

2. *Knowledge.* Because the ideal gas law is used, the pressure and temperature have to be in absolute values.

Continuity Equation for Flow in a Conduit

A conduit is a pipe or duct or channel that is completely filled with a flowing fluid. Because flow in conduits is common, it is useful to derive an equation that applies to this case. To begin the derivation, recognize that in a conduit (see Fig. 5.13), there is no place for mass to accumulate, so Eq. (5.28) simplifies to

$$\int_{cs} \rho \mathbf{V} \cdot \mathbf{dA} = 0 \tag{5.31}$$

Mass is crossing the control surface at sections 1 and 2, so Eq. (5.31) simplifies to

$$\int_{\text{section 2}} \rho V dA - \int_{\text{section 1}} \rho V dA = 0 \tag{5.32}$$

If density is assumed to be constant across each section, Eq. (5.32) simplifies to

$$\rho_1 A_1 V_1 = \rho_2 A_2 V_2 \tag{5.33}$$

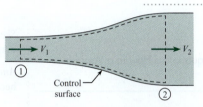

FIGURE 5.13

Flow through a conduit.

Eq. (5.33), which is called the *pipe flow form* of the continuity equation, is the final result. The meaning of this equation is (rate of inflow of mass at section 1) = (rate of outflow of mass at section 2).

There are other useful ways of writing the continuity equation. For example, Eq. (5.33) can be written in several equivalent forms:

$$\rho_1 Q_1 = \rho_2 Q_2 \tag{5.34}$$

$$\dot{m}_1 = \dot{m}_2 \tag{5.35}$$

If density is assumed to be constant, then Eq. (5.34) reduces to

$$Q_2 = Q_1 \tag{5.36}$$

Eq. (5.34) is valid for both steady and unsteady incompressible flow in a pipe. If there are more than two ports and the accumulation term is zero, then Eq. (5.29) can be reduced to

$$\sum_{cs} \dot{m}_i = \sum_{cs} \dot{m}_o \tag{5.37}$$

If the flow is assumed to have constant density, Eq. (5.37) can be written in terms of discharge:

$$\sum_{cs} Q_i = \sum_{cs} Q_o \tag{5.38}$$

Summary Depending on the assumptions of the problem, there are many ways to write the continuity equation. However, one can analyze any problem using the three equations summarized in Table 5.5. Thus, we recommend starting with one of these three equations because this is simpler than remembering many different equations.

✔ CHECKPOINT PROBLEM 5.3

Water and alcohol mix in a tank. Can the continuity equation be used to show that the outlet flow rate is 2 liters per second?

    a. yes

    b. no

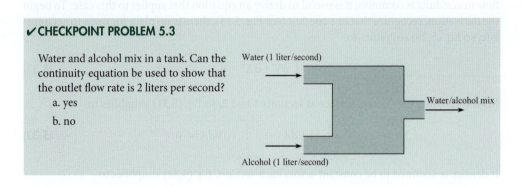

Water (1 liter/second)

Water/alcohol mix

Alcohol (1 liter/second)

Example 5.8 shows how to apply continuity to flow in a pipe.

EXAMPLE 5.8

Applying the Continuity Equation to Flow in a Variable Area Pipe

Problem Statement

A 120 cm pipe is in series with a 60 cm pipe. The speed of the water in the 120 cm pipe is 2 m/s. What is the water speed in the 60 cm pipe?

$V = 2$ m/s   120 cm   60 cm

Define the Situation

Water flows through a contraction in a pipe.

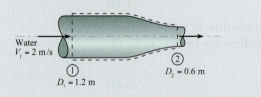

Water
$V_1 = 2$ m/s

①   ②
$D_1 = 1.2$ m $D_2 = 0.6$ m

State the Goal

V_2(m/s) ⬅ Mean velocity at section 2

Generate Ideas and Make a Plan

Selection. Select the continuity equation because the problem variables are velocity and pipe diameter.

Sketch. Select a fixed CV. Sketch this CV on the situation diagram. Label the inlet as section 1 and outlet as section 2.

Analysis. Select the pipe flow form of continuity (i.e., Eq. 5.33) because the problem involves flow in a pipe.

$$\rho A_1 V_1 = \rho A_2 V_2 \tag{a}$$

Assume density is constant (this is standard practice for steady flow of a liquid). The continuity equation reduces to

$$A_1 V_1 = A_2 V_2 \tag{b}$$

Validate. To validate Eq (b), notice that the primary dimensions of each term are L³/T. Also, this equation makes physical sense because it can be interpreted as (inflow) = (outflow).

Plan. Eq (b) contains the goal (V_2) and all other variables are known. Thus, the plan is to substitute numbers into this equation.

Take Action (Execute the Plan)

Continuity Equation:

$$V_2 = V_1 \frac{A_1}{A_2} = V_1 \left(\frac{D_1}{D_2}\right)^2$$

$$V_2 = (2 \text{ m/s}) \left(\frac{1.2 \text{ m}}{0.6 \text{ m}}\right)^2 = \boxed{8 \text{ m/s}}$$

Example 5.9 shows how the continuity equation can be applied together with the Bernoulli equation

EXAMPLE 5.9

Applying the Bernoulli and Continuity Equations to Flow through a Venturi

Problem Statement

Water with a density of 1000 kg/m³ flows through a vertical venturimeter as shown. A pressure gage is connected across two taps in the pipe (station 1) and the throat (station 2). The area ratio $A_{\text{throat}}/A_{\text{pipe}}$ is 0.5. The velocity in the pipe is 10 m/s. Find the pressure difference recorded by the pressure gage. Assume the flow has a uniform velocity distribution and that viscous effects are not important.

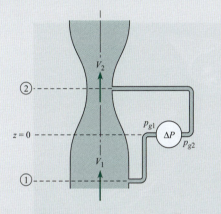

Define the Situation

Water flows in venturimeter. Area ratio = 0.5. V_1 = 10 m/s.

Assumptions:

1. Velocity distribution is uniform.
2. Viscous effects are unimportant.

Properties: ρ = 1000 kg/m³.

State the Goal

Find: Pressure difference measured by gage.

Generate Ideas and Make a Plan

1. Because viscous effects are unimportant, apply the Bernoulli equation between stations 1 and 2.
2. Combine the continuity equation (5.33) with the results of step 1.
3. Find the pressure on the gage by applying the hydrostatic equation.

Take Action (Execute the Plan)

1. The Bernoulli equation

$$p_1 + \gamma z_1 + \rho \frac{V_1^2}{2} = p_2 + \gamma z_2 + \rho \frac{V_2^2}{2}$$

Rewrite the equation in terms of piezometric pressure.

$$p_{z_1} - p_{z_2} = \frac{\rho}{2}\left(V_2^2 - V_1^2\right)$$

$$= \frac{\rho V_1^2}{2}\left(\frac{V_2^2}{V_1^2} - 1\right)$$

2. Continuity equation $V_2/V_1 = A_1/A_2$

$$p_{z_1} - p_{z_2} = \frac{\rho V_1^2}{2}\left(\frac{A_1^2}{A_2^2} - 1\right)$$

$$= \frac{1000 \text{ kg/m}^3}{2} \times (10 \text{ m/s})^2 \times (2^2 - 1)$$

$$= 150 \text{ kPa}$$

3. Apply the hydrostatic equation between the gage attachment point where the pressure is p_{g_1} and station 1 where the gage line is tapped into the pipe,

$$p_{z_1} = p_{g_1}$$

Also $p_{z_2} = p_{g_2}$ so

$$\Delta p_{\text{gage}} = p_{g_1} - p_{g_2} = p_{z_1} - p_{z_2} = \boxed{150 \text{ kPa}}$$

5.5 Predicting Cavitation

Designers can encounter a phenomenon, called cavitation, in which a liquid starts to boil due to low pressure. This situation is beneficial for some applications, but it is usually a problem that should be avoided by thoughtful design. Thus, this section describes cavitation and discusses how to design systems to minimize the possibility of harmful cavitation.

Description of Cavitation

Cavitation is when fluid pressure at a given point in a system drops to the vapor pressure and boiling occurs.

EXAMPLE. Consider water flowing at 15°C in a piping system. If the pressure of the water drops to the vapor pressure, the water will boil, and engineers will say that the system is cavitating. Because the vapor pressure of water at 15°C, which can be looked up in Appendix A.5, is $p_v = 1.7$ kPa abs, the condition required for cavitation is known. To avoid cavitation, the designer can configure the system so that pressures at all locations are above 1.7 kPa absolute.

Cavitation can damage equipment and degrade performance. Boiling causes vapor bubbles to form, grow, and then collapse, producing shock waves, noise, and dynamic effects that lead to decreased equipment performance and, frequently, equipment failure. Cavitation damage to a propeller (see Fig. 5.14) occurs because the spinning propeller creates low pressures near the tips of the blades where the velocity is high. Serious erosion produced by cavitation in a spillway tunnel of Hoover Dam is shown in Fig. 5.15.

FIGURE 5.14

Cavitation damage to a propeller. (Photo by Erik Axdahl)

Cavitation degrades materials because of the high pressures associated with the collapse of vapor bubbles. Experimental studies reveal that very high intermittent pressure, as high as 800 MPa, develops in the vicinity of the bubbles when they collapse (1). Therefore, if bubbles collapse close to boundaries such as pipe walls, pump impellers, valve casings, and dam slipway floors, they can cause considerable damage. Usually this damage occurs in the form of fatigue failure brought about by the action of millions of bubbles impacting (in effect, imploding) against the material surface over a long period of time, thus producing a material pitting in the zone of cavitation.

In some applications, cavitation is beneficial. Cavitation is responsible for the effectiveness of ultrasonic cleaning. Supercavitating torpedoes have been developed in which a large bubble envelops the torpedo, significantly reducing the contact area with the water and leading to significantly faster speeds. Cavitation plays a medical role in shock wave lithotripsy for the destruction of kidney stones.

FIGURE 5.15

Cavitation damage to a hydroelectric power dam spillway tunnel. (U.S. Bureau of Reclamation)

The world's largest and most technically advanced water tunnel for studying cavitation is located in Memphis, Tennessee—the William P. Morgan Large Cavitation Tunnel. This facility is used to test large-scale models of submarine systems and full-scale torpedoes as well as applications in the maritime shipping industry. More detailed discussions of cavitation can be found in Brennen (2) and Young (3).

Identifying Cavitation Sites

To predict cavitation, engineers looks for locations with low pressures. For example, when water flows through a pipe restriction (Fig. 5.16), the velocity increases according to the continuity equation, and in turn, the pressure decreases as dictated by the Bernoulli equation. For low flow rates, there is a relatively small drop in pressure at the restriction, so the water remains well above the vapor pressure, and boiling does not occur. However, as the flow rate increases, the pressure at the restriction becomes progressively lower until a flow rate is reached where the pressure is equal to the vapor pressure as shown in Fig. 5.16. At this point, the liquid boils to form bubbles, and cavitation ensues. The onset of cavitation can also be affected by the presence of contaminant gases, turbulence and by viscous effects.

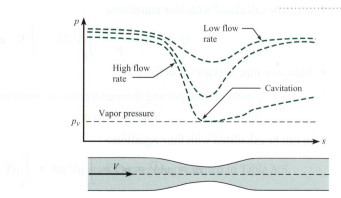

FIGURE 5.16

Flow through pipe restriction: variation of pressure for three different flow rates.

- Cavitation is usually undesirable because it can cause reduced performance Cavitation can cause erosion or pitting of solid materials, noise, vibrations, and structural failures.
- Cavitation is most likely to occur in regions of high velocity, in inlet regions of centrifugal pumps, and at locations of high elevations.
- To reduce the probability of cavitation, designers can specify that components that are susceptible to cavitation (e.g., values and centrifugal pumps) be situated at low elevations.

REFERENCES

1. Knapp, R.T., J. W. Daily, and F. G. Hammitt. *Cavitation*. New York: McGraw-Hill, 1970.

2. Brennen, C. E. *Cavitation and Bubble Dynamics*. New York: Oxford University Press, 1995.

3. Young, F. R. *Cavitation*. New York: McGraw-Hill, 1989.

PROBLEMS

PLUS Problem available in *WileyPLUS* at instructor's discretion.

GO Guided Online (GO) Problem, available in *WileyPLUS* at instructor's discretion.

Characterizing Flow Rates (§5.1)

5.1 Consider filling the gasoline tank of an automobile at a gas station. (a) Estimate the discharge in L/min. (b) Using the same nozzle, estimate the time to put 190 L in the tank. (c) Estimate the cross-sectional area of the nozzle and calculate the velocity at the nozzle exit.

5.2 The average flow rate (release) through Grand Coulee Dam is 3100 m^3/s. The width of the river downstream of the dam is 90 m. Making a reasonable estimate of the river velocity, estimate the river depth.

5.3 Taking a jar of known volume, fill with water from your household tap and measure the time to fill. Calculate the discharge from the tap. Estimate the cross-sectional area of the faucet outlet, and calculate the water velocity issuing from the tap.

5.4 **PLUS** Another name for the volume flow rate equation could be:

   a. the discharge equation

   b. the mass flow rate equation

   c. either a or b

5.5 A liquid flows through a pipe with a constant velocity. If a pipe twice the size is used with the same velocity, will the flow rate be (a) halved, (b) doubled, (c) quadrupled? Explain.

5.6 **PLUS** For flow of a gas in a pipe, which form of the continuity equation is more general?

   a. $V_1 A_1 = V_2 A_2$

   b. $\rho_1 V_1 A_1 = \rho_2 V_2 A_2$

   c. both are equally applicable

5.7 **PLUS** The discharge of water in a 35-cm-diameter pipe is 0.06 m^3/s. What is the mean velocity?

5.8 **PLUS** A pipe with a 46 cm diameter carries water having a velocity of 1.2 m/s. What is the discharge in cubic meters per second and in liters per minute?

5.9 A pipe with a 2 m diameter carries water having a velocity of 4 m/s. What is the discharge in cubic meters per second?

5.10 **PLUS** A pipe whose diameter is 6 cm transports air with a temperature of 20°C and pressure of 180 kPa absolute at 19 m/s. Determine the mass flow rate.

5.11 **PLUS** Natural gas (methane) flows at 25 m/s through a pipe with a 0.84 m diameter. The temperature of the methane is 15°C, and the pressure is 160 kPa gage. Determine the mass flow rate.

5.12 An aircraft engine test pipe is capable of providing a flow rate of 180 kg/s at altitude conditions corresponding to an absolute pressure of 50 kPa and a temperature of −18°C. The velocity of air through the duct attached to the engine is 255 m/s. Calculate the diameter of the duct.

5.13 A heating and air-conditioning engineer is designing a system to move 1000 m^3 of air per hour at 100 kPa abs, and 30°C. The duct is rectangular with cross-sectional dimensions of 1 m by 20 cm. What will be the air velocity in the duct?

5.14 The hypothetical velocity distribution in a circular duct is

$$\frac{V}{V_0} = 1 - \frac{r}{R}$$

where r is the radial location in the duct, R is the duct radius, and V_0 is the velocity on the axis. Find the ratio of the mean velocity to the velocity on the axis.

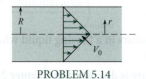

PROBLEM 5.14

5.15 Water flows in a two-dimensional channel of width W and depth D as shown in the diagram. The hypothetical velocity profile for the water is

$$V(x, y) = V_s\left(1 - \frac{4x^2}{W^2}\right)\left(1 - \frac{y^2}{D^2}\right)$$

where V_s is the velocity at the water surface midway between the channel walls. The coordinate system is as shown; x is measured from the center plane of the channel and y downward from the water surface. Find the discharge in the channel in terms of V_s, D, and W.

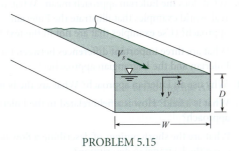

PROBLEM 5.15

5.16 ^{WILEY} **GO** Water flows in a pipe that has a 1.2 m diameter and the following hypothetical velocity distribution: The velocity is maximum at the centerline and decreases linearly with r to a minimum at the pipe wall. If $V_{max} = 4.5$ m/s and $V_{min} = 3.6$ m/s, what is the discharge in cubic meters per second and in liters per minute?

5.17 In Prob. 5.16, if $V_{max} = 8$ m/s, $V_{min} = 6$ m/s, and $D = 2$ m, what is the discharge in cubic meters per second and the mean velocity?

5.18 ^{WILEY} **GO** Air enters this square duct at section 1 with the velocity distribution as shown. Note that the velocity varies in the y direction only (for a given value of y, the velocity is the same for all values of z).

  a. What is the volume rate of flow?

  b. What is the mean velocity in the duct?

  c. What is the mass rate of flow if the mass density of the air is 1.2 kg/m³?

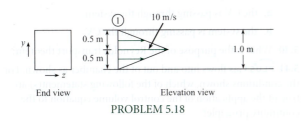

PROBLEM 5.18

5.19 ^{WILEY} **PLUS** The velocity at section A-A is 4.5 m/s, and the vertical depth y at the same section is 1.2 m. If the width of the channel is 8.5 m, what is the discharge in cubic meter per second?

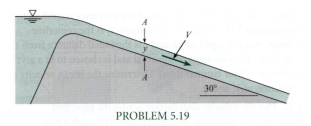

PROBLEM 5.19

5.20 ^{WILEY} **PLUS** The rectangular channel shown is 1.2 m wide. What is the discharge in the channel?

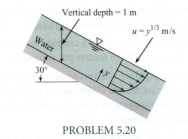

PROBLEM 5.20

5.21 If the velocity in the channel of Prob. 5.20 is given as $u = 8[\exp(y) - 1]$ m/s and the channel width is 2 m, what is the discharge in the channel and what is the mean velocity?

5.22 ^{WILEY} **PLUS** Water from a pipe is diverted into a weigh tank for exactly 20 min. The increased weight in the tank is 20 kN. What is the discharge in cubic meters per second? Assume $T = 20°$C.

5.23 Water enters the lock of a ship canal through 180 ports, each port having a 0.6 m by 0.6 m cross section. The lock is 275 m long and 32 m wide. The lock is designed so that the water surface in it will rise at a maximum rate of 1.8 m/min. For this condition, what will be the mean velocity in each port?

5.24 ^{WILEY} **GO** An empirical equation for the velocity distribution in a horizontal, rectangular, open channel is given by $u = u_{max}(y/d)^n$, where u is the velocity at a distance y meters above the floor of the channel. If the depth d of flow is 1.2 m, $u_{max} = 3$ m/s, and $n = 1/6$, what is the discharge in cubic meters per second per meter of width of channel? What is the mean velocity?

5.25 The hypothetical water velocity in a V-shaped channel (see the accompanying figure) varies linearly with depth from zero at the bottom to maximum at the water surface. Determine the discharge if the maximum velocity is 1.8 m/s.

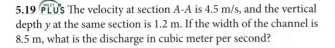

PROBLEM 5.25

5.51 Two parallel disks of diameter D are brought together, each with a normal speed of V. When their spacing is h, what is the radial component of convective acceleration at the section just inside the edge of the disk (section A) in terms of V, h, and D? Assume uniform velocity distribution across the section.

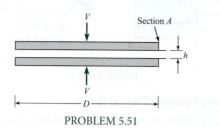

PROBLEM 5.51

5.52 PLUS Two streams discharge into a pipe as shown. The flows are incompressible. The volume flow rate of stream A into the pipe is given by $Q_A = 0.04t$ m³/s and that of stream B by $Q_B = 0.006\,t^2$ m³/s, where t is in seconds. The exit area of the pipe is 0.01 m². Find the velocity and acceleration of the flow at the exit at $t = 1$ s.

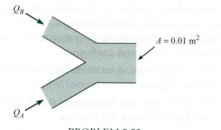

PROBLEM 5.52

5.53 Air discharges downward in the pipe and then outward between the parallel disks. Assuming negligible density change in the air, derive a formula for the acceleration of air at point A, which is a distance r from the center of the disks. Express the acceleration in terms of the constant air discharge Q, the radial distance r, and the disk spacing h. If $D = 10$ cm, $h = 0.6$ cm, and $Q = 0.380$ m³/s, what are the velocity in the pipe and the acceleration at point A where $r = 20$ cm?

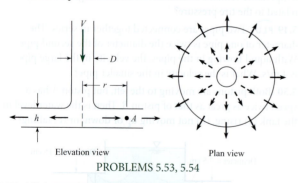

Elevation view   Plan view

PROBLEMS 5.53, 5.54

5.54 All the conditions of Prob. 5.53 are the same except that $h = 1$ cm and the discharge is given as $Q = Q_0(t/t_0)$, where $Q_0 = 0.1$ m³/s and $t_0 = 1$ s. For the additional conditions, what will be the acceleration at point A when $t = 2$ s and $t = 3$ s?

5.55 GO A tank has a hole in the bottom with a cross-sectional area of 0.0025 m² and an inlet line on the side with a cross-sectional area of 0.0025 m², as shown. The cross-sectional area of the tank is 0.1 m². The velocity of the liquid flowing out the bottom hole is $V = \sqrt{2gh}$, where h is the height of the water surface in the tank above the outlet. At a certain time the surface level in the tank is 1 m and rising at the rate of 0.1 cm/s. The liquid is incompressible. Find the velocity of the liquid through the inlet.

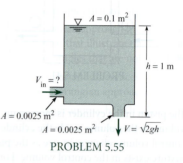

PROBLEM 5.55

5.56 PLUS A mechanical pump is used to pressurize a bicycle tire. The inflow to the pump is 0.02 m³/min. The density of the air entering the pump is 1.2 kg/m³. The inflated volume of a bicycle tire is 0.0009 m³. The density of air in the inflated tire is 6.4 kg/m³. How many seconds does it take to pressurize the tire if there initially was no air in the tire?

5.57 A 15-cm cylinder falls at a rate of 1.2 m/s in a 20-cm-diameter tube containing an incompressible liquid. What is the mean velocity of the liquid (with respect to the tube) in the space between the cylinder and the tube wall?

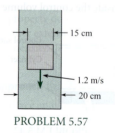

PROBLEM 5.57

5.58 PLUS This circular tank of water is being filled from a pipe as shown. The velocity of flow of water from the pipe is 3 m/s. What will be the rate of rise of the water surface in the tank?

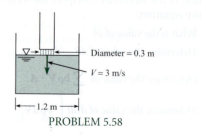

PROBLEM 5.58

5.59 A sphere 20 cm in diameter falls at 1.2 m/s downward axially through water in a 30 cm-diameter container. Find the upward speed of the water with respect to the container wall at the midsection of the sphere.

5.60 PLUS A rectangular air duct 20 cm by 60 cm carries a flow of 1.44 m³/s. Determine the velocity in the duct. If the duct tapers to 10 cm by 40 cm, what is the velocity in the latter section? Assume constant air density.

5.61 PLUS A 30 cm pipe divides into a 20 cm branch and a 18 cm branch. If the total discharge is 0.40 m³/s and if the same mean velocity occurs in each branch, what is the discharge in each branch?

5.62 The conditions are the same as in Prob. 5.61 except that the discharge in the 20 cm branch is twice that in the 15 cm branch. What is the mean velocity in each branch?

5.63 PLUS Water flows in a 25 cm pipe that is connected in series with a 15 cm pipe. If the rate of flow is 3400 liters per minute, what is the mean velocity in each pipe?

5.64 What is the velocity of the flow of water in leg B of the tee shown in the figure?

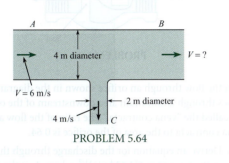

PROBLEM 5.64

5.65 PLUS For a steady flow of gas in the conduit shown, what is the mean velocity at section 2?

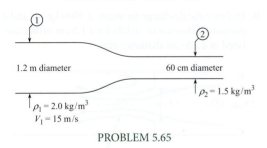

PROBLEM 5.65

5.66 Two pipes, A and B, are connected to an open water tank. The water is entering the bottom of the tank from pipe A at 0.28 m³/min. The water level in the tank is rising at 2.5 cm/min, and the surface area of the tank is 7.4 m². Calculate the discharge in a second pipe, pipe B, that is also connected to the bottom of the tank. Is the flow entering or leaving the tank from pipe B?

5.67 Is the tank in the figure filling or emptying? At what rate is the water level rising or falling in the tank?

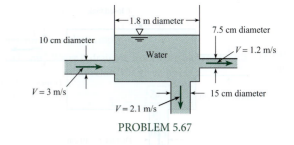

PROBLEM 5.67

5.68 GO Given: Flow velocities as shown in the figure and water surface elevation (as shown) at $t = 0$ s. At the end of 22 s, will the water surface in the tank be rising or falling, and at what speed?

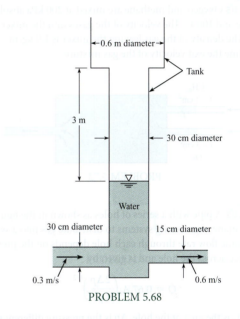

PROBLEM 5.68

5.69 GO A lake with no outlet is fed by a river with a constant flow of 3.4 m³/s. Water evaporates from the surface at a constant rate of 0.037 m³/s per square kilometer surface area. The area varies with depth h (meter) as A (square kilometers) $= 4.5 + 5.5h$. What is the equilibrium depth of the lake? Below what river discharge will the lake dry up?

5.70 A stationary nozzle discharges water against a plate moving toward the nozzle at half the jet velocity. When the discharge from the nozzle is 0.14 m³/s, at what rate will the plate deflect water?

5.71 An open tank has a constant inflow of 0.57 m³/s. A 30 cm-diameter drain provides a variable outflow velocity V_{out} equal to $\sqrt{(2gh)}$ m/s. What is the equilibrium height h_{eq} of the liquid in the tank?

5.72 Assuming that complete mixing occurs between the two inflows before the mixture discharges from the pipe at C, find the mass rate of flow, the velocity, and the specific gravity of the mixture in the pipe at C.

5.84 WILEY PLUS A spherical tank with a diameter of 1 m is half filled with water. A port at the bottom of the tank is opened to drain the tank. The hole diameter is 1 cm, and the velocity of the water draining from the hole is $V_e = \sqrt{2gh}$, where h is the elevation of the water surface above the hole. Find the time required for the tank to empty.

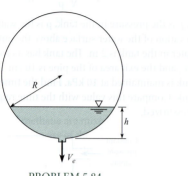

PROBLEM 5.84

5.85 A tank containing oil is to be pressurized to decrease the draining time. The tank, shown in the figure, is 2 m in diameter and 6 m high. The oil is originally at a level of 5 m. The oil has a density of 880 kg/m³. The outlet port has a diameter of 2 cm, and the velocity at the outlet is given by

$$V_e = \sqrt{2gh + \frac{2p}{\rho}}$$

where p is the gage pressure in the tank, ρ is the density of the oil, and h is the elevation of the surface above the hole. Assume during the emptying operation that the temperature of the air in the tank is constant. The pressure will vary as

$$p = (p_0 + p_{atm})\frac{(L - h_0)}{(L - h)} - p_{atm}$$

where L is the height of the tank, p_{atm} is the atmospheric pressure, and the subscript 0 refers to the initial conditions. The initial pressure in the tank is 300 kPa gage, and the atmospheric pressure is 100 kPa.

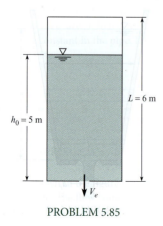

PROBLEM 5.85

Applying the continuity equation to this problem, one finds

$$\frac{dh}{dt} = -\frac{A_e}{A_T}\sqrt{2gh + \frac{2p}{\rho}}$$

Integrate this equation to predict the depth of the oil with time for a period of one hour.

5.86 *Rocket Propulsion.* To prepare for Problems 5.87, 5.88, and 5.89, use the Internet or other resources and define the following terms in the context of rocket propulsion: (a) solid fuel, (b) grain, and (c) surface regression. Also explain how a solid-fuel rocket engine works.

5.87 WILEY PLUS An end-burning rocket motor has a chamber diameter of 10 cm and a nozzle exit diameter of 8 cm. The density of the solid propellant is 1800 kg/m³, and the propellant surface regresses at the rate of 1.5 cm/s. The gases crossing the nozzle exit plane have a pressure of 10 kPa abs and a temperature of 2200°C. The gas constant of the exhaust gases is 415 J/kg K. Calculate the gas velocity at the nozzle exit plane.

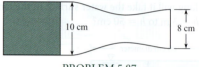

PROBLEM 5.87

5.88 A cylindrical-port rocket motor has a grain design consisting of a cylindrical shape as shown. The curved internal surface and both ends burn. The solid propellant surface regresses uniformly at 1 cm/s. The propellant density is 2000 kg/m³. The inside diameter of the motor is 20 cm. The propellant grain is 40 cm long and has an inside diameter of 12 cm. The diameter of the nozzle exit plane is 20 cm. The gas velocity at the exit plane is 1800 m/s. Determine the gas density at the exit plane.

PROBLEM 5.88

5.89 The mass flow rate through a rocket nozzle (shown) is given by

$$\dot{m} = 0.65\frac{p_c A_t}{\sqrt{RT_c}}$$

where p_c and T_c are the pressure and temperature in the rocket chamber and R is the gas constant of the gases in the chamber. The propellant burning rate (surface regression rate) can be expressed as $\dot{r} = ap_c^n$, where a and n are two empirical constants. Show, by application of the continuity equation, that the chamber pressure can be expressed as

$$p_c = \left(\frac{a\rho_p}{0.65}\right)^{1/(1-n)}\left(\frac{A_g}{A_t}\right)^{1/(1-n)}(RT_c)^{1/[2(1-n)]}$$

where ρ_p is the propellant density and A_g is the grain surface burning area. If the operating chamber pressure of a rocket motor is 3.5 MPa and $n = 0.3$, how much will the chamber pressure increase if a crack develops in the grain, increasing the burning area by 20%?

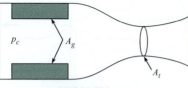

PROBLEM 5.89

5.90 The piston shown is moving up during the exhaust stroke of a four-cycle engine. Mass escapes through the exhaust port at a rate given by

$$\dot{m} = 0.65 \frac{p_c A_v}{\sqrt{RT_c}}$$

where p_c and T_c are the cylinder pressure and temperature, A_v is the valve opening area, and R is the gas constant of the exhaust gases. The bore of the cylinder is 10 cm, and the piston is moving upward at 30 m/s. The distance between the piston and the head is 10 cm. The valve opening area is 1 cm^2, the chamber pressure is 300 kPa abs, the chamber temperature is 600°C, and the gas constant is 350 J/kg K. Applying the continuity equation, determine the rate at which the gas density is changing in the cylinder. Assume the density and pressure are uniform in the cylinder and the gas is ideal.

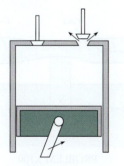

PROBLEM 5.90

5.91 [WILEY PLUS] Gas is flowing from Location 1 to 2 in the pipe expansion shown. The inlet density, diameter and velocity are ρ_1, D_1, and V_1 respectively. If D_2 is $2D_1$ and V_2 is half of V_1, what is the magnitude of ρ_2?

 a. $\rho_2 = 4\,\rho_1$
 b. $\rho_2 = 2\,\rho_1$
 c. $\rho_2 = \frac{1}{2}\,\rho_1$
 d. $\rho_2 = \rho_1$

5.92 [WILEY PLUS] Air is flowing from a ventilation duct (cross section 1) as shown, and is expanding to be released into a room at cross section 2.

The area at cross section 2, A_2, is 3 times A_1. Assume that the density is constant. The relation between Q_1 and Q_2 is:

 a. $Q_2 = \frac{1}{3}\,Q_1$
 b. $Q_2 = Q_1$
 c. $Q_2 = 3\,Q_1$
 d. $Q_2 = 9\,Q_1$

5.93 [WILEY PLUS] Water is flowing from Location 1 to 2 in this pipe expansion. D_1 and V_1 are known at the inlet. D_2 and P_2 are known at the outlet. What equation(s) do you need to solve for the inlet pressure P_1? Neglect viscous effects.

 a. The continuity equation
 b. The continuity equation and the flow rate equation
 c. The continuity equation, the flow rate equation, and the Bernoulli equation
 d. There is insufficient information to solve the problem

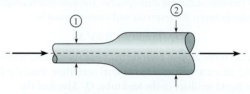

PROBLEMS 5.91, 5.92, 5.93

5.94 The flow pattern through the pipe contraction is as shown, and the Q of water is 1.7 m^3/s. For $d = 0.6$ m and $D = 1.8$ m, what is the pressure at point B if the pressure at point C is 153 kPa?

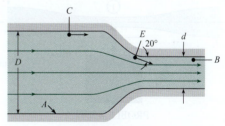

PROBLEM 5.94

5.95 Water flows through a rigid contraction section of circular pipe in which the outlet diameter is one-half the inlet diameter. The velocity of the water at the inlet varies with time as $V_{in} = (10\ \text{m/s})$ $[1 - \exp(-t/10)]$. How will the velocity vary with time at the outlet?

5.96 [WILEY PLUS] The annular venturimeter is useful for metering flows in pipe systems for which upstream calming distances are limited. The annular venturimeter consists of a cylindrical section mounted inside a pipe as shown. The pressure difference is measured between the upstream pipe and at the region adjacent to the cylindrical section. Air at standard conditions flows in the system. The pipe diameter is 15 cm. The ratio of the cylindrical section diameter to the inside pipe diameter is 0.8. A pressure difference of 5 cm of water is measured. Find the volume flow rate. Assume the flow is incompressible, inviscid, and steady and that the velocity is uniformly distributed across the pipe.

PROBLEM 5.96

5.97 Venturi-type applicators are frequently used to spray liquid fertilizers. Water flowing through the venturi creates a subatmospheric pressure at the throat, which in turn causes the liquid fertilizer to flow up the feed tube and mix with the water in the throat region. The venturi applicator shown uses water at 20°C to spray a liquid fertilizer with the same density. The venturi exhausts to the atmosphere, and the exit diameter is 1 cm. The ratio of exit area to throat area (A_2/A_1) is 2. The flow rate of water through the venturi is 8 L/m (liters/min). The bottom of the feed tube in the reservoir is 5 cm below the liquid fertilizer surface and 10 cm below the centerline of the venturi. The pressure at the liquid fertilizer surface is atmospheric. The flow rate through the feed tube between the reservoir and venturi throat is

$$Q_1(\text{L/min}) = 0.5\sqrt{\Delta h}$$

where Δh is the drop in piezometric head (in meters) between the feed tube entrance and the venturi centerline. Find the flow rate of liquid fertilizer in the feed tube, Q_l. Also find the concentration of liquid fertilizer in the mixture, $[Q_l/(Q_l + Q_w)]$, at the end of the sprayer.

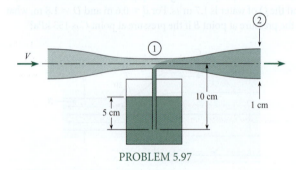

PROBLEM 5.97

5.98 ⓌILEY PLUS Air with a density of 1 kg/m³ is flowing upward in the vertical duct, as shown. The velocity at the inlet (station 1) is 24 m/s, and the area ratio between stations 1 and 2 is 0.5 $(A_2/A_1 = 0.5)$. Two pressure taps, 3 m apart, are connected to a manometer, as shown. The specific weight of the manometer liquid is 18.9 kN/m³. Find the deflection, Δh, of the manometer.

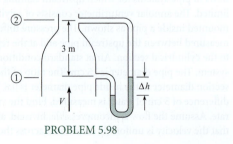

PROBLEM 5.98

5.99 An atomizer utilizes a constriction in an air duct as shown. Design an operable atomizer making your own assumptions regarding the air source.

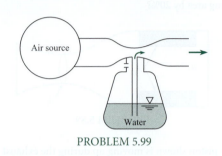

PROBLEM 5.99

5.100 ⓌILEY PLUS A suction device is being designed based on the venturi principle to lift objects submerged in water. The operating water temperature is 15°C. The suction cup is located 1 m below the water surface, and the venturi throat is located 1 m above the water. The atmospheric pressure is 100 kPa. The ratio of the throat area to the exit area is 1/4, and the exit area is 0.001 m². The area of the suction cup is 0.1 m².

   a. Find the velocity of the water at the exit for maximum lift condition.

   b. Find the discharge through the system for maximum lift condition.

   c. Find the maximum load the suction cup can support.

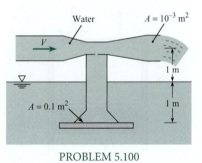

PROBLEM 5.100

5.101 ⓌILEY PLUS A design for a hovercraft is shown in the figure. A fan brings air at 15.5°C into a chamber, and the air is exhausted between the skirts and the ground. The pressure inside the chamber is responsible for the lift. The hovercraft is 4.5 m long and 2.1 m wide. The weight of the craft including crew, fuel, and load is 8.9 kN. Assume that the pressure in the chamber is the stagnation pressure (zero velocity) and the pressure where the air exits around the skirt is atmospheric. Assume the air is incompressible, the flow is steady, and viscous effects are negligible. Find the airflow rate necessary to maintain the skirts at a height of 7.5 cm above the ground.

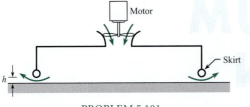

PROBLEM 5.101

5.102 Water is forced out of this cylinder by the piston. If the piston is driven at a speed of 1.8 m/s, what will be the speed of efflux of the water from the nozzle if $d = 5$ cm and $D = 10$ cm? Neglecting friction and assuming irrotational flow, determine the force F that will be required to drive the piston. The exit pressure is atmospheric pressure.

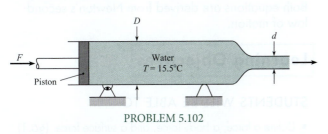

PROBLEM 5.102

5.103 Air flows through a constant-area heated pipe. At the entrance, the velocity is 10 m/s, the pressure is 100 kPa absolute, and the temperature is 20°C. At the outlet, the pressure is 80 kPa absolute, and the temperature is 50°C. What is the velocity at the outlet? Can the Bernoulli equation be used to relate the pressure and velocity changes? Explain.

Predicting Cavitation (§5.5)

5.104 Sometimes driving your car on a hot day, you may encounter a problem with the fuel pump called pump cavitation. What is happening to the gasoline? How does this affect the operation of the pump?

5.105 What is cavitation? Why does the tendency for cavitation in a liquid increase with increased temperatures?

5.106 PLUS The following questions have to do with cavitation.

   a. Is it more correct to say that cavitation has to do with (i) vacuum pressures, or (ii) vapor pressures?

   b. Is cavitation more likely to occur on the low pressure (suction) side of a pump, or the high pressure (discharge) side? Why?

   c. What does the word cavitation have to do with cavities, like the ones we get in our teeth? Is this aspect of cavitation the (i) cause, or the (ii) result of the phenomenon?

   d. When water goes over a waterfall, and one can see lots of bubbles in the water, is that due to cavitation? Why, or why not?

5.107 GO When gage A indicates a pressure of 130 kPa gage, then cavitation just starts to occur in the venturi meter. If $D = 50$ cm

and $d = 10$ cm, what is the water discharge in the system for this condition of incipient cavitation? The atmospheric pressure is 100 kPa gage, and the water temperature is 10°C. Neglect gravitational effects.

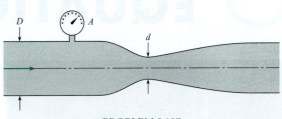

PROBLEM 5.107

5.108 A sphere 30 cm in diameter is moving horizontally at a depth of 3.6 m below a water surface where the water temperature is 10°C. $V_{max} = 1.5\ V_o$, where V_o is the free stream velocity and occurs at the maximum sphere width. At what speed in still water will cavitation first occur?

5.109 GO When the hydrofoil shown was tested, the minimum pressure on the surface of the foil was found to be 70 kPa absolute when the foil was submerged 1.80 m and towed at a speed of 8 m/s. At the same depth, at what speed will cavitation first occur? Assume irrotational flow for both cases and $T = 10°C$.

5.110 For the hydrofoil of Prob. 5.109, at what speed will cavitation begin if the depth is increased to 3 m?

5.111 PLUS When the hydrofoil shown was tested, the minimum pressure on the surface of the foil was found to be 17 kPa vacuum when the foil was submerged 1.2 m and towed at a speed of 7.5 m/s. At the same depth, at what speed will cavitation first occur? Assume irrotational flow for both cases and $T = 10°C$.

5.112 For the conditions of Prob. 5.111, at what speed will cavitation begin if the depth is increased to 3 m?

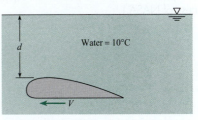

PROBLEMS 5.109, 5.110, 5.111, 5.112

5.113 A sphere is moving in water at a depth where the absolute pressure is 124 kPa abs. The maximum velocity on a sphere occurs 90° from the forward stagnation point and is 1.5 times the free-stream velocity. The density of water is 1000 kg/m³. Calculate the speed of the sphere at which cavitation will occur. $T = 10°C$.

5.114 The minimum pressure on a cylinder moving horizontally in water ($T = 10°C$) at 5 m/s at a depth of 1 m is 80 kPa absolute. At what velocity will cavitation begin? Atmospheric pressure is 100 kPa absolute.

6 MOMENTUM EQUATION

FIGURE 6.1
Engineers design systems by using a small set of fundamental equations such as the momentum equation. (Photo courtesy of NASA.)

Chapter Road Map

This chapter presents (a) the linear momentum equation and the (b) angular momentum equation. Both equations are derived from Newton's second law of motion.

Learning Objectives

STUDENTS WILL BE ABLE TO

- Define a force, a body force, and a surface force. (§6.1)
- Explain Newton's second law (particle or system of particles). (§6.1)
- Solve a vector equation with the VSM (Visual Solution Method). (§6.1)
- List the steps to derive the linear momentum equation. (§6.2)
- Describe or calculate (a) momentum flow and (b) momentum accumulation. (§6.2)
- Sketch a force diagram. Sketch a momentum diagram. (§6.3)
- Describe the physics of the momentum equation and the meaning of the variables that appear in the equation. (§6.2, §6.3)
- Describe the process for applying the momentum equation. (§6.3)
- Apply the linear momentum equation to problems involving jets, vanes, pipe bends, nozzles, and other stationary objects. (§6.4)
- Apply the linear momentum equations to moving objects such as carts and rockets. (§6.5)
- Apply the angular momentum equation to analyze rotating machinery such as pumps and turbines. (§6.6)

6.1 Understanding Newton's Second Law of Motion

Because Newton's second law is the theoretical foundation of the momentum equation, this section reviews relevant concepts.

Body and Surface Forces

A **force** is an interaction between two bodies that can be idealized as a push or pull of one body on other body. A push/pull interaction is one that can cause acceleration.

Newton's third law tells us that forces must involve the interaction of *two bodies* and that *forces occur in pairs*. The two forces are equal in magnitude, opposite in direction, and colinear.

EXAMPLE. To give examples of force, consider an airplane that is flying in a straight path at constant speed (Fig. 6.2). Select the airplane as the *system* for analysis. Idealize the airplane as a *particle*. Newton's first law (i.e., force equilibrium) tells us that the sum of forces must balance. There are four forces on the airplane.

- The *lift force* is the net upward push of the air (body 1) on the airplane (body 2).
- The *weight* is the pull of the earth (body 1) on the airplane (body 2) through the action of gravity.
- The *drag force* is the net resistive force of the air (body 1) on the airplane (body 2).
- The *thrust force* is the net horizontal push of the air (body 1) on the surfaces of the propeller (body 2).

Notice that each of the four interactions just described can be classified as a force because: (a) they involve a push or pull, and (b) they involve the interaction of two bodies of matter.

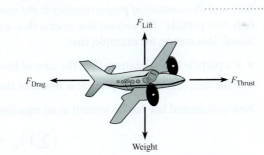

FIGURE 6.2

When an airplane is flying in straight and level flight, the forces sum to zero.

F_{Lift}

F_{Drag}

F_{Thrust}

Weight

Forces can be classified into two categories: body force and surface force. A **surface force** (also known as a contact force) is a force that requires physical contact or touching between the two interacting bodies. The lift force (Fig. 6.2) is a surface force because the air (body 1) must touch the wing (body 2) to create the lift force. Similarly, the thrust and drag forces are surface forces.

A **body force** is a force that can act without physical contact. For example, the weight force is a body force because the airplane (body 1) does not need to touch the earth (body 2) for the weight force to act.

A body force acts on every particle within a system. In contrast, a surface force acts only on the particles that are in physical contact with the other interacting body. For example, consider a system comprised of a glass of water sitting on a table. The weight force is pulling on every particle within the system, and we represent this force as a vector that passes through the center of gravity of the system. In contrast, the normal force on the bottom of the cup acts only on the particles of glass that are touching the table.

Summary Forces can be classified in two categories: body forces and surface forces (see Fig. 6.3). Most forces are surface forces.

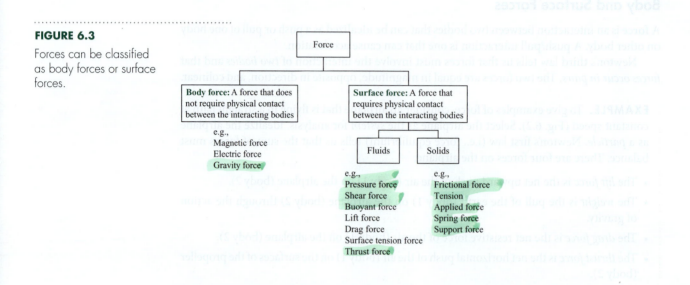

Newton's Second Law of Motion

In words, Newtons' second law is: *The sum of forces on a particle is proportional to the acceleration, and the constant of proportionality is the mass of the particle.* Notice that this law applies only to a particle. The second law asserts that *acceleration and unbalanced forces are proportional.* This means, for example, that

- If a particle is accelerating, then the sum of forces on the particle is nonzero.
- If the sum of forces on a particle is nonzero, then the particle will be accelerating.

Newton's second law can be written as an equation:

$$\left(\sum \mathbf{F} \right)_{ext} = m\mathbf{a} \tag{6.1}$$

where the subscript "ext" is a reminder to sum only external forces.

EXAMPLE. To illustrate the relationship between unbalanced forces and acceleration, consider an airplane that is turning left while flying at a constant speed in a horizontal plane (Fig. 6.4a). Select the airplane as a *system*. Idealize the airplane as a *particle*. Because the airplane is traveling in a circular path at constant speed, the acceleration vector must point inward. Fig. 6.4b shows the vectors that appear in Newton's second law. For Newton's second law of motion to be satisfied, the sum of the force vectors (Fig. 6.4c) must be equal to the *m**a*** vector.

The airplane example illustrates a method for visualizing and solving a vector equation called the *Visual Solution Method* (VSM). This method was adapted from Hibbler (1). This method is presented in the next subsection. Checkpoint Problem 6.1 gives you a chance to test your understanding of Newton's second law.

FIGURE 6.4

An airplane flying with a steady speed on curved path in a horizontal plane (a) Top view, (b) Front view, (c) Sketch showing how the $\Sigma\mathbf{F}$ vectors balance the $\mathbf{ma}$ vector.

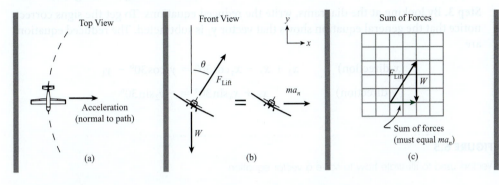

(a)   (b)   (c)

✔ **CHECKPOINT PROBLEM 6.1**

A disk in a horizontal plane is rotating in a counterclockwise direction and the speed of rotation is decreasing. A penny stays in place on the disk due to friction. Which letter (a to h) best represents the direction of acceleration of the penny? Which letter best represents the direction of the sum of forces vector?

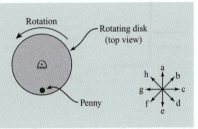

Solving a Vector Equation with the Visual Solution Method (VSM)

The VSM is an approach for solving a vector equation that reveals the physics while also showing visually how the equation can be solved. Thus, the VSM simplifies problem solving. The VSM has three steps.

Step 1: Identify the vector equation in its general form.

Step 2: Draw a diagram that shows the vectors that appear in the left side of the equation. Then, draw a second diagram that shows the vectors that appear on the right side of the equation. Add the equal sign between the diagrams.

Step 3: From the diagrams, apply the general equation and simplify the results to create the reduced equation(s). The reduced equation(s) can be written as a vector equation or as one or more scaler equations.

EXAMPLE. This example shows how to apply the VSM to the airplane problem (see Fig. 6.4).

Step 1: The general equation is Newton's second law ($\Sigma\mathbf{F})_{\text{ext}} = m\mathbf{a}$.

Step 2: The two diagrams separated by an equal sign are shown in Fig. 6.4b.

Step 3: By looking at the diagrams, one can write the reduced equation using scalar equations:

$$(x\text{-direction}) \qquad F_{\text{lift}} \sin \theta = ma_n$$

$$(y\text{-direction}) \qquad -W + F_{\text{lift}} \cos \theta = 0$$

Alternatively, one can look at the diagrams and then write the reduced equation using a vector equation.

$$F_{\text{Lift}}(\sin \theta \mathbf{i} + \cos \theta \mathbf{j}) - W\mathbf{j} = (ma_n)\mathbf{i}$$

EXAMPLE. This example shows how to apply the VSM to a generic vector equation.

Step 1. Suppose the general equation is $\Sigma \mathbf{x} = \mathbf{y}_2 - \mathbf{y}_1$.

Step 2. Suppose the vectors are known. Then, one can sketch the diagrams (Fig. 6.5).

Step 3. By looking at the diagrams, write the reduced equations. To get the signs correct, notice that the general equation shows that vector $\mathbf{y}_1$ is subtracted. The reduced equations are

$$\text{(x-direction)} \qquad x_2 + x_3 - x_4 \cos 30° = y_2 \cos 30° - y_1$$

$$\text{(y-direction)} \qquad x_1 + x_4 \sin 30° = -y_2 \sin 30°$$

FIGURE 6.5

Vectors used to illustrate how to solve a vector equation.

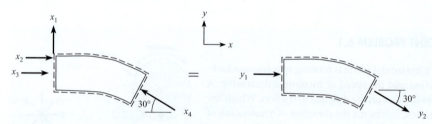

Newton's Second Law (System of Particles)

Newton's second law (Eq. 6.1) applies to one particle. Because a flowing fluid involves many particles, the next step is to modify the second law so that it applies to a system of particles. To begin the derivation, note that the mass of a particle must be constant. Then, modify Eq. (6.1) to give

$$\left(\sum \mathbf{F} \right)_{\text{ext}} = \frac{d(m\mathbf{v})}{dt} \tag{6.2}$$

Where $m\mathbf{v}$ is the momentum of one particle.

To extend Eq. (6.2) to multiple particles, apply Newton's second law to each particle, and then add the equations together. Internal forces, which are defined as forces between the particles of the system, cancel out, and the result is

$$\left(\sum \mathbf{F} \right)_{\text{ext}} = \frac{d}{dt} \sum_{i=1}^{N} (m_i \mathbf{v}_i) \tag{6.3}$$

where $m_i \mathbf{v}_i$ is the momentum of the ith particle, and $(\Sigma \mathbf{F})_{\text{ext}}$ are forces that are external to the system. Next, let

$$\text{(Total momentum of the system)} \equiv \mathbf{M} = \sum_{i=1}^{N} (m_i \mathbf{v}_i) \tag{6.4}$$

Combine Eqs. (6.3) and (6.4).

$$\left(\sum \mathbf{F} \right)_{\text{ext}} = \frac{d(\mathbf{M})}{dt} \Big|_{\text{closed system}} \tag{6.5}$$

The subscript "closed system" reminds us that Eq. (6.5) is for a closed system.

6.2 The Linear Momentum Equation: Theory

This section shows how to derive the linear momentum equation and explains the physics.

Derivation

Start with Newton's second law for a system of particles (Eq. 6.5). Next, apply the Reynolds transport theorem (Eq. 5.23) to the right side of the equation. The extensive property is momentum, and the corresponding intensive property is the momentum per unit mass which ends up being the velocity. Thus, Reynolds transport theorem gives

$$\frac{d\mathbf{M}}{dt}\bigg|_{\text{closed system}} = \frac{d}{dt}\int_{cv} \mathbf{v}\rho \, d\mathcal{V} + \int_{cs} \mathbf{v}\rho \, \mathbf{V} \cdot \mathbf{dA} \tag{6.6}$$

Combining Eqs. (6.5) and (6.6) gives the *general form* of the *momentum equation*.

$$\left(\sum \mathbf{F}\right)_{\text{ext}} = \frac{d}{dt}\int_{cv} \mathbf{v}\rho \, d\mathcal{V} + \int_{cs} \rho \mathbf{v}(\mathbf{V} \cdot \mathbf{dA}) \tag{6.7}$$

where $(\Sigma \mathbf{F})_{\text{ext}}$ is the sum of external forces acting on the matter in the control volume, $\mathbf{v}$ is fluid velocity relative to an inertial reference frame, and $\mathbf{V}$ is velocity relative to the control surface.

Eq. (6.7) can be simplified. To begin, assume that each particle inside the CV has the same velocity. Thus, the first term on the right side of Eq. (6.7) can be written as

$$\frac{d}{dt}\int_{cv} \mathbf{v}\rho \, d\mathcal{V} = \frac{d}{dt}\left[\mathbf{v}\int_{cv} \rho \, d\mathcal{V}\right] = \frac{d(m_{cv}\mathbf{v}_{cv})}{dt} \tag{6.8}$$

Next, assume that velocity is uniformly distributed as it crosses the control surface. Then, the last term in Eq. (6.7) can be written as

$$\int_{cs} \mathbf{v}\rho \mathbf{V} \cdot \mathbf{dA} = \mathbf{v}\int_{cs} \rho \mathbf{V} \cdot \mathbf{dA} = \sum_{cs} \dot{m}_o \mathbf{v}_o - \sum_{cs} \dot{m}_i \mathbf{v}_i \tag{6.9}$$

Combining Eqs. (6.7) to (6.9) gives the final result:

$$\left(\sum \mathbf{F}\right)_{\text{ext}} = \frac{d(m_{cv}\mathbf{v}_{cv})}{dt} + \sum_{cs} \dot{m}_o \mathbf{v}_o - \sum_{cs} \dot{m}_i \mathbf{v}_i \tag{6.10}$$

where m_{cv} is the mass of the matter that is inside the control volume. The subscripts o and i refer to the outlet and inlet ports, respectively. Eq. (6.10) is the *simplified form* of the momentum equation.

Physical Interpretation of the Momentum Equation

The momentum equation asserts that the sum of forces is exactly balanced by the momentum terms; see Fig. 6.6.

Momentum Flow (Physical Interpretation)

To understand what momentum flow means, select a cylindrical fluid particle passing across a CS (see Fig. 6.7). Let the particle be long enough so that it travels across the CS during a time interval Δt. Then, the particle's length is

$$L = (\text{length}) = \left(\frac{\text{length}}{\text{time}}\right)(\text{time}) = (\text{speed})(\text{time}) = v\Delta t$$

FIGURE 6.6

The conceptual meaning of the momentum equation.

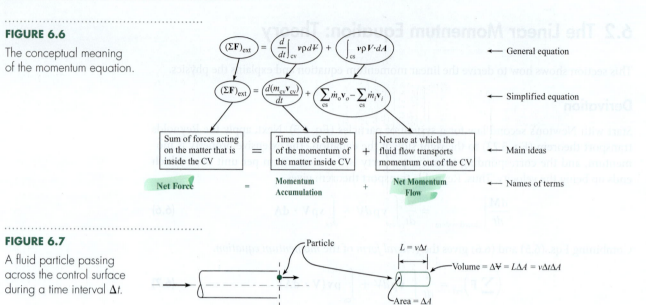

$$\left(\sum \mathbf{F}\right)_{ext} = \left(\frac{d}{dt}\int_{cv} \mathbf{v}\rho\, d\mathcal{V}\right) + \int_{cs} \mathbf{v}\rho V \cdot dA \qquad \leftarrow \text{General equation}$$

$$\left(\sum \mathbf{F}\right)_{ext} = \frac{d(m_{cv}\mathbf{v}_{cv})}{dt} + \sum_{cs}\dot{m}_o\mathbf{v}_o - \sum_{cs}\dot{m}_i\mathbf{v}_i \qquad \leftarrow \text{Simplified equation}$$

| Sum of forces acting on the matter that is inside the CV | = | Time rate of change of the momentum of the matter inside CV | + | Net rate at which the fluid flow transports momentum out of the CV | $\leftarrow$ Main ideas |

Net Force = **Momentum Accumulation** + **Net Momentum Flow** $\leftarrow$ Names of terms

FIGURE 6.7

A fluid particle passing across the control surface during a time interval Δt.

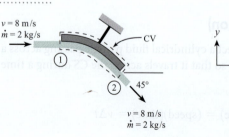

and the particle's volume is $\mathcal{V} = (v\Delta t)\Delta A$. The momentum of the particle is

$$\text{momentum of one particle} = (\text{mass})(\text{velocity}) = (\rho\Delta\mathcal{V})\mathbf{v} = (\rho v\Delta t\Delta A)\mathbf{v}$$

Next, add up the momentum of all particles that are crossing the control surface through a given face.

$$\text{momentum of all particles} = \sum_{cs}(\rho v\Delta t\Delta A)\mathbf{v} \qquad (6.11)$$

Now, let the time interval Δt and the area ΔA approach zero and replace the sum with the integral. Eq. (6.11) becomes

$$\left(\frac{\text{momentum of all particles crossing the CS}}{\text{interval of time}}\right)_{\text{instant in time}} = \int_{cs}(\rho v)\mathbf{v}\, dA$$

Summary Momentum flow describes the rate at which the flowing fluid transports momentum across the control surface.

Momentum Flow (Calculations)

When fluid crosses the control surface, it transports momentum across the CS. At section 1 (Fig. 6.8), momentum is transporting into the CV. At section 2, momentum is transported out of the CV.

FIGURE 6.8

A fluid jet striking a flat vane.

$v = 8$ m/s
$\dot{m} = 2$ kg/s

CV

①

② 45°

y

x

$v = 8$ m/s
$\dot{m} = 2$ kg/s

When the velocity is uniformly distributed across the CS, Eq. (6.10) indicates that the

$$\left(\begin{array}{c}\text{magnitude of}\\\text{momentum flow}\end{array}\right) = \dot{m}v = \rho A v^2 \qquad (6.12)$$

Thus, at section 1, the momentum flow has a magnitude of

$$\dot{m}v = (2\ \text{kg/s})(8\ \text{m/s}) = 16\ \text{kg} \cdot \text{m/s}^2 = 16\ \text{N}$$

and the direction of vector is to the right. Similarly at section 2, the momentum flow has a magnitude of 16 newtons and a direction of 45° below horizontal. From Eq. (6.10), the net momentum flow term is:

$$\dot{m}\mathbf{v}_2 - \dot{m}\mathbf{v}_1 = \{(16\ \text{N})\cos(45°\mathbf{i} - \sin 45°\mathbf{j})\} - \{(16\ \text{N})\mathbf{i}\}$$

Summary For uniform velocity, momentum flow terms have a magnitude $\dot{m}v = \rho A v^2$ and a direction parallel to the velocity vector. The net momentum flow is calculated by subtracting the inlet momentum flow vector(s) from the outlet momentum flow vector(s).

✔ **CHECKPOINT PROBLEM 6.2**

Pressurized air forces water out of a tank. If the air pressure is increased so that the exit speed increases from V to $2V$, what happens to the rate of momentum flow out the bottom of the tank? The rate

  a. Stays the same
  b. Increases by 1x
  c. Increases by 2x
  d. Increases by 3x
  e. Increases by 4x
  f. Increases by 8x

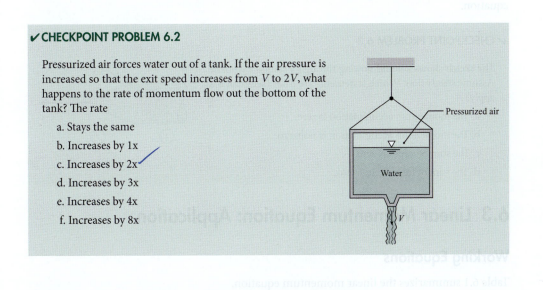

Momentum Accumulation (Physical Interpretation)

To understand what accumulation means, consider a control volume around a nozzle (Fig. 6.9). Then, divide the control volume into many small volumes. Pick one of these small volumes, and note that the momentum inside this volume is $(\rho\Delta\forall)\mathbf{v}$.

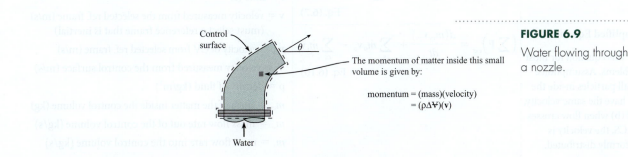

FIGURE 6.9

Water flowing through a nozzle.

The momentum of matter inside this small volume is given by:

$$\text{momentum} = (\text{mass})(\text{velocity})$$
$$= (\rho\Delta\forall)(\mathbf{v})$$

To find the total momentum inside the CV add up the momentum for all the small volumes that comprise the CV. Then, let $\Delta V \to 0$, and use the fact that an integral is the sum of many small terms.

$$\begin{pmatrix} \text{Total momentum} \\ \text{inside the CV} \end{pmatrix} = \sum (\rho \Delta V)\mathbf{v} = \sum \mathbf{v}\rho\Delta V = \int_{cv} \mathbf{v}\rho \, dV \qquad (6.13)$$

Taking the time derivative of Eq. (6.13) gives the final result:

$$\begin{pmatrix} \text{Momentum} \\ \text{Accumulation} \end{pmatrix} = \begin{pmatrix} \text{Rate of change of the} \\ \text{total momentum} \\ \text{inside the CV} \end{pmatrix} = \frac{d}{dt}\int_{cv} \mathbf{v}\rho \, dV \qquad (6.14)$$

Summary Momentum accumulation describes the time rate of change of the momentum inside the CV. For most problems, the accumulation term is zero or negligible. To analyze the momentum accumulation term, one can ask two questions. *Is the momentum of the matter inside the CV changing with time? Is this change significant?* If the answers to both questions are yes, then the momentum accumulation term should be analyzed. Otherwise, the accumulation term can be set to zero.

Checkpoint Problem 6.3 gives you a chance to test your understanding of the momentum equation.

✔ CHECKPOINT PROBLEM 6.3

The sketch shows a liquid flowing through a stationary nozzle. Assume steady flow. Which statements are true? (select all that apply)

   a. The momentum accumulation is zero.

   b. The momentum accumulation is nonzero.

   c. The sum of forces is zero.

   d. The sum of forces is nonzero.

6.3 Linear Momentum Equation: Application

Working Equations

Table 6.1 summarizes the linear momentum equation.

TABLE 6.1 Summary of the Linear Momentum Equation

Description	Equation	Terms
General Equation	$$\left(\sum \mathbf{F}\right)_{ext} = \frac{d}{dt}\int_{cv} \mathbf{v}\rho \, dV + \int_{cs} \rho\mathbf{v}(\mathbf{V}\cdot d\mathbf{A})$$ Eq. (6.7)	$(\sum\mathbf{F})_{ext}$ = sum of external forces (N) t = time (s) $\mathbf{v}$ = velocity measured from the selected ref. frame (m/s) (must select a reference frame that is inertial)
Simplified Equation Use this equation for most problems. Assumptions: (a) all particles inside the CV have the same velocity, and (b) when flow crosses the CS, the velocity is uniformly distributed.	$$\left(\sum \mathbf{F}\right)_{ext} = \frac{d(m_{cv}\mathbf{v}_{cv})}{dt} + \sum_{cs} \dot{m}_o\mathbf{v}_o - \sum_{cs} \dot{m}_i\mathbf{v}_i$$ Eq. (6.10)	$\mathbf{v}_{cv}$ = velocity of CV from selected ref. frame (m/s) $\mathbf{V}$ = velocity measured from the control surface (m/s) ρ = density of fluid (kg/m³) m_{cv} = mass of the matter inside the control volume (kg) $\dot{m}_o$ = mass flow rate out of the control volume (kg/s) $\dot{m}_i$ = mass flow rate into the control volume (kg/s)

Force and Momentum Diagram

The recommended method for applying the momentum equation, the VSM (visual solution method), is illustrated in the next example.

EXAMPLE. This example explains how to apply the VSM for water flowing out a nozzle (Fig. 6.10a). The water enters at section 1 and jets out at section 2.

FIGURE 6.10

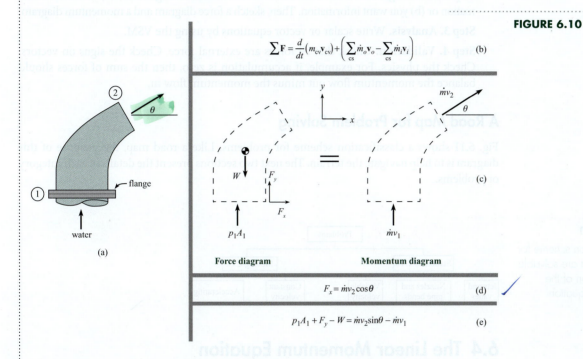

$$\sum \mathbf{F} = \frac{d}{dt}\left(m_{cv}\mathbf{v}_{cv}\right) + \left(\sum_{cs}\dot{m}_o\mathbf{v}_o - \sum_{cs}\dot{m}_i\mathbf{v}_i\right) \tag{b}$$

(c)

Force diagram   Momentum diagram

$$F_x = \dot{m}v_2\cos\theta \tag{d}$$

$$p_1A_1 + F_y - W = \dot{m}v_2\sin\theta - \dot{m}v_1 \tag{e}$$

Step 1. Write the momentum equation (see Fig. 6.10b). Select a control volume that surrounds the nozzle.

Step 2a. To represent the force terms, sketch a *force diagram* (Fig. 6.10c). A **force diagram** illustrates the forces that are acting on the matter that is inside the CV. A force diagram is similar to a *free body diagram* in terms of how it is drawn and how it looks. However, a free-body diagram is an Lagrangian idea, whereas a force diagram is an Eulerian idea. This is why different names are used.

To draw the force diagram, sketch the CV, then sketch the external forces acting on the CV. In Fig. 6.10c, the weight vector, W, represents the weight of the water plus the weight of the nozzle material. The pressure vector, symbolized with p_1A_1, represents the water in the pipe pushing the water through the nozzle. The force vector, symbolized with F_x and F_y, represents the force of the support that is holding the nozzle stationary.

Step 2b. To represent the momentum terms, sketch a **momentum diagram** (Fig. 6.10c). This diagram shows the momentum terms from the right side of the momentum equation. The momentum outflow is represented with $\dot{m}v_2$ and momentum inflow is represented with $\dot{m}v_1$. The momentum accumulation term is zero because the total momentum inside the CV is constant with time.

Step 3. Using the diagrams, write the reduced equations (see Figs. 6.10d and 6.10e).

Summary The *force diagram* shows forces on the CV, and the *momentum diagram* shows the momentum terms. We recommend drawing these diagrams and using the VSM.

A Process for Applying the Momentum Equation

Step 1. Selection. Select the linear momentum equation when the problem involves forces, accelerating fluid particles, and torque does not need to be considered.

Step 2. Sketching. Select a CV so that control surfaces cut through where (a) you know information or (b) you want information. Then, sketch a force diagram and a momentum diagram.

Step 3. Analysis. Write scalar or vector equations by using the VSM.

Step 4. Validation. Check that all forces are external force. Check the signs on vectors. Check the physics. For example, if accumulation is zero, then the sum of forces should balance the momentum flow out minus the momentum flow in.

A Road Map for Problem Solving

Fig. 6.11 shows a classification scheme for problems. Like a road map, the purpose of this diagram is to help navigate the terrain. The next two sections present the details of each category of problems.

FIGURE 6.11

A classification scheme for problems that are solvable by application of the momentum equation.

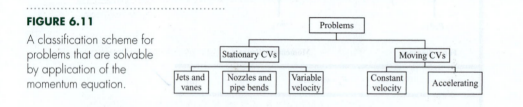

6.4 The Linear Momentum Equation for a Stationary Control Volume

When a CV is stationary with respect to the earth, then the accumulation term is nearly always zero or negligible. Thus, the momentum equations simplifies to

$$\text{(sum of forces)} = \text{(rate of momentum out)} - \text{(rate of momentum in)}$$

Fluid Jets

Problems in the category of **fluid jet** involve a free jet leaving a nozzle. However, analysis of the nozzle itself is not part of the problem. An example of a fluid jet problem is shown in Fig. 6.12. This problem shown involves a water cannon on a cart. The water leaves the nozzle with velocity V, and the goal is to find the tension in the cable.

FIGURE 6.12

A problem involving a fluid jet.

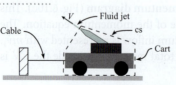

Each category of problems has certain facts that make problem solving easier. These facts will be presented in the form of tips. **Tips** for fluid jet problems are

- When a free jet crosses the control surface, the jet does not exert a force. Thus, do not draw a force on the force diagram. The reason is that the pressure in the jet is ambient pressure, so there is no net force. This can be proven by applying Euler's equation.
- The momentum flow of the fluid jet is $\dot{m}\mathbf{v}$.

Example 6.1 shows a problem in the "fluid jet" category.

EXAMPLE 6.1

Momentum Equation Applied to a Stationary Rocket

Problem Statement

The following sketch shows a 40 g rocket, of the type used for model rocketry, being fired on a test stand to evaluate thrust. The exhaust jet from the rocket motor has a diameter of $d = 1$ cm, a speed of $v = 450$ m/s, and a density of $\rho = 0.5$ kg/m^3. Assume the pressure in the exhaust jet equals ambient pressure. Find the force F_s acting on the support that holds the rocket stationary.

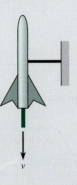

Define the Situation

A small rocket is fired on a test stand.

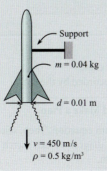

Assumptions. Pressure is 0.0 kPa gage at the nozzle exit plane.

State the Goal

F_s (N) ← Force that acts on the support

Generate Ideas and Make a Plan

Selection. Select the momentum equation because fluid particles are accelerating due to pressures generated by combustion and because force is the goal.

Sketching. Select a CV surrounding the rocket because the control surface cuts

- through the support (where we want information), and
- across the rocket nozzle (where information is known).

Then, sketch a *force diagram* and a *momentum diagram*. Notice that the diagrams include an arrow to indicate the positive y-direction. This is important because the momentum equation is a vector equation.

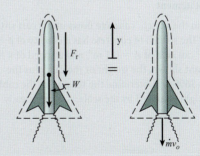

In the force diagram, the body force is the weight (W). The force (F_r) represents the downward push of the support on the rocket. There is no pressure force at the nozzle exit plane because pressure is atmospheric.

Analysis. Apply the momentum equation in vertical direction by selecting terms off the diagrams.

$$F_r + W = \dot{m}v_o \qquad \text{(a)}$$

In Eq. (a), the only unknown is F_r. Thus, the plan is

1. Calculate momentum flow: $\dot{m}v_o = \rho A v_o^2$.
2. Calculate weight.
3. Solve for force F_r. Then, apply Newton's third law.

Take Action (Execute the Plan)

1. Momentum flow.

$$\rho A v^2 = (0.5 \text{ kg/m}^3)(\pi \times 0.01^2 \text{ m}^2/4)(450^2 \text{ m}^2/\text{s}^2)$$

$$= 7.952 \text{ N}$$

2. Weight

$$W = mg = (0.04 \text{ kg})(9.81 \text{ m/s}^2) = 0.3924 \text{ N}$$

3. Force on the rocket (from Eq. (a))

$$F_r = \rho A v_o^2 - W = (7.952 \text{ N}) - (0.3924 \text{ N}) = 7.56 \text{ N}$$

By Newton's third law, the force on the support is equal in magnitude to F_r and opposite in direction.

$$\boxed{F_s = 7.56 \text{ N (upward)}}$$

Review

1. *Knowledge.* Notice that forces acting on the rocket do not sum to zero. This is because the fluid is accelerating.

2. *Knowledge.* For a rocket, the term $\dot{m}v$ is sometimes called a "thrust force." For this example $\dot{m}v = 7.95$ N; this value is typical of a small motor used for model rocketry.

3. *Knowledge.* Newton's third law tells us that forces always occur in pairs, equal in magnitude and opposite in direction. In the sketch below, F_r and F_s are equal in magnitude and opposite in direction.

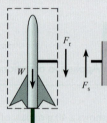

Example 6.2 gives another problem in the category of "fluid jet."

EXAMPLE 6.2

Momentum Equation Applied to a Fluid Jet

Problem Statement

As shown in the sketch, concrete flows into a cart sitting on a scale. The stream of concrete has a density of $\rho = 2400 \text{ kg/m}^3$, an area of $A = 0.1 \text{ m}^2$, and a speed of $v = 3$ m/s. At the instant shown, the weight of the cart plus the concrete is 3600 N. Determine the tension in the cable and the weight recorded by the scale. Assume steady flow.

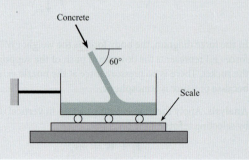

Define the Situation

Concrete is flowing into a cart that is being weighed.

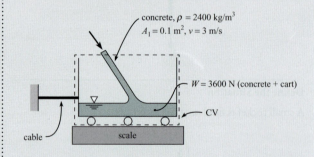

State the Goal

T(N) ← Tension in cable
W_s(N) ← Weight recorded by the scale

Generate Ideas and Make a Plan

Select the momentum equation. Then, select a CV and sketch this in the situation diagram. Next, sketch a force diagram and momentum diagram.

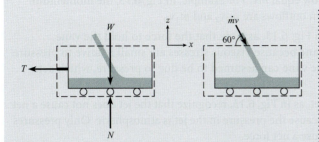

Notice in the force diagram that the liquid jet does not exert a force at the control surface. This is because the pressure in the jet equals atmospheric pressure.

To apply the momentum equation, use the force and momentum diagrams to visualize the vectors.

$$\sum \mathbf{F} = \dot{m}_o \mathbf{v}_o - \dot{m}_i \mathbf{v}_i$$

$$-T\mathbf{i} + (N - W)\mathbf{k} = -\dot{m}v((\cos 60°)\mathbf{i} - (\sin 60°)\mathbf{j})$$

Next, write scalar equations

$$-T = -\dot{m}v \cos 60° \qquad \text{(a)}$$

$$(N - W) = \dot{m}v \sin 60° \qquad \text{(b)}$$

Now, the goals can be solved for. The plan is to:

1. Calculate T using Eq. (a).
2. Calculate N using Eq. (b). Then let $W_s = -N$.

Take Action (Execute the Plan)

1. Momentum equation (horizontal direction)

$$T = \dot{m}v \cos 60° = \rho A v^2 \cos 60°$$

$$T = (2400 \text{ kg/m}^3)(0.1 \text{ m}^2)(3 \text{ m/s})^2 \cos 60°$$

$$= \boxed{1080 \text{ N}}$$

2. Momentum equation (vertical direction)

$$N - W = \dot{m}v \sin 60° = \rho A v^2 \sin 60°$$

$$N = W + \rho A v^2 \sin 60°$$

$$= 3600 \text{ N} + 1871 \text{ N} = \boxed{5471 \text{ N}}$$

Review

1. *Discussion.* The weight recorded by the scale is larger than the weight of the cart because of the momentum carried by the fluid jet.

2. *Discussion.* The momentum accumulation term in this problem is nonzero. However, it was assumed to be small and was neglected.

Vanes

A **vane** is a structural component, typically thin, that is used to turn a fluid jet (Fig. 6.13). A vane is used to idealize many components of engineering interest. Examples include a blade in a turbine, a sail on a ship, and a thrust reverser on an aircraft engine.

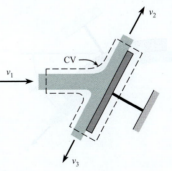

FIGURE 6.13

A fluid jet striking a flat vane.

To make solving of vane problems easier, we offer the following **Tips**.

- **Tip 1.** Assume that $v_1 = v_2 = v_3$. This assumption can be justified with the Bernoulli equation. In particular, assume inviscid flow and neglect elevation changes, and the Bernoulli equation can be used to prove that the velocity of the fluid jet is constant.

- **Tip 2.** Let each momentum flow equal $\dot{m}\mathbf{v}$. For example, in Fig. 6.13, the momentum inflow is $\dot{m}_1\mathbf{v}_1$. The momentum outflows are $\dot{m}_2\mathbf{v}_2$ and $\dot{m}_3\mathbf{v}_3$.
- **Tip 3.** If the vane is flat, as in Fig. 6.13, assume that the force to hold the vane stationary is normal to the vane because viscous stresses are small relative to pressure stresses. Thus, the load on the vane can assumed to be due to pressure, which acts normal to the vane.
- **Tip 4.** When the jet is a free jet, as in Fig. 6.13, recognize that the jet does not cause a net force at the control surface because the pressure in the jet is atmospheric. Only pressures different than atmospheric cause a net force.

EXAMPLE 6.3

Momentum Equation Applied to a Vane

Problem Statement

A water jet ($\rho = 1000 \text{ kg/m}^3$) is deflected 60° by a stationary vane as shown in the figure. The incoming jet has a speed of 30 m/s and a diameter of 3 cm. Find the force exerted by the jet on the vane.

Define the Situation

A water jet is deflected by a vane.

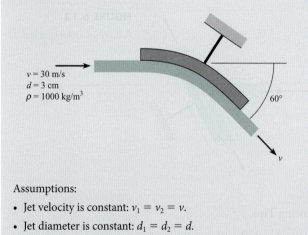

$v = 30$ m/s
$d = 3$ cm
$\rho = 1000$ kg/m³

Assumptions:

- Jet velocity is constant: $v_1 = v_2 = v$.
- Jet diameter is constant: $d_1 = d_2 = d$.
- Neglect gravitational effects.

State the Goal

$\mathbf{F}_{jet}(\text{N}) \leftarrow$ Force of the fluid jet on the vane

---

Generate Ideas and Make a Plan

Select. Because force is a parameter and fluid particles accelerate as the jet turns, select the linear momentum equation.

Sketch. Select a CV that cuts through support so that the force of the support can be found. Then, sketch a force diagram and a momentum diagram.

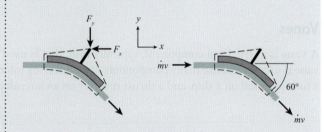

In the force and momentum diagrams, notice that

- Pressure forces are zero because pressures in the water jet at the control surface are zero gage.
- Each momentum flow is represented with $\dot{m}v$.

Analysis. To apply the momentum equation, use the force and momentum diagrams to write a vector equation.

$$\sum \mathbf{F} = \dot{m}_o\mathbf{v}_o - \dot{m}_i\mathbf{v}_i$$

$$(-F_x)\mathbf{i} + (-F_y)\mathbf{j} = \dot{m}v(\cos 60°\,\mathbf{i} - \sin 60°\,\mathbf{j}) - \dot{m}v\,\mathbf{i}$$

Now, write scalar equations

$$-F_x = \dot{m}v(\cos 60° - 1) \tag{a}$$

$$-F_y = -\dot{m}v(\sin 60°) \tag{b}$$

Because there is enough information to solve Eqs. (a) and (b), the problem is cracked. The plan is

1. Calculate $\dot{m}v$.
2. Apply Eq. (a) to calculate F_x.
3. Apply Eq. (b) to calculate F_y.
4. Apply Newton's third law to find the force of the jet.

Take Action (Execute the Plan)

1. Momentum flow rate.

$$\dot{m}v = (\rho A v)v$$
$$= (1000 \text{ kg/m}^3)(3.14 \times 0.015^2 \text{ m}^2)(30 \text{ m/s})^2$$
$$= 636.3 \text{ N}$$

2. Linear momentum equation (x-direction)

$$F_x = \dot{m}v(1 - \cos 60°)$$
$$= (636.3 \text{ N})(1 - \cos 60°)$$
$$F_x = 318.5 \text{ N}$$

3. Linear momentum equation (y-direction)

$$F_y = \dot{m}v \sin 60°$$
$$= (636.3 \text{ N}) \sin 60°$$
$$F_y = 551 \text{ N}$$

4. Newton's third law

The force of the jet on the vane ($\mathbf{F}_{jet}$) is opposite in direction to the force required to hold the vane stationary ($\mathbf{F}$). Therefore,

$$\mathbf{F}_{jet} = (318.5 \text{ N})\mathbf{i} + (551 \text{ N})\mathbf{j}$$

Review

1. *Discussion.* Notice that the problem goal was specified as a vector. Thus, the answer was given as a vector.

2. *Skill.* Notice how the common assumptions for a vane were applied in the "define the situation" portion.

Nozzles

Nozzles are flow devices used to accelerate a fluid stream by reducing the cross-sectional area of the flow (Fig. 6.14). Problems in this category involve analysis of the nozzle itself, not analysis of the free jet.

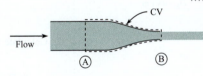

FIGURE 6.14

A fluid jet exiting a nozzle.

To make solving of nozzle problems easier, we offer the following **Tips**.

- **Tip 1.** Let each momentum flow equal $\dot{m}\mathbf{v}$. For the nozzle in Fig. 6.14, the momentum inflow is $\dot{m}\mathbf{v}_A$ and the outflow is $\dot{m}\mathbf{v}_B$.
- **Tip 2.** Include a pressure force where the nozzle connects to a pipe. For the nozzle in Fig. 6.14, include a pressure force of magnitude $p_A A_A$ on the force diagram. This pressure force, like all pressure forces, is compressive.
- **Tip 3.** To find p_A, apply the Bernoulli equation between A and B.
- **Tip 4.** To relate v_A and v_B, apply the continuity equation.
- **Tip 5.** When the CS cuts through a support structure (e.g., a pipe wall, a flange), represent the associated force on the force diagram. For the nozzle shown in Fig. 6.14, add a force F_{Ax} and F_{Ay} to the force diagram.

EXAMPLE 6.4

Momentum Equation Applied to a Nozzle

Problem Statement

The sketch shows air flowing through a nozzle. The inlet pressure is $p_1 = 105$ kPa abs, and the air exhausts into the atmosphere, where the pressure is 101.3 kPa abs. The nozzle has an inlet diameter of 60 mm and an exit diameter of 10 mm, and the nozzle is connected to the supply pipe by flanges. Find the force required to hold the nozzle stationary. Assume the air has a constant density of 1.22 kg/m³. Neglect the weight of the nozzle.

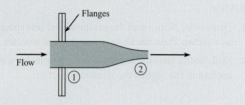

Define the Situation

Air flows through a nozzle

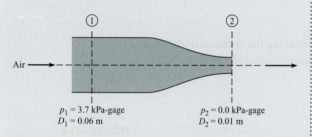

$p_1 = 3.7$ kPa-gage
$D_1 = 0.06$ m

$p_2 = 0.0$ kPa-gage
$D_2 = 0.01$ m

Properties. $\rho = 1.22$ kg/m³.

Assumptions

- Weight of nozzle is negligible.
- Steady flow, constant density flow, inviscid flow.

State the Goals

$\mathbf{F}(N)$ ⬅ Force required to hold nozzle stationary

Generate Ideas and Make a Plan

Select. Because force is a parameter and fluid particles are accelerating in the nozzle, select the momentum equation.

Sketch. Sketch a force and momentum diagram.

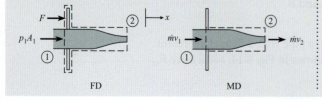

FD   MD

Write the momentum equation (x-direction)

$$F + p_1A_1 = \dot{m}(v_2 - v_1) \qquad (a)$$

To solve for F, we need v_2 and v_1, which can be found using the Bernoulli equation. Thus, the plan is

1. Derive an equation for v_2 by applying the Bernoulli equation and the continuity equation.
2. Calculate v_2 and v_1.
3. Calculate F by applying Eq. (a).

Take Action (Execute the Plan)

1. *Bernoulli Equation* (apply between 1 and 2)

$$p_1 + \gamma z_1 + \frac{1}{2}\rho v_1^2 = p_2 + \gamma z_2 + \frac{1}{2}\rho v_2^2$$

Term-by-term analysis

- $z_1 = z_2 = 0$
- $p_1 = 3.7$ kPa; $p_2 = 0.0$

The Bernoulli equation reduces to

$$p_1 + \rho v_1^2/2 = \rho v_2^2/2$$

Continuity Equation. Select a CV that cuts through sections 1 and 2. Neglect the mass accumulation terms. Continuity simplifies to

$$v_1A_1 = v_2A_2$$
$$v_1d_1^2 = v_2d_2^2$$

Substitute into the Bernoulli equation and solve for v_2:

$$v_2 = \sqrt{\frac{2p_1}{\rho(1 - (d_2/d_1)^4)}}$$

2. Calculate v_2 and v_1.

$$v_2 = \sqrt{\frac{2 \times 3.7 \times 1000 \text{ Pa}}{(1.22 \text{ kg/m}^3)(1 - (10/60)^4)}} = 77.9 \text{ m/s}$$

$$v_1 = v_2\left(\frac{d_2}{d_1}\right)^2$$

$$= 77.9 \text{ m/s} \times \left(\frac{1}{6}\right)^2 = 2.16 \text{ m/s}$$

3. Momentum equation

$$F + p_1A_1 = \dot{m}(v_2 - v_1)$$
$$F = \rho A_1 v_1(v_2 - v_1) - p_1A_1$$

$$= (1.22 \text{ kg/m}^3)\left(\frac{\pi}{4}\right)(0.06 \text{ m})^2(2.16 \text{ m/s})$$

$$\times (77.9 - 2.16)(\text{m/s})$$

$$- 3.7 \times 1000 \text{ N/m}^2 \times \left(\frac{\pi}{4}\right)(0.06 \text{ m})^2$$

$$= 0.564 \text{ N} - 10.46 \text{ N} = -9.90 \text{ N}$$

Because *F* is negative, the direction is opposite to the direction assumed on the force diagram. Thus,

> Force to hold nozzle = 9.90 N (← direction)

Review

1. *Knowledge.* The direction initially assumed for the force on a force diagram is arbitrary. If the answer for the force is negative, then the force acts in a direction opposite the chosen direction.

2. *Knowledge.* Pressures were changed to gage pressure in the "define the situation" operation because it is the pressures differences as compared to atmospheric pressure that cause net pressure forces.

Pipe Bends

A **pipe bend** is a structural component that is used to turn through an angle (Fig. 6.15). A pipe bend is often connected to straight runs of pipe by flanges. A flange is round disk with a hole in the center that slides over a pipe and is often welded in place. Flanges are bolted together to connect sections of pipe.

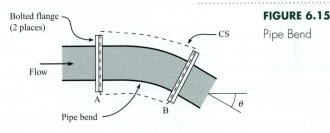

FIGURE 6.15

Pipe Bend

To make solving of nozzle problems easier, we offer the following **Tips**.

- **Tip 1.** Let each momentum flow equal $\dot{m}\mathbf{v}$. For the bend in Fig. 6.15, the momentum inflow is $\dot{m}\mathbf{v}_A$ and the outflow is $\dot{m}\mathbf{v}_B$.

- **Tip 2.** Include pressure forces where the CS cuts through a pipe. In Fig. 6.15, there is a pressure force at section A: $F_A = p_A A_A$ and at section B: $F_B = p_B A_B$. As always, both pressure forces are compressive.

- **Tip 3.** To relate p_A and p_B, it is most correct to apply the *energy equation* from Chapter 7 and include head loss. An alternative is to assume that pressure is constant or to assume inviscid flow and apply the Bernoulli equation.

- **Tip 4.** To relate v_A and v_B, apply the continuity equation.

- **Tip 5.** When the CS cuts through a support structure (pipe wall, flange), include the loads caused by the support on the force diagram.

EXAMPLE 6.5

Momentum Equation Applied to a Pipe Bend

Problem Statement

A 1-m-diameter pipe bend shown in the diagram is carrying crude oil (*S* = 0.94) with a steady flow rate of 2 m³/s. The bend has an angle of 30° and lies in a horizontal plane. The volume of oil in the bend is 1.2 m³, and the empty weight of the bend is 4 kN. Assume the pressure along the centerline of the bend is constant with a value of 75 kPa gage. Find the force required to hold the bend in place.

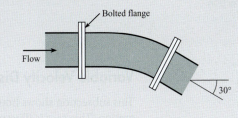

Define the Situation

Crude oil flows through a pipe bend.

- Bend lies in a horizontal plane.
- $\mathcal{V}_{oil} = 1.2 \text{ m}^3$ = volume of oil in bend.
- $W_{bend} = 4000 \text{ N}$ = empty weight of bend.
- $p = 75 \text{ kPa-gage}$ = pressure along centerline.

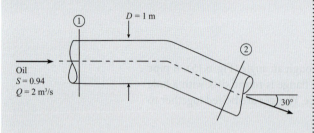

State the Goal

$F(N)$ ◀— Force to hold the bend stationary.

Generate Ideas and Make a Plan

Select. Because force is a parameter and fluid particles accelerate in the pipe bend, select the momentum equation.

Sketch. Select a CV that cuts through the support structure and through sections 1 and 2. Then, sketch the force and momentum diagrams.

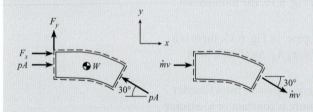

Analysis. Using the diagrams as guides, write the momentum equation in each direction:

- x-direction

$$F_x + p_1 A_1 - p_2 A_2 \cos 30° = \dot{m} v_2 \cos 30° - \dot{m} v_1 \quad \text{(a)}$$

- y-direction

$$F_y - p_2 A_2 \sin 30° = -\dot{m} v_2 \sin 30° \quad \text{(b)}$$

- z-direction

$$-F_z - W_{total} = 0 \quad \text{(c)}$$

Review these equations and notice that there is enough information to solve for the goals F_x, F_y, and F_z. Thus, create a plan.

1. Calculate the momentum flux $\dot{m}v$.
2. Calculate the pressure force pA.
3. Solve Eq. (a) for F_x.
4. Solve Eq. (b) for F_y.
5. Solve Eq. (c) for F_z.

Take Action (Execute the Plan)

1. Momentum Flow

Example 6.6

- Apply the volume flow rate equation

$$v = Q/A = \frac{(2 \text{ m}^3/s)}{(\pi \times 0.5^2 \text{ m}^2)} = 2.55 \text{ m/s}$$

- Next, calculate the momentum flow

$$\dot{m}v = \rho Q v = (0.94 \times 1000 \text{ kg/m}^3)(2 \text{ m}^3/s)(2.55 \text{ m/s})$$
$$= 4.79 \text{ kN}$$

2. Pressure Force

$$pA = (75 \text{ kN/m}^2)(\pi \times 0.5^2 \text{ m}^2) = 58.9 \text{ kN}$$

3. Momentum Equation (x-direction)

$$F_x + p_1 A_1 - p_2 A_2 \cos 30° = \dot{m} v_2 \cos 30° - \dot{m} v_1$$
$$F_x = -pA(1 - \cos 30°) - \dot{m}v(1 - \cos 30°)$$
$$= -(pA + \dot{m}v)(1 - \cos 30°)$$
$$= -(58.9 + 4.79)(\text{kN})(1 - \cos 30°)$$
$$= -8.53 \text{ kN}$$

4. Momentum Equation (y-direction)

$$F_y + p_2 A_2 \sin 30° = -\dot{m} v_2 \sin 30°$$
$$F_y = -(pA + \dot{m}v) \sin 30°$$
$$= -(58.9 + 4.79)(\text{kN})(\sin 30°) = -31.8 \text{ kN}$$

Reaction force in z-direction. (The bend weight includes the oil plus the empty pipe).

$$-F_z - W_{total} = 0$$

$$W = \gamma \mathcal{V} + 4 \text{ kN}$$
$$= (0.94 \times 9.81 \text{ kN/m}^3)(1.2 \text{ m}^3) + 4 \text{ kN} = 15.1 \text{ kN}$$

Force to hold the bend

$$\boxed{\mathbf{F} = (-8.53 \text{ kN})\mathbf{i} + (-31.8 \text{ kN})\mathbf{j} + (15.1 \text{ kN})\mathbf{k}}$$

Variable Velocity Distribution

This subsection shows how to solve a problem when the momentum flow is evaluated by integration. This case is illustrated by Example 6.6.

EXAMPLE 6.6

Momentum Equation Applied with a Variable Velocity Distribution

Problem Statement

The drag force of a bullet-shaped device may be measured using a wind tunnel. The tunnel is round with a diameter of 1 m, the pressure at section 1 is 1.5 kPa gage, the pressure at section 2 is 1.0 kPa gage, and air density is 1.0 kg/m³. At the inlet, the velocity is uniform with a magnitude of 30 m/s. At the exit, the velocity varies linearly as shown in the sketch. Determine the drag force on the device and support vanes. Neglect viscous resistance at the wall, and assume pressure is uniform across sections 1 and 2.

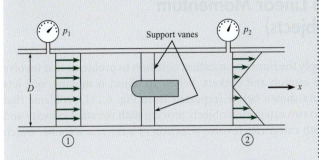

Define the Situation

Data is supplied for wind tunnel test (see above).
Assume. Steady flow.
Air: $\rho = 1.0$ kg/m³.

State the Goal

Find: Drag force (in newtons) on model

Make a Plan

1. Select a control volume that encloses the model.
2. Sketch the force diagram.
3. Sketch the momentum diagram.
4. The downstream velocity profile is not uniformly distributed. Apply the integral form of the momentum equation, Eq. (6.7).
5. Evaluate the sum of forces.
6. Determine velocity profile at section 2 by application of continuity equation.
7. Evaluate the momentum terms.
8. Calculate drag force on model.

Take Action (Execute the Plan)

1. The control volume selected is shown. The control volume is stationary.

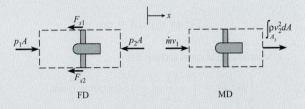

FD   MD

2. The forces consist of the pressure forces and the force on the model support struts cut by the control surface. The drag force on the model is equal and opposite to the force on the support struts: $F_D = F_{s1} + F_{s2}$.

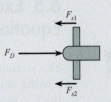

3. There is inlet and outlet momentum flux.
4. Integral form of momentum equation in x-direction

$$\sum F_x = \frac{d}{dt}\int_{cv} \rho v_x \, d\Psi + \int_{cs} \rho v_x (\mathbf{V} \cdot d\mathbf{A})$$

On cross section 1, $\mathbf{V} \cdot d\mathbf{A} = -v_x dA$, and on cross section 2, $\mathbf{V} \cdot d\mathbf{A} = v_x dA$, so

$$\sum F_x = \frac{d}{dt}\int_{cv} \rho v_x \, d\Psi - \int_1 \rho v_x^2 \, dA + \int_2 \rho v_x^2 \, dA$$

5. Evaluation of force terms.

$$\sum F_x = p_1 A - p_2 A - (F_{s1} + F_{s2})$$
$$= p_1 A - p_2 A - F_D$$

6. Velocity profile at section 2.

Velocity is linear in radius, so choose $v_2 = v_1 K(r/r_o)$, where r_o is the tunnel radius and K is a proportionality factor to be determined.

$$Q_1 = Q_2$$

$$A_1 v_1 = \int_{A_2} v_2(r) \, dA = \int_0^{r_o} v_1 K(r/r_o) 2\pi r \, dr$$

$$\pi r_o^2 v_1 = 2\pi v_1 K \frac{1}{3} r_o^2$$

$$K = \frac{3}{2}$$

7. Evaluation of momentum terms

- Accumulation term for steady flow is $\dfrac{d}{dt}\displaystyle\int_{cv}\rho v_x\,d\mathcal{V}=0$

- Momentum at cross section 1 with $v_x = v_1$ is

$$\int_1 \rho v_x^2\,dA = \rho v_1^2 A = \dot{m}v_1$$

- Momentum at cross section 2 is

$$\int_2 \rho v_x^2\,dA = \int_0^{r_o}\rho\left[\frac{3}{2}v_1\left(\frac{r}{r_o}\right)\right]^2 2\pi r\,dr = \frac{9}{8}\dot{m}v_1$$

8. Drag force

$$p_1 A - p_2 A - F_D = \dot{m}v_1\left(\frac{9}{8}-1\right)$$

$$F_D = (p_1 - p_2)A - \frac{1}{8}\rho A v_1^2$$

$$= (\pi \times 0.5^2\ \text{m}^2)(1.5 - 1.0)(10^3)\text{N/m}^2$$

$$-\frac{1}{8}(1\ \text{kg/m}^3)(\pi \times 0.5^2\ \text{m}^2)(30\ \text{m/s})^2$$

$$F_D = \boxed{304\ \text{N}}$$

6.5 Examples of the Linear Momentum Equation (Moving Objects)

This section describes how to apply the linear momentum equation to problems that involve moving objects such as carts in motion and rockets. When an object is moving, one lets the CV move with the object. As shown below (repeated from Fig. 6.11), problems that involve moving CVs classify into two categories: objects moving with constant velocity and objects that are accelerating. Both categories involve selection of a reference frame, which is the next topic.

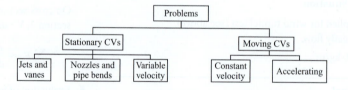

Reference Frame

When an object is moving, it is necessary to specify a reference frame. A **reference frame** is a three-dimensional framework from which an observer takes measurements. For example, Fig. 6.16 shows a rocket in flight. For this situation, one possible reference frame is fixed to the earth. Another possible reference frame is fixed to the rocket. Observers in these two frames of reference would report different values of the rocket velocity V_{Rocket} and the velocity of the fluid jet V_{jet}. The ground-based reference frame is inertial. An **inertial reference frame** is any reference frame that is stationary or moving with constant velocity with respect to the earth. Thus, an inertial reference frame is a nonaccelerating reference frame. Alternatively, a **noninertial reference frame** is any reference frame that is accelerating.

Regarding the linear momentum equation as presented in this text, this equation is only valid for an inertial frame. Thus, when objects are moving, the engineer should specify an inertial reference frame.

FIGURE 6.16

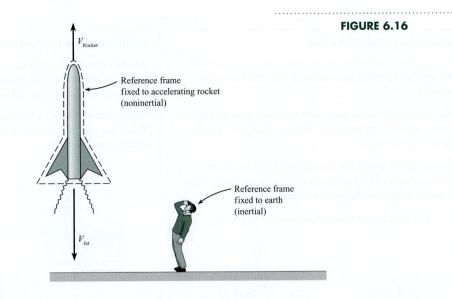

Analyzing a Moving Body (Constant Velocity)

When an object is moving with constant velocity, then the reference frame can be placed on the moving object or fixed to the earth. However, most problems are simpler if the frame is fixed to the moving object. Example 6.7 shows how to solve a problem involving an object moving with constant velocity.

EXAMPLE 6.7

Momentum Equation Applied to a Moving CV

Problem Statement

A stationary nozzle produces a water jet with a speed of 50 m/s and a cross-sectional area of 5 cm². The jet strikes a moving block and is deflected 90° relative to the block. The block is sliding with a constant speed of 25 m/s on a surface with friction. The density of the water is 1000 kg/m³. Find the frictional force F acting on the block.

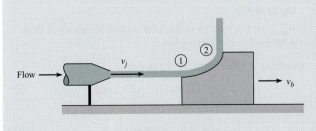

Define the Situation

A block slides at constant velocity due to a fluid jet.

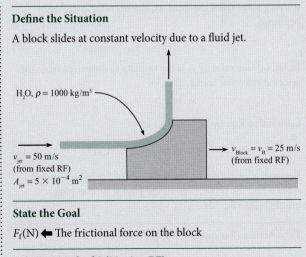

H_2O, $\rho = 1000 \text{ kg/m}^3$

$v_{jet} = 50$ m/s
(from fixed RF)
$A_{jet} = 5 \times 10^{-4} \text{ m}^2$

$v_{Block} = v_B = 25$ m/s
(from fixed RF)

State the Goal

$F_f(\text{N})$ ◀ The frictional force on the block

Solution Method I (Moving RF)

When a body is moving at constant velocity, the easiest way to solve the problem is to put the RF on the moving body. This solution method is shown first.

Generate Ideas and Make a Plan

Select the linear momentum equation because force is the goal and fluid particles accelerate as they interact with the block.

Select a moving CV that surrounds the block because this CV involves known parameters (i.e., the two fluid jets) and the goal (frictional force).

Because the CV is moving at a constant velocity, select a reference frame (RF) that is fixed to the moving block. This RF makes analysis of the problem simpler.

Sketch the force and momentum diagrams and the RF.

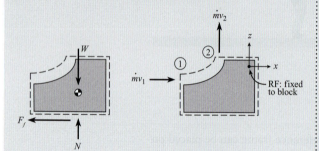

To apply the momentum equation, use the force and momentum diagrams to visualize the vectors. The momentum equation in the x-direction is

$$-F_f = -\dot{m}v_1 \tag{a}$$

In Eq. (a), the mass flow rate describes the rate at which mass is crossing the control surface. Because the CS is moving away from the fluid jet, the mass flow rate term becomes

$$\dot{m} = \rho A V = \rho A_{\text{jet}}(v_{\text{jet}} - v_{\text{block}}) \tag{b}$$

In Eq. (a), the velocity v_1 is the velocity as measured from the selected reference frame. Thus,

$$v_1 = v_{\text{jet}} - v_{\text{block}} \tag{c}$$

Combining Eqs. (a), (b), and (c) gives

$$F_f = \dot{m}v_1 = \rho A_{\text{jet}}(v_{\text{jet}} - v_{\text{block}})^2 \tag{d}$$

Because, all variables on the right side of Eq. (d) are known, we can find the problem goal. The plan is simple: plug numbers into Eq. (d).

Take Action (Execute the Plan)

$$F_f = \rho A_{\text{jet}}(v_{\text{jet}} - v_{\text{block}})^2$$
$$F_f = (1000 \text{ kg/m}^2)(5 \times 10^{-4} \text{ m}^2)(50 - 25)^2 (\text{m/s})^2$$
$$\boxed{F_f = 312 \text{ N}}$$

Solution Method II (Fixed RF)

Another way to solve this problem is to use a fixed reference frame. To implement this approach, sketch the force diagram, the momentum diagram, and the selected RF.

Notice that $\dot{m}v_2$ shows a vertical and horizontal component. This is because an observer in the selected RF would see these velocity components.

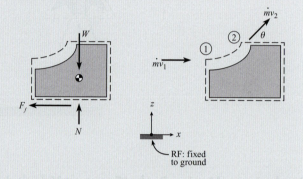

From the diagrams, one can write the momentum equation in the x-direction:

$$-F_f = \dot{m}v_2 \cos\theta - \dot{m}v_1$$
$$F_f = \dot{m}(v_1 - v_2 \cos\theta) \tag{e}$$

In the momentum equation, the mass flow rate is measured relative to the control surface. Thus, $\dot{m}$ is independent of the RF, and one can use Eq. (b), which is repeated below:

$$\dot{m} = \rho A V = \rho A_{\text{jet}}(v_{\text{jet}} - v_{\text{block}}) \tag{f}$$

In Eq. (e), the velocity v_1 is the velocity as measured from the selected reference frame. Thus,

$$v_1 = v_{\text{jet}} \tag{g}$$

To analyze v_2, relate velocities by using a relative-velocity equation from a Dynamics Text:

$$v_{\text{jet}} = v_{\text{block}} + v_{\text{jet/block}} \tag{h}$$

where

• $v_2 = v_{\text{jet}}$ is the velocity of the jet at section 2 as measured from the fixed RF.

• v_{block} is the velocity of the moving block as measured from the fixed RF.

• $v_{\text{jet/block}}$ is the velocity of the jet at section as measured from a RF fixed to the moving block.

Substitute numbers into Eq. (h) to give

$$\mathbf{v}_2 = (25 \text{ m/s})\mathbf{i} + (25 \text{ m/s})\mathbf{j} \tag{i}$$

Thus

$$v_2 \cos\theta = v_{2x} = 25 \text{ m/s} = v_{\text{block}} \tag{j}$$

Substitute Eqs. (f), (g), and (j) into Eq. (e).

$$F_f = \{\dot{m}\}(v_1 - v_2 \cos\theta)$$
$$= \{\rho A_{jet}(v_{jet} - v_{block})\}(v_{jet} - v_{block}) \qquad \text{(k)}$$
$$= \rho A_{jet}(v_{jet} - v_{block})^2$$

Eq. (k) is identical to Eq. (d). Thus, *Solution Method I* is equivalent to *Solution Method II*.

Review the Solution and the Process

1. *Knowledge.* When an object moves with constant velocity, select an RF fixed to the moving object because this is much easier than selecting an RF fixed to the earth.

2. *Knowledge.* Specifying the control volume and the reference frame are independent decisions.

Analyzing a Moving Body (Accelerating)

This section presents an example of an accelerating object, namely the analysis of a rocket (Fig. 6.17). To begin, sketch a control volume around the rocket. Note that the reference frame cannot be fixed to the rocket because the rocket is accelerating.

Assume the rocket is moving vertically upward with a speed v_r measured with respect to the ground. Exhaust gases leave the engine nozzle (area A_e) at a speed V_e relative to the rocket nozzle with a gage pressure of p_e. The goal is to obtain the equation of motion of the rocket.

The control volume is drawn around and accelerates with the rocket. The force and momentum diagrams are shown in Fig. 6.18. There is a drag force of D and a weight of W acting downward. There is a pressure force of $p_e A_e$ on the nozzle exit plane because the pressure in a supersonic jet is greater than ambient pressure. The summation of the forces in the z-direction is

$$\sum F_z = p_e A_e - W - D \qquad \text{(6.15)}$$

FIGURE 6.17

Vertical launch of rocket.

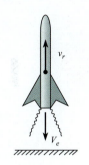

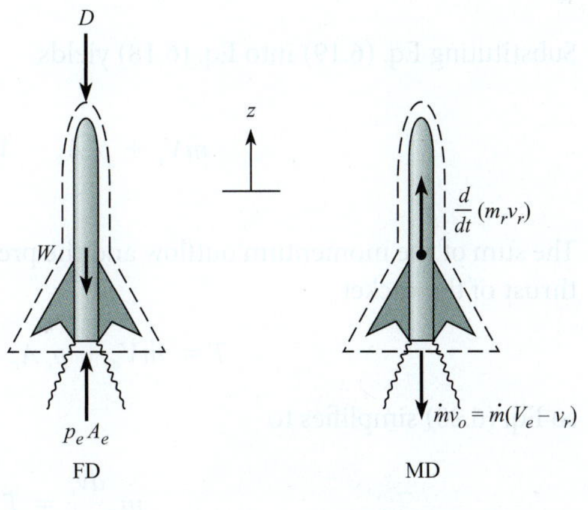

FIGURE 6.18

Force and momentum diagrams for rocket.

There is only one momentum flux out of the rocket nozzle, $\dot{m}v_o$. The speed v_o must be referenced to an inertial reference frame, which in this case is chosen as the ground. The speed of the exit gases with respect to the ground is

$$v_o = (V_e - v_r) \qquad \text{(6.16)}$$

because the rocket is moving upward with speed v_r with respect to the ground, and the exit gases are moving downward at speed V_e with respect to the rocket.

The momentum equation, in the z-direction is

$$\sum F_z = \frac{d}{dt}\int_{cv} v_z \rho\, d\forall + \sum_{cs} \dot{m}_o v_{oz} - \sum_{cs} \dot{m}_i v_{iz}$$

The velocity inside the control volume is the speed of the rocket, v_r, so the accumulation term becomes

$$\frac{d}{dt}\left(\int_{cv} v_z \rho\, d\forall\right) = \frac{d}{dt}\left[v_r \int_{cv} \rho\, d\forall\right] = \frac{d}{dt}(m_r v_r)$$

Substituting the sum of the forces and momentum terms into the momentum equation gives

$$p_e A_e - W - D = \frac{d}{dt}(m_r v_r) - \dot{m}(V_e - v_r) \tag{6.17}$$

Next, apply the product rule to the accumulation term. This gives

$$p_e A_e - W - D = m_r \frac{dv_r}{dt} + v_r\left(\frac{dm_r}{dt} + \dot{m}\right) - \dot{m}V_e \tag{6.18}$$

The continuity equation can now be used to eliminate the second term on the right. Applying the continuity equation to the control surface around the rocket leads to

$$\frac{d}{dt}\int_{cv} \rho\, d\forall + \sum \dot{m}_o - \sum \dot{m}_i = 0$$

$$\frac{dm_r}{dt} + \dot{m} = 0 \tag{6.19}$$

Substituting Eq. (6.19) into Eq. (6.18) yields

$$\dot{m}V_e + p_e A_e - W - D = m_r \frac{dv_r}{dt} \tag{6.20}$$

The sum of the momentum outflow and the pressure force at the nozzle exit is identified as the thrust of the rocket

$$T = \dot{m}V_e + p_e A_e = \rho_e A_e V_e^2 + p_e A_e$$

so Eq. (6.20) simplifies to

$$m_r \frac{dv_r}{dt} = T - D - W \tag{6.21}$$

which is the equation used to predict and analyze rocket performance.

Integration of Eq. (6.21) leads to one of the fundamental equations for rocketry: the burnout velocity or the velocity achieved when all the fuel is burned. Neglecting the drag and weight, the equation of motion reduces to

$$T = m_r \frac{dv_r}{dt} \tag{6.22}$$

The instantaneous mass of the rocket is given by $m_r = m_i - \dot{m}t$, where m_i is the initial rocket mass and t is the time from ignition. Substituting the expression for mass into Eq. (6.22) and integrating with the initial condition $v_r(0) = 0$ results in

$$v_{bo} = \frac{T}{\dot{m}} \ln \frac{m_i}{m_f} \tag{6.23}$$

where v_{bo} is the burnout velocity and m_f is the final (or payload) mass. The ratio $T/\dot{m}$ is known as the specific impulse, I_{sp}, and has units of velocity.

6.6 The Angular Momentum Equation

This section presents the *angular momentum equation*, which is also called the *moment-of-momentum equation*. The angular momentum equation is very useful for situations that involve torques. Examples include analyses of rotating machinery such as pumps, turbines, fans, and blowers.

Derivation of the Equation

Newton's second law of motion can be used to derive an equation for the rotational motion of a system of particles:

$$\sum \mathbf{M} = \frac{d(\mathbf{H}_{sys})}{dt} \tag{6.24}$$

where $\mathbf{M}$ is a moment and $\mathbf{H}_{sys}$ is the total angular momentum of all mass forming the system.

To convert Eq. (6.24) to an Eulerian equation, apply the Reynolds transport theorem, Eq. (5.23). The extensive property B_{sys} becomes the angular momentum of the system: $B_{sys} = \mathbf{H}_{sys}$. The intensive property b becomes the angular momentum per unit mass. The angular momentum of an element is $\mathbf{r} \times m\mathbf{v}$, and so $b = \mathbf{r} \times \mathbf{v}$. Substituting for B_{sys} and b in Eq. (5.23) gives

$$\frac{d(\mathbf{H}_{sys})}{dt} = \frac{d}{dt} \int_{cv} (\mathbf{r} \times \mathbf{v}) \rho \, d\mathcal{V} + \int_{cs} (\mathbf{r} \times \mathbf{v}) \rho \mathbf{V} \cdot d\mathbf{A} \tag{6.25}$$

Combining Eqs. (6.24) and (6.25) gives the integral form of the *moment-of-momentum equation*:

$$\sum \mathbf{M} = \frac{d}{dt} \int_{cv} (\mathbf{r} \times \mathbf{v}) \rho \, d\mathcal{V} + \int_{cs} (\mathbf{r} \times \mathbf{v}) \rho \mathbf{V} \cdot d\mathbf{A} \tag{6.26}$$

where $\mathbf{r}$ is a position vector that extends from the moment center, $\mathbf{V}$ is flow velocity relative to the control surface, and $\mathbf{v}$ is flow velocity relative to the inertial reference frame selected.

If the mass crosses the control surface through a series of inlet and outlet ports with uniformly distributed properties across each port, the moment-of-momentum equation becomes

$$\sum \mathbf{M} = \frac{d}{dt} \int_{cv} (\mathbf{r} \times \mathbf{v}) \rho \, d\mathcal{V} + \sum_{cs} \mathbf{r}_o \times (\dot{m}_o \mathbf{v}_o) - \sum_{cs} \mathbf{r}_i \times (\dot{m}_i \mathbf{v}_i) \tag{6.27}$$

The moment-of-momentum equation has the following physical interpretation:

$$\begin{pmatrix} \text{sum of} \\ \text{moments} \end{pmatrix} = \begin{pmatrix} \text{angular momentum} \\ \text{accumulation} \end{pmatrix} + \begin{pmatrix} \text{angular momentum} \\ \text{outflow} \end{pmatrix} - \begin{pmatrix} \text{angular momentum} \\ \text{inflow} \end{pmatrix}$$

Application

The process for applying the angular momentum equation is similar to the process for applying the linear momentum equation. To illustrate this process, Example 6.8 shows how to apply the angular momentum equation to a pipe bend.

EXAMPLE 6.8

Applying the Angular Momentum Equation to Calculate the Moment on a Reducing Bend

Problem Statement

The reducing bend shown in the figure is supported on a horizontal axis through point A. Water (20°C) flows through the bend at 0.25 m³/s. The inlet pressure at cross section 1 is 150 kPa gage, and the outlet pressure at section 2 is 59.3 kPa gage. A weight of 1420 N acts 20 cm to the right of point A. Find the moment the support system must resist. The diameters of the inlet and outlet pipes are 30 cm and 10 cm, respectively.

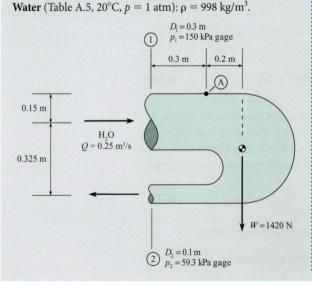

Define the Situation

Water flows through a pipe bend.
Assume steady flow.
Water (Table A.5, 20°C, $p = 1$ atm): $\rho = 998$ kg/m³.

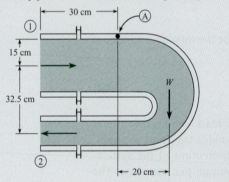

$D_1 = 0.3$ m
$p_1 = 150$ kPa gage
0.3 m | 0.2 m
0.15 m
H₂O
$Q = 0.25$ m³/s
0.325 m
$W = 1420$ N
$D_2 = 0.1$ m
$p_2 = 59.3$ kPa gage

State the Goal

M_A(N) ◄ Moment acting on the support structure

Generate Ideas and Make a Plan

Select the moment-of-momentum equation (Eq. 6.27) because (a) torque is a parameter and (b) fluid particles are accelerating as they pass through the pipe bend.

Select a control volume surrounding the reducing bend. The reason is that this CV cuts through point A (where we want to know the moment) and also cuts through sections 1 and 2 where information is known.

Sketch the force and momentum diagrams. Add dimensions to the sketches so that it is easier to evaluate cross products.

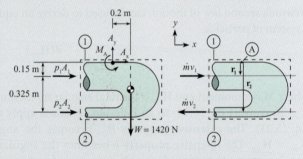

Select point "A" to sum moments about. Because the flow is steady, the accumulation of momentum term is zero. Also, there is one inflow of angular momentum and one outflow. Thus, the angular momentum equation (Eq. 6.27) simplifies to:

$$\sum \mathbf{M}_A = \{\mathbf{r}_2 \times (\dot{m}\mathbf{v}_2)\} - \{\mathbf{r}_1 \times (\dot{m}\mathbf{v}_1)\} \quad \text{(a)}$$

Sum moments in the z-direction

$$\sum M_{A,z} = (p_1 A_1)(0.15 \text{ m}) + (p_2 A_2)(0.475 \text{ m}) + M_A - W(0.2 \text{ m}) \quad \text{(b)}$$

Next, analyze the momentum terms in Eq. (a).

$$\{\mathbf{r}_2 \times (\dot{m}\mathbf{v}_2)\} - \{\mathbf{r}_1 \times (\dot{m}\mathbf{v}_1)\}_z = \{-r_2 \dot{m} v_2\} - \{r_1 \dot{m} v_1\} \quad \text{(c)}$$

Substitute Eqs. (b) and (c) into Eq. (a)

$$(p_1 A_1)(0.15 \text{ m}) + (p_2 A_2)(0.475 \text{ m}) + M_A - W(0.2 \text{ m}) = \{-r_2 \dot{m} v_2\} - \{r_1 \dot{m} v_1\} \quad \text{(d)}$$

All the terms in Eq. (d) are known, so M_A can be calculated. Thus, the plan is

1. Calculate torques to due to pressure: $r_1 p_1 A_1$ and $r_2 p_2 A_2$.
2. Calculate momentum flow terms: $r_2 \dot{m} v_2 + r_1 \dot{m} v_1$.
3. Calculate M_A.

Take Action (Execute the Plan)

1. Torques due to pressure

$$r_1 p_1 A_1 = (0.15 \text{ m})(150 \times 1000 \text{ N/m}^2)(\pi \times 0.3^2/4 \text{ m}^2)$$
$$= 1590 \text{ N} \cdot \text{m}$$
$$r_2 p_2 A_2 = (0.475 \text{ m})(59.3 \times 1000 \text{ N/m}^2)(\pi \times 0.15^2/4 \text{ m}^2)$$
$$= 498 \text{ N} \cdot \text{m}$$

2. Momentum flow terms

$$\dot{m} = \rho Q = (998 \text{ kg/m}^3)(0.25 \text{ m}^3/\text{s})$$
$$= 250 \text{ kg/s}$$

$$v_1 = \frac{Q}{A_1} = \frac{0.25 \text{ m}^3/\text{s}}{\pi \times 0.15^2 \text{ m}^2} = 3.54 \text{ m/s}$$

$$v_2 = \frac{Q}{A_2} = \frac{0.25 \text{ m}^3/\text{s}}{\pi \times 0.075^2 \text{ m}^2} = 14.15 \text{ m/s}$$

$$\dot{m}(r_2 v_2 + r_1 v_1) = (250 \text{ kg/s})$$
$$\times (0.475 \times 14.15 + 0.15 \times 3.54)(\text{m}^2/\text{s})$$
$$= 1813 \text{ N} \cdot \text{m}$$

3. Moment exerted by support

$$M_A = -0.15 p_1 A_1 - 0.475 p_2 A_2 + 0.2W - \dot{m}(r_2 v_2 + r_1 v_1)$$
$$= -(1590 \text{ N} \cdot \text{m}) - (498 \text{ N} \cdot \text{m})$$
$$+ (0.2 \text{ m} \times 1420 \text{ N}) - (1813 \text{ N} \cdot \text{m})$$
$$M_A = -3.62 \text{ kN} \cdot \text{m}$$

Thus, a moment of 3.62 kN · m acting in the clockwise, direction is needed to hold the bend stationary.

> By Newton's third law, the moment acting on the support structure is $M_A = 3.62 \text{ kN} \cdot \text{m}$ (counterclockwise).

Review the Solution and the Process

Tip. Use the "right-hand-rule" to find the correct direction of moments.

Example 6.9 illustrates how to apply the angular momentum equation to predict the power delivered by a turbine. This analysis can be applied to both power-producing machines (turbines) and power-absorbing machines (pumps and compressors). Additional information is presented in Chapter 14.

EXAMPLE 6.9

Applying the Angular Momentum Equation to Predict the Power Delivered by a Francis Turbine

Problem Statement

A Francis turbine is shown in the diagram. Water is directed by guide vanes into the rotating wheel (runner) of the turbine. The guide vanes have a 70° angle from the radial direction. The water exits with only a radial component of velocity with respect to the environment. The outer diameter of the wheel is 1 m, and the inner diameter is 0.5 m. The distance across the runner is 4 cm. The discharge is 0.5 m³/s, and the rotational rate of the wheel is 1200 rpm. The water density is 1000 kg/m³. Find the power (kW) produced by the turbine.

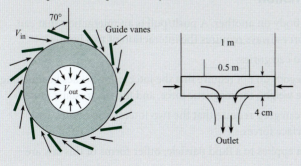

Define the Situation

A Francis turbine generates power.

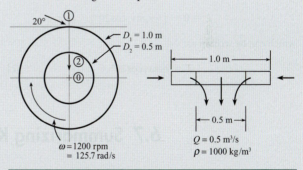

State the Goal

$P(\text{W}) \Leftarrow$ Power generated by the turbine

Generate Ideas and Make a Plan

Because power is the goal, select the *power equation*.

$$P = T\omega \qquad \text{(a)}$$

where T is torque acting on the turbine, and ω is turbine angular speed. In Eq. (a), torque is unknown, so it becomes the new goal. Torque can be found using the angular momentum equation.

Sketch. To apply the angular momentum equation, select a control volume surrounding the turbine. Then, sketch a force and momentum diagram

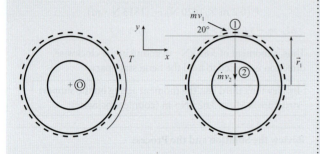

In the force diagram, the torque T is the external torque from the generator. Because this torque opposes angular acceleration, its direction is counterclockwise. The flow is idealized by using one inlet momentum flow at section 1 and one outlet momentum flow at section 2.

Select point "O" to sum moments about. Because the flow is steady, the accumulation of momentum is zero. Thus, the angular momentum equation (Eq. 6.26) simplifies to:

$$\sum \mathbf{M}_A = \{ \mathbf{r}_2 \times (\dot{m}\mathbf{v}_2) \} - \{ \mathbf{r}_1 \times (\dot{m}\mathbf{v}_1) \} \qquad \text{(b)}$$

Apply Eq. (b) in the z-direction. Also, recognize that the flow at section 2 has no angular momentum. That is, $\{ \mathbf{r}_2 \times (\dot{m}\mathbf{v}_2) \} = 0$. Thus, Eq. (b) simplifies to

$$T = 0 - \{ -r_1 \dot{m} v_1 \cos 20° \}$$

which can be written as:

$$T = r_1 \dot{m} v_1 \cos 20° \qquad \text{(c)}$$

In Eq. (c), the velocity v_1 can be calculated using the flow rate equation. Because velocity is not perpendicular to area, use the dot product.

$$Q_1 = \mathbf{V}_1 \cdot \mathbf{A}_1$$

$$Q = v_1 A_1 \sin 20°$$

which can be rewritten as

$$v_1 = \frac{Q}{A_1 \sin 20°} \qquad \text{(d)}$$

Now, the number of equations equals the number of unknowns. Thus, the plan is to

1. Calculate inlet velocity v_1 using Eq. (d).
2. Calculate mass flow rate using $\dot{m} = \rho Q$.
3. Calculate torque using Eq. (c).
4. Calculate power using Eq. (a).

Take Action (Execute the Plan)

1. Volume flow rate equation:

$$v_1 = \frac{Q}{A_1 \sin 20°} = \frac{(0.5 \text{ m}^3/\text{s})}{\pi (1.0 \text{ m})(0.04 \text{ m}) \sin 20°} = 11.63 \text{ m/s}$$

2. Mass flow rate equation:

$$\dot{m} = \rho Q = (1000 \text{ kg/m}^3)(0.5 \text{ m}^3/\text{s}) = 500 \text{ kg/s}$$

3. Angular momentum equation:

$$T = r_1 \dot{m} v_1 \cos 20°$$
$$= (0.5 \text{ m})(500 \text{ kg/s})(11.63 \text{ m/s}) \cos 20°$$
$$= 2732 \text{ N} \cdot \text{m}$$

4. Power equation:

$$P = T\omega = (2732 \text{ N} \cdot \text{m})(125.7 \text{ rad/s})$$

$$\boxed{P = 343 \text{ kW}}$$

6.7 Summarizing Key Knowledge

Newton's Second Law of Motion

- A *force* is a push or pull of one body on another. A push/pull is an interaction that can cause a body to accelerate. A force always requires the interaction of two bodies.
- Forces can be classified into two categories:
  - *Body forces.* Forces in this category do not require that the interacting bodies be touching. Common body forces include weight, the magnetic force, and the electrostatic force.
  - *Surface forces.* Forces in this category require that the two interacting bodies are touching. Most forces are surface forces.
- Newton's second law $\sum \mathbf{F} = m\mathbf{a}$ applies to a fluid particle; other forms of this law are derived from this equation.

- Newton's second law asserts that forces are related to accelerations:
  ▸ Thus, if $\sum \mathbf{F} > \mathbf{0}$, the particle must accelerate.
  ▸ Thus, if $\mathbf{a} > \mathbf{0}$, the sum of forces must be nonzero.

Solving Vector Equations

- A vector equation is one whose terms are vectors.
- A vector equation can be written as one or more equivalent scalar equations.
- The Visual Solution Method (VSM) is an approach for solving a vector equation that makes problem solving easier. The process for the VSM is
  ▸ **Step 1:** Identify the vector equation in its general form.
  ▸ **Step 2:** Sketch a diagram that shows the vectors on the left side of the equation. Sketch an equal sign. Sketch a diagram that shows the vectors on the right side of the equation.
  ▸ **Step 3:** From the diagrams, apply the general equation, write the final results, and simplify the results to create the reduced equation(s).

The Linear Momentum Equation

- The linear momentum equation is Newton's second law in a form that is useful for solving problems in fluid mechanics
- To derive the momentum equation
  ▸ Begin with Newton's second law for a single particle.
  ▸ Derive Newton's second law for a system of particles.
  ▸ Apply the Reynolds transport theorem to give the final result.
- Physical Interpretation

$$\begin{pmatrix} \text{sum of} \\ \text{forces} \end{pmatrix} = \begin{pmatrix} \text{momentum} \\ \text{accumulation} \end{pmatrix} + \begin{pmatrix} \text{momentum} \\ \text{outflow} \end{pmatrix} - \begin{pmatrix} \text{momentum} \\ \text{inflow} \end{pmatrix}$$

- The *momentum accumulation* term gives the rate at which the momentum inside the control volume is changing with time.
- The *momentum flow* terms give the rate at which momentum is being transported across the control surfaces.

The Angular Momentum Equation

- The angular momentum equation is the rotational analog to the linear momentum equation.
  ▸ This equation is useful for problems involving torques (i.e., moments)
  ▸ This equation is commonly applied to rotating machinery such as pumps, fans, and turbines.
- The physics of the angular momentum equation are

$$\begin{pmatrix} \text{sum of} \\ \text{moments} \end{pmatrix} = \begin{pmatrix} \text{angular momentum} \\ \text{accumulation} \end{pmatrix} + \begin{pmatrix} \text{angular momentum} \\ \text{outflow} \end{pmatrix} - \begin{pmatrix} \text{angular momentum} \\ \text{inflow} \end{pmatrix}$$

- To apply the angular momentum equation, use the same process as that used for the linear momentum equation.

REFERENCES

1. Hibbeler, R.C. *Dynamics*. Englewood Cliffs, NJ: Prentice Hall, 1995.

PROBLEMS

PLUS Problem available in *WileyPLUS* at instructor's discretion.

GO Guided Online (GO) Problem, available in *WileyPLUS* at instructor's discretion.

Newton's Second Law of Motion (§6.1)

6.1 Identify the surface and body forces acting on a glider in flight. Also, sketch a free body diagram and explain how Newton's laws of motion apply.

6.2 Newton's second law can be stated that the force is equal to the rate of change of momentum, $F = d(mv)/dt$. Taking the derivative by parts yields $F = m(dv)/(dt) + v(dm)/(dt)$. This does not correspond to $F = ma$. What is the source of the discrepancy?

The Linear Momentum Equation: Theory (§6.2)

6.3 **PLUS** Which of the following are correct with respect to the derivation of the momentum equation? (Select all that apply.)

   a. Reynold's transport theorem is applied to Fick's law.

   b. The extensive property is momentum.

   c. The intensive property is mass.

   d. The velocity is assumed to be uniformly distributed across each inlet and outlet.

   e. The net momentum flow is the "ins" minus the "outs."

   f. The net force is the sum of forces acting on the matter inside the CV

The Linear Momentum Equation: Application (§6.3)

6.4 **PLUS** When making a force diagram (FD) and its partner momentum diagram (MD) to set up the equations for a momentum equation problem (see Fig. 6.10 on p. 217 in §6.3), which of the following elements should be in the FD, and which should be in the MD? (Classify all below as either FD or MD.)

   a. Each mass stream with product $\dot{m}_o \mathbf{v}_o$ or product $\dot{m}_i \mathbf{v}_i$ crossing a control surface boundary.

   b. Reaction forces required to hold walls, vanes, or pipes in place.

   c. Weight of a solid body that contains or contacts the fluid.

   d. Weight of the fluid.

   e. Pressure force caused by a fluid flowing across a control surface boundary.

Applying the Momentum Equation to Fluid Jets (§6.4)

6.5 Give five examples of jets and how they are used in practice.

6.6 **PLUS** A "balloon rocket" is a balloon suspended from a taut wire by a hollow tube (drinking straw) and string. The nozzle is

formed of a 0.8-cm-diameter tube, and an air jet exits the nozzle with a speed of 45 m/s and a density of 1.2 kg/m³. Find the force F needed to hold the balloon stationary. Neglect friction.

6.7 **PLUS** The balloon rocket is held in place by a force F. The pressure inside the balloon is 20 cm-H₂O, the nozzle diameter is 1.0 cm, and the air density is 1.2 kg/m³. Find the exit velocity v and the force F. Neglect friction and assume the air flow is inviscid and irrotational.

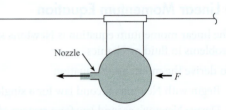

PROBLEMS 6.6, 6.7

6.8 **PLUS** For Example 6.2 in §6.4, the situation diagram shows concrete being "shot" at an angle into a cart that is tethered by a cable, and sitting on a scale. Determine whether the following two statements are "true" or "false."

   a. Mass is being accumulated in the cart.

   b. Momentum is being accumulated in the cart.

6.9 **PLUS** A water jet of diameter 30 mm and speed $v = 25$ m/s is filling a tank. The tank has a mass of 25 kg and contains 25 liters of water at the instant shown. The water temperature is 15°C. Find the force acting on the bottom of the tank and the force acting on the stop block. Neglect friction.

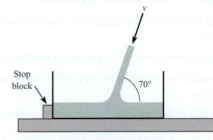

PROBLEMS 6.9, 6.10

6.10 **GO** A water jet of diameter 5 cm and speed $v = 18$ m/s is filling a tank. The tank has a mass of 12 kg and contains 20 liters of water at the instant shown. The water temperature is 21°C. Find the minimum coefficient of friction such that the force acting on the stop block is zero.

6.11 A design contest features a submarine that will travel at a steady speed of $V_{sub} = 1$ m/s in 15°C water. The sub is powered by a water jet. This jet is created by drawing water from an inlet of diameter 25 mm, passing this water through a pump and then accelerating the water through a nozzle of diameter 5 mm to a speed of V_{jet}. The hydrodynamic drag force (F_D) can be calculated using

$$F_D = C_D\left(\frac{\rho V_{sub}^2}{2}\right)A_p$$

where the coefficient of drag is $C_D = 0.3$ and the projected area is $A_p = 0.28$ m². Specify an acceptable value of V_{jet}.

PROBLEM 6.11

6.12 A horizontal water jet at 21°C impinges on a vertical-perpendicular plate. The discharge is 0.05 m³/s. If the external force required to hold the plate in place is 900 N, what is the velocity of the water?

6.13 WILEY PLUS A horizontal water jet at 21°C issues from a circular orifice in a large tank. The jet strikes a vertical plate that is normal to the axis of the jet. A force of 2.7 kN is needed to hold the plate in place against the action of the jet. If the pressure in the tank is 170 kPa gage at point A, what is the diameter of the jet just downstream of the orifice?

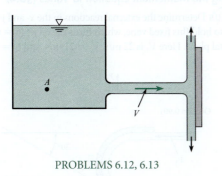

PROBLEMS 6.12, 6.13

6.14 WILEY PLUS An engineer, who is designing a water toy, is making preliminary calculations. A user of the product will apply a force F_1 that moves a piston ($D = 80$ mm) at a speed of $V_{piston} = 300$ mm/s. Water at 20°C jets out of a converging nozzle

of diameter $d = 15$ mm. To hold the toy stationary, the user applies a force F_2 to the handle. Which force (F_1 versus F_2) is larger? Explain your answer using concepts of the momentum principle. Then calculate F_1 and F_2. Neglect friction between the piston and the walls.

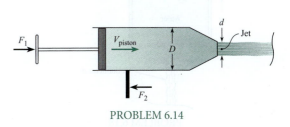

PROBLEM 6.14

6.15 A firehose on a boat is producing a 10 cm-diameter water jet with a speed of $V =100$ km/hr. The boat is held stationary by a cable attached to a pier, and the water temperature is 10°C. Calculate the tension in the cable.

6.16 WILEY PLUS A boat is held stationary by a cable attached to a pier. A firehose directs a spray of 5°C water at a speed of $V = 50$ m/s. If the allowable load on the cable is 5 kN, calculate the mass flow rate of the water jet. What is the corresponding diameter of the water jet?

PROBLEMS 6.15, 6.16

6.17 WILEY GO A group of friends regularly enjoys white-water rafting, and they bring piston water guns to shoot water from one raft to another. One summer they notice that when on placid slack water (no current), after just a few volleys at each other, they are drifting apart. They wonder whether the jet being ejected out of a piston gun has enough momentum to force the shooter and raft backward. To answer this question,

    a. Sketch a CV, an FD, and an MD for this system.

    b. Calculate the momentum flux (N) generated by ejecting water with a flow rate of 3.8 L/s from a cross section of 4 cm.

6.18 WILEY GO A tank of water (15°C) with a total weight of 200 N (water plus the container) is suspended by a vertical cable. Pressurized air drives a water jet ($d = 12$ mm) out the bottom of the tank such that the tension in the vertical cable is 10 N. If $H = 425$ mm, find the required air pressure in units of atmospheres (gage). Assume the flow of water is irrotational.

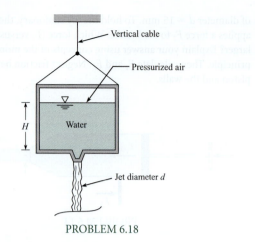

PROBLEM 6.18

6.19 WILEY PLUS A jet of water (15.5°C) is discharging at a constant rate of 0.05 m³/s from the upper tank. If the jet diameter at section 1 is 10 cm, what forces will be measured by scales A and B? Assume the empty tank weighs 1.4 kN, the cross-sectional area of the tank is 0.37 m², $h = 0.3$ m, and $H = 2.7$ m.

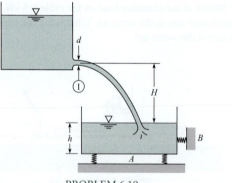

PROBLEM 6.19

6.20 A conveyor belt discharges gravel into a barge as shown at a rate of 38 m³/min. If the gravel weighs 18.9 kN/m³, what is the tension in the hawser that secures the barge to the dock?

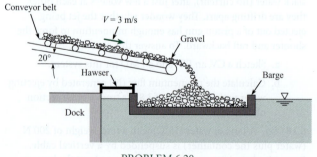

PROBLEM 6.20

6.21 The semicircular nozzle sprays a sheet of liquid through 180° of arc as shown. The velocity is V at the efflux section where

the sheet thickness is t. Derive a formula for the external force F (in the y-direction) required to hold the nozzle system in place. This force should be a function of ρ, V, r, and t.

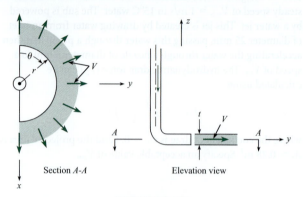

PROBLEM 6.21

6.22 The expansion section of a rocket nozzle is often conical in shape, and because the flow diverges, the thrust derived from the nozzle is less than it would be if the exit velocity were everywhere parallel to the nozzle axis. By considering the flow through the spherical section suspended by the cone and assuming that the exit pressure is equal to the atmospheric pressure, show that the thrust is given by

$$T = \dot{m}V_e \frac{(1 + \cos\alpha)}{2}$$

where $\dot{m}$ is the mass flow through the nozzle, V_e is the exit velocity, and α is the nozzle half-angle.

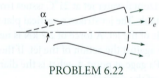

PROBLEM 6.22

Applying the Momentum Equation to Vanes (§6.4)

6.23 WILEY PLUS Determine the external reactions in the x- and y-directions needed to hold this fixed vane, which turns the oil jet ($S = 0.9$) in a horizontal plane. Here V_1 is 22 m/s, $V_2 = 21$ m/s, and $Q = 0.15$ m³/s.

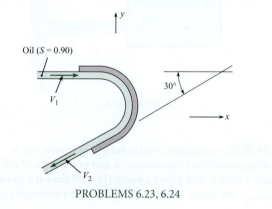

PROBLEMS 6.23, 6.24

6.24 Solve Prob. 6.23 for $V_1 = 21$ m/s, $V_2 = 19.8$ m/s, and $Q = 0.043$ m³/s.

6.25 PLUS This planar water jet (15.5°C) is deflected by a fixed vane. What are the x- and y-components of force per unit width needed to hold the vane stationary? Neglect gravity.

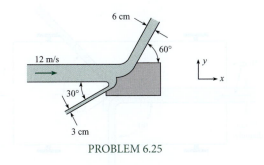

PROBLEM 6.25

6.26 PLUS A water jet with a speed of 9 m/s and a mass flow rate of 1.6 kg/s is turned 30° by a fixed vane. Find the force of the water jet on the vane. Neglect gravity.

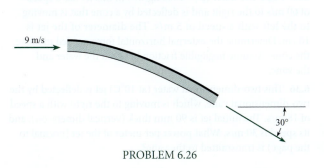

PROBLEM 6.26

6.27 GO Water ($\rho = 1000$ kg/m³) strikes a block as shown and is deflected 30°. The flow rate of the water is 1.5 kg/s, and the inlet velocity is $V = 10$ m/s. The mass of the block is 1 kg. The coefficient of static friction between the block and the surface is 0.1 (friction force/normal force). If the force parallel to the surface exceeds the frictional force, the block will move. Determine the force on the block and whether the block will move. Neglect the weight of the water.

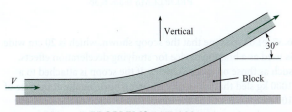

PROBLEMS 6.27, 6.28

6.28 For the situation described in Prob. 6.27, find the maximum inlet velocity (V) such that the block will not slip.

6.29 PLUS Plate A is 50 cm in diameter and has a sharp-edged orifice at its center. A water jet (at 10°C) strikes the plate concentrically with a speed of 90 m/s. What external force is needed to hold the plate in place if the jet issuing from the orifice also has a speed of 90 m/s? The diameters of the jets are $D = 10$ cm and $d = 3.5$ cm.

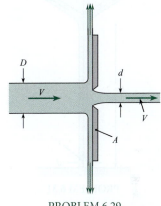

PROBLEM 6.29

6.30 A two-dimensional liquid jet impinges on a vertical wall. Assuming that the incoming jet speed is the same as the exiting jet speed ($V_1 = V_2$), derive an expression for the force per unit width of jet exerted on the wall. What form do you think the upper liquid surface will take next to the wall? Sketch the shape you think it will take, and explain your reasons for drawing it that way.

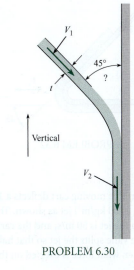

PROBLEM 6.30

6.31 PLUS A cone that is held stable by a wire is free to move in the vertical direction and has a jet of water (at 10°C) striking it from below. The cone weighs 30 N. The initial speed of the jet as it comes from the orifice is 15 m/s, and the initial jet diameter is 2 cm. Find the height to which the cone will rise and remain stationary. *Note:* The wire is only for stability and should not enter into your calculations.

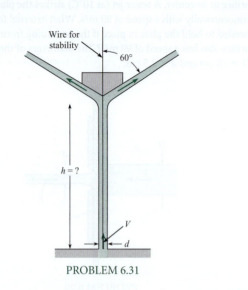

PROBLEM 6.31

6.32 A horizontal jet of water (at 10°C) that is 6 cm in diameter and has a velocity of 20 m/s is deflected by the vane as shown. If the vane is moving at a rate of 7 m/s in the x-direction, what components of force are exerted on the vane by the water in the x- and y-directions? Assume negligible friction between the water and the vane.

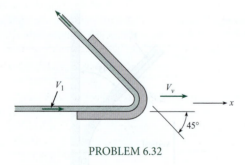

PROBLEM 6.32

6.33 PLUS A vane on this moving cart deflects a 15-cm-diameter water ($\rho = 1000$ kg/m³) jet as shown. The initial speed of the water in the jet is 50 m/s, and the cart moves at a speed of 3 m/s. If the vane splits the jet so that half goes one way and half the other, what force is exerted on the vane by the water?

6.34 Refer to the cart of Prob. 6.33. If the cart speed is constant at 1.5 m/s, and if the initial jet speed is 18 m/s, and jet diameter = 45 mm, what is the rolling resistance of the cart? ($\rho = 1000$ kg/m³)

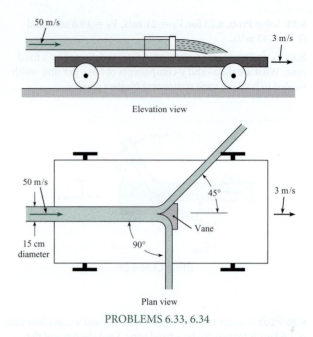

Elevation view

Plan view

PROBLEMS 6.33, 6.34

6.35 PLUS The water ($\rho = 1000$ kg/m³) in this jet has a speed of 60 m/s to the right and is deflected by a cone that is moving to the left with a speed of 5 m/s. The diameter of the jet is 10 cm. Determine the external horizontal force needed to move the cone. Assume negligible friction between the water and the vane.

6.36 This two-dimensional water (at 10°C) jet is deflected by the two-dimensional vane, which is moving to the right with a speed of 18 m/s. The initial jet is 90 mm thick (vertical dimension), and its speed is 30 m/s. What power per meter of the jet (normal to the page) is transmitted to the vane?

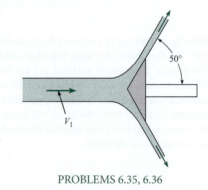

PROBLEMS 6.35, 6.36

6.37 PLUS Assume that the scoop shown, which is 20 cm wide, is used as a braking device for studying deceleration effects, such as those on space vehicles. If the scoop is attached to a 1000 kg sled that is initially traveling horizontally at the rate of 100 m/s, what will be the initial deceleration of the sled? The scoop dips into the water 8 cm ($d = 8$ cm). ($T = 10$°C.)

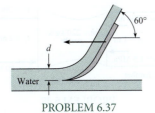

PROBLEM 6.37

6.38 This snowplow "cleans" a swath of snow that is 10 cm deep ($d = 10$ cm) and 0.6 m wide ($B = 0.6$ m). The snow leaves the blade in the direction indicated in the sketches. Neglecting friction between the snow and the blade, estimate the power required for just the snow removal if the speed of the snowplow is 12 m/s.

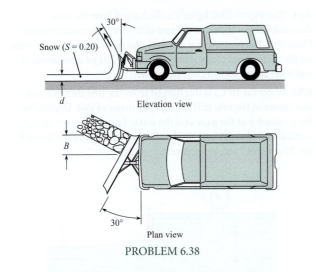

Elevation view

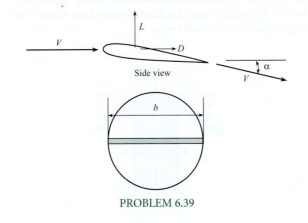

Plan view

PROBLEM 6.38

6.39 WILEY GO A finite span airfoil can be regarded as a vane as shown in the figure. The cross section of air affected is equal to the circle with the diameter of the wing span, b. The wing deflects the air by an angle α and produces a force normal to the free-stream velocity, the lift L, and in the free-stream direction, the drag D. The airspeed is unchanged. Calculate the lift and drag for a 9 m wing span in a 90 m/s airstream at 101.3 kPa abs psia and 15.5°C for flow deflection of 2°.

V → → D

Side view

PROBLEM 6.39

6.40 The "clam shell" thrust reverser sketched in the figure is often used to decelerate aircraft on landing. The sketch shows normal operation (a) and when deployed (b). The vanes are oriented 20° with respect to the vertical. The mass flow through the engine is 68 kg/s, the inlet velocity is 90 m/s, and the exit velocity is 425 m/s. Assume that when the thrust reverser is deployed, the exit velocity of the exhaust is unchanged. Assume the engine is stationary. Calculate the thrust under normal operation (N) and when the thrust reverser is deployed.

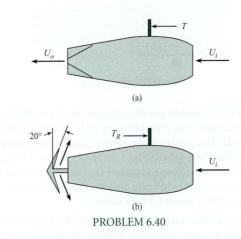

PROBLEM 6.40

Applying the Momentum Equation to Nozzles (§6.4)

6.41 Firehoses are fitted with special nozzles. Use the Internet or contact your local fire department to find information on operational conditions and typical hose and nozzle sizes used.

6.42 WILEY PLUS High-speed water jets are used for speciality cutting applications. The pressure in the chamber is approximately 4.14×10^5 kPa gage. Using the Bernoulli equation, estimate the water speed exiting the nozzle exhausting to atmospheric pressure. Neglect compressibility effects and assume a water temperature of 15.5°C.

6.43 WILEY PLUS Water at 15.5°C flows through a nozzle that contracts from a diameter of 7.5 cm to 2.5 cm. The pressure at section 1 is 120 kPa, and atmospheric pressure prevails at the exit of the jet. Calculate the speed of the flow at the nozzle exit and the force required to hold the nozzle stationary. Neglect weight.

6.44 WILEY GO Water at 15°C flows through a nozzle that contracts from a diameter of 10 cm to 2 cm. The exit speed is $v_2 = 25$ m/s, and atmospheric pressure prevails at the exit of the jet. Calculate the pressure at section 1 and the force required to hold the nozzle stationary. Neglect weight.

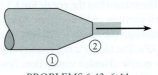

PROBLEMS 6.43, 6.44

6.45 $\overset{\text{WILEY}}{\text{PLUS}}$ Water (at 10°C) flows through this nozzle at a rate of 0.56 m³/s and discharges into the atmosphere. $D_1 = 65$ cm, and $D_2 = 22.5$ cm. Determine the force required at the flange to hold the nozzle in place. Assume irrotational flow. Neglect gravitational forces.

6.46 Solve Prob. 6.45 using the following values: $Q = 0.30$ m³/s, $D_1 = 30$ cm, and $D_2 = 10$ cm. ($\rho = 1000$ kg/m³.)

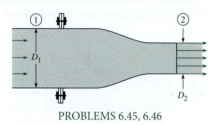

PROBLEMS 6.45, 6.46

6.47 $\overset{\text{WILEY}}{\text{PLUS}}$ This "double" nozzle discharges water ($\rho = 1000$ kg/m³) into the atmosphere at a rate of 0.45 m³/s. If the nozzle is lying in a horizontal plane, what x-component of force acting through the flange bolts is required to hold the nozzle in place? *Note:* Assume irrotational flow, and assume the water speed in each jet to be the same. Jet A is 10 cm in diameter, jet B is 12 cm in diameter, and the pipe is 0.3 m in diameter.

6.48 This "double" nozzle discharges water (at 10°C) into the atmosphere at a rate of 0.65 m³/s. If the nozzle is lying in a horizontal plane, what x-component of force acting through the flange bolts is required to hold the nozzle in place? *Note:* Assume irrotational flow, and assume the water speed in each jet to be the same. Jet A is 8 cm in diameter, jet B is 9 cm in diameter, and the pipe is 30 cm in diameter.

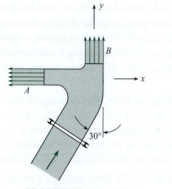

PROBLEMS 6.47, 6.48

6.49 $\overset{\text{WILEY}}{\text{PLUS}}$ A rocket-nozzle designer is concerned about the force required to hold the nozzle section on the body of a rocket. The nozzle section is shaped as shown in the figure. The pressure and velocity at the entrance to the nozzle are 1.5 MPa and 100 m/s. The exit pressure and velocity are 80 kPa and 2000 m/s. The mass flow through the nozzle is 220 kg/s. The atmospheric pressure is 100 kPa. The rocket is not accelerating. Calculate the force on the nozzle-chamber connection. *Note:* The given pressures are absolute.

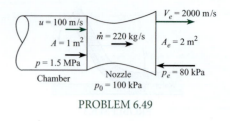

PROBLEM 6.49

6.50 A 15 cm nozzle is bolted with six bolts to the flange of a 30 cm pipe. If water ($\rho = 1000$ kg/m³) discharges from the nozzle into the atmosphere, calculate the tension load in each bolt when the pressure in the pipe is 200 kPa. Assume irrotational flow.

6.51 Water ($\rho = 1000$ kg/m³) is discharged from the two-dimensional slot shown at the rate of 0.23 m³/s per meter of slot. Determine the pressure p at the gage and the water force per meter on the vertical end plates A and C. The slot and jet dimensions B and b are 20 cm and 10 cm, respectively.

6.52 Water (at 10°C) is discharged from the two-dimensional slot shown at the rate of 0.40 m³/s per meter of slot. Determine the pressure p at the gage and the water force per meter on the vertical end plates A and C. The slot and jet dimensions B and b are 20 cm and 7 cm, respectively.

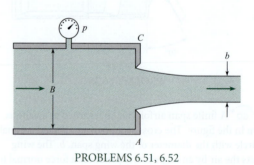

PROBLEMS 6.51, 6.52

6.53 This spray head discharges water ($\rho = 1000$ kg/m³) at a rate of 0.12 m³/s. Assuming irrotational flow and an efflux speed of 20 m/s in the free jet, determine what force acting through the bolts of the flange is needed to keep the spray head on the 15 cm pipe. Neglect gravitational forces.

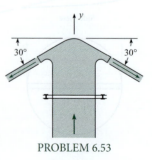

PROBLEM 6.53

6.54 Two circular water ($\rho = 1000$ kg/m³) jets of 12 mm diameter ($d = 12$ mm) issue from this unusual nozzle. If the efflux speed is 25 m/s, what force is required at the flange to hold the nozzle in place? The pressure in the 10 cm pipe ($D = 9$ cm) is 345 kPa gage.

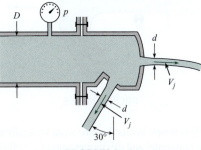

PROBLEM 6.54

6.55 Liquid (S = 1.2) enters the "black sphere" through a 5 cm pipe with velocity of 15 m/s and a pressure of 400 kPa gage. It leaves the sphere through two jets as shown. The velocity in the vertical jet is 30 m/s, and its diameter is 2.5 cm. The other jet's diameter is also 2.5 cm. What force through the 5 cm pipe wall is required in the x- and y-directions to hold the sphere in place? Assume the sphere plus the liquid inside it weighs 900 N.

6.56 Liquid (S = 1.5) enters the "black sphere" through a 5 cm pipe with a velocity of 10 m/s and a pressure of 400 kPa. It leaves the sphere through two jets as shown. The velocity in the vertical jet is 30 m/s, and its diameter is 25 mm. The other jet's diameter is also 25 mm. What force through the 5 cm pipe wall is required in the x- and y-directions to hold the sphere in place? Assume the sphere plus the liquid inside it weighs 600 N.

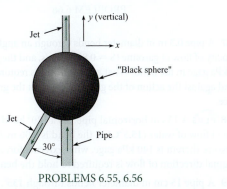

PROBLEMS 6.55, 6.56

Applying the Momentum Equation to Pipe Bends (§6.4)

6.57 PLUS A hot gas stream enters a uniform-diameter return bend as shown. The entrance velocity is 90 m/s, the gas density is 0.32 kg/m³, and the mass flow rate is 0.45 kg/s. Water is sprayed into the duct to cool the gas down. The gas exits with a density of 1 kg/m³. The mass flow of water into the gas is negligible. The pressures at the entrance and exit are the same and equal to the atmospheric pressure. Find the force required to hold the bend.

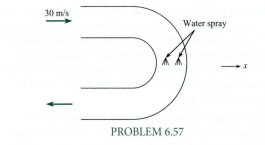

PROBLEM 6.57

6.58 Assume that the gage pressure p is the same at sections 1 and 5 cm the horizontal bend shown in the figure. The fluid flowing in the bend has density ρ, discharge Q, and velocity V. The cross-sectional area of the pipe is A. Then the magnitude of the force (neglecting gravity) required at the flanges to hold the bend in place will be (a) pA, (b) $pA + \rho QV$, (c) $2pA + \rho QV$, or (d) $2pA + 2\rho QV$.

6.59 PLUS The pipe shown has a 180° vertical bend in it. The diameter D is 0.3m, and the pressure at the center of the upper pipe is 105 kPa gage. If the flow in the bend is 0.6 m³/s, what external force will be required to hold the bend in place against the action of the water? The bend weighs 900 N, and the volume of the bend is 0.085 m³. Assume the Bernoulli equation applies. ($\rho = 1000$ kg/m³)

6.60 The pipe shown has a 180° horizontal bend in it as shown, and D is 20 cm. The discharge of water ($\rho = 1000$ kg/m³) in the pipe and bend is 0.35 m³/s, and the pressure in the pipe and bend is 100 kPa gage. If the bend volume is 0.10 m³, and the bend itself weighs 400 N, what force must be applied at the flanges to hold the bend in place?

6.61 Set up the solution for Problem 6.60, and answer the following questions:

   a. Do the two pressure forces from the inlet and exit act in the same direction, or in opposite directions?

   b. For the data given, which term has the larger magnitude (in N), the pressure force term, or the net momentum flux term?

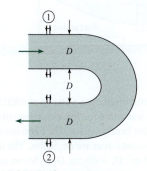

PROBLEMS 6.58, 6.59, 6.60, 6.61

6.62 Water (at 10°C) flows in the 90° horizontal bend at a rate of 0.34 m³/s and discharges into the atmosphere past the downstream flange. The pipe diameter is 0.3 m. What force must be applied at the upstream flange to hold the bend in place? Assume that the volume of water downstream of the upstream flange is 0.1 m³ and that the bend and pipe weigh 445 N. Assume the pressure at the inlet section is 28 kPa gage.

6.63 PLUS The gage pressure throughout the horizontal 90° pipe bend is 300 kPa. If the pipe diameter is 1 m and the water (at 10°C) flow rate is 10 m³/s, what x-component of force must be applied to the bend to hold it in place against the water action?

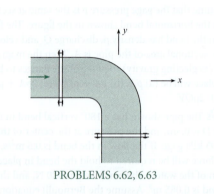

PROBLEMS 6.62, 6.63

6.64 This 30° vertical bend in a pipe with a 0.6 m diameter carries water ($\rho = 1000$ kg/m³) at a rate of 0.9 m³/s. If the pressure p_1 is 70 kPa at the lower end of the bend, where the elevation is 30 m, and p_2 is 60 kPa at the upper end, where the elevation is 31 m, what will be the vertical component of force that must be exerted by the "anchor" on the bend to hold it in position? The bend itself weighs 1350 N, and the length L is 1.2 m.

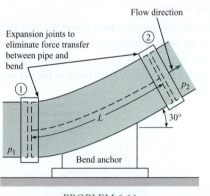

PROBLEM 6.64

6.65 GO This bend discharges water ($\rho = 1000$ kg/m³) into the atmosphere. Determine the force components at the flange required to hold the bend in place. The bend lies in a horizontal plane. Assume viscous forces are negligible. The interior volume of the bend is 0.25 m³, $D_1 = 60$ cm, $D_2 = 30$ cm, and $V_2 = 10$ m/s. The mass of the bend material is 250 kg.

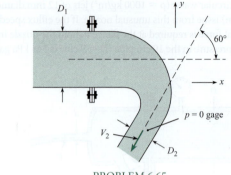

PROBLEM 6.65

6.66 PLUS This nozzle bends the flow from vertically upward to 30° with the horizontal and discharges water ($\gamma = 9810$ N/m³) at a speed of $V = 40$ m/s. The volume within the nozzle itself is 0.05 m³, and the weight of the nozzle is 445 N. For these conditions, what *vertical* force must be applied to the nozzle at the flange to hold it in place?

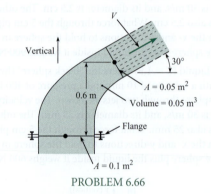

PROBLEM 6.66

6.67 A pipe 0.3 m in diameter bends through an angle of 135°. The velocity of flow of gasoline (S = 0.8) is 6 m/s, and the pressure is 70 kPa gage in the bend. What external force is required to hold the bend against the action of the gasoline? Neglect the gravitational force.

6.68 PLUS A 15 cm horizontal pipe has a 180° bend in it. If the rate of flow of water (15.5°C) in the bend is 0.06 m³/s and the pressure therein is 140 kPa gage, what external force in the original direction of flow is required to hold the bend in place?

6.69 A pipe 15 cm in diameter bends through 135°. The velocity of flow of gasoline (S = 0.8) is 8 m/s, and the pressure is 100 kPa gage throughout the bend. Neglecting gravitational force, determine the external force required to hold the bend against the action of the gasoline.

6.70 A horizontal reducing bend turns the flow of water ($\rho = 1000$ kg/m³) through 60°. The inlet area is 0.001 m², and the outlet area is 0.0001 m². The water from the outlet discharges into the atmosphere with a velocity of 50 m/s. What horizontal force (parallel to the initial flow direction) acting through the metal of the bend at the inlet is required to hold the bend in place?

6.71 Water (at 10°C) flows in a duct as shown. The inlet water velocity is 10 m/s. The cross-sectional area of the duct is 0.1 m². Water is injected normal to the duct wall at the rate of 500 kg/s midway between stations 1 and 2. Neglect frictional forces on the duct wall. Calculate the pressure difference ($p_1 - p_2$) between stations 1 and 2.

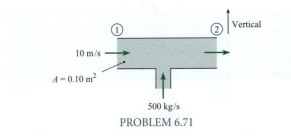

PROBLEM 6.71

6.72 PLUS For this wye fitting, which lies in a horizontal plane, the cross-sectional areas at sections 1, 2, and 3 are 0.1 m², 0.1 m², and 0.025 m², respectively. At these same respective sections the pressures are 48 kPa gage, 43 kPa gage, and 0 kPa gage, and the water discharges are 0.6 m³/s to the right, 0.35 m³/s to the right, and exits to atmosphere at 0.23 m³/s. What x-component of force would have to be applied to the wye to hold it in place?

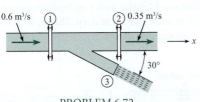

PROBLEM 6.72

6.73 Water ($\rho = 1000$ kg/m³) flows through a horizontal bend and T section as shown. The mass flow rate entering at section a is 6 kg/s, and those exiting at sections b and c are 3 kg/s each. The pressure at section a is 35 kPa gage. The pressure at the two outlets is atmospheric. The cross-sectional areas of the pipes are the same: 32 cm². Find the x-component of force necessary to restrain the section.

6.74 Water ($\rho = 1000$ kg/m³) flows through a horizontal bend and T section as shown. At section a the flow enters with a velocity of 6 m/s, and the pressure is 4.8 kPa. At both sections b and c the flow exits the device with a velocity of 3 m/s, and the pressure at these sections is atmospheric ($p = 0$). The cross-sectional areas at a, b, and c are all the same: 0.20 m². Find the x- and y-components of force necessary to restrain the section.

PROBLEMS 6.73, 6.74

6.75 For this horizontal T through which water ($\rho = 1000$ kg/m³) is flowing, the following data are given: $Q_1 = 0.25$ m³/s, $Q_2 = 0.10$ m³/s, $p_1 = 100$ kPa, $p_2 = 70$ kPa, $p_3 = 80$ kPa, $D_1 = 15$ cm, $D_2 = 7$ cm, and $D_3 = 15$ cm. For these conditions, what external force in the x–y plane (through the bolts or other supporting devices) is needed to hold the T in place?

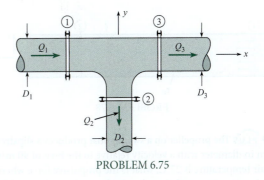

PROBLEM 6.75

Applying Momentum Equation: Other Situations (§6.4)

6.76 GO Firehoses can break windows. A 0.2-m diameter (D_1) firehose is attached to a nozzle with a 0.1 m diameter (d_2) outlet. The free jet from the nozzle is deflected by 90° when it hits the window as shown. Find the force the window must withstand due to the impact of the jet when water flows through the firehose at a rate of 0.15 m³/s.

6.77 PLUS A fireman is soaking a home that is dangerously close to a burning building. To prevent water damage to the inside of the neighboring home, he throttles down his flow rate so that it will not break windows. Assuming the typical window should be able to withstand a force up to 110 N, what is the largest volumetric flow rate he should allow (liter/min.), given a 20-cm diameter (D_1) firehose discharging through a nozzle with 10 cm diameter (d_2) outlet. The free jet from the nozzle is deflected by 90° when it hits the window as shown.

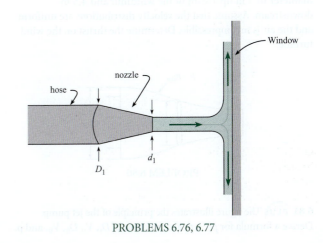

PROBLEMS 6.76, 6.77

6.78 For laminar flow in a pipe, wall shear stress (τ_0) causes the velocity distribution to change from uniform to parabolic as shown. At the fully developed section (section 2), the velocity is distributed as follows: $u = u_{max}[1 - (r/r_0)^2]$. Derive a formula for the force on the wall due to shear stress, F_τ, between 1 and 2 as a function of U (the mean velocity in the pipe), ρ, p_1, p_2, and D (the pipe diameter).

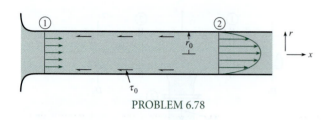

PROBLEM 6.78

6.79 (WILEY PLUS) The propeller on a swamp boat produces a slipstream 0.9 m in diameter with a velocity relative to the boat of 90 m/s. If the air temperature is 27°C, what is the propulsive force when the boat is not moving and also when its forward speed is 9 m/s? *Hint:* Assume that the pressure, except in the immediate vicinity of the propeller, is atmospheric.

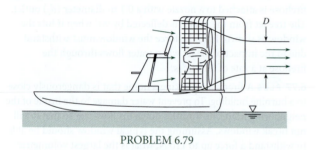

PROBLEM 6.79

6.80 (WILEY PLUS) A wind turbine is operating in a 12 m/s wind that has a density of 1.2 kg/m³. The diameter of the turbine silhouette is 4 m. The constant-pressure (atmospheric) streamline has a diameter of 3 m upstream of the windmill and 4.5 m downstream. Assume that the velocity distributions are uniform and the air is incompressible. Determine the thrust on the wind turbine.

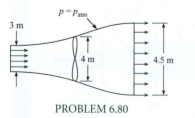

PROBLEM 6.80

6.81 (WILEY PLUS) The figure illustrates the principle of the jet pump. Derive a formula for $p_2 - p_1$ as a function of D_j, V_j, D_0, V_0, and ρ.

Assume that the fluid from the jet and the fluid initially flowing in the pipe are the same, and assume that they are completely mixed at section 2, so that the velocity is uniform across that section. Also assume that the pressures are uniform across both sections 1 and 2. What is $p_2 - p_1$ if the fluid is water, $A_j/A_0 = 1/3$, $V_j = 15$ m/s, and $V_0 = 2$ m/s? Neglect shear stress.

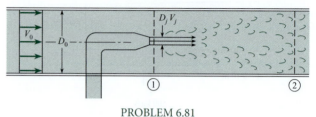

PROBLEM 6.81

6.82 Jet-type pumps are sometimes used to circulate the flow in basins in which fish are being reared. The use of a jet-type pump eliminates the need for mechanical machinery that might be injurious to the fish. The accompanying figure shows the basic concept for this type of application. For this type of basin the jets would have to increase the water surface elevation by an amount equal to $6V^2/2g$, where V is the average velocity in the basin (0.3 m/s as shown in this example). Propose a basic design for a jet system that would make such a recirculating system work for a channel 2.4 m wide and 1.2 m deep. That is, determine the speed, size, and number of jets.

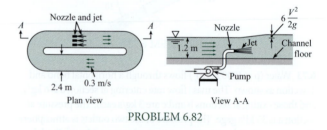

PROBLEM 6.82

6.83 An engineer is measuring the lift and drag on a wind turbine blade section mounted in a two-dimensional wind tunnel. The wind tunnel is 0.5 m high and 0.5 m deep (into the paper). The upstream wind velocity is uniform at 10 m/s, and the downstream velocity is 12 m/s and 8 m/s as shown. The vertical component of velocity is zero at both stations. The test section is 1 m long. The engineer measures the pressure distribution in the tunnel along the upper and lower walls and finds

$$p_u = 100 - 10x - 20x(1 - x)\,(\text{Pa gage})$$
$$p_l = 100 - 10x + 20x(1 - x)\,(\text{Pa gage})$$

where x is the distance in meters measured from the beginning of the test section. The gas density is homogeneous throughout and equal to 1.2 kg/m³. The lift and drag are the vectors indicated on the figure. The forces acting on the fluid are in the opposite direction to these vectors. Find the lift and drag forces acting on the wind turbine blade section.

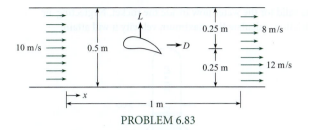

PROBLEM 6.83

6.84 🅟🅛🅤🅢 A torpedolike device is tested in a wind tunnel with an air density of 1.35 kg/m³. The tunnel is 0.9 m in diameter, the upstream pressure is 1.65 kPa gage, and the downstream pressure is 0.7 kPa gage. If the mean air velocity V is 36 m/s, what are the mass rate of flow and the maximum velocity at the downstream section at C? If the pressure is assumed to be uniform across the sections at A and C, what is the drag of the device and support vanes? Assume viscous resistance at the walls is negligible.

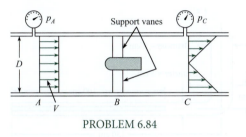

PROBLEM 6.84

6.85 A ramjet operates by taking in air at the inlet, providing fuel for combustion, and exhausting the hot air through the exit. The mass flow at the inlet and outlet of the ramjet is 60 kg/s (the mass flow rate of fuel is negligible). The inlet velocity is 225 m/s. The density of the gases at the exit is 0.25 kg/m³, and the exit area is 0.5 m². Calculate the thrust delivered by the ramjet. The ramjet is not accelerating, and the flow within the ramjet is steady.

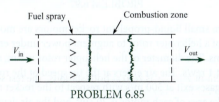

PROBLEM 6.85

6.86 🅟🅛🅤🅢 A modern turbofan engine in a commercial jet takes in air, part of which passes through the compressors, combustion chambers, and turbine, and the rest of which bypasses the compressor and is accelerated by the fans. The mass flow rate of bypass air to the mass flow rate through the compressor-combustor-turbine path is called the "bypass ratio." The total flow rate of air entering a turbofan is 300 kg/s with a velocity of 300 m/s. The engine has a bypass ratio of 2.5. The bypass air exits at 600 m/s, whereas the air through the compressor–combustor–turbine path exits at 1000 m/s. What is the thrust of the turbofan engine? Clearly show your control volume and application of momentum equation.

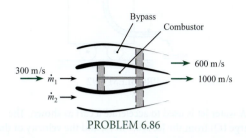

PROBLEM 6.86

Applying Momentum Equation to Moving CVs (§6.5)

6.87 Using the Internet or some other source as reference, define in your own words the meaning of "inertial reference frame."

6.88 The surface of the earth is not a true inertial reference frame because there is a centripetal acceleration due to the earth's rotation. The earth rotates once every 24 hours and has a diameter of 12,900 kilometers. What is the centripetal acceleration on the surface of the earth, and how does it compare to the gravitational acceleration?

6.89 A large tank of liquid is resting on a frictionless plane as shown. Explain in a qualitative way what will happen after the cap is removed from the short pipe.

6.90 🅟🅛🅤🅢 The open water tank shown is resting on a frictionless plane. The capped orifice on the side has a 4-cm diameter exit pipe that is located 3 m below the surface of the water. Ignore all friction effects, and determine the force necessary to keep the tank from moving when the cap is removed.

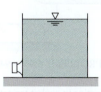

PROBLEMS 6.89, 6.90

6.91 Consider a tank of water ($\rho = 1000$ kg/m³) in a container that rests on a sled. A high pressure is maintained by a compressor so that a jet of water leaving the tank horizontally from an orifice does so at a constant speed of 25 m/s relative to the tank. If there is 0.10 m³ of water in the tank at time t and the diameter of the jet is 15 mm, what will be the acceleration of the sled at time t if the empty tank and compressor have a weight of 350 N and the coefficient of friction between the sled and the ice is 0.05?

6.92 🅟🅛🅤🅢 A cart is moving along a railroad track at a constant velocity of 5 m/s as shown. Water ($\rho = 1000$ kg/m³) issues from a nozzle at 10 m/s and is deflected through 180° by a vane on the cart. The cross-sectional area of the nozzle is 0.002 m². Calculate the resistive force on the cart.

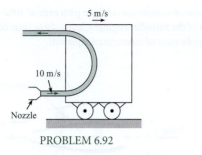

PROBLEM 6.92

6.93 A water jet is used to accelerate a cart as shown. The discharge (Q) from the jet is 0.1 m³/s, and the velocity of the jet (V_j) is 10 m/s. When the water hits the cart, it is deflected normally as shown. The mass of the cart (M) is 10 kg. The density of water (ρ) is 1000 kg/m³. There is no resistance on the cart, and the initial velocity of the cart is zero. The mass of the water in the jet is much less than the mass of the cart. Derive an equation for the acceleration of the cart as a function of Q, ρ, V_c, M, and V_j. Evaluate the acceleration of the cart when the velocity is 5 m/s.

6.94 PLUS A water jet strikes a cart as shown. After striking the cart, the water is deflected vertically with respect to the cart. The cart is initially at rest and is accelerated by the water jet. The mass in the water jet is much less than that of the cart. There is no resistance on the cart. The mass flow rate from the jet is 45 kg/s. The mass of the cart is 100 kg. Find the time required for the cart to achieve a speed one-half of the jet speed.

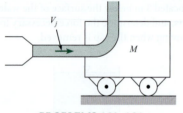

PROBLEMS 6.93, 6.94

6.95 It is common practice in rocket trajectory analyses to neglect the body-force term and drag, so the velocity at burnout is given by

$$v_{bo} = \frac{T}{\lambda} \ln \frac{M_0}{M_f}$$

Assuming a thrust-to-mass-flow ratio of 3000 N · s/kg and a final mass of 50 kg, calculate the initial mass needed to establish the rocket in an earth orbit at a velocity of 7200 m/s.

6.96 A very popular toy on the market several years ago was the water rocket. Water (at 10°C) was loaded into a plastic rocket and pressurized with a hand pump. The rocket was released and would travel a considerable distance in the air. Assume that a water rocket has a mass of 50 g and is charged with 100 g of water. The pressure inside the rocket is 100 kPa gage. The exit area is one-tenth of the chamber cross-sectional area. The inside diameter of the rocket is 5 cm. Assume that Bernoulli's equation

is valid for the water flow inside the rocket. Neglecting air friction, calculate the maximum velocity it will attain.

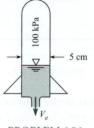

PROBLEM 6.96

The Angular Momentum Equation (§6.6)

6.97 PLUS Water ($\rho = 1000$ kg/m³) is discharged from the slot in the pipe as shown. If the resulting two-dimensional jet is 100 cm long and 15 mm thick, and if the pressure at section A-A is 30 kPa, what is the reaction at section A-A? In this calculation, do not consider the weight of the pipe.

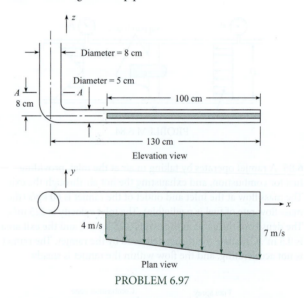

PROBLEM 6.97

6.98 Two small liquid-propellant rocket motors are mounted at the tips of a helicopter rotor to augment power under emergency conditions. The diameter of the helicopter rotor is 7 m, and it rotates at 1 rev/s. The air enters at the tip speed of the rotor, and exhaust gases exit at 500 m/s with respect to the rocket motor. The intake area of each motor is 20 cm², and the air density is 1.2 kg/m³. Calculate the power provided by the rocket motors. Neglect the mass rate of flow of fuel in this calculation.

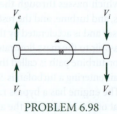

PROBLEM 6.98

6.99 Design a rotating lawn sprinkler to deliver 0.625 m of water per hour over a circle of 15 m radius. Make the simplifying assumptions that the pressure to the sprinkler is 350 kPa gage and that frictional effects involving the flow of water through the sprinkler flow passages are negligible (the Bernoulli equation is applicable). However, do not neglect the friction between the rotating element and the fixed base of the sprinkler.

6.100 PLUS What is the force and moment reaction at section 1? Water (at 10°C) is flowing in the system. Neglect gravitational forces.

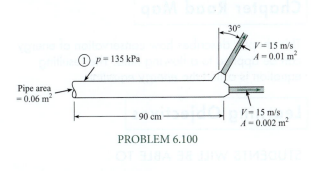

PROBLEM 6.100

6.101 What is the reaction at section 1? Water ($\rho = 1000$ kg/m³) is flowing, and the axes of the two jets lie in a vertical plane. The pipe and nozzle system weighs 90 N.

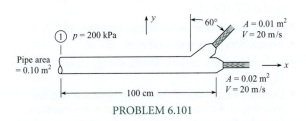

PROBLEM 6.101

6.102 A reducing pipe bend is held in place by a pedestal as shown. There are expansion joints at sections 1 and 2, so no force

is transmitted through the pipe past these sections. The pressure at section 1 is 140 kPa gage, and the rate of flow of water ($\rho = 1000$ kg/m³) is 0.06 m³/s. Find the force and moment that must be applied at section 3 to hold the bend stationary. Assume the flow is irrotational, and neglect the influence of gravity.

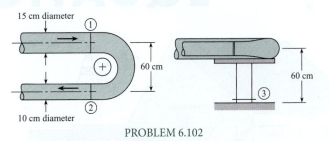

PROBLEM 6.102

6.103 A centrifugal fan is used to pump air. The fan rotor is 0.3 m in diameter, and the blade spacing is 5 cm. The air enters with no angular momentum and exits radially with respect to the fan rotor. The discharge is 0.7 m³/s. The rotor spins at 3600 rev/min. The air is at atmospheric pressure and a temperature of 15.5°C. Neglect the compressibility of the air. Calculate the power (kW) required to operate the fan.

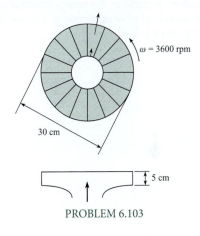

PROBLEM 6.103

7 THE ENERGY EQUATION

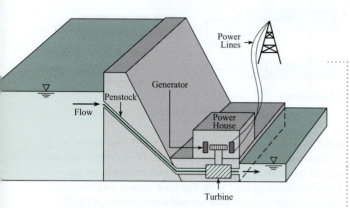

FIGURE 7.1

The energy equation can be applied to hydroelectric power generation. In addition, the energy equation can be applied to thousands of other applications. It is one of the most useful equations in fluid mechanics.

Chapter Road Map

This chapter describes how conservation of energy can be applied to a flowing fluid. The resulting equation is called the *energy equation*.

Learning Objectives

STUDENTS WILL BE ABLE TO

- Explain the meaning of energy, work, and power. (§7.1)
- Classify energy into categories. (§7.1)
- Define a pump and a turbine. (§7.1)
- Explain conservation of energy for a closed system and a CV. (§7.2)
- List the steps to derive the energy equation. (§7.3)
- Explain flow work and shaft work. (§7.3)
- Define head loss and the kinetic energy correction factor. (§7.3)
- Describe the physics of the energy equation and the meaning of the variables that appear in the equation. Describe the process for applying the energy equation. Apply the energy equation. (§7.3)
- Apply the power equation. (§7.4)
- Define mechanical efficiency and apply this concept. (§7.5)
- Contrast the energy equation and the Bernoulli equation. (§7.6)
- Calculate head loss for a sudden expansion. (§7.7)
- Explain the conceptual foundations of the energy grade line and hydraulic grade line. Sketch these lines. (§7.8)

7.1 Energy Concepts

The energy equation is built on foundational concepts that are introduced in this section.

Energy

Energy is the property of a system that characterizes the amount of work that this system can do on its environment. In simple terms, if matter (i.e., the system) can be used to lift a weight, then that matter has energy.

Examples

- Water behind a dam has energy because the water can be directed through a pipe (i.e., a penstock), then used to rotate a wheel (i.e., a water turbine) that lifts a weight. Of course this work can also rotate the shaft of an electrical generator, which is used to produce electrical power.

- Wind has energy because the wind can pass across a set of blades (e.g., a windmill), rotate the blades, and lift a weight that is attached to a rotating shaft. This shaft can also do work to rotate the shaft of an electrical generator.

- Gasoline has energy because it can be placed into a cylinder (e.g., a gas engine), burned and expanded to move a piston in a cylinder. This moving cylinder can then be connected to a mechanism that is used to lift a weight.

The SI unit of energy, the *joule*, is the energy associated with a force of one newton acting through a distance of one meter. For example, if a person with a weight of 700 newtons travels up a 10-meter flight of stairs, their gravitational potential energy has changed by $\Delta PE = (700 \text{ N})(10 \text{ m}) = 700 \text{ N} \cdot \text{m} = 700 \text{ J}$. In traditional units, the unit of energy, the *foot-pound-force* (lbf) is defined as energy associated with a force of 1.0 lbf moving through a distance of 1.0 foot.

Another way to define a unit of energy is describe the heating of water. A *small calorie* (cal) is the amount of energy required to increase the temperature of 1.0 gram of water by 1°C. The unit conversion between small calories and joules is 1.0 cal = 4.187 J. The *large calorie* (Cal), is the amount of energy to raise 1.0 kg of water by 1°C. Thus, 1.0 Cal = 4187 J.

Energy can be classified into categories.

- **Mechanical Energy.** This is the energy associated with motion (i.e., *kinetic energy*) plus the energy associated with position in a field. Regarding position in a field, this refers to position in a gravitational field (i.e., gravitational potential energy) and to deflection of an elastic object such as a spring (i.e., spring potential energy).

- **Thermal Energy.** This is energy associated with temperature changes and phase changes. For example, select a system comprised of 1 kg of ice (about 1 liter). The energy to melt the ice is 334 kJ. The energy to raise the temperature of the liquid water from 0°C to 100°C is 419 kJ.

- **Chemical Energy.** This is the energy associated with chemical bonds between elements. For example, when methane (CH_4) is burned, there is a chemical reaction that involves the breaking of the bonds in the methane and formation of new bonds to produce CO_2 and water. This chemical reaction releases heat, which is another way of saying the chemical energy is converted to thermal energy during combustion.

- **Electrical Energy.** This is the energy associated with electrical change. For example, a charged capacitor contains the amount of electrical energy $\Delta E = 1/2\ CV^2$ where C is capacitance and V is voltage.
- **Nuclear Energy.** This is energy associated with the binding of the particles in the nucleus of an atom. For example, when the uranium atom divides into two other atoms during fission, energy is released.

Work

Work is done on a system when a force acts on the system over a distance. In the absence of completing effects, the effect of work is to increase the energy of the system.

> **EXAMPLE.** For the spray bottle in Fig. 7.2a, work is done when a finger acts through a distance as the trigger is displaced. Work is also done by the piston as it exerts a force on the liquid as the piston is displaced. The magnitude of work done W can be evaluated using.

$$W = \int_{s_1}^{s_2} \mathbf{F} \cdot \mathbf{ds} \tag{7.1}$$

where $\mathbf{s}$ is position, and $\mathbf{F}$ is force. The effect of this work is to increase the energy of the system in several ways: water is lifted through a elevation, thereby increasing its potential energy, and water is sprayed out the nozzle, thereby increasing its kinetic energy.

FIGURE 7.2

(a) In a spray bottle, a piston pump does work on the fluid, thereby increasing the energy of the liquid. (b) For a wind turbine, the air does work on the blade, thereby allowing the wind turbine to be used to produce electrical power.

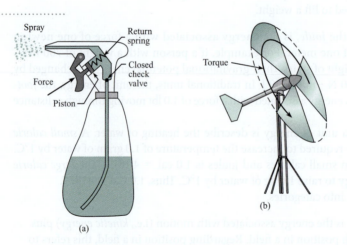

Spray

Return spring

Closed check valve

Force

Piston

Torque

(a)

(b)

> **EXAMPLE.** For the wind turbine in Fig 7.2b, work is done by air that exerts a force on the blades and causes the blades to rotate through a distance.

Work has the same units as energy: joules or newton-meters in SI and ft-lbf in traditional units.

Power

Power, which expresses a rate of work or energy, is defined by

$$P \equiv \frac{\text{quantity of work (or energy)}}{\text{interval of time}} = \lim_{\Delta t \to 0} \frac{\Delta W}{\Delta t} = \dot{W} \tag{7.2}$$

Equation (7.2) is defined at an instant in time because power can vary with time. To calculate power, engineers use several different equations. For rectilinear motion, such as a car or bicycle, the amount of work is the product of force and displacement $\Delta W = F \Delta x$: Then, power can be found using

$$P = \lim_{\Delta t \to 0} \frac{F \Delta x}{\Delta t} = FV \tag{7.3a}$$

where V is the velocity of the moving body.

When a shaft is rotating (Fig. 7.2b), the amount of work is given by the product of torque and angular displacement $\Delta W = T \Delta \theta$. In this case, the power equation is

$$P = \lim_{\Delta t \to 0} \frac{T \Delta \theta}{\Delta t} = T \omega \tag{7.3b}$$

where ω is the angular speed. The SI units of angular speed are rad/s.

Because power has units of energy per time, the SI unit is a joule/second, which is called a watt. Common units for power are the watt (W), horsepower (hp), and the ft-lbf/s. Some typical values of power:

- A incandescent lightbulb can use 60 to 100 J/s of energy.

- A well-conditioned athlete can sustain a power output of about 300 J/s for an hour.

- A typical midsize car (2011 Toyota Camry) has a rated power of 126 kW.

- A large hydroelectric facility (i.e., Bonneville Dam on the Columbia River 40 miles east of Portland, Oregon) has a rated power of 1080 MW.

Pumps and Turbines

A **turbine** is a machine that is used to extract energy from a flowing fluid.* Examples of turbines include the horizontal-axis wind turbine shown in Fig. 7.2b, the gas turbine, the Kaplan turbine, the Francis turbine, and the Pelton wheel.

A **pump** is a machine that is used to provide energy to a flowing fluid. Examples of pumps include the piston pump shown in Fig. 7.2a, the centrifugal pump, the diaphragm pump, and the gear pump.

7.2 Conservation of Energy

When James Prescott Joule died, his obituary in *The Electrical Engineer* (1) stated that

Very few indeed who read this announcement will realize how great of a man has passed away; and yet it must be admitted by those most competent to judge that his name must be classed among the greatest original workers in science.

Joule was a brewer who did science as a hobby, yet he formulated one of the most important scientific laws ever developed. However, Joule's theory of conservation of energy was so

---

*The engine on a jet, which is called a gas turbine, is a notable exception. The jet engine adds energy to a flowing fluid, thereby increasing the momentum of a fluid jet and producing thrust.

controversial that he could not get a scientific journal to publish it. So his theory first appeared in a local Manchester newspaper (2). What a fine example of persistence! Nowadays, Joule's ideas about work and energy are foundational to engineering. This section introduces Joule's theory.

Joule's Theory of Energy Conservation

Joule recognized that the energy of a *closed system* can be changed in only two ways.

- **Work.** The energy of the system can be changed by work interactions at the boundary.
- **Heat Transfer.** The energy of the system can change by heat transfer across the boundary. **Heat transfer** can be defined as the transfer of thermal energy from hot to cold by mechanisms of conduction, convection, and radiation.

Joule's idea of energy conservation is illustrated in Fig. 7.3. The system is represented by the blue box. The scale on the left side of the figure represents the quantity of energy in the system. The arrows on the right side illustrate that energy can increase or decrease via work or heat transfer interactions. Note that energy is a property of a system, whereas work and heat transfer are interactions that occur on system boundaries.

FIGURE 7.3

The law of conservation of energy for a closed system.

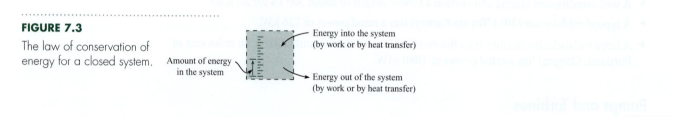

Amount of energy in the system

Energy into the system
(by work or by heat transfer)

Energy out of the system
(by work or by heat transfer)

The work and energy balance proposed by Joule is captured with an equation called the first law of thermodynamics:

$$\Delta E = Q - W$$

$$\left\{ \begin{array}{c} \text{increase in} \\ \text{energy stored} \\ \text{in the system} \end{array} \right\} = \left\{ \begin{array}{c} \text{amount of energy} \\ \text{that entered system} \\ \text{by heat transfer} \end{array} \right\} - \left\{ \begin{array}{c} \text{amount of energy} \\ \text{that left system} \\ \text{due to work} \end{array} \right\} \qquad (7.4)$$

Terms in Eq. (7.4) have units of joules, and the equation is applied during a time interval when the system undergoes a process to move from state 1 to state 2. To modify Eq. (7.4) so that it applies at an instant in time, take the derivative to give

$$\frac{dE}{dt} = \dot{Q} - \dot{W} \qquad (7.5)$$

Eq. (7.5) applies at an instant in time and has units of joules per second or watts. The work and time terms have sign conventions:

- W and $\dot{W}$ are positive if work is done by the system on the surroundings.
- W and $\dot{W}$ are negative if work is done by the surroundings on the system.
- Q and $\dot{Q}$ are positive if heat (i.e., thermal energy) is transferred into the system.
- Q and $\dot{Q}$ are negative if heat (i.e., thermal energy) is transferred out of the system.

✔ **CHECKPOINT PROBLEM 7.1**

A battery is used to power a DC motor, which is then used to drive a centrifugal pump. For the indicated system, which statements are true? Circle all that apply. Assume steady state operation.

a. $\dot{W} > 0$

b. $\dot{W} < 0$

c. $\dot{Q} > 0$

d. $\dot{Q} < 0$

e. $\dfrac{dE}{dt} > 0$

f. $\dfrac{dE}{dt} < 0$

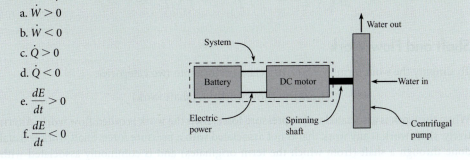

Control Volume (Open System)

Eq. (7.5) applies to a *closed system*. To extend it to a CV, apply the Reynolds transport theorem Eq. (5.23). Let the extensive property be energy ($B_{sys} = E$), and let $b = e$ to obtain

$$\dot{Q} - \dot{W} = \frac{d}{dt} \int_{cv} e\rho \, d\mathcal{V} + \int_{cs} e\rho \mathbf{V} \cdot d\mathbf{A} \tag{7.6}$$

where e is energy per mass in the fluid. Eq. (7.6) is the general form of conservation of energy for a control volume. However, most problems in fluid mechanics can be solved with a simpler form of this equation. This simpler equation will be derived in the next section.

7.3 The Energy Equation

This section shows how to simplify Eq. (7.6) to a form that is convenient for problems that occur in fluid mechanics.

Select Eq. (7.6). Then, let $e = e_k + e_p + u$ where e_k is the kinetic energy per unit mass, e_p is the gravitational potential energy per unit mass, and u is the internal energy* per unit mass.

$$\dot{Q} - \dot{W} = \frac{d}{dt} \int_{cv} (e_k + e_p + u)\rho \, d\mathcal{V} + \int_{cs} (e_k + e_p + u)\rho \mathbf{V} \cdot d\mathbf{A} \tag{7.7}$$

Next, let[†]

$$e_k = \frac{\text{kinetic energy of a fluid particle}}{\text{mass of this fluid particle}} = \frac{mV^2/2}{m} = \frac{V^2}{2} \tag{7.8}$$

Similarly, let

$$e_p = \frac{\text{gravitational potential energy of a fluid particle}}{\text{mass of this fluid particle}} = \frac{mgz}{m} = gz \tag{7.9}$$

---

*By definition, internal energy contains all forms of energy that are not kinetic energy or gravitational potential energy.
[†]It is assumed that the control surface is not accelerating, so V, which is referenced to the control surface, is also referenced to an inertial reference frame.

where z is the elevation measured relative to a datum. When Eqs. (7.8) and (7.9) are substituted into Eq. (7.7), the result is

$$\dot{Q} - \dot{W} = \frac{d}{dt} \int_{cv} \left(\frac{V^2}{2} + gz + u \right) \rho \, dV + \int_{cs} \left(\frac{V^2}{2} + gz + u \right) \rho \mathbf{V} \cdot d\mathbf{A} \qquad \textbf{(7.10)}$$

Shaft and Flow Work

To simplify the work term in Eq. (7.10), classify work into two categories:

$$(\text{work}) = (\text{flow work}) + (\text{shaft work})$$

When the work is associated with a pressure force, then the work is called **flow work**. Alternatively, **shaft work** is any work that is not associated with a pressure force. Shaft work is usually done through a shaft (from which the term originates) and is commonly associated with a pump or turbine. According to the sign convention for work, pump work is negative. Similarly, turbine work is positive. Thus,

$$\dot{W}_{\text{shaft}} = \dot{W}_{\text{turbines}} - \dot{W}_{\text{pumps}} = \dot{W}_t - \dot{W}_p \qquad \textbf{(7.11)}$$

To derive an equation for flow work, use the idea that work equals force times distance. Begin the derivation by defining a control volume situated inside a converging pipe (Fig. 7.4). At section 2, the fluid that is inside the control volume will push on the fluid that is outside the control volume. The magnitude of the pushing force is $p_2 A_2$. During a time interval Δt, the displacement of the fluid at section 2 is $\Delta x_2 = V_2 \Delta t$. Thus, the amount of work is

$$\Delta W_2 = (F_2)(\Delta x_2) = (p_2 A_2)(V_2 \Delta t) \qquad \textbf{(7.12)}$$

FIGURE 7.4

Sketch for deriving flow work.

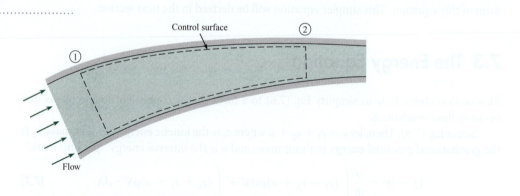

Convert the amount of work given by Eq. (7.12) into a rate of work:

$$\dot{W}_2 = \lim_{\Delta t \to 0} \frac{\Delta W_2}{\Delta t} = p_2 A_2 V_2 = \left(\frac{p_2}{\rho} \right) (\rho A_2 V_2) = \dot{m} \left(\frac{p_2}{\rho} \right) \qquad \textbf{(7.13)}$$

This work is positive because the fluid inside the control volume is doing work on the environment. In a similar manner, the flow work at section 1 is negative and is given by

$$\dot{W}_1 = -\dot{m} \left(\frac{p_1}{\rho} \right)$$

The net flow work for the situation pictured in Fig. 7.4 is

$$\dot{W}_{\text{flow}} = \dot{W}_2 + \dot{W}_1 = \dot{m} \left(\frac{p_2}{\rho} \right) - \dot{m} \left(\frac{p_1}{\rho} \right) \qquad \textbf{(7.14)}$$

Equation (7.14) can be generalized to a situation involving multiple streams of fluid passing across a control surface:

$$\dot{W}_{\text{flow}} = \sum_{\text{outlets}} \dot{m}_{\text{out}}\left(\frac{p_{\text{out}}}{\rho}\right) - \sum_{\text{inlets}} \dot{m}_{\text{in}}\left(\frac{p_{\text{in}}}{\rho}\right) \tag{7.15}$$

To develop a general equation for flow work, use integrals to account for velocity and pressure variations on the control surface. Also, use the dot product to account for flow direction. The general equation for flow work is

$$\dot{W}_{\text{flow}} = \int_{\text{cs}} \left(\frac{p}{\rho}\right)\rho\mathbf{V} \cdot d\mathbf{A} \tag{7.16}$$

In summary, the work term is the sum of flow work [Eq. (7.16)] and shaft work [Eq. (7.11)]:

$$\dot{W} = \dot{W}_{\text{flow}} + \dot{W}_{\text{shaft}} = \left(\int_{\text{cs}} \left(\frac{p}{\rho}\right)\rho\mathbf{V} \cdot d\mathbf{A}\right) + \dot{W}_{\text{shaft}} \tag{7.17}$$

Introduce the work term from Eq. (7.17) into Eq. (7.10) and let $\dot{W}_{\text{shaft}} = \dot{W}_s$

$$\dot{Q} - \dot{W}_s - \int_{\text{cs}} \frac{p}{\rho}\rho\mathbf{V} \cdot d\mathbf{A}$$

$$= \frac{d}{dt}\int_{\text{cv}}\left(\frac{V^2}{2} + gz + u\right)\rho\,d\forall + \int_{\text{cs}}\left(\frac{V^2}{2} + gz + u\right)\rho\mathbf{V} \cdot d\mathbf{A} \tag{7.18}$$

In Eq. (7.18), combine the last term on the left side with the last term on the right side:

$$\dot{Q} - \dot{W}_s = \frac{d}{dt}\int_{\text{cv}}\left(\frac{V^2}{2} + gz + u\right)\rho\,d\forall + \int_{\text{cs}}\left(\frac{V^2}{2} + gz + u + \frac{p}{\rho}\right)\rho\mathbf{V} \cdot d\mathbf{A} \tag{7.19}$$

Replace $p/\rho + u$ by the specific enthalpy, h. The integral form of the energy principle is

$$\dot{Q} - \dot{W}_s = \frac{d}{dt}\int_{\text{cv}}\left(\frac{V^2}{2} + gz + u\right)\rho\,d\forall + \int_{\text{cs}}\left(\frac{V^2}{2} + gz + h\right)\rho\mathbf{V} \cdot d\mathbf{A} \tag{7.20}$$

Kinetic Energy Correction Factor

The next simplification is to extract the velocity terms out of the integrals on the right side of Eq. (7.20). This is done by introducing the kinetic energy correction factor.

Figure 7.5 shows fluid that is pumped through a pipe. At sections 1 and 2, kinetic energy is transported across the control surface by the flowing fluid. To derive an equation for this kinetic energy, start with the mass flow rate equation.

$$\dot{m} = \rho A \overline{V} = \int_A \rho V dA$$

This integral can be conceptualized as adding up the mass of each fluid particle that is crossing the section area and then dividing by the time interval associated with this crossing. To convert this integral to kinetic energy (KE), multiply the mass of each fluid particle by $(V^2/2)$.

$$\left\{\begin{array}{c} \text{Rate of KE} \\ \text{transported} \\ \text{across a section} \end{array}\right\} = \int_A \rho V\left(\frac{V^2}{2}\right)dA = \int_A \frac{\rho V^3 dA}{2}$$

FIGURE 7.5

Flow carries kinetic energy into and out of a control volume.

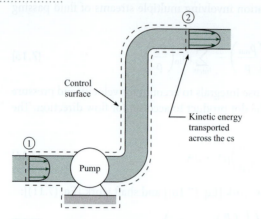

The **kinetic energy correction factor** is defined as

$$\alpha = \frac{\text{actual KE/time that crosses a section}}{\text{KE/time by assuming a uniform velocity distribution}} = \frac{\displaystyle\int_A \frac{\rho V^3 dA}{2}}{\displaystyle\frac{\overline{V}^3}{2}\int_A \rho\,dA}$$

For a constant density fluid, this equation simplifies to

$$\alpha = \frac{1}{A}\int_A \left(\frac{V}{\overline{V}}\right)^3 dA \qquad (7.21)$$

For theoretical development, α is found by integrating the velocity profile using Eq. (7.21). This approach, illustrated in Example 7.1, is a lot of work. Thus in application, engineers commonly estimate a value of α. Some *guidelines* are listed here.

- For fully developed laminar flow in a pipe, the velocity distribution is parabolic. Use $\alpha = 2.0$ because this is the correct value as shown by Example 7.1.
- For fully developed turbulent flow in a pipe, $\alpha \approx 1.05$ because the velocity profile is pluglike. Use $\alpha = 1.0$ for this case.
- For flow at the exit of a nozzle or converging section, use $\alpha = 1.0$ because converging flow leads to a uniform velocity profile. This is why wind tunnels use converging sections.
- For a uniform flow such as air flow in a wind tunnel or air flow incident on a wind turbine, use $\alpha = 1.0$.

EXAMPLE 7.1

Calculating the Kinetic Energy Correction Factor for Laminar Flow

Problem Statement

The velocity distribution for laminar flow in a pipe is given by the equation

$$V(r) = V_{\max}\left[1 - \left(\frac{r}{r_0}\right)^2\right]$$

where $V_{\max}$ is the velocity in the center of the pipe, r_0 is the radius of the pipe, and r is the radial distance from the center. Find the kinetic-energy correction factor α.

Define the Situation

There is laminar flow in a round pipe.

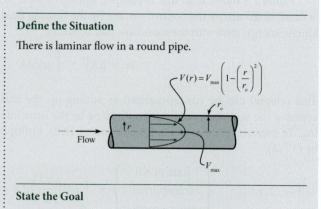

State the Goal

α ◄ Find the kinetic-energy correction factor (no units)

Generate Ideas and Make a Plan

Because the goal is α, apply the definition given by Eq. (7.21).

$$\alpha = \frac{1}{A}\int_A \left(\frac{V(r)}{\bar{V}}\right)^3 dA \tag{a}$$

Eq. (a) has one known (A) and two unknowns (dA, $\bar{V}$). To find dA use Fig. 5.3 (see page 175).

$$dA = 2\pi r dr \tag{b}$$

To find $\bar{V}$, apply the *flow rate equation*,

$$\bar{V} = \frac{1}{A}\int_A V(r)dA = \frac{1}{\pi r_o^2}\int_{r=0}^{r=r_o} V(r)2\pi r dr \tag{c}$$

Now the problem is cracked. There are three equations and three unknowns. The plan is:

1. Find the mean velocity $\bar{V}$ using Eq. (c)
2. Plug $\bar{V}$ into Eq. (a) and integrate

Take Action (Execute the Plan)

1. Flow Rate Equation (find mean velocity)

$$\bar{V} = \frac{1}{\pi r_o^2}\left[\int_0^{r_o} V_{max}\left(1 - \frac{r^2}{r_o^2}\right)2\pi r\, dr\right]$$

$$= \frac{2V_{max}}{r_o^2}\left[\int_0^{r_o}\left(1 - \frac{r^2}{r_o^2}\right)r\, dr\right] = \frac{2V_{max}}{r_o^2}\left[\int_0^{r_o}\left(r - \frac{r^3}{r_o^2}\right)dr\right]$$

$$= \frac{2V_{max}}{r_o^2}\left[\left(\frac{r^2}{2} - \frac{r^4}{4r_o^2}\right)\Big|_0^{r_o}\right] = \frac{2V_{max}}{r_o^2}\left[\frac{r_o^2}{2} - \frac{r_o^2}{4}\right] = V_{max}/2$$

---

2. Definition of α

$$\alpha = \frac{1}{A}\left[\int_A \left(\frac{V(r)}{\bar{V}}\right)^3 dA\right] = \frac{1}{\pi r_o^2 \bar{V}^3}\left[\int_0^{r_o} V(r)^3 2\pi r\, dr\right]$$

$$= \frac{1}{\pi r_o^2 (V_{max}/2)^3}\left[\int_0^{r_o}\left[V_{max}\left(1 - \frac{r^2}{r_o^2}\right)\right]^3 2\pi r\, dr\right]$$

$$= \frac{16}{r_o^2}\left[\int_0^{r_o}\left(1 - \frac{r^2}{r_o^2}\right)^3 r\, dr\right]$$

To evaluate the integral, make a change of variable by letting $u = (1 - r^2/r_o^2)$. The integral becomes

$$\alpha = \left(\frac{16}{r_o^2}\right)\left(-\frac{r_o^2}{2}\right)\left(\int_1^0 u^3\, du\right) = 8\left(\int_0^1 u^3\, du\right)$$

$$= 8\left(\frac{u^4}{4}\Big|_0^1\right) = 8\left(\frac{1}{4}\right)$$

$$\boxed{\alpha = 2}$$

Review the Solution and the Process

1. *Knowledge.* Laminar fully developed flow in a round pipe is called Poiseuille flow. Useful facts:

   - The velocity profile is parabolic.
   - The mean velocity is one-half of the maximum (centerline) velocity: $\bar{V} = V_{max}/2$.
   - The kinetic energy correction factor is $\alpha = 2$.

2. *Knowledge.* In practice, engineers commonly estimate α. The purpose of this example is to illustrate how to calculate α.

Last Steps of the Derivation

Now that the KE correction factor is available, the derivation of the energy equation may be completed. Begin by applying Eq. (7.20) to the control volume shown in Fig. 7.5. Assume steady flow and that velocity is normal to the control surfaces. Then, Eq. (7.20) simplifies to:

$$\dot{Q} - \dot{W}_s + \int_{A_1}\left(\frac{p_1}{\rho} + gz_1 + u_1\right)\rho V_1 dA_1 + \int_{A_1}\frac{\rho V_1^3}{2}dA_1$$

$$= \int_{A_2}\left(\frac{p_2}{\rho} + gz_2 + u_2\right)\rho V_2 dA_2 + \int_{A_2}\frac{\rho V_2^3}{2}dA_2 \tag{7.22}$$

Assume that piezometric head $p/\gamma + z$ is constant across sections 1 and 2.[*] If temperature is also assumed constant across each section, then $p/\rho + gz + u$ can be taken outside the integral to yield

$$\dot{Q} - \dot{W}_s + \left(\frac{p_1}{\rho} + gz_1 + u_1\right)\int_{A_1}\rho V_1 dA_1 + \int_{A_1}\rho\frac{V_1^3}{2}dA_1$$

$$= \left(\frac{p_2}{\rho} + gz_2 + u_2\right)\int_{A_2}\rho V_2 dA_2 + \int_{A_2}\rho\frac{V_2^3}{2}dA_2 \tag{7.23}$$

---

[*]Euler's equation can be used to show that pressure variation normal to rectilinear streamlines is hydrostatic.

Next, factor out $\int \rho V \, dA = \rho \overline{V} A = \dot{m}$ from each term in Eq. (7.23). Because $\dot{m}$ does not appear as a factor of $\int (\rho V^3/2) dA$, express $\int (\rho V^3/2) dA$ as $\alpha(\rho \overline{V}^3/2)A$, where α is the kinetic energy correction factor:

$$\dot{Q} - \dot{W}_s + \left(\frac{p_1}{\rho} + gz_1 + u_1 + \alpha_1 \frac{\overline{V}_1^2}{2} \right) \dot{m} = \left(\frac{p_2}{\rho} + gz_2 + u_2 + \alpha_2 \frac{\overline{V}_2^2}{2} \right) \dot{m} \qquad \textbf{(7.24)}$$

Divide through by $\dot{m}$:

$$\frac{1}{\dot{m}}(\dot{Q} - \dot{W}_s) + \frac{p_1}{\rho} + gz_1 + u_1 + \alpha_1 \frac{\overline{V}_1^2}{2} = \frac{p_2}{\rho} + gz_2 + u_2 + \alpha_2 \frac{\overline{V}_2^2}{2} \qquad \textbf{(7.25)}$$

Introduce Eq. (7.11) into Eq. (7.25):

$$\frac{\dot{W}_p}{\dot{m}g} + \frac{p_1}{\gamma} + z_1 + \alpha_1 \frac{\overline{V}_1^2}{2g} = \frac{\dot{W}_t}{\dot{m}g} + \frac{p_2}{\gamma} + z_2 + \alpha_2 \frac{\overline{V}_2^2}{2g} + \frac{u_2 - u_1}{g} - \frac{\dot{Q}}{\dot{m}g} \qquad \textbf{(7.26)}$$

Introduce **pump head** and **turbine head**:

$$Pump\ head = h_p = \frac{\dot{W}_p}{\dot{m}g} = \frac{\text{work/time done by pump on flow}}{\text{weight/time of flowing fluid}}$$

$$Turbine\ head = h_t = \frac{\dot{W}_t}{\dot{m}g} = \frac{\text{work/time done by flow on turbine}}{\text{weight/time of flowing fluid}}$$

$$\textbf{(7.27)}$$

Equation (7.26) becomes:

$$\frac{p_1}{\gamma} = \alpha_1 \frac{\overline{V}_1^2}{2g} + z_1 + h_p = \frac{p_2}{\gamma} + \alpha_2 \frac{\overline{V}_2^2}{2g} + z_2 + h_t + \left[\frac{1}{g}(u_2 - u_1) - \frac{\dot{Q}}{\dot{m}g} \right] \qquad \textbf{(7.28)}$$

Equation (7.28) is separated into terms that represent mechanical energy (nonbracketed terms) and terms that represent thermal energy (the bracketed term). This bracketed term is always positive because of the second law of thermodynamics. This term is called head loss and is represented by h_L. **Head loss** is the conversion of useful mechanical energy to waste thermal energy through viscous action. Head loss is analogous to thermal energy (heat) that is produced by Coulomb friction. When the bracketed term is replaced by head loss h_L, Eq. (7.28) becomes the *energy equation.*

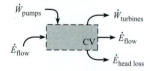

Energy into CV by flow and pumps $=$ Energy out of CV by flow, turbines, and head loss

$$\left(\frac{p_1}{\gamma} + \alpha_1 \frac{\overline{V}_1^2}{2g} + z_1 \right) + h_p = \left(\frac{p_2}{\gamma} + \alpha_2 \frac{\overline{V}_2^2}{2g} + z_2 \right) + h_t + h_L \qquad \textbf{(7.29)}$$

Physical Interpretation of the Energy Equation

The energy equation describes an energy balance for a control volume (Fig. 7.6). The inflows of energy are balanced with the outflows of energy.* Regarding inflows, energy can be transported across the control surface by the flowing fluid or a pump can do work on the fluid and thereby add energy to the fluid. Regarding outflows, energy within the flow can be used to do work on a turbine, energy can be transported across the control surface by the flowing fluid, or mechanical energy can be converted to waste thermal heat via head loss.

*The term "$\dot{E}_{\text{flow}}$" includes a work term, namely flow work. Remember that energy is a property of a system, whereas work and heat transfer are interactions that occur on system boundaries. Here, we are using the term "energy balance" to describe (energy terms) + (work terms) + (heat transfer terms).

The energy balance can also be expressed using head:

$$\left(\frac{p_1}{\gamma} + \alpha_1 \frac{\overline{V}_1^2}{2g} + z_1\right) + h_p = \left(\frac{p_2}{\gamma} + \alpha_2 \frac{\overline{V}_2^2}{2g} + z_2\right) + h_t + h_L$$

$$\begin{pmatrix} \text{pressure head} \\ \text{velocity head} \\ \text{elevation head} \end{pmatrix}_1 + \begin{pmatrix} \text{pump} \\ \text{head} \end{pmatrix} = \begin{pmatrix} \text{pressure head} \\ \text{velocity head} \\ \text{elevation head} \end{pmatrix}_2 + \begin{pmatrix} \text{turbine} \\ \text{head} \end{pmatrix} + \begin{pmatrix} \text{head} \\ \text{loss} \end{pmatrix}$$

Head can be thought of as the *ratio of energy to weight for a fluid particle.* Or, head can describe the *energy per time that is passing across a section* because head and power are related by $P = \dot{m}gh$.

✔ CHECKPOINT PROBLEM 7.2

As shown, a pump moves water from a lower reservoir to a higher reservoir. The pipe has a constant diameter. Which statements are true? (circle all that apply)

 a. Pressure head at 1 is zero.

 b. Pressure head at 2 is zero.

 c. Velocity head at 1 > velocity head at 2.

 d. Velocity head at 2 < velocity head at 1.

 e. Pump head is negative.

 f. Pump head is positive.

 g. Head loss is positive.

 h. Head loss is negative.

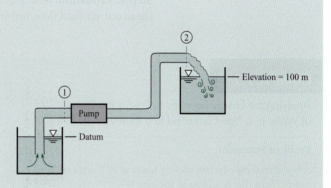

Working Equations

Table 7.1 summarizes the energy equation, its variables, and the main assumptions.

TABLE 7.1 Summary of the Energy Equation

Description	Equation	Terms

The energy equation has only one form.
 	
Major assumptions
 	
• Steady state; no energy accumulation in CV
• CV has one inlet and one outlet
• Constant density flow
• All thermal energy terms (except for head loss) can be neglected.
• Streamlines are straight and parallel at each section
• Temperature is constant across each section.
	$$\left(\frac{p_1}{\gamma} + \alpha_1 \frac{\overline{V}_1^2}{2g} + z_1\right) + h_p =$$ $$\left(\frac{p_2}{\gamma} + \alpha_2 \frac{\overline{V}_2^2}{2g} + z_2\right) + h_t + h_L$$ Eq. (7.29)	$$\left(\frac{p}{\gamma} + \alpha \frac{\overline{V}^2}{2g} + z\right) = \begin{pmatrix} \text{energy/weight transported} \\ \text{into or out of cv} \\ \text{by fluid flow} \end{pmatrix}$$ p/γ = pressure head at cs (m)
		$\alpha \dfrac{\overline{V}^2}{2g}$ = velocity head at cs (m)
		   (α = kinetic energy (KE) correction factor at cs)
		   ($\alpha \approx 1.0$ for turbulent flow)
		   ($\alpha \approx 1.0$ for nozzles)
		   ($\alpha \approx 2.0$ for full-developed laminar flow in round pipe)
		z = elevation head at cs (m)
		
		h_p = head added by a pump (m)
		h_t = head removed by a turbine (m)
		h_L = head loss (m)
		   (to predict head loss, apply Eq. (10.45))

The process for applying the energy equation is

Step 1. Selection. Select the continuity equation when the problem involves pumps, turbine, or head loss. Check to ensure that the assumptions used to derive the energy equation are satisfied. The assumptions are steady flow, one inlet port and one outlet port, constant density, and negligible thermal energy terms (except for head loss).

Step 2. CV Selection. Select and label section 1 (inlet port) and section 2 (outlet port). Locate sections 1 and 2 where (a) you know information or (b) where you want information. By convention, engineers usually do not sketch a CV when applying the energy equation.

Step 3. Analysis. Write the general form of the energy equation. Conduct a term-by-term analysis. Simplify the general equation to the reduced equation.

Step 4. Validation. Check units. Check the physics: (head in via fluid flow and pump) = (head out via fluid flow, turbine, and head loss)

EXAMPLE 7.2

Applying the Energy Equation to Predict the Speed of Water in a Pipe Connected to a Reservoir

---

Problem Statement

A horizontal pipe carries cooling water at 10°C for a thermal power plant. The head loss in the pipe is

$$h_L = \frac{0.02(L/D)V^2}{2g}$$

where L is the length of the pipe from the reservoir to the point in question, V is the mean velocity in the pipe, and D is the diameter of the pipe. If the pipe diameter is 20 cm and the rate of flow is 0.06 m³/s, what is the pressure in the pipe at $L = 2000$ m. Assume $\alpha_2 = 1$.

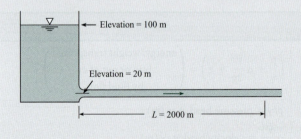

Define the Situation

Water flows in a system.

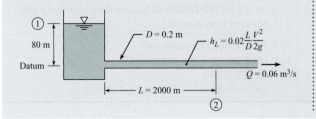

Assumptions:
- $\alpha_2 = 1.0$
- Steady Flow

Water (10°C, 1 atm., Table A.5): $\gamma = 9810$ N/m³

State the Goal

p_2(kPa) ⬅ Pressure at section 2

Generate Ideas and Make a Plan

Select the energy equation because (a) the situation involves water flowing through a pipe, and (b) the energy equation contains the goal (p_2). Locate section 1 at the surface and section 2 at the location where we want to know pressure. The plan is to:

1. Write the general form of the energy equation (7.29)
2. Analyze each term in the energy equation.
3. Solve for p_2.

Take Action (Execute the Plan)

1. Energy equation (general form)

$$\frac{p_1}{\gamma} + \alpha_1 \frac{\overline{V}_1^2}{2g} + z_1 + h_p = \frac{p_2}{\gamma} + \alpha_2 \frac{\overline{V}_2^2}{2g} + z_2 + h_t + h_L$$

2. Term-by-term analysis
- $p_1 = 0$ because the pressure at top of a reservoir is $p_{atm} = 0$ gage.
- $V_1 \approx 0$ because the level of the reservoir is constant or changing very slowly.
- $z_1 = 100$ m; $z_2 = 20$ m.
- $h_p = h_t = 0$ because there are no pumps or turbines in the system.
- Find V_2 using the flow rate equation (5.3).

$$V_2 = \frac{Q}{A} = \frac{0.06 \text{ m}^3/\text{s}}{(\pi/4)(0.2 \text{ m})^2} = 1.910 \text{ m/s}$$

- Head loss is

$$h_L = \frac{0.02(L/D)V^2}{2g} = \frac{0.02(2000 \text{ m}/0.2 \text{ m})(1.910 \text{ m/s})^2}{2(9.81 \text{ m/s}^2)}$$

$$= 37.2 \text{ m}$$

3. Combine steps 1 and 2.

$$(z_1 - z_2) = \frac{p_2}{\gamma} + \alpha_2 \frac{\overline{V}_2^2}{2g} + h_L$$

$$80 \text{ m} = \frac{p_2}{\gamma} + 1.0 \frac{(1.910 \text{ m/s})^2}{2(9.81 \text{ m/s}^2)} + 37.2 \text{ m}$$

$$80 \text{ m} = \frac{p_2}{\gamma} + (0.186 \text{ m}) + (37.2 \text{ m})$$

$$p_2 = \gamma(42.6 \text{ m}) = (9810 \text{ N/m}^3)(42.6 \text{ m}) = \boxed{418 \text{ kPa}}$$

Review the Solution and the Process

1. *Skill.* Notice that section 1 was set at the free surface because properties are known there. Section 2 was set where we want to find information.

2. *Knowledge.* Regarding selection of an equation, we could have chosen the Bernoulli equation. However, it would have been a lousy choice because the Bernoulli equation assumes inviscid flow.

- *Key Idea*: Select the Bernoulli equation if viscous effects can be neglected; select the energy equation if viscous effects are significant.

- *Rule of Thumb*: When fluid is flowing through a pipe that is more than about five diameters long, i.e., ($L/D > 5$), viscous effects are significant.

7.4 The Power Equation

Depending on context, engineers use various equations for calculating power. This section shows how to calculate power associated with pumps and turbines. An equation for pump power follows from the definition of pump head given in Eq. (7.27):

$$\dot{W}_p = \gamma Q h_p = \dot{m} g h_p \tag{7.30a}$$

Similarly, the power delivered from a flow to a turbine is

$$\dot{W}_t = \gamma Q h_t = \dot{m} g h_t \tag{7.30b}$$

Equations (7.30a) and (7.30b) can be generalized to give an equation for calculating power associated with a pump or turbine.

$$P = \dot{m} g h = \gamma Q h \tag{7.31}$$

Equations for calculating power are summarized in Table 7.2.

You can try out the equations in Table 7.2 in the next checkpoint problem.

TABLE 7.2 Summary of the Power Equation

Description	Equation	Terms
Rectilinear motion of an object such as an airplane, a submarine, or a car	$P = FV$ (7.3a)	P = power (W) F = force doing work (N) V = speed of object (m/s)
Rotational motion such as a shaft driving a pump or an output shaft from a turbine	$P = T\omega$ (7.3b)	T = torque (N · m) ω = angular speed (rad/s)
Power supplied from a pump to a flowing fluid Power supplied from a flowing fluid to a turbine	$P = \dot{m}gh = \gamma Qh$ (7.31)	$\dot{m}$ = mass flow rate through machine (kg/s) g = gravitational constant = 9.81 (m/s^2) h = head of pump or head of turbine (m) γ = specific weight (N/m^3) Q = volume flow rate (m^3/s)

EXAMPLE 7.3

Applying the Energy Equation to Calculate the Power Required by a Pump

Problem Statement

A pipe 50 cm in diameter carries water (10°C) at a rate of 0.5 m³/s. A pump in the pipe is used to move the water from an elevation of 30 m to 40 m. The pressure at section 1 is 70 kPa gage, and the pressure at section 2 is 350 kPa gage. What power in kilowatts must be supplied to the flow by the pump? Assume $h_L = 3$ m of water and $\alpha_1 = \alpha_2 = 1$.

Define the Situation

Water is being pumped through a system.

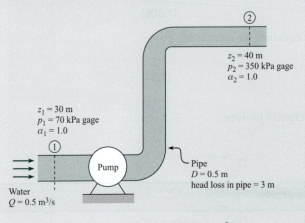

$z_2 = 40$ m
$p_2 = 350$ kPa gage
$\alpha_2 = 1.0$

$z_1 = 30$ m
$p_1 = 70$ kPa gage
$\alpha_1 = 1.0$

Pump

Pipe
$D = 0.5$ m
head loss in pipe = 3 m

Water
$Q = 0.5$ m³/s

Water (10°C, 1 atm., Table A.5): $\gamma = 9810$ N/m³

State the Goal

$P(W) \Leftarrow$ Power the pump is supplying to the water in watts.

Generate Ideas and Make a Plan

Because this problem involves water being pumped through a system, it is an energy equation problem. However, the goal is to find power, so the power equation will also be needed. The steps are

1. Write the energy equation between section 1 and section 2.

2. Analyze each term in the energy equation.

3. Calculate the head of the pump h_p.

4. Find the power by applying the power equation (7.30a).

Take Action (Execute the Plan)

1. Energy equation (general form)

$$\frac{p_1}{\gamma} + \alpha_1 \frac{\overline{V}_1^2}{2g} + z_1 + h_p = \frac{p_2}{\gamma} + \alpha_2 \frac{\overline{V}_2^2}{2g} + z_2 + h_t + h_L$$

2. Term-by-term analysis

   • Velocity head cancels because $V_1 = V_2$.

   • $h_t = 0$ because there are no turbines in the system.

   • All other head terms are given.

   • Inserting terms into the general equation gives

$$\frac{p_1}{\gamma} + z_1 + h_p = \frac{p_2}{\gamma} + z_2 + h_L$$

3. Pump head (from step 2)

$$h_p = \left(\frac{p_2 - p_1}{\gamma}\right) + (z_2 - z_1) + h_L$$

$$= \left(\frac{(350{,}000 - 70{,}000)\,\text{N/m}^2}{9810\,\text{N/m}^3}\right) + (10\,\text{m}) + (3\,\text{m})$$

$$= (28.5\,\text{m}) + (10\,\text{m}) + (3\,\text{m}) = 41.5\,\text{m}$$

Physics: The head provided by the pump (41.5 m) is balanced by the increase in pressure head (28.5 m) plus the increase in elevation head (10 m) plus the head loss (3 m).

4. Power equation

$$P = \gamma Q h_p$$

$$= (9810\,\text{N/m}^3)(0.5\,\text{m}^3/\text{s})(41.5\,\text{m})$$

$$= \boxed{204\,\text{kW}}$$

7.5 Mechanical Efficiency

Fig. 7.7 shows an electric motor connected to a centrifugal pump. Motors, pump, turbines, and similar device have energy losses. In pump and turbines, energy losses are due to factors such as mechanical friction, viscous dissipation, and leakage. Energy losses are accounted for by using efficiency.

Mechanical efficiency is defined as the ratio of power output to power input:

$$\eta \equiv \frac{\text{power output from a machine or system}}{\text{power input to a machine or system}} = \frac{P_{\text{output}}}{P_{\text{input}}} \qquad (7.32)$$

The symbol for mechanical efficiency is the Greek letter η, which is pronounced as "eta." In addition to mechanical efficiency, engineers also use *thermal efficiency,* which is defined using thermal energy input into a system. In this text, only mechanical efficiency is used, and we sometimes use the label "efficiency" instead of "mechanical efficiency."

FIGURE 7.7

CAD drawing of a centrifugal pump and electric motor. (Image courtesy of Ted Kyte; www.ted-kyte.com.)

EXAMPLE. Suppose an electric motor like the one shown in Fig. 7.7 is drawing 1000 W of electrical power from a wall circuit. As shown in Fig. 7.8, the motor provides 750 J/s of power to its output shaft. This power drives the pump, and the pump supplies 450 J/s to the fluid.

In this example, the efficiency of the electric motor is

$$\eta_{\text{motor}} = (750 \text{ J/s})/(1000 \text{ J/s}) = 0.75 = 75\%$$

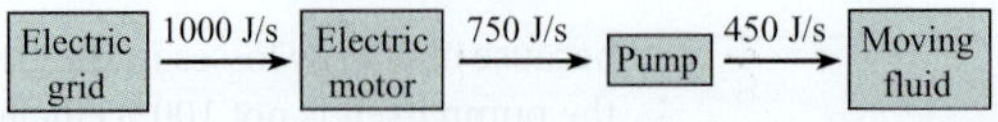

FIGURE 7.8

The energy flow through a pump that is powered by an electric motor.

Similarly, the efficiency of the pump is

$$\eta_{\text{pump}} = (450 \text{ J/s})/(750 \text{ J/s}) = 0.60 = 60\%$$

and the combined efficiency is

$$\eta_{\text{combined}} = (450 \text{ J/s})/(1000 \text{ J/s}) = 0.45 = 45\%$$

EXAMPLE. Suppose that wind incident on a wind turbine contains 1000 J/s of energy as shown in Fig. 7.9. Because a wind turbine cannot extract all the energy and because of losses, the work that the wind turbine does on its output shaft is 360 J/s. This power drives an electric generator, and the generator produces 324 J/s of electrical power, which is supplied to the power grid. Calculate the system efficiency and the efficiency of the components.

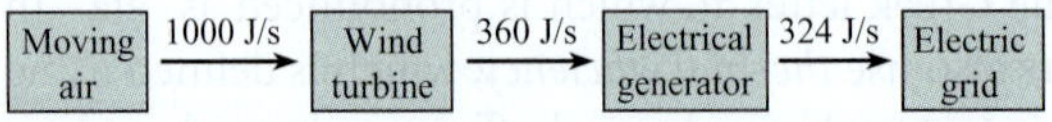

FIGURE 7.9

The energy flow associated with generating electrical power from a wind turbine.

The efficiency of the wind turbine is

$$\eta_{\text{wind turbine}} = (360 \text{ J/s})/(1000 \text{ J/s}) = 0.36 = 36\%$$

The efficiency of the electric generator is

$$\eta_{\text{electric generator}} = (324 \text{ J/s})/(360 \text{ J/s}) = 0.90 = 90\%$$

The combined efficiency is

$$\eta_{\text{combined}} = (324 \text{ J/s})/(1000 \text{ J/s}) = 0.324 = 32.4\%$$

We can generalize the results of the last two examples to summarize the efficiency equations (Table 7.3). Example 7.4 shows how efficiency enters into a calculation of power.

TABLE 7.3 Summary of the Efficiency Equation

Description	Equation		Terms
Pump	$P_{\text{pump}} = \eta_{\text{pump}} P_{\text{shaft}}$	(7.33a)	P_{pump} = power that the pump supplies to the fluid (W) $\quad [P_{\text{pump}} = \dot{m}gh_p = \gamma Qh_p]$
			η_{pump} = efficiency of pump ()
			P_{shaft} = power that is supplied to the pump shaft (W)
Turbine	$P_{\text{shaft}} = \eta_{\text{turbine}} P_{\text{turbine}}$	(7.33b)	P_{turbine} = power that the fluid supplies to a turbine (W) $\quad [P_{\text{turbine}} = \dot{m}gh_t = \gamma Qh_t]$
			η_{turbine} = efficiency of turbine ()
			P_{shaft} = power that is supplied by the turbine shaft (W)

EXAMPLE 7.4

Applying the Energy Equation to Predict the Power Produced by a Turbine

Problem Statement

At the maximum rate of power generation, a small hydroelectric power plant takes a discharge of 14.1 m³/s through an elevation drop of 61 m. The head loss through the intakes, penstock, and outlet works is 1.5 m. The combined efficiency of the turbine and electrical generator is 87%. What is the rate of power generation?

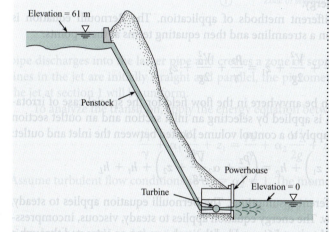

Define the Situation

A small hydroelectric plant is producing electrical power
- Combined head loss: $h_L = 1.5$ m
- Combined efficiency (turbine/generator): $\eta = 0.87$
- **Water** (10°C, 1 atm, Table A.5): $\gamma = 9810$ N/m³

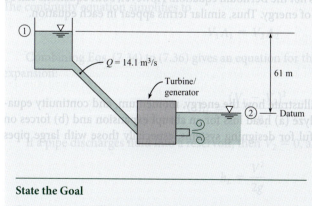

State the Goal

$P_{\text{output from generator}}$ (MW) ⬅ Power produced by generator

Generate Ideas and Make a Plan

Because this problem involves a fluid system for producing power, select the energy equation. Because power is the goal, also select the power equation. The plan is

1. Write the energy equation (7.29) between section 1 and section 2.
2. Analyze each term in the energy equation.
3. Solve for the head of the turbine h_t.
4. Find the input power to the turbine using the power equation (7.30b).
5. Find the output power from the generator by using the efficiency equation (7.33b).

Take Action (Execute the Plan)

1. Energy equation (general form)

$$\frac{p_1}{\gamma} + \alpha_1\frac{\overline{V}_1^2}{2g} + z_1 + h_p = \frac{p_2}{\gamma} + \alpha_2\frac{\overline{V}_2^2}{2g} + z_2 + h_t + h_L$$

2. Term-by-term analysis
   - Velocity heads are negligible because $V_1 \approx 0$ and $V_2 \approx 0$.
   - Pressure heads are zero because $p_1 = p_2 = 0$ gage.
   - $h_p = 0$ because there is no pump in the system.
   - Elevation head terms are given.

3. Combine steps 1 and 2:

$$h_t = (z_1 - z_2) - h_L$$
$$= (61 \text{ m}) - (1.5 \text{ m}) = 59.5 \text{ m}$$

Physics: Head supplied to the turbine (59.5 m) is equal to the net elevation change of the dam (61 m) minus the head loss (1.5 m).

4. Power equation

$$P_{\text{input to turbine}} = \gamma Q h_t = (9810 \text{ N/m}^3)(14.1 \text{ m}^3/\text{s})(59.5 \text{ m})$$
$$= 8.23 \text{ MW}$$

5. Efficiency equation

$$P_{\text{output from generator}} = \eta P_{\text{input to turbine}} = 0.87(8.23 \text{ MW})$$
$$= \boxed{7.16 \text{ MW}}$$

Review the Solution and the Process

1. *Knowledge.* Notice that sections 1 and 2 were located on the free surfaces. This is because information is known at these locations.

2. *Discussion.* The maximum power that can be generated is a function of the elevation head and the flow rate. This maximum power is decreased by head loss and by energy losses in the turbine and the generator.

Forces on Transitions

To find forces on transitions in pipes, apply the momentum equation in combination with the energy equation, the flow rate equation, and the head loss equation. This approach is illustrated by Example 7.5.

EXAMPLE 7.5

Applying the Energy and Momentum Equations to Find Force on a Pipe Contraction

Problem Statement

A pipe 30 cm in diameter carries water (10°C, 250 kPa) at a rate of 0.707 m³/s. The pipe contracts to a diameter of 20 cm. The head loss through the contraction is given by

$$h_L = 0.1 \frac{V_2^2}{2g}$$

where V_2 is the velocity in the 20 cm pipe. What horizontal force is required to hold the transition in place? Assume the kinetic energy correction factor is 1.0 at both the inlet and exit

Define the Situation

Water flows through a contraction.

- $\alpha_1 = \alpha_2 = 1.0$
- $h_L = 0.1 \, (V_2^2/(2g))$

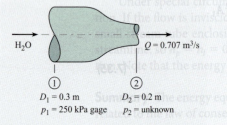

$D_1 = 0.3$ m $D_2 = 0.2$ m
$p_1 = 250$ kPa gage $p_2 =$ unknown

Properties: Water (10°C, 1 atm., Table A.5):
$\gamma = 9810$ N/m³

State the Goal

F_x(N) ◀ Horizontal force acting on the contraction

Generate Ideas and Make a Plan

Because force is the goal, start with the momentum equation. To solve the momentum equation, we need p_2. Find this with the energy equation. The step-by-step plan is

1. Derive an equation for F_x by applying the momentum eqn.
2. Derive an equation for p_2 by applying the energy eqn.

3. Calculate p_2.
4. Calculate F_x.

Take Action (Execute the Plan)

1. Momentum equation
   - Sketch a force diagram and a momentum diagram

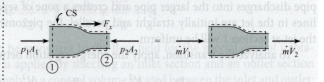

   - Write the x-direction momentum equation.

$$p_1 A_1 - p_2 A_2 + F_x = \dot{m} V_2 - \dot{m} V_1$$

   - Rearrange to give

$$F_x = \rho Q(V_2 - V_1) + p_2 A_2 - p_1 A_1$$

2. Energy equation (from section 1 to section 2)
   - Let $\alpha_1 = \alpha_2 = 1$, $z_1 = z_2$, and $h_p = h_t = 0$
   - Eq. (7.29) simplifies to

$$\frac{p_1}{\gamma} + \frac{V_1^2}{2g} = \frac{p_2}{\gamma} + \frac{V_2^2}{2g} + h_L$$

   - Rearrange to give

$$p_2 = p_1 - \gamma \left(\frac{V_2^2}{2g} - \frac{V_1^2}{2g} + h_L \right)$$

3. Pressure at section 2.
   - Find velocities using the flow rate equation.

$$V_1 = \frac{Q}{A_1} = \frac{0.707 \text{ m}^3/\text{s}}{(\pi/4) \times (0.3 \text{ m})^2} = 10 \text{ m/s}$$

$$V_2 = \frac{Q}{A_2} = \frac{0.707 \text{ m}^3/\text{s}}{(\pi/4) \times (0.2 \text{ m})^2} = 22.5 \text{ m/s}$$

   - Calculate head loss.

$$h_L = \frac{0.1 \, V_2^2}{2g} = \frac{0.1 \times (22.5 \text{ m/s})^2}{2 \times (9.81 \text{ m/s}^2)} = 2.58 \text{ m}$$

- Calculate pressure.

$$p_2 = p_1 - \gamma\left(\frac{V_2^2}{2g} - \frac{V_1^2}{2g} + h_L\right)$$

$$= 250 \text{ kPa} - 9.81 \text{ kN/m}^3$$

$$\times \left(\frac{(22.5 \text{ m/s})^2}{2(9.81 \text{ m/s}^2)} - \frac{(10 \text{ m/s})^2}{2(9.81 \text{ m/s}^2)} + 2.58 \text{ m}\right)$$

$$= 21.6 \text{ kPa}$$

4. Calculate F_x.

$$F_x = \rho Q(V_2 - V_1) + p_2 A_2 - p_1 A_1$$

$$= (1000 \text{ kg/m}^3)(0.707 \text{ m}^3/\text{s})(22.5 - 10)(\text{m/s})$$

$$+ (21,600 \text{ Pa})\left(\frac{\pi(0.2 \text{ m})^2}{4}\right) - (250,000 \text{ Pa})$$

$$\times \left(\frac{\pi(0.3 \text{ m})^2}{4}\right)$$

$$= (8837 + 677 - 17,670)\text{N} = -8.16 \text{ kN}$$

$$\boxed{F_x = 8.16 \text{ kN acting to the left}}$$

7.8 Hydraulic and Energy Grade Lines

This section introduces the hydraulic grade line (HGL) and the energy grade line (EGL), which are graphical representations that show head in a system. This visual approach provides insights and helps one locate and correct trouble spots in the system (usually points of low pressure).

The **EGL**, shown in Fig. 7.11, is a line that indicates the total head at each location in a system. The *EGL* is related to terms in the energy equation by

$$\text{EGL} = \left(\begin{array}{c}\text{velocity}\\\text{head}\end{array}\right) + \left(\begin{array}{c}\text{pressure}\\\text{head}\end{array}\right) + \left(\begin{array}{c}\text{elevation}\\\text{head}\end{array}\right) = \alpha\frac{V^2}{2g} + \frac{p}{\gamma} + z = \left(\begin{array}{c}\text{total}\\\text{head}\end{array}\right) \quad \textbf{(7.38)}$$

Notice that **total head**, which characterizes the energy that is carried by a flowing fluid, is the sum of velocity head, the pressure head, and the elevation head.

FIGURE 7.11

EGL and HGL in a straight pipe.

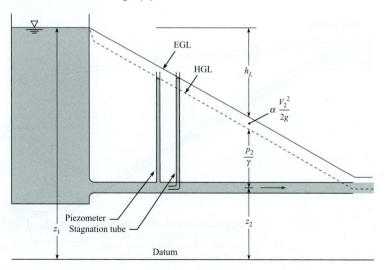

The **HGL,** shown in Fig. 7.11, is a line that indicates the piezometric head at each location in a system:

$$\text{HGL} = \left(\begin{array}{c}\text{pressure}\\\text{head}\end{array}\right) + \left(\begin{array}{c}\text{elevation}\\\text{head}\end{array}\right) = \frac{p}{\gamma} + z = \left(\begin{array}{c}\text{piezometric}\\\text{head}\end{array}\right) \qquad \textbf{(7.39)}$$

Because the HGL gives piezometric head, the HGL will be coincident with the liquid surface in a piezometer as shown in Fig. 7.11. Similarly, the EGL will be coincident with the liquid surface in a stagnation tube.

Tips for Drawing HGLs and EGLs

1. In a lake or reservoir, the HGL and EGL will coincide with the liquid surface. Also, both the HGL and EGL will indicate piezometric head.
2. A pump causes an abrupt rise in the EGL and HGL by adding energy to the flow. For example, see Fig. 7.12.
3. For steady flow in a pipe of constant diameter and wall roughness, the slope ($\Delta h_L/\Delta L$) of the EGL and the HGL will be constant. For example, see Fig. 7.11.

FIGURE 7.12

Rise in EGL and HGL due to pump.

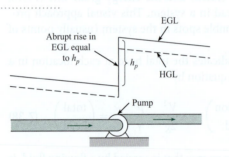

4. Locate the HGL below the EGL by a distance of the velocity head ($\alpha V^2/2g$).
5. Height of the EGL decreases in the flow direction unless a pump is present.
6. A turbine causes an abrupt drop in the EGL and HGL by removing energy from the flow. For example, see Fig. 7.13.

FIGURE 7.13

Drop in EGL and HGL due to turbine.

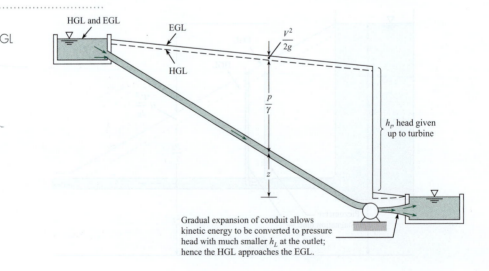

Gradual expansion of conduit allows kinetic energy to be converted to pressure head with much smaller h_L at the outlet; hence the HGL approaches the EGL.

7. Power generated by a turbine can be increased by using a gradual expansion at the turbine outlet. As shown in Fig. 7.13, the expansion converts kinetic energy to pressure. If the outlet to a reservoir is an abrupt expansion, as in Fig. 7.15, this kinetic energy is lost.

8. When a pipe discharges into the atmosphere the HGL is coincident with the system because $p/\gamma = 0$ at these points. For example, in Figures 7.14 and 7.16, the HGL in the liquid jet is drawn through the centerline of the jet.

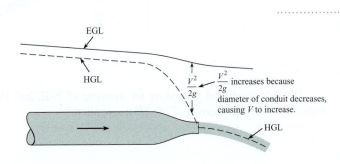

FIGURE 7.14

Change in HGL and EGL due to flow through a nozzle.

9. When a flow passage changes diameter, the distance between the EGL and the HGL will change (see Fig. 7.14 and Fig. 7.15) because velocity changes. In addition, the slope on the EGL will change because the head loss per length will be larger in the conduit with the larger velocity (see Fig. 7.15).

FIGURE 7.15

Change in EGL and HGL due to change in diameter of pipe.

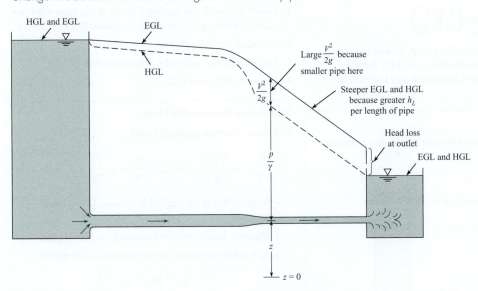

10. If the HGL falls below the pipe, then p/γ is negative, indicating subatmospheric pressure (see Fig. 7.16) and a potential location of cavitation.

FIGURE 7.16

Subatmospheric pressure when pipe is above HGL.

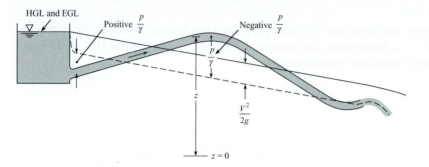

The recommended procedure for drawing an EGL and HGL is shown in Example 7.6. Notice how the tips from pp. 274–275 are applied.

EXAMPLE 7.6

Sketching the EGL and HGL for a Piping System

Problem Statement

A pump draws water (10°C) from a reservoir, where the water-surface elevation is 160 m, and forces the water through a pipe 1525 m long and 0.3 m in diameter. This pipe then discharges the water into a reservoir with water-surface elevation of 190 m. The flow rate is 0.2 m³/s, and the head loss in the pipe is given by

$$h_L = 0.01\left(\frac{L}{D}\right)\left(\frac{V^2}{2g}\right)$$

Determine the head supplied by the pump, h_p, and the power supplied to the flow, and draw the HGL and EGL for the system. Assume that the pipe is horizontal and is 155 m in elevation.

Define the Situation

Water is pumped from a lower reservoir to a higher reservoir.

- $h_L = 0.01\left(\frac{L}{D}\right)\left(\frac{V^2}{2g}\right)$

- Water (10°C, 1 atm, Table A.5): $\gamma = 9810$ N/m³.

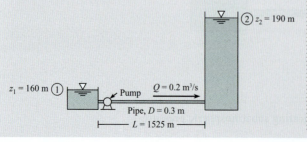

State the Goals

1. h_p(m) ⬅ pump head
2. P(kW) ⬅ power supplied by the pump
3. Draw the HGL and the EGL.

Generate Ideas and Make a Plan

Because pump head and power are goals, apply the energy equation and the power equation, respectively. The step-by-step plan is

1. Locate section 1 and section 2 at top of the reservoirs (see sketch). Then, apply the energy equation (7.29).
2. Calculate terms in the energy equation.
3. Calculate power using the power equation (7.30a).
4. Draw the HGL and EGL.

Take Action (Execute the Plan)

1. Energy equation (general form)

$$\frac{p_1}{\gamma} + \alpha_1\frac{\overline{V}_1^2}{2g} + z_1 + h_p = \frac{p_2}{\gamma} + \alpha_2\frac{\overline{V}_2^2}{2g} + z_2 + h_t + h_L$$

- Velocity heads are negligible because $V_1 \approx 0$ and $V_2 \approx 0$.
- Pressure heads are zero because $p_1 = p_2 = 0$ gage.
- $h_t = 0$ because there are no turbines in the system.

$$h_p = (z_2 - z_1) + h_L$$

Interpretation: Head supplied by the pump provides the energy to lift the fluid to a higher elevation plus the energy to overcome head loss.

2. Calculations.

- Calculate V using the flow rate equation.

$$V = \frac{Q}{A} = \frac{0.2 \text{ m}^3/\text{s}}{(\pi/4)(0.3 \text{ m})^2} = 2.83 \text{ m/s}$$

- Calculate head loss.

$$h_L = 0.01\left(\frac{L}{D}\right)\left(\frac{V^2}{2g}\right) = 0.01\left(\frac{1525 \text{ m}}{0.3 \text{ m}}\right)\left(\frac{(2.83 \text{ m/s})^2}{2 \times (9.81 \text{ m/s}^2)}\right)$$

$$= 20.75 \text{ m}$$

- Calculate h_p.

$$h_p = (z_2 - z_1) + h_L = (190 \text{ m} - 160 \text{ m}) + 20.75 \text{ m} = \boxed{50.75 \text{ m}}$$

3. Power

$$\dot{W}_p = \gamma Q h_p = (9810 \text{ N/m}^3)(0.2 \text{ m}^3/\text{s})(50.75 \text{ m})$$

$$= \boxed{99.5 \text{ kW}}$$

4. HGL and EGL

- From Tip 1 on p. 274, locate the HGL and EGL along the reservoir surfaces.

- From Tip 2, sketch in a head rise of 50.75 m corresponding to the pump.

- From Tip 3, sketch the EGL from the pump outlet to the reservoir surface. Use the fact that the head loss is 20.75 m. Also, sketch EGL from the reservoir on the left to the pump inlet. Show a small head loss.

- From Tip 4, sketch the HGL below the EGL by a distance of $V^2/2g \approx 0.5$ m.

- From Tip 5, check the sketches to ensure that EGL and HGL are decreasing in the direction of flow (except at the pump).

HGL (dashed black line) and EGL (solid blue line)

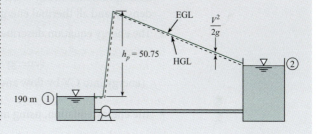

7.9 Summarizing Key Knowledge

Foundational Concepts

- *Energy* is a property of a system that allows the system to do work on its surroundings. Energy can be classified into five categories: mechanical energy, thermal energy, chemical energy, electrical energy, and nuclear energy.

- *Mechanical work* is done by a force that acts through a distance. A more general definition of work is that *work* is an interaction of a system with the surroundings in such a way that the sole effect on the surroundings could have been the lifting of a weight.

- *Power* is the ratio of work to time or energy to time at an instant in time. Note the key difference between energy and power

  ▸ Energy (and work) describe an *amount* (e.g., how many joules).

  ▸ Power describes an *amount/time* or *rate* (e.g., how many joules/second or watts).

- Machines can be classified into two categories:

  ▸ A *pump* is any machine that adds energy to a flowing fluid.

  ▸ A *turbine* is any machine that extracts energy from a flowing fluid.

Conservation of Energy and Derivation of the Energy Equation

- The law of conservation of energy asserts that work and energy balance.

  ▸ The balance for a closed system is (Energy changes of the system) = (Energy increases due to heat transfer) − (Energy decreases due to the system doing work).

▸ The balance for a CV is (Energy changes in the CV) = (Energy increases in the CV due to heat transfer) − (Energy out of CV via work done on the surrounding) + (Energy transported into the CV by fluid flow)

• Work can be classified into two categories

▸ *Flow work* is work that is done by the pressure force in a flowing fluid

▸ *Shaft work* is any work that is not flow work.

The Energy Equation

• The energy equation is the law of conservation of energy simplified so that it applies to common situations that occur in fluid mechanics. Some of the most important assumptions are steady state, one inflow and one outflow port to the CV, constant density, and all thermal energy terms (except for head loss) are neglected.

• The energy equation describes an energy balance for a control volume (CV).

$$(\text{energy into CV}) = (\text{energy out of CV})$$

$$(\text{energy into CV by flow and pumps}) = (\text{energy out by flow, turbines, and head loss})$$

• The energy equation, using math symbols, is

$$\left(\frac{p_1}{\gamma} + \alpha_1 \frac{V_1^2}{2g} + z_1\right) + h_p = \left(\frac{p_2}{\gamma} + \alpha_2 \frac{V_2^2}{2g} + z_2\right) + h_t + h_L$$

$$\begin{pmatrix}\text{pressure head}\\\text{velocity head}\\\text{elevation head}\end{pmatrix}_1 + \begin{pmatrix}\text{pump}\\\text{head}\end{pmatrix} = \begin{pmatrix}\text{pressure head}\\\text{velocity head}\\\text{elevation head}\end{pmatrix}_2 + \begin{pmatrix}\text{turbine}\\\text{head}\end{pmatrix} + \begin{pmatrix}\text{head}\\\text{loss}\end{pmatrix}$$

• Regarding head

▸ Head can be thought of as the ratio of energy to weight for a fluid particle.

▸ Head can also describe the energy per time that is passing across a section because head and power are related by $P = \dot{m}gh$

• Regarding head loss (h_L)

▸ Head loss represents an irreversible conversion of mechanical energy to thermal energy through the action of viscosity.

▸ Head loss is always positive and is analogous to frictional heating.

▸ Head loss for a sudden expansion is given by

$$h_L = \frac{(V_1 - V_2)^2}{2g}$$

• Regarding the kinetic energy correction factor α

▸ This factor accounts for the distribution of kinetic energy in a flowing fluid. It is defined as the ratio of (actual KE/time that crosses a surface) to (KE/time that would cross if the velocity was uniform).

▸ For most situations, engineers set $\alpha = 1$. If the flow is known to be fully developed and laminar, then engineers use $\alpha = 2$. In other cases, one can go back to the mathematical definition and calculate a value of α.

Power and Mechanical Efficiency

- Mechanical efficiency is the ratio of (power output) to (power input) for a machine or system.
- There are several equations that engineers use to calculate power.
  - ▸ For translational motion such as a car or an airplane $P = FV$
  - ▸ For rotational motion such as the shaft on a pump $P = T\omega$
  - ▸ For the pump, the power added to the flow is: $P = \gamma Q h_p$
  - ▸ For a turbine, the power extracted from the flow is $P = \gamma Q h_t$

The HGL and EGL

- The hydraulic grade line (HGL) is a profile of the piezometric head, $p/\gamma + z$, along a pipe.
- The energy grade line (EGL) is a profile of the total head, $V^2/2g + p/\gamma + z$, along a pipe.
- If the hydraulic grade line falls below the elevation of a pipe, subatmospheric pressure exists in the pipe at that location, giving rise to the possibility of cavitation.

REFERENCES

1. Electrical Engineer, October 18, 1889, p. 311–312, accessed 1/23/11, http://books.google.com/books?id=PQsAAAAAMAAJ&pg=PA311&lpg=PA311&dq=James+Joule+obituary&hl=en#v=onepage&q=James%20Joule%20obituary&f=false

2. "James Prescott Joule (1818–1889): A Manchester Son and the Father of the International Unit of Energy," Winhoven, S.H. &

Gibbs, N.K., accessed on 1/23/11, http://www.bad.org.uk/Portals/_Bad/History/Historical%20poster%2006.pdf

3. Cengel, Y. A., and M. A. Boles. *Thermodynamics: An Engineering Approach*. New York: McGraw-Hill, 1998.

4. Moran, M. J., and H. N. Shapiro. *Fundamentals of Engineering Thermodynamics*. New York: John Wiley, 1992.

PROBLEMS

PLUS Problem available in *WileyPLUS* at instructor's discretion.

GO Guided Online (GO) Problem, available in *WileyPLUS* at instructor's discretion.

Energy Concepts (§7.1)

7.1 From the list below, select one topic that is interesting to you. Then, use references such as the Internet to research your topic and prepare one page of documentation that you could use to present your topic to your peers.

  a. Explain how hydroelectric power is produced.

  b. Explain how a Kaplan turbine works, how a Francis turbine works, and the differences between these two types of turbines.

  c. Explain how a horizontal-axis wind turbine is used to produce electrical power.

  d. Explain how a steam turbine is used to produce electrical power.

7.2 PLUS Using Section 7.1 and other resources, answer the following questions. Strive for depth, clarity, and accuracy. Also, strive for effective use of sketches, words, and equations.

  a. What are the common forms of energy? Which of these forms are relevant to fluid mechanics?

  b. What is work? Describe three example of work that are relevant to fluid mechanics.

  c. List three significant differences between power and energy.

7.3 (WILEY PLUS) Apply the grid method to each situation.

a. Calculate the energy in joules used by a 746 W pump that is operating for 6 hours. Also, calculate the cost of electricity for this time period. Assume that electricity costs $0.15 per kW-hr.

b. A motor is being to used to turn the shaft of a centrifugal pump. Apply Eq. (7.3b) on p. 255 of §7.2 to calculate the power in watts corresponding to a torque of 11 N · m and a rotation speed of 89.3 rad/s.

c. A turbine produces a power of 10.2 kW. Calculate the power in hp.

7.4 (WILEY PLUS) Energy (select all that are correct):

a. has same units as work

b. has same units as power

c. has same units work/time

d. can have units of Joule

e. can have units of Watt

7.5 (WILEY PLUS) Power (select all that are correct)

a. has same units as energy

b. has same units as energy/time

c. has same units as work/time

d. can have units of Joule

e. can have units of Watt

7.6 Estimate the power required to spray water out of the spray bottle that is pictured in Fig. 7.2a on p. 254 of §7.2. *Hint:* Make appropriate assumptions about the number of sprays per unit time and the force exerted by the finger.

7.7 (WILEY PLUS) The sketch shows a common consumer product called the Water Pik. This device uses a motor to drive a piston pump that produces a jet of water ($d = 1$ mm, $T = 10°C$) with a speed of 27 m/s. Estimate the minimum electrical power in watts that is required by the device. *Hints:* (a) Assume that the power is used only to produce the kinetic energy of the water in the jet; and (b) in a time interval Δt, the amount of mass that flows out the

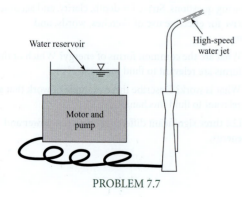

PROBLEM 7.7

nozzle is Δm, and the corresponding amount of kinetic energy is ($\Delta m V^2/2$).

7.8 An engineer is considering the development of a small wind turbine ($D = 1.25$ m) for home applications. The design wind speed is 24 km/hr at $T = 10°C$ and $p = 90$ kPa. The efficiency of the turbine is $\eta = 20\%$, meaning that 20% of the kinetic energy in the wind can be extracted. Estimate the power in watts that can be produced by the turbine. *Hint:* In a time interval Δt, the amount of mass that flows through the rotor is $\Delta m = \dot{m}\Delta t$, and the corresponding amount of kinetic energy in this flow is ($\Delta m V^2/2$).

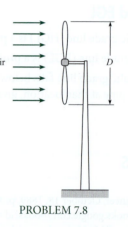

PROBLEM 7.8

Conservation of Energy (§7.2)

7.9 (WILEY PLUS) The first law of thermodynamics for a closed system can be characterized in words as

a. (change in energy in a system) = (thermal energy in) − (work done on surroundings)

b. (change in energy in a system) = (thermal energy out) − (work done by surroundings)

c. either of the above

7.10 (WILEY PLUS) The application of Reynolds transport theorem to the first law of thermodynamics (select all that are correct)

a. refers to the increase of energy stored in a closed system

b. extends the applicability of the first law from a closed system to an open system (control volume)

c. refers only to heat transfer, and not to work

The Kinetic Energy Correction Factor (§7.3)

7.11 (WILEY PLUS) Using Section 7.3 and other resources, answer the questions below. Strive for depth, clarity, and accuracy while also combining sketches, words, and equations in ways that enhance the effectiveness of your communication.

a. What is the kinetic-energy correction factor? Why do engineers use this term?

b. What is the meaning of each variable (α, A, V, $\overline{V}$) that appears in Eq. (7.21) on p. 260 of §7.3?

c. What values of α are commonly used?

7.12 For this hypothetical velocity distribution in a wide rectangular channel, evaluate the kinetic-energy correction factor α.

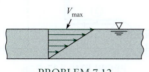

PROBLEM 7.12

7.13 PLUS For these velocity distributions in a round pipe, indicate whether the kinetic-energy correction factor α is greater than, equal to, or less than unity.

7.14 Calculate α for case (*c*).

7.15 Calculate α for case (*d*).

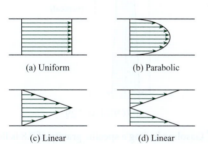

(a) Uniform   (b) Parabolic

(c) Linear   (d) Linear

PROBLEMS 7.13, 7.14, 7.15

7.16 An approximate equation for the velocity distribution in a pipe with turbulent flow is

$$\frac{V}{V_{\max}} = \left(\frac{y}{r_0}\right)^n$$

where $V_{\max}$ is the centerline velocity, y is the distance from the wall of the pipe, r_0 is the radius of the pipe, and n is an exponent that depends on the Reynolds number and varies between 1/6 and 1/8 for most applications. Derive a formula for α as a function of n. What is α if $n = 1/7$?

7.17 An approximate equation for the velocity distribution in a rectangular channel with turbulent flow is

$$\frac{u}{u_{\max}} = \left(\frac{y}{d}\right)^n$$

where $u_{\max}$ is the velocity at the surface, y is the distance from the floor of the channel, d is the depth of flow, and n is an exponent that varies from about 1/6 to 1/8 depending on the Reynolds number. Derive a formula for α as a function of n. What is the value of α for $n = 1/7$?

7.18 The following data were taken for turbulent flow in a circular pipe with a radius of 3.5 cm. Evaluate the kinetic energy correction factor. The velocity at the pipe wall is zero.

r (cm)	V (m/s)	r (cm)	V (m/s)
0.0	32.5	2.8	22.03
0.5	32.44	2.9	21.24
1.0	32.27	3.0	20.49
1.5	31.22	3.1	19.6
2.0	28.21	3.2	18.69
2.25	26.51	3.25	18.16
2.5	24.38	3.3	17.54
2.6	23.7	3.35	17.02
2.7	22.88	3.4	16.14

The Energy Equation (§7.3)

7.19 Using Section 7.3 and other resources, answer the questions below. Strive for depth, clarity, and accuracy. Also, strive for effective use of sketches, words, and equations.

a. What is conceptual meaning of the first law of thermodynamics for a system?

b. What is flow work? How is the equation for flow work (Eq. 7.16) on p. 259 of §7.3 derived?

c. What is shaft work? How is shaft work different than flow work?

7.20 Using Section 7.3 and other resources, answer the questions below. Strive for depth, clarity, and accuracy. Also, strive for effective use of sketches, words, and equations.

a. What is head? How is head related to energy? To power?

b. What is head of a turbine?

c. How is head of a pump related to power? To energy?

d. What is head loss?

7.21 PLUS **part (a) only** Using Sections 7.3 and 7.7 and using other resources, answer the following questions. Strive for depth, clarity, and accuracy. Also, strive for effective use of sketches, words and equations.

a. What are the five main terms in the energy equation (7.29) on p. 262 of §7.3? What does each term mean?

b. How are terms in the energy equation related to energy? To power?

c. What assumptions are required for using the energy equation (7.29) on p. 262 of §7.3?

7.22 Using the energy equation (7.29 on p. 262 of §7.3), prove that fluid in a pipe will flow from a location with high piezometric head to a location with low piezometric head. Assume there are no pumps or turbines and that the pipe has a constant diameter.

7.23 PLUS Water flows at a steady rate in this vertical pipe. The pressure at A is 10 kPa, and at B it is 98.1 kPa. Then the flow in the pipe is (a) upward, (b) downward, or (c) no flow. (*Hint:* See problem 7.23.)

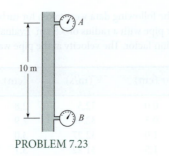

PROBLEM 7.23

7.24 Determine the discharge in the pipe and the pressure at point B. Neglect head losses. Assume $\alpha = 1.0$ at all locations.

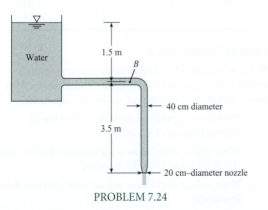

PROBLEM 7.24

7.25 PLUS A pipe drains a tank as shown. If $x = 4.25$ m, $y = 1.2$ m, and head losses are neglected, what is the pressure at point A and what is the velocity at the exit? Assume $\alpha = 1.0$ at all locations.

7.26 PLUS A pipe drains a tank as shown. If $x = 6$ m, $y = 4$ m, and head losses are neglected, what is the pressure at point A and what is the velocity at the exit? Assume $\alpha = 1.0$ at all locations.

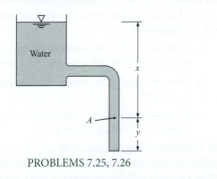

PROBLEMS 7.25, 7.26

7.27 For this system, the discharge of water is 0.1 m³/s, $x = 1.0$ m, $y = 1.5$ m, $z = 6.0$ m, and the pipe diameter is 30 cm. Assuming a head loss of 0.5 m, what is the pressure head at point 2 if the jet from the nozzle is 10 cm in diameter? Assume $\alpha = 1.0$ at all locations.

7.28 PLUS For this diagram of an industrial pressure washer system, $x = 0.3$ m, $y = 0.9$ m, $z = 3$ m, $Q = 0.1$ m³/s, and the hose diameter is 10 cm. Assuming a head loss of 0.3 m is derived over the distance from point 2 to the jet, what is the pressure at point 2 if the jet from the nozzle is 2.5 cm in diameter? Assume $\alpha = 1.0$ throughout.

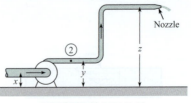

PROBLEMS 7.27, 7.28

7.29 PLUS For this refinery pipe, $D_A = 20$ cm, $D_B = 14$ cm, and $L = 1$ m. If crude oil (S = 0.90) is flowing at a rate of 0.05 m³/s, determine the difference in pressure between sections A and B. Neglect head losses.

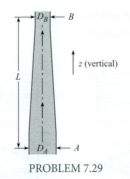

PROBLEM 7.29

7.30 GO Gasoline having a specific gravity of 0.8 is flowing in the pipe shown at a rate of 0.15 m³/s. What is the pressure at section 2 when the pressure at section 1 is 124 kPa gage and the head loss is 1.8 m between the two sections? Assume $\alpha = 1.0$ at all locations.

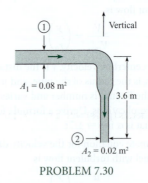

PROBLEM 7.30

7.31 GO Water flows from a pressurized tank as shown. The pressure in the tank above the water surface is 100 kPa gage, and the water surface level is 8 m above the outlet. The water exit velocity is 10 m/s. The head loss in the system varies as $h_L = K_L V^2 / 2g$, where K_L is the minor-loss coefficient. Find the value for K_L. Assume $\alpha = 1.0$ at all locations.

7.32 PLUS A reservoir with water is pressurized as shown. The pipe diameter is 2.5 cm. The head loss in the system is given by $h_L = 5V^2/2g$. The height between the water surface and the pipe outlet is 3 m. A discharge of 0.003 m³/s is needed. What must the pressure in the tank be to achieve such a flow rate? Assume $\alpha = 1.0$ at all locations.

7.33 In the figure shown, suppose that the reservoir is open to the atmosphere at the top. The valve is used to control the flow rate from the reservoir. The head loss across the valve is given as $h_L = 4V^2/2g$, where V is the velocity in the pipe. The cross-sectional area of the pipe is 8 cm². The head loss due to friction in the pipe is negligible. The elevation of the water level in the reservoir above the pipe outlet is 9 m. Find the discharge in the pipe. Assume $\alpha = 1.0$ at all locations.

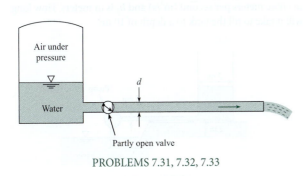

PROBLEMS 7.31, 7.32, 7.33

7.34 PLUS A minor artery in the human arm, diameter $D = 3$ mm, tapers gradually over a distance of 20 cm to a diameter of $d = 1.6$ mm. The blood pressure at D is 110 mm Hg, and at d is 85 mm Hg. What is the head loss (m) that occurs over this 20-cm distance if the blood ($S = 1.06$) is moving with a flowrate of 300 milliliters/min, and the arm is being held horizontally? Idealize the flow in the artery as steady, the fluid as Newtonian, and the walls of the artery as rigid.

7.35 PLUS As shown, a microchannel is being designed to transfer fluid in a MEMS (microelectrical mechanical system) application. The channel is 200 micrometers in diameter and is 5 cm long. Ethyl alcohol is driven through the system at the rate of 0.1 microliters/s (μL/s) with a syringe pump, which is essentially a moving piston. The pressure at the exit of the channel is atmospheric. The flow is laminar, so $\alpha = 2$. The head loss in the channel is given by

$$h_L = \frac{32\mu LV}{\gamma D^2}$$

where L is the channel length, D the diameter, V the mean velocity, μ the viscosity of the fluid, and γ the specific weight of the fluid. Find the pressure in the syringe pump. The velocity head associated with the motion of the piston in the syringe pump is negligible.

PROBLEM 7.35

7.36 Firefighting equipment requires that the exit velocity of the firehose be 30 m/s at an elevation of 45 m above the hydrant. The nozzle at the end of the hose has a contraction ratio of 4:1 ($A_e/A_{hose} = 1/4$). The head loss in the hose is $8V^2/2g$, where V is the velocity in the hose. What must the pressure be at the hydrant to meet this requirement? The pipe supplying the hydrant is much larger than the firehose.

7.37 GO The discharge in the siphon is 0.08 m³/s, $D = 20$ cm, $L_1 = 0.9$ m, and $L_2 = 0.9$ m. Determine the head loss between the reservoir surface and point C. Determine the pressure at point B if three-quarters of the head loss (as found above) occurs between the reservoir surface and point B. Assume $\alpha = 1.0$ at all locations.

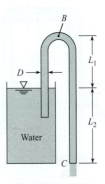

PROBLEM 7.37

7.38 GO For this siphon the elevations at A, B, C, and D are 30 m, 32 m, 27 m, and 26 m, respectively. The head loss between the inlet and point B is three-quarters of the velocity head, and the head loss in the pipe itself between point B and the end of the pipe is one-quarter of the velocity head. For these conditions, what is the discharge and what is the pressure at point B? The pipe diameter = 25 cm. Assume $\alpha = 1.0$ at all locations.

7.39 PLUS For this system, point B is 10 m above the bottom of the upper reservoir. The head loss from A to B is $1.1V^2/2g$, and the pipe area is 8×10^{-4} m². Assume a constant discharge of 8×10^{-4} m³/s. For these conditions, what will be the depth of water in the upper reservoir for which cavitation will begin at point B? Vapor pressure = 1.23 kPa and atmospheric pressure = 100 kPa. Assume $\alpha = 1.0$ at all locations.

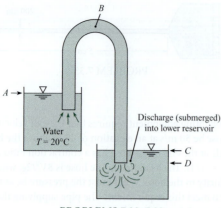

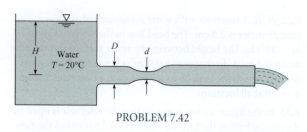

PROBLEM 7.42

7.40 In this system, $d = 15$ cm, $D = 30$ cm, $\Delta z_1 = 1.8$ m, and $\Delta z_2 = 3.6$ m. The discharge of water in the system is 0.3 m³/s. Is the machine a pump or a turbine? What are the pressures at points A and B? Neglect head losses. Assume $\alpha = 1.0$ at all locations.

7.43 (WILEY GO) A pump is used to fill a tank 5 m in diameter from a river as shown. The water surface in the river is 2 m below the bottom of the tank. The pipe diameter is 5 cm, and the head loss in the pipe is given by $h_L = 10\, V^2/2g$, where V is the mean velocity in the pipe. The flow in the pipe is turbulent, so $\alpha = 1$. The head provided by the pump varies with discharge through the pump as $h_p = 20 - 4 \times 10^4\, Q^2$, where the discharge is given in cubic meters per second (m³/s) and h_p is in meters. How long will it take to fill the tank to a depth of 10 m?

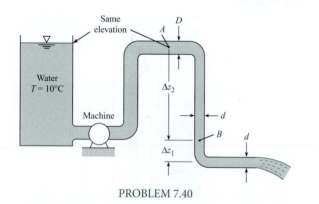

PROBLEM 7.40

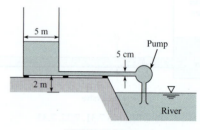

PROBLEM 7.43

7.41 (WILEY GO) The pipe diameter D is 30 cm, d is 15 cm, and the atmospheric pressure is 100 kPa. What is the maximum allowable discharge before cavitation occurs at the throat of the venturi meter if $H = 5$ m? Assume $\alpha = 1.0$ at all locations.

7.44 A pump is used to transfer SAE-30 oil from tank A to tank B as shown. The tanks have a diameter of 12 m. The initial depth of the oil in tank A is 20 m, and in tank B the depth is 1 m. The pump delivers a constant head of 60 m. The connecting pipe has a diameter of 20 cm, and the head loss due to friction in the pipe is 20 $V^2/2g$. Find the time required to transfer the oil from tank A to B; that is, the time required to fill tank B to 20 m depth.

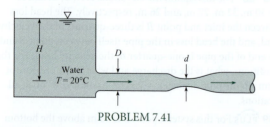

PROBLEM 7.41

7.42 (WILEY GO) In this system $d = 15$ cm, $D = 35$ cm, and the head loss from the venturi meter to the end of the pipe is given by $h_L = 1.5\, V^2/2g$, where V is the velocity in the pipe. Neglecting all other head losses, determine what head H will first initiate cavitation if the atmospheric pressure is 100 kPa absolute. What will be the discharge at incipient cavitation? Assume $\alpha = 1.0$ at all locations.

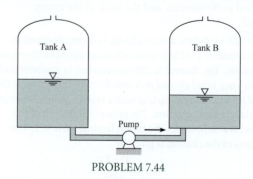

PROBLEM 7.44

7.45 A pump is used to pressurize a tank to 300 kPa abs. The tank has a diameter of 2 m and a height of 4 m. The initial level of water in the tank is 1 m, and the pressure at the water surface is 0 kPa gage. The atmospheric pressure is 100 kPa. The pump operates with a constant head of 50 m. The water is drawn from a source that is 4 m below the tank bottom. The pipe connecting the source and the tank is 4 cm in diameter and the head loss, including the expansion loss at the tank, is $10 \ V^2/2g$. The flow is turbulent.

Assume the compression of the air in the tank takes place isothermally, so the tank pressure is given by

$$p_T = \frac{3}{4 - z_t} p_0$$

where z_t is the depth of fluid in the tank in meters. Write a computer program that will show how the pressure varies in the tank with time, and find the time to pressurize the tank to 300 kPa abs.

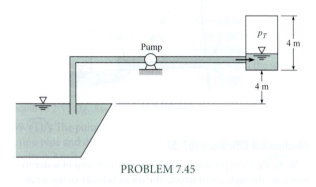

PROBLEM 7.45

The Power Equation (§7.4)

7.46 PLUS As shown, water at 15°C is flowing in a 15-cm-diameter by 60-m-long run of pipe that is situated horizontally. The mean velocity is 2 m/s, and the head loss is 2 m. Determine the pressure drop and the required pumping power to overcome head loss in the pipe.

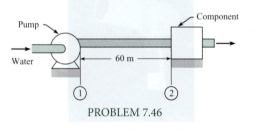

PROBLEM 7.46

7.47 PLUS The pump shown in the figure supplies energy to the flow such that the upstream pressure (30 cm pipe) is 35 kPa and the downstream pressure (15 cm pipe) is 380 kPa when the flow of water is 0.085 m³/s. What kilowatt power is delivered by the pump to the flow? Assume $\alpha = 1.0$ at all locations.

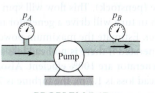

PROBLEM 7.47

7.48 GO A water discharge of 8 m³/s is to flow through this horizontal pipe, which is 1 m in diameter. If the head loss is given by $7 \ V^2/2g$ (V is velocity in the pipe), how much power will have to be supplied to the flow by the pump to produce this discharge? Assume $\alpha = 1.0$ at all locations.

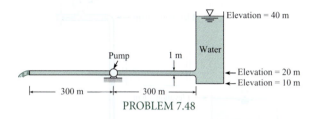

PROBLEM 7.48

7.49 An engineer is designing a subsonic wind tunnel. The test section is to have a cross-sectional area of 4 m² and an airspeed of 60 m/s. The air density is 1.2 kg/m³. The area of the tunnel exit is 10 m². The head loss through the tunnel is given by $h_L = (0.025)(V_T^2/2g)$, where V_T is the airspeed in the test section. Calculate the power needed to operate the wind tunnel. *Hint:* Assume negligible energy loss for the flow approaching the tunnel in region A, and assume atmospheric pressure at the outlet section of the tunnel. Assume $\alpha = 1.0$ at all locations.

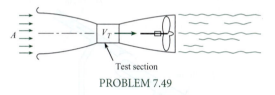

Test section

PROBLEM 7.49

7.50 PLUS Neglecting head losses, determine how many kilowatts of power the pump must deliver to produce the flow as shown. Here the elevations at points A, B, C, and D are 35 m, 47 m, 33 m, and 27 m, respectively. The nozzle area is 0.01 m².

7.51 PLUS Neglecting head losses, determine what power the pump must deliver to produce the flow as shown. Here the elevations at points A, B, C, and D are 40 m, 65 m, 35 m, and 30 m, respectively. The nozzle area is 25 cm².

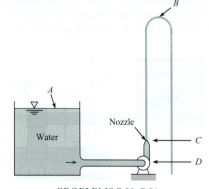

PROBLEMS 7.50, 7.51

7.69 This abrupt expansion is to be used to dissipate the high-energy flow of water in the 1.5 m–diameter penstock. Assume $\alpha = 1.0$ at all locations.

    a. What power (in kW) is lost through the expansion?

    b. If the pressure at section 1 is 35 kPa gage, what is the pressure at section 2?

    c. What force is needed to hold the expansion in place?

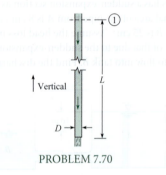

PROBLEM 7.69

7.70 This rough aluminum pipe is 15 cm in diameter. It weighs 22 N per meter of length, and the length L is 15 m. If the discharge of water is 0.17 m³/s and the head loss due to friction from section 1 to the end of the pipe is 3 m, what is the longitudinal force transmitted across section 1 through the pipe wall?

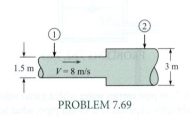

PROBLEM 7.70

7.71 Water flows in this bend at a rate of 5 m³/s, and the pressure at the inlet is 650 kPa. If the head loss in the bend is 10 m, what will the pressure be at the outlet of the bend? Also estimate the force of the anchor block on the bend in the x direction required to hold the bend in place. Assume $\alpha = 1.0$ at all locations.

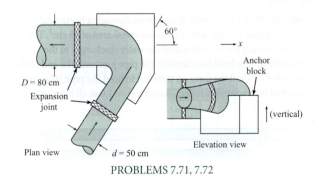

PROBLEMS 7.71, 7.72

7.72 ⓦ**PLUS** In a local water treatment plant, water flows in this bend at a rate of 7 m³/s, and the pressure at the inlet is 800 kPa. If the head loss in the bend is 13 m, what will the pressure be at the outlet of the bend? Also estimate the force of the anchor block on the bend in the x direction required to hold the bend in place. Assume $\alpha = 1.0$ at all locations.

7.73 Fluid flowing along a pipe of diameter D accelerates around a disk of diameter d as shown in the figure. The velocity far upstream of the disk is U, and the fluid density is ρ. Assuming incompressible flow and that the pressure downstream of the disk is the same as that at the plane of separation, develop an expression for the force required to hold the disk in place in terms of U, D, d, and ρ. Using the expression you developed, determine the force when $U = 10$ m/s, $D = 5$ cm, $d = 4$ cm, and $\rho = 1.2$ kg/m³. Assume $\alpha = 1.0$ at all locations.

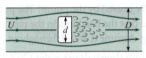

PROBLEM 7.73

EGL and HGL (§7.8)

7.74 ⓦ**PLUS part (b) only** Using Section 7.8 and other resources, answer the following questions. Strive for depth, clarity, and accuracy while also combining sketches, words, and equations in ways that enhance the effectiveness of your communication.

    a. What are three important reasons that engineers use the HGL and the EGL?

    b. What factors influence the magnitude of the HGL? What factors influence the magnitude of the EGL?

    c. How are the EGL and HGL related to the piezometer? To the stagnation tube?

    d. How is the EGL related to the energy equation?

    e. How can you use an HGL or an EGL to determine the direction of flow?

7.75 ⓦ**PLUS** The energy grade line for steady flow in a uniform-diameter pipe is shown. Which of the following could be in the "black box"? (a) a pump, (b) a partially closed valve, (c) an abrupt

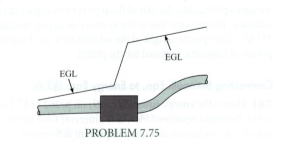

PROBLEM 7.75

expansion, or (d) a turbine. Choose all valid answer(s) and state your rationale.

7.76 If the pipe shown has constant diameter, is this type of HGL possible? If so, under what additional conditions? If not, why not?

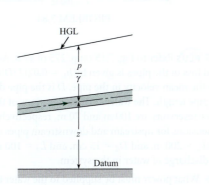

PROBLEM 7.76

7.77 PLUS For the system shown,

  a. What is the flow direction?

  b. What kind of machine is at A?

  c. Do you think both pipes, AB and CA, are the same diameter?

  d. Sketch in the EGL for the system.

  e. Is there a vacuum at any point or region of the pipes? If so, identify the location.

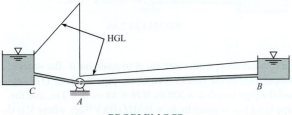

PROBLEM 7.77

7.78 The HGL and the EGL are as shown for a certain flow system.

  a. Is flow from A to E or from E to A?

  b. Does it appear that a reservoir exists in the system?

  c. Does the pipe at E have a uniform or a variable diameter?

  d. Is there a pump in the system?

  e. Sketch the physical setup that could yield the conditions shown between C and D.

  f. Is anything else revealed by the sketch?

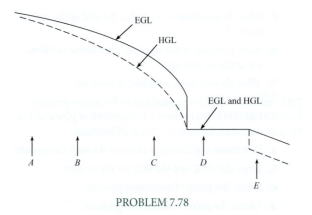

PROBLEM 7.78

7.79 Sketch the HGL and the EGL for this conduit, which tapers uniformly from the left end to the right end.

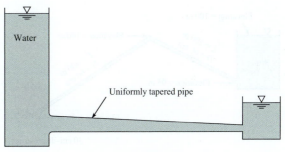

PROBLEM 7.79

7.80 PLUS The HGL and the EGL for a pipeline are shown in the figure.

  a. Indicate which is the HGL and which is the EGL.

  b. Are all pipes the same size? If not, which is the smallest?

  c. Is there any region in the pipes where the pressure is below atmospheric pressure? If so, where?

  d. Where is the point of maximum pressure in the system?

  e. Where is the point of minimum pressure in the system?

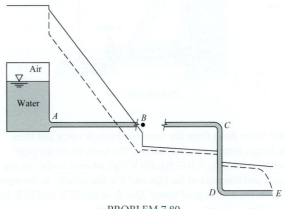

PROBLEM 7.80

f. What do you think is located at the end of the pipe at point *E*?

g. Is the pressure in the air in the tank above or below atmospheric pressure?

h. What do you think is located at point *B*?

7.81 GO Assume that the head loss in the pipe is given by $h_L = 0.014(L/D)(V^2/2g)$, where *L* is the length of pipe and *D* is the pipe diameter. Assume $\alpha = 1.0$ at all locations.

a. Determine the discharge of water through this system.

b. Draw the HGL and the EGL for the system.

c. Locate the point of maximum pressure.

d. Locate the point of minimum pressure.

e. Calculate the maximum and minimum pressures in the system.

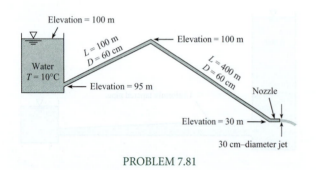

PROBLEM 7.81

7.82 Sketch the HGL and the EGL for the reservoir and pipe of Example 7.2.

7.83 The discharge of water through this turbine is 28.3 m³/s. What power is generated if the turbine efficiency is 85% and the total head loss is 1.2 m? *H* = 30 m. Also, carefully sketch the EGL and the HGL.

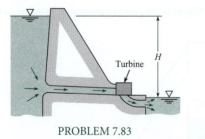

PROBLEM 7.83

7.84 Water flows from the reservoir through a pipe and then discharges from a nozzle as shown. The head loss in the pipe itself is given as $h_L = 0.025(L/D)(V^2/2g)$, where *L* and *D* are the length and diameter of the pipe and *V* is the velocity in the pipe. What is the discharge of water? Also draw the HGL and EGL for the system. Assume $\alpha = 1.0$ at all locations.

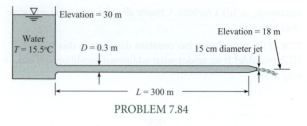

PROBLEM 7.84

7.85 PLUS Refer to Fig. 7.15 on p. 275 of §7.8. Assume that the head loss in the pipes is given by $h_L = 0.02(L/D)(V^2/2g)$, where *V* is the mean velocity in the pipe, *D* is the pipe diameter, and *L* is the pipe length. The water surface elevations of the upper and lower reservoirs are 100 m and 70 m, respectively. The respective dimensions for upstream and downstream pipes are $D_u = 30$ cm, and $L_u = 200$ m, and $D_d = 15$ cm, and $L_d = 100$ m. Determine the discharge of water in the system.

7.86 What power must be supplied to the water to pump 0.1 m³/s at 20°C from the lower to the upper reservoir? Assume that the head loss in the pipes is given by $h_L = 0.018(L/D)(V^2/2g)$, where *L* is the length of the pipe in meters and *D* is the pipe diameter in meters. Sketch the HGL and the EGL.

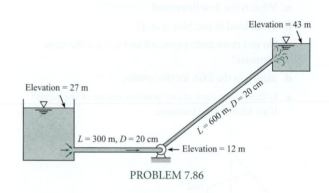

PROBLEM 7.86

7.87 Water flows from reservoir *A* to reservoir *B*. The water temperature in the system is 10°C, the pipe diameter *D* is 1 m, and the pipe length *L* is 300 m. If *H* = 16 m, *h* = 2 m, and the pipe head loss is given by $h_L = 0.01(L/D)(V^2/2g)$, where *V* is the velocity in the pipe, what will be the discharge in the pipe? In your solution, include the head loss at the pipe outlet, and sketch the HGL and the EGL. What will be the pressure at point *P* halfway between the two reservoirs? Assume $\alpha = 1.0$ at all locations.

7.88 GO Water flows from reservoir *A* to reservoir *B* in a desert retirement community. The water temperature in the system is 37.8°C, the pipe diameter *D* is 1.2 m, and the pipe length *L* is 60 m. If *H* = 10 m, *h* = 3 m, and the pipe head loss is given by $h_L = 0.01(L/D)(V^2/2g)$, where *V* is the velocity in the pipe, what will be the discharge in the pipe? In your solution, include the head loss at the pipe outlet. What will be the pressure at point *P* halfway between the two reservoirs? Assume $\alpha = 1.0$ at all locations.

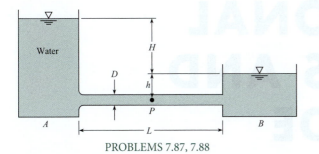

PROBLEMS 7.87, 7.88

7.89 Water flows from the reservoir on the left to the reservoir on the right at a rate of 0.45 m³/s. The formula for the head losses in the pipes is $h_L = 0.02(L/D)(V^2/2g)$. What elevation in the left reservoir is required to produce this flow? Also carefully sketch the HGL and the EGL for the system. *Note:* Assume the head-loss formula can be used for the smaller pipe as well as for the larger pipe. Assume $\alpha = 1.0$ at all locations.

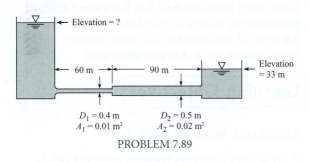

PROBLEM 7.89

7.90 What power is required to pump water at a rate of 3 m³/s from the lower to the upper reservoir? Assume the pipe head loss

is given by $h_L = 0.018(L/D)(V^2/2g)$, where L is the length of pipe, D is the pipe diameter, and V is the velocity in the pipe. The water temperature is 10°C, the water surface elevation in the lower reservoir is 150 m, and the surface elevation in the upper reservoir is 250 m. The pump elevation is 100 m, $L_1 = 100$ m, $L_2 = 1000$ m, $D_1 = 1$ m, and $D_2 = 50$ cm. Assume the pump and motor efficiency is 74%. In your solution, include the head loss at the pipe outlet and sketch the HGL and the EGL. Assume $\alpha = 1.0$ at all locations.

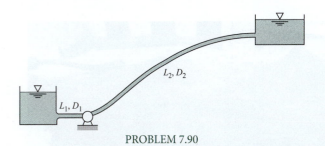

PROBLEM 7.90

7.91 Refer to Fig. 7.16 on p. 276 of §7.8. Assume that the head loss in the pipe is given by $h_L = 0.02(L/D)(V^2/2g)$, where V is the mean velocity in the pipe, D is the pipe diameter, and L is the pipe length. The elevations of the reservoir water surface, the highest point in the pipe, and the pipe outlet are 250 m, 250 m, and 210 m, respectively. The pipe diameter is 30 cm, and the pipe length is 200 m. Determine the water discharge in the pipe, and, assuming that the highest point in the pipe is halfway along the pipe, determine the pressure in the pipe at that point. Assume $\alpha = 1.0$ at all locations.

8 DIMENSIONAL ANALYSIS AND SIMILITUDE

FIGURE 8.1
The photo shows a model of a formula racing car that was built out of clay for testing in a small wind tunnel. The purpose of the testing was to assess the drag characteristics. The work was done by Josh Hartung, while he was an undergraduate engineering student. (Photo courtesy of Josh Hartung.)

Chapter Road Map

Because of the complexity of flows, designs are often based on experimental results, which are commonly done using scale models. The theoretical basis of experimental testing is called dimensional analysis, the topic of this chapter. This topic is also used to simplify analysis and to present results.

Learning Objectives

STUDENTS WILL BE ABLE TO

- Explain why dimensional analysis is needed. (§8.1)
- Explain or apply the Buckingham Π theorem. (§8.2)
- Find π-groups using the step-by-step method. (§8.3)
- Find π-groups using the exponent method. (§8.3)
- Define and describe common π-groups. (§8.4).
- Define a model and a prototype. (§8.5)
- Explain what similitude means, including geometric similitude and dynamic similitude. Describe the criteria for acheiving similitude. (§8.5)

8.1 Need for Dimensional Analysis

Fluid mechanics is more heavily involved with experimental testing than other disciplines because the analytical tools currently available to solve the momentum and energy equations are not capable of providing accurate results. This is particularly evident in turbulent, separating flows. The solutions obtained by utilizing techniques from computational fluid dynamics with the largest computers available yield only fair approximations for turbulent flow problems—hence the need for experimental evaluation and verification.

For analyzing model studies and for correlating the results of experimental research, it is essential that researchers employ dimensionless groups. To appreciate the advantages of using dimensionless groups, consider the flow of water through the unusual orifice illustrated in Fig. 8.2. Actually, this is much like a nozzle used for flow metering except that the flow is in the opposite direction. An orifice operating in this flow condition will have a much different performance than one operating in the normal mode. However, it is not unlikely that a firm or city water department might have such a situation where the flow may occur the "right way" most of the time and the "wrong way" part of the time—hence the need for such knowledge.

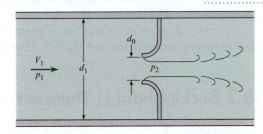

FIGURE 8.2

Flow through inverted flow nozzle.

Because of size and expense it is not always feasible to carry out tests on a full-scale prototype. Thus engineers will test a subscale model and measure the pressure drop across the model. The test procedure may involve testing several orifices, each with a different throat diameter d_0. For purposes of discussion, assume that three nozzles are to be tested. The Bernoulli equation, introduced in Chapter 4, suggests that the pressure drop will depend on flow velocity and fluid density. It may also depend on the fluid viscosity.

The test program may be carried out with a range of velocities and possibly with fluids of different density (and viscosity). The pressure drop, $p_1 - p_2$, is a function of the velocity V_1, density ρ, and diameter d_0. By carrying out numerous measurements at different values of V_1 and ρ for the three different nozzles, the data could be plotted as shown in Fig. 8.3a for tests using water. In addition, further tests could be planned with different fluids at considerably more expense.

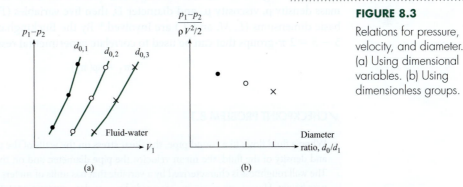

FIGURE 8.3

Relations for pressure, velocity, and diameter. (a) Using dimensional variables. (b) Using dimensionless groups.

The material introduced in this chapter leads to a much better approach. Through dimensional analysis it can be shown that the pressure drop can be expressed as

$$\frac{p_1 - p_2}{(\rho V^2)/2} = f\left(\frac{d_0}{d_1}, \frac{\rho V_1 d_0}{\mu}\right) \tag{8.1}$$

which means that dimensionless group for pressure, $(p_1 - p_2)/(\rho V^2/2)$, is a function of the dimensionless throat/pipe diameter ratio d_0/d_1 and the dimensionless group, $(\rho V_1 d_0)/\mu$, which will be identified later as the Reynolds number. The purpose of the experimental program is to establish the functional relationship. As will be shown later, if the Reynolds number is sufficiently large, the results are independent of Reynolds number. Then

$$\frac{p_1 - p_2}{(\rho V^2)/2} = f\left(\frac{d_0}{d_1}\right) \qquad (8.2)$$

Thus for any specific orifice design (same d_0/d_1) the pressure drop, $p_1 - p_2$, divided by $\rho V_1^2/2$ for the model is same for the prototype. Therefore the data collected from the model tests can be applied directly to the prototype. Only one test is needed for each orifice design. Consequently only three tests are needed, as shown in Fig. 8.2b. The fewer tests result in considerable savings in effort and expense.

The identification of dimensionless groups that provide correspondence between model and prototype data is carried out through **dimensional analysis**.

8.2 Buckingham Π Theorem

In 1915 Buckingham (1) showed that the number of independent dimensionless groups of variables (dimensionless parameters) needed to correlate the variables in a given process is equal to $n - m$, where n is the number of variables involved and m is the number of basic dimensions included in the variables.

Buckingham referred to the dimensionless groups as Π, which is the reason the theorem is called the Π theorem. Henceforth dimensionless groups will be referred to as **π-groups**. If the equation describing a physical system has n dimensional variables and is expressed as

$$y_1 = f(y_2, y_3, \ldots y_n)$$

then it can be rearranged and expressed in terms of $(n - m)$ π-groups as

$$\pi_1 = \varphi(\pi_2, \pi_3, \ldots \pi_{n-m})$$

Thus if the drag force F of a fluid flowing past a sphere is known to be a function of the velocity V, mass density ρ, viscosity μ, and diameter D, then five variables (F, V, ρ, μ, and D) and three basic dimensions (L, M, and T) are involved.* By the Buckingham Π theorem there will be $5 - 3 = 2$ π-groups that can be used to correlate experimental results in the form

$$\pi_1 = \varphi(\pi_2)$$

✔ CHECKPOINT PROBLEM 8.1

When a fluid flows in a round pipe, the shear stress on the walls of the pipe depends on the viscosity and density of the fluid, the mean velocity, the pipe diameter, and on the roughness of the pipe wall. The wall roughness is characterized by a variable that has units of meters that is called the sand roughness height. How many π-groups are needed to correlate experimental data?

   a. 1
   b. 2
   c. 3
   d. 4
   e. 5

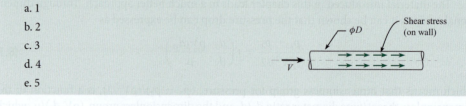

---

*Note that only three basic dimensions will be considered here. Temperature will not be included.

8.3 Dimensional Analysis

Dimensional analysis is the process for applying π-groups to analysis, experiment design, and the presentation of results. This section presents two methods for finding π-groups:

The Step-by-Step Method

Several methods may be used to carry out the process of finding the π-groups, but the step-by-step approach, very clearly presented by Ipsen (2), is one of the easiest and reveals much about the process. The process for the step-by-step method follows in Table 8.1.

The final result can be expressed as a functional relationship of the form

$$\pi_1 = f(\pi_2, \pi_2, \ldots \pi_n) \tag{8.3}$$

The selection of the dependent and independent π-groups depends on the application. Also the selection of variables used to eliminate dimensions is arbitrary.

TABLE 8.1 The Step-by-Step Approach

Step	Action Taken during This Step
1	Identify the significant dimensional variables and write out the primary dimensions of each.
2	Apply the Buckingham Π theorem to find the number of π-groups.*
3	Set up table with the number of rows equal to the number of dimensional variables and the number of columns equal to the number of basic dimensions plus one ($m + 1$).
4	List all the dimensional variables in the first column with primary dimensions.
5	Select a dimension to be eliminated, choose a variable with that dimension in the first column, and combine with remaining variables to eliminate the dimension. List combined variables in the second column with remaining primary dimensions.
6	Select another dimension to be eliminated, choose from variables in the second column that have that dimension, and combine with the remaining variables. List the new combinations with remaining primary dimensions in the third column
7	Repeat Step 6 until all dimensions are eliminated. The remaining dimensionless groups are the π-groups. List the π-groups in the last column

*Note that, in rare instances, the number of π-groups may be one more than predicted by the Buckingham Π theorem. This anomaly can occur because it is possible that two-dimensional categories can be eliminated when dividing (or multiplying) by a given variable. See Ipsen (2) for an example of this.

Example 8.1 shows how to use the step-by-step method to find the π-groups for a body falling in a vacuum.

EXAMPLE 8.1

Finding π-Group for a Body Falling in a Vacuum

Problem Statement

There are three significant dimensional variables for a body falling in a vacuum (no viscous effects): the velocity V; the acceleration due to gravity, g; and the distance through which the body falls, h. Find the π-groups using the step-by-step method.

Define the Situation

A body is falling in a vacuum, $V = f(g, h)$.

State the Goal

Find the π-groups.

Generate Ideas and Make a Plan

Apply the step-by-step method in Table 8.1.

Take Action (Execute the Plan)

1. Significant variables and dimensions

$$[V] = L/T$$
$$[g] = L/T^2$$
$$[h] = L$$

There are only two dimensions, L and T.

2. From the Buckingham Π theorem, there is only one (three variables–two dimensions) π-group.

3. Set up table with three rows (number of variables) and three (dimensions + 1) columns.

4. List variables and primary dimensions in first column.

Variable	[]	Variable	[]	Variable	[]
V	$\dfrac{L}{T}$	$\dfrac{V}{h}$	$\dfrac{1}{T}$	$\dfrac{V}{\sqrt{gh}}$	0
g	$\dfrac{L}{T^2}$	$\dfrac{g}{h}$	$\dfrac{1}{T^2}$		
h	L				

5. Select h to eliminate L. Divide g by h, enter in second column with dimension $1/T^2$. Divide V by h, enter in second column with dimension $1/T$.

6. Select g/h to eliminate T. Divide V/h by $\sqrt{g/h}$ and enter in third column.

As expected, there is only one π-group,

$$\pi = \frac{V}{\sqrt{gh}}$$

The final functional form of equation of the equation is

$$\frac{V}{\sqrt{gh}} = C$$

Review the Solution and the Process

1. *Knowledge.* From physics, one can show that $C = \sqrt{2}$.

2. *Knowledge.* The proper relationship between V, h, and g was found with dimensionless analysis. If the value of C was not known, it could be determined from experiment.

Example 8.2 illustrates the application of the step-by-step method for finding π-groups for a problem with five variables and three primary dimensions.

EXAMPLE 8.2

Finding π-Groups for Drag on a Sphere Using Step-by-Step Method

Problem Statement

The drag F_D of a sphere in a fluid flowing past the sphere is a function of the viscosity μ, the mass density ρ, the velocity of flow V, and the diameter of the sphere D. Use the step-by-step method to find the π-groups.

Define the Situation

The functional relationship is $F_D = f(V, \rho, \mu, D)$.

State the Goal

Find the π-groups using the step-by-step method.

Generate Ideas and Make a Plan

Apply the step-by-step procedure from Table 8.1.

Take Action (Execute the Plan)

1. Dimensions of significant variables

$$F = \frac{ML}{T^2}, V = \frac{L}{T}, \rho = \frac{M}{L^3}, \mu = \frac{M}{LT}, D = L$$

2. Number of π-groups, $5 - 3 = 2$.

3. Set up table with five rows and four columns.

4. Write variables and dimensions in first column.

Variable	[]	Variable	[]	Variable	[]	Variable	[]
F_D	$\dfrac{ML}{T^2}$	$\dfrac{F_D}{D}$	$\dfrac{M}{T^2}$	$\dfrac{F_D}{\rho D^4}$	$\dfrac{1}{T^2}$	$\dfrac{F_D}{\rho V^2 D^2}$	0
V	$\dfrac{L}{T}$	$\dfrac{V}{D}$	$\dfrac{1}{T}$	$\dfrac{V}{D}$	$\dfrac{1}{T}$		
ρ	$\dfrac{M}{L^3}$	ρD^3	M				
μ	$\dfrac{M}{LT}$	μD	$\dfrac{M}{T}$	$\dfrac{\mu}{\rho D^2}$	$\dfrac{1}{T}$	$\dfrac{\mu}{\rho VD}$	0
D	L						

5. Eliminate L using D and write new variable combinations with corresponding dimensions in the second column.

6. Eliminate M using ρD^3 and write new variable combinations with dimensions in the third column.

7. Eliminate T using V/D and write new combinations in the fourth column.

The final two π-groups are

$$\pi_1 = \frac{F_D}{\rho V^2 D^2} \quad \text{and} \quad \pi_2 = \frac{\mu}{\rho VD}$$

The functional equation can be written as

$$\frac{F_D}{\rho V^2 D^2} = f\left(\frac{\mu}{\rho VD}\right)$$

The form of the π-groups obtained will depend on the variables selected to eliminate dimensions. For example, if in Example 8.2, $\mu/\rho D^2$ had been used to eliminate the time dimension, the two π-groups would have been

$$\pi_1 = \frac{\rho F_D}{\mu^2} \quad \text{and} \quad \pi_2 = \frac{\mu}{\rho VD}$$

The result is still valid but may not be convenient to use. The form of any π-group can be altered by multiplying or dividing by another π-group. Multiplying the π_1 by the square of π_2 yields the original π_1 in Example 8.2.

$$\frac{\rho F_D}{\mu^2} \times \left(\frac{\mu}{\rho VD}\right)^2 = \frac{F_D}{\rho V^2 D^2}$$

By so doing the two π-groups would be the same as in Example 8.2.

The Exponent Method

An alternative method for finding the π-groups is the exponent method. This method involves solving a set of algebraic equations to satisfy dimensional homogeneity. The process for the exponent method is listed in Table 8.2.

TABLE 8.2 The Exponent Method

Step	Action Taken During This Step
1	Identify the significant dimensional variables, y_i, and write out the primary dimensions of each, $[y_i]$.
2	Apply the Buckingham Π theorem to find the number of π-groups.
3	Write out the product of the primary dimensions in the form $$[y_1] = [y_2]^a \times [y_3]^b \times \cdots \times [y_n]^k$$ where n is the number of dimensional variables and a, b, etc. are exponents.
4	Find the algebraic equations for the exponents that satisfy dimensional homogeneity (same power for dimensions on each side of equation).
5	Solve the equations for the exponents.
6	Express the dimensional equation in the form $y_1 = y_2^a y_3^b \ldots y_n^k$ and identify the π-groups.

Example 8.3 illustrates how to apply the exponent method to find the π-groups of the same problem addressed in Example 8.2.

EXAMPLE 8.3

Finding π-Groups for Drag on a Sphere Using Exponent Method

Problem Statement

The drag of a sphere, F_D, in a flowing fluid is a function of the velocity V, the fluid density ρ, the fluid viscosity μ, and the sphere diameter D. Find the π-groups using the exponent method.

Define the Situation

The functional equation is $F_D = f(V, \rho, \mu, D)$.

State the Goal

Find the π-groups using the exponent method.

Generate Ideas and Make a Plan

Apply the process for the exponent method from Table 8.2.

Take Action (Execute the Plan)

1. Dimensions of significant variables are

$$[F] = \frac{ML}{T^2}, [V] = \frac{L}{T}, [\rho] = \frac{M}{L^3}, [\mu] = \frac{M}{LT}, [D] = L$$

2. Number of π-groups is $5 - 3 = 2$.

3. Form product with dimensions.

$$\frac{ML}{T^2} = \left[\frac{L}{T}\right]^a \times \left[\frac{M}{L^3}\right]^b \times \left[\frac{M}{LT}\right]^c \times [L]^d$$

$$= \frac{L^{a-3b-c+d} M^{b+c}}{T^{a+c}}$$

4. Dimensional homogeneity. Equate powers of dimensions on each side.

$$L: \quad a - 3b - c + d = 1$$
$$M: \quad b + c = 1$$
$$T: \quad a + c = 2$$

5. Solve for exponents a, b, and c in terms of d.

$$\begin{pmatrix} 1 & -3 & -1 \\ 0 & 1 & 1 \\ 1 & 0 & 1 \end{pmatrix} \begin{pmatrix} a \\ b \\ c \end{pmatrix} = \begin{pmatrix} 1 - d \\ 1 \\ 2 \end{pmatrix}$$

The value of the determinant is -1 so a unique solution is achievable. Solution is $a = d, b = d - 1, c = 2 - d$

6. Write dimensional equation with exponents.

$$F = V^d \rho^{d-1} \mu^{2-d} D^d$$

$$F = \frac{\mu^2}{\rho} \left(\frac{\rho VD}{\mu}\right)^d$$

$$\frac{F\rho}{\mu^2} = \left(\frac{\rho VD}{\mu}\right)^d$$

There are two π-groups:

$$\pi_1 = \frac{F\rho}{\mu^2} \quad \text{and} \quad \pi_2 = \frac{\rho VD}{\mu}$$

By dividing π_1 by the square of π_2, the π_1 group can be written as $F_D/(\rho V^2 D^2)$, so the functional form of the equation can be written as

$$\frac{F}{\rho V^2 D^2} = f\left(\frac{\rho VD}{\mu}\right)$$

Review the Solution and the Process

Discussion. The functional relationship between the two π-groups can be obtained from experiments.

Selection of Significant Variables

All the foregoing procedures deal with straightforward situations. However, some problems do occur. To apply dimensional analysis one must first decide which variables are significant. If the problem is not sufficiently well understood to make a good choice of the significant variables, dimensional analysis seldom provides clarification.

A serious shortcoming might be the omission of a significant variable. If this is done, one of the significant π-groups will likewise be missing. In this regard, it is often best to identify a list of variables that one regards as significant to a problem and to determine if only one dimensional category (such as M or L or T) occurs. When this happens, it is likely that there is an error in choice of significant variables because it is not possible to combine two variables to eliminate the lone dimension. Either the variable with the lone dimension should not have been included in the first place (it is not significant), or another variable should have been included.

How does one know if a variable is significant for a given problem? Probably the truest answer is by experience. After working in the field of fluid mechanics for several years, one develops a feel for the significance of variables to certain kinds of applications. However, even the inexperienced engineer will appreciate the fact that free-surface effects have no significance in closed-conduit flow; consequently, surface tension, σ, would not be included as a variable. In closed-conduit flow, if the velocity is less than approximately one-third the speed of sound, compressibility effects are usually negligible. Such guidelines, which have been observed by previous experimenters, help the novice engineer develop confidence in her or his application of dimensional analysis and similitude.

8.4 Common π-Groups

The most common π-groups can be found by applying dimensional analysis to the variables that might be significant in a general flow situation, The purpose of this section is to develop these common π-groups and discuss their significance.

Variables that have significance in a general flow field are the velocity V, the density ρ, the viscosity μ, and the acceleration due to gravity g. In addition, if fluid compressibility were likely, then the bulk modulus of elasticity, E_v, should be included. If there is a liquid-gas interface, the surface tension effects may also be significant. Finally the flow field will be affected by a general length, L, such as the width of a building or the diameter of a pipe. These variables will be regarded as the independent variables. The primary dimensions of the significant independent variables are

$$[V] = L/T \quad [\rho] = M/L^3 \quad [\mu] = M/LT$$

$$[g] = L/T^2 \quad [E_v] = M/LT^2 \quad [\sigma] = M/T^2 \quad [L] = L$$

There are several other independent variables that could be identified for thermal effects, such as temperature, specific heat, and thermal conductivity. Inclusion of these variables is beyond the scope of this text.

Products that result from a flowing fluid are pressure distributions (p), shear stress distributions (τ), and forces on surfaces and objects (F) in the flow field. These will be identified as the dependent variables. The primary dimensions of the dependent variables are

$$[p] = M/LT^2 \quad [\tau] = [\Delta p] = M/LT^2 \quad [F] = (ML)/T^2$$

There are other dependent variables not included here, but they will be encountered and introduced for specific applications.

Altogether there are 10 significant variables, which, by application of the Buckingham Π theorem, means there are seven π-groups. Utilizing either the step-by-step method or the exponent method yields

$$\frac{p}{\rho V^2} \quad \frac{\tau}{\rho V^2} \quad \frac{F}{\rho V^2 L^2}$$

$$\frac{\rho V L}{\mu} \quad \frac{V}{\sqrt{E_v/\rho}} \quad \frac{\rho L V^2}{\sigma} \quad \frac{V^2}{gL}$$

The first three groups, the dependent π-groups, are identified by specific names. For these groups it is common practice to use the kinetic pressure, $\rho V^2/2$, instead of ρV^2. In most applications one is concerned with a pressure difference, so the pressure π-group is expressed as

$$C_p = \frac{p - p_0}{\frac{1}{2}\rho V^2}$$

where C_p is called the pressure coefficient and p_0 is a reference pressure. The pressure coefficient was introduced earlier in Chapter 4 and Section 8.1. The π-group associated with shear stress is called the shear-stress coefficient and defined as

$$c_f = \frac{\tau}{\frac{1}{2}\rho V^2}$$

where the subscript f denotes "friction." The π-group associated with force is referred to, here, as a force coefficient and defined as

$$C_F = \frac{F}{\frac{1}{2}\rho V^2 L^2}$$

This coefficient will be used extensively in Chapter 11 for lift and drag forces on airfoils and hydrofoils.

The independent π-groups are named after earlier contributors to fluid mechanics. The π-group $VL\rho/\mu$ is called the Reynolds number, after Osborne Reynolds, and designated by Re. The group $V/(\sqrt{E_v/\rho})$ is rewritten as (V/c) because $\sqrt{E_v/\rho}$ is the speed of sound, c. This π-group is called the Mach number and designated by M. The π-group $\rho L V^2/\sigma$ is called the Weber number and designated by We. The remaining π-group is usually expressed as $V/\sqrt{gL}$ and identified as the Froude (rhymes with "food") number* and written as Fr.

The general functional form for all the π-groups is

$$C_p, c_f, C_F = f(\text{Re, M, We, Fr}) \tag{8.4}$$

which means that either of the three dependent π-groups are functions of the four independent π-groups; that is, the pressure coefficient, the shear-stress coefficient, or the force coefficient are functions of the Reynolds number, Mach number, Weber number, and Froude number.

The π-groups, their symbols, and their names are summarized in Table 8.3. Each independent π-group has an important physical interpretation as indicated by the ratio column. The Reynolds number can be viewed as the ratio of kinetic to viscous forces. The kinetic forces are the forces associated with fluid motion. The Bernoulli equation indicates that the pressure difference required to bring a moving fluid to rest is the kinetic pressure, $\rho V^2/2$, so the kinetic forces,[†] F_k, should be proportional to

$$F_k \propto \rho V^2 L^2$$

The shear force due to viscous effects, F_v, is proportional to the shear stress and area

$$F_v \propto \tau A \propto \tau L^2$$

and the shear stress is proportional to

$$\tau \propto \mu \frac{dV}{dy} \propto \frac{\mu V}{L}$$

---

*Sometimes the Froude number is written as $V/\sqrt{(\Delta \gamma g L)/\gamma}$ and called the densimetric Froude number. It has application in studying the motion of fluids in which there is density stratification, such as between saltwater and freshwater in an estuary or heated-water effluents associated with thermal power plants.

[†]Traditionally the kinetic force has been identified as the "inertial" force.

TABLE 8.3 Common Π-Groups

π-Group	Symbol	Name	Ratio
$\dfrac{p - p_0}{(\rho V^2)/2}$	C_p	Pressure coefficient	$\dfrac{\text{Pressure differenece}}{\text{Kinetic pressure}}$
$\dfrac{\tau}{(\rho V^2)/2}$	c_f	Shear-stress coefficient	$\dfrac{\text{Shear stress}}{\text{Kinetic pressure}}$
$\dfrac{F}{(\rho V^2 L^2)/2}$	C_F	Force coefficient	$\dfrac{\text{Force}}{\text{Kinetic force}}$
$\dfrac{\rho L V}{\mu}$	Re	Reynolds number	$\dfrac{\text{Kinetic force}}{\text{Viscous force}}$
$\dfrac{V}{c}$	M	Mach number	$\sqrt{\dfrac{\text{Kinetic force}}{\text{Compressive force}}}$
$\dfrac{\rho L V^2}{\sigma}$	We	Weber number	$\dfrac{\text{Kinetic force}}{\text{Surface-tension force}}$
$\dfrac{V}{\sqrt{gL}}$	Fr	Froude number	$\sqrt{\dfrac{\text{Kinetic force}}{\text{Gravitational force}}}$

so $F_v \propto \mu VL$. Taking the ratio of the kinetic to the viscous forces

$$\frac{F_k}{F_v} \propto \frac{\rho VL}{\mu} = \text{Re}$$

yields the Reynolds number. The magnitude of the Reynolds number provides important information about the flow. A low Reynolds number implies viscous effects are important; a high Reynolds number implies kinetic forces predominate. The Reynolds number is one of the most widely used π-groups in fluid mechanics. It is also often written using kinematic viscosity, $\text{Re} = \rho VL/\mu = VL/\nu$.

The ratios of the other independent π-groups have similar significance. The Mach number is an indicator of how important compressibility effects are in a fluid flow. If the Mach number is small, then the kinetic force associated with the fluid motion does not cause a significant density change, and the flow can be treated as incompressible (constant density). On the other hand, if the Mach number is large, there are often appreciable density changes that must be considered in model studies.

The Weber number is an important parameter in liquid atomization. The surface tension of the liquid at the surface of a droplet is responsible for maintaining the droplet's shape. If a droplet is subjected to an air jet and there is a relative velocity between the droplet and the gas, kinetic forces due to this relative velocity cause the droplet to deform. If the Weber number is too large, the kinetic force overcomes the surface-tension force to the point that the droplet shatters into even smaller droplets. Thus a Weber-number criterion can be useful in predicting the droplet size to be expected in liquid atomization. The size of the droplets resulting from liquid atomization is a very significant parameter in gas-turbine and rocket combustion.

The Froude number is unimportant when gravity causes only a hydrostatic pressure distribution, such as in a closed conduit. However, if the gravitational force influences the pattern of flow, such as in flow over a spillway or in the formation of waves created by a ship as it cruises over the sea, the Froude number is a most significant parameter.

8.5 Similitude

Scope of Similitude

Similitude is the theory and art of predicting prototype performance from model observations. Whenever it is necessary to perform tests on a model to obtain information that cannot be obtained by analytical means alone, the rules of similitude must be applied. The theory of similitude involves the application of π-groups, such as the Reynolds number or the Froude number, to predict prototype performance from model tests. The art of similitude enters the problem when the engineer must make decisions about model design, model construction, performance of tests, or analysis of results that are not included in the basic theory.

Present engineering practice makes use of model tests more frequently than most people realize. For example, whenever a new airplane is being designed, tests are made not only on the general scale model of the prototype airplane but also on various components of the plane. Numerous tests are made on individual wing sections as well as on the engine pods and tail sections.

Models of automobiles and high-speed trains are also tested in wind tunnels to predict the drag and flow patterns for the prototype. Information derived from these model studies often indicates potential problems that can be corrected before the prototype is built, thereby saving considerable time and expense in development of the prototype.

In civil engineering, model tests are always used to predict flow conditions for the spillways of large dams. In addition, river models assist the engineer in the design of flood-control structures as well as in the analysis of sediment movement in the river. Marine engineers make extensive tests on model ship hulls to predict the drag of the ships. Much of this type of testing is done at the David Taylor Model Basin, Naval Surface Warfare Center, Carderock Division, near Washington, D.C. (see Fig. 8.4). Tests are also regularly performed on models of tall buildings

FIGURE 8.4

Ship-model test at the David Taylor Model Basin, Naval Surface Warfare Center, Carderock Division. (Naval Surface Warfare Center Carderock Division)

to help predict the wind loads on the buildings, the stability characteristics of the buildings, and the airflow patterns in their vicinity. The latter information is used by the architects to design walkways and passageways that are safer and more comfortable for pedestrians to use.

Geometric Similitude

Geometric similitude means that the model is an exact geometric replica of the prototype.[*] Consequently, if a 1:10 scale model is specified, all linear dimensions of the model must be 1/10 of those of the prototype. In Fig. 8.5 if the model and prototype are geometrically similar, the following equalities hold:

$$\frac{\ell_m}{\ell_p} = \frac{w_m}{w_p} = \frac{c_m}{c_p} = L_r \tag{8.5}$$

Here ℓ, w, and c are specific linear dimensions associated with the model and prototype, and L_r is the scale ratio between model and prototype. It follows that the ratio of corresponding areas between model and prototype will be the square of the length ratio: $A_r = L_r^2$. The ratio of corresponding volumes will be given by $\Psi_m/\Psi_p = L_r^3$.

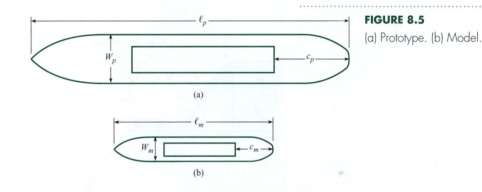

(a)

(b)

FIGURE 8.5

(a) Prototype. (b) Model.

Dynamic Similitude

Dynamic similitude means that the forces that act on corresponding masses in the model and prototype are in the same ratio (F_m/F_p = constant) throughout the entire flow field. For example, the ratio of the kinetic to viscous forces must be the same for the model and the prototype. Because the forces acting on the fluid elements control the motion of those elements, it follows that dynamic similarity will yield similarity of flow patterns. Consequently, the flow patterns for the model and the prototype will be the same if geometric similitude is satisfied and if the relative forces acting on the fluid are the same in the model as in the prototype. This latter condition requires that the appropriate π-groups introduced in Section 8.4 be the same for the model and prototype because these π-groups are indicators of relative forces within the fluid.

A more physical interpretation of the force ratios can be illustrated by considering the flow over the spillway shown in Fig. 8.6a. Here corresponding masses of fluid in the model and prototype are acted on by corresponding forces. These forces are the force of gravity F_g, the pressure force F_p, and the viscous resistance force F_v. These forces add vectorially as shown in Fig. 8.6 to yield a resultant force F_R, which will in turn produce an acceleration of the volume of fluid in accordance with Newton's second law of motion. Hence, because the force polygons in the

---

[*]For most model studies this is a basic requirement. However, for certain types of problems, such as river models, distortion of the vertical scale is often necessary to obtain meaningful results.

FIGURE 8.6

Model-prototype relations: prototype view (a) and model view (b).

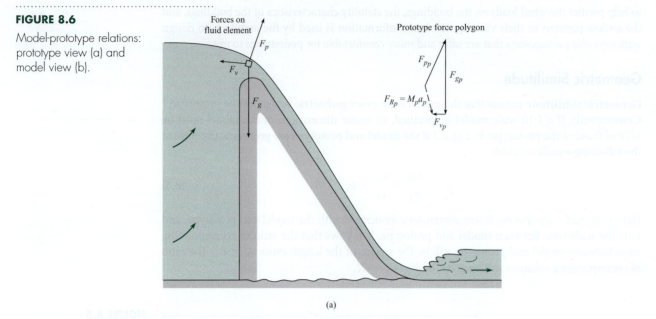

(a)

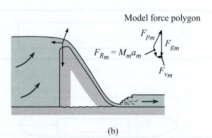

(b)

prototype and model are similar, the magnitudes of the forces in the prototype and model will be in the same ratio as the magnitude of the vectors representing mass times acceleration:

$$\frac{m_m a_m}{m_p a_p} = \frac{F_{gm}}{F_{gp}}$$

or

$$\frac{\rho_m L_m^3 (V_m/t_m)}{\rho_p L_p^3 (V_p/t_p)} = \frac{\gamma_m L_m^3}{\gamma_p L_p^3}$$

which reduces to

$$\frac{V_m}{g_m t_m} = \frac{V_p}{g_p t_p}$$

But

$$\frac{t_m}{t_p} = \frac{L_m/V_m}{L_p/V_p}$$

so

$$\frac{V_m^2}{g_m L_m} = \frac{V_p^2}{g_p L_p} \tag{8.6}$$

Taking the square root of each side of Eq. (8.6) gives

$$\frac{V_m}{\sqrt{g_m L_m}} = \frac{V_p}{\sqrt{g_p L_p}} \qquad \text{or} \qquad \text{Fr}_m = \text{Fr}_p \qquad (8.7)$$

Thus the Froude number for the model must be equal to the Froude number for the prototype to have the same ratio of forces on the model and the prototype.

Equating the ratio of the forces producing acceleration to the ratio of viscous forces,

$$\frac{m_m a_m}{m_p a_p} = \frac{F_{vm}}{F_{vp}} \qquad (8.8)$$

where $F_v \propto \mu V L$ leads to

$$\text{Re}_m = \text{Re}_p$$

The same analysis can be carried out for the Mach number and the Weber number. To summarize, if the independent π-groups for the model and prototype are equal, then the condition for dynamic similitude is satisfied.

Referring back to Eq. (8.4) for the general functional relationship,

$$C_p, c_f, C_F = f(\text{Re}, \text{M}, \text{We}, \text{Fr})$$

if the independent π-groups are the same for the model and the prototype, then dependent π-groups must also be equal so

$$C_{p,m} = C_{p,p} \qquad c_{f,m} = c_{f,p} \qquad C_{F,m} = C_{F,p} \qquad (8.9)$$

To have complete similitude between the model and the prototype, it is necessary to have both geometric and dynamic similitude.

In many situations it may not be possible nor necessary to have all the independent π-groups the same for the model and the prototype to carry out useful model studies. For the flow of a liquid in a horizontal pipe, for example, in which the fluid completely fills the pipe (no free surface), there would be no surface tension effects, so the Weber number would be inappropriate. Compressibility effects would not be important, so the Mach number would not be needed. In addition, gravity would not be responsible for the flow, so the Froude number would not have to be considered. The only significant π-group would be the Reynolds number; thus dynamic similitude would be achieved by matching the Reynolds number between the model and the prototype.

On the other hand if a model test were to be done for the flow over a spillway, the Froude number would be a significant π-group because gravity is responsible for the motion of the fluid. Also, the action of viscous stresses due to the spillway surface could possibly affect the flow pattern, so the Reynolds number may be a significant π-group. In this situation, dynamic similitude may require that both the Froude number and the Reynolds number be the same for the model and prototype.

The choice of significant π-groups for dynamic similitude and their actual use in predicting prototype performance are considered in the next two sections.

8.6 Model Studies for Flows without Free-Surface Effects

Free-surface effects are absent in the flow of liquids or gases in closed conduits, including control devices such as valves, or in the flow about bodies (e.g., aircraft) that travel through air or are deeply submerged in a liquid such as water (submarines). Free-surface effects are also absent where a structure such as a building is stationary and wind flows past it. In all these cases, given relatively low Mach numbers, the Reynolds-number criterion is the most significant for dynamic similarity. That is, the Reynolds number for the model must equal the Reynolds number for the prototype.

Example 8.4 illustrates the application of Reynolds-number similitude for the flow over a blimp.

EXAMPLE 8.4

Reynolds-Number Similitude

Problem Statement

The drag characteristics of a blimp 5 m in diameter and 60 m long are to be studied in a wind tunnel. If the speed of the blimp through still air is 10 m/s, and if a 1/10 scale model is to be tested, what airspeed in the wind tunnel is needed for dynamically similar conditions? Assume the same air pressure and temperature for both model and prototype.

Define the Situation

A 1/10 scale model blimp is being testing in a wind tunnel. Prototype speed is 10 m/s.

Assumptions: Same air pressure and temperature for model and prototype, therefore $\nu_m = \nu_p$.

State the Goal

Find the air speed (m/s) in the wind tunnel for dynamic similitude.

Generate Ideas and Make a Plan

The only π-group that is appropriate is the Reynolds number (there are no compressibility effects, free-surface effects, or gravitation effects). Thus equating the model and prototype Reynolds number satisfies dynamic similitude.

1. Equate the Reynolds number of the model and the prototype.
2. Calculate model speed.

Take Action (Execute the Plan)

1. Reynolds-number similitude

$$\mathrm{Re}_m = \mathrm{Re}_p$$
$$\frac{V_m L_m}{\nu_m} = \frac{V_p L_p}{\nu_p}$$

2. Model velocity

$$V_m = V_p \frac{L_p}{L_m} \frac{\nu_m}{\nu_p} = 10 \text{ m/s} \times 10 \times 1 = \boxed{100 \text{ m/s}}$$

Example 8.4 shows that the airspeed in the wind tunnel must be 100 m/s for true Reynolds-number similitude. This speed is quite large, and in fact Mach-number effects may start to become important at such a speed. However, it will be shown in Section 8.8 that it is not always necessary to operate models at true Reynolds-number similitude to obtain useful results.

If the engineer feels that it is essential to maintain Reynolds-number similitude, then only a few alternatives are available. One way to produce high Reynolds numbers at nominal airspeeds is to increase the density of the air. A NASA wind tunnel at the Ames Research Center at Moffett Field in California is one such facility. It has a 3.6 m-diameter test section, it can be pressurized up to 620 kPa, it can be operated to yield a Reynolds number per foot up to 1.2×10^7, and the maximum Mach number at which a model can be tested in this wind tunnel is 0.6. The airflow in this wind tunnel is produced by a single-stage, 20-blade axial-flow fan, which is powered by a 1120 MW, variable-speed, synchronous electric motor (3). Several problems are peculiar to a pressurized tunnel. First, a shell (essentially a pressurized bottle) must surround the entire tunnel and its components, adding to the cost of the tunnel. Second, it takes a long time to pressurize the tunnel in preparation for operation, increasing the time from the start to the finish of runs. In this regard it should be noted that the original pressurized wind tunnel at the Ames Research Center was built in 1946; however, because of extensive use, the tunnel's pressure shell began to deteriorate, so a new facility (the one previously described) was built and put in operation in 1995. Improvements over the old facility include a better data collection system, very low turbulence, and capability of depressurizing only the test section instead of the entire 17,600 m³ wind tunnel circuit when installing and removing models. The original pressurized wind tunnel was used to test most models of U.S. commercial aircraft over the past half-century, including the Boeing 737, 757, and 767; Lockheed L-1011; and McDonnell Douglas DC-9 and DC-10.

The Boeing 777 was tested in the low-speed, pressurized 5 m-by-5 m tunnel in Farnborough, England. This tunnel, operated by the Defence Evaluation and Research Agency (DERA) of Great Britain, can operate at three atmospheres with Mach numbers up to 0.2. Approximately 15,000 hours of total testing time was required for the Boeing 777 (4).

Another method of obtaining high Reynolds numbers is to build a tunnel in which the test medium (gas) is at a very low temperature, thus producing a relatively high-density–low-viscosity fluid. NASA has built such a tunnel and operates it at the Langley Research Center. This tunnel, called the National Transonic Facility, can be pressurized up to 9 atmospheres. The test medium is nitrogen, which is cooled by injecting liquid nitrogen into the system. In this wind tunnel it is possible to reach Reynolds numbers of 10^8 based on a model size of 0.25 m (5). Because of its sophisticated design, its initial cost was approximately $100,000,000 (6), and its operating expenses are high.

Another modern approach in wind-tunnel technology is the development of magnetic or electrostatic suspension of models. The use of the magnetic suspension with model airplanes has been studied (6), and the electrostatic suspension for the study of single-particle aerodynamics has been reported (7).

The use of wind tunnels for aircraft design has grown significantly as the size and sophistication of aircraft have increased. For example, in the 1930s the DC-3 and B-17 each had about 100 hours of wind-tunnel tests at a rate of $100 per hour of run time. By contrast the F-15 fighter required about 20,000 hours of tests at a cost of $20,000 per hour (6). The latter test time is even more staggering when one realizes that a much greater volume of data per hour at higher accuracy is obtained from the modern wind tunnels because of the high-speed data acquisition made possible by computers.

Example 8.5 illustrates the use of Reynolds-number similitude to design a test for a valve.

EXAMPLE 8.5

Reynolds-Number Similitude of a Valve

Problem Statement

The valve shown is the type used in the control of water in large conduits. Model tests are to be done, using water as the fluid, to determine how the valve will operate under wide-open conditions. The prototype size is 1.8 m in diameter at the inlet. What flow rate is required for the model if the prototype flow is 20 m³/s? Assume that the temperature for model and prototype is 15°C and that the model inlet diameter is 0.3 m.

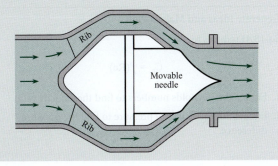

Rib

Movable needle

Rib

Define the Situations

A 1/6 scale model of a valve will be tested in a water tunnel. Prototype flow rate is 20 m³/s.

Assumptions:

1. No compressibility, free surface or gravitational effects.

2. Temperature of water in model and prototype is the same. Therefore kinematic viscosity for model and prototype are equal.

State the Goal

Find the flow rate through the model in m³/s.

Generate Ideas and Make a Plan

Dynamic similitude is obtained by equating the model and prototype Reynolds number. The model/prototype area ratio is the square of the scale ratio.

1. Equate Reynolds number of model and prototype.

2. Calculate the velocity ratio.

3. Calculate the discharge ratio using model/prototype area ratio.

Take Action (Execute the Plan)

1. Reynolds-number similitude

$$\text{Re}_m = \text{Re}_p$$

$$\frac{V_m L_m}{\nu_m} = \frac{V_p L_p}{\nu_p}$$

2. Velocity ratio

$$\frac{V_m}{V_p} = \frac{L_p}{L_m}\frac{\nu_m}{\nu_p}$$

Since $\nu_p = \nu_m$,

$$\frac{V_m}{V_p} = \frac{L_p}{L_m}$$

3. Discharge

$$\frac{Q_m}{Q_p} = \frac{V_m}{V_p}\frac{A_m}{A_p} = \frac{L_p}{L_m}\left(\frac{L_m}{L_p}\right)^2 = \frac{L_m}{L_p}$$

$$Q_m = 20 \text{ m}^3/\text{s} \times \frac{1}{6} = \boxed{3.3 \text{ m}^3/\text{s}}$$

Review the Solution and the Process

Discussion. This discharge is very large and serves to emphasize that very few model studies are made that completely satisfy the Reynolds-number criterion. This subject will be discussed further in the next sections.

8.7 Model-Prototype Performance

Geometric (scale model) and dynamic (same π-groups) similitude mean that the dependent π-groups are the same for both the model and the prototype. For this reason, measurements made with the model can be applied directly to the prototype. Such correspondence is illustrated in this section.

Example 8.6 shows how the pressure difference measured in a model test can be used to find the pressure difference between the corresponding two points on the prototype.

EXAMPLE 8.6

Application of Pressure Coefficient

Problem Statement

A 1/10 scale model of a blimp is tested in a wind tunnel under dynamically similar conditions. The speed of the blimp through still air is 10 m/s. A 17.8 kPa pressure difference is measured between two points on the model. What will be the pressure difference between the two corresponding points on the prototype? The temperature and pressure in the wind tunnel is the same as the prototype.

Define the Situation

A 1/10 scale of a blimp is tested in a wind tunnel under dynamically similar conditions. A pressure difference of 17.8 kPa is measured on the model.

Properties: Pressure and temperature are the same for wind tunnel test and prototype, so $\nu_m = \nu_p$.

State the Goal

Find the corresponding pressure difference (Pa) on prototype.

Generate Ideas and Make a Plan

Eq. (8.4) reduces to

$$C_p = f(\text{Re})$$

1. Equate the Reynolds numbers to find the velocity ratio.
2. Equate the coefficient of pressure to find the pressure difference.

Take Action (Execute the Plan)

1. Reynolds-number similitude

$$\text{Re}_m = \text{Re}_p$$

$$\frac{V_m L_m}{\nu_m} = \frac{V_p L_p}{\nu_p}$$

$$\frac{V_p}{V_m} = \frac{L_m}{L_p} = \frac{1}{10}$$

2. Pressure coefficient correspondence

$$\frac{\Delta p_m}{\frac{1}{2}\rho_m V_m^2} = \frac{\Delta p_p}{\frac{1}{2}\rho_p V_p^2}$$

$$\frac{\Delta p_p}{\Delta p_m} = \left(\frac{V_p}{V_m}\right)^2 = \left(\frac{L_m}{L_p}\right)^2 = \frac{1}{100}$$

Pressure difference on prototype

$$\Delta p_p = \frac{\Delta p_m}{100} = \frac{17.8 \text{ kPa}}{100} = \boxed{178 \text{ Pa}}$$

Example 8.7 illustrates calculating the fluid dynamic force on a prototype blimp from wind tunnel data using similitude.

EXAMPLE 8.7

Drag Force from Wind Tunnel Testing

Problem Statement

A 1/10 scale of a blimp is tested in a wind tunnel under dynamically similar conditions. If the drag force on the model blimp is measured to be 1530 N, what corresponding force could be expected on the prototype? The air pressure and temperature are the same for both model and prototype.

Define the Situation

A 1/10 scale model of blimp is tested in a wind tunnel, and a drag force of 1530 N is measured.

Properties: Pressure and temperature are the same, $\nu_m = \nu_p$.

State the Goal

Find the drag force (in newtons) on the prototype.

Generate Ideas and Make a Plan

Reynolds number is the only significant π-group, so Eq. (8.4) reduces to $C_F = f(\text{Re})$.

1. Find velocity ratio by equating Reynolds numbers.
2. Find the force by equating the force coefficients.

Take Action (Execute the Plan)

1. Reynolds-number similitude

$$\text{Re}_m = \text{Re}_p$$

$$\frac{V_m L_m}{\nu_m} = \frac{V_p L_p}{\nu_p}$$

$$\frac{V_p}{V_m} = \frac{V_m}{L_p} = \frac{1}{10}$$

2. Force coefficient correspondence

$$\frac{F_p}{\frac{1}{2}\rho_p V_p^2 L_p^2} = \frac{F_m}{\frac{1}{2}\rho_m V_m^2 L_m^2}$$

$$\frac{F_p}{F_m} = \frac{V_p^2}{V_m^2}\frac{L_p^2}{L_m^2} = \frac{L_m^2}{L_p^2}\frac{L_p^2}{L_m^2} = 1$$

Therefore

$$F_p = 1530 \text{ N}$$

Review the Solution and the Process

Discussion. The result that the model force is the same as the prototype force is interesting. When Reynolds-number similitude is used, and the fluid properties are the same, the forces on the model will always be the same as the forces on the prototype.

8.8 Approximate Similitude at High Reynolds Numbers

The primary justification for model tests is that it is more economical to get answers needed for engineering design by such tests than by any other means. However, as revealed by Examples 8.3, 8.4, and 8.6, Reynolds-number similitude requires expensive model tests (high-pressure facilities, large test sections, or using different fluids). This section shows that approximate similitude is achievable even though high Reynolds numbers cannot be reached in model tests.

Consider the size and power required for wind-tunnel tests of the blimp in Example 8.4. The wind tunnel would probably require a section at least 2 m by 2 m to accommodate the model blimp. With a 100 m/s airspeed in the tunnel, the power required for producing continuously a stream of air of this size and velocity is in the order of 4 MW. Such a test is not prohibitive, but it is very expensive. It is also conceivable that the 100 m/s airspeed would introduce Mach-number effects not encountered with the prototype, thus generating concern over the validity of the model data. Furthermore, a force of 1530 N is generally larger than that usually associated with model tests. Therefore, especially in the study of problems involving non-free-surface flows, it is desirable to perform model tests in such a way that large magnitudes of forces or pressures are not encountered.

For many cases, it is possible to obtain all the needed information from abbreviated tests. Often the Reynolds-number effect (relative viscous effect) either becomes insignificant at high Reynolds numbers or becomes independent of the Reynolds number. The point where testing can be stopped often can be detected by inspection of a graph of the pressure coefficient C_p versus the Reynolds number Re. Such a graph for a venturi meter in a pipe is shown in Fig. 8.7. In this meter, Δp is the pressure difference between the points shown, and V is the velocity in the restricted section of the venturi meter. Here it is seen that viscous forces affect the value of C_p below a Reynolds number of approximately 50,000. However, for higher Reynolds numbers, C_p is virtually constant. Physically this means that at low Reynolds numbers (relatively high viscous forces), a significant part of the change in pressure comes from viscous resistance, and the remainder comes from the acceleration (change in kinetic energy) of the fluid as it passes through the venturi meter. However, with high Reynolds numbers (resulting from either small viscosity or a large product of V, D, and ρ), the viscous resistance is negligible compared with the force required to accelerate the fluid. Because the ratio of Δp to the kinetic pressure does not change (constant C_p) for high Reynolds numbers, there is no need to carry out tests at higher Reynolds numbers. This is true in general, so long as the flow pattern does not change with the Reynolds number.

FIGURE 8.7

C_p for a venturi meter as a function of the Reynolds numbers.

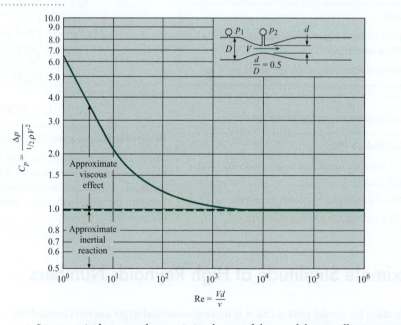

In a practical sense, whoever is in charge of the model test will try to predict from previous works approximately what maximum Reynolds number will be needed to reach the point of insignificant Reynolds-number effect and then will design the model accordingly. After a series of tests has been made on the model, C_p versus Re will be plotted to see whether the range of constant C_p has indeed been reached. If so, then no more data are needed to predict the prototype performance. However, if C_p has not reached a constant value, the test program has to be expanded or results extrapolated. Thus the results of some model tests can be used to predict prototype performance, even though the Reynolds numbers are not the same for the model and the prototype. This is especially valid for angular-shaped bodies, such as model buildings, tested in wind tunnels.

In addition, the results of model testing can be combined with analytic results. Computational fluid dynamics (CFD) may predict the change in performance with Reynolds number but may not be reliable to predict the performance level. In this case, the model testing would be used to establish the level and of performance, and the trends predicted by CFD would be used to extrapolate the results to other conditions.

Example 8.8 is an illustration on the approximate similitude at high Reynolds number for flow through a constriction.

EXAMPLE 8.8

Measuring Head Loss in a Nozzle in Reverse Flow

Problem Statement

Tests are to be performed to determine the head loss in a nozzle under a reverse-flow situation. The prototype operates with water at 10°C and with a nominal reverse-flow velocity of 1.5 m/s. The diameter of the prototype is 1 m. The tests are done in a 1/12 scale model facility with water at 15°C. A head loss (pressure drop) of 7.0 kPa is measured with a velocity of 6 m/s. What will be the head loss in the actual nozzle?

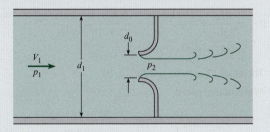

Define the Situation

A 1/12 scale model tests for head loss in a reverse-flow nozzle. A pressure difference of 7.0 kPa is measured with model at 6 m/s.

Properties: Table F.5.: Water at 10°C, $\rho = 1000 \text{ kg/m}^3$, $\nu = 1.31 \times 10^{-6} \text{ m}^2/\text{s}$; water at 15°C, $\rho = 999 \text{ kg/m}^3$, $\nu = 1.14 \times 10^{-6} \text{ m}^2/\text{s}$

State the Goal

Find the pressure drop (Pa) for the prototype nozzle.

Generate Ideas and Make a Plan

The only significant π-group is the Reynolds number, so Eq. (8.4) reduces to $C_p = f(\text{Re})$. Dynamic similitude

achieved if $\text{Re}_m = \text{Re}_p$, then $C_{p,m} = C_{p,p}$. From Fig. 8.7, if $\text{Re}_m, \text{Re}_p > 10^3$, then $C_{p,m} = C_{p,p}$.

1. Calculate Reynolds number for model and prototype.
2. Check if both exceed 10^3. If not, model tests need to be reevaluated.
3. Calculate pressure coefficient.
4. Evaluate pressure drop in prototype.

Take Action (Execute the Plan)

1. Reynolds numbers

$$\text{Re}_m = \frac{VD}{\nu} = \frac{(6 \text{ m/s} \times (1/12) \text{ m})}{(1.14 \times 10^{-6} \text{ m}^2/\text{s})} = 4.4 \times 10^5$$

$$\text{Re}_p = \frac{(1.5 \text{ m} \times 0.9 \text{ m})}{(1.31 \times 10^{-6} \text{ m}^2/\text{s})} = 1.04 \times 10^6$$

2. Both Reynolds numbers exceed 10^3. Therefore $C_{p,m} = C_{p,p}$. The test is valid.

3. Pressure coefficient from model tests

$$C_{p,m} = \frac{\Delta p}{\frac{1}{2}\rho V^2} = \frac{(7000 \text{ Pa})}{\left(\frac{1}{2} \times 999 \text{ kg/m}^3 \times (6 \text{ m/s})^2\right)} = 0.389$$

4. Pressure drop in prototype

$$\Delta p_p = 0.389 \times \frac{1}{2}\rho V^2 = 0.389 \times 0.5 \times 1000 \text{ kg/m}^3 \times (1.5 \text{ m/s})^2$$

$$= \boxed{292 \text{ Pa}}$$

Review the Solution and the Process

1. *Knowledge.* Because the Reynolds numbers are so much greater than 10^3, the equation for pressure drop is valid over a wide range of velocities.

2. *Discussion.* This example justifies the independence of Reynolds number referred to in Section 8.1.

In some situations viscous and compressibility effects may both be important, but it is not possible to have dynamic similitude with both π-groups. Which π-group is chosen for similitude depends a great deal on what information the engineer is seeking. If the engineer is interested in the viscous motion of fluid near a wall in shock-free supersonic flow, then the Reynolds number should be selected as the significant π-group. However, if the shock wave pattern over a body is of interest, then the Mach number should be selected for similitude. A useful rule of thumb is that compressibility effects are unimportant for M < 0.3.

Example 8.9 shows the difficulty in having Reynolds-number similitude and avoiding Mach-number effects in wind tunnel tests of an automobile.

EXAMPLE 8.9

Model Tests for Drag Force on an Automobile

Problem Statement

A 1/10 scale of an automobile is tested in a wind tunnel with air at atmospheric pressure and 20°C. The automobile is 4 m long and travels at a velocity of 100 km/hr in air at the same conditions. What should the wind-tunnel speed be such that the measured drag can be related to the drag of the prototype? Experience shows that the dependent π-groups are independent of Reynolds numbers for values exceeding 10^5. The speed of sound is 1235 km/hr.

Define the Situation

A 1/10 scale model of a 4 m–long automobile moving at 100 km/hr is tested in wind tunnel.

Properties: Air (20°C), Table A.3,
$\rho = 1.2$ kg/m^3, $\nu = 1.51 \times 10^{-5}$ N $\cdot$ s/m^2

State the Goal

Find the wind tunnel speed to achieve similitude.

Generate Ideas and Make a Plan

Mach number of the prototype is about 0.08 (100/1235), so Mach-number effects are unimportant. Dynamic similitude is achieved with Reynolds numbers, $\text{Re}_m = \text{Re}_p$. With dynamic similitude, $C_{F,m} = C_{F,p}$, and model measurements can be applied to prototype.

1. Determine the model speed for dynamic similitude.

2. Evaluate the model speed. If it is not feasible, continue to next step.

3. Calculate the prototype Reynolds number. If $\text{Re}_p > 10^5$, then $\text{Re}_m \geq 10^5$, for $C_{F,m} = C_{F,p}$.

4. Find the speed for which $\text{Re}_m \geq 10^5$.

Take Action (Execute the Plan)

1. Velocity from Reynolds-number similitude

$$\left(\frac{VL}{\nu}\right)_m = \left(\frac{VL}{\nu}\right)_p$$

$$\frac{V_m}{V_p} = \frac{L_p}{L_m} = 10$$

$$V_m = 10 \times 100 \text{ km/hr} = 1000 \text{ km/hr}$$

2. With this velocity, M = 1000/1235 = 0.81. This is too high for model tests because it would introduce unwanted compressibility effects.

3. Reynolds number of prototype

$$\text{Re}_p = \frac{VL\rho}{\mu} = \frac{100 \text{ km/hr} \times 0.278 \text{ (m/s)(km/hr)} \times 4 \text{ m}}{1.51 \times 10^{-5} \text{ m}^2/\text{s}}$$

$$= 7.4 \times 10^6$$

Therefore $C_{F,m} = C_{F,p}$, if $\text{Re}_m \geq 10^5$.

4. Wind tunnel speed

$$V_m \geq \text{Re}_m \frac{\nu_m}{L_m} = 10^5 \times \frac{1.51 \times 10^{-5} \text{ m}^2/\text{s}}{0.4 \text{ m}}$$

$$\geq \boxed{3.8 \text{ m/s}}$$

Review the Solution and the Process

Discussion. The wind-tunnel speed must exceed 3.8 m/s. From a practical point of view, the speed will be chosen to provide sufficiently large forces for reliable and accurate measurements.

8.9 Free-Surface Model Studies

Spillway Models

The flow over a spillway is a classic case of a free-surface flow. The major influence, besides the spillway geometry itself, on the flow of water over a spillway is the action of gravity. Hence the Froude-number similarity criterion is used for such model studies. It can be appreciated for large spillways with depths of water on the order of 3 m or 4 m and velocities on the order of 10 m/s or more, that the Reynolds number is very large. At high values of the Reynolds number, the relative viscous forces are often independent of the Reynolds number, as noted in the foregoing section (Sec. 8.8). However, if the reduced-scale model is made too small, the viscous forces as well as the surface-tension forces would have a larger relative effect on the flow in the model than in the prototype. Therefore, in practice, spillway models are made large enough so that the viscous effects have about the same relative effect in the model as in the prototype (i.e., the viscous effects are nearly independent of the Reynolds number). Then the Froude number is the significant π-group. Most model spillways are made at least 1 m high, and for precise studies, such as calibration of individual spillway bays, it is not uncommon to design and construct model spillway sections that are 2 m or 3 m high. Figures 8.8 and 8.9 show a comprehensive model and spillway model for Hell's Canyon Dam in Idaho.

FIGURE 8.8

Comprehensive model for Hell's Canyon Dam. Tests were made at the Albrook Hydraulic Laboratory, Washington State University. (Photo courtesy of Albrook Hydraulic Laboratory, Washington State University)

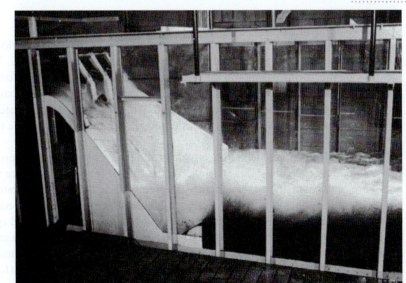

FIGURE 8.9

Spillway model for Hell's Canyon Dam. Tests were made at the Albrook Hydraulic Laboratory, Washington State University. (Photo courtesy of Albrook Hydraulic Laboratory, Washington State University)

Example 8.10 is an application of Froude-number similitude in modeling discharge over a spillway.

EXAMPLE 8.10

Modeling Flood Discharge Over a Spillway

Problem Statement

A 1/49 scale model of a proposed dam is used to predict prototype flow conditions. If the design flood discharge over the spillway is 15,000 m³/s, what water flow rate should be established in the model to simulate this flow? If a velocity of 1.2 m/s is measured at a point in the model, what is the velocity at a corresponding point in the prototype?

Define the Situation

A 1/49 scale model of spillway will be tested.
Prototype discharge is 15,000 m³/s.

State the Goal

Find:

1. Flow rate over model.

2. Velocity on prototype at point where velocity is 1.2 m/s on model.

Generate Ideas and Make a Plan

Gravity is responsible for the flow, so the significant π-group is the Froude number. For dynamic similitude, $\mathrm{Fr}_m = \mathrm{Fr}_p$.

1. Calculate velocity ratio from Froude-number similitude.

2. Calculate discharge ratio using scale ratio and calculate model discharge.

3. Use velocity ratio from step 1 to find velocity at point on prototype.

Take Action (Execute the Plan)

1. Froude-number similitude

$$\mathrm{Fr}_m = \mathrm{Fr}_p$$

$$\frac{V_m}{\sqrt{g_m L_m}} = \frac{V_p}{\sqrt{g_p L_p}}$$

The acceleration due to gravity is the same, so

$$\frac{V_m}{V_p} = \sqrt{\frac{L_m}{L_p}}$$

2. Discharge ratio

$$\frac{Q_m}{Q_p} = \frac{A_m}{A_p}\frac{V_m}{V_p} = \frac{L_m^2}{L_p^2}\sqrt{\frac{L_m}{L_p}} = \left(\frac{L_m}{L_p}\right)^{5/2}$$

Discharge for model

$$Q_m = Q_p\left(\frac{1}{49}\right)^{5/2} = 15{,}000\,\frac{\text{m}^3}{\text{s}} \times \frac{1}{16{,}800} = \boxed{0.89\ \text{m}^3/\text{s}}$$

3. Velocity on prototype

$$\frac{V_p}{V_m} = \sqrt{\frac{L_p}{L_m}}$$

$$V_p = \sqrt{49} \times 1.2\ \text{m/s} = \boxed{8.4\ \text{m/s}}$$

Ship Model Tests

The largest facility for ship testing in the United States is the David Taylor Model Basin, Naval Surface Warfare Center, Carderock Division, near Washington, D.C. Two of the core facilities are the towing basins and the rotating arm facility. In the rotating arm facility, models are suspended from the end of a rotating arm in a larger circular basin. Forces and moments can be measured on ship models up to 9 m in length at steady-state speeds as high as 15.4 m/s. In the high-speed towing basin, models 1.2 m to 6.1 m can be towed at speeds up to 16.5 m/s.

The aim of the ship model testing is to determine the resistance that the propulsion system of the ship must overcome. This resistance is the sum of the wave resistance and the surface resistance of the hull. The wave resistance is a free-surface, or Froude-number, phenomenon, and the hull resistance is a viscous, or Reynolds-number, phenomenon. Because both wave and viscous effects contribute significantly to the overall resistance, it would appear that both the Froude and Reynolds criteria should be used. However, it is impossible to satisfy both if the model liquid is water (the only practical test liquid), because the Reynolds-number similitude dictates a higher velocity for the model than for the prototype [equal to $V_p(L_p/L_m)$], whereas the Froude-number similitude dictates a lower velocity for the model [equal to $V_p(\sqrt{L_m}/\sqrt{L_p})$]. To circumvent such a dilemma, the procedure is to model for the phenomenon that is the most difficult to predict analytically and to account for the other resistance by analytical means. Because the wave resistance is the most difficult problem, the model is operated according to the Froude-number similitude, and the hull resistance is accounted for analytically.

To illustrate how the test results and the analytical solutions for surface resistance are merged to yield design data, the following necessary sequential steps are indicated.

1. Make model tests according to Froude-number similitude, and the total model resistance is measured. This total model resistance will be equal to the wave resistance plus the surface resistance of the hull of the model.

2. Estimate the surface resistance of the model by analytical calculations.

3. Subtract the surface resistance calculated in step 2 from the total model resistance of step 1 to yield the wave resistance of the model.

4. Using the Froude-number similitude, scale the wave resistance of the model up to yield the wave resistance of the prototype.

5. Estimate the surface resistance of the hull of the prototype by analytical means.

6. The sum of the wave resistance of the prototype from step 4 and the surface resistance of the prototype from step 5 yields the total prototype resistance, or drag.

8.10 Summarizing Key Knowledge

Rationale and Description of Dimensional Analysis

- Dimensional analysis involves combining dimensional variables to form dimensionless groups. These groups, called π-groups, can be regarded as the scaling parameters for fluid flow. Dimensional analysis is applied to analysis, experiment design, and to the presentation of results.

- The *Buckingham* Π *theorem* states that the number of independent π-groups is $n - m$, where n is the number of dimensional variables and m is the number of basic dimensions included in the variables.

Rationale and Description of Dimensional Analysis

- The π-groups can be found by either the *step-by-step method* or the *exponent method*.

  ▶ In the *step-by-step method* each dimension is removed by successively using a dimensional variable until the π-groups are obtained.

  ▶ In the *exponent method*, each variable is raised to a power, they are multiplied together, and three simultaneous algebraic equations formulated for dimensional homogeneity, are solved to yield the π-groups.

Common π-groups

- Four common *independent* π-groups are

$$\text{Reynolds number } \mathrm{Re} = \frac{\rho V L}{\mu} \qquad \text{Mach number } \mathrm{M} = \frac{V}{c}$$

$$\text{Weber number } \mathrm{We} = \frac{\rho V^2 L}{\sigma} \qquad \text{Froude number } \mathrm{Fr} = \frac{V}{\sqrt{gL}}$$

- Three common *dependent* π-groups are

$$\text{Pressure coefficient, } C_p = \frac{\Delta p}{(\rho V^2)/2}$$

$$\text{Shear stress coefficient, } c_f = \frac{\tau}{(\rho V^2)/2}$$

$$\text{Force coefficient, } C_F = \frac{F}{(\rho V^2 L^2)/2}$$

- The general functional form of the common π-groups is

$$C_F, c_f, C_p = f(\text{Re, M, We, Fr})$$

Dimensional Analysis in Experimental Testing

- Experimental testing is often performed with a small-scale replica (*model*) of the full-scale structure (*prototype*).
- *Similitude* is the art and theory of predicting prototype performance from model observations. To achieve exact similitude:
  ▸ The model must be a scale model of the prototype (*geometric similitude*).
  ▸ Values of the π-groups must be the same for the model and the prototype (*dynamic similitude*).
- In practice, it is not always possible to have complete dynamic similitude, so *only the most important π-groups are matched*.

REFERENCES

1. Buckingham, E. "Model Experiments and the Forms of Empirical Equations." *Trans. ASME,* 37 (1915), 263.

2. Ipsen, D. C. *Units, Dimensions and Dimensionless Numbers.* New York: McGraw-Hill, 1960.

3. NASA publication available from the U.S. Government Printing Office: No. 1995-685-893.

4. Personal communication. Mark Goldhammer, Manager, Aerodynamic Design of the 777.

5. Kilgore, R. A., and D. A. Dress. "The Application of Cryogenics to High Reynolds-Number Testing in Wind Tunnels, Part 2. Development and Application of the Cryogenic Wind Tunnel Concept." *Cryogenics,* Vol. 24, no. 9, September 1984.

6. Baals, D. D., and W. R. Corliss. *Wind Tunnels of NASA.* Washington, DC: U.S. Govt. Printing Office, 1981.

7. Kale, S., et al. "An Experimental Study of Single-Particle Aerodynamics." *Proc. of First Nat. Congress on Fluid Dynamics,* Cincinnati, Ohio, July 1988.

PROBLEMS

PLUS Problem available in *WileyPLUS* at instructor's discretion.

GO Guided Online (GO) Problem, available in *WileyPLUS* at instructor's discretion.

Dimensional Analysis (§8.2)

8.1 Find the primary dimensions of density ρ, viscosity μ, and pressure p.

8.2 According to the Buckingham Π theorem, if there are six dimensional variables and three primary dimensions, how many dimensionless variables will there be?

8.3 Explain what is meant by dimensional homogeneity.

8.4 PLUS Determine which of the following equations are dimensionally homogeneous:

$$\textbf{a. } Q = \tfrac{2}{3} CL \sqrt{2g} H^{3/2}$$

where Q is discharge, C is a pure number, L is length, g is acceleration due to gravity, and H is head.

b. $V = \dfrac{1.49}{n} R^{2/3} S^{1/2}$

where V is velocity, n is length to the one-sixth power, R is length, and S is slope.

c. $h_f = f \dfrac{L}{D} \dfrac{V^2}{2g}$

where h_f is head loss, f is a dimensionless resistance coefficient, L is length, D is diameter, V is velocity, and g is acceleration due to gravity.

d. $D = \dfrac{0.074}{Re^{0.2}} \dfrac{Bx\rho V^2}{2}$

where D is drag force, Re is Vx/v, B is width, x is length, ρ is mass density, v is the kinematic viscosity, and V is velocity.

8.5 Determine the dimensions of the following variables and combinations of variables in terms of primary dimensions.

 a. T (torque)

 b. $\rho V^2/2$, where V is velocity and ρ is mass density

 c. $\sqrt{\tau/\rho}$, where τ is shear stress

 d. Q/ND^3, where Q is discharge, D is diameter, and N is angular speed of a pump

8.6 It takes a certain length of time for the liquid level in a tank of diameter D to drop from position h_1 to position h_2 as the tank is being drained through an orifice of diameter d at the bottom. Determine the π-groups that apply to this problem. Assume that the liquid is nonviscous. Express your answer in the functional form.

$$\dfrac{\Delta h}{d} = f(\pi_1, \pi_2, \pi_3)$$

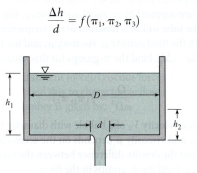

PROBLEM 8.6

8.7 The maximum rise of a liquid in a small capillary tube is a function of the diameter of the tube, the surface tension, and the specific weight of the liquid. What are the significant π-groups for the problem?

8.8 For very low velocities it is known that the drag force F_D of a small sphere is a function solely of the velocity V of flow past the sphere, the diameter d of the sphere, and the viscosity μ of the fluid. Determine the π-groups involving these variables.

8.9 Observations show that the side thrust F, for a rough spinning ball in a fluid is a function of the ball diameter D, the free-stream velocity V_0, the density ρ, the viscosity μ, the

roughness height k_s, and the angular velocity of spin ω. Determine the dimensionless parameter(s) that would be used to correlate the experimental results of a study involving the variables noted above. Express your answer in the functional form

$$\dfrac{F}{\rho V_0^2 D^2} = f(\pi_1, \pi_2, \pi_3)$$

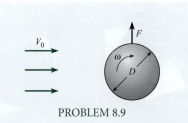

PROBLEM 8.9

8.10 Consider steady viscous flow through a small horizontal tube. For this type of flow, the pressure gradient along the tube, $\Delta p/\Delta \ell$ should be a function of the viscosity μ, the mean velocity V, and the diameter D. By dimensional analysis, derive a functional relationship relating these variables.

8.11 A flow-metering device, called a vortex meter, consists of a square element mounted inside a pipe. Vortices are generated by the element, which gives rise to an oscillatory pressure measured on the leeward side of the element. The fluctuation frequency is related to the flow velocity. The discharge in the pipe is a function of the frequency of the oscillating pressure ω, the pipe diameter D, the size of the element l, the density ρ, and the viscosity μ. Thus

$$Q = f(\omega, D, l, \rho, \mu)$$

Find the π-groups in the form

$$\dfrac{Q}{\omega D^3} = f(\pi_1, \pi_2)$$

8.12 It is known that the pressure developed by a centrifugal pump, Δp, is a function of the diameter D of the impeller, the speed of rotation n, the discharge Q, and the fluid density ρ. By dimensional analysis, determine the π-groups relating these variables.

8.13 The force on a satellite in the earth's upper atmosphere depends on the mean path of the molecules λ (a length), the density ρ, the diameter of the body D, and the molecular speed c: $F = f(\lambda, \rho, D, c)$. Find the nondimensional form of this equation.

8.14 A general study is to be made of the height of rise of liquid in a capillary tube as a function of time after the start of a test. Other significant variables include surface tension, mass density, specific weight, viscosity, and diameter of the tube. Determine the dimensionless parameters that apply to the problem. Express your answer in the functional form

$$\dfrac{h}{d} = f(\pi_1, \pi_2, \pi_3)$$

8.36 ⌀PLUS A large venturi meter is calibrated by means of a 1/10 scale model using the prototype liquid. What is the discharge ratio Q_m/Q_p for dynamic similarity? If a pressure difference of 400 kPa is measured across ports in the model for a given discharge, what pressure difference will occur between similar ports in the prototype for dynamically similar conditions?

8.37 ⌀PLUS A 1/5 scale model of an experimental deep sea submersible that will operate at great depths is to be tested to determine its drag characteristic by towing it behind a submarine. For true similitude, what should be the towing speed relative to the speed of the prototype?

8.38 ⌀PLUS A spherical balloon that is to be used in air at 15.5°C and atmospheric presssure is tested by towing a 1/12 scale model in a lake. The model is 0.43 m in diameter, and a drag of 165 N is measured when the model is being towed in deep water at 1.5 m/s. What drag (in newtons and pound force) can be expected for the prototype in air under dynamically similar conditions? Assume that the water temperature is 15°C.

8.39 ⌀PLUS An engineer needs a value of lift force for an airplane that has a coefficient of lift (C_L) of 0.4. The π-group is defined as

$$C_L = 2\frac{F_L}{\rho V^2 S}$$

where F_L is the lift force, ρ is the density of ambient air, V is the speed of the air relative to the airplane, and S is the area of the wings from a top view. Estimate the lift force in newtons for a speed of 80 m/s, an air density of 1.1 kg/m^3, and a wing area (planform area) of 15 m^2.

PROBLEM 8.39 (© Daniel Karlsson/Stocktrek Images, Inc.)

8.40 ⌀PLUS An airplane travels in air ($p = 100$ kPa, $T = 10°C$) at 150 m/s. If a 1/8 scale model of the plane is tested in a wind tunnel at 25°C, what must the density of the air in the tunnel be so that both the Reynolds-number and the Mach-number criteria are satisfied? The speed of sound varies with the square root of the absolute temperature. (*Note:* The dynamic viscosity is independent of pressure.)

8.41 The Airbus A380-300 has a wing span of 79.8 m. The cruise altitude is 10,000 m in a standard atmosphere. Assume you are designing a wind tunnel to operate with air at 20°C. The

span of the scale model A380 in the wind tunnel is 1 m. Assume Mach-number correspondence between model and prototype. Both the speed of sound and the dynamic viscosity vary linearly with the square root of the absolute temperature. What would the pressure of the air in the wind tunnel have to be to have Reynolds-number similitude? Use the properties for a standard atmosphere in Chapter 3 to find properties at 10,000 m altitude.

8.42 ⌀PLUS The Boeing 787-3 Dreamliner has a wing span of 52 m. It flies at a cruise Mach number of 0.85, which corresponds to a velocity of 945 km/hr at an altitude of 10,000 m. You are going to estimate the drag on the prototype by measuring the drag on a 1 m wing span scale model in a wind tunnel with air where the speed of sound is 340 m/s and the density is 0.98 kg/m^3. What is the ratio of the force on the prototype to the force on the model? Only Mach-number similitude is considered. Use the properties of the standard atmosphere in Chapter 3 to evaluate the density of air for the prototype.

8.43 ⌀GO Flow in a given pipe is to be tested with air and then with water. Assume that the velocities (V_A and V_W) are such that the flow with air is dynamically similar to the flow with water. Then for this condition, the magnitude of the ratio of the velocities, V_A/V_W, will be (a) less than unity, (b) equal to unity, or (c) greater than unity.

8.44 ⌀PLUS A smooth pipe designed to carry crude oil (D = 120 cm, $\rho = 900$ kg/m^3, and $\mu = 0.02$ Pa-s) is to be modeled with a smooth pipe 10 cm in diameter carrying water ($T = 15°C$). If the mean velocity in the prototype is 0.6 m/s, what should be the mean velocity of water in the model to ensure dynamically similar conditions?

8.45 ⌀GO A student is competing in a contest to design a radio-controlled blimp. The drag force acting on the blimp depends on the Reynolds number, Re = $(\rho VD)/\mu$, where V is the speed of the blimp, D is the maximum diameter, ρ is the density of air, and μ is the viscosity of air. This blimp has a coefficient of drag (C_D) of 0.3. This π-group is defined as

$$C_D = 2\frac{F_D}{\rho V^2 A_p}$$

where F_D is the drag force ρ is the density of ambient air, V is the speed of the blimp, and $A_p = \pi D^2/4$ is the maximum section area of the blimp from a front view. Calculate the Reynolds number, the drag force in newtons, and the power in watts required to move the blimp through the air. Blimp speed is 800 mm/s, and the maximum diameter is 475 mm. Assume that ambient air is at 20°C.

PROBLEM 8.45

8.46 **WILEY PLUS** Colonization of the moon will require an improved understanding of fluid flow under reduced gravitational forces. The gravitational force on the moon is 1/5 that on the surface of the earth. An engineer is designing a model experiment for flow in a conduit on the moon. The important scaling parameters are the Froude number and the Reynolds number. The model will be full scale. The kinematic viscosity of the fluid to be used on the moon is 0.5×10^{-5} m²/s. What should be the kinematic viscosity of the fluid to be used for the model on earth?

8.47 A drying tower at an industrial site is 10 m in diameter. The air inside the tower has a kinematic viscosity of 4×10^{-5} m²/s and enters at 12 m/s. A 1/15 scale model of this tower is fabricated to operate with water that has a kinematic viscosity of 10^{-6} m²/s. What should be the entry velocity of the water be to achieve Reynolds-number scaling?

8.48 **WILEY PLUS** A flow meter to be used in a 40 cm pipeline carrying oil ($\nu = 10^{-5}$ m²/s, $\rho = 860$ kg/m³) is to be calibrated by means of a model (1/9 scale) carrying water ($T = 20°C$ and standard atmospheric pressure). If the model is operated with a velocity of 1.6 m/s, find the velocity for the prototype based on Reynolds-number scaling. For the given conditions, if the pressure difference in the model was measured as 3.0 kPa, what pressure difference would you expect for the discharge meter in the oil pipeline?

8.49 Water at 10°C flowing through a rough pipe 10 cm in diameter is to be simulated by air (20°C) flowing through the same pipe. If the velocity of the water is 1.5 m/s, what will the air velocity have to be to achieve dynamic similarity? Assume the absolute air pressure in the pipe to be 150 kPa. If the pressure difference between two sections of the pipe during air flow was measured as 780 Pa, what pressure difference occurs between these two sections when water is flowing under dynamically similar conditions?

8.50 **WILEY GO** The "noisemaker" B is towed behind the minesweeper A to set off enemy acoustic mines such as that shown at C. The drag force of the "noisemaker" is to be studied in a water tunnel at a 1/5 scale (the model is 1/5 the size of the full scale). If the full-scale towing speed is 5 m/s, what should be the water velocity in the water tunnel for the two tests to be exactly similar? What will be the prototype drag force if the model drag force is found to be 2400 N? Assume that seawater at the same temperature is used in both the full-scale and the model tests.

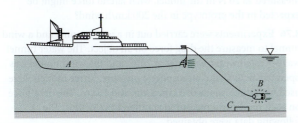

PROBLEM 8.50

8.51 **WILEY PLUS** An experiment is being designed to measure aerodynamic forces on a building. The model is a 1/500 scale replica of the prototype. The wind velocity on the prototype is 14.5 m/s, and the density is 1.24 kg/m³. The maximum velocity in the wind tunnel is 90 m/s. The viscosity of the air flowing for the model and the prototype is the same. Find the density needed in the wind tunnel for dynamic similarity. A force of 200 N is measured on the model. What will the force be on the prototype?

8.52 A 60 cm valve is designed for control of flow in a petroleum pipeline. A 1/3 scale model of the full-size valve is to be tested with water in the laboratory. If the prototype flow rate is to be 0.5 m³/s, what flow rate should be established in the laboratory test for dynamic similitude to be established? Also, if the pressure coefficient C_p in the model is found to be 1.07, what will be the corresponding C_p in the full-scale valve? The relevant fluid properties for the petroleum are S = 0.82 and $\mu = 3 \times 10^{-3}$ N · s/m². The viscosity of water is 10^{-3} N · s/m².

8.53 **WILEY PLUS** The moment acting on a submarine rudder is studied by a 1/40 scale model. If the test is made in a water tunnel and if the moment measured on the model is 2 N · m when the freshwater speed in the tunnel is 6.6 m/s, what are the corresponding moment and speed for the prototype? Assume the prototype operates in sea water. Assume $T = 10°C$ for both the freshwater and the seawater.

8.54 **WILEY PLUS** A model hydrofoil is tested in a water tunnel. For a given angle of attack, the lift of the hydrofoil is measured to be 25 kN when the water velocity is 15 m/s in the tunnel. If the prototype hydrofoil is to be twice the size of the model, what lift force would be expected for the prototype for dynamically similar conditions? Assume a water temperature of 20°C for both model and prototype.

8.55 A 1/10 scale model of an automobile is tested in a pressurized wind tunnel. The test is to simulate the automobile traveling at 100 km/h in air at atmospheric pressure and 25°C. The wind tunnel operates with air at 25°C. At what pressure in the test section must the tunnel operate to have the same Mach and Reynolds numbers? The speed of sound in air at 25°C is 345 m/s.

8.56 If the tunnel in Prob. 8.55 were to operate at atmospheric pressure and 25°C, what speed would be needed to achieve the same Reynolds number for the prototype? At this speed, would you conclude that Mach-number effects were important?

8.57 **WILEY PLUS** Experimental studies have shown that the condition for breakup of a droplet in a gas stream is

$$We/Re^{1/2} = 0.5$$

where Re is the Reynolds number and We is the Weber number based on the droplet diameter. What diameter water droplet would break up in a 12 m/s airstream at 20°C and standard atmospheric pressure? The surface tension of water is 0.041 N/m.

8.58 Water is sprayed from a nozzle at 30 m/s into air at atmospheric pressure and 20°C. Estimate the size of the droplets produced if the Weber number for breakup is 6.0 based on the droplet diameter.

8.59 Determine the relationship between the kinematic viscosity ratio v_m/v_p and the scale ratio if both the Reynolds-number and the Froude-number criteria are to be satisfied in a given model test.

8.60 GO A hydraulic model, 1/20 scale, is built to simulate the flow conditions of a spillway of a dam. For a particular run, the waves downstream were observed to be 8 cm high. How high would be similar waves on the full-scale dam operating under the same conditions? If the wave period in the model is 2 s, what would the wave period in the prototype be?

8.61 The scale ratio between a model dam and its prototype is 1/25. In the model test, the velocity of flow near the crest of the spillway was measured to be 2.5 m/s. What is the corresponding prototype velocity? If the model discharge is 0.10 m³/s, what is the prototype discharge?

8.62 PLUS A seaplane model is built at a 1/6 scale. To simulate takeoff conditions at 117 km/h, what should be the corresponding model speed to achieve Froude-number scaling?

8.63 If the scale ratio between a model spillway and its prototype is 1/36, what velocity and discharge ratio will prevail between model and prototype? If the prototype discharge is 3000 m³/s, what is the model discharge?

8.64 The depth and velocity at a point in a river are measured to be 6 m and 4.5 m/s, respectively. If a 1/64 scale model of this river is constructed and the model is operated under dynamically similar conditions to simulate the free-surface conditions, then what velocity and depth can be expected in the model at the corresponding point?

8.65 PLUS A 1/40 scale model of a spillway is tested in a laboratory. If the model velocity and discharge are 1 m/s and 0.1 m³/s, respectively, what are the corresponding values for the prototype?

8.66 Flow around a bridge pier is studied using a model at 1/12 scale. When the velocity in the model is 0.9 m/s, the standing wave at the pier nose is observed to be 2.5 cm in height. What are the corresponding values of velocity and wave height in the prototype?

8.67 A 1/25 scale model of a spillway is tested. The discharge in the model is 0.1 m³/s. To what prototype discharge does this correspond? If it takes 1 min for a particle to float from one point to another in the model, how long would it take a similar particle to traverse the corresponding path in the prototype?

8.68 PLUS A tidal estuary is to be modeled at 1/600 scale. In the actual estuary, the maximum water velocity is expected to be 3.6 m/s, and the tidal period is approximately 12.5 h. What corresponding velocity and period would be observed in the model?

8.69 PLUS The maximum wave force on a 1/36 model seawall was found to be 80 N. For a corresponding wave in the full-scale wall, what full-scale force would you expect? Assume freshwater

is used in the model study. Assume $T = 10°C$ for both model and prototype water.

8.70 PLUS A model of a spillway is to be built at 1/80 scale. If the prototype has a discharge of 800 m³/s, what must be the water discharge in the model to ensure dynamic similarity? The total force on part of the model is found to be 51 N. To what prototype force does this correspond?

8.71 PLUS A newly designed dam is to be modeled in the laboratory. The prime objective of the general model study is to determine the adequacy of the spillway design and to observe the water velocities, elevations, and pressures at critical points of the structure. The reach of the river to be modeled is 1200 m long, the width of the dam (also the maximum width of the reservoir upstream) is to be 300 m, and the maximum flood discharge to be modeled is 5000 m³/s. The maximum laboratory discharge is limited to 0.90 m³/s, and the floor space available for the model construction is 50 m long and 20 m wide. Determine the largest feasible scale ratio (model/prototype) for such a study.

8.72 A ship model 1.2 m long is tested in a towing tank at a speed that will produce waves that are dynamically similar to those observed around the prototype. The test speed is 1.5 m/s. What should the prototype speed be, given that the prototype length is 30 m? Assume both the model and the prototype are to operate in freshwater.

8.73 PLUS The wave resistance of a model of a ship at 1/25 scale is 9 N at a model speed of 1.5 m/s. What are the corresponding velocity and wave resistance of the prototype?

8.74 A 1/20 scale model building that is rectangular in plan view and is three times as high as it is wide is tested in a wind tunnel. If the drag of the model in the wind tunnel is measured to be 200 N for a wind speed of 20 m/s, then the prototype building in a 40 m/s wind (same temperature) should have a drag of about (a) 40 kN, (b) 80 kN, (c) 230 kN, or (d) 320 kN.

8.75 PLUS A model of a high-rise office building at 1/550 scale is tested in a wind tunnel to estimate the pressures and forces on the full-scale structure. The wind-tunnel air speed is 20 m/s at 20°C and atmospheric pressure, and the full-scale structure is expected to withstand winds of 200 km/h (10°C). If the extreme values of the pressure coefficient are found to be 1.0, −2.7, and −0.8 on the windward wall, side wall, and leeward wall of the model, respectively, what corresponding pressures could be expected to act on the prototype? If the lateral wind force (wind force on building normal to wind direction) was measured as 20 N in the model, what lateral force might be expected in the prototype in the 200 km/h wind?

8.76 Experiments were carried out in a water tunnel and a wind tunnel to measure the drag force on an object. The water tunnel was operated with freshwater at 20°C, and the wind tunnel was operated at 20°C and atmospheric pressure. Three models were used with dimensions of 5 cm, 8 cm, and 15 cm. The drag force on each model was measured at different velocities. The following data were obtained.

Data for the water tunnel

Model Size, cm	Velocity, m/s	Force, N
5	1.0	0.064
5	4.0	0.69
5	8.0	2.20
8	1.0	0.135
8	4.0	1.52
8	8.0	4.52

Data for the wind tunnel

Model Size, cm	Velocity, m/s	Force, N
8	10	0.025
8	40	0.21
8	80	0.64
15	10	0.06
15	40	0.59
15	80	1.82

The drag force is a function of the density, viscosity, velocity, and model size,

$$F_D = f(\rho, \mu, V, D)$$

Using dimensional analysis, express this equation using π-groups and then write a computer program or use a spreadsheet to reduce the data. Plot the data using the dimensionless parameters.

8.77 Experiments are performed to measure the pressure drop in a pipe with water at 20°C and crude oil at the same temperature. Data are gathered with pipes of two diameters, 5 cm and 10 cm. The following data were obtained for pressure drop per unit length.

For water

Pipe Diameter, cm	Velocity, m/s	Pressure Drop, N/m^3
5	1	210
5	2	730
5	5	3750
10	1	86
10	2	320
10	5	1650

For crude oil

Pipe Diameter, cm	Velocity, m/s	Pressure Drop, N/m^3
5	1	310
5	2	1040
5	5	5300
10	1	130
10	2	450
10	5	2210

The pressure drop per unit length is assumed to be a function of the pipe diameter, liquid density and viscosity, and the velocity,

$$\frac{\Delta p}{L} = f(\rho, \mu, V, D)$$

Perform a dimensional analysis to obtain the π-groups and then write a computer program or use a spreadsheet to reduce the data. Plot the results using the dimensionless parameters.

9 PREDICTING SHEAR FORCE

FIGURE 9.1

When engineers design sailboats for racing, they consider the drag force on the hull. This drag force is caused by the pressure and shear stress distributions. This chapter is concerned with the shear stress and the shear force. (Foucras G./StockImage/Getty Images.)

Chapter Road Map

This chapter describes how to predict shear stress and shear force on a flat surface. The emphasis is on the theory because this theory provides the foundation for more advanced study in fluid mechanics.

Learning Objectives

STUDENTS WILL BE ABLE TO

- Describe Couette flow. Show how to derive and apply the working equations. (§9.1)
- Describe Hele-Shaw flow. Show how to derive and apply the working equations. (§9.1)
- Sketch the development of the boundary layer on a flat plate. Label and explain the main features. (§9.2)
- Define the local shear stress coefficient, c_f, and the average shear stress coefficient, C_f. (§9.3)
- Define or calculate Re_x and Re_L. (§9.3)
- For the laminar boundary, calculate the boundary layer thickness, the shear stress, and the shear force using suitable correlations. (§9.3)
- Describe the transition Reynolds number. (§9.4)
- Describe or apply the power law equation for the turbulent boundary layer. (§9.5)
- Sketch a turbulent boundary layer. Label and describe the three zones of flow. (§9.5)
- For the turbulent boundary layer, calculate the boundary layer thickness, the shear stress, and the shear force using suitable correlations. (§9.5)

9.1 Uniform Laminar Flow

In this section, Newton's second law of motion is used to derive a differential equation that governs a 1-D, steady, viscous flow. Then, the equation is solved for two classic problems: Couette flow (see Section 2.6) and Hele-Shaw flow (fully developed laminar flow between two parallel plates). The rationale for this section is to introduce fundamentals that are useful for analyzing viscous flows.

The equation derived in this section is a special case of the Navier-Stokes equation. The Navier-Stokes equation is probably the single most important equation in fluid mechanics.

The Equation of Motion for Steady and Uniform Laminar Flow

Consider a CV (Fig. 9.2), which is aligned with the flow direction s. The streamlines are inclined at an angle θ with respect to the horizontal plane. The control volume has dimensions $\Delta s \times \Delta y \times$ unity; that is, the control volume has a unit length into the page. By application of the momentum equation, the sum of the forces acting in the s-direction is equal to the net outflow of momentum from the control volume. The flow is uniform, so the outflow of momentum is equal to the inflow and the momentum equation reduces to

$$\sum F_s = 0 \tag{9.1}$$

FIGURE 9.2

Control volume for analysis of uniform flow with parallel streamlines.

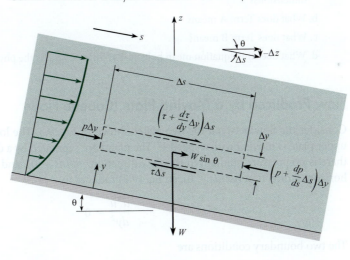

There are three forces acting on the matter in the control volume: the forces due to pressure, shear stress, and gravity. The net pressure force is

$$p\Delta y - \left(p + \frac{dp}{ds}\Delta s\right)\Delta y = -\frac{dp}{ds}\Delta s \Delta y$$

The net force due to shear stress is

$$\left(\tau + \frac{d\tau}{dy}\Delta y\right)\Delta s - \tau\Delta s = \frac{d\tau}{dy}\Delta y \Delta s$$

The component of gravitational force is $\rho g \Delta s \Delta y \sin \theta$. However, $\sin \theta$ can be related to the rate at which the elevation, z, decreases with increasing s and is given by $-dz/ds$. Thus the gravitational force becomes

$$\rho g \Delta s \Delta y \sin \theta = -\gamma \Delta y \Delta s \frac{dz}{ds}$$

Summing the forces and dividing by volume ($\Delta s \Delta y$) results in

$$\frac{d\tau}{dy} = \frac{d}{ds}(p + \gamma z) \tag{9.2}$$

where it is noted that the gradient of the shear stress is equal to the gradient in piezometric pressure in the flow direction. The shear stress is equal to $\mu \, du/dy$, so the basic equation becomes

$$\frac{d^2u}{dy^2} = \frac{1}{\mu}\frac{d}{ds}(p + \gamma z) \tag{9.3}$$

where μ is constant. Eq. (9.3) is the Navier-Stokes equation applied to a uniform and steady flow. The general form of this equation is introduced in Chapter 16. This equation is now applied to the two flow configurations.

✔ CHECKPOINT PROBLEM 9.1

The sketch identifies terms that appear in the Navier-Stokes equation.

a. What are the secondary dimensions of each term? Primary dimensions?

b. What does Term A mean?

c. What does Term B mean?

d. What does this equation mean holistically? That is, what is the physical interpretation?

$$\underbrace{\mu \frac{d^2u}{dy^2}}_{\text{Term A}} = \underbrace{\frac{d(p + \gamma z)}{ds}}_{\text{Term B}}$$

Flow Produced by a Moving Plate (Couette Flow)

Consider the flow between the two plates shown in Fig. 9.3. The lower plate is fixed, and the upper plate is moving with a speed U. The plates are separated by a distance L. In this problem there is no pressure gradient in the flow direction ($dp/ds = 0$), and the streamlines are in the horizontal direction ($dz/ds = 0$), so Eq. (9.3) reduces to

$$\frac{d^2u}{dy^2} = 0$$

The two boundary conditions are

$$u = 0 \quad \text{at} \quad y = 0$$
$$u = U \quad \text{at} \quad y = L$$

FIGURE 9.3

Flow generated by a moving plate (Couette flow).

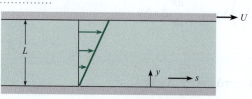

Integrating this equation twice gives

$$u = C_1 y + C_2$$

Applying the boundary conditions results in

$$u = \frac{y}{L} U \tag{9.4}$$

which shows that the velocity profile is linear between the two plates. The shear stress is constant and equal to

$$\tau = \mu \frac{du}{dy} = \mu \frac{U}{L} \tag{9.5}$$

This flow is known as a **Couette flow** after a French scientist, M. Couette, who did pioneering work on the flow between parallel plates and rotating cylinders. It has application in the design of lubrication systems.

Example 9.1 illustrates the application of Couette flow in calculating shear stress.

EXAMPLE 9.1

Shear Stress in Couette Flow

Problem Statement

SAE 30 lubricating oil at $T = 38°C$ flows between two parallel plates, one fixed and the other moving at 1.0 m/s. Plates are spaced 0.3 mm apart. What is the shear stress on the plates?

Define the Situation

SAE 30 lubricating oil is flowing between parallel plates

Properties: From Table A.4, $\mu = 1.0 \times 10^{-1}$ N · s/m^2

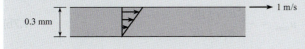

State the Goal

Find: Shear stress (in N/m^2) on top plate.

Generate Ideas and Make a Plan

Calculate shear stress using Eq. (9.5).

Take Action (Execute the Plan)

$$\tau = \mu \frac{du}{dy} = \mu \frac{U}{L}$$

$$= (1.0 \times 10^{-1} \text{ N} \cdot \text{s/m}^2)(1.0 \text{ m/s})/(3 \times 10^{-4} \text{ m})$$

$$\tau = \boxed{333 \text{ N/m}^2}$$

Review the Solution and the Process

Knowledge. Because the velocity gradient is constant, the shear stress is constant throughout the flow. Thus, the magnitude of the shear stress is the same for the bottom plate as the top plate.

Flow Between Stationary Parallel Plates (Hele-Shaw Flow)

Consider the two parallel plates separated by a distance B in Fig. 9.4. In this case, the flow velocity is zero at the surface of both plates, so the boundary conditions for Eq. (9.3) are

$$u = 0 \quad \text{at} \quad y = 0$$
$$u = 0 \quad \text{at} \quad y = B$$

Because the flow is uniform (i.e., there is no change in velocity in the streamwise direction), u is a function of y only. Therefore, d^2u/dy^2 in Eq. (9.3), as well as the gradient in piezometric

FIGURE 9.4

Uniform flow between
two stationary plates
(Hele-Shaw flow).

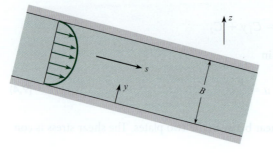

pressure, must also be equal to a constant in the streamwise direction. Integrating Eq. (9.3) twice gives

$$u = \frac{y^2}{2\mu} \frac{d}{ds}(p + \gamma z) + C_1 y + C_2$$

To satisfy the boundary condition at $y = 0$, set $C_2 = 0$. Applying the boundary condition at $y = B$ requires that C_1 be

$$C_1 = -\frac{B}{2\mu} \frac{d}{ds}(p + \gamma z)$$

so the final equation for the velocity is

$$u = -\frac{1}{2\mu} \frac{d}{ds}(p + \gamma z)(By - y^2) = -\frac{1}{2\mu}(By - y^2)\frac{d(p + \gamma z)}{ds} \quad (9.6)$$

which is a parabolic profile with the maximum velocity occurring on the centerline between the plates, as shown in Fig. 9.3. The maximum velocity is

$$u_{max} = -\left(\frac{B^2}{8\mu}\right)\frac{d}{ds}(p + \gamma z) \quad (9.7a)$$

or in terms of piezometric head

$$u_{max} = -\left(\frac{B^2 \gamma}{8\mu}\right)\frac{dh}{ds} \quad (9.7b)$$

The fluid always flows in the direction of decreasing piezometric pressure or piezometric head, so dh/ds is negative, giving a positive value for u_{max}.

The discharge per unit width, q, is obtained by integrating the velocity over the distance between the plates:

$$q = \int_0^B u\, dy = -\left(\frac{B^3}{12\mu}\right)\frac{d}{ds}(p + \gamma z) = -\left(\frac{B^3 \gamma}{12\mu}\right)\frac{dh}{ds} \quad (9.8)$$

The average velocity is

$$V = \frac{q}{B} = -\left(\frac{B^2}{12\mu}\right)\frac{d}{ds}(p + \gamma z) = \frac{2}{3}u_{max} \quad (9.9)$$

Note that flow is the result of a change of the piezometric head, not just a change of p or z alone. Experiments reveal that if the Reynolds number (VB/ν) is less than 1000, the flow is laminar. For a Reynolds number greater than 1000, the flow may be turbulent, and the equations developed in this section are invalid.

The flow between parallel plates is often called **Hele-Shaw flow**. It has application in flow visualization studies and in microchannel flows.

A significant difference between Couette flow and Hele-Shaw flow is that the motion of a plate is responsible for Couette flow, whereas a gradient in piezometric pressure provides the force to move a Hele-Shaw flow.

Example 9.2 illustrates how to calculate the pressure gradient required for flow between two parallel plates.

EXAMPLE 9.2

Pressure Gradient for Flow Between Parallel Plates

Problem Statement

Oil having a specific gravity of 0.8 and a viscosity of 2×10^{-2} N·s/m² flows downward between two vertical smooth plates spaced 10 mm apart. If the discharge per meter of width is 0.01 m²/s, what is the pressure gradient dp/ds for this flow?

Define the Situation

Oil flows downward between two vertical smooth plates spaced 10 mm apart. The discharge per meter of width is 0.01 (m²/s).

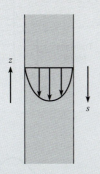

State the Goal

Find: Pressure gradient dp/ds (in Pa/m) for this flow.

Properties: S = 0.8, $\mu = 2 \times 10^{-2}$ N·s/m².

Generate Ideas and Make a Plan

1. Check to see if the flow is laminar using $VB/\nu < 1000$. If it is laminar, continue.

2. Calculate piezometric head gradient using Eq. (9.8).

3. Subtract elevation gradient to obtain the pressure gradient.

Take Action (Execute the Plan)

1. Check for laminar flow

$$\text{Re} = \frac{VB}{\nu} = \frac{VB\rho}{\mu} = \frac{q\rho}{\mu}$$

$$= \frac{(0.01 \text{ m}^2/\text{s}) \times 800 \text{ kg/m}^3}{0.02 \text{ N} \cdot \text{s/m}^2} = 400$$

$VB/\nu < 1000$. Flow is laminar, equations apply.

2. Kinematic viscosity:

$$\nu = \mu/\rho = \frac{2 \times 10^{-2} \text{ N} \cdot \text{s/m}^2}{0.8 \times 1000 \text{ kg/m}^3} = 2.5 \times 10^{-5} \text{ m}^2/\text{s}$$

Piezometric head gradient is

$$\frac{dh}{ds} = -\frac{12\mu}{B^3\gamma}q = -\frac{12\nu}{B^3 g}q$$

$$\frac{dh}{ds} = -\frac{12 \times 2.5 \times 10^{-5} \text{ m}^2/\text{s}}{(0.01 \text{ m})^3 \times 9.81 \text{ m}^4/\text{s}^2} \times 0.01 \text{ m}^2/\text{s} = -0.306$$

3. Plates are oriented vertically, s is positive downward, so $dz/ds = -1$. Thus

$$\frac{dh}{ds} = \frac{d}{ds}\left(\frac{p}{\gamma}\right) + \frac{dz}{ds}$$

$$\frac{d}{ds}\left(\frac{p}{\gamma}\right) = \frac{dh}{ds} - \frac{dz}{ds} = -0.306 + 1 = 0.694$$

or

$$\frac{dp}{ds} = (0.8 \times 9810 \text{ N/m}^3) \times 0.694 = \boxed{5450 \text{ N/m}^2 \text{ per meter}}$$

Review the Solution and the Process

Note that the pressure increases in the downward direction, which means that the pressure, in part, supports the weight of the fluid.

9.2 Qualitative Description of the Boundary Layer

The purpose of this section is to provide a qualitative description of the **boundary layer**, which is the region adjacent to a surface over which the velocity changes from the free-stream value (with respect to the object) to zero at the surface. This region, which is generally very thin, occurs because of the viscosity of the fluid. The velocity gradient at the surface is responsible for the viscous shear stress and shear force.

The boundary-layer development for flow past a thin plate oriented parallel to the flow direction shown in Fig. 9.5a. The thickness of the boundary layer, δ, is defined as the distance from the surface where the velocity is 99% of the free-stream velocity. The actual thickness of a boundary layer may be 2% to 3% of the plate length, so the boundary-layer thickness shown in Fig. 9.5a is exaggerated at least by a factor of five to show details of the flow field. Fluid passes over the top and underneath the plate, so two boundary layers are depicted (one above and one below the plate). For convenience, the surface is assumed to be stationary, and the free-stream fluid is moving at a velocity U_0.

FIGURE 9.5

Development of boundary layer and shear stress along a thin, flat plate.
(a) Flow pattern above and below the plate.
(b) Shear-stress distribution on either side of plate.

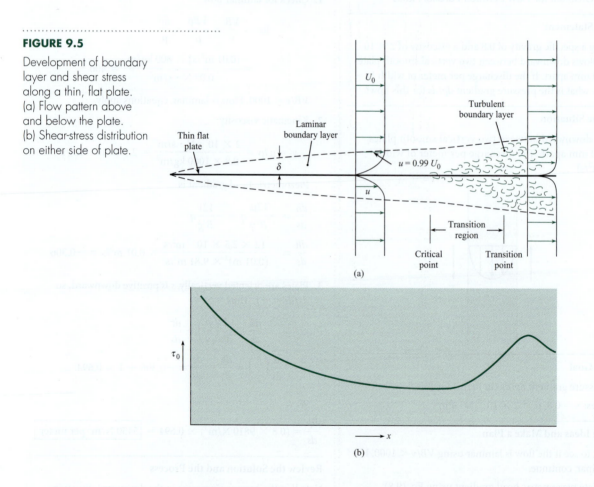

The development and growth of the boundary layer occurs because of the "no-slip" condition at the surface; that is, the fluid velocity at the surface must be zero. As the fluid particles next to the plate pass close to the leading edge of the plate, a retarding force (from the shear stress) begins to act on the particles to slow them down. As these particles progress farther downstream, they continue to be subjected to shear stress from the plate, so they continue to decelerate. In addition, these particles (because of their lower velocity) retard other particles adjacent to them but farther out from the plate. Thus the boundary layer becomes thicker, or "grows," in the downstream direction. The broken line in Fig. 9.5a identifies the outer limit of the boundary layer. As the boundary layer becomes thicker, the velocity gradient at the wall becomes smaller and the local shear stress is reduced.

The initial section of the boundary layer is the laminar boundary layer. In this region the flow is smooth and steady. Thickening of the laminar boundary layer continues smoothly in the downstream direction until a point is reached where the boundary layer becomes unstable. Beyond this point, the critical point, small disturbances in the flow will grow and spread, leading to turbulence. The boundary becomes fully turbulent at the transition point. The region between the critical point and the transition point is called the transition region.

In most problems of practical interest, the extent of the laminar boundary layer is small and contributes little to the total drag force on a body. Still it is important for flow of very viscous liquids and for flow problems with small length scales.

The turbulent boundary layer is characterized by intense cross-stream mixing as turbulent eddies transport high-velocity fluid from the boundary layer edge to the region close to the wall. This cross-stream mixing gives rise to a high effective viscosity, which can be three orders of magnitude higher than the actual viscosity of the fluid itself. The effective viscosity, due to turbulent mixing is not a property of the fluid but rather a property of the flow, namely, the mixing process. Because of this intense mixing, the velocity profile is much "fuller" than the laminar-flow velocity profile as shown in Fig. 9.5a. This situation leads to an increased velocity gradient at the surface and a larger shear stress.

The shear-stress distribution along the plate is shown in Fig. 9.4b. It is easy to visualize that the shear stress must be relatively large near the leading edge of the plate where the velocity gradient is steep, and that it becomes progressively smaller as the boundary layer thickens in the downstream direction. At the point where the boundary layer becomes turbulent, the shear stress at the boundary increases because the velocity profile changes producing a steeper gradient at the surface.

These qualitative aspects of the boundary layer serve as a foundation for the quantitative relations presented in the next section.

9.3 Laminar Boundary Layer

This section summarizes the equations for the velocity profile and shear stress in a laminar boundary layer and describes how to calculate shear stress and shear forces on a surface. This information can be used to estimate drag forces on surfaces in low Reynolds-number flows.

Boundary-Layer Equations

In 1904 Prandtl (1) first stated the essence of the boundary-layer hypothesis, which is that viscous effects are concentrated in a thin layer of fluid (the boundary layer) next to solid boundaries. Along with his discussion of the qualitative aspects of the boundary layer, he also simplified the general equations of motion of a fluid (Navier-Stokes equations) for application to the boundary layer.

In 1908, Blasius, one of Prandtl's students, obtained a solution for the flow in a laminar boundary layer (2) on a flat plate with a constant free-stream velocity. One of Blasius's key assumptions was that the shape of the nondimensional velocity distribution did not vary from section to section along the plate. That is, he assumed that a plot of the relative velocity, u/U_0, versus the relative distance from the boundary, y/δ, would be the same at each section. With this assumption and with Prandtl's equations of motion for boundary layers, Blasius obtained a numerical solution for the relative velocity distribution, shown in Fig. 9.6.* In this plot, x is the distance from the leading edge of the plate, and Re_x is the Reynolds number based on the free-stream velocity and the length along the plate ($\text{Re}_x = U_0 x/\nu$). In Fig. 9.6 the outer limit of the boundary layer ($u/U_0 = 0.99$) occurs at approximately $y\text{Re}_x^{1/2}/x = 5$. Because $y = \delta$ at this point, the following relationship is derived for the *boundary-layer thickness* in laminar flow on a flat plate:

$$\frac{\delta}{x}\text{Re}_x^{1/2} = 5 \quad \text{or} \quad \delta = \frac{5x}{\text{Re}_x^{1/2}} \tag{9.10}$$

FIGURE 9.6

Velocity distribution in laminar boundary layer. [After Blasius (2).]

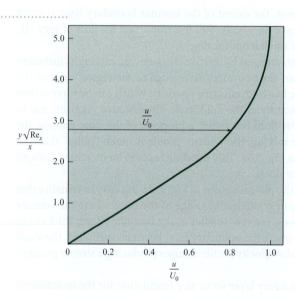

The Blasius solution also showed that

$$\left.\frac{d(u/U_0)}{d[(y/x)\text{Re}_x^{1/2}]}\right|_{y=0} = 0.332$$

which can be used to find the shear stress at the surface. The velocity gradient at the boundary becomes

$$\left.\frac{du}{dy}\right|_{y=0} = 0.332 \frac{U_0}{x}\text{Re}_x^{1/2}$$

$$\left.\frac{du}{dy}\right|_{y=0} = 0.332 \frac{U_0^{3/2}}{x^{1/2}\nu^{1/2}} \tag{9.11}$$

*Experimental evidence indicates that the Blasius solution is valid except very near the leading edge of the plate. In the vicinity of the leading edge, an error results because of certain simplifying assumptions. However, the discrepancy is not significant for most engineering problems.

Equation (9.11) shows that the velocity gradient (and shear stress) decreases with increasing distance x along the plate.

Shear Stress

The shear stress at the boundary is obtained from

$$\tau_0 = \mu \left. \frac{du}{dy} \right|_{y=0} = 0.332\mu \frac{U_0}{x} \mathrm{Re}_x^{1/2} \tag{9.12}$$

Equation (9.12) is used to obtain the local shear stress at any given section (any given value of x) for the laminar boundary layer.

Example 9.3 illustrates the application of the laminar boundary layer equations for calculating boundary layer thickness and shear stress.

Shear Force

Consider one side of a flat plate with width B and length L. Because the shear stress at the boundary, τ_0, varies along the plate, it is necessary to integrate this stress over the entire surface to obtain the total shear force, F_s.

$$F_s = \int_0^L \tau_0 B \, dx \tag{9.13}$$

EXAMPLE 9.3

Laminar Boundary-Layer Thickness and Shear Stress

Problem Statement

Crude oil at 20°C ($v = 10^{-5}\,\mathrm{m^2/s}$, $S = 0.86$) with a free-stream velocity of 0.3 m/s flows past a thin, flat plate that is 1.2 m wide and 1.8 m long in a direction parallel to the flow. The flow is laminar. Determine and plot the boundary-layer thickness and the shear stress distribution along the plate.

Define the Situation

Crude oil flows past a thin, flat plate. Free-stream velocity is 0.3 m/s.

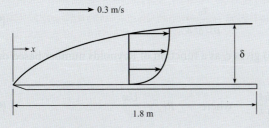

Oil. $v = 10^{-5}\,\mathrm{m^2/s}$, $S = 0.86$

Assumptions:

1. Plate is smooth, flat with sharp leading edge.
2. Boundary layer is laminar.

State the Goal

Surface shear stress, τ_0, as function of x.
Boundary-layer thickness, δ, as function of x.

Generate Ideas and Make a Plan

1. Calculate boundary-layer thickness with Eq. (9.10).
2. Calculate shear-stress distribution with Eq. (9.12).
3. Summarize results using a table and a plot.

Take Action (Execute the Plan)

1. Reynolds-number variation with distance

$$\mathrm{Re}_x = \frac{U_0 x}{v} = \frac{0.3 \times x}{10^{-5}} = 3 \times 10^4 x$$

Boundary-layer thickness

$$\delta = \frac{5x}{\mathrm{Re}_x^{1/2}} = \frac{5x}{(3 \times 10^4)^{1/2} x^{1/2}} = 8.66 \times 10^{-2} x^{1/2}\,\mathrm{m}$$

2. Shear-stress distribution

$$\tau_0 = 0.332\mu \frac{U_0}{x} \mathrm{Re}_x^{1/2}$$

$$\mu = \rho v = 998\,\mathrm{kg/m^3} \times 0.86 \times 10^{-5}\,\mathrm{m^2/s}$$

$$= 8.58 \times 10^{-3}\,\mathrm{N \cdot s/m^2}$$

$$\tau_0 = 0.332(8.58 \times 10^{-3})(0.3/x)(3 \times 10^2 x)^{1/2}$$

$$= \frac{0.493\,\mathrm{N/m^2}}{x^{1/2}}$$

3. Summary (make a plot and build a table).

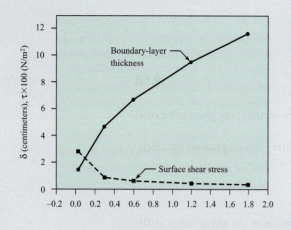

TABLE 9.1 Results: δ and τ_0 for Different Values of x

	$x =$ 0.03 m	$x =$ 0.3 m	$x =$ 0.6 m	$x =$ 1.2 m	$x =$ 1.8 m
$x^{1/2}$	0.173	0.548	0.775	1.095	1.342
τ_0, N/m^2	2.85	0.899	0.639	0.450	0.367
δ, m	0.015	0.047	0.067	0.095	0.116
δ, cm	1.5	4.7	6.7	9.5	11.6

Review the Solution and the Process

1. Notice that the boundary-layer thickness increases with distance. At the end of the plate $\delta/x = 0.02$, or the boundary-layer thickness is 2% of the distance from leading edge.

2. Notice also that shear stress decreases with distance from leading edge of the plate.

Substituting in Eq. (9.12) for τ_0 and integrating gives

$$F_s = \int_0^L 0.332 B \mu \frac{U_0 U_0^{1/2} x^{1/2}}{x \nu^{1/2}} \, dx$$

$$= 0.664 B \mu U_0 \frac{U_0^{1/2} L^{1/2}}{\nu^{1/2}} \tag{9.14}$$

$$= 0.664 B \mu U_0 \mathrm{Re}_L^{1/2}$$

where Re_L is the Reynolds number based on the approach velocity and the length of the plate.

Shear-Stress Coefficients

It is convenient to express the shear stress at the boundary, τ_0, and the total shearing force F_s in terms of π-groups involving the kinetic pressure of the free stream, $\rho U_0^2/2$. The **local shear-stress coefficient**, c_f, is defined as

$$c_f = \frac{\tau_0}{\rho U_0^2/2} \tag{9.15}$$

Substituting Eq. (9.12) into Eq. (9.15) gives c_f as a function of Reynolds number based on the distance from the leading edge.

$$c_f = \frac{0.664}{\mathrm{Re}_x^{1/2}} \qquad \text{where} \qquad Re_x = \frac{Ux}{\nu} \tag{9.16}$$

The total shearing force, as given by Eq. (9.13), can also be expressed as a π-group

$$C_f = \frac{F_s}{(\rho U_0^2/2)A} \tag{9.17}$$

where A is the plate area. This π-group is called the average shear-stress coefficient. Substituting Eq. (9.14) into Eq. (9.17) gives C_f:

$$C_f = \frac{1.33}{\mathrm{Re}_L^{1/2}} \qquad \text{where} \qquad \mathrm{Re}_L = \frac{UL}{\nu} \qquad \textbf{(9.18)}$$

Example 9.4 shows how to calculate the total shear force for a laminar boundary layer on a flat plate.

EXAMPLE 9.4

Resistance Calculation for Laminar Boundary Layer on a Flat Plate

Problem Statement

Crude oil at 21°C ($\nu = 10^{-5}$ m²/s, S = 0.86.) with a free-stream velocity of 0.3 m/s flows past a thin, flat plate that is 1 m wide and 2 m long in a direction parallel to the flow. The flow is laminar. Determine the resistance on one side of the plate.

Define the Situation

Crude oil flows past a thin, flat plate. Free-stream velocity is 0.3 m/s.

Properties: For oil, $\nu = 10^{-5}$ m²/s, S = 0.86.

Assumptions: Flow is laminar.

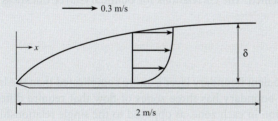

State the Goal

Find: Shear force (in N) on one side of plate.

Generate Ideas and Make a Plan

1. Calculate the Reynolds number based on plate length.
2. Evaluate C_f using Eq. (9.18).
3. Calculate the shear force using Eq. (9.17).

Take Action (Execute the Plan)

1. Reynolds number.

$$\mathrm{Re}_L = \frac{U_0 L}{\nu} = \frac{0.3 \text{ m/s} \times 2 \text{ m}}{10^{-5} \text{ m}^2/\text{s}} = 6 \times 10^4$$

2. Value for C_f:

$$C_f = \frac{1.33}{\mathrm{Re}_L^{1/2}} = \frac{1.33}{(6 \times 10^4)^{1/2}} = 0.0054$$

3. Shear force.

$$F_s = \frac{C_f B L \rho U_0^2}{2}$$

$$= 0.0054 \times 1 \text{ m} \times 2 \text{ m} \times 0.86$$
$$\times 998 \text{ kg}/\text{m}^3 \times \frac{(0.3 \text{ m/s})^2}{2} = \boxed{0.42 \text{ N}}$$

9.4 Boundary Layer Transition

Transition is the zone where the laminar boundary layer changes into a turbulent boundary layer as shown in Fig. 9.5a. As the laminar boundary layer continues to grow, the viscous stresses are less capable of damping disturbances in the flow. A point is then reached where disturbances occurring in the flow are amplified, leading to turbulence. The critical point occurs at a Reynolds number of about 10^5 ($\mathrm{Re}_{cr} \cong 10^5$) based on the distance from the leading edge. Vortices created near the wall grow and mutually interact, ultimately leading to a fully turbulent boundary layer at the transition point, which nominally occurs at a Reynolds number of 3×10^6 ($\mathrm{Re}_{tr} \cong 3 \times 10^6$). For purposes of simplicity in this text, it will be assumed that the boundary layer changes from laminar to turbulent flow at a Reynolds number 500,000. The details of the transition region can be found in White (3).

Transition to a turbulent boundary layer can be influenced by several other flow conditions, such as free-stream turbulence, pressure gradient, wall roughness, wall heating, and wall cooling. With appropriate roughness elements at the leading edge, the boundary layer can become turbulent at the very beginning of the plate. In this case it is said that the boundary layer is "**tripped**" at the leading edge.

✔ **CHECKPOINT PROBLEM 9.2**

Suppose the roof of an automobile is idealized as a flat plate. Given the data in the figure, what is the speed V of the car in mph? Assume $T = 20°C$ and $p = 1$ atm.

a. 12.6

b. 14.1

c. 16.9

d. 28.1

e. 34.7

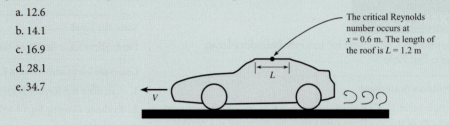

The critical Reynolds number occurs at $x = 0.6$ m. The length of the roof is $L = 1.2$ m

---

9.5 Turbulent Boundary Layer

Understanding the mechanics of the turbulent boundary layer is important because in the majority of practical problems, the turbulent boundary layer is primarily responsible for shear force. In this section the velocity distribution in the turbulent boundary layer on a flat plate oriented parallel to the flow is presented. The correlations for boundary-layer thickness and shear stress are also included.

Velocity Distribution

The velocity distribution in the turbulent boundary layer is more complicated than the laminar boundary layer. The turbulent boundary has three zones of flow that require different equations for the velocity distribution in each zone, as opposed to the single relationship of the laminar boundary layer. Figure 9.7 shows a portion of a turbulent boundary layer in which the three different zones of flow are identified. The zone adjacent to the wall is the viscous sublayer; the zone immediately above the viscous sublayer is the logarithmic region; and

FIGURE 9.7

Sketch of zones in turbulent boundary layer.

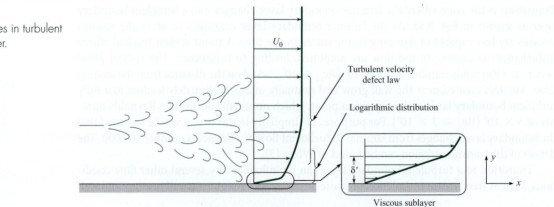

finally, beyond that region is the velocity defect region. Each of these velocity zones will be discussed separately.

Viscous Sublayer The zone immediately adjacent to the wall is a layer of fluid that is essentially laminar because the presence of the wall dampens the cross-stream mixing and turbulent fluctuations. This very thin layer is called the **viscous sublayer**. This thin layer behaves as a Couette flow introduced in Section 9.1. In the viscous sublayer, τ is virtually constant and equal to the shear stress at the wall, τ_0. Thus $du/dy = \tau_0/\mu$, which on integration yields

$$u = \frac{\tau_0 \, y}{\mu} \tag{9.19}$$

Dividing the numerator and denominator by ρ gives

$$u = \frac{\tau_0/\rho}{\mu/\rho} y \tag{9.20}$$

$$\frac{u}{\sqrt{\tau_0/\rho}} = \frac{\sqrt{\tau_0/\rho}}{\nu} y$$

The combination of variables $\sqrt{\tau_0/\rho}$ has the dimensions of velocity and recurs again and again in derivations involving boundary-layer theory. It has been given the special name **shear velocity**. The shear velocity (which is also sometimes called **friction velocity**) is symbolized as u_*. Thus, by definition,

$$u_* = \sqrt{\frac{\tau_0}{\rho}} \tag{9.21}$$

Now, substituting u_* for $\sqrt{\tau_0/\rho}$ in Eq. (9.20), yields the nondimensional velocity distribution in the viscous sublayer:

$$\frac{u}{u_*} = \frac{y}{\nu/u_*} \tag{9.22}$$

Experimental results show that the limit of viscous sublayer occurs when yu_*/ν is approximately 5. Consequently, the thickness of the viscous sublayer, identified by δ', is given as

$$\delta' = \frac{5\nu}{u_*} \tag{9.23}$$

The thickness of the viscous sublayer is very small (typically less than one-tenth the thickness of a dime). The thickness of the viscous sublayer increases as the wall shear stress decreases in the downstream direction.

The Logarithmic Velocity Distribution The flow zone outside the viscous sublayer is turbulent; therefore, a completely different type of flow is involved. The mixing action of turbulence causes small fluid masses to be swept back and forth in a direction transverse to the mean flow direction. A small mass of fluid swept from a low-velocity zone next to the viscous sublayer into a higher-velocity zone farther out in the stream has a retarding effect on the higher-velocity stream. Similarly, a small mass of fluid that originates farther out in the boundary layer in a high-velocity flow zone and is swept into a region of low velocity has the effect of accelerating the lower-velocity fluid. Although the process just described is primarily a momentum exchange phenomenon, it has the same effect as applying a shear stress to the fluid; thus in turbulent flow these "stresses" are termed **apparent shear stresses**, or **Reynolds stresses** after the British scientist–engineer who first did extensive research in turbulent flow in the late l800s.

The mixing action of turbulence causes the velocities at a given point in a flow to fluctuate with time. If one places a velocity-sensing device, such as a hot-wire anemometer, in a turbulent flow, one can measure a fluctuating velocity, as illustrated in Fig. 9.8. It is convenient to think of the velocity as composed of two parts: a mean value, $\bar{u}$, plus a fluctuating part, u'. The fluctuating part of the velocity is responsible for the mixing action and the momentum exchange, which manifests itself as an apparent shear stress as noted previously. In fact, the apparent shear stress is related to the fluctuating part of the velocity by

$$\tau_{\text{app}} = -\rho \overline{u'v'} \tag{9.24}$$

FIGURE 9.8

Velocity fluctuations in turbulent flow.

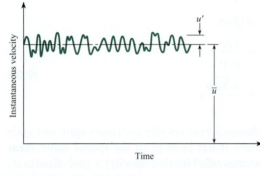

where u' and v' refer to the x and y components of the velocity fluctuations, respectively, and the bar over these terms denotes the product of $u'v'$ averaged over a period of time.* The expression for apparent shear stress is not very useful in this form, so Prandtl developed a theory to relate the apparent shear stress to the temporal mean velocity distribution.

The theory developed by Prandtl is analogous to the idea of molecular transport creating shear stress presented in Chapter 2. In the turbulent boundary layer, the principal flow is parallel to the boundary. However, because of turbulent eddies, there are fluctuating components transverse to the principal flow direction. These fluctuating velocity components are associated with small masses of fluid, as shown in Fig. 9.8, that move across the boundary layer. As the mass moves from the lower-velocity region to the higher-velocity region, it tends to retain its original velocity. The difference in velocity between the surrounding fluid and the transported mass is identified as the fluctuating velocity component u'. For the mass shown in Fig. 9.8, u' would be negative and approximated by[†]

$$u' \approx \ell \frac{du}{dy}$$

where du/dy is the mean velocity gradient and ℓ is the distance the small fluid mass travels in the transverse direction. Prandtl identified this distance as the "mixing length." Prandtl assumed that the magnitude of the transverse fluctuating velocity component is proportional to the magnitude of the fluctuating component in the principal flow direction: $|v'| \cong |u'|$, which seems to be a reasonable assumption because both components arise from the same set

---

*Equation (9.24) can be derived by considering the momentum exchange that results when the transverse component of turbulent flow passes through an area parallel to the x-z plane. Or, by including the fluctuating velocity components in the Navier-Stokes equations, one can obtain the apparent shear stress terms, one of which is Eq. (9.24). Details of these derivations appear in Chapter 18 of Schlichting (4).

[†]For convenience, the bar used to denote time-averaged velocity is deleted.

of eddies. Also, it should be noted that a positive v' will be associated with a negative u', so the product $\overline{u'v'}$ will be negative. Thus the apparent shear stress can be expressed as

$$\tau_{\text{app}} = -\rho\overline{u'v'} = \rho\ell^2\left(\frac{du}{dy}\right)^2 \tag{9.25}$$

A more general form of Eq. (9.25) is

$$\tau_{\text{app}} = \rho\ell^2\left|\frac{du}{dy}\right|\frac{du}{dy}$$

which ensures that the sign for the apparent shear stress is correct.

The theory leading to Eq. (9.25) is called Prandtl's mixing-length theory and is used extensively in analyses involving turbulent flow.* Prandtl also made the important and clever assumption that the mixing length is proportional to the distance from the wall ($\ell = \kappa y$) for the region close to the wall. If one considers the velocity distribution in a boundary layer where du/dy is positive, as is shown in Fig. 9.9, and substitutes κy for ℓ, then Eq. (9.25) reduces to

$$\tau_{\text{app}} = \rho\kappa^2 y^2\left(\frac{du}{dy}\right)^2$$

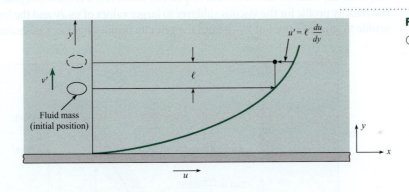

FIGURE 9.9

Concept of mixing length.

For the zone of flow near the boundary, it is assumed that the shear stress is uniform and approximately equal to the shear stress at the wall. Thus the foregoing equation becomes

$$\tau_0 = \rho\kappa^2 y^2\left(\frac{du}{dy}\right)^2 \tag{9.26}$$

Taking the square root of each side of Eq. (9.26) and rearranging yields

$$du = \frac{\sqrt{\tau_0/\rho}}{\kappa}\frac{dy}{y}$$

Integrating the above equation and substituting u_* for $\sqrt{\tau_0/\rho}$ gives

$$\frac{u}{u_*} = \frac{1}{\kappa}\ln y + C \tag{9.27}$$

*Prandtl published an account of his mixing-length concept in 1925. G. I. Taylor (5) published a similar concept in 1915, but the idea has been traditionally attributed to Prandtl.

Experiments on smooth boundaries indicate that the constant of integration C can be given in terms of u_*, ν, and a pure number as

$$C = 5.56 - \frac{1}{\kappa}\ln\frac{\nu}{u_*}$$

When this expression for C is substituted into Eq. (9.27), the result is

$$\frac{u}{u_*} = \frac{1}{\kappa}\ln\frac{yu_*}{\nu} + 5.56 \tag{9.28}$$

In Eq. (9.28), κ has sometimes been called the universal turbulence constant, or Karman's constant. Experiments show that this constant is approximately 0.41 (3) for the turbulent zone next to the viscous sublayer. Introducing this value for κ into Eq. (9.28) gives the *logarithmic velocity distribution*

$$\frac{u}{u_*} = 2.44\ln\frac{yu_*}{\nu} + 5.56 \tag{9.29}$$

Obviously the region where this model is valid is limited because the mixing length cannot continuously increase to the boundary layer edge. This distribution is valid for values of yu_*/ν ranging from approximately 30 to 500.

The region between the viscous sublayer and the logarithmic velocity distribution is the buffer zone. There is no equation for the velocity distribution in this zone, although various empirical expressions have been developed (6). However, it is common practice to extrapolate the velocity profile for the viscous sublayer to larger values of yu_*/ν and the logarithmic velocity profile to smaller values of yu_*/ν until the velocity profiles intersect as shown in Fig. 9.10.

FIGURE 9.10

Velocity distribution in a turbulent boundary layer.

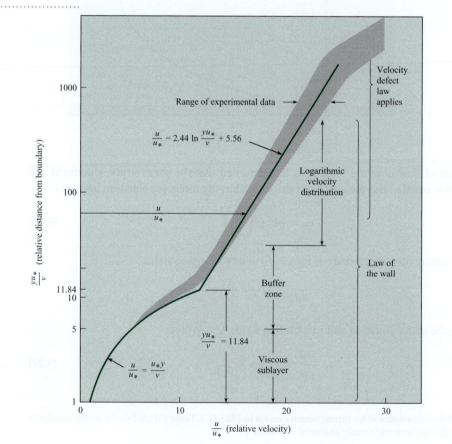

The intersection occurs at $yu_*/\nu = 11.84$ and is regarded as the demarcation between the viscous sublayer and the logarithmic profile. The "nominal" thickness of the viscous sublayer is

$$\delta'_N = 11.84 \frac{\nu}{u_*} \tag{9.30}$$

The combination of the viscous and logarithmic velocity profile for the range of yu_*/ν from 0 to approximately 500 is called the **law of the wall**.

Making a semilogarithmic plot of the velocity distribution in a turbulent boundary layer, as shown in Fig. 9.10, makes it straightforward to identify the velocity distribution in the viscous sublayer and in the region where the logarithmic equation applies. However, the logarithmic nature of this plot accentuates the nondimensional distance yu_*/ν near the wall. A better perspective of the relative extent of the regions is obtained by plotting the graph on a linear scale, as shown in Fig. 9.11. From this plot one notes that the laminar sublayer and buffer zone are a very small part of the thickness of the turbulent boundary layer.

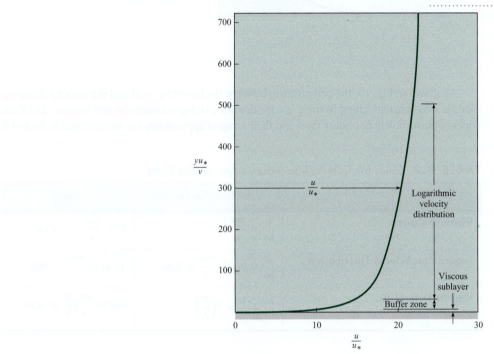

FIGURE 9.11

Velocity distribution in a turbulent boundary layer—linear scales.

Velocity Defect Region For $y/\delta > 0.15$ and $yu_*/\nu > 500$ the velocity profile corresponding to the law of the wall becomes increasingly inadequate to match experimental data, so a third zone, called the velocity defect region, is identified. The velocity in this region is represented by the **velocity defect law**, which for a flat plate with zero pressure gradient is simply expressed as

$$\frac{U_0 - u}{u_*} = f\left(\frac{y}{\delta}\right) \tag{9.31}$$

and the correlation with experimental data is plotted in Fig. 9.12. At the edge of the boundary layer $y = \delta$ and $(U_0 - u)/u_* = 0$, so $u = U_0$, or the free-stream velocity. This law applies to rough as well as smooth surfaces. However, the functional relationship has to be modified for flows with free-stream pressure gradients.

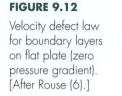

FIGURE 9.12

Velocity defect law for boundary layers on flat plate (zero pressure gradient). [After Rouse (6).]

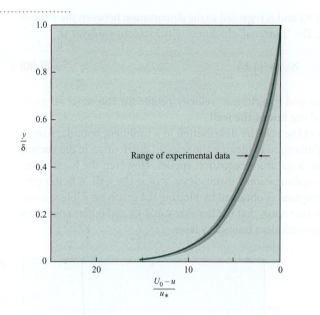

As shown in Fig. 9.9, the demarcation between the law of the wall and the velocity defect regions is somewhat arbitrary, so there is considerable overlap between the two regions. The three zones of the turbulent boundary layer and their range of applicability are summarized in Table 9.2.

TABLE 9.2 Zones for Turbulent Boundary Layer on Flat Plate

Zone	Velocity Distribution	Range
Viscous Sublayer	$\dfrac{u}{u_*} = \dfrac{yu_*}{\nu}$	$0 < \dfrac{yu_*}{\nu} < 11.84$
Logarithmic Velocity Distribution	$\dfrac{u}{u_*} = 2.44\ln\dfrac{yu_*}{\nu} + 5.56$	$11.84 \le \dfrac{yu_*}{\nu} < 500$
Velocity Defect Law	$\dfrac{U_0 - u}{u_*} = f\left(\dfrac{y}{\delta}\right)$	$500 \le \dfrac{yu_*}{\nu}, \dfrac{y}{\delta} > 0.15$

Power-Law Formula for Velocity Distribution Analyses have shown that for a wide range of Reynolds numbers ($10^5 < \text{Re} < 10^7$), the velocity profile in the turbulent boundary layer on a flat plate is approximated reasonably by the *power-law* equation

$$\frac{u}{U_0} = \left(\frac{y}{\delta}\right)^{1/7} \tag{9.32}$$

Comparisons with experimental results show that this formula conforms to those results very closely over about 90% of the boundary layer ($0.1 < y/\delta < 1$). Obviously it is not valid at the surface because $(du)/(dy)\big|_{y=0} \to \infty$, which implies infinite surface shear stress. For the inner 10% of the boundary layer, one must resort to equations for the law of the wall (see Fig. 9.10) to obtain a more precise prediction of velocity. Because Eq. (9.32) is valid over the major portion of the boundary layer, it is used to advantage in deriving the overall thickness of the boundary layer as well as other relations for the turbulent boundary layer.

Example 9.5 illustrates the application of various equations to calculate the velocity in the turbulent boundary layer.

EXAMPLE 9.5

Turbulent Boundary-Layer Properties

Problem Statement

Water (15°C) flows with a velocity of 6 m/s past a flat plate. The plate is oriented parallel to the flow. At a particular section downstream of the leading edge of the plate, the boundary layer is turbulent, the shear stress on the plate is 43 N/m², and the boundary-layer thickness is 0.02 mm. Find the velocity of the water at a distance of 0.002 mm from the plate as determined by

a. The logarithmic velocity distribution

b. The velocity defect law

c. The power-law formula

Also, what is the nominal thickness of the viscous sublayer?

Define the Situation

Water flows past a flat plate oriented parallel to the flow. At a point downstream of the leading edge of the plate, shear stress on the plate is 43 N/m², and boundary layer thickness is 0.02 mm.

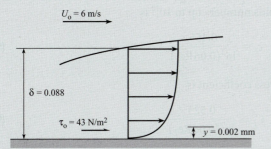

Properties:

From Table A.5, $\rho = 999$ kg/m³, $\nu = 1.14 \times 10^{-6}$ m²/s

State the Goal

1. V(m/s) ◀ Velocity at $y = 0.002$ mm using:

a. Logarithmic velocity distribution

b. Velocity defect law

c. Power-law formula

2. Calculate the nominal thickness of the viscous sublayer

Generate Ideas and Make a Plan

1. Calculate shear velocity, u_*, from Eq. (9.21).

2. Calculate u using Eq. (9.29) for logarithmic profile.

3. Calculate y/δ and find $(U_0 - u)/u_*$ from Fig. 9.12.

4. Calculate u from $(U_0 - u)/u_*$ for velocity defect law.

5. Calculate u from Eq. (9.32) for power law.

6. Calculate δ'_N from Eq. (9.30).

Take Action (Execute the Plan)

1. Shear velocity

$$u_* = (\tau_0/\rho)^{1/2}$$
$$= [(43 \text{ N/m}^2)/(999 \text{ kg/m}^3)]^{1/2} = 0.207 \text{ m/s}$$

2. Logarithmic velocity distribution

$$yu_*/\nu = (0.002 \text{ mm})(0.207 \text{ m/s})/(1.14 \times 10^{-6} \text{ m}^2/\text{s}) = 363$$
$$u/u_* = 2.44 \ln(yu_*/\nu) + 5.56$$
$$= 2.44 \times \ln(363) + 5.56 = 20$$
$$u = 20 \times 0.207 \text{ m/s} = \boxed{4.14 \text{ m/s}}$$

3. Nondimensional distance

$$y/\delta = 0.002 \text{ mm}/0.02 \text{ mm} = 0.10$$

From Fig. 9.12

$$\frac{U_0 - u}{u_*} = 8.2$$

4. Velocity from defect law

$$u = U_0 - 8.2u_*$$
$$= 6 \text{ m/s} - (8.2)(0.207) \text{ m/s}$$
$$= \boxed{4.30 \text{ m/s}}$$

5. Power-law formula

$$u/U_0 = (y/\delta)^{1/7}$$
$$u = (U_0)(0.10)^{1/7}$$
$$= (6 \text{ m/s})(0.7197)$$
$$= \boxed{4.32 \text{ m/s}}$$

6. Nominal sublayer thickness

$$\delta'_N = 11.84\nu/u_* = (11.84)(1.14 \times 10^{-6} \text{ m}^2/\text{s})/(0.207 \text{ m/s})$$
$$= \boxed{6.52 \times 10^{-5} \text{ m}}$$

Review the Solution and the Process

Notice that the velocity obtained using logarithmic distribution and defect law are nearly the same, which indicates that the point is in the overlap region.

Boundary-Layer Thickness and Shear-Stress Correlations

Unlike the laminar boundary layer, there is no analytically derived equation for the thickness of the turbulent boundary layer. There is a way to obtain an equation by using momentum principles and empirical data for the local shear stress and by assuming the 1/7 power velocity profile (3). The result is

$$\delta = \frac{0.16x}{\text{Re}_x^{1/7}} \tag{9.33}$$

where x is the distance from the leading edge of the plate and Re_x is $U_0 x / \nu$.

Many empirical expressions have been proposed for the local shear-stress distribution for the turbulent boundary layer on a flat plate. One of the simplest correlations is

$$c_f = \frac{\tau_0}{\rho U_0^2 / 2} = \frac{0.027}{\text{Re}_x^{1/7}} \tag{9.34a}$$

and the corresponding average shear-stress coefficient is

$$C_f = \frac{0.032}{\text{Re}_L^{1/7}} \tag{9.34b}$$

where Re_L is the Reynolds number of the plate based on the length of the plate in the streamwise direction.

Even though the variation of c_f with Reynolds number given by Eq. (9.34a) provides a reasonably good fit with experimental data for Reynolds numbers less than 10^7, it tends to underpredict the skin friction at higher Reynolds numbers. Several correlations have been proposed in the literature; see the review by Schlichting (4). A correlation proposed by White (3) that fits the data for turbulent Reynolds numbers up to 10^{10} is

$$c_f = \frac{0.455}{\ln^2(0.06\,\text{Re}_x)} \tag{9.35}$$

The corresponding average shear-stress coefficient is

$$C_f = \frac{0.523}{\ln^2(0.06\,\text{Re}_L)} \tag{9.36}$$

These are the correlations for shear-stress coefficients recommended here.

The boundary layer on a flat plate is composed of both a laminar and turbulent part. The purpose here is to develop a correlation valid for the combined boundary layer. As noted in Section 9.3, the boundary layer on a flat plate consists first of a laminar boundary layer that grows in thickness, develops instability, and becomes turbulent. A turbulent boundary layer develops over the remainder of the plate. As discussed earlier in Section 9.4, the transition from a laminar to turbulent boundary layer is not immediate but takes place over a transition length. However for the purposes of analysis here it is assumed that transition occurs at a point corresponding to a transition Reynolds number, Re_{tr}, of about 500,000.

The idea here is to take the turbulent shear force for length L, $F_{s,\,turb}(L)$, assuming the boundary layer is turbulent from the leading edge, subtract the portion up to the transition point, $F_{s,\,turb}(L_{tr})$ and replace it with the laminar shear force up to the transition point $F_{s,\,lam}(L_{tr})$. Thus the composite shear force on the plate is

$$F_s = F_{s,\,turb}(L) - F_{s,\,turb}(L_{tr}) + F_{s,\,lam}(L_{tr})$$

Substituting in Eq. (9.18) for laminar flow and Eq. (9.36) for turbulent flow over a plate of width B gives

$$F_s = \left(\frac{0.523}{\ln^2(0.06\,\mathrm{Re}_L)} BL - \frac{0.523}{\ln^2(0.06\,\mathrm{Re}_{tr})} BL_{tr} + \frac{1.33}{\mathrm{Re}_{tr}^{1/2}} BL_{tr} \right) \rho \frac{U_0^2}{2} \qquad \textbf{(9.37)}$$

where Re_{tr} is the Reynolds number at the transition, Re_L is the Reynolds number at the end of the plate, and L_{tr} is the distance from the leading edge of the plate to the transition zone.

Expressing the resistance force in terms of the average shear-stress coefficient, $C_f = F_s/(BL\rho U_0^2/2)$, gives

$$C_f = \frac{0.523}{\ln^2(0.06\mathrm{Re}_L)} + \frac{L_{tr}}{L}\left(\frac{1.33}{\mathrm{Re}_{tr}^{1/2}} - \frac{0.523}{\ln^2(0.06\mathrm{Re}_{tr})} \right)$$

Here $L_{tr}/L = \mathrm{Re}_{tr}/\mathrm{Re}_L$. Therefore,

$$C_f = \frac{0.523}{\ln^2(0.06\mathrm{Re}_L)} + \frac{\mathrm{Re}_{tr}}{\mathrm{Re}_L}\left(\frac{1.33}{\mathrm{Re}_{tr}^{1/2}} - \frac{0.523}{\ln^2(0.06\mathrm{Re}_{tr})} \right)$$

Finally, for $\mathrm{Re}_{tr} = 500{,}000$, the equation for average shear-stress coefficient becomes

$$C_f = \frac{0.523}{\ln^2(0.06\mathrm{Re}_L)} - \frac{1520}{\mathrm{Re}_L} \qquad \textbf{(9.38)}$$

The variation of C_f with Reynolds number is shown by the solid line in Fig. 9.13. This curve corresponds to a boundary layer that begins as a laminar boundary layer and then changes to a turbulent boundary layer after the transition Reynolds number. This is the normal condition for a flat-plate boundary layer. Table 9.3 summarizes the equations for boundary-layer thickness, and for local shear-stress and average shear-stress coefficients for the boundary layer on a flat plate.

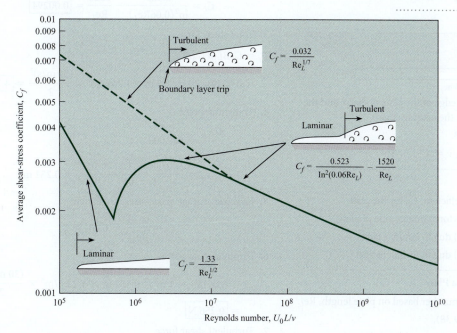

FIGURE 9.13

Average shear-stress coefficients.

Example 9.6 shows the calculation of shear force due to a boundary layer on a flat plate.

TABLE 9.3 Summary of Equations for Boundary Layer on a Flat Plate

	Laminar Flow $\mathrm{Re}_x, \mathrm{Re}_L < 5 \times 10^5$	Turbulent Flow $\mathrm{Re}_x, \mathrm{Re}_L \geq 5 \times 10^5$
Boundary-Layer Thickness, δ	$\delta = \dfrac{5x}{\mathrm{Re}_x^{1/2}}$	$\delta = \dfrac{0.16x}{\mathrm{Re}_x^{1/7}}$
Local Shear-Stress Coefficient, c_f	$c_f = \dfrac{0.664}{\mathrm{Re}_x^{1/2}}$	$c_f = \dfrac{0.455}{\ln^2(0.06\mathrm{Re}_x)}$
Average Shear-Stress Coefficient, C_f (mixed boundary layer)	$C_f = \dfrac{1.33}{\mathrm{Re}_L^{1/2}}$	$C_f = \dfrac{0.523}{\ln^2(0.06\mathrm{Re}_L)} - \dfrac{1520}{\mathrm{Re}_L}$
Average Shear-Stress Coefficient, C_f (tripped boundary layer)		$C_f = \dfrac{0.032}{\mathrm{Re}_L^{1/7}}$

EXAMPLE 9.6

Calculating Shear Force on a Flat Plate

Problem Statement

Assume that air 20°C and normal atmospheric pressure flows over a smooth, flat plate with a velocity of 30 m/s. The initial boundary layer is laminar and then becomes turbulent at a transitional Reynolds number of 5×10^5. The plate is 3 m long and 1 m wide. What will be the average resistance coefficient C_f for the plate? Also, what is the total shearing resistance of one side of the plate, and what will be the resistance due to the turbulent part and the laminar part of the boundary layer?

Define the Situation

Air flows past a flat plate

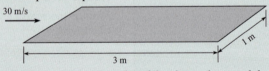

30 m/s

1 m

3 m

Assumptions: The leading edge of the plate is sharp, and the boundary is not tripped on the leading edge.

Properties: From Table A.3,

$$\rho = 1.2 \text{ kg/m}^3, \nu = 1.51 \times 10^{-5} \text{ m}^2/\text{s.}$$

State the Goal

1. Average shear-stress coefficient, C_f, for the plate
2. Total shear force (in newtons) on one side of plate
3. Shear force (in newtons) due to laminar part
4. Shear force (in newtons) due to turbulent part

Generate Ideas and Make a Plan

1. Calculate the Reynolds number based on plate length, Re_L.
2. Calculate C_f using Eq. (9.38).
3. Calculate the shear force on one side of plate using $F_s = (1/2)\rho U_0^2 C_f B L$.

4. Using value for transition Reynolds number, find transition point.
5. Use Eq. (9.18) to find average shear-stress coefficient for laminar portion.
6. Calculate shear force for laminar portion.
7. Subtract laminar portion from total shear force.

Take Action (Execute the Plan)

1. Reynolds number based on plate length

$$\mathrm{Re}_L = \frac{30 \text{ m/s} \times 3 \text{ m}}{1.51 \times 10^{-5} \text{ m}^2/\text{s}} = 5.96 \times 10^6$$

2. Average shear-stress coefficient

$$C_f = \frac{0.523}{\ln^2(0.06\mathrm{Re}_L)} - \frac{1520}{\mathrm{Re}_L} = \boxed{0.00294}$$

3. Total shear force

$$F_s = C_f B L \rho (U_0^2/2)$$

$$= 0.00294 \times 1 \text{ m} \times 3 \text{ m} \times 1.2 \text{ kg/m}^3 \times \frac{(30 \text{ m/s})^2}{2} = \boxed{4.76 \text{ N}}$$

4. Transition point

$$\frac{U x_{\mathrm{tr}}}{\nu} = 500{,}000$$

$$x_{\mathrm{tr}} = \frac{500{,}000 \times 1.51 \times 10^{-5}}{30} = 0.252 \text{ m}$$

5. Laminar average shear-stress coefficient

$$C_f = \frac{1.33}{\mathrm{Re}_{\mathrm{tr}}^{1/2}} = 0.00188$$

6. Laminar shear force

$$F_{s, \text{lam}} = 0.00188 \times 1 \text{ m} \times 0.252 \text{ m} \times 1.2 \text{ kg/m}^3 \times \frac{(30 \text{ m/s})^2}{2}$$

$$= \boxed{0.256 \text{ N}}$$

7. Turbulent shear force

$$F_{s, \text{turb}} = 4.76 \text{ N} - 0.26 \text{ N} = \boxed{4.50 \text{ N}}$$

If the boundary layer is "tripped" by some roughness or leading-edge disturbance (such as a wire across the leading edge), the boundary layer is turbulent from the leading edge. This is shown by the dashed line in Fig. 9.13. For this condition the boundary layer thickness, local shear-stress coefficient, and average shear-stress coefficient are fit reasonably well by Eqs. (9.33), (9.34a), and (9.34b).

$$\delta = \frac{0.16x}{\mathrm{Re}_x^{1/7}} \qquad c_f = \frac{0.027}{\mathrm{Re}_x^{1/7}} \qquad C_f = \frac{0.032}{\mathrm{Re}_L^{1/7}} \qquad \textbf{(9.39)}$$

which are valid up to a Reynolds number of 10^7. For Reynolds numbers beyond 10^7, the average shear-stress coefficient given by Eq. (9.36) can be used. It is of interest to note that marine engineers incorporate tripping mechanisms for the boundary layer on ship models to produce a boundary layer that can be predicted more precisely than a combination of laminar and turbulent boundary layers.

Example 9.7 illustrates calculating shear force with a tripped boundary layer.

EXAMPLE 9.7

Shear Force with a Tripped Boundary Layer

Problem Statement

Air at 20°C flows past a smooth, thin plate with a free-stream velocity of 20 m/s. Plate is 3 m wide and 6 m long in the direction of flow, and boundary layer is tripped at the leading edge.

Define the Situation

Air flows past a smooth, thin plate. Boundary layer is tripped at leading edge.

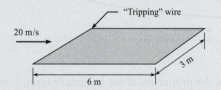

Properties: From Table A.3,

$$\rho = 1.2 \text{ kg/m}^3, \quad \mu = 1.81 \times 10^{-5} \text{ N} \cdot \text{s/m}^2.$$

State the Goal

Find: Total shear force (in newtons) on both sides of plate.

Generate Ideas and Make a Plan

1. Calculate the Reynolds number based on plate length.
2. Find average shear-stress coefficient from Eq. (9.39).
3. Calculate shear force for both sides of plate.

Take Action (Execute the Plan)

1. Reynolds number

$$\mathrm{Re}_L = \frac{\rho U L}{\mu} = \frac{1.2 \times 20 \times 6}{1.81 \times 10^{-5}} = 7.96 \times 10^6$$

Reynolds number is less than 10^7.

2. Average shear-stress coefficient

$$C_f = \frac{0.032}{\mathrm{Re}_L^{1/7}}$$

$$= \frac{0.032}{(7.96 \times 10^6)^{1/7}} = 0.0033$$

3. Shear force

$$F_s = 2 \times C_f A \frac{\rho U_0^2}{2}$$

$$= 0.0033 \times 3 \text{ m} \times 6 \text{ m} \times 1.2 \text{ kg/m}^3 \times (20 \text{ m/s})^2$$

$$= \boxed{28.5 \text{ N}}$$

Even though the equations in this chapter have been developed for flat plates, they are useful for engineering estimates for some surfaces that are not truly flat plates. For example, the skin friction drag of the submerged part of the hull of a ship can be estimated with Eq. (9.38).

9.6 Pressure Gradient Effects on Boundary Layers

In the preceding sections the features of a boundary layer on a flat plate where the external pressure gradient is zero have been presented. The boundary layer begins as laminar, goes through transition, and becomes turbulent with a "fuller" velocity profile and an increase in local shear stress. The purpose of this section is to present some features of the boundary layer over a curved surface where the external pressure gradient is not zero.

The flow over an airfoil section is shown in Fig. 9.14. The variation in static pressure with distance, s, along the surface is also shown on the figure. The point corresponding to $s = 0$ is the forward stagnation point where the pressure is equal to the stagnation pressure. The pressure then decreases toward a minimum value at the midsection. This minimum pressure corresponds to the location of maximum speed as predicted by the Bernoulli equation. The pressure then rises again as the flow decelerates toward the trailing edge. When the pressure decreases with increasing distance ($dp/ds < 0$), the pressure gradient is referred to as a favorable pressure gradient as introduced in Chapter 4. This means that the direction of the force due to the pressure gradient is in the flow direction. In other words, the effect of the pressure gradient is to accelerate the flow. This is the condition between the forward stagnation point and the point of minimum pressure. A rise in pressure with distance ($dp/ds > 0$) is called an adverse pressure gradient and occurs between the point of minimum pressure and the trailing edge. The pressure force due to the adverse pressure gradient acts in the direction opposite to the flow direction and tends to decelerate the flow.

FIGURE 9.14

Surface pressure distribution on airfoil section.

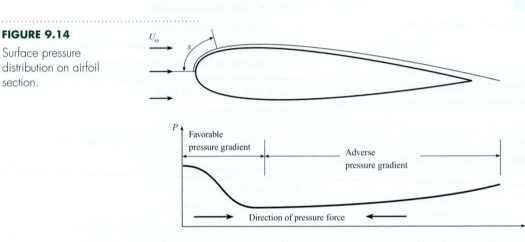

The external pressure gradient effects the properties of the boundary layer. Compared to a flat plate, the laminar boundary layer in a favorable pressure gradient grows more slowly and is more stable. This means that the boundary-layer thickness is less and the local shear stress is increased. Also the transition region is moved downstream, so the boundary layer becomes turbulent somewhat later. Of course, free-stream turbulence and surface roughness will still promote the early transition to a fully turbulent boundary layer.

The effect of external pressure gradient on the boundary layer is most pronounced for the adverse pressure gradient. The development of the velocity profiles for the laminar and turbulent boundary layers in an adverse pressure gradient are shown in Fig. 9.15. The retarding force associated with the adverse pressure gradient decelerates the flow, especially near the surface, where the velocities are the lowest. Ultimately there is a reversal of flow at the wall, which gives rise to a recirculatory pattern and the formation of an eddy. This phenomenon is called **boundary-layer separation**. The point of separation is defined where the velocity gradient $\partial u/\partial y$ becomes zero as indicated on the figure. The separation point for the turbulent boundary layer occurs farther downstream because the velocity profile is much fuller (higher velocities persist closer to the wall) than the laminar profile, and it takes longer for the adverse pressure gradient to decelerate the flow. Thus the turbulent boundary layer is less affected by the adverse pressure gradient.

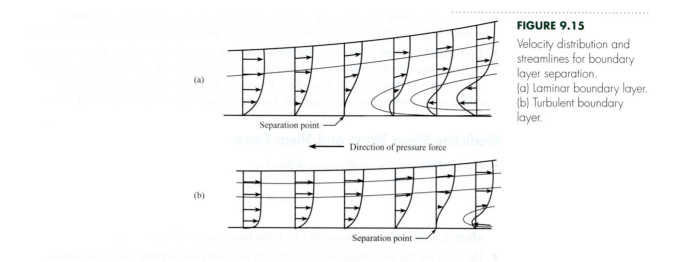

FIGURE 9.15
Velocity distribution and streamlines for boundary layer separation.
(a) Laminar boundary layer.
(b) Turbulent boundary layer.

Even though shear stresses on a body in a flow may not contribute significantly to the total drag force, the effect of boundary-layer separation can be very important. When boundary-layer separation takes place on airfoils at a high angle of attack, "stall" occurs, which means the airfoil loses its capability to provide lift. A photograph illustrating boundary-layer separation on an airfoil section is shown in Fig. 4.26. Boundary-layer separation on a cylinder was discussed and illustrated in Section 4.8. Understanding and controlling boundary-layer separation is important in the design of fluid dynamic shapes for maximum performance.

9.7 Summarizing Key Knowledge

Uniform Laminar Flow

- The variation in velocity for a planar, viscous, steady flow with parallel streamlines is governed by the equation

$$\frac{d^2u}{dy^2} = \frac{1}{\mu}\frac{d}{ds}(p + \gamma z)$$

  where the distance y is normal to the streamlines and the distance s is along the streamlines.
- In this chapter, this equation is used to analyze two flow configurations:
  ▸ Couette flow (flow generated by a moving plate)
  ▸ Hele-Shaw flow (flow between stationary parallel plates).

Boundary Layer

- The boundary layer is the region where the viscous stresses are responsible for the velocity change between the wall and the free stream.
- The boundary-layer thickness is the distance from the wall to the location where the velocity is 99% of the free-stream velocity.
- The laminar boundary layer is characterized by smooth (nonturbulent) flow where the momentum transfer between fluid layers occurs because of viscosity.
- As the boundary layer thickness grows, the laminar boundary layer becomes unstable, and a turbulent boundary layer ensues.

9.10 (WILEY GO) A flat plate is pulled to the right at a speed of 30 cm/s. Oil with a viscosity of 4 N · s/m² fills the space between the plate and the solid boundary. The plate is 1 m long ($L = 1$ m) by 30 cm wide, and the spacing between the plate and boundary is 2.0 mm.

a. Express the velocity mathematically in terms of the coordinate system shown.

b. By mathematical means, determine whether this flow is rotational or irrotational.

c. Determine whether continuity is satisfied, using the differential form of the continuity equation.

d. Calculate the force required to produce this plate motion.

PROBLEM 9.10

9.11 (WILEY PLUS) The velocity distribution that is shown represents laminar flow. Indicate which of the following statements are true.

a. The velocity gradient at the boundary is infinitely large.

b. The maximum shear stress in the liquid occurs midway between the walls.

c. The maximum shear stress in the liquid occurs next to the boundary.

d. The flow is irrotational.

e. The flow is rotational.

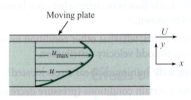

PROBLEM 9.11

9.12 The upper plate shown is moving to the right with a velocity V, and the lower plate is free to move laterally under the action of the viscous forces applied to it. For steady-state conditions, derive an equation for the velocity of the lower plate. Assume that the area of oil contact is the same for the upper plate, each side of the lower plate, and the fixed boundary.

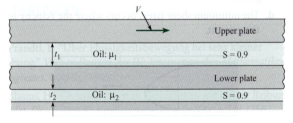

PROBLEM 9.12

9.13 (WILEY PLUS) A circular horizontal disk with a 27 cm diameter has a clearance of 3.0 mm from a horizontal plate. What torque is required to rotate the disk about its center at an angular speed of 31 rad/s when the clearance space contains oil ($\mu = 8$ N · s/m²)?

9.14 (WILEY PLUS) A plate 2 mm thick and 1 m wide (normal to the page) is pulled between the walls shown in the figure at a speed of 0.40 m/s. Note that the space that is not occupied by the plate is filled with glycerine at a temperature of 20°C. Also, the plate is positioned midway between the walls. Sketch the velocity distribution of the glycerine at section A-A. Neglecting the weight of the plate, estimate the force required to pull the plate at the speed given.

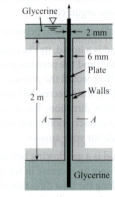

PROBLEM 9.14

9.15 (WILEY PLUS) A bearing uses SAE 30 oil with a viscosity of 0.1 N · s/m². The bearing is 30 mm in diameter, and the gap between the shaft and the casing is 1 mm. The bearing has a length of 1 cm. The shaft turns at $\omega = 200$ rad/s. Assuming that the flow between the shaft and the casing is a Couette flow, find the torque required to turn the bearing.

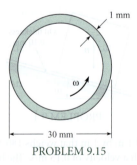

PROBLEM 9.15

9.16 An important application of viscous flow is found in lubrication theory. Consider a shaft that turns inside a stationary cylinder, with a lubricating fluid in the annular region. By considering a system consisting of an annulus of fluid of radius r and width Δr, and realizing that under steady-state operation the net torque on this ring is zero, show that $d(r^2 \tau)/dr = 0$, where τ is the viscous shear stress. For a flow that has a tangential component of velocity only, the shear stress is related to the velocity by $\tau = \mu r d(V/r)/dr$. Show that the torque per unit length acting on the inner cylinder is given by $T = 4\pi\mu\omega r_s^2/ (1 - r_s^2/r_0^2)$, where ω is the angular velocity of the shaft.

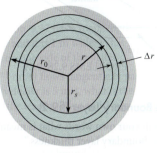

PROBLEM 9.16

9.17 Using the equation developed in Prob. 9.16, find the power necessary to rotate a 2 cm shaft at 60 rad/s if the inside diameter of the casing is 2.2 cm, the bearing is 3 cm long, and SAE 30 oil at 38°C is the lubricating fluid.

9.18 The analysis developed in Prob. 9.16 applies to a device used to measure the viscosity of a fluid. By applying a known torque to the inner cylinder and measuring the angular velocity achieved, one can calculate the viscosity of the fluid. Assume you have a 4 cm inner cylinder and a 4.5 cm outer cylinder. The cylinders are 10 cm long. When a force of 0.6 N is applied to the tangent of the inner cylinder, it rotates at 20 rpm. Calculate the viscosity of the fluid.

9.19 PLUS Two horizontal parallel plates are spaced 0.0045 m apart. The pressure decreases at a rate of 3650 Pa/m in the horizontal x-direction in the fluid between the plates. What is the maximum fluid velocity in the x direction? The fluid has a dynamic viscosity of 0.05 N · s/m² and a specific gravity of 0.80.

9.20 A viscous fluid fills the space between these two plates, and the pressures at A and B are 7 kPa and 5 kPa, respectively. The fluid is not accelerating. If the specific weight of the fluid is 15.7 kN/m³, then one must conclude that (a) flow is downward, (b) flow is upward, or (c) there is no flow.

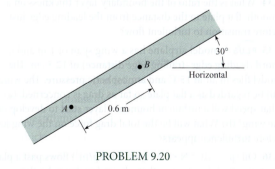

PROBLEM 9.20

9.21 Glycerine at 20°C flows downward between two vertical parallel plates separated by a distance of 0.4 cm. The ends are open, so there is no pressure gradient. Calculate the discharge per unit width, q, in m²/s.

9.22 PLUS Two vertical parallel plates are spaced 3 mm apart. If the pressure decreases at a rate of 9400 Pa/m in the vertical z-direction in the fluid between the plates, what is the maximum fluid velocity in the z-direction? The fluid has a viscosity of 0.05 Pa · s and a specific gravity of 0.80.

9.23 GO Two parallel plates are spaced 2 mm apart, and motor oil (SAE 30) with a temperature of 38°C flows at a rate of 0.00025 m³/s per meter of width between the plates. What is the pressure gradient in the direction of flow if the plates are inclined at 60° with the horizontal and if the flow is downward between the plates?

9.24 GO Glycerin at 20°C flows downward in the annular region between two cylinders. The internal diameter of the outer cylinder is 3 cm, and the external diameter of the inner cylinder is 2.8 cm. The pressure is constant along the flow direction. The flow is laminar. Calculate the discharge. (*Hint:* The flow between the two cylinders can be treated as the flow between two flat plates.)

PROBLEM 9.24

9.25 PLUS One type of bearing that can be used to support very large structures is shown in the accompanying figure. Here fluid under pressure is forced from the bearing midpoint (slot A) to the exterior zone B. Thus a pressure distribution occurs as shown. For this bearing, which is 43 cm wide, what discharge of oil from slot A per meter of length of bearing is required? Assume a 190 kN load per meter of bearing length with a clearance space t between the floor and the bearing surface of 1.5 mm. Assume an oil viscosity of 0.20 N · s/m². How much oil per hour would have to be pumped per meter of bearing length for the given conditions?

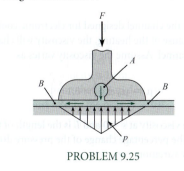

PROBLEM 9.25

9.52 Assume that the velocity profile in a boundary layer is replaced by a step profile, as shown in the figure, where the velocity is zero adjacent to the surface and equal to the free-stream velocity (U) at a distance greater than δ_* from the surface. Assume also that the density is uniform and equal to the free-stream density (ρ_∞). The distance δ_* (displacement thickness) is so chosen that the mass flux corresponding to the step profile is equal to the mass flux through the actual boundary layer. Derive an integral expression for the displacement thickness as a function of u, U, y, and δ.

9.53 Because of the reduction of velocity associated with the boundary layer, the streamlines outside the boundary layer are shifted away from the boundary. This amount of displacement of the streamlines is defined as the displacement thickness δ_*. Using the expression developed in Prob. 9.52, evaluate the displacement thickness of the boundary layer at the downstream edge of the plate (point A) in Prob. 9.51.

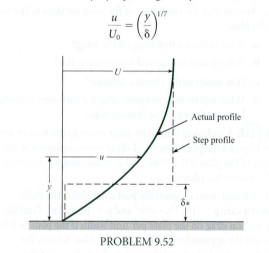

PROBLEMS 9.51, 9.53

9.54 Use the expression developed in Prob. 9.52 to find the ratio of the displacement thickness to the boundary layer thickness for the turbulent boundary layer profile given by

$$\frac{u}{U_0} = \left(\frac{y}{\delta}\right)^{1/7}$$

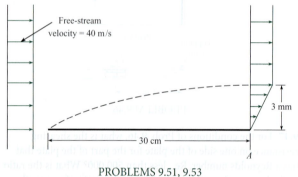

PROBLEM 9.52

9.55 PLUS What is the ratio of the skin friction drag of a plate 30 m long and 5 m wide to that of a plate 10 m long and 5 m wide if both plates are towed lengthwise through water ($T = 20°C$) at 10 m/s?

9.56 PLUS Estimate the power required to pull the sign shown if it is towed at 41 m/s and if it is assumed that the sign has the same resistance characteristics as a flat plate. Assume standard atmospheric pressure and a temperature of 10°C.

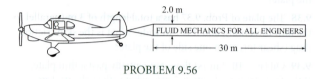

PROBLEM 9.56

9.57 GO A thin plastic panel (3 mm thick) is lowered from a ship to a construction site on the ocean floor. The plastic panel weighs 300 N in air and is lowered at a rate of 3 m/s. Assuming that the panel remains vertically oriented, calculate the tension in the cable.

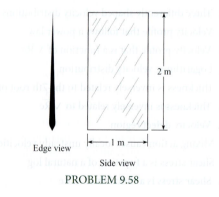

PROBLEM 9.57

9.58 The plate shown in the figure is weighted at the bottom so it will fall stably and steadily in a liquid. The weight of the plate in air is 23.5 N, and the plate has a volume of 0.002 m³. Estimate its falling speed in freshwater at 20°C. The boundary layer is normal; that is, it is not tripped at the leading edge.

In this problem, the final falling speed (terminal velocity) occurs when the weight is equal to the sum of the skin friction and buoyancy.

$$W = B + F_s = \gamma\Psi + \frac{1}{2}C_f\rho U_0^2 S$$

Hints: Find the final falling speed. This problem requires an iterative solution.

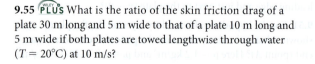

Edge view   Side view

PROBLEM 9.58

9.59 WILEY **PLUS** A turbulent boundary layer develops from the leading edge of a flat plate with water at 20°C flowing tangentially past the plate with a free-stream velocity of 7.7 m/s. Determine the thickness of the viscous sublayer, δ', at a distance 7.8 m downstream from the leading edge.

9.60 A model airplane descends in a vertical dive through air at standard conditions (1 atmosphere and 20°C). The majority of the drag is due to skin friction on the wing (like that on a flat plate). The wing has a span of 1 m (tip to tip) and a chord length (leading edge to trailing edge distance) of 10 cm. The leading edge is rough, so the turbulent boundary layer is "tripped." The model weighs 3 N. Determine the speed (in meters per second) at which the model will fall.

9.61 WILEY **PLUS** A flat plate is oriented parallel to a 24-m/s airflow at 20°C and atmospheric pressure. The plate is $L = 3$ m in the flow direction and 0.5 m wide. On one side of the plate, the boundary layer is tripped at the leading edge, and on the other side there is no tripping device. Find the total drag force on the plate.

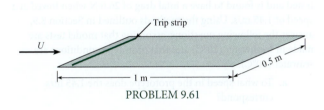

PROBLEM 9.61

9.62 An engineer is designing a horizontal, rectangular conduit that will be part of a system that allows fish to bypass a dam. Inside the conduit, a flow of water at 5°C will be divided into two streams by a flat, rectangular metal plate. Calculate the viscous drag force on this plate, assuming boundary-layer flow with free-stream velocity of 3 m/s and plate dimensions of $L = 1.8$ m and $W = 1.2$ m.

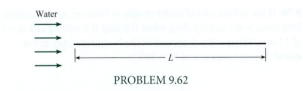

PROBLEM 9.62

9.63 A model is being developed for the entrance region between two flat plates. As shown in the figure, it is assumed that the region is approximated by a turbulent boundary layer originating at the leading edge. The system is designed such that the plates end where the boundary layers merge. The spacing between the plates is 4 mm, and the entrance velocity is 10 m/s. The fluid is water at 20°C. Roughness at the leading edge trips the boundary layers. Find the length L where the boundary layers merge, and find the force per unit depth (into the paper) due to shear stress on both plates.

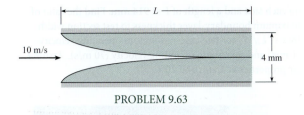

PROBLEM 9.63

9.64 An outboard racing boat "planes" at 110 km/hr over water at 15°C. The part of the hull in contact with the water has an average width of 1 m and a length of 2.5 m. Estimate the power required to overcome its shear force.

9.65 A motor boat pulls a long, smooth, water-soaked log (0.5 m in diameter and 50 m long) at a speed of 1.7 m/s. Assuming total submergence, estimate the force required to overcome the shear force of the log. Assume a water temperature of 10°C and that the boundary layer is tripped at the front of the log.

9.66 WILEY **PLUS** High-speed passenger trains are streamlined to reduce shear force. The cross section of a passenger car of one such train is shown. For a train 81 m long, (a) estimate the shear force for a speed of 81.1 km/hr and (b) for one of 204 km/hr. What power is required for just the shear force at these speeds? These two power calculations will be answers (c) and (d) respectively. Assume $T = 10$°C and that the boundary layer is tripped at the front of the train.

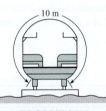

PROBLEM 9.66

9.67 Consider the boundary layer next to the smooth hull of a ship. The ship is cruising at a speed of 15 m/s in 15°C freshwater. Assuming that the boundary layer on the ship hull develops the same as on a flat plate, determine

 a. The thickness of the boundary layer at a distance $x = 30$ m downstream from the bow.

 b. The velocity of the water at a point in the boundary layer at $x = 30$ m and $y/\delta = 0.50$.

 c. The shear stress, τ_0, adjacent to the hull at $x = 30$ m.

9.68 A wind tunnel operates by drawing air through a contraction, passing this air through a test section, and then exhausting the air using a large axial fan. Experimental data are recorded in the test section, which is typically a rectangular section of duct that is made of clear plastic (usually acrylic). In the test section, the velocity should have a very uniform distribution; thus, it is important that the boundary layer be very thin at the end of the test section. For the pictured wind tunnel, the test section is square with a dimension of $W = 457$ mm on each side and a length of $L = 914$ mm. Find the ratio of maximum boundary-layer thickness to test section width $[\delta(x = L)/W]$ for two cases: minimum operating velocity (1 m/s) and maximum operating velocity (70 m/s). Assume air properties at 1 atm and 20°C.

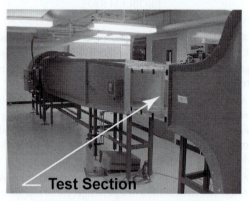

Test Section

PROBLEM 9.68 (Photo by Donald Elger)

9.69 A ship 180 m long steams at a rate of 8 m/s through still freshwater ($T = 10$°C). If the submerged area of the ship is 4600 m², what is the skin friction drag of this ship?

9.70 A river barge has the dimensions shown. It draws 0.6 m of water when empty. Estimate the skin friction drag of the barge when it is being towed at a speed of 3 m/s through still freshwater at 15°C.

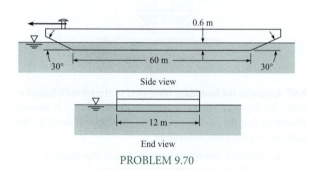

Side view

End view

PROBLEM 9.70

9.71 PLUS A supertanker has length, breadth, and draught (fully loaded) dimensions of 325 m, 48 m, and 19 m, respectively. In open seas the tanker normally operates at a speed of 9.27 m/s. For these conditions, and assuming that flat-plate boundary-layer conditions are approximated, estimate the skin friction drag of such a ship steaming in 10°C water. What power is required to overcome the skin friction drag? What is the boundary-layer thickness at 300 m from the bow?

9.72 A model test is needed to predict the wave drag on a ship. The ship is 150 m long and operates at 9 m/s in seawater at 10°C. The wetted area of the prototype is 2300 m². The model/prototype scale ratio is 1/100. Modeling is done in freshwater at 15°C to match the Froude number. The viscous drag can be calculated by assuming a flat plate with the wetted area of the model and a length corresponding to the length of the model. A total drag of 0.5 N is measured in model tests. Calculate the wave drag on the actual ship.

9.73 A ship is designed so that it is 250 m long, its beam measures 30 m, and its draft is 12 m. The surface area of the ship below the water line is 8800 m². A 1/40 scale model of the ship is tested and is found to have a total drag of 26.0 N when towed at a speed of 1.45 m/s. Using the methods outlined in Section 8.9, answer the following questions, assuming that model tests are made in freshwater (20°C) and that prototype conditions are seawater (10°C).

 a. To what speed in the prototype does the 1.45 m/s correspond?

 b. What are the model skin friction drag and wave drag?

 c. What would the ship drag be in saltwater corresponding to the model test conditions in freshwater?

9.74 A hydroplane 3 m long skims across a very calm lake ($T = 20$°C) at a speed of 15 m/s. For this condition, what will be the minimum shear stress along the smooth bottom?

9.75 Estimate the power required to overcome the shear force of a water skier if he or she is towed at 50 km/hr and each ski is 1.2 m by 15 cm. Assume the water temperature is 15°C.

9.76 If the wetted area of an 80-m ship is 1500 m², approximately how great is the surface drag when the ship is traveling at a speed of 15 m/s. What is the thickness of the boundary layer at the stern? Assume seawater at $T = 10$°C.

FLOW IN CONDUITS 10

Chapter Road Map

This chapter explains how to analyze flow in conduits. The primary tool, the energy equation, was presented in Chapter 7. This chapter expands on this knowledge by describing how to calculate head loss. In addition, this chapter explains how to design pumps into systems and how to analyze a network of pipes.

Learning Objectives

STUDENTS WILL BE ABLE TO

- Define a conduit. Classify a flow as laminar or turbulent. Define or calculate the Reynolds number. (§10.1)
- Describe developing flow and fully developed flow. Classify a flow into these categories. (§10.1)
- Specify a pipe size using Diameter Normal (DN). (§10.2)
- Describe total head loss, pipe head loss, and component head loss. (§10.3)
- Define the friction factor f. List the steps to derive the Darcy-Weisbach equation. (§10.3)
- Describe the physics of the Darcy-Weisbach equation and the meaning of the variables that appear in the equation. Apply this equation. (§10.3)
- Calculate h_f or f for laminar flow. (§10.5)
- Describe the main features of the Moody diagram. Calculate f for turbulent flow using the Moody diagram or the Swamee-Jain correlation. (§10.6)
- Solve turbulent flow problems when the equations cannot be solved by algebra alone. (§10.7)
- Define the minor loss coefficient. Describe and apply the combined head loss equation. (§10.8)
- Define hydraulic diameter and hydraulic radius and solve relevant problems. (§10.9)
- Solve problems that involve pumps and pipe networks. (§10.10)

FIGURE 10.1
The Alaskan pipeline, a significant accomplishment of the engineering profession, transports oil 1286 km across the state of Alaska. The pipe diameter is 1.2 m, and 44 pumps are used to drive the flow. This chapter presents information for designing systems involving pipes, pumps, and turbines. (© Eastcott/Momatiuk/The Image Works.)

A **conduit** is any pipe, tube, or duct that is completely filled with a flowing fluid. Examples include a pipeline transporting liquefied natural gas, a microchannel transporting hydrogen in a fuel cell, and a duct transporting air for heating of a building. A pipe that is partially filled with a flowing fluid, for example a drainage pipe, is classified as an open-channel flow and is analyzed using ideas from Chapter 15.

10.1 Classifying Flow

This section describes how to classify flow in a conduit by considering (a) whether the flow is laminar or turbulent, and (b) whether the flow is developing or fully developed. Classifying flow is essential for selecting the proper equation for calculating head loss.

Laminar Flow and Turbulent Flow

Flow in a conduit is classified as being either laminar or turbulent, depending on the magnitude of the Reynolds number. The original research involved visualizing flow in a glass tube as shown in Fig. 10.2a. Reynolds (1) in the 1880s injected dye into the center of the tube and observed the following:

- When the velocity was low, the streak of dye flowed down the tube with little expansion, as shown in Fig. 10.2b. However, if the water in the tank was disturbed, the streak would shift about in the tube.

- If velocity was increased, at some point in the tube, the dye would all at once mix with the water as shown in Fig. 10.2c.

- When the dye exhibited rapid mixing (Fig. 10.2c), illumination with an electric spark revealed eddies in the mixed fluid as shown in Fig. 10.2d.

The flow regimes shown in Fig. 10.2 are laminar flow (Fig. 10.2b) and turbulent flow (Figs. 10.2c and 10.2d). Reynolds showed that the onset of turbulence was related to a π-group that is now called the Reynolds number (Re = $\rho VD/\mu$) in honor of Reynolds' pioneering work.

The Reynolds number is often written as Re_D, where the subscript "D" denotes that diameter is used in the formula. This subscript is called a *length scale*. Indicating the length scale for Reynolds number is good practice because muliple values are used. For example, Chapter 9 introduced Re_x and Re_L.

FIGURE 10.2

Reynolds' experiment.
(a) Apparatus.
(b) Laminar flow of dye in tube.
(c) Turbulent flow of dye in tube.
(d) Eddies in turbulent flow.

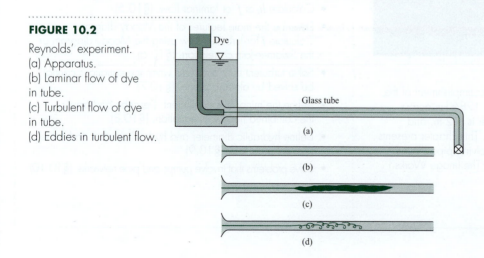

Reynolds number can be calculated with four different equations. These equations are equivalent because one can start with one formula and derive the others. The formulas are

$$\mathrm{Re}_D = \frac{VD}{\nu} = \frac{\rho VD}{\mu} = \frac{4Q}{\pi D \nu} = \frac{4\dot{m}}{\pi D \mu} \tag{10.1}$$

Reynolds discovered that if the fluid in the upstream reservoir was not completely still or if the pipe had some vibrations, then the change from laminar to turbulent flow occurred at $\mathrm{Re}_D \sim 2100$. However, if conditions were ideal, it was possible to reach a much higher Reynolds number before the flow became turbulent. Reynolds also found that, when going from high velocity to low velocity, the change back to laminar flow occurred at $\mathrm{Re}_D \sim 2000$. Based on Reynolds' experiments, engineers use guidelines to establish whether or not flow in a conduit will be laminar or turbulent. The guidelines used in this text are as follows:

$$
\begin{array}{ll}
\mathrm{Re}_D \le 2000 & \text{laminar flow} \\
2000 \le \mathrm{Re}_D \le 3000 & \text{unpredictable} \\
\mathrm{Re}_D \ge 3000 & \text{turbulent flow}
\end{array} \tag{10.2}
$$

In Eq. (10.2), the middle range ($2000 \le \mathrm{Re}_D \le 3000$) corresponds to a type of flow that is unpredictable because it can change back and forth between laminar and turbulent states. Recognize that precise values of Reynolds number versus flow regime do not exist. Thus, the guidelines given in Eq. (10.2) are approximate, and other references may give different values. For example, some references use $\mathrm{Re}_D = 2300$ as the criteria for turbulence.

Developing Flow and Fully Developed Flow

Flow in a conduit is classified as either developing flow or fully developed flow. For example, consider laminar fluid entering a pipe from a reservoir as shown in Fig. 10.3. As the fluid moves down the pipe, the velocity profile changes in the streamwise direction as viscous effects cause the plug-type profile to gradually change into a parabolic profile. This region of changing velocity profile is called **developing flow**. After the parabolic distribution is achieved, the flow profile remains unchanged in the streamwise direction, and flow is called **fully developed flow**.

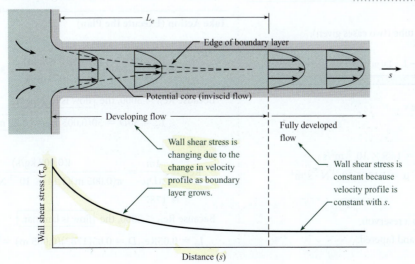

FIGURE 10.3

In developing flow, the wall shear stress is changing. In fully developed flow, the wall shear stress is constant.

The distance required for flow to develop is called the **entry or entrance length** (L_e). In the entry length, the wall shear stress is decreasing in the streamwise (i.e. s) direction. For laminar flow, the wall shear-stress distribution is shown in Fig. 10.3. Near the pipe entrance, the radial velocity gradient (change in velocity with distance from the wall) is high, so the shear stress is large. As the velocity profile progresses to a parabolic shape, the velocity gradient and the wall shear stress decrease until a constant value is achieved. The entry length is defined as the distance at which the shear stress reaches 2% of the fully developed value. Correlations for entry length are

$$\frac{L_e}{D} = 0.05 \, \text{Re}_D \qquad \text{(laminar flow: Re}_D \leq 2000) \tag{10.3a}$$

$$\frac{L_e}{D} = 50 \qquad \text{(turbulent flow: Re}_D \geq 3000) \tag{10.3b}$$

Eq. (10.3) is valid for flow entering a circular pipe from a reservoir under quiescent conditions. Other upstream components such as valves, elbows, and pumps produce complex flow fields that require different lengths to achieve fully developing flow.

In summary, flow in a conduit is classified into four categories: laminar developing, laminar fully developed, turbulent developing, or turbulent fully developed. The key to classification is to calculate the Reynolds number as shown by Example 10.1.

EXAMPLE 10.1

Classifying Flow in Conduits

Problem Statement

Consider fluid flowing in a round tube of length 1 m and diameter 5 mm. Classify the flow as laminar or turbulent and calculate the entrance length for (a) air (50°C) with a speed of 12 m/s and (b) water (15°C) with a mass flow rate of $\dot{m} = 8$ g/s.

Define the Situation

Fluid is flowing in a round tube (two cases given).

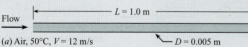

(a) Air, 50°C, $V = 12$ m/s
(b) Water, 15°C, $\dot{m} = 0.008$ kg/s
$L = 1.0$ m
$D = 0.005$ m

Properties:

1. Air (50°C), Table A.3, $\nu = 1.79 \times 10^{-5} \, \text{m}^2/\text{s}$

2. Water (15°C), Table A.5, $\mu = 1.14 \times 10^{-3} \, \text{N} \cdot \text{s/m}^2$

Assumptions:

1. The pipe is connected to a reservoir.

2. The entrance is smooth and tapered.

State the Goal

- Determine whether each flow is laminar or turbulent.
- Calculate the entrance length (in meters) for each case.

Generate Ideas and Make a Plan

- Calculate the Reynolds number using Eq. (10.1).
- Establish whether the flow is laminar or turbulent using Eq. (10.2).
- Calculate the entrance length using Eq. (10.3).

Take Action (Execute the Plan)

a. Air

$$\text{Re}_D = \frac{VD}{\nu} = \frac{(12 \text{ m/s})(0.005 \text{ m})}{1.79 \times 10^{-5} \text{ m}^2/\text{s}} = 3350$$

Because $\text{Re}_D > 3000$, the $\boxed{\text{flow is turbulent.}}$

$$L_e = 50D = 50(0.005 \text{ m}) = \boxed{0.25 \text{ m}}$$

b. Water

$$\text{Re}_D = \frac{4\dot{m}}{\pi D \mu} = \frac{4(0.008 \text{ kg/s})}{\pi(0.005 \text{ m})(1.14 \times 10^{-3} \text{ N} \cdot \text{s/m}^2)}$$

$$= 1787$$

Because $\text{Re}_D < 2000$, the $\boxed{\text{flow is laminar.}}$

$$L_e = 0.05 \text{Re}_D D = 0.05(1787)(0.005 \text{ m}) = \boxed{0.447 \text{ m}}$$

10.2 Specifying Pipe Sizes

This section describes how to specify pipes using the Diameter Normal (DN). This information is useful for specifying a size of pipe that is available commercially.

Standard Sizes for Pipes (DN)

One of the most common standards for pipe sizes is called the Diameter Nominal (DN) system. The terms used in the DN system are introduced in Fig. 10.4. The ID (pronounced "eye dee") indicates the inner pipe diameter, and the OD ("oh dee") indicates the outer pipe diameter. As shown in Table 10.1, an DN pipe is specified using two values: a diameter nominal (DN) and a schedule. The diameter determines the outside diameter or OD. For example, pipes with a nominal size of 50 mm have an OD of 60.3 mm.

Pipe schedule is related to the thickness of the wall. The original meaning of schedule was the ability of a pipe to withstand pressure, thus pipe schedule correlates with wall thickness. Each nominal pipe size has many possible schedules that range from schedule 5 to schedule 160. The data in Table 10.1 show representative ODs and schedules; more pipe sizes are specified in engineering handbooks and on the Internet.

FIGURE 10.4

Section view of a pipe.

A larger schedule indicates thicker walls. A schedule 40 pipe has thicker walls than a schedule 10 pipe.

ID (Inside diameter)

OD (Outside diameter)

TABLE 10.1 Diameter Nominal

NPS (mm)	OD (mm)	Schedule	Wall Thickness (mm)	ID (mm)
15	21.3	40	2.77	15.8
		80	3.73	13.9
25	33.4	40	3.38	26.6
		80	4.55	24.3
50	60.3	40	3.91	52.5
		80	5.54	49.2
100	114.3	40	6.02	102.3
		80	8.56	97.2
200	219.1	40	8.18	202.7
		80	12.7	193.7
350	355.6	10	6.35	342.9
		40	11.13	333.3
		80	19.05	317.5
		120	27.79	300
600	610	10	6.35	596.9
		40	17.4	576.6
		80	30.96	547.7
		120	46.02	517.6

10.3 Pipe Head Loss

This section presents the Darcy-Weisbach equation, which is used for calculating head loss in a straight run of pipe. This equation is one of the most useful equations in fluid mechanics.

Combined (Total) Head Loss

Pipe head loss is one type of head loss; the other type is called component head loss. All head loss is classified using these two categories:

$$\text{(Total head loss)} = \text{(Pipe head loss)} + \text{(Component head loss)} \qquad (10.4)$$

Component head loss is associated with flow through devices such as valves, bends, and tees. **Pipe head loss** is associated with fully developed flow in conduits, and it is caused by shear stresses that act on the flowing fluid. Note that pipe head loss is sometimes called major head loss, and component head loss is sometimes called minor head loss. Pipe head loss is predicted with the Darcy-Weisbach equation.

Derivation of the Darcy-Weisbach Equation

To derive the Darcy-Weisbach equation, start with the situation shown in Fig. 10.5. Assume fully developed and steady flow in a round tube of constant diameter D. Situate a cylindrical control volume of diameter D and length ΔL inside the pipe. Define a coordinate system with an axial coordinate in the streamwise direction (s direction) and a radial coordinate in the r direction.

Apply the momentum equation to the control volume shown in Fig. 10.5.

$$\sum \mathbf{F} = \frac{d}{dt} \int_{cv} \mathbf{v}\rho \, d\Psi + \int_{cs} \mathbf{v}\rho \mathbf{V} \cdot d\mathbf{A} \qquad (10.5)$$

$$\text{(Net forces)} = \text{(Momentum accumulation rate)} + \text{(Net efflux of momentum)}$$

FIGURE 10.5

Initial situation for the derivation of the Darcy-Weisbach equation.

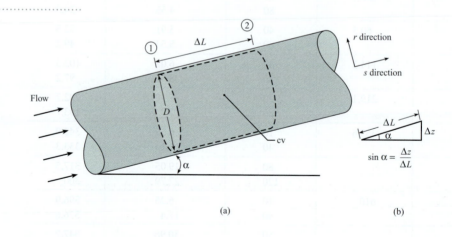

(a)   (b)

Select the streamwise direction and analyze each of the three terms in Eq. (10.5). The net efflux of momentum is zero because the velocity distribution at section 2 is identical to the velocity distribution at section 1. The momentum accumulation term is also zero because the flow is steady. Thus, Eq. (10.5) simplifies to $\Sigma F = 0$. Forces are shown in Fig. 10.6. Summing of forces in the streamwise direction gives

$$F_{\text{pressure}} + F_{\text{shear}} + F_{\text{weight}} = 0$$

$$(p_1 - p_2)\left(\frac{\pi D^2}{4}\right) - \tau_0(\pi D \Delta L) - \gamma \left[\left(\frac{\pi D^2}{4}\right)\Delta L\right] \sin \alpha = 0 \qquad (10.6)$$

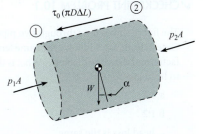

FIGURE 10.6
Force diagram.

Figure 10.5b shows that $\sin \alpha = (\Delta z / \Delta L)$. Equation (10.6) becomes

$$(p_1 + \gamma z_1) - (p_2 + \gamma z_2) = \frac{4 \Delta L \tau_0}{D} \tag{10.7}$$

Next, apply the energy equation to the control volume shown in Fig. 10.5. Recognize that $h_p = h_t = 0$, $V_1 = V_2$, and $\alpha_1 = \alpha_2$. Thus, the energy equation reduces to

$$\frac{p_1}{\gamma} + z_1 = \frac{p_2}{\gamma} + z_2 + h_L \tag{10.8}$$
$$(p_1 + \gamma z_1) - (p_2 + \gamma z_2) = \gamma h_L$$

Combine Eqs. (10.7) and (10.8) and replace ΔL by L. Also, introduce a new symbol h_f to represent head loss in pipe.

$$h_f = \left(\frac{\text{head loss}}{\text{in a pipe}} \right) = \frac{4 L \tau_0}{D \gamma} \tag{10.9}$$

Rearrange the right side of Eq. (10.9).

$$h_f = \left(\frac{L}{D} \right) \left\{ \frac{4 \tau_0}{\rho V^2 / 2} \right\} \left\{ \frac{\rho V^2 / 2}{\gamma} \right\} = \left\{ \frac{4 \tau_0}{\rho V^2 / 2} \right\} \left(\frac{L}{D} \right) \left\{ \frac{V^2}{2g} \right\} \tag{10.10}$$

Define a new π-group called the **friction factor** f that gives the ratio of wall shear stress (τ_0) to kinetic pressure ($\rho V^2 / 2$):

$$f \equiv \frac{(4 \cdot \tau_0)}{(\rho V^2 / 2)} \approx \frac{\text{shear stress acting at the wall}}{\text{kinetic pressure}} \tag{10.11}$$

In the technical literature, the friction factor is identified by several different labels that are synonymous: friction factor, Darcy friction factor, Darcy-Weisbach friction factor, and the resistance coefficient. There is also another coefficient called the Fanning friction factor, often used by chemical engineers, which is related to the Darcy-Weisbach friction factor by a factor of 4.

$$f_{\text{Darcy}} = 4 f_{\text{Fanning}}$$

This text uses only the Darcy-Weisbach friction factor. Combining Eqs. (10.10) and (10.11) gives the Darcy-Weisbach equation:

$$h_f = f \frac{L}{D} \frac{V^2}{2g} \quad \rightarrow \text{major loss} \tag{10.12}$$

To use the Darcy-Weisbach equation, the flow should be fully developed and steady. The Darcy-Weisbach equation is used for either laminar flow or turbulent flow and for either round pipes or nonround conduits such as a rectangular duct.

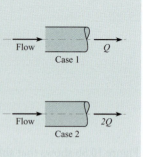

The Darcy-Weisbach equation shows that head loss depends on the friction factor, the
pipe length-to-diameter ratio, and the mean velocity squared. The key to using the Darcy-
Weisbach equation is calculating a value of the friction factor f. This topic is addressed in the
next sections of this text.

10.4 Stress Distributions in Pipe Flow

This section derives equations for the stress distributions on a plane that is oriented normal to
stream lines. These equations, which apply to both laminar and turbulent flow, provide insights
about the nature of the flow. Also, these equations are used for subsequent derivations.

In pipe flow the pressure acting on a plane that is normal to the direction of flow is hydrostatic. This means that the pressure distribution varies linearly as shown in Fig. 10.7. The reason
that the pressure distribution is hydrostatic can be explained with Euler's equation (see p. 130).

FIGURE 10.7

For fully developed flow
in a pipe, the pressure
distribution on an area
normal to streamlines is
hydrostatic.

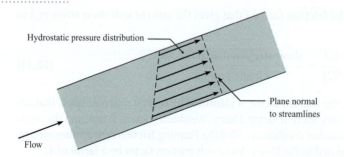

To derive an equation for the shear-stress variation, consider flow of a Newtonian fluid in
a round tube that is inclined at an angle α with respect to the horizontal as shown in Fig. 10.8.
Assume that the flow is fully developed, steady, and laminar. Define a cylindrical control volume of length ΔL and radius r.

Apply the momentum equation in the s direction. The net momentum efflux is zero
because the flow is fully developed; that is, the velocity distribution at the inlet is the same as
the velocity distribution at the exit. The momentum accumulation is also zero because the flow
is steady. The momentum equation simplifies to force equilibrium.

$$\sum F_s = F_{\text{pressure}} + F_{\text{weight}} + F_{\text{shear}} = 0 \qquad \text{(10.13)}$$

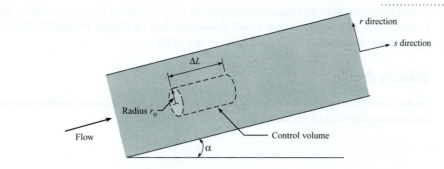

FIGURE 10.8

Sketch for derivation of an equation for shear stress.

Analyze each term in Eq. (10.13) using the force diagram shown in Fig. 10.9:

$$pA - \left(p + \frac{dp}{ds}\Delta L\right)A - W\sin\alpha - \tau(2\pi r)\Delta L = 0 \qquad \textbf{(10.14)}$$

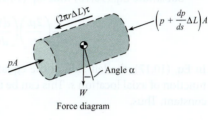

FIGURE 10.9

Force diagram corresponding to the control volume defined in Fig. 10.8.

Let $W = \gamma A\Delta L$, and let $\sin\alpha = \Delta z/\Delta L$ as shown in Fig. 10.5b. Next, divide Eq. (10.14) by $A\Delta L$:

$$\tau = \frac{r}{2}\left[-\frac{d}{ds}(p + \gamma z)\right] \qquad \textbf{(10.15)}$$

Equation (10.15) shows that the shear-stress distribution varies linearly with r as shown in Fig. 10.10. Notice that the shear stress is zero at the centerline, it reaches a maximum value of τ_0 at the wall, and the variation is linear in between. This linear shear stress variation applies to both laminar and turbulent flow.

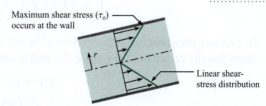

FIGURE 10.10

In fully developed flow (laminar or turbulent), the shear-stress distribution on an area that is normal to streamlines is linear.

10.5 Laminar Flow in a Round Tube

This section describes laminar flow and derives relevant equations. Laminar flow is important for flow in small conduits called microchannels, for lubrication flow, and for analyzing other flows in which viscous forces are dominant. Also, knowledge of laminar flow provides a foundation for the study of advanced topics.

Laminar flow is a flow regime in which fluid motion is smooth, the flow occurs in layers (laminae), and the mixing between layers occurs by molecular diffusion, a process that is much slower than turbulent mixing. According to Eq. (10.2), laminar flow occurs when $\mathrm{Re}_D \leq 2000$. Laminar flow in a round tube is called **Poiseuille flow** or **Hagen-Poiseuille flow** in honor of researchers who studied low-speed flows in the 1840s.

Velocity Profile

To derive an equation for the velocity profile in laminar flow, begin by relating stress to rate-of-strain using the viscosity equation:

$$\tau = \mu \frac{dV}{dy}$$

where y is the distance from the pipe wall. Change variables by letting $y = r_0 - r$, where r_0 is pipe radius and r is the radial coordinate. Next, use the chain rule of calculus:

$$\tau = \mu \left(\frac{dV}{dy} \right) = \mu \left(\frac{dV}{dr} \right) \left(\frac{dr}{dy} \right) = -\left(\mu \frac{dV}{dr} \right) \tag{10.16}$$

Substitute Eq. (10.16) into Eq. (10.15).

$$-\left(\frac{2\mu}{r} \right) \left(\frac{dV}{dr} \right) = \frac{d}{ds}(p + \gamma z) \tag{10.17}$$

In Eq. (10.17), the left side of the equation is a function of radius r, and the right side is a function of axial location s. This can be true if and only if each side of Eq. (10.17) is equal to a constant. Thus,

$$\text{constant} = \frac{d}{ds}(p + \gamma z) = \left(\frac{\Delta(p + \gamma z)}{\Delta L} \right) = \left(\frac{\gamma \Delta h}{\Delta L} \right) \tag{10.18}$$

where Δh is the change in piezometric head over a length ΔL of conduit. Combine Eqs. (10.17) and (10.18):

$$\frac{dV}{dr} = -\left(\frac{r}{2\mu} \right) \left(\frac{\gamma \Delta h}{\Delta L} \right) \tag{10.19}$$

Integrate Eq. (10.19):

$$V = -\left(\frac{r^2}{4\mu} \right) \left(\frac{\gamma \Delta h}{\Delta L} \right) + C \tag{10.20}$$

To evaluate the constant of integration C in Eq. (10.20), apply the no-slip condition, which states that the velocity of the fluid at the wall is zero. Thus,

$$V(r = r_0) = 0$$

$$0 = -\frac{r_0^2}{4\mu} \left(\frac{\gamma \Delta h}{\Delta L} \right) + C$$

Solve for C and substitute the result into Eq. (10.20):

$$V = \frac{r_0^2 - r^2}{4\mu} \left[-\frac{d}{ds}(p + \gamma z) \right] = -\left(\frac{r_0^2 - r^2}{4\mu} \right) \left(\frac{\gamma \Delta h}{\Delta L} \right) \tag{10.21}$$

The maximum velocity occurs at $r = r_0$:

$$V_{\max} = -\left(\frac{r_0^2}{4\mu} \right) \left(\frac{\gamma \Delta h}{\Delta L} \right) \tag{10.22}$$

Combine Eqs. (10.21) and (10.22):

$$V(r) = -\left(\frac{r_0^2 - r^2}{4\mu}\right)\left(\frac{\gamma\Delta h}{\Delta L}\right) = V_{max}\left(1 - \left(\frac{r}{r_0}\right)^2\right) \tag{10.23}$$

Equation (10.23) shows that velocity varies as radius squared ($V \sim r^2$), meaning that the velocity distribution in laminar flow is parabolic as plotted in Fig. 10.11.

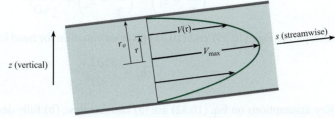

FIGURE 10.11

The velocity profile in Poiseuille flow is parabolic.

Discharge and Mean Velocity V

To derive an equation for discharge Q, introduce the velocity profile from Eq. (10.23) into the flow rate equation.

$$Q = \int V \, dA$$

$$= -\int_0^{r_0} \frac{(r_0^2 - r^2)}{4\mu}\left(\frac{\gamma\Delta h}{\Delta L}\right)(2\pi r \, dr) \tag{10.24}$$

Integrate Eq. (10.24):

$$Q = -\left(\frac{\pi}{4\mu}\right)\left(\frac{\gamma\Delta h}{\Delta L}\right)\frac{(r^2 - r_0^2)^2}{2}\Bigg|_0^{r_0} = -\left(\frac{\pi r_0^4}{8\mu}\right)\left(\frac{\gamma\Delta h}{\Delta L}\right) \tag{10.25}$$

To derive an equation for mean velocity, apply $Q = \overline{V}A$ and use Eq. (10.25).

$$\overline{V} = -\left(\frac{r_0^2}{8\mu}\right)\left(\frac{\gamma\Delta h}{\Delta L}\right) \tag{10.26}$$

Comparing Eqs. (10.26) and (10.22) reveals that $\overline{V} = V_{max}/2$. Next, substitute $D/2$ for r_0 in Eq. (10.26). The final result is an equation for mean velocity in a round tube.

$$\overline{V} = -\left(\frac{D^2}{32\mu}\right)\left(\frac{\gamma\Delta h}{\Delta L}\right) = \frac{V_{max}}{2} \tag{10.27}$$

Head Loss and Friction Factor f

To derive an equation for head loss in a round tube, assume fully developed flow in the pipe shown in Fig. 10.12. Apply the energy equation from section 1 to 2 and simplify to give

$$\left(\frac{p_1}{\gamma} + z_1\right) = \left(\frac{p_2}{\gamma} + z_2\right) + h_L \tag{10.28}$$

Let $h_L = h_f$ and then Eq. (10.28) becomes

$$\left(\frac{p_1}{\gamma} + z_1\right) = \left(\frac{p_2}{\gamma} + z_2\right) + h_f \tag{10.29}$$

FIGURE 10.12

Flow in a pipe.

Expand Eq. (10.27).

$$\overline{V} = -\left(\frac{\gamma D^2}{32\mu}\right)\left(\frac{\Delta h}{\Delta L}\right) = -\left(\frac{\gamma D^2}{32\mu}\right)\frac{\left(\dfrac{p_2}{\gamma} + z_2\right) - \left(\dfrac{p_1}{\gamma} + z_1\right)}{\Delta L} \tag{10.30}$$

Reorganize Eq. (10.30) and replace ΔL with L.

$$\left(\frac{p_1}{\gamma} + z_1\right) = \left(\frac{p_2}{\gamma} + z_2\right) + \frac{32\mu \overline{V} L}{\gamma D^2} \tag{10.31}$$

Comparing Eqs. (10.29) and (10.31) gives an equation for head loss in a pipe.

$$h_f = \frac{32\mu L \overline{V}}{\gamma D^2} \tag{10.32}$$

Key assumptions on Eq. (10.32) are (a) laminar flow, (b) fully developed flow, (c) steady flow, and (d) Newtonian fluid.

Equation (10.32) shows that head loss in laminar flow varies linearly with velocity. Also, head loss is influenced by viscosity, pipe length, specific weight, and pipe diameter squared.

To derive an equation for the friction factor f, combine Eq. (10.32) with the Darcy-Weisbach equation (10.12).

$$h_f = \frac{32\,\mu L V}{\gamma D^2} = f\frac{L}{D}\frac{V^2}{2g} \tag{10.33}$$

$$\text{or } f = \left(\frac{32\mu L V}{\gamma D^2}\right)\left(\frac{D}{L}\right)\left(\frac{2g}{V^2}\right) = \frac{64\mu}{\rho D V} = \boxed{\frac{64}{\mathrm{Re}_D}} \tag{10.34}$$

Equation (10.34) shows that the friction factor for laminar flow depends only on Reynolds number. Example 10.2 illustrates how to calculate head loss.

EXAMPLE 10.2

Head Loss for Laminar Flow

Problem Statement

Oil (S = 0.85) with a kinematic viscosity of $6 \times 10^{-4}\ \text{m}^2/\text{s}$ flows in a 15 cm diameter pipe at a rate of 0.020 m³/s. What is the head loss for a 100 m length of pipe?

Define the Situation

- Oil is flowing in a pipe at a flow rate of $Q = 0.02\ \text{m}^3/\text{s}$.
- Pipe diameter is $D = 0.15$ m.

Assumptions: Fully developed, steady flow

Properties: Oil: S = 0.85, $\nu = 6 \times 10^{-4}\ \text{m}^2/\text{s}$

State the Goal

Calculate head loss (in meters) for a pipe length of 100 m.

Generate Ideas and Make a Plan

1. Calculate the mean velocity using the flow rate equation.
2. Calculate the Reynolds number using Eq. (10.1).
3. Check whether the flow is laminar or turbulent using Eq. (10.2).
4. Calculate head loss using Eq. (10.32).

Take Action (Execute the Plan)

1. Mean velocity

$$V = \frac{Q}{A} = \frac{0.020\ \text{m}^3/\text{s}}{(\pi D^2)/4} = \frac{0.020\ \text{m}^3/\text{s}}{\pi((0.15\ \text{m})^2/4)} = 1.13\ \text{m/s}$$

2. Reynolds number

$$\mathrm{Re}_D = \frac{VD}{\nu} = \frac{(1.13\ \text{m/s})(0.15\ \text{m})}{6 \times 10^{-4}\ \text{m}^2/\text{s}} = 283$$

3. Because $Re_D < 2000$, the flow is laminar.

4. Head loss (laminar flow).

$$h_f = \frac{32\mu LV}{\gamma D^2} = \frac{32\rho \nu LV}{\rho g D^2} = \frac{32\nu LV}{gD^2}$$

$$= \frac{32(6 \times 10^{-4} \text{ m}^2/\text{s})(100 \text{ m})(1.13 \text{ m/s})}{(9.81 \text{ m/s}^2)(0.15 \text{ m})^2}$$

$$= \boxed{9.83 \text{ m}}$$

Review the Solution and the Process

Knowledge. An alternative way to calculate head loss for laminar flow is to use the Darcy-Weisbach equation (10.12) as follows:

$$f = \frac{64}{Re_D} = \frac{64}{283} = 0.226$$

$$h_f = f\left(\frac{L}{D}\right)\left(\frac{V^2}{2g}\right) = 0.226\left(\frac{100 \text{ m}}{0.15 \text{ m}}\right)\left(\frac{(1.13 \text{ m/s})}{2 \times 9.81 \text{ m/s}^2}\right)^2$$

$$= 9.83 \text{ m}$$

10.6 Turbulent Flow and the Moody Diagram

This section describes the characteristics of turbulent flow, presents equations for calculating the friction factor f, and presents a famous graph called the Moody diagram. This information is important because most flows in conduits are turbulent.

Qualitative Description of Turbulent Flow

Turbulent flow is a flow regime in which the movement of fluid particles is chaotic, eddying, and unsteady, with significant movement of particles in directions transverse to the flow direction. Because of the chaotic motion of fluid particles, turbulent flow produces high levels of mixing and has a velocity profile that is more uniform or flatter than the corresponding laminar velocity profile. According to Eq. (10.2), turbulent flow occurs when $Re \geq 3000$.

Engineers and scientists model turbulent flow by using an empirical approach. This is because the complex nature of turbulent flow has prevented researchers from establishing a mathematical solution of general utility. Still, the empirical information has been used successfully and extensively in system design. Over the years, researchers have proposed many equations for shear stress and head loss in turbulent pipe flow. The empirical equations that have proven to be the most reliable and accurate for engineering use are presented in the next section.

Equations for the Velocity Distribution

The time-average velocity distribution is often described using an equation called the power-law formula.

$$\frac{u(r)}{u_{max}} = \left(\frac{r_0 - r}{r_0}\right)^m \tag{10.35}$$

where u_{max} is velocity in the center of the pipe, r_0 is the pipe radius, and m is an empirically determined variable that depends on Re as shown in Table 10.2. Notice in Table 10.2 that the velocity in the center of the pipe is typically about 20% higher than the mean velocity V. Although Eq. (10.35) provides an accurate representation of the velocity profile, it does not predict an accurate value of wall shear stress.

An alternative approach to Eq. (10.35) is to use the turbulent boundary-layer equations presented in Chapter 9. The most significant of these equations, called the logarithmic velocity distribution, is given by Eq. (9.29) and repeated here:

$$\frac{u(r)}{u_*} = 2.44 \ln \frac{u_*(r_0 - r)}{\nu} + 5.56 \tag{10.36}$$

where u_*, the shear velocity, is given by $u_* = \sqrt{\tau_0/\rho}$.

TABLE 10.2 Exponents for Power-Law Equation and Ratio of Mean to Maximum Velocity

Re	4×10^3	2.3×10^4	1.1×10^5	1.1×10^6	3.2×10^6
m	$\dfrac{1}{6.0}$	$\dfrac{1}{6.6}$	$\dfrac{1}{7.0}$	$\dfrac{1}{8.8}$	$\dfrac{1}{10.0}$
u_{max}/V	1.26	1.24	1.22	1.18	1.16

Source: Schlichting (2).

Equations for the Friction Factor, f

To derive an equation for f *in turbulent flow*, substitute the log law in Eq. (10.36) into the definition of mean velocity given by Eq. (5.10):

$$V = \frac{Q}{A} = \left(\frac{1}{\pi r_0^2} \right) \int_0^{r_0} u(r) 2\pi r\, dr = \left(\frac{1}{\pi r_0^2} \right) \int_0^{r_0} u_* \left[2.44 \ln \frac{u_*(r_0 - r)}{\nu} + 5.56 \right] 2\pi r\, dr$$

After integration, algebra, and tweaking the constants to better fit experimental data, the result is

$$\frac{1}{\sqrt{f}} = 2.0 \log_{10}(\text{Re}\sqrt{f}) - 0.8 \tag{10.37}$$

Equation (10.37), first derived by Prandtl in 1935, gives the friction factor for turbulent flow in tubes that have smooth walls. The details of the derivation of Eq. (10.37) are presented by White (21). To determine the influence of roughness on the walls, Nikuradse (4), one of Prandtl's graduate students, glued uniform-sized grains of sand to the inner walls of a tube and then measured pressure drops and flow rates.

Nikuradse's data, Fig. 10.13, shows the friction factor f plotted as function of Reynolds number for various sizes of sand grains. To characterize the size of sand grains, Nikuradse used

FIGURE 10.13

Resistance coefficient f versus Reynolds number for sand-roughened pipe. [After Nikuradse (4)].

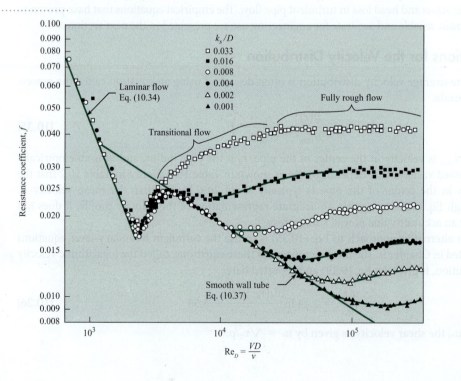

a variable called the **sand roughness height** with the symbol k_s. The π-group, k_s/D, is given the name **relative roughness**.

In laminar flow, the data in Fig. 10.13 show that wall roughness does not influence f. In particular, notice how the data corresponding to various values of k_s/D collapse into a single blue line that is labeled "laminar flow."

In turbulent flow, the data in Fig. 10.13 show that wall roughness has a major impact on f. When $k_s/D = 0.033$, then values of f are about 0.04. As the relative roughness drops to 0.002, values of f decrease by a factor of about 3. Eventually wall roughness does not matter, and the value of f can be predicted by assuming that the tube has a smooth wall. This latter case corresponds to the blue curve in Fig. 10.13 that is labeled "smooth wall tube." The effects of roughness are summarized by White (5) and presented in Table 10.3. These regions are also labeled in Fig. 10.13.

TABLE 10.3 Effects of Wall Roughness

Type of Flow	Parameter Ranges		Influence of Parameters on f
Laminar Flow	$Re_D < 2000$	NA	f depends on Reynolds number f is independent of wall roughness (k_s/D)
Turbulent Flow, Smooth Tube	$Re_D > 3000$	$\left(\dfrac{k_s}{D}\right) Re_D < 10$	f depends on Reynolds number f is independent of wall roughness (k_s/D)
Transitional Turbulent Flow	$Re_D > 3000$	$10 < \left(\dfrac{k_s}{D}\right) Re_D < 1000$	f depends on Reynolds number f depends on wall roughness (k_s/D)
Fully Rough Turbulent Flow	$Re_D > 3000$	$\left(\dfrac{k_s}{D}\right) Re_D > 1000$	f is independent of Reynolds number f depends on wall roughness (k_s/D)

Moody Diagram

Colebrook (6) advanced Nikuradse's work by acquiring data for commercial pipes and then developing an empirical equation, called the Colebrook-White formula, for the friction factor. Moody (3) used the Colebrook-White formula to generate a design chart similar to that shown in Fig. 10.14. This chart is now known as the **Moody diagram** for commercial pipes.

In the Moody diagram, Fig. 10.14, the variable k_s denotes the **equivalent sand roughness**. That is, a pipe that has the same resistance characteristics at high Re values as a sand-roughened pipe is said to have a roughness equivalent to that of the sand-roughened pipe. Table 10.4 gives the equivalent sand roughness for various kinds of pipes. This table can be used to calculate the relative roughness for a given pipe diameter, which, in turn, is used in Fig. 10.14, to find the friction factor.

In the Moody diagram, Fig. 10.14, the abscissa is the Reynolds number Re, and the ordinate is the resistance coefficient f. Each blue curve is for a constant relative roughness k_s/D, and the values of k_s/D are given on the right at the end of each curve. To find f, given Re and k_s/D, one goes to the right to find the correct relative roughness curve. Then one looks at the bottom of the chart to find the given value of Re and, with this value of Re, moves vertically upward until the given k_s/D curve is reached. Finally, from this point one moves horizontally to the left scale to read the value of f. If the curve for the given value of k_s/D is not plotted in Fig. 10.14, then one simply finds the proper position on the graph by interpolation between the k_s/D curves that bracket the given k_s/D.

To provide a more convenient solution to some types of problems, the top of the Moody diagram presents a scale based on the parameter $Re\, f^{1/2}$. This parameter is useful when h_f and k_s/D are known but the velocity V is not. Using the Darcy-Weisbach equation given in Eq. (10.12) and the definition of Reynolds number, one can show that

$$\mathrm{Re}\, f^{1/2} = \frac{D^{3/2}}{\nu}(2gh_f/L)^{1/2} \tag{10.38}$$

FIGURE 10.14

Friction Factor f versus Reynolds number. Reprinted with minor variations. [After Moody (3). Reprinted with permission from the ASME.]

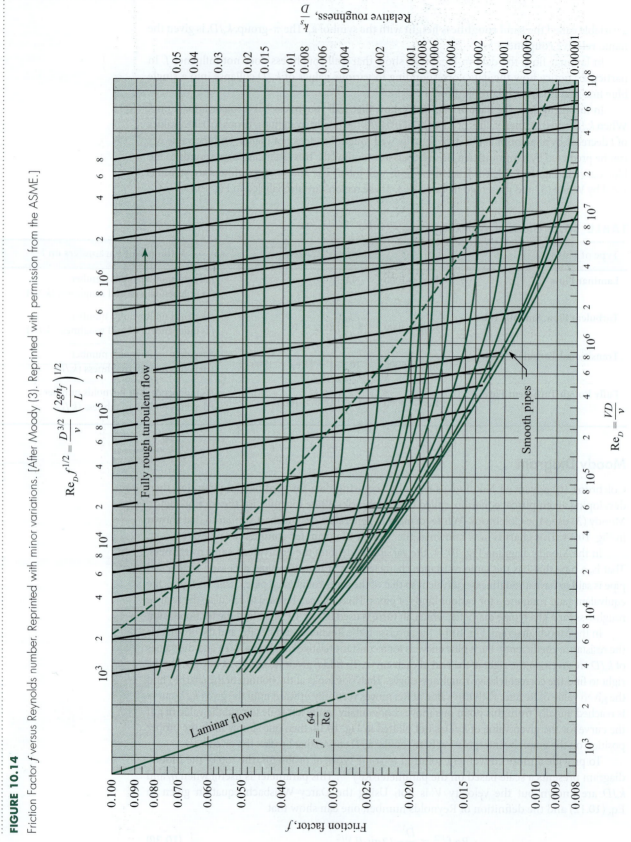

TABLE 10.4 Equivalent Sand-Grain Roughness, (k_s), for Various Pipe Materials

Boundary Material	k_s, Millimeters	k_s, Inches
Glass, plastic	Smooth	Smooth
Copper or brass tubing	0.0015	6×10^{-5}
Wrought iron, steel	0.046	0.002
Asphalted cast iron	0.12	0.005
Galvanized iron	0.15	0.006
Cast iron	0.26	0.010
Concrete	0.3 to 3.0	0.012–0.12
Riveted steel	0.9–9	0.035–0.35
Rubber pipe (straight)	0.025	0.001

In the Moody diagram, Fig. 10.14, curves of constant $\mathrm{Re}\, f^{1/2}$ are plotted using heavy black lines that slant from the left to right. For example, when $\mathrm{Re}\, f^{1/2} = 10^5$ and $k_s/D = 0.004$, then $f = 0.029$. When using computers to carry out pipe-flow calculations, it is much more convenient to have an equation for the friction factor as a function of Reynolds number and relative roughness. By using the Colebrook-White formula, Swamee and Jain (7) developed an explicit equation for friction factor, namely

$$f = \frac{0.25}{\left[\log_{10} \left(\dfrac{k_s}{3.7D} + \dfrac{5.74}{\mathrm{Re}_D^{0.9}} \right) \right]^2} \qquad (10.39)$$

It is reported that this equation predicts friction factors that differ by less than 3% from those on the Moody diagram for $4 \times 10^3 < \mathrm{Re}_D < 10^8$ and $10^{-5} < k_s/D < 2 \times 10^{-2}$.

✔ **CHECKPOINT PROBLEM 10.2**

Water (15°C) flows in a 100 m length of cast iron pipe. The pipe inside diameter is 0.15 m, and the mean velocity is 0.6 m/s.

   a. What is the value of Reynolds number?

   b. What is the value of k_s/D ?

   c. What is the value of f from the Moody diagram?

   d. What is the value of f from the Swamee-Jain correlation?

   e. What is the value of head loss?

10.7 Strategy for Solving Problems

Analyzing flow in conduits can be challenging because the equations often cannot be solved with algebra. Thus, this section presents a strategy.

   Conduit problems are solved with the energy equation together with equations for head-loss. Thus, the next checkpoint problem allows you to test your understanding of the energy equation.

✔ **CHECKPOINT PROBLEM 10.3**

The sketch shows an idealization of a garden hose of diameter D and length L connected to a pipe bib at a residence. Assume that the supply pressure p_s upstream of the valve is constant. Assume that the faucet valve has no head loss because it is fully open. Thus, the only head loss is in the garden hose.

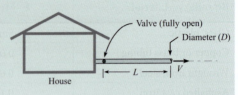

a. Derive an equation for the mean velocity V of the water in terms of the friction factor and other relevant variables.

b. How much will V change if L is doubled? Assume f remains constant.

Fig. 10.15 provides a strategy for problem solving. When flow is laminar, solutions are straightforward because head loss is linear with velocity V and the equations are simple enough to solve with algebra. When flow is turbulent, head loss is nonlinear with V and the equations are too complex to solve with algebra. Thus for turbulent flow, engineers use computer solutions or the traditional approach.

FIGURE 10.15

A strategy for solving conduit flow problems.

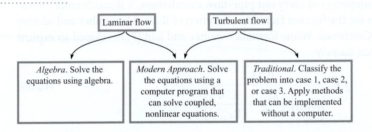

To solve a turbulent flow problem using the *traditional approach*, one classifies the problems into three cases:

Case 1 is when the goal is to find the *head loss*, given the pipe length, pipe diameter, and flow rate. This problem is straightforward because it can be solved using algebra; see Example 10.3.

Case 2 is when the goal is to find the *flow rate*, given the head loss (or pressure drop), the pipe length, and the pipe diameter. This problem usually requires an iterative approach. See Examples 10.4 and 10.5.

Case 3 is when the goal is to find the *pipe diameter*, given the flow rate, length of pipe, and head loss (or pressure drop). This problem usually requires an iterative approach; see Example 10.6.

There are several approaches that sometimes eliminate the need for an iterative approach. For case 2, an iterative approach can sometimes be avoided by using an explicit equation developed by Swamee and Jain (7):

$$Q = -2.22\, D^{5/2} \sqrt{gh_f/L} \, \log\left(\frac{k_s}{3.7\,D} + \frac{1.78\,\nu}{D^{3/2} \sqrt{gh_f/L}} \right) \tag{10.40}$$

Using Eq. (10.40) is equivalent to using the top of the Moody diagram, which presents a scale for $\text{Re}\, f^{1/2}$. For case 3, one can sometimes use an explicit equation developed by Swamee and Jain (7) and modified by Streeter and Wylie (8):

$$D = 0.66 \left[k_s^{1.25} \left(\frac{LQ^2}{gh_f} \right)^{4.75} + \nu Q^{9.4} \left(\frac{L}{gh_f} \right)^{5.2} \right]^{0.04} \tag{10.41}$$

Example 10.3 shows an example of a case 1 problem.

EXAMPLE 10.3

Head Loss in a Pipe (Case 1)

Problem Statement

Water ($T = 20°C$) flows at a rate of 0.05 m³/s in a 20 cm asphalted cast-iron pipe. What is the head loss per kilometer of pipe?

Define the Situation

Water is flowing in a pipe.

$L = 1000$ m

Water, 20°C
$Q = 0.05$ m³/s
$D = 0.2$ m
Asphalted, cast-iron

Assumptions: Fully developed flow

Properties: Water (20°C), Table A.5: $v = 1 \times 10^{-6}$ m²/s

State the Goal

Calculate the head loss (in meters) for $L = 1000$ m.

Generate Ideas and Make a Plan

Because this is a case 1 problem (head loss is the goal), the solution is straightforward.

1. Calculate the mean velocity using the flow rate equation.
2. Calculate the Reynolds number using Eq. (10.1).

3. Calculate the relative roughness and then look up f on the Moody diagram.
4. Find head loss by applying the Darcy-Weisbach equation (10.1).

Take Action (Execute the Plan)

1. Mean velocity

$$V = \frac{Q}{A} = \frac{0.05 \text{ m}^3/\text{s}}{(\pi/4)(0.02 \text{ m})^2} = 1.59 \text{ m/s}$$

2. Reynolds number

$$\text{Re}_D = \frac{VD}{v} = \frac{(1.59 \text{ m/s})(0.20 \text{ m})}{10^{-6} \text{ m}^2/\text{s}} = 3.18 \times 10^5$$

3. Resistance coefficient
   - Equivalent sand roughness (Table 10.4):
     $$k_s = 0.12 \text{ mm}$$
   - Relative roughness:
     $$k_s/D = (0.00012 \text{ m})/(0.2 \text{ m}) = 0.0006$$
   - Look up f on the Moody diagram for Re = 3.18×10^5 and $k_s/D = 0.0006$:
     $$f = 0.019$$

4. Darcy-Weisbach equation

$$h_f = f\left(\frac{L}{D}\right)\left(\frac{V^2}{2g}\right) = 0.019\left(\frac{1000 \text{ m}}{0.20 \text{ m}}\right)\left(\frac{1.59^2 \text{ m}^2/\text{s}^2}{2(9.81 \text{ m/s}^2)}\right)$$

$$= \boxed{12.2 \text{ m}}$$

Example 10.4 shows an example of a case 2 problem. Notice that the solution involves application of the scale on the top of the Moody diagram, thereby avoiding an iterative solution.

EXAMPLE 10.4

Flow Rate in a Pipe (Case 2)

Problem Statement

The head loss per kilometer of 20 cm asphalted cast-iron pipe is 12.2 m. What is the flow rate of water through the pipe?

Define the Situation

This is the same situation as Example 10.3 except that the head loss is now specified and the discharge is unknown.

State the Goal

Calculate the discharge (m³/s) in the pipe.

Generate Ideas and Make a Plan

This is a case 2 problem because flow rate is the goal. However, a direct (i.e., noniterative) solution is possible because head loss is specified. The strategy will be to use the horizontal scale on the top of the Moody diagram.

1. Calculate the parameter on the top of the Moody diagram.

2. Using the Moody diagram, find the friction factor f.
3. Calculate mean velocity using the Darcy-Weisbach equation (10.12).
4. Find discharge using the flow rate equation.

Take Action (Execute the Plan)

1. Compute the parameter $D^{3/2}\sqrt{2gh_f/L}/v$.

$$D^{3/2}\frac{\sqrt{2gh_f/L}}{v} = (0.20 \text{ m})^{3/2}$$

$$\times \frac{[2(9.81 \text{ m/s}^2)(12.2 \text{ m}/1000 \text{ m})]^{1/2}}{1.0 \times 10^{-6} \text{ m}^2/\text{s}}$$

$$= 4.38 \times 10^4$$

2. Determine resistance coefficient.
   - Relative roughness:
     $k_s/D = (0.00012 \text{ m})/(0.2 \text{ m}) = 0.0006$
   - Look up f on the Moody diagram for
     $D^{3/2}\sqrt{2gh_f/L}/v = 4.4 \times 10^4$ and $k_s/D = 0.0006$:
     $$f = 0.019$$

3. Find V using the Darcy-Weisbach equation.

$$h_f = f\left(\frac{L}{D}\right)\left(\frac{V^2}{2g}\right)$$

$$12.2 \text{ m} = 0.019\left(\frac{1000 \text{ m}}{0.20 \text{ m}}\right)\left(\frac{V^2}{2(9.81 \text{ m/s}^2)}\right)$$

$$V = 1.59 \text{ m/s}$$

4. Use flow rate equation to find discharge.

$$Q = VA = (1.59 \text{ m/s})(\pi/4)(0.2 \text{ m})^2 = \boxed{0.05 \text{ m}^3/\text{s}}$$

Review the Solution and the Process

Validation. The calculated flow rate matches the value from Example 10.3. This is expected because the data are the same.

When case 2 problems require iteration, several methods can be used to find a solution. One of the easiest ways is a method called "successive substitution," which is illustrated in Example 10.5.

EXAMPLE 10.5

Flow Rate in a Pipe (Case 2)

Problem Statement

Water ($T = 20°C$) flows from a tank through a 50 cm diameter steel pipe. Determine the discharge of water.

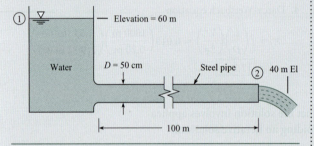

Define the Situation

Water is draining from a tank through a steel pipe.

Assumptions:

• Flow is fully developed.
• Include only the head loss in the pipe.

Properties:

• Water (20°C), Table A.5: $\nu = 1 \times 10^{-6} \text{ m}^2/\text{s}$.
• Steel pipe, Table 10.4, equivalent sand roughness: $k_s = 0.046 \text{ mm}$. Relative roughness (k_s/D) is 9.2×10^{-5}.

State the Goal

Find: Discharge (m^3/s) for the system.

Generate Ideas and Make a Plan

This is a case 2 problem because flow rate is the goal. An iterative solution is used because V is unknown, so there is no direct way to use the Moody diagram.

1. Apply the energy equation from section 1 to section 2.
2. First trial. Guess a value of f and then solve for V.

3. Second trial. Using V from the first trial, calculate a new value of f.

4. Convergence. If the value of f is constant within a few percent between trials, then stop. Otherwise, continue with more iterations.

5. Calculate flow rate using the flow rate equation.

Take Action (Execute the Plan)

1. Energy equation (reservoir surface to outlet)

$$\frac{p_1}{\gamma} + \frac{V_1^2}{2g} + z_1 = \frac{p_2}{\gamma} + \frac{V_2^2}{2g} + z_2 + h_L$$

$$0 + 0 + 60 = 0 + \frac{V_2^2}{2g} + 40 + f\frac{L}{D}\frac{V_2^2}{2g}$$

or

$$V = \left(\frac{2g \times 20}{1 + 200f}\right)^{1/2} \tag{a}$$

2. First trial (iteration 1)
   • Guess a value of $f = 0.020$.
   • Use Eq. (a) to calculate $V = 8.86$ m/s.
   • Use $V = 8.86$ m/s to calculate Re $= 4.43 \times 10^6$.
   • Use Re $= 4.43 \times 10^6$ and $k_s/D = 9.2 \times 10^{-5}$ on the Moody diagram to find that $f = 0.012$.
   • Use Eq. (a) with $f = 0.012$ to calculate $V = 10.7$ m/s.

3. Second trial (iteration 2)
   • Use $V = 10.7$ m/s to calculate $\text{Re}_D = 5.35 \times 10^6$.
   • Use $\text{Re}_D = 5.35 \times 10^6$ and $k_s/D = 9.2 \times 10^{-5}$ on the Moody diagram to find that $f = 0.012$.

4. Convergence. The value of $f = 0.012$ is unchanged between the first and second trials. Therefore, there is no need for more iterations.

5. Flow rate

$$Q = VA = (10.7 \text{ m/s}) \times (\pi/4) \times (0.50)^2 \text{ m}^2 = 2.10 \text{ m}^3/\text{s}$$

In a case 3 problem, derive an equation for diameter D and then use the method of successive substitution to find a solution. Iterative approaches, as illustrated in Example 10.6, can employ a spreadsheet program to perform the calculations.

EXAMPLE 10.6

Finding Pipe Diameter (Case 3)

Problem Statement

What size of asphalted cast-iron pipe is required to carry water (15°C) at a discharge of 0.08 m³/s and with a head loss of 1 m per 300 m of pipe?

Define the Situation

Water is flowing in a asphalted cast-iron pipe. $Q = 0.08$ m³/s.

Assumptions: Fully developed flow

Properties:

- Water (15°C), Table A.5: $\nu = 1.14 \times 10^{-6}$ m²/s
- Asphalted cast-iron pipe, Table 10.4, equivalent sand roughness: $k_s = 0.12$ mm

---

State the Goal

Calculate the pipe diameter so that head loss is 1 m per 300 m of pipe length.

Generate Ideas and Make a Plan

Because this is a case 3 problem (pipe diameter is the goal), use an iterative approach.

1. Derive an equation for pipe diameter by using the Darcy-Weisbach equation.

2. For iteration 1, guess f, solve for pipe diameter, and then recalculate f.

3. To complete the problem, build a table in a spreadsheet program.

Take Action (Execute the Solution)

1. Develop an equation to use for iteration.

- Darcy-Weisbach equation

$$h_f = f\left(\frac{L}{D}\right)\left(\frac{V^2}{2g}\right) = f\left(\frac{L}{D}\right)\left(\frac{Q^2/A^2}{2g}\right) = \frac{fLQ^2}{2g(\pi/4)^2 D^5}$$

- Solve for pipe diameter

$$D^5 = \frac{fLQ^2}{0.785^2(2gh_f)} \tag{a}$$

2. Iteration 1

- Guess $f = 0.015$.
- Solve for diameter using Eq. (a):

$$D^5 = \frac{0.015(300 \text{ m})(0.08 \text{ m}^3/\text{s})^2}{0.785^2(19.62 \text{ m/s}^2)(1 \text{ m})} = 2.38 \times 10^{-3} \text{ m}^5$$

$$D = 0.299 \text{ m}$$

- Find parameters needed for calculating f:

$$V = \frac{Q}{A} = \frac{0.08 \text{ m}^3/\text{s}}{(\pi/4)(0.299 \text{ m})^2} = 1.14 \text{ m/s}$$

$$\text{Re} = \frac{VD}{\nu} = \frac{(1.14 \text{ m/s})(0.299 \text{ m})}{1.14 \times 10^{-6} \text{ m}^2/\text{s}} = 3 \times 10^5$$

$$k_s/D = 0.12/(0.299 \times 10^3) = 0.0004$$

- Calculate f using Eq. (10.39): $f = 0.0178$.

3. In the table below, the first row contains the values from iteration 1. The value of $f = 0.0178$ from iteration 1 is used for the initial value for iteration 2. Notice how the solution has converged by iteration 2.

Iteration #	Initial f	D	V	Re	k_s/D	New f
		(m)	(m/s)			
1	0.0150	0.299	1.14	2.99E+05	4.00E−04	0.0177
2	0.0177	0.309	1.07	2.99E+05	3.88E−04	0.0176
3	0.0176	0.309	1.07	2.99E+05	3.88E−04	0.0176
4	0.0176	0.309	1.07	2.99E+05	3.88E−04	0.0176

Specify a pipe with a 30 cm inside diameter.

10.8 Combined Head Loss

Previous sections have described how to calculate head loss in pipes. This section completes the story by describing how to calculate head loss in components. This knowledge is essential for modeling and design of systems.

The Minor Loss Coefficient, K

When fluid flows through a component such as a partially open value or a bend in a pipe, viscous effects cause the flowing fluid to lose mechanical energy. For example, Fig. 10.16 shows flow through a "generic component." At section 2, the head of the flow will be less than at section 1. To characterize component head loss, engineers use a π-group called the **minor loss coefficient** K

$$K \equiv \frac{(\Delta h)}{(V^2/2g)} = \frac{(\Delta p_z)}{(\rho V^2/2)} \qquad (10.42)$$

where Δh is the drop in piezometric head that is caused by a component, Δp_z is the drop in pizeometric pressure, and V is mean velocity. As shown in Eq. (10.42), the minor loss coefficient has two useful interpretations:

$$K = \frac{\text{drop in piezometric head across component}}{\text{velocity head}} = \frac{\text{pressure drop due to component}}{\text{kinetic pressure}}$$

Thus, the head loss across a single component or transition is $h_L = K(V^2/(2g))$, where K is the minor loss coefficient for that component or transition.

Most values of K are found by experiment. For example, consider the setup shown in Fig. 10.17. To find K, flow rate is measured and mean velocity is calculated using $V = (Q/A)$. Pressure and elevation measurements are used to calculate the change in piezometric head.

$$\Delta h = h_2 - h_1 = \left(\frac{p_2}{\gamma} + z_2\right) - \left(\frac{p_1}{\gamma} + z_1\right) \qquad (10.43)$$

Then, values of V and Δh are used in Eq. (10.42) to calculate K. The next section presents typical data for K.

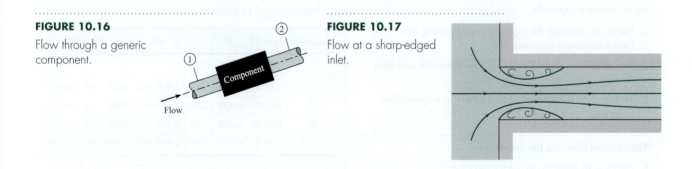

FIGURE 10.16

Flow through a generic component.

FIGURE 10.17

Flow at a sharp-edged inlet.

Data for the Minor Loss Coefficient

This section presents K data and relates these data to flow separation and wall shear stress. This information is useful for system modeling.

Pipe inlet. Near the entrance to a pipe when the entrance is rounded, flow is developing as shown in Fig. 10.3 and the wall shear stress is higher than that found in fully developed flow. Alternatively, if the pipe inlet is abrupt, or sharp-edged, as in Fig. 10.17, separation occurs just downstream of the entrance. Hence the streamlines converge and then diverge with consequent turbulence and relatively high head loss. The loss coefficient for the abrupt inlet is $K_e = 0.5$. This value is found in Table 10.5 using the row labeled "Pipe entrance" and the criteria of $r/d = 0.0$. Other values of head loss are summarized in Table 10.5.

TABLE 10.5 Loss Coefficients for Various Transitions and Fittings

Description	Sketch	Additional Data			K	Source
Pipe entrance $h_L = K_e V^2/2g$		r/d 0.0 0.1 >0.2			K_e 0.50 0.12 0.03	(10)[†]
Contraction $h_L = K_C V_2^2/2g$		D_2/D_1 0.00 0.20 0.40 0.60 0.80 0.90	K_C $\theta = 60°$ 0.08 0.08 0.07 0.06 0.06 0.06		K_C $\theta = 180°$ 0.50 0.49 0.42 0.27 0.20 0.10	(10)
Expansion $h_L = K_E V_1^2/2g$		D_1/D_2 0.00 0.20 0.40 0.60 0.80	K_E $\theta = 20°$ 0.30 0.25 0.15 0.10		K_E $\theta = 180°$ 1.00 0.87 0.70 0.41 0.15	(9)
90° miter bend		Without vanes			$K_b = 1.1$	(15)
90° smooth bend		With vanes r/d 1 2 4 6 8 10			$K_b = 0.2$ $K_b = 0.35$ 0.19 0.16 0.21 0.28 0.32	(15) (16) and (9)

Threaded pipe fittings	Globe valve—wide open
	Angle valve—wide open
	Gate valve—wide open
	Gate valve—half open
	Return bend
	Tee
	   Straight-through flow
	   Side-outlet flow
	90° elbow
	45° elbow
					$K_v = 10.0$
					$K_v = 5.0$
					$K_v = 0.2$
					$K_v = 5.6$
					$K_b = 2.2$
					
					$K_t = 0.4$
					$K_t = 1.8$
					$K_b = 0.9$
					$K_b = 0.4$
						(15)

[†]Reprinted by permission of the American Society of Heating, Refrigerating and Air Conditioning Engineers, Atlanta, Georgia, from the 1981 *ASHRAE Handbook—Fundamentals.*

FIGURE 10.18

Flow pattern in an elbow.

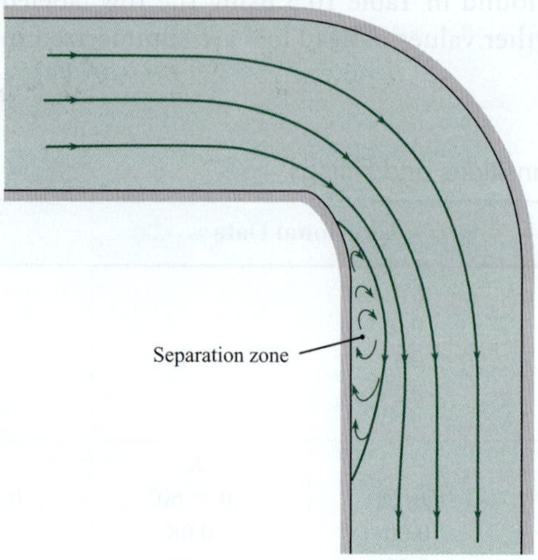

Separation zone

Flow in an elbow. In an elbow (90° smooth bend), considerable head loss is produced by secondary flows and by separation that occurs near the inside of the bend and downstream of the midsection as shown in Fig. 10.18.

The loss coefficient for an elbow at high Reynolds numbers depends primarily on the shape of the elbow. For a very short-radius elbow, the loss coefficient is quite high. For larger-radius elbows, the coefficient decreases until a minimum value is found at an r/d value of about 4 (see Table 10.5). However, for still larger values of r/d, an increase in loss coefficient occurs because the elbow itself is significantly longer.

Other components. The loss coefficients for a number of other fittings and flow transitions are given in Table 10.5. This table is representative of engineering practice. For more extensive tables, see references (10–15).

In Table 10.5, the K was found by experiment, so one must be careful to ensure that Reynolds number values in the application correspond to the Reynolds number values used to acquire the data.

Combined Head Loss Equation

The total head loss is given by Eq. (10.4), which is repeated here:

$$\{\text{Total head loss}\} = \{\text{Pipe head loss}\} + \{\text{Component head loss}\} \tag{10.44}$$

To develop an equation for the combined head loss, substitute Eqs. (10.12) and (10.42) in Eq. (10.44):

$$h_L = \sum_{\text{pipes}} f \frac{L}{D} \frac{V^2}{2g} + \sum_{\text{components}} K \frac{V^2}{2g} \tag{10.45}$$

Equation (10.45) is called the *combined head loss equation*. To apply this equation, follow the same approaches that were used for solving pipe problems. That is, classify the flow as case 1, 2, or 3 and apply the usual equations: the energy, Darcy-Weisbach, and flow rate equations. Example 10.7 illustrates this approach for a case 1 problem.

EXAMPLE 10.7

Pipe System with Combined Head Loss

Problem Statement

If oil ($v = 4 \times 10^{-5}$ m^2/s; $S = 0.9$) flows from the upper to the lower reservoir at a rate of 0.028 m^3/s in the 15 cm smooth pipe, what is the elevation of the oil surface in the upper reservoir?

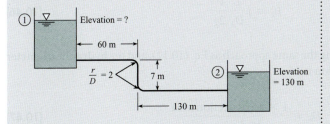

Define the Situation

Oil is flowing from a upper reservoir to a lower reservoir.

Properties:

- Oil: $v = 4 \times 10^{-5}$ m^2/s, $S = 0.9$
- Minor head loss coefficients, Table 10.5: entrance = $K_e = 0.5$; bend = $K_b = 0.19$; outlet = $K_E = 1.0$

State the Goal

Calculate the elevation (in meters) of the free surface of the upper reservoir.

Generate Ideas and Make a Plan

This is a case 1 problem because flow rate and pipe dimensions are known. Thus, the solution is straightforward.

1. Apply the energy equation from 1 to 2.
2. Apply the combined head loss equation (10.45).
3. Develop an equation for z_1 by combining results from steps 1 and 2.
4. Calculate the resistance coefficient f.
5. Solve for z_1 using the equation from step 3.

Take Action (Execute the Plan)

1. Energy equation and term-by-term analysis

$$\frac{p_1}{\gamma} + \alpha_1 \frac{\overline{V}_1^2}{2g} + z_1 + h_p = \frac{p_2}{\gamma} + \alpha_2 \frac{\overline{V}_2^2}{2g} + z_2 + h_t + h_L$$

$$0 + 0 + z_1 + 0 = 0 + 0 + z_2 + 0 + h_L$$

$$z_1 = z_2 + h_L$$

Interpretation: Change in elevation head is balanced by the total head loss.

2. Combined head loss equation

$$h_L = \sum_{pipes} f \frac{L}{D} \frac{V^2}{2g} + \sum_{components} K \frac{V^2}{2g}$$

$$h_L = f \frac{L}{D} \frac{V^2}{2g} + \left(2K_b \frac{V^2}{2g} + K_e \frac{V^2}{2g} + K_E \frac{V^2}{2g} \right)$$

$$= \frac{V^2}{2g} \left(f \frac{L}{D} + 2K_b + K_e + K_E \right)$$

3. Combine eqs. (1) and (2).

$$z_1 = z_2 + \frac{V^2}{2g} \left(f \frac{L}{D} + 2K_b + K_e + K_E \right)$$

4. Resistance coefficient
- Flow rate equation

$$V = \frac{Q}{A} = \frac{(0.028 \text{ m}^3/\text{s})}{(\pi/4)(0.15 \text{ m})^2} = 1.58 \text{ m/s}$$

- Reynolds number

$$\text{Re}_D = \frac{VD}{v} = \frac{1.58 \text{ m/s}(0.15 \text{ m})}{4 \times 10^{-5} \text{ m}^2/\text{s}} = 5.93 \times 10^3$$

Thus, flow is turbulent.
- Swamee-Jain equation (10.39)

$$f = \frac{0.25}{\left[\log_{10} \left(\frac{k_s}{3.7D} + \frac{5.74}{\text{Re}^{0.9}} \right) \right]^2} = \frac{0.25}{\left[\log_{10} \left(0 + \frac{5.74}{5930^{0.9}} \right) \right]^2} = 0.036$$

5. Calculate z_1 using the equation from step (3):

$$z_1 = (130 \text{ m}) + \frac{(1.58 \text{ m/s})^2}{2(9.81) \text{m/s}^2}$$

$$\left(0.036 \frac{(197 \text{ m})}{(0.15 \text{ m})} + 2(0.19) + 0.5 + 1.0 \right)$$

$$\boxed{z_1 = 136 \text{ m}}$$

Review the Solution and the Process

1. *Discussion.* Notice the difference is the magnitude of the pipe head loss versus the magnitude of the component head loss:

$$\text{Pipe head loss} \sim \Sigma f \frac{L}{D} = 0.036 \frac{(197 \text{ m})}{(0.15 \text{ m})} = 47.2$$

Component head loss $\sim \Sigma K = 2(0.19) + 0.5 + 1.0 = 1.88$

Thus pipe losses $\gg$ component losses for this problem.

2. *Skill.* When pipe head loss is dominant, make simple estimates of K because these estimates will not impact the prediction very much.

10.9 Nonround Conduits

Previous sections have considered round pipes. This section extends this information by describing how to account for conduits that are square, triangular, or any other nonround shape. This information is important for applications such as sizing ventilation ducts in buildings and for modeling flow in open channels.

When a conduit is noncircular, then engineers modify the Darcy-Weisbach equation, Eq. (10.12), to use hydraulic diameter D_h in place of diameter.

$$h_L = f \frac{L}{D_h} \frac{V^2}{2g} \tag{10.46}$$

Equation (10.46) is derived using the same approach as Eq. (10.12), and the **hydraulic diameter** that emerges from this derivation is

$$D_h \equiv \frac{4 \times \text{cross-section area}}{\text{wetted perimeter}} \tag{10.47}$$

where the "wetted perimeter" is that portion of the perimeter that is physically touching the fluid. The wetted perimeter of a rectangular duct of dimension $L \times w$ is $2L + 2w$. Thus, the hydraulic diameter of this duct is:

$$D_h \equiv \frac{4 \times Lw}{2L + 2w} = \frac{2Lw}{L + w}$$

Using Eq. (10.47), the hydraulic diameter of a round pipe is the pipe's diameter D. When Eq. (10.46) is used to calculate head loss, the resistance coefficient f is found using a Reynolds number based on hydraulic diameter. Use of hydraulic diameter is an approximation. According to White (20), this approximation introduces an uncertainty of 40% for laminar flow and 15% for turbulent flow.

$$f = \left(\frac{64}{\text{Re}_{D_h}} \right) \pm 40\% \text{ (laminar flow)}$$

$$f = \frac{0.25}{\left[\log_{10} \left(\frac{k_s}{3.7 D_h} + \frac{5.74}{\text{Re}_{D_h}^{0.9}} \right) \right]^2} \pm 15\% \text{ (turbulent flow)} \tag{10.48}$$

In addition to hydraulic diameter, engineers also use **hydraulic radius**, which is defined as

$$R_h \equiv \frac{\text{section area}}{\text{wetted perimeter}} = \frac{D_h}{4} \tag{10.49}$$

Notice that the ratio of R_h to D_h is 1/4 instead of 1/2. Although this ratio is not logical, it is the convention used in the literature and is useful to remember. Chapter 15, which focuses on open-channel flow, will present examples of hydraulic radius.

Summary. To model flow in a nonround conduit, the approaches of the previous sections are followed with the only difference being the use of hydraulic diameter in place of diameter. This is illustrated by Example 10.8.

EXAMPLE 10.8

Pressure Drop in an HVAC Duct

Problem Statement

Air ($T = 20°C$ and $p = 101$ kPa absolute) flows at a rate of 2.5 m³/s in a horizontal, commercial steel, HVAC duct. (Note that HVAC is an acronym for heating, ventilating, and air conditioning.) What is the pressure drop in millimeters of water per 50 m of duct?

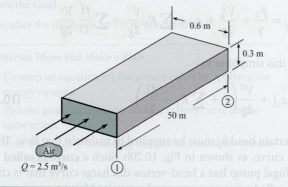

Define the Situation

Air is flowing through a duct.

Assumptions:

- Fully developed flow, meaning that $V_1 = V_2$. Thus, the velocity head terms in the energy equation cancel out.
- No sources of component head loss.

Properties:

- Air (20°C, 1 atm, Table A.2:) $\rho = 1.2$ kg/m³, $\nu = 15.1 \times 10^{-6}$ m²/s
- Steel pipe, Table 10.4: $k_s = 0.046$ mm

State the Goal

Find: Pressure drop (mm H_2O) in a length of 50 m.

Generate Ideas and Make a Plan

This is a case 1 problem because flow rate and duct dimensions are known. Thus, the solution is straightforward.

1. Derive an equation for pressure drop by using the energy equation.
2. Calculate parameters needed to find head loss.
3. Calculate head loss by using the Darcy-Weisbach equation (10.12).
4. Calculate pressure drop Δp by combining steps 1, 2, and 3.

Take Action (Execute the Plan)

1. Energy equation (after term-by-term analysis)

$$p_1 - p_2 = \rho g h_L$$

2. Intermediate calculations
   - Flow rate equation

$$V = \frac{Q}{A} = \frac{2.5 \text{ m}^3/\text{s}}{(0.3 \text{ m})(0.6 \text{ m})} = 13.9 \text{ m/s}$$

   - Hydraulic diameter

$$D_h \equiv \frac{4 \times \text{section area}}{\text{wetted perimeter}} = \frac{4(0.3 \text{ m})(0.6 \text{ m})}{(2 \times 0.3 \text{ m}) + (2 \times 0.6 \text{ m})} = 0.4 \text{ m}$$

   - Reynolds number

$$\text{Re} = \frac{VD_h}{\nu} = \frac{(13.9 \text{ m/s})(0.4 \text{ m})}{(15.1 \times 10^{-6} \text{ m}^2/\text{s})} = 368,000$$

   Thus, flow is turbulent.
   - Relative roughness

$$k_s/D_h = (0.000046 \text{ m})/(0.4 \text{ m}) = 0.000115$$

   - Resistance coefficient (Moody diagram): $f = 0.015$

3. Darcy-Weisbach equation

$$h_f = f\left(\frac{L}{D_h}\right)\left(\frac{V^2}{2g}\right) = 0.015\left(\frac{50 \text{ m}}{0.4 \text{ m}}\right)\left\{\frac{(13.9 \text{ m/s})^2}{2(9.81 \text{ m/s}^2)}\right\}$$

$$= 18.6 \text{ m}$$

4. Pressure drop (from step 1)

$$p_1 - p_2 = \rho g h_L = (1.2 \text{ kg/m}^3)(9.81 \text{ m/s}^2)(18.6 \text{ m}) = 220 \text{ Pa}$$

$$\boxed{p_1 - p_2 = 22.4 \text{ mm } H_2O}$$

10.10 Pumps and Systems of Pipes

This section explains how to model flow in a network of pipes and how to incorporate performance data for a centrifugal pump. These topics are important because pumps and pipe networks are common.

No matter which pipe is involved, the pressure difference between the two junction points is the same. Also, the elevation difference between the two junction points is the same. Because $h_L = (p_1/\gamma + z_1) - (p_2/\gamma + z_2)$, it follows that h_L between the two junction points is the same in both of the pipes of the parallel pipe system. Thus,

$$h_{L_1} = h_{L_2}$$

$$f_1 \frac{L_1}{D_1} \frac{V_1^2}{2g} = f_2 \frac{L_2}{D_2} \frac{V_2^2}{2g}$$

Then

$$\left(\frac{V_1}{V_2}\right)^2 = \frac{f_2 L_2 D_1}{f_1 L_1 D_2} \quad \text{or} \quad \frac{V_1}{V_2} = \left(\frac{f_2 L_2 D_1}{f_1 L_1 D_2}\right)^{1/2}$$

If f_1 and f_2 are known, the division of flow can be easily determined. However, some trial-and-error analysis may be required if f_1 and f_2 are in the range where they are functions of the Reynolds number.

Pipe Networks

The most common pipe networks are the water distribution systems for municipalities. These systems have one or more sources (discharges of water into the system) and numerous loads: one for each household and commercial establishment. For purposes of simplification, the loads are usually lumped throughout the system. Figure 10.22 shows a simplified distribution system with two sources and seven loads.

FIGURE 10.22

Pipe network.

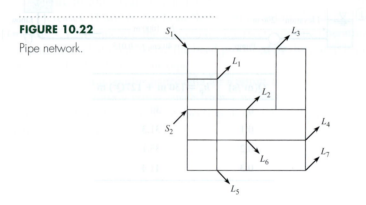

The engineer is often engaged to design the original system or to recommend an economical expansion to the network. An expansion may involve additional housing or commercial developments, or it may be to handle increased loads within the existing area.

In the design of such a system, the engineer will have to estimate the future loads for the system and will need to have sources (wells or direct pumping from streams or lakes) to satisfy the loads. Also, the layout of the pipe network must be made (usually parallel to streets), and pipe sizes will have to be determined. The object of the design is to arrive at a network of pipes that will deliver the design flow at the design pressure for minimum cost. The cost will include first costs (materials and construction) as well as maintenance and operating costs. The design process usually involves a number of iterations on pipe sizes and layouts before the optimum design (minimum cost) is achieved.

So far as the fluid mechanics of the problem are concerned, the engineer must determine pressures throughout the network for various conditions—that is, for various combinations of pipe sizes, sources, and loads. The solution of a problem for a given layout and a given set of sources and loads requires that two conditions be satisfied:

1. The continuity equation must be satisfied. That is, the flow into a junction of the network must equal the flow out of the junction. This must be satisfied for all junctions.

2. The head loss between any two junctions must be the same regardless of the path in the series of pipes taken to get from one junction point to the other. This requirement results because pressure must be continuous throughout the network (pressure cannot have two values at a given point). This condition leads to the conclusion that the algebraic sum of head losses around a given loop must be equal to zero. Here the sign (positive or negative) for the head loss in a given pipe is given by the sense of the flow with respect to the loop, that is, whether the flow has a clockwise or counterclockwise direction.

At one time, these solutions were made by trial-and-error hand computation, but computers have made the older methods obsolete. Even with these advances, however, the engineer charged with the design or analysis of such a system must understand the basic fluid mechanics of the system to be able to interpret the results properly and to make good engineering decisions based on the results. Therefore, an understanding of the original method of solution by Hardy Cross (17) may help you to gain this basic insight. The Hardy Cross method is as follows.

The engineer first distributes the flow throughout the network so that loads at various nodes are satisfied. In the process of distributing the flow through the pipes of the network, the engineer must be certain that continuity is satisfied at all junctions (flow into a junction equals flow out of the junction), thus satisfying requirement 1. The first guess at the flow distribution obviously will not satisfy requirement 2 regarding head loss; therefore, corrections are applied. For each loop of the network, a discharge correction is applied to yield a zero net head loss around the loop. For example, consider the isolated loop in Fig. 10.23. In this loop, the loss of head in the clockwise direction will be given by

$$\sum h_{L_c} = h_{L_{AB}} + h_{L_{BC}}$$
$$= \sum k Q_c^n \qquad \text{(10.51)}$$

FIGURE 10.23

A typical loop of a pipe network.

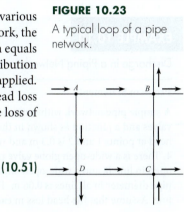

The loss of head for the loop in the counterclockwise direction is

$$\sum h_{L_{cc}} = \sum_{cc} k Q_{cc}^n \qquad \text{(10.52)}$$

For a solution, the clockwise and counterclockwise head losses have to be equal, or

$$\sum h_{L_c} = \sum h_{L_{cc}}$$
$$\sum k Q_c^n = \sum k Q_{cc}^n$$

As noted, the first guess for flow in the network will undoubtedly be in error; therefore, a correction in discharge, ΔQ, will have to be applied to satisfy the head loss requirement. If the clockwise head loss is greater than the counterclockwise head loss, ΔQ will have to be applied in the counterclockwise direction. That is, subtract ΔQ from the clockwise flows and add it to the counterclockwise flows:

$$\sum k(Q_c - \Delta Q)^n = \sum k(Q_{cc} + \Delta Q)^n \qquad \text{(10.53)}$$

Expand the summation on either side of Eq. (10.53) and include only two terms of the expansion:

$$\sum k(Q_c^n - nQ_c^{n-1}\Delta Q) = \sum k(Q_{cc}^n + nQ_{cc}^{n-1}\Delta Q)$$

Solve for ΔQ:

$$\Delta Q = \frac{\sum kQ_c^n - \sum kQ_{cc}^n}{\sum nkQ_c^{n-1} + \sum nkQ_{cc}^{n-1}} \tag{10.54}$$

Thus if ΔQ as computed from Eq. (10.54) is positive, the correction is applied in a counterclockwise sense (add ΔQ to counterclockwise flows and subtract it from clockwise flows).

A different ΔQ is computed for each loop of the network and applied to the pipes. Some pipes will have two ΔQs applied because they will be common to two loops. The first set of corrections usually will not yield the final desired result because the solution is approached only by successive approximations. Thus the corrections are applied successively until the corrections are negligible. Experience has shown that for most loop configurations, applying ΔQ as computed by Eq. (10.54) produces too large a correction. Fewer trials are required to solve for Qs if approximately 0.6 of the computed ΔQ is used.

More information on methods of solution of pipe networks is available in references (18) and (19). A search of the Internet under "pipe networks" yields information on software available from various sources.

EXAMPLE 10.10

Discharge in a Piping Network

Problem Statement

A simple pipe network with water flow consists of three valves and a junction as shown in the figure. The piezometric head at points 1 and 2 is 0.3 m and reduces to zero at point 4. There is a wide-open globe valve in line A, a gate valve half open in line B, and a wide-open angle valve in line C. The pipe diameter in all lines is 0.06 m. Find the flow rate in each line. Assume that the head loss in each line is due only to the valves.

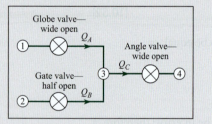

Define the Situation

Water flows through a network of pipes.
- $h_1 = h_2 = 0.3$ m.
- $h_4 = 0$ m.
- Pipe diameter (all pipes) is 0.06 m.

Assumptions: Head loss is due to valves only.

State the Goal

Find the flow rate (in m³/s) in each pipe.

Generate Ideas and Make a Plan

1. Let $h_{L,1\to3} = h_{L,2\to3}$.
2. Let $h_{L,2\to4} = 0.3$ m.
3. Solve equations using the Hardy Cross approach.

Take Action (Execute the Plan)

The piezometric heads at points 1 and 2 are equal, so

$$h_{L,1\to3} + h_{L,3\to2} = 0$$

The head loss between points 2 and 4 is 0.3 m, so

$$h_{L,2\to3} + h_{L,3\to4} = 0$$

Continuity must be satisfied at point 3, so

$$Q_A + Q_B = Q_C$$

The head loss through a valve is given by

$$h_L = K_V \frac{V^2}{2g}$$

$$= K_V \frac{1}{2g}\left(\frac{Q}{A}\right)^2$$

where K_V is the loss coefficient. For a 0.06 m pipe, the head loss becomes

$$h_L = 6375.5 K_v Q^2$$

where h_L is in meter and Q is in m³/s.

The head loss equation between points 1 and 2 expressed in term of discharge is

$$6375.5 K_A Q_A^2 - 6375.5 K_B Q_B^2 = 0$$

or

$$K_A Q_A^2 - K_B Q_B^2 = 0$$

where K_A is the loss coefficient for the wide-open globe valve ($K_A = 10$) and K_B is the loss coefficient for the half-open gate valve ($K_B = 5.6$). The head loss equation between points 2 and 4 is

$$6375.5 K_B Q_B^2 + 6375.5 K_C Q_C^2 = 0.3$$

or

$$K_B Q_B^2 + K_C Q_C^2 = 4.7 \times 10^{-5}$$

where K_C is the loss coefficient for the wide-open angle valve ($K_C = 5$). The two head loss equations and the continuity equation comprise three equations for Q_A, Q_B, and Q_C. However, the equations are nonlinear and require linearization and solution by iteration (Hardy Cross approach). The discharge is written as

$$Q = Q_0 + \Delta Q$$

where Q_0 is the starting value and ΔQ is the change. Then

$$Q^2 \cdot Q_0^2 + 2Q_0 \Delta Q$$

where the $(\Delta Q)^2$ term is neglected. The equations in terms of ΔQ become

$$2K_A Q_{0,A} \Delta Q_A - 2K_B Q_{0,B} \Delta Q_B = K_B Q_{0,B}^2 - K_A Q_{0,A}^2$$

$$2K_C Q_{0,C} \Delta Q_C - 2K_B Q_{0,B} \Delta Q_B = 0.0307 - K_B Q_{0,B}^2 - K_C Q_{0,C}^2$$

$$\Delta Q_A + \Delta Q_B = \Delta Q_C$$

which can be expressed in matrix form as

$$\begin{bmatrix} 2K_A Q_{0,A} & -2K_B Q_{0,B} & 0 \\ 0 & 2K_B Q_{0,B} & 2K_C Q_{0,C} \\ 1 & 1 & -1 \end{bmatrix} \begin{Bmatrix} \Delta Q_A \\ \Delta Q_B \\ \Delta Q_C \end{Bmatrix}$$

$$= \begin{bmatrix} K_B Q_{0,B}^2 - K_A Q_{0,A}^2 \\ 4.7 \times 10^{-5} - K_B Q_{0,B}^2 - K_C Q_{0,C}^2 \\ 0 \end{bmatrix}$$

The procedure begins by selecting values for $Q_{0,A}$, $Q_{0,B}$, and $Q_{0,C}$. Assume $Q_{0,A} = Q_{0,B}$ and $Q_{0,C} = 2Q_{0,A}$. Then from the head loss equation from points 2 to 4

$$K_B Q_{0,B}^2 + K_C Q_{0,C}^2 = 4.7 \times 10^{-5}$$

$$(K_B + 4K_C)Q_{0,B}^2 = 4.7 \times 10^{-5}$$

$$(5.6 + 4 \times 5)Q_{0,B}^2 = 4.7 \times 10^{-5}$$

$$Q_{0,B} = 1.35 \times 10^{-3}$$

and $Q_{0,A} = 1.35 \times 10^{-3}$ and $Q_{0,C} = 2.7 \times 10^{-3}$. These values are substituted into the matrix equation to solve for the ΔQ's. The discharges are corrected by $Q_0^{new} = Q_0^{old} + \Delta Q$ and substituted into the matrix equation again to yield new ΔQ's. The iterations are continued until sufficient accuracy is obtained. The accuracy is judged by how close the column matrix on the right approaches zero.

Review the Solution and the Process

Knowledge. This solution technique is called the Newton-Raphson method. This method is useful for nonlinear systems of algebraic equations. It can be implemented easily on a computer. The solution procedure for more complex systems is the same.

10.11 Key Knowledge

Classifying Flow in Conduits

- A *conduit* is any pipe, tube, or duct that is filled with a flowing fluid.

- Flow in a conduit is characterized using the Reynolds number based on pipe diameter. This π-group is given by several equivalent formulas

$$\mathrm{Re}_D = \frac{VD}{\nu} = \frac{\rho VD}{\mu} = \frac{4Q}{\pi D \nu} = \frac{4\dot{m}}{\pi D \mu}$$

- To classify a flow as *laminar* or *turbulent*, calculate the Reynolds number

$$\mathrm{Re}_D \leq 2000 \qquad \text{laminar flow}$$

$$\mathrm{Re}_D \geq 3000 \qquad \text{turbulent flow}$$

- Flow in a conduit can be developing or fully developed

  ▶ *Developing flow* occurs near an entrance or after the flow is disrupted (i.e., downstream of a valve, a bend, an orifice). *Developing flow* means that the velocity profile and wall shear stress are changing with axial location.

  ▶ *Fully developed flow* occurs in straight runs of pipe that are long enough to allow the flow to develop. Fully developed flow means that the velocity profile and the shear stress are constant with axial location *x*. In fully developed flow, the flow is uniform, and the pressure gradient (*dp/dx*) is constant.

- To classify a flow at a pipe inlet as developing or fully developed, calculate the *entrance length* (L_e). At any axial location greater than L_e, the flow will be fully developed. The equations for entrance length are

$$\frac{L_e}{D} = 0.05 \text{Re}_D \qquad \text{(laminar flow: Re}_D \leq 2000)$$

$$\frac{L_e}{D} = 50 \qquad \text{(turbulent flow: Re}_D \geq 3000)$$

- To describe commercial pipe in the NPS system, specify a nominal diameter in inches and a schedule number. The schedule number characterizes the wall thickness. Actual dimensions need to be looked up.

Head Loss (Pipe Head Loss)

- The sum of head losses in a piping system is called *total head loss*. Sources of head loss classify into two categories:

  ▶ *Pipe Head Loss.* Head loss in straight runs of pipe with fully developed flow

  ▶ *Component Head Loss.* Head loss in components and transitions such as valves, elbows, and bends

- To characterize *pipe head loss*, engineers use a π-group called the *friction factor*. The friction factor *f* gives the ratio of wall shear stress ($4\tau_0$) to kinetic pressure ($\rho V^2/2$).

- Pipe head loss has two symbols that are used: h_L or h_f. To predict pipe head loss, apply the Darcy-Weisbach equation (DWE)

$$h_L = h_f = f \frac{L}{D} \frac{V^2}{2g}$$

There are three methods for using the DWE:

▶ *Method 1 (laminar flow).* Apply the DWE in this form

$$h_f = \frac{32 \mu L V}{\gamma D^2}$$

▶ *Method 2 (laminar or turbulent flow).* Apply the DWE and use a formula for *f*.

$$f = \frac{64}{\text{Re}} \qquad\qquad \text{Laminar flow}$$

$$f = \frac{0.25}{\left[\log_{10} \left(\dfrac{k_s}{3.7D} + \dfrac{5.74}{\text{Re}_D^{0.9}} \right) \right]^2} \qquad\qquad \text{Turbulent flow}$$

▶ *Method 3 (laminar or turbulent flow).* Apply the DWE; and look up *f* on the Moody diagram.

- The *roughness* of a pipe wall sometimes affects the friction factor
  - ▶ *Laminar Flow.* The roughness does not matter; the friction factor f is independent of roughness.
  - ▶ *Turbulent Flow.* The roughness is characterized by looking up an equivalent sand roughness height k_s and then finding f as a function of Reynolds number and k_s/D. When the flow is *fully turbulent,* then f is independent of Reynolds number.

Head Loss (Component Head Loss)

- To characterize the head loss in a component, engineers use a π-group called the *minor loss coefficient, K,* which gives the ratio of head loss to velocity head. Values of K, which come from experimental studies, are tabulated in engineering references. Each component has a specific value of K, which is looked up. The head loss for a component is

$$h_L = K_{\text{component}} \frac{V^2}{2g}$$

- The *total head loss* in a pipe is given by

Overall (total) head loss $= \sum (\text{Pipe head losses}) + \sum (\text{Component head losses})$

$$h_L = \sum_{\text{pipes}} f \frac{L}{D} \frac{V^2}{2g} + \sum_{\text{components}} K \frac{V^2}{2g}$$

Additional Useful Results

- Noncircular conduits can be analyzed using the hydraulic diameter D_h or the hydraulic radius (R_h). To analyze a noncircular conduit, apply the same equations that are used for round conduits and replace D with D_h in the formulas. The equations for D_h and R_h are

$$D_h = 4R_h = \frac{4 \times \text{section area}}{\text{wetted perimeter}}$$

- To find the operating point of a centrifugal pump in a system, the traditional approach is a graphical solution. One plots a system curve that is derived using the energy equation, and one plots the head versus flow rate curve of the centrifugal pump. The intersection of these two curves gives the operating point of the system.

- The analysis of pipe networks is based on the continuity equation being satisfied at each junction and the head loss between any two junctions being independent of pipe path between the two junctions. A series of equations based on these principles are solved iteratively to obtain the flow rate in each pipe and the pressure at each junction in the network.

REFERENCES

1. Reynolds, O. "An Experimental Investigation of the Circumstances Which Determine Whether the Motion of Water Shall Be Direct or Sinuous and of the Law of Resistance in Parallel Channels." *Phil. Trans. Roy. Soc. London,* 174, part III (1883).

2. Schlichting, Hermann. *Boundary Layer Theory,* 7th ed. New York: McGraw-Hill, 1979.

3. Moody, Lewis F. "Friction Factors for Pipe Flow." *Trans. ASME,* 671 (November 1944).

4. Nikuradse, J. "Strömungsgesetze in rauhen Rohren." *VDI-Forschungsh.,* no. 361 (1933). Also translated in *NACA Tech. Memo,* 1292.

5. White, F. M. *Viscous Fluid Flow.* New York: McGraw-Hill, 1991.

6. Colebrook, F. "Turbulent Flow in Pipes with Particular Reference to the Transition Region between the Smooth and Rough Pipe Laws." *J. Inst. Civ. Eng.,* vol. 11, 133–156 (1939).

7. Swamee, P. K., and A. K. Jain. "Explicit Equations for Pipe-Flow Problems." *J. Hydraulic Division of the ASCE,* vol. 102, no. HY5 (May 1976).

8. Streeter, V. L., and E. B. Wylie. *Fluid Mechanics,* 7th ed. New York: McGraw-Hill, 1979.

9. Barbin, A. R., and J. B. Jones. "Turbulent Flow in the Inlet Region of a Smooth Pipe." *Trans. ASME, Ser. D: J. Basic Eng.,* vol. 85, no. 1 (March 1963).

10. ASHRAE. *ASHRAE Handbook—1977 Fundamentals.* New York: Am. Soc. of Heating, Refrigerating and Air Conditioning Engineers, Inc., 1977.

11. Crane Co. "Flow of Fluids Through Valves, Fittings and Pipe." Technical Paper No. 410, Crane Co. (1988), 104 N. Chicago St., Joliet, IL 60434.

12. Fried, Irwin, and I. E. Idelchik. *Flow Resistance: A Design Guide for Engineers.* New York: Hemisphere, 1989.

13. Hydraulic Institute. *Engineering Data Book,* 2nd ed., Hydraulic Institute, 30200 Detroit Road, Cleveland, OH 44145.

14. Miller, D. S. *Internal Flow—A Guide to Losses in Pipe and Duct Systems.* British Hydrodynamic and Research Association (BHRA), Cranfield, England (1971).

15. Streeter, V. L. (ed.) *Handbook of Fluid Dynamics.* New York: McGraw-Hill, 1961.

16. Beij, K. H. "Pressure Losses for Fluid Flow in 90% Pipe Bends." *J. Res. Nat. Bur. Std.,* 21 (1938). Information cited in Streeter (20).

17. Cross, Hardy. "Analysis of Flow in Networks of Conduits or Conductors." *Univ. Illinois Bull.,* 286 (November 1936).

18. Hoag, Lyle N., and Gerald Weinberg. "Pipeline Network Analysis by Digital Computers." *J. Am. Water Works Assoc.,* 49 (1957).

19. Jeppson, Roland W. *Analysis of Flow in Pipe Networks.* Ann Arbor, MI: Ann Arbor Science Publishers, 1976.

20. White, F. M. *Fluid Mechanics,* 5th Ed. New York: McGraw-Hill, 2003.

PROBLEMS

PLUS Problem available in *WileyPLUS* at instructor's discretion.

GO Guided Online (GO) Problem, available in *WileyPLUS* at instructor's discretion.

Notes on Pipe Diameter for Chapter 10 Problems

When a pipe diameter is given using the label "NPS" or "nominal," find the dimensions using Table 10.1 on p. 363 of §10.2. Otherwise, assume the specified diameter is an inside diameter (ID).

Classifying Flow (§10.1)

10.1 **PLUS** Kerosene (20°C) flows at a rate of 0.02 m³/s in a 17.7-cm-diameter pipe. Would you expect the flow to be laminar or turbulent? Calculate the entrance length.

10.2 **PLUS** A compressor draws 0.3 m³/s of ambient air (20°C) in from the outside through a round duct that is 10 m long and 150 mm in diameter. Determine the entrance length and establish whether the flow is laminar or turbulent.

10.3 Design a lab demo for laminar flow. Specify the diameter and length for a tube that carries SAE 10W-30 oil at 38°C so that the system demonstrates laminar flow, and fully developed flow, with a discharge of $Q = 0.1$ L/s.

Darcy-Weisbach Equation (§10.3)

10.4 Using §10.3 and other resources, answer the following questions. Strive for depth, clarity, and accuracy while also combining sketches, words, and equations in ways that enhance the effectiveness of your communication.

   a. What is pipe head loss? How is pipe head loss related to total head loss?

   b. What is the friction factor f? How is f related to wall shear stress?

   c. What assumptions need to be satisfied to apply the Darcy-Weisbach equation?

10.5 **PLUS** For each case that follows, apply the Darcy-Weisbach equation from Eq. (10.12) in §10.3 to calculate the head loss in a pipe. Apply the grid method to carry and cancel units.

   a. Water flows at a rate of 75 lpm and a mean velocity of 55 m/min in a pipe of length 60 m. For a resistance coefficient of $f = 0.02$, find the head loss in meters.

   b. The head loss in a section of PVC pipe is 0.8 m, the resistance coefficient is $f = 0.012$, the length is 15 m, and the flow rate is 0.028 m³/s. Find the pipe diameter in meters.

10.6 **PLUS** As shown, air (20°C) is flowing from a large tank, through a horizontal pipe, and then discharging to ambient. The pipe length is $L = 50$ m, and the pipe is schedule 40 PVC with a nominal diameter of 25 mm. The mean velocity in the pipe is 10 m/s, and $f = 0.015$. Determine the pressure (in Pa) that needs to be maintained in the tank.

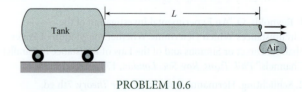

PROBLEM 10.6

10.7 **PLUS** Water (15°C) flows through a garden hose (ID = 22 mm) with a mean velocity of 2 m/s. Find the pressure drop for a section of hose that is 20 meters long and situated horizontally. Assume that $f = 0.012$.

10.8 As shown, water (15°C) is flowing from a tank through a tube and then discharging to ambient. The tube has an ID of 8 mm and a length of $L = 6$ m, and the resistance coefficient is $f = 0.015$. The water level is $H = 3$ m. Find the exit velocity in m/s. Find the discharge in L/s. Sketch the HGL and the EGL. Assume that the only head loss occurs in the tube.

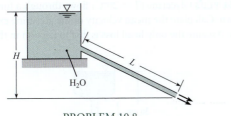

PROBLEM 10.8

10.9 Water flows in the pipe shown, and the manometer deflects 90 cm. What is f for the pipe if $V = 3$ m/s?

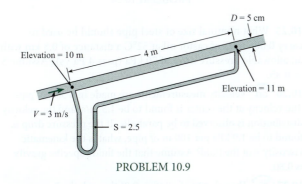

PROBLEM 10.9

Laminar Flow in Pipes (§10.5)

10.10 Using §10.5 and other resources, answer the questions that follow. Strive for depth, clarity, and accuracy while also combining sketches, words, and equations in ways that enhance the effectiveness of your communication.

  a. What are the main characteristics of laminar flow?

  b. What is the meaning of each variable that appears in Eq. (10.27) in §10.5?

  c. In Eq. (13.33) of §10.5, what is the meaning of h_f?

10.11 A fluid ($\mu = 10^{-2}$ N · s/m^2; $\rho = 800$ kg/m^3) flows with a mean velocity of 4 cm/s in a 10 cm smooth pipe.

  a. What is the value of Reynolds number?

  b. What is the magnitude of the maximum velocity in the pipe?

  c. What is the magnitude of the friction factor f?

  d. What is the shear stress at the wall?

  e. What is the shear stress at a radial distance of 25 mm from the center of the pipe?

10.12 Water (15°C) flows in a horizontal schedule 40 pipe that has a nominal diameter of 15 mm. The Reynolds number is Re $= 1000$. Work in SI units.

  a. What is mass flow rate?

  b. What is the magnitude of the friction factor f?

  c. What is the head loss per meter of pipe length?

  d. What is the pressure drop per meter of pipe length?

10.13 Flow of a liquid in a smooth 2.5 cm pipe yields a head loss of 2 m per meter of pipe length when the mean velocity is 0.5 m/s. Calculate f and the Reynolds number. Prove that doubling the flow rate will double the head loss. Assume fully developed flow.

10.14 As shown, a round tube of diameter 0.5 mm and length 750 mm is connected to plenum. A fan produces a negative gage pressure of -373 Pa in the plenum and draws air (20°C) into the microchannel. What is the mean velocity of air in the microchannel? Assume that the only head loss is in the tube.

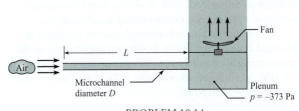

PROBLEM 10.14

10.15 Liquid ($\gamma = 10$ kN/m^3) is flowing in a pipe at a steady rate, but the direction of flow is unknown. Is the liquid moving upward or moving downward in the pipe? If the pipe diameter is 8 mm and the liquid viscosity is 3.0×10^{-3} N · s/m^2, what is the magnitude of the mean velocity in the pipe?

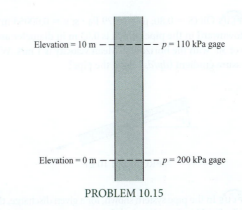

PROBLEM 10.15

10.16 Oil (S $= 0.97$, $\mu = 0.479$ Pa · s) is pumped through a nominal 25 mm, schedule 80 pipe at the rate of 1×10^{-4} m^3/s. What is the head loss per 30 m of level pipe?

10.17 A liquid ($\rho = 1000$ kg/m^3; $\mu = 10^{-1}$ N · s/2 m^2; $v = 10^{-4}$ m^2/s) flows uniformly with a mean velocity of 1.5 m/s in a pipe with a diameter of 100 mm. Show that the flow is laminar. Also, find the friction factor f and the head loss per meter of pipe length.

10.18 ^{WILEY} GO ° Kerosene ($S = 0.80$ and $T = 20°C$) flows from the tank shown and through the 6 mm–diameter (ID) tube. Determine the mean velocity in the tube and the discharge. Assume the only head loss is in the tube.

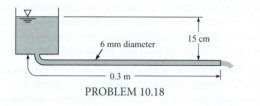

PROBLEM 10.18

10.19 WILEY PLUS Oil ($S = 0.94$; $\mu = 0.048$ N · s/m²) is pumped through a horizontal 5 cm pipe. Mean velocity is 0.5 m/s. What is the pressure drop per 10 m of pipe?

10.20 WILEY PLUS As shown, SAE 10W-30 oil is pumped through an 8 m length of 1-cm-diameter drawn tubing at a discharge of 7.85×10^{-4} m³/s. The pipe is horizontal, and the pressures at points 1 and 2 are equal. Find the power necessary to operate the pump, assuming the pump has an efficiency of 100%. Properties of SAE 10W-30 oil: kinematic viscosity $= 7.6 \times 10^{-5}$ m²/s; specific weight $= 8630$ N/m³.

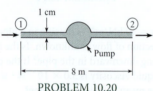

PROBLEM 10.20

10.21 WILEY PLUS Oil ($S = 0.80$; $\mu = 0.479$ Pa · s; $\nu = 0.00053$ m²/s) flows downward in the pipe, which is 0.03 m in diameter and has a slope of 30° with the horizontal. Mean velocity is 1 m/s. What is the pressure gradient (dp/ds) along the pipe?

PROBLEM 10.21

10.22 WILEY PLUS In the pipe system shown, for a given discharge, the ratio of the head loss in a given length of the 1 m pipe to the head loss in the same length of the 2 m pipe is (a) 2, (b) 4, (c) 16, or (d) 32.

10.23 Glycerine ($T = 20°C$) flows in a pipe with a 150-mm diameter at a mean velocity of 0.45 m/s. Is the flow laminar or turbulent? Plot the velocity distribution across the flow section, in 12-mm increments of radius.

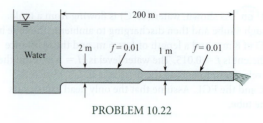

PROBLEM 10.22

10.24 WILEY PLUS Glycerine ($T = 20°C$) flows through a funnel as shown. Calculate the mean velocity of the glycerine exiting the tube. Assume the only head loss is due to friction in the tube.

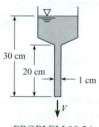

PROBLEM 10.24

10.25 What nominal size of steel pipe should be used to carry 0.006 m³/s of castor oil at 32°C a distance of 0.8 km with an allowable pressure drop of 70 kPa ($\mu = 4$ Pa · s)? Assume S $= 0.85$.

10.26 ^{WILEY} GO ° Velocity measurements are made in a 30-cm pipe. The velocity at the center is found to be 1.5 m/s, and the velocity distribution is observed to be parabolic. If the pressure drop is found to be 1.9 kPa per 100 m of pipe, what is the kinematic viscosity ν of the fluid? Assume that the fluid's specific gravity is 0.80.

10.27 ^{WILEY} GO ° The velocity of oil ($S = 0.8$) through the 5-cm smooth pipe is 1.2 m/s. Here $L = 12$ m, $z_1 = 1$ m, $z_2 = 2$ m, and the manometer deflection is 10 cm. Determine the flow direction, the resistance coefficient f, whether the flow is laminar or turbulent, and the viscosity of the oil.

10.28 The velocity of oil ($S = 0.8$) through the 50-mm smooth pipe is 1.5 m/s. Here $L = 9$ m, $z_1 = 0.6$ m, $z_2 = 1.2$ m, and the manometer deflection is 100 mm. Determine the flow direction, the resistance coefficient f, whether the flow is laminar or turbulent, and the viscosity of the oil.

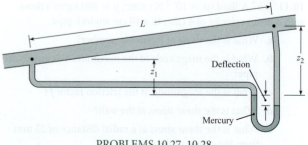

PROBLEMS 10.27, 10.28

10.29 Glycerine at 20°C flows at 0.6 m/s in the 2-cm commercial steel pipe. Two piezometers are used as shown to measure the piezometric head. The distance along the pipe between the standpipes is 1 m. The inclination of the pipe is 20°. What is the height difference Δh between the glycerine in the two standpipes?

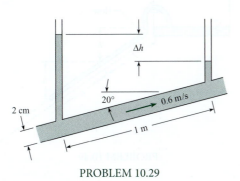

PROBLEM 10.29

10.30 PLUS Water is pumped through a heat exchanger consisting of tubes 6 mm in diameter and 6 m long. The velocity in each tube is 12 cm/s. The water temperature increases from 20°C at the entrance to 30°C at the exit. Calculate the pressure difference across the heat exchanger, neglecting entrance losses but accounting for the effect of temperature change by using properties at average temperatures.

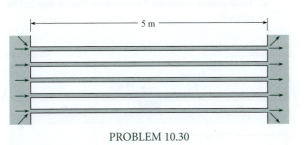

PROBLEM 10.30

Turbulent Flow in Pipes (§10.6)

10.31 PLUS Use Figure 10.14, Table 10.3, and Table 10.4 (in §10.6) to assess the following statements as True or False:

a. If k_s/D is 0.05 or larger, and the flow is turbulent, the value of f is not dependent on Re_D.

b. For smooth pipes and turbulent flow, f depends on k_s/D and not Re_D.

c. For laminar flow, f is always given by $f = 64/Re_D$.

d. If $Re_D = 2 \times 10^7$ and $k_s/D = 0.00005$, then $f = 0.012$.

e. If $Re_D = 1000$ and the pipe is smooth, $f = 0.04$.

f. The sand roughness height k_s for wrought iron is 0.002 mm.

10.32 PLUS Water (21°C) flows through a nominal 100-mm, schedule 40, PVC pipe at the rate of 0.03 m³/s. What is the resistance coefficient f? Use the Swamee-Jain Eq. (10.39) in §10.6.

10.33 PLUS Water at 20°C flows through a 2-cm ID smooth brass tube at a rate of 0.003 m³/s. What is f for this flow? Use the Swamee-Jain Eq. (10.39) in §10.6.

10.34 Water (10°C) flows through a 25-cm smooth pipe at a rate of 0.05 m³/s. What is the resistance coefficient f?

10.35 PLUS What is f for the flow of water at 10°C through a 10-cm cast-iron pipe with a mean velocity of 4 m/s?

10.36 GO A fluid ($\mu = 10^{-2}$ N · s/m²; $\rho = 800$ kg/m³) flows with a mean velocity of 500 mm/s in a 100-mm-diameter smooth pipe. Answer the following questions relating to the given flow conditions.

a. What is the magnitude of the maximum velocity in the pipe?

b. What is the magnitude of the resistance coefficient f?

c. What is the shear velocity?

d. What is the shear stress at a radial distance of 25 mm from the center of the pipe?

e. If the discharge is doubled, will the head loss per length of pipe also be doubled?

10.37 PLUS Water (20°C) flows in a 16-cm cast-iron pipe at a rate of 0.1 m³/s. For these conditions, determine or estimate the following:

a. Reynolds number

b. Friction factor f (use Swamee-Jain Eq. (10.39) in §10.6.)

c. Shear stress at the wall, τ_0

10.38 In a 100-mm uncoated cast-iron pipe, 0.0006 m³/s of water flows at 15°C. Determine f from Fig. 10.14.

10.39 PLUS Determine the head loss in 270 m of a concrete pipe with a 150-mm diameter ($k_s = 0.06$ mm) carrying 0.085 m³/s of fluid. The properties of the fluid are $v = 3 \times 10^{-4}$ m²/s and $\rho = 773$ kg/m³.

10.40 PLUS Points A and B are 1.5 km apart along a 15-cm new steel pipe ($k_s = 4.6 \times 10^{-5}$ m). Point B is 20 m higher than A. With a flow from A to B of 0.03 m³/s of crude oil ($S = 0.82$) at 10°C ($\mu = 10^{-2}$ N · s/m²), what pressure must be maintained at A if the pressure at B is to be 300 kPa?

10.41 A pipe can be used to measure the viscosity of a fluid. A liquid flows in a 1.5-cm smooth pipe 1 m long with an average velocity of 4 m/s. A head loss of 50 cm is measured. Estimate the kinematic viscosity.

10.42 GO For a 40-cm pipe, the resistance coefficient f was found to be 0.06 when the mean velocity was 3 m/s and the kinematic viscosity was 10^{-5} m²/s. If the velocity were doubled, would you expect the head loss per meter of length of pipe to double, triple, or quadruple?

10.43 GO Water (10°C) flows with a speed of 1.5 m/s through a horizontal run of PVC pipe. The length of the pipe is 30 m, and the pipe is schedule 40 with a nominal diameter of 65 mm. Calculate (a) the pressure drop in kPa, (b) the head loss in meters, and (c) the power in watts needed to overcome the head loss.

10.44 Water (10°C) flows with a speed of 2 m/s through a horizontal run of PVC pipe. The length of the pipe is 50 m, and the pipe is schedule 40 with a nominal diameter of 65 mm. Calculate (a) the pressure drop in kilopascals, (b) the head loss in meters, and (c) the power in watts needed to overcome the head loss.

10.45 ⓦ Air flows in a 3-cm smooth tube at a rate of 0.015 m³/s. If $T = 20°C$ and $p = 110$ kPa absolute, what is the pressure drop per meter of length of tube?

10.46 Points A and B are 4.8 km apart along a 600-mm new cast-iron pipe carrying water ($T = 10°C$). Point A is 9 m higher than B. The pressure at B is 140 kPa greater than that at A. Determine the direction and rate of flow.

10.47 ⓦ Air flows in a 25-mm smooth tube at a rate of 1.42 × 10^{-2} m³/s. If $T = 25°C$ and $p = 103.4$ kPa, what is the pressure drop per meter of length of tube?

10.48 ⓦ Water is pumped through a vertical 10-cm new steel pipe to an elevated tank on the roof of a building. The pressure on the discharge side of the pump is 1.6 MPa. What pressure can be expected at a point in the pipe 110 m above the pump when the flow is 0.02 m³/s? Assume $T = 20°C$.

10.49 ⓖⓞ The house located on a hill as shown is flooded by a broken waterline. The frantic owners siphon water out of the basement window and down the hill in the backyard, with one hose, of length L, and thus an elevation difference of h to drive the siphon. Water drains from the siphon, but too slowly for the desperate home owners. They reason that with a larger head difference, they can generate more flow. So they get another hose, same length as the first, and connect the 2 hoses for total length 2L. The backyard has a constant slope, so that a hose length of 2L correlates to a head difference of 2h.

  a. Assume no head loss, and calculate whether the flow rate doubles when the hose length is doubled from Case 1 (length L and height h) to Case 2 (length 2L and height 2h).

  b. Assume $h_L = 0.025(L/D)(V^2/2g)$, and calculate the flow rate for Cases 1 and 2, where $D = 0.025$ m, $L = 15$ m, and $h = 6$ m. How much of an improvement in flow rate is accomplished in Case 2 as compared to Case 1?

  c. Both the husband and wife of this couple took fluid mechanics in college. They review with new appreciation the energy equation and the form of the head loss term and realize that they should use a larger diameter hose. Calculate the flow rate for Case 3, where $L = 15$ m, $h = 6$ m, and $D = 50$ mm. Use the same expression for h_L as in part (b). How much of an improvement in flow rate is accomplished in Case 3 as compared to Case 1 in part (b)?

10.50 A train travels through a tunnel as shown. The train and tunnel are circular in cross section. Clearance is small, causing all air (15°C) to be pushed from the front of the train and

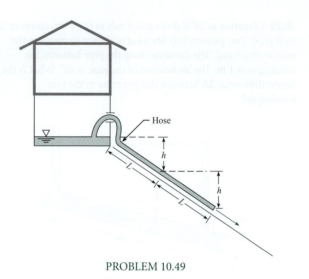

PROBLEM 10.49

discharged from the tunnel. The tunnel is 3 m in diameter and is concrete. The train speed is 15 m/s. Assume the concrete is very rough ($k_s = 0.015$ m).

  a. Determine the change in pressure between the front and rear of the train that is due to *pipe friction* effects.

  b. Sketch the energy and hydraulic grade lines for the train position shown.

  c. What power is required to produce the air flow in the tunnel?

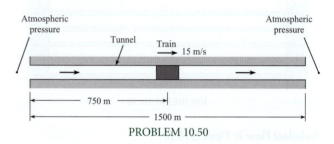

PROBLEM 10.50

10.51 Water (15°C) is pumped from a reservoir to a large, pressurized tank as shown. The steel pipe is 100 mm in diameter and 90 m long. The discharge is 0.03 m³/s. The initial water levels in the tanks are the same, but the pressure in tank B is 70 kPa gage, and tank A is open to the atmosphere. The pump efficiency is 90%. Find the power necessary to operate the pump for the given conditions.

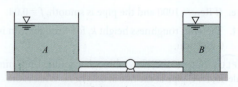

PROBLEM 10.51

Solving Turbulent Flow Problems (§10.7)

10.52 PLUS Using the information at the beginning of §10.7, classify each problem given below as case 1, case 2, or case 3. For each of your choices, state your rationale.

    a. Problem 10.51

    b. Problem 10.54

    c. Problem 10.57

10.53 A plastic siphon hose with $D = 1.2$ cm and $L = 5.5$ m is used to drain water ($15°C$) out a tank. Calculate the velocity in the tube for the two situations given below. Use $H = 3$ m and $h = 1$ m.

    a. Assume the Bernoulli equation applies (neglect all head loss).

    b. Assume the component head loss is zero, and the pipe head loss is nonzero.

10.54 GO A plastic siphon hose of length 7 m is used to drain water ($15°C$) out of a tank. For a flow rate of 1.5 L/s, what hose diameter is needed? Use $H = 5$ m and $h = 0.5$ m. Assume all head loss occurs in the tube.

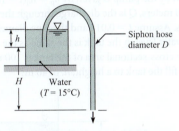

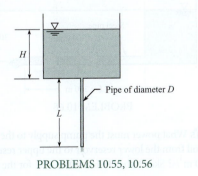

PROBLEMS 10.53, 10.54

10.55 PLUS As shown, water ($21°C$) is draining from a tank through a galvanized iron pipe. The pipe length is $L = 3$ m, the tank depth is $H = 1.2$ m, and the pipe is 25-mm DN schedule 40. Calculate the velocity in the pipe and the flow rate. Neglect component head loss.

10.56 As shown, water ($15°C$) is draining from a tank through a galvanized iron pipe. The pipe length is $L = 2$ m, the tank depth is $H = 1$ m, and the pipe is a 15-mm DN schedule 40. Calculate the velocity in the pipe. Neglect component head loss.

PROBLEMS 10.55, 10.56

10.57 Air ($40°C$, 1 atm) will be transported in a straight horizontal copper tube over a distance of 150 m at a rate of 0.1 m³/s. If the pressure drop in the tube should not exceed 6 in H_2O, what is the minimum pipe diameter?

10.58 GO A fluid with $\nu = 10^{-6}$ m²/s and $\rho = 800$ kg/m³ flows through the 8-cm galvanized iron pipe. Estimate the flow rate for the conditions shown in the figure.

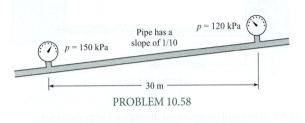

PROBLEM 10.58

10.59 Determine the diameter of commercial steel pipe required to convey 8.5 m³/s of water at $15°C$ with a head loss of 0.3 m per 300 m of pipe. Assume pipes are available in the multiple on 50 mm sizes when the diameters are expressed in millimeter (that is, 250 mm, 300 mm etc.).

10.60 A pipeline is to be designed to carry crude oil ($S = 0.93$, $\nu = 10^{-5}$ m²/s) with a discharge of 0.10 m³/s and a head loss per kilometer of 50 m. What diameter of steel pipe is needed? What power output from a pump is required to maintain this flow? Available pipe diameters are 20, 22, and 24 cm.

Combined Head Loss in Systems (§10.8)

10.61 PLUS Use Table 10.5 (on p. 381 in §10.8) to select loss coefficients, K, for the following transitions and fittings.

    a. A threaded pipe $90°$ elbow

    b. A $90°$ smooth bend with $r/d = 2$

    c. A pipe entrance with r/d of 0.3

    d. An abrupt contraction, with $\theta = 180°$, and $D_2/D_1 = 0.60$

    e. A gate valve, wide open

10.62 PLUS The sketch shows a test of an electrostatic air filter. The pressure drop for the filter is 75 mm of water when the airspeed is 10 m/s. What is the minor loss coefficient for the filter? Assume air properties at $20°C$.

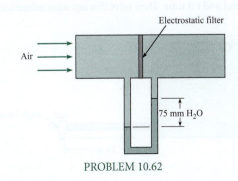

PROBLEM 10.62

10.63 PLUS If the flow of 0.10 m³/s of water is to be maintained in the system shown, what power must be added to the water by the pump? The pipe is made of steel and is 15 cm in diameter.

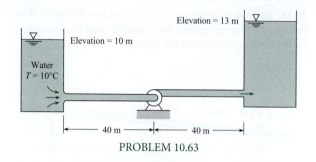

PROBLEM 10.63

10.64 Water will be siphoned through a 5 mm-diameter, 1.2-m-long Tygon tube from a jug on an upside-down wastebasket into a graduated cylinder as shown. The initial level of the water in the jug is 530 mm above the tabletop. The graduated cylinder is a 500 mL cylinder, and the water surface in the cylinder is 300 mm above the tabletop when the cylinder is full. The bottom of the cylinder is 12 mm above the table. The inside diameter of the jug is 180 mm. Calculate the time it will take to fill the cylinder from an initial depth of 50 mm of water in the cylinder.

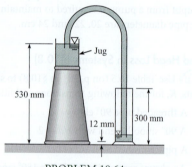

PROBLEM 10.64

10.65 Water flows from a tank through a 2.6-m length of galvanized iron pipe 26 mm in diameter. At the end of the pipe is an angle valve that is wide open. The tank is 2 m in diameter. Calculate the time required for the level in the tank to change from 10 m to 2 m. *Hint:* Develop an equation for dh/dt where h is the level and t if time. Then solve this equation numerically.

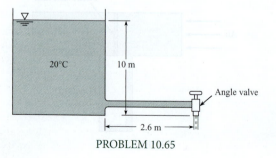

PROBLEM 10.65

10.66 PLUS A tank and piping system is shown. The galvanized pipe diameter is 1.5 cm, and the total length of pipe is 10 m. The two 90° elbows are threaded fittings. The vertical distance from the water surface to the pipe outlet is 5 m. The velocity of the water in the tank is negligible. Find (a) the exit velocity of the water and (b) the height (h) the water jet would rise on exiting the pipe. The water temperature is 20°C.

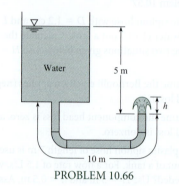

PROBLEM 10.66

10.67 A pump is used to fill a tank from a reservoir as shown. The head provided by the pump is given by $h_p = h_0(1 - (Q^2/Q_{max}^2))$ where h_0 is 50 meters, Q is the discharge through the pump, and Q_{max} is 2 m³/s. Assume $f = 0.018$ and the pipe diameter is 90 cm. Initially the water level in the tank is the same as the level in the reservoir. The cross-sectional area of the tank is 100 m². How long will it take to fill the tank to a height, h, of 40 m?

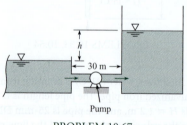

PROBLEM 10.67

10.68 GO A water turbine is connected to a reservoir as shown. The flow rate in this system is 0.14 m³/s. What power can be delivered by the turbine if its efficiency is 80%? Assume a temperature of 21°C.

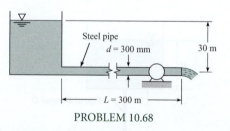

PROBLEM 10.68

10.69 PLUS What power must the pump supply to the system to pump the oil from the lower reservoir to the upper reservoir at a rate of 0.20 m³/s? Sketch the HGL and the EGL for the system.

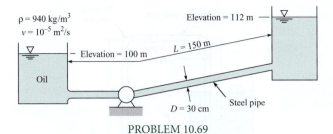

$\rho = 940 \text{ kg/m}^3$
$\nu = 10^{-5} \text{ m}^2/\text{s}$
Elevation = 112 m
Elevation = 100 m
$L = 150 \text{ m}$
Oil
$D = 30 \text{ cm}$
Steel pipe

PROBLEM 10.69

10.70 GO A cast-iron pipe 300 mm in diameter and 60 m long joins two water (15°C) reservoirs. The upper reservoir has a water-surface elevation of 45 m, and the lower on has a water-surface elevation of 12 m. The pipe exits from the side of the upper reservoir at an elevation of 36 m and enters the lower reservoir at an elevation of 9 m. There are two wide-open gate valves in the pipe. (a) List all sources of h_L and the quantitative factors associated with each. (b) Draw the EGL and the HGL for the system, and (c) determine the discharge in the pipe.

10.71 GO An engineer is making an estimate of hydroelectric power for a home owner. This owner has a small stream ($Q = 0.06$ m³/s, $T = 5$°C) that is located at an elevation $H = 10$ m above the owner's residence. The owner is proposing to divert the stream and operate a water turbine connected to an electric generator to supply electrical power to the residence. The maximum acceptable head loss in the penstock (a penstock is a conduit that supplies a turbine) is 0.9 m. The penstock has a length of 26 m. If the penstock is going to be fabricated from commercial-grade, plastic pipe, find the minimum diameter that can be used. Neglect component head losses. Assume that pipes are available in following sizes—that is, like 50 mm, 100 mm, 150 mm, etc.

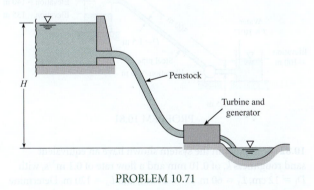

H
Penstock
Turbine and generator

PROBLEM 10.71

10.72 The water-surface elevation in a reservoir is 45 m. A straight pipe 30 m long and 150 mm in diameter conveys water from the reservoir to an open drain. The pipe entrance (it is abrupt) is at elevation 30 m, and the pipe outlet is at elevation 18 m. At the outlet the water discharges freely into the air. The water temperature is 10°C. If the pipe is asphalted cast iron, what will be the discharge rate in the pipe? Consider all head losses. Also draw the HGL and the EGL for this system.

10.73 PLUS A heat exchanger is being designed as a component of a geothermal power system in which heat is transferred from the geothermal brine to a "clean" fluid in a closed-loop power cycle. The heat exchanger, a shell-and-tube type, consists of 100 galvanized-iron tubes 2 cm in diameter and 5 m long, as shown. The temperature of the fluid is 200°C, the density is 860 kg/m³, and the viscosity is 1.35×10^{-4} N · s/m². The total mass flow rate through the exchanger is 50 kg/s.

    a. Calculate the power required to operate the heat exchanger, neglecting entrance and outlet losses.

    b. After continued use, 2 mm of scale develops on the inside surfaces of the tubes. This scale has an equivalent roughness of 0.5 mm. Calculate the power required under these conditions.

5 m
2 cm
Side view

PROBLEM 10.73

10.74 The heat exchanger shown consists of 10 m of drawn tubing 2 cm in diameter with 19 return bends. The flow rate is 3×10^{-4} m³/s. Water enters at 20°C and exits at 80°C. The elevation difference between the entrance and the exit is 0.8 m. Calculate the pump power required to operate the heat exchanger if the pressure at 1 equals the pressure at 2. Use the viscosity corresponding to the average temperature in the heat exchanger.

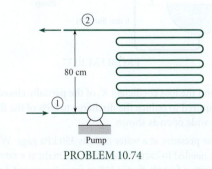

②
80 cm
①
Pump

PROBLEM 10.74

10.75 PLUS A heat exchanger consists of a closed system with a series of parallel tubes connected by 180° elbows as shown in the figure. There are a total of 14 return elbows. The pipe diameter is 2 cm, and the total pipe length is 10 m. The head loss coefficient for each return elbow is 2.2. The tube is copper. Water with an average temperature of 40°C flows through the system with a mean velocity of 8 m/s. Find the power required to operate the pump if the pump is 85% efficient.

10.76 A heat exchanger consists of 15 m of copper tubing with an internal diameter of 15 mm. There are 14 return elbows in the system with a loss coefficient of 2.2 for each elbow. The pump in the system has a pump curve given by

$$h_p = h_{p0}\left[1 - \left(\frac{Q}{Q_{max}}\right)^3\right]$$

where h_{p0} is head provided by the pump at zero discharge and Q_{max} is 10^{-3} m³/s. Water at 40°C flows through the system. Find the system operating point for values of h_{p0} of 2 m, 10 m, and 20 m.

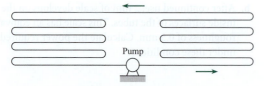

Pump

PROBLEMS 10.75, 10.76

10.77 Ⓦ🅟🅛🅤🅢 Gasoline ($T = 10°C$) is pumped from the gas tank of an automobile to the carburetor through a 6-mm fuel line of drawn tubing 3 m long. The line has five 90° smooth bends with an r/d of 6. The gasoline discharges through a 0.75-mm jet in the carburetor to a pressure of 96.5 kPa abs. The pressure in the tank is 101.3 kPa abs. The pump is 80% efficient. What power must be supplied to the pump if the automobile is consuming fuel at the rate of 0.45 lpm? Obtain gasoline properties from Figs. A.2 and A.3.

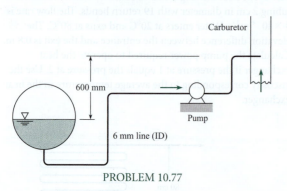

Carburetor

600 mm

Pump

6 mm line (ID)

PROBLEM 10.77

10.78 Find the loss coefficient K_v of the partially closed valve that is required to reduce the discharge to 50% of the flow with the valve wide open as shown.

10.79 The pressure at a water main is 350 kPa gage. What size of pipe is needed to carry water from the main at a rate of 0.025 m³/s to a factory that is 160 m from the main? Assume that galvanized-steel pipe is to be used and that the pressure required at the factory is 70 kPa gage at a point 8 m above the main connection.

10.80 The 12-cm galvanized-steel pipe shown is 800 m long and discharges water into the atmosphere. The pipeline has an open globe valve and four threaded elbows; $h_1 = 3$ m and $h_2 = 15$ m. What is the discharge, and what is the pressure at A, the midpoint of the line?

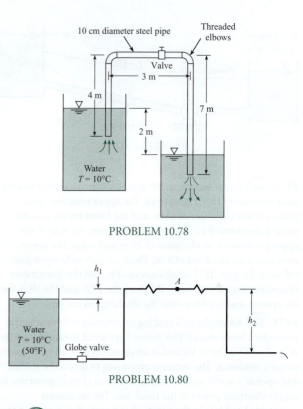

10 cm diameter steel pipe

Threaded elbows

Valve

3 m

4 m

7 m

2 m

Water
$T = 10°C$

PROBLEM 10.78

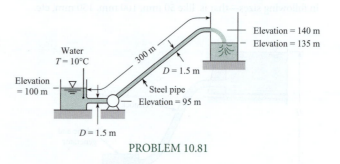

h_1

A

Water
$T = 10°C$
(50°F)

Globe valve

h_2

PROBLEM 10.80

10.81 Ⓦ🅟🅛🅤🅢 Water is pumped at a rate of 25 m³/s from the reservoir and out through the pipe, which has a diameter of 1.50 m. What power must be supplied to the water to effect this discharge?

Elevation = 140 m
Elevation = 135 m

Water
$T = 10°C$

300 m

$D = 1.5$ m

Elevation
= 100 m

Steel pipe
Elevation = 95 m

$D = 1.5$ m

PROBLEM 10.81

10.82 Both pipes in the system shown have an equivalent sand roughness k_s of 0.10 mm and a flow rate of 0.1 m³/s, with $D_1 = 12$ cm, $L_1 = 60$ m, $D_2 = 24$ cm, and $L_2 = 120$ m. Determine the difference in the water-surface elevation between the two reservoirs.

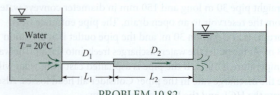

Water
$T = 20°C$

D_1

D_2

L_1

L_2

PROBLEM 10.82

10.83 Liquid discharges from a tank through the piping system shown. There is a venturi section at A and a sudden contraction at B. The liquid discharges to the atmosphere. Sketch the energy and hydraulic gradelines. Where might cavitation occur?

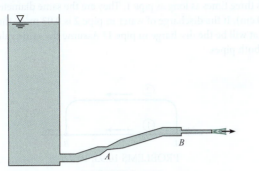

PROBLEM 10.83

10.84 The steel pipe shown carries water from the main pipe A to the reservoir and is 50 mm in diameter and 72 m long. What must be the pressure in pipe A to provide a flow of 190 lpm?

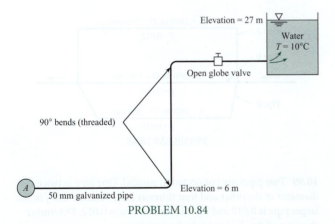

PROBLEM 10.84

10.85 If the water surface elevation in reservoir B is 110 m, what must be the water surface elevation in reservoir A if a flow of 0.03 m³/s is to occur in the cast-iron pipe? Draw the HGL and the EGL, including relative slopes and changes in slope.

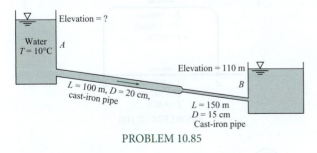

PROBLEM 10.85

Nonround Conduits (§10.9)

10.86 PLUS Air at 60°F and atmospheric pressure flows in a horizontal duct with a cross section corresponding to an

equilateral triangle (all sides equal). The duct is 30 m long, and the dimension of a side is 150 mm. The duct is constructed of galvanized iron ($k_s = 0.15$ mm). The mean velocity in the duct is 3.6 m/s. What is the pressure drop over the 30-m length?

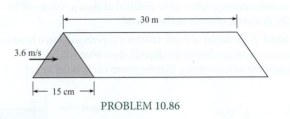

PROBLEM 10.86

10.87 PLUS A cold-air duct 100 cm by 15 cm in cross section is 100 m long and made of galvanized iron. This duct is to carry air at a rate of 6 m³/s at a temperature of 15°C and atmospheric pressure. What is the power loss in the duct?

10.88 PLUS Air (20°C) flows with a speed of 10 m/s through a horizontal rectangular air-conditioning duct. The duct is 20 m long and has a cross section of 1. Calculate (a) the pressure drop in meters of water and (b) the power in watts needed to overcome head loss. Assume the roughness of the duct is $k_s = 0.004$ mm. Neglect component head losses.

10.89 An air-conditioning system is designed to have a duct with a rectangular cross section 30 cm by 60 cm, as shown. During construction, a truck driver backed into the duct and made it a trapezoidal section, as shown. The contractor, behind schedule, installed it anyway. For the same pressure drop along the pipe, what will be the ratio of the velocity in the trapezoidal duct to that in the rectangular duct? Assume the Darcy-Weisbach resistance coefficient is the same for both ducts.

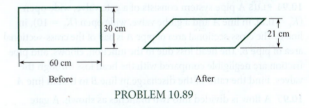

PROBLEM 10.89

Modeling Pumps in Systems (§10.10)

10.90 What power must be supplied by the pump to the flow if water ($T = 20°C$) is pumped through the 300-mm steel pipe from the lower tank to the upper one at a rate of 0.4 m³/s?

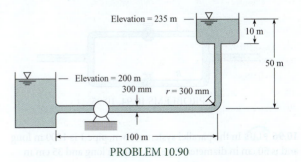

PROBLEM 10.90

10.91 If the pump for Fig. 10.20b is installed in the system of Prob. 10.90, what will be the rate of discharge of water from the lower tank to the upper one?

10.92 A pump that has the characteristic curve shown in the accompanying graph is to be installed as shown. What will be the discharge of water in the system?

10.93 If the liquid of Prob. 10.92 is a superliquid (zero head loss occurs with the flow of this liquid), then what will be the pumping rate, assuming that the pump curve is the same?

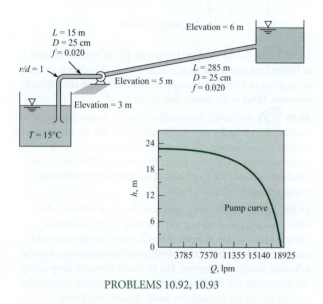

PROBLEMS 10.92, 10.93

Pipes in Parallel and in Networks (§10.10)

10.94 PLUS A pipe system consists of a gate valve, wide open ($K_v = 0.2$), in line A and a globe valve, wide open ($K_v = 10$), in line B. The cross-sectional area of pipe A is half of the cross-sectional area of pipe B. The head loss due to the junction, elbows, and pipe friction are negligible compared with the head loss through the valves. Find the ratio of the discharge in line B to that in line A.

10.95 A flow is divided into two branches as shown. A gate valve, half open, is installed in line A, and a globe valve, fully open, is installed in line B. The head loss due to friction in each branch is negligible compared with the head loss across the valves. Find the ratio of the velocity in line A to that in line B (include elbow losses for threaded pipe fittings).

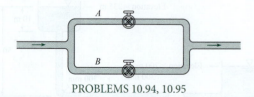

PROBLEMS 10.94, 10.95

10.96 PLUS In the parallel system shown, pipe 1 is 1200 m long and is 50 cm in diameter. Pipe 2 is 1500 m long and 35 cm in

diameter. Assume f is the same in both pipes. What is the division of the flow of water at 10°C if the flow rate will be 1.2 m³/s?

10.97 Pipes 1 and 2 are the same kind (cast-iron pipe), but pipe 2 is three times as long as pipe 1. They are the same diameter (30 cm). If the discharge of water in pipe 2 is 0.03 m³/s, then what will be the discharge in pipe 1? Assume the same value of f in both pipes.

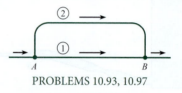

PROBLEMS 10.93, 10.97

10.98 Water flows from left to right in this parallel pipe system. The pipe having the greatest velocity is (a) pipe A, (b) pipe B, or (c) pipe C.

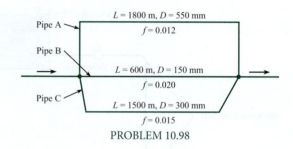

PROBLEM 10.98

10.99 Two pipes are connected in parallel. One pipe is twice the diameter of the other and four times as long. Assume that f in the larger pipe is 0.010 and f in the smaller one is 0.012. Determine the ratio of the discharges in the two pipes.

10.100 PLUS With a total flow of 0.4 m³/s, determine the division of flow and the head loss from A to B.

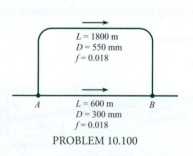

PROBLEM 10.100

10.101 PLUS The pipes shown in the system are all concrete. With a flow of 0.7 m³/s of water, find the head loss and the division of flow in the pipes from A to B. Assume $f = 0.030$ for all pipes.

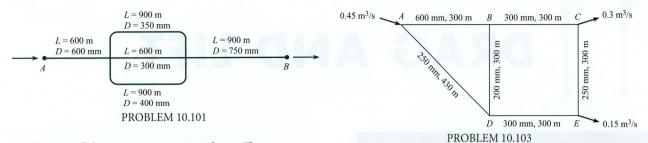

PROBLEM 10.101

PROBLEM 10.103

10.102 A parallel pipe system is set up as shown. Flow occurs from A to B. To augment the flow, a pump having the characteristics shown in Fig. 10.20b is installed at point C. For a total discharge of 0.60 m³/s, what will be the division of flow between the pipes and what will be the head loss between A and B? Assume commercial steel pipe.

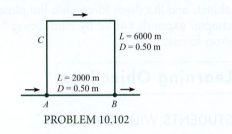

PROBLEM 10.102

10.103 For the given source and loads shown, how will the flow be distributed in the simple network, and what will be the pressures at the load points if the pressure at the source is 400 kPa? Assume horizontal pipes and $f = 0.012$ for all pipes.

10.104 Frequently in the design of pump systems, a bypass line will be installed in parallel to the pump so that some of the fluid can recirculate as shown. The bypass valve then controls the flow rate in the system. Assume that the head-versus-discharge curve for the pump is given by $h_p = 100 - 100Q$, where h_p is in meters and Q is in m³/s. The bypass line is 10 cm in diameter. Assume the only head loss is that due to the valve, which has a head-loss coefficient of 0.2. The discharge leaving the system is 0.2 m³/s. Find the discharge through the pump and bypass line.

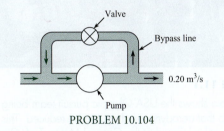

PROBLEM 10.104

11 DRAG AND LIFT

FIGURE 11.1

This photo shows the USA Olympic pursuit team being tested so that aerodynamic drag can be reduced. This wind tunnel is located at the General Motors Tech Center in Warren, Michigan. (Andy Sacks/Photodisc/Getty Images)

Chapter Road Map

Previous chapters have described the hydrostatic force on a panel, the buoyant force on a submerged object, and the shear force on a flat plate. This chapter expands this list by introducing the lift and drag forces.

Learning Objectives

STUDENTS WILL BE ABLE TO

- Define lift and drag. Explain how lift and drag are related to shear stress and pressure distributions. (§11.1)
- Define form drag. Define friction drag. (§11.1)
- For flow over a circular cylinder, describe the three drag regimes and the drag crisis. (§11.2)
- Define the coefficient of drag and find C_D values. Calculate the drag force. (§11.2, §11.3)
- Describe how to calculate the power required to overcome drag. Solve relevant problems. (§11.3)
- Explain how to calculate terminal velocity. Solve relevant problems. (§11.4)
- Describe vortex shedding. (§11.5)
- Explain what streamlining means. (§11.6)
- Define circulation and describe the circulation theory of lift. (§11.8)
- Define the coefficient of lift and calculate the lift force. (§11.8)
- Calculate the lift and drag on an airfoil. (§11.9)

When a body moves through a stationary fluid or when a fluid flows past a body, the fluid exerts a resultant force. The component of this resultant force that is parallel to the free-stream velocity is called the **drag force**. Similarly, the **lift force** is the component of the resultant force that is perpendicular to the free stream. For example, as air flows over a kite, it creates a resultant force that can be resolved in lift and drag components as shown in Fig. 11.2. By definition, lift and drag forces are limited to those forces produced by a flowing fluid.

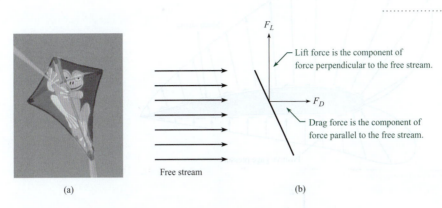

(a)   (b)

FIGURE 11.2

(a) A kite. [Photo by Donald Elger]
(b) Forces acting on the kite due to the air flowing over the kite.

11.1 Relating Lift and Drag to Stress Distributions

This section explains how lift and drag forces are related to stress distributions. This section also introduces the concepts of form and friction drag. These ideas are fundamental to understanding lift and drag.

Integrating a Stress Distribution to Yield Force

Lift and drag forces are related to the stress distribution on a body through integration. For example, consider the stress acting on the airfoil shown in Fig. 11.3. As shown, there is a pressure distribution and a shear-stress distribution. To relate stress to force, select a differential area as shown in Fig. 11.4. The magnitude of the pressure force is $dF_p = p\,dA$, and the magnitude of the viscous force is $dF_v = \tau\,dA$.* The differential lift force is normal to the free-stream direction

$$dF_L = -p\,dA\sin\theta - \tau\,dA\cos\theta$$

and the differential drag is parallel to the free-stream direction

$$dF_D = -p\,dA\cos\theta + \tau\,dA\sin\theta$$

Integration over the surface of the airfoil gives lift force (F_L) and drag force (F_D):

$$F_L = \int(-p\sin\theta - \tau\cos\theta)dA \qquad \textbf{(11.1)}$$

$$F_D = \int(-p\cos\theta + \tau\sin\theta)dA \qquad \textbf{(11.2)}$$

Equations (11.1) and (11.2) show that the lift and drag forces are related to the stress distributions through integration.

---

*The sign convention on τ is such that a clockwise sense of $\tau\,dA$ on the surface of the foil signifies a positive sign for τ.

FIGURE 11.3

Pressure and shear stress
acting on an airfoil.

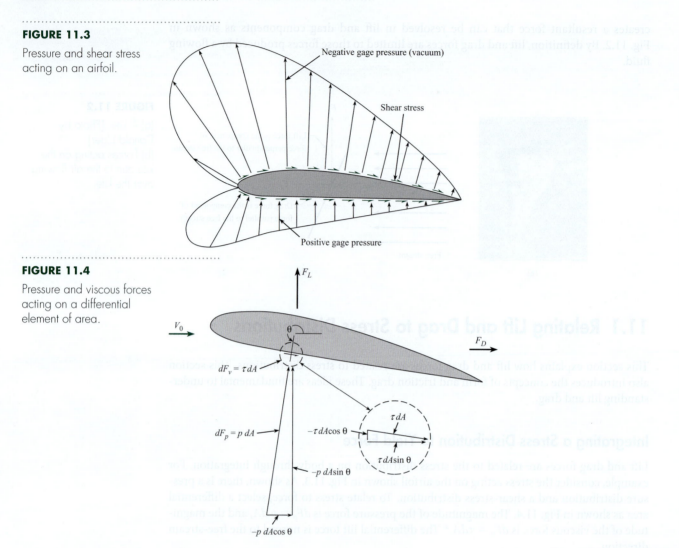

FIGURE 11.4

Pressure and viscous forces
acting on a differential
element of area.

Form Drag and Friction Drag

Notice that Eq. (11.2) can be written as the sum of two integrals.

$$F_D = \underbrace{\int (-p \cos\theta)\, dA}_{\text{form drag}} + \underbrace{\int (\tau \sin\theta)\, dA}_{\text{friction drag}} \tag{11.3}$$

Form drag is the portion of the total drag force that is associated with the pressure distribution. **Friction drag** is the portion of the total drag force that is associated with the viscous shear-stress distribution. The drag force on any body is the sum of form drag and friction drag. In words, Eq. (11.3) can be written as

$$(\text{total drag force}) = (\text{form drag}) + (\text{friction drag}) \tag{11.4}$$

11.2 Calculating Drag Force

This section introduces the drag force equation, the coefficient of drag, and presents data for two-dimensional bodies. This information is used to calculate drag force on objects.

Drag Force Equation

The drag force F_D on a body is found by using the drag force equation:

$$F_D = C_D A\left(\frac{\rho V_0^2}{2}\right)$$ (11.5)

where C_D is called the coefficient of drag, A is a reference area of the body, ρ is the fluid density, and V_0 is the free-stream velocity measured relative to the body.

The reference area A depends on the type of body. One common reference area, called **projected area** and given the symbol A_p, is the silhouetted area that would be seen by a person looking at the body from the direction of flow. For example, the projected area of a plate normal to the flow is $b\ell$, and the projected area of a cylinder with its axis normal to the flow is $d\ell$. Other geometries use different reference areas; for example, the reference area for an airplane wing is the planform area, which is the area observed when the wing is viewed from above.

The **coefficient of drag** C_D is a parameter that characterizes the drag force associated with a given body shape. For example, an airplane might have $C_D = 0.03$, and a baseball might have $C_D = 0.4$. The coefficient of drag is a π-group that is defined by

$$C_D \equiv \frac{F_D}{A_{\text{Ref}}(\rho V_0^2/2)} = \frac{\text{(drag force)}}{\text{(reference area)(kinetic pressure)}}$$ (11.6)

Values of the coefficient of drag C_D are usually found by experiment. For example, drag force F_D can be measured using a force balance in a wind tunnel. Then C_D can be calculated using Eq. (11.6). For this calculation, speed of the air in the wind tunnel V_0 can be measured using a Pitot-static tube or similar device, and air density can be calculated by applying the ideal gas law using measured values of temperature and pressure.

Equation (11.5) shows that drag force is related to four variables. Drag is related to the shape of an object because shape is characterized by the value of C_D. Drag is related to the size of the object because size is characterized by the reference area. Drag is related to the density of ambient fluid. Finally, drag is related to the speed of the fluid squared. This means that if the wind velocity doubles and C_D is constant, then the wind load on a building goes up by a factor of four.

✔ CHECKPOINT PROBLEM 11.1

Consider a car that is traveling in a straight line at constant speed.

  Case A: The car speed is 40 km/h. There is no wind.
  Case B: The car speed is 80 km/h. There is no wind.
  Case C: The car speed is 65 km/h. There is a 15 km/h steady headwind.

The coefficient of drag is the same in all three cases.

Which statement(s) are true? (Select all that apply).

  a. (Drag in Case B) = 2(Drag in Case A).
  b. (Drag in Case B) = 4(Drag in Case A).
  c. (Drag in Case B) = 8(Drag in Case A).
  d. (Drag in Case C) < (Drag in Case B).
  e. (Drag in Case C) > (Drag in Case B).
  f. (Drag in Case C) = (Drag in Case B).

Coefficient of Drag (Two-Dimensional Bodies)

This section presents C_D data and describes how C_D varies with the Reynolds number for objects that can be classified as two dimensional. A **two-dimensional body** is a body with a uniform section area and a flow pattern that is independent of the ends of the body. Examples of two-dimensional bodies are shown in Fig. 11.5. In the aerodynamics literature, C_D values for two-dimensional bodies are called **sectional drag coefficients**. Two-dimensional bodies can be visualized as objects that are infinitely long in the direction normal to the flow.

FIGURE 11.5

Coefficient of drag versus Reynolds number for two-dimensional bodies. [Data sources: Bullivant (1), DeFoe (2), Goett and Bullivant (3), Jacobs (4), Jones (5), and Lindsey (6).]

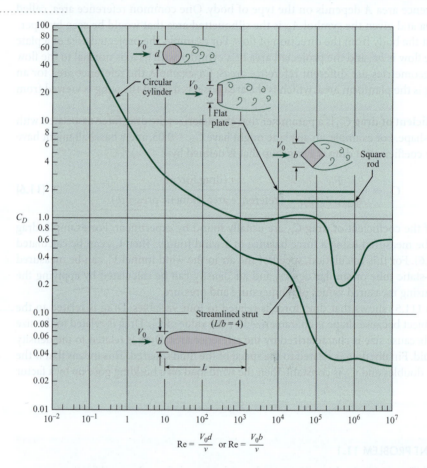

$$\text{Re} = \frac{V_0 d}{\nu} \text{ or Re} = \frac{V_0 b}{\nu}$$

The sectional drag coefficient can be used to estimate C_D for real objects. For example, C_D for a cylinder with a length to diameter ratio of 20 (e.g., $L/D \geq 20$) approaches the sectional drag coefficient because the end effects have an insignificant contribution to the total drag force. Alternatively, the sectional drag coefficient would be inaccurate for a cylinder with a small L/D ratio (e.g., $L/D \approx 1$) because the end effects would be important.

As shown in Fig. 11.5, the Reynolds number sometimes, but not always, influences the sectional drag coefficient. The value of C_D for the flat plate and square rod are independent of Re. The sharp edges of these bodies produce flow separation, and the drag force is due to the pressure distribution (form drag) and not on the shear-stress distribution (friction drag, which depends on Re). Alternatively, C_D for the cylinder and the streamlined strut show strong Re dependence because both form and friction drag are significant.

To calculate drag force on an object, find a suitable coefficient of drag, and then apply the drag force equation. This approach is illustrated by Example 11.1.

EXAMPLE 11.1

Drag Force on a Cylinder

Problem Statement

A vertical cylinder that is 30 m high and 30 cm in diameter is being used to support a television transmitting antenna. Find the drag force acting on the cylinder and the bending moment at its base. The wind speed is 35 m/s, the air pressure is 1 atm, and temperature is 20°C.

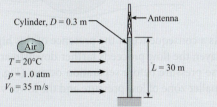

Define the Situation

Wind is blowing across a tall cylinder.

Assumptions:

- Wind speed is steady.
- Effects associated with the ends of the cylinder are negligible because $L/D = 100$.
- Neglect drag force on the antenna because the frontal area is much less than the frontal area of the cylinder.
- The line of action of the drag force is at an elevation of 15 m.

Properties: Air (20°C), Table A.5: $\rho = 1.2$ kg/m³, and $\mu = 1.81 \times 10^{-5}$ N · s/m²

State the Goals

Calculate

- Drag force (in N) on the cylinder
- Bending moment (in N · m) at the base of the cylinder

Generate Ideas and Make a Plan

1. Calculate the Reynolds number.
2. Find coefficient of drag using Fig. 11.5
3. Calculate drag force using Eq. (11.5).
4. Calculate bending moment using $M = F_D \cdot L/2$.

Take Action (Execute the Plan)

1. Reynolds number

$$\text{Re}_D = \frac{V_0 D \rho}{\mu} = \frac{35 \text{ m/s} \times 0.30 \text{ m} \times 1.20 \text{ kg/m}^3}{1.81 \times 10^{-5} \text{ N} \cdot \text{s/m}^2} = 7.0 \times 10^5$$

2. From Fig. 11.5, the coefficient of drag is $C_D = 0.20$.

3. Drag force

$$F_D = \frac{C_D A_p \rho V_0^2}{2}$$

$$= \frac{(0.2)(30 \text{ m})(0.3 \text{ m})(1.20 \text{ kg/m}^3)(35^2 \text{ m}^2/\text{s}^2)}{2}$$

$$= \boxed{1323 \text{ N}}$$

4. Moment at the base

$$M = F_D \left(\frac{L}{2}\right) = (1323 \text{ N})\left(\frac{30}{2} \text{ m}\right) = \boxed{19,800 \text{ N} \cdot \text{m}}$$

Discussion of C_D for a Circular Cylinder

Drag Regimes The coefficient of drag C_D, as shown in Fig. 11.4, can be described in terms of three regimes.

Regime I (Re < 10^3). In this regime, C_D depends on both form drag and friction drag. As shown, C_D decreases with increasing Re.

Regime II (10^3 < Re < 10^5). In this regime, C_D has a nearly constant value. The reason is that form drag, which is associated with the pressure distribution, is the dominant cause of drag. Over this range of Reynolds numbers, the flow pattern around the cylinder remains virtually unchanged, thereby producing very similar pressure distributions. This characteristic, the constancy of C_D at high values of Re, is representative of most bodies that have angular form.

Regime III (10^5 < Re < 5×10^5). In this regime, C_D decreases by about 80%, a remarkable change! This change occurs because the boundary layer on the circular cylinder changes. For Reynolds numbers less than 10^5, the boundary layer is laminar, and separation occurs about midway between the upstream side and downstream side of the cylinder (Fig. 11.6). Hence the entire downstream half of the cylinder is exposed to a relatively low pressure, which in turn produces a relatively high value for C_D. When the Reynolds number is increased to about 10^5, the boundary layer becomes turbulent, which causes higher-velocity fluid to be mixed into the

FIGURE 11.6

Flow pattern around a cylinder for $10^3 < Re < 10^5$.

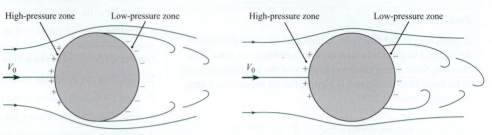

High-pressure zone   Low-pressure zone

V_0

FIGURE 11.7

Flow pattern around a cylinder for $Re > 5 \times 10^5$.

High-pressure zone   Low-pressure zone

V_0

region close to the wall of the cylinder. As a consequence of the presence of this high-velocity, high-momentum fluid in the boundary layer, the flow proceeds farther downstream along the surface of the cylinder against the adverse pressure before separation occurs (Fig. 11.7). This change in separation produces a much smaller zone of low pressure and the lower value of C_D.

Surface Roughness

Surface roughness has a major influence on drag. For example, if the surface of the cylinder is slightly roughened upstream of the midsection, the boundary layer will be forced to become turbulent at lower Reynolds numbers than those for a smooth cylinder surface. The same trend can also be produced by creating abnormal turbulence in the approach flow. The effects of roughness are shown in Fig. 11.8 for cylinders that were roughened with sand grains of size k. A small to medium size of roughness ($10^{-3} < k/d < 10^{-2}$) on a cylinder triggers an early onset of reduction of C_D. However, when the relative roughness is quite large ($10^{-2} < k/d$), the characteristic dip in C_D is absent.

FIGURE 11.8

Effects of roughness on C_D for a cylinder. [After Miller et al. (7).]

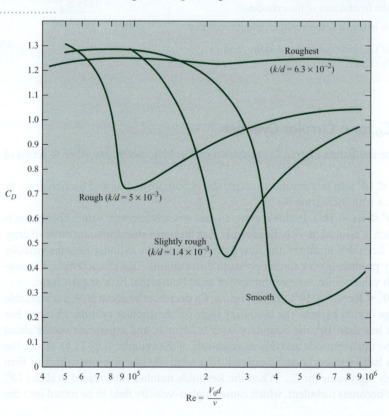

Roughest

($k/d = 6.3 \times 10^{-2}$)

Rough ($k/d = 5 \times 10^{-3}$)

Slightly rough

($k/d = 1.4 \times 10^{-3}$)

Smooth

$$Re = \frac{V_0 d}{\nu}$$

11.3 Drag of Axisymmetric and 3-D Bodies

Section 11.2 described drag for two-dimensional bodies. Drag on other body shapes is presented in this section. This section also describes power and rolling resistance.

Drag Data

An object is classified as an **axisymmetric body** when the flow direction is parallel to an axis of symmetry of the body and the resulting flow is also symmetric about its axis. Examples of axisymmetric bodies include a sphere, a bullet, and a javelin. When flow is not aligned with an axis of symmetry, the flow field is three dimensional (3D), and the body is classified as a three-dimensional or **3-D body**. Examples of 3-D bodies include a tree, a building, and an automobile.

The principles that apply to two-dimensional flow over a body also apply to axisymmetric flows. For example, at very low values of the Reynolds number, the coefficient of drag is given by exact equations relating C_D and Re. At high values of Re, the coefficient of drag becomes constant for angular bodies, whereas rather abrupt changes in C_D occur for rounded bodies. All these characteristics can be seen in Fig. 11.9, where C_D is plotted against Re for several axisymmetric bodies.

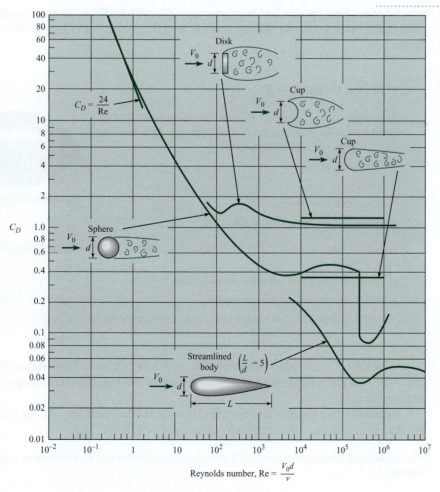

FIGURE 11.9

Coefficient of drag versus Reynolds number for axisymmetric bodies. [Data sources: Abbott (9), Brevoort and Joyner (10), Freeman (11), and Rouse (12).]

The drag coefficient of a sphere is of special interest because many applications involve the drag of spherical or near-spherical objects, such as particles and droplets. Also, the drag of a sphere is often used as a standard of comparison for other shapes. For Reynolds numbers less than 0.5, the flow around the sphere is laminar and amenable to analytical solutions. An exact solution by Stokes yielded the following equation, which is called Stokes's equation, for the drag of a sphere:

$$F_D = 3\pi\mu V_0 d \tag{11.7}$$

Note that the drag for this laminar flow condition varies directly with the first power of V_0. This is characteristic of all laminar flow processes. For completely turbulent flow, the drag is a function of the velocity to the second power. When the drag force given by Eq. (11.7) is substituted into Eq. (11.6), the result is the drag coefficient corresponding to Stokes's equation:

$$C_D = \frac{24}{Re} \tag{11.8}$$

Thus for flow past a sphere, when Re $\leq$ 0.5, one may use the direct relation for C_D given in Eq. (11.8).

Several correlations for the drag coefficient of a sphere are available (13). One such correlation has been proposed by Clift and Gauvin (14):

$$C_D = \frac{24}{Re}(1 + 0.15Re^{0.687}) + \frac{0.42}{1 + 4.25 \times 10^4 Re^{-1.16}} \tag{11.9}$$

which deviates from the *standard drag curve** by −4% to 6% for Reynolds numbers up to 3×10^5. Note that as the Reynolds number approaches zero, this correlation reduces to the equation for Stokes flow.

✔ CHECKPOINT PROBLEM 11.2

Suppose you are estimating C_D for an American football oriented so that its long axis is into the wind. You have available Fig. 11.9. Which choice would you make?

I would idealize the football

(Stockbyte/Getty Images)

    a. As a sphere

    b. As a streamlined body

    c. As one of the other bodies on the figure.

Do you think that a spinning football (about its long axis) has a different value of drag than a nonspinning football?

    d. Yes, the spinning football will have lower drag.

    e. Yes, the spinning football has higher drag.

    f. No, the spinning football has the same drag as a nonspinning football.

Values for C_D for other axisymmetric and 3-D bodies at high Reynolds numbers (Re $> 10^4$) are given in Table 11.1. Extensive data on the drag of various shapes is available in Hoerner (15).

To find the drag force on an object, find or estimate the coefficient of drag and then apply the drag force equation. This approach is illustrated by Example 11.2.

---

*The *standard drag curve* represents the best fit of the cumulative data that have been obtained for drag coefficient of a sphere.

TABLE 11.1 Approximate C_D Values for Various Bodies

Type of Body		Length Ratio	Re	C_D
	Rectangular plate	$l/b = 1$	$>10^4$	1.18
		$l/b = 5$	$>10^4$	1.20
		$l/b = 10$	$>10^4$	1.30
		$l/b = 20$	$>10^4$	1.50
		$l/b = \infty$	$>10^4$	1.98
	Circular cylinder with axis parallel to flow	$l/d = 0$ (disk)	$>10^4$	1.17
		$l/d = 0.5$	$>10^4$	1.15
		$l/d = 1$	$>10^4$	0.90
		$l/d = 2$	$>10^4$	0.85
		$l/d = 4$	$>10^4$	0.87
		$l/d = 8$	$>10^4$	0.99
	Square rod	∞	$>10^4$	2.00
	Square rod	∞	$>10^4$	1.50
	Triangular cylinder	∞	$>10^4$	1.39
	Semicircular shell	∞	$>10^4$	1.20
	Semicircular shell	∞	$>10^4$	2.30
	Hemispherical shell		$>10^4$	0.39
	Hemispherical shell		$>10^4$	1.40
	Cube		$>10^4$	1.10
	Cube		$>10^4$	0.81
	Cone—60° vertex		$>10^4$	0.49
	Parachute		$\approx 3 \times 10^7$	1.20

Sources: Brevoort and Joyner (10), Lindsey (6), Morrison (16), Roberson et al. (17), Rouse (12), and Scher and Gale (18).

EXAMPLE 11.2

Drag on a Sphere

Problem Statement

What is the drag of a 12-mm sphere that drops at a rate of 8 cm/s in oil ($\mu = 10^{-1}$ N·s/m², S = 0.85)?

Define the Situation

A sphere ($d = 0.012$ m) is falling in oil.

Speed of the sphere is $V = 0.08$ m/s.

Assumptions: Sphere is moving at a steady speed (terminal velocity).

Properties:

Oil: $\mu = 10^{-1}$ N·s/m², S = 0.85, $\rho = 850$ kg/m³

State the Goal

Find: Drag force (in newtons) on the sphere.

Generate Ideas and Make a Plan

1. Calculate the Reynolds number.
2. Find the coefficient of drag using Fig. 11.9.
3. Calculate drag force using Eq. (11.5).

Take Action (Execute the Plan)

1. Reynolds number

$$Re = \frac{Vd\rho}{\mu} = \frac{(0.08 \text{ m/s})(0.012 \text{ m})(850 \text{ kg/m}^3)}{10^{-1} \text{ N} \cdot \text{s/m}^2} = 8.16$$

2. Coefficient of drag (from Fig. 11.9) is $C_D = 5.3$.

3. Drag force

$$F_D = \frac{C_D A_p \rho V_0^2}{2}$$

$$F_D = \frac{(5.3)(\pi/4)(0.012^2 \text{ m}^2)(850 \text{ kg/m}^3)(0.08 \text{ m/s})^2}{2}$$

$$= \boxed{1.63 \times 10^{-3} \text{ N}}$$

Power and Rolling Resistance

Before reading this section, you can try out your knowledge with the Checkpoint Problem. The knowledge you need has been previously covered in this text.

✔ **CHECKPOINT PROBLEM 11.3**

Consider a bicycle racer that is traveling in a straight line at constant speed.

Case A: The speed is 20 km/h. There is no wind.
Case B: The speed is 40 km/h. There is no wind.

For both cases, C_D is the same, and rolling resistance is negligible.

Which statement is true?

  a. (Power in Case B) = (Power in Case A).
  b. (Power in Case B) = 2(Power in Case A).
  c. (Power in Case B) = 4(Power in Case A).
  d. (Power in Case B) = 8(Power in Case A).

When power is involved in a problem, the power equation from Chapter 7 is applied. For example, consider a car moving at a steady speed on a level road. Because the car is not accelerating, the horizontal forces are balanced as shown in Fig. 11.10. Force equilibrium gives

$$F_{\text{Drive}} = F_{\text{Drag}} + F_{\text{Rolling resistance}}$$

The driving force (F_{Drive}) is the frictional force between the driving wheels and the road. The drag force is the resistance of the air on the car. The rolling resistance is the frictional force that occurs when an object such as a ball or tire rolls. It is related to the deformation and types of the materials that are in contact. For example, a rubber tire on asphalt will have a larger rolling resistance than a steel train wheel on a steel rail. The rolling resistance is calculated using

$$F_{\text{Rolling resistance}} = F_r = C_r N \tag{11.10}$$

where C_r is the coefficient of rolling resistance and N is the normal force.

FIGURE 11.10

Horizontal forces acting on car that is moving at a steady speed.

The power required to move the car shown in Fig. 11.10 at a constant speed is given by Eq. (7.2a)

$$P = FV = F_{Drive}V_{Car} = (F_{Drag} + F_{Rolling\ resistance})V_{Car} \qquad \textbf{(11.11)}$$

Thus, when power is involved in a problem, apply the equation $P = FV$ while concurrently using a free-body diagram to determine the appropriate force. This approach is illustrated in Example 11.3.

EXAMPLE 11.3

Speed of a Bicycle Rider

Problem Statement

A bicyclist of mass 70 kg supplies 300 watts of power while riding into a 3 m/s headwind. The frontal area of the cyclist and bicycle together is 0.362 m^2, the drag coefficient is 0.88, and the coefficient of rolling resistance is 0.007. Determine the speed V_c of the cyclist. Express your answer in mph and in m/s.

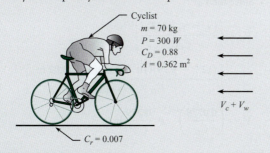

Cyclist
$m = 70$ kg
$P = 300$ W
$C_D = 0.88$
$A = 0.362$ m^2

$V_c + V_w$

$C_r = 0.007$

Define the Situation

A bicycle rider is cycling into a headwind of magnitude $V_w = 3$ m/s.

Assumptions:

1. The path is level, with no hills.
2. Mechanical losses in the bike gear train are zero.

Properties: Air (20°C, 1 atm), Table A.2: $\rho = 1.2$ kg/m^3

State the Goal

Find the speed (m/s) of the rider.

Generate Ideas and Make a Plan

1. Relate bike speed (V_c) to power using Eq. (11.11).
2. Calculate rolling resistance.
3. Develop an equation for drag force using Eq. (11.5).

4. Combine steps 1 to 3.
5. Solve for V_c.

Take Action (Execute the Plan)

1. Power equation
   - The power from the bike rider is being used to overcome drag and rolling resistance. Thus,
   $$P = (F_D + F_r)V_c$$

2. Rolling resistance
   $$F_r = C_r N = C_r mg = 0.007(70\ \text{kg})(9.81\ \text{m/s}^2) = 4.81\ \text{N}$$

3. Drag force
   - $V_0 =$ speed of the air relative to the bike rider
   $$V_0 = V_c + 3\ \text{m/s}$$

   - Drag force
   $$F_D = C_D A\left(\frac{\rho V_0^2}{2}\right) = \frac{0.88(0.362\ \text{m}^2)(1.2\ \text{kg/m}^3)}{2}$$
   $$\times (V_c + 3\ \text{m/s})^2$$
   $$= 0.1911(V_c + 3\ \text{m/s})^2$$

4. Combine results:
   $$P = (F_D + F_r)V_c$$
   $$300\ \text{W} = (0.1911(V_c + 3)^2 + 4.81)V_c$$

5. Because the equation is cubic, use a spreadsheet program as shown. In this spreadsheet, let V_c vary and then search for the value of V_c that causes the right side of the equation to equal 300. The result is

   $$V_c = \boxed{9.12\ \text{m/s}}$$

V_c	RHS
(m/s)	(W)
0	0.0
5	85.2
8	223.5
9	291.0
9.1	298.4
9.11	299.1
9.12	299.9
9.13	300.6

11.4 Terminal Velocity

Another common application of the drag force equation is finding the steady-state speed of a body that is falling through a fluid. When a body is dropped, it accelerates under the action of gravity. As the speed of the falling body increases, the drag increases until the upward force (drag) equals the net downward force (weight minus buoyant force). Once the forces are balanced, the body moves at a constant speed called the **terminal velocity**, which is identified as the maximum velocity attained by a falling body.

To find terminal velocity, balance the forces acting on the object, and then solve the resulting equation. In general this process is iterative, as illustrated by Example 11.4.

EXAMPLE 11.4

Terminal Velocity of a Sphere in Water

Problem Statement

A 20 mm plastic sphere (S = 1.3) is dropped in water. Determine its terminal velocity. Assume $T = 20°C$.

Define the Situation

A smooth sphere ($D = 0.02$ m, S = 1.3) is falling in water.

Properties: Water (20°C), Table A.5, $\nu = 1 \times 10^{-6}$ m²/s, $\rho = 998$ kg/m³, and $\gamma = 9790$ N/m³

State the Goal

Find the terminal velocity (m/s) of the sphere.

Generate Ideas and Make a Plan

This problem requires an iterative solution because the terminal velocity equation is implicit. The plan steps are

1. Apply force equilibrium.
2. Develop an equation for terminal velocity.
3. To solve the terminal velocity equation, set up a procedure for iteration.
4. To implement the iterative solution, build a table in a spreadsheet program.

Take Action (Execute the Plan)

1. Force equilibrium
   - Sketch a free-body diagram.

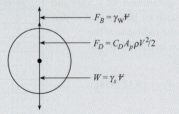

- Apply force equilibrium (vertical direction):

$$F_{\text{Drag}} + F_{\text{Buoyancy}} = W$$

2. Terminal velocity equation
   - Analyze terms in the equilibrium equation:

$$C_D A \left(\frac{\rho V_0^2}{2} \right) + \gamma_w \mathcal{V} = \gamma_s \mathcal{V}$$

$$C_D \left(\frac{\pi d^2}{4} \right) \left(\frac{\rho V_0^2}{2} \right) + \gamma_w \left(\frac{\pi d^3}{6} \right) = \gamma_s \left(\frac{\pi d^3}{6} \right)$$

   - Solve for V_0

$$V_0 = \left[\frac{(\gamma_s - \gamma_w)(4/3)d}{C_D \rho_w} \right]^{1/2}$$

$$= \left[\frac{(12.7 - 9.79)(10^3 \text{ N/m}^3)(4/3)(0.02 \text{ m})}{C_D \times 998 \text{ kg/m}^3} \right]^{1/2}$$

$$V_0 = \left(\frac{0.0778}{C_D} \right)^{1/2} = \frac{0.279}{C_D^{1/2}} \text{ m/s}$$

3. Iteration 1
   - Initial guess: $V_0 = 1.0$ m/s
   - Calculate Re:

$$\text{Re} = \frac{Vd}{\nu} = \frac{(1.0 \text{ m/s})(0.02 \text{ m})}{1 \times 10^{-6} \text{ m}^2/\text{s}} = 20000$$

   - Calculate C_D using Eq. (11.9):

$$C_D = \frac{24}{20000}(1 + 0.15(20000^{0.687}))$$

$$+ \frac{0.42}{1 + 4.25 \times 10^4 (20000)^{-1.16}} = 0.456$$

   - Find new value of V_0 (use equation from step 2):

$$V_0 = \left(\frac{0.0778}{C_D} \right)^{1/2} = \frac{0.279}{0.456^{0.5}} = 0.413 \text{ m/s}$$

4. Iterative solution

- As shown, use a spreadsheet program to build a table. The first row shows the results of iteration 1.
- The terminal velocity from iteration 1 $V_0 = 0.413$ m/s is used as the initial velocity for iteration 2.
- The iteration process is repeated until the terminal velocity reaches a constant value of $V_0 = 0.44$ m/s. Notice that convergence is reached in two iterations.

Iteration #	Initial V_0	Re	C_D	New V_0
	(m/s)			(m/s)
1	1.000	20000	0.456	0.413
2	0.413	8264	0.406	0.438
3	0.438	8752	0.409	0.436
4	0.436	8721	0.409	0.436
5	0.436	8723	0.409	0.436
6	0.436	8722	0.409	0.436

$$V_0 = 0.44 \text{ m/s}$$

11.5 Vortex Shedding

This section introduces vortex shedding, which is important for two reasons: It can be used to enhance heat transfer and mixing, and it can cause unwanted vibrations and failures of structures.

Flow past a bluff body generally produces a series of vortices that are shed alternatively from each side, thereby producing a series of alternating vortices in the wake. This phenomenon is call **vortex shedding**. Vortex shedding for a cylinder occurs for Re $\geqslant$ 50 and gives the flow pattern sketched in Fig. 11.11. In this figure, a vortex is in the process of formation near the top of the cylinder. Below and to the right of the first vortex is another vortex, which was formed and shed a short time before. Thus the flow process in the wake of a cylinder involves the formation and shedding of vortices alternately from one side and then the other. This alternate formation and shedding of vortices creates a cyclic change in pressure with consequent periodicity in side thrust on the cylinder. Vortex shedding was the primary cause of failure of the Tacoma Narrows suspension bridge in the state of Washington in 1940.

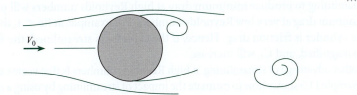

FIGURE 11.11

Formation of a vortex behind a cylinder.

Experiments reveal that the frequency of shedding can be represented by plotting Strouhal number (St) as a function of Reynolds number. The Strouhal number is a π-group defined as

$$\text{St} = \frac{nd}{V_0} \tag{11.12}$$

where n is the frequency of shedding of vortices from one side of the cylinder, in Hz, d is the diameter of the cylinder, and V_0 is the free-stream velocity. The Strouhal number for vortex shedding from a circular cylinder is given in Fig. 11.12. Other cylindrical and two-dimensional bodies also shed vortices. Consequently, the engineer should always be alert to vibration problems when designing structures that are exposed to wind or water flow.

FIGURE 11.12

Strouhal number versus
Reynolds number for flow
past a circular cylinder.
[After Jones (5) and
Roshko (8)]

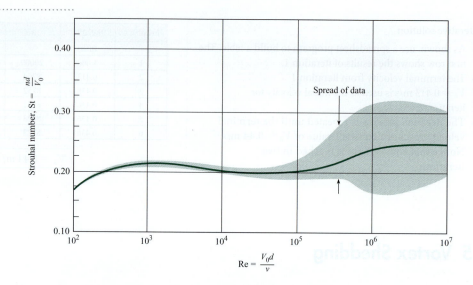

11.6 Reducing Drag by Streamlining

An engineer can design a body shape to minimize the drag force. This process is called **streamlining** and is often focused on reducing form drag. The reason for focusing on form drag is that drag on most bluff objects (e.g., a cylindrical body at Re > 1000) is predominantly due to the pressure variation associated with flow separation. In this case, streamlining involves modifying the body shape to reduce or eliminate separation. The impacts of streamlining can be dramatic. For example, Fig. 11.5 shows that C_D for the streamlined shape is about 1/6 of C_D for the circular cylinder when Re $\approx 5 \times 10^5$.

While streamlining reduces form drag, friction drag is typically increased. This is because there is more surface area on a streamlined body as compared to a nonstreamlined body. Consequently, when a body is streamlined the optimum condition results when the sum of form drag and friction drag is minimum.

Streamlining to produce minimum drag at high Reynolds numbers will probably not produce minimum drag at very low Reynolds numbers. For example, at Re < 1, the majority of the drag of a cylinder is friction drag. Hence, if the cylinder is streamlined, the friction drag will likely be magnified, and C_D will increase.

Another advantage of streamlining at high Reynolds numbers is that vortex shedding is eliminated. Example 11.5 shows how to estimate the impact of streamlining by using a ratio of C_D values.

EXAMPLE 11.5

Comparing Drag on Bluff and Streamlined Shapes

Problem Statement

Compare the drag of the cylinder of Example 11.1 with the drag of the streamlined shape shown in Fig. 11.5. Assume that both shapes have the same projected area.

Define the Situation

The cylinder from Example 11.1 is being compared to a streamlined shape.

Assumptions:

1. The cylinder and the streamlined body have the same projected area.

2. Both objects are two-dimensional bodies (neglect end effects).

State the Goal

Find the ratio of drag force on the streamlined body to drag force on the cylinder.

Generate Ideas and Make a Plan

1. Retrieve Re and C_D from Example 11.1.
2. Find the coefficient of drag for the streamlined shape using Fig. 11.5
3. Calculate the ratio of drag forces using Eq. (11.5).

Take Action (Execute the Plan)

1. From Example 11.1, Re $= 7 \times 10^5$ and C_D(cylinder) $= 0.2$.
2. Using this Re and Fig. 11.5 gives C_D(streamlined shape) $= 0.034$.

3. Drag force ratio (derived from Eq. 11.5) is

$$\frac{F_D(\text{streamlined shape})}{F_D(\text{cylinder})} = \frac{C_D(\text{streamlined shape})}{C_D(\text{cylinder})}$$

$$\times \left(\frac{A_p(\rho V_0^2/2)}{A_p(\rho V_0^2/2)} \right)$$

$$\frac{F_D(\text{streamlined shape})}{F_D(\text{cylinder})} = \frac{0.034}{0.2} = \boxed{0.17}$$

Review the Results and the Process

Discussion. The streamlining provided nearly a sixfold reduction in drag!

11.7 Drag in Compressible Flow

So far, this chapter has described drag for flows with constant density. This section describes drag when the density of a gas is changing due to pressure variations. These types of flow are called *compressible flows*. This information is important for modeling of projectiles such as bullets and rockets.

In steady flow, the influence of compressibility depends on the ratio of fluid velocity to the speed of sound. This ratio is a π-group called the Mach number.

The variation of drag coefficient with Mach number for three axisymmetric bodies is shown in Fig. 11.13. In each case, the drag coefficient increases only slightly with the Mach number at low Mach numbers and then increases sharply as transonic flow (M $\approx$ 1) is approached. Note that the rapid increase in drag coefficient occurs at a higher Mach number (closer to unity) if the body is slender with a pointed nose. The drag coefficient reaches a maximum at a Mach number somewhat larger than unity and then decreases as the Mach number is further increased.

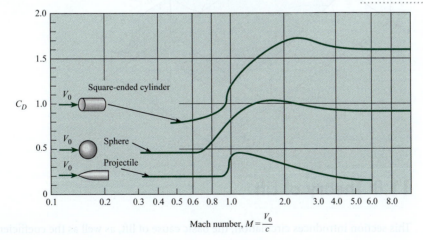

FIGURE 11.13

Drag characteristics of projectile, sphere, and cylinder with compressibility effects. [After Rouse (12)]

The slight increase in drag coefficient with low Mach numbers is attributed to an increase in form drag due to compressibility effects on the pressure distribution. However, as the flow velocity is increased, the maximum velocity on the body finally becomes sonic. The Mach number of the free-stream flow at which sonic flow first appears on the body is called the *critical Mach number*. Further increases in flow velocity result in local regions of supersonic flow (M > 1), which lead to wave drag due to shock wave formation and an appreciable increase in drag coefficient.

The critical Mach number for a sphere is approximately 0.6. Note in Fig. 11.13 that the drag coefficient begins to rise sharply at about this Mach number. The critical Mach number for the pointed body is larger, and correspondingly, the rise in drag coefficient occurs at a Mach number closer to unity.

The drag coefficient data for the sphere shown in Fig. 11.13 are for a Reynolds number of the order of 10^4. The data for the sphere shown in Fig. 11.9, on the other hand, are for very low Mach numbers. The question then arises about the general variation of the drag coefficient of a sphere with both Mach number and Reynolds number. Information of this nature is often needed to predict the trajectory of a body through the upper atmosphere or to model the motion of a nanoparticle.

A contour plot of the drag coefficient of a sphere versus both Reynolds and Mach numbers based on available data (19) is shown in Fig. 11.14. Notice the C_D-versus-Re curve from Fig. 11.9 in the M = 0 plane. Correspondingly, notice the C_D-versus-M curve from Fig. 11.13 in the Re = 10^4 plane. At low Reynolds numbers C_D decreases with an increasing Mach number, whereas at high Reynolds numbers the opposite trend is observed. Using this figure, the engineer can determine the drag coefficient of a sphere at any combination of Re and M. Of course, corresponding C_D contour plots can be generated for any body, provided the data are available.

FIGURE 11.14

Contour plot of the drag coefficient of the sphere versus Reynolds and Mach numbers.

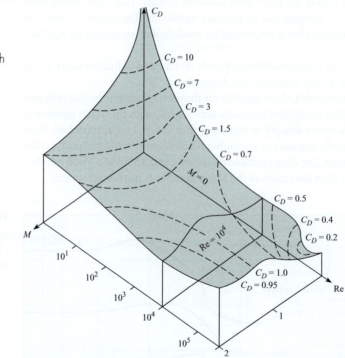

11.8 Theory of Lift

This section introduces circulation, the basic cause of lift, as well as the coefficient of lift.

Circulation

Circulation, a characteristic of a flow field, gives a measure of the average rate of rotation of fluid particles that are situated in an area that is bounded by a closed curve. Circulation is

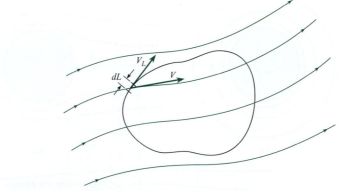

FIGURE 11.15
Concept of circulation.

defined by the path integral as shown in Fig. 11.15. Along any differential segment of the path, the velocity can be resolved into components that are tangent and normal to the path. Signify the tangential component of velocity as V_L. Integrate $V_L\, dL$ around the curve; the resulting quantity is called circulation, which is represented by the Greek letter Γ (capital gamma). Hence

$$\Gamma = \oint V_L\, dL \qquad (11.13)$$

Sign convention dictates that in applying Eq. (11.13), one uses tangential velocity vectors that have a counterclockwise sense around the curve as negative and take those that have a clockwise direction as having a positive contribution.[*] For example, consider finding the circulation for an irrotational vortex. The tangential velocity at any radius is C/r, where a positive C means a clockwise rotation. Therefore, if circulation is evaluated about a curve with radius r, the differential circulation is

$$d\Gamma = V_L\, dL = \frac{C}{r_1} r_1\, d\theta = C\, d\theta \qquad (11.14)$$

Integrate this around the entire circle:

$$\Gamma = \int_0^{2\pi} C\, d\theta = 2\pi C \qquad (11.15)$$

One way to induce circulation physically is to rotate a cylinder about its axis. Fig. 11.16a shows the flow pattern produced by such action. The velocity of the fluid next to the surface of the cylinder is equal to the velocity of the cylinder surface itself because of the no-slip condition that must prevail between the fluid and solid. At some distance from the cylinder, however, the velocity decreases with r, much like it does for the irrotational vortex. The next section shows how circulation produces lift.

Combination of Circulation and Uniform Flow around a Cylinder

Superpose the velocity field produced for uniform flow around a cylinder, Fig. 11.16b, onto a velocity field with circulation around a cylinder, Fig. 11.16a. Observe that the velocity is reinforced on the top side of the cylinder and reduced on the other side (Fig. 11.16c). Also observe that the stagnation points have both moved toward the low-velocity side of the cylinder. Consistent with the Bernoulli equation (assuming irrotational flow throughout), the pressure on the high-velocity side is lower than the pressure on the low-velocity side. Hence a pressure

[*]The sign convention is the opposite of that for the mathematical definition of a line integral.

FIGURE 11.16

Ideal flow around
a cylinder.
(a) Circulation.
(b) Uniform flow.
(c) Combination of
circulation and uniform
flow.

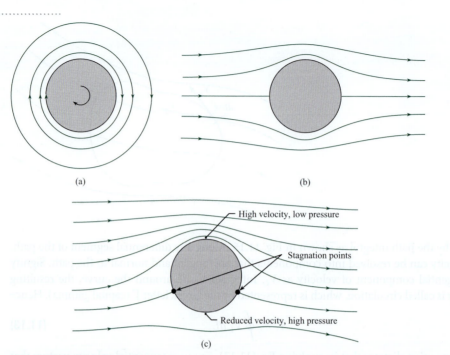

(a)

(b)

High velocity, low pressure

Stagnation points

Reduced velocity, high pressure

(c)

differential exists that causes a side thrust, or lift, on the cylinder. According to ideal flow theory, the lift per unit length of an infinitely long cylinder is given by $F_L/\ell = \rho V_0 \Gamma$, where F_L is the lift on the segment of length ℓ. For this ideal irrotational flow there is no drag on the cylinder. For the real-flow case, separation and viscous stresses do produce drag, and the same viscous effects will reduce the lift somewhat. Even so, the lift is significant when flow occurs past a rotating body or when a body is translating and rotating through a fluid. Hence the reason for the "curve" on a pitched baseball or the "drop" on a Ping-Pong ball is a fore spin. This phenomenon of lift produced by rotation of a solid body is called the **Magnus effect** after a nineteenth-century German scientist who made early studies of the lift on rotating bodies. A paper by Mehta (28) offers an interesting account of the motion of rotating sports balls.

Coefficients of lift and drag for the rotating cylinder with end plates are shown in Fig. 11.17. In this figure, the parameter $r\omega/V_0$ is the ratio of cylinder surface speed to the free-stream velocity, where r is the radius of the cylinder and ω is the angular speed in radians per second. The corresponding curves for the rotating sphere are given in Fig. 11.18.

Coefficient of Lift

The **coefficient of lift** is a parameter that characterizes the lift that is associated with a body. For example, a wing at a high angle of attack will have a high coefficient of lift, and a wing that has a zero angle of attack will have a low or zero coefficient of lift. The coefficient of lift is defined using a π-group:

$$C_L \equiv \frac{F_L}{A(\rho V_0^2/2)} = \frac{\text{lift force}}{(\text{reference area})(\text{dynamic pressure})} \tag{11.16}$$

To calculate lift force, engineers use the lift equation:

$$F_L = C_L A \left(\frac{\rho V_0^2}{2} \right) \tag{11.17}$$

where the reference area for a rotating cylinder or sphere is the projected area A_p.

FIGURE 11.17

Coefficients of lift and drag for a rotating cylinder. [After Rouse (12).]

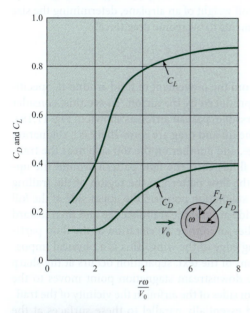

FIGURE 11.18

Coefficients of lift and drag for a rotating sphere. [After Barkla et al. (20). Reprinted with the permission of Cambridge University Press.]

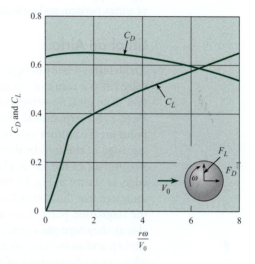

EXAMPLE 11.6

Lift on a Rotating Sphere

Problem Statement

A Ping-Pong ball is moving at 10 m/s in air and is spinning at 100 revolutions per second in the clockwise direction. The diameter of the ball is 3 cm. Calculate the lift and drag force and indicate the direction of the lift (up or down). The density of air is 1.2 kg/m³.

Define the Situation

A Ping-Pong ball is moving horizontally and rotating.

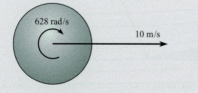

Properties: Air: $\rho = 1.2$ kg/m³

State the Goal

Find

1. Drag force (in newtons) on the ball
2. Lift force (in newtons) on the ball
3. The direction of lift (up or down?)

Generate Ideas and Make a Plan

1. Calculate the value of $r\omega/V_0$.
2. Use the value of $r\omega/V_0$ to look up the coefficients of lift and drag on Fig. 11.7.
3. Calculate lift force using Eq. (11.8).
4. Calculate drag force using Eq. (11.5).

Take Action (Execute the Plan)

The rotation rate in rad/s is

$$\omega = (100 \text{ rev/s})(2\pi \text{ rad/rev}) = 628 \text{ rad/s}$$

The rotational parameter is

$$\frac{\omega r}{V_0} = \frac{(628 \text{ rad/s})(0.015 \text{ m})}{10 \text{ m/s}} = 0.942$$

From Fig. 11.18, the lift coefficient is approximately 0.26, and the drag coefficient is 0.64. The lift force is

$$F_L = \frac{1}{2}\rho V_0^2 C_L A_p$$

$$= \frac{1}{2}(1.2 \text{ kg/m}^3)(10 \text{ m/s})^2(0.26)\frac{\pi}{4}(0.03 \text{ m})^2$$

$$= \boxed{1.10 \times 10^{-2} \text{ N}}$$

The lift force is downward. The drag force is

$$F_D = \frac{1}{2}\rho V_0^2 C_D A_p$$

$$= \boxed{27.1 \times 10^{-3} \text{ N}}$$

11.9 Lift and Drag on Airfoils

This section presents information on how to calculate lift and drag on winglike objects. Some typical applications include calculating the takeoff weight of an airplane, determining the size of wings needed, and estimating power requirements to overcome drag force.

Lift of an Airfoil

An **airfoil** is a body designed to produce lift from the movement of fluid around it. Specifically, lift is a result of circulation in the flow produced by the airfoil. To see this, consider flow of an ideal flow (nonviscous and incompressible) past an airfoil as shown in Fig. 11.19a. Here, as for irrotational flow past a cylinder, the lift and drag are zero. There is a stagnation point on the bottom side near the leading edge, and another on the top side near the trailing edge of the foil. In the real flow (viscous fluid) case, the flow pattern around the upstream half of the foil is plausible. However, the flow pattern in the region of the trailing edge, as shown in Fig. 11.19a, cannot occur. A stagnation point on the upper side of the foil indicates that fluid must flow from the lower side around the trailing edge and then toward the stagnation point. Such a flow pattern implies an infinite acceleration of the fluid particles as they turn the corner around the trailing edge of the wing. This is a physical impossibility, and as we have seen in previous sections of the text, separation occurs at the sharp edge. As a consequence of the separation, the downstream stagnation point moves to the trailing edge. Flow from both the top and bottom sides of the airfoil in the vicinity of the trailing edge then leaves the airfoil smoothly and essentially parallel to these surfaces at the trailing edge (Fig. 11.19b).

FIGURE 11.19

Patterns of flow around an airfoil.
(a) Ideal flow—no circulation.
(b) Real flow—circulation.

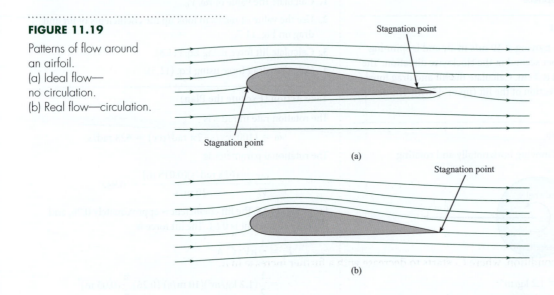

Stagnation point

Stagnation point

(a)

Stagnation point

(b)

To bring theory into line with the physically observed phenomenon, it was hypothesized that a circulation around the airfoil must be induced in just the right amount so that the downstream stagnation point is moved all the way back to the trailing edge of the airfoil, thus allowing the flow to leave the airfoil smoothly at the trailing edge. This is called the *Kutta condition* (21), named after a pioneer in aerodynamic theory. When analyses are made with this simple assumption concerning the magnitude of the circulation, very good agreement occurs between theory and experiment for the flow pattern and the pressure distribution, as well as for the lift on a two-dimensional airfoil section (no end effects). Ideal flow theory then shows that the magnitude of the circulation required to maintain the rear stagnation point at the trailing edge (the Kutta condition) of a symmetric airfoil with a small angle of attack is given by

$$\Gamma = \pi c V_0 \alpha \qquad \text{(11.18)}$$

where Γ is the circulation, c is the chord length of the airfoil, and α is the angle of attack of the chord of the airfoil with the free-stream direction (see Fig. 11.20 for a definition sketch).

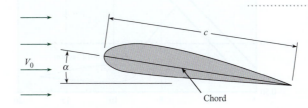

FIGURE 11.20

Definition sketch for an airfoil section.

Like that for the cylinder, the lift per unit length for an infinitely long wing is

$$F_L/\ell = \rho V_0 \Gamma$$

The planform area for the length segment ℓ is ℓc. Hence the lift on segment ℓ is

$$F_L = \rho V_0^2 \pi c \ell \alpha \qquad \text{(11.19)}$$

For an airfoil the coefficient of lift is

$$C_L = \frac{F_L}{S\rho V_0^2/2} \qquad \text{(11.20)}$$

where the reference area S is the planform area of the wing—that is, the area seen from the plan view. On combining Eqs. (11.18) and (11.19) and identifying S as the area associated with length segment ℓ, one finds that C_L for irrotational flow past a two-dimensional airfoil is given by

$$C_L = 2\pi\alpha \qquad \text{(11.21)}$$

Equations (11.19) and (11.21) are the theoretical lift equations for an infinitely long airfoil at a small angle of attack. Flow separation near the leading edge of the airfoil produces deviations (high drag and low lift) from the ideal flow predictions at high angles of attack. Hence experimental wind-tunnel tests are always made to evaluate the performance of a given type of airfoil section. For example, the experimentally determined values of lift coefficient versus α for two NACA airfoils are shown in Fig. 11.21. Note in this figure that the coefficient of lift increases with the angle of attack, α, to a maximum value and then decreases with further increase in α. This condition, where C_L starts to decrease with a further increase in α, is called **stall**. Stall occurs because of the onset of separation over the top of the airfoil, which changes the pressure distribution so that it not only decreases lift but also increases drag. Data for many other airfoil sections are given by Abbott and Von Doenhoff (22).

Airfoils of Finite Length—Effect on Drag and Lift

The drag of a two-dimensional foil at a low angle of attack (no end effects) is primarily viscous drag. However, wings of finite length also have an added drag and a reduced lift associated with

FIGURE 11.21

Values of C_l for two NACA airfoil sections. [After Abbott and Van Doenhoff (22).]

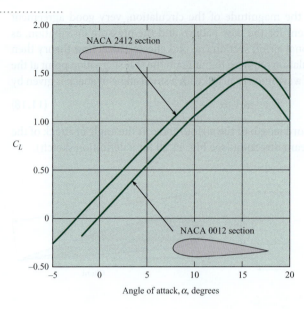

vortices generated at the wing tips. These vortices occur because the high pressure below the wing and the low pressure on top cause fluid to circulate around the end of the wing from the high-pressure zone to the low-pressure zone, as shown in Fig. 11.22. This induced flow has the effect of adding a downward component of velocity, w, to the approach velocity V_0. Hence, the "effective" free-stream velocity is now at an angle ($\phi \approx w/V_0$) to the direction of the original free-stream velocity, and the resultant force is tilted back as shown in Fig. 11.23. Thus the effective lift is smaller than the lift for the infinitely long wing because the effective angle of incidence is smaller. This resultant force has a component parallel to V_0 that is called the *induced drag* and is given by $F_L\phi$. Prandtl (23) showed that the induced velocity w for an elliptical spanwise lift distribution is given by the following equation:

$$w = \frac{2F_L}{\pi \rho V_0 b^2} \tag{11.22}$$

where b is the total length (or span) of the finite wing. Hence

$$F_{Di} = F_L\phi = \frac{2F_L^2}{\pi \rho V_0^2 b^2} = \frac{C_L^2}{\pi} \frac{S^2}{b^2} \frac{\rho V_0^2}{2} \tag{11.23}$$

From Eq. (11.23) it can be easily shown that the coefficient of induced drag, C_{Di}, is given by

$$C_{Di} = \frac{C_L^2}{\pi(b^2/S)} = \frac{C_L^2}{\pi \Lambda} \tag{11.24}$$

FIGURE 11.22

Formation of tip vortices.

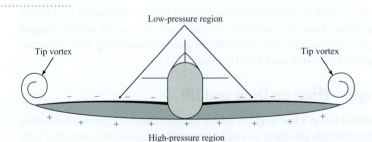

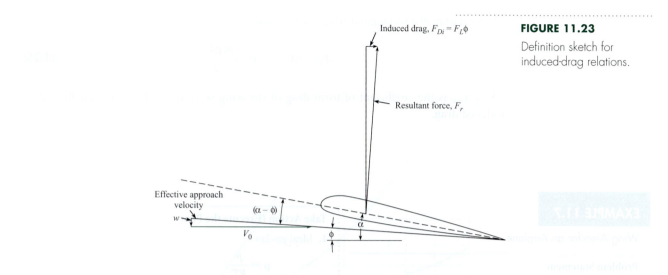

Induced drag, $F_{Di} = F_L \phi$

Resultant force, F_r

Effective approach velocity

w

V_0

$(\alpha - \phi)$

α

ϕ

FIGURE 11.23

Definition sketch for induced-drag relations.

which happens to represent the minimum induced drag for any wing planform. Here the ratio b^2/S is called the aspect ratio Λ of the wing, and S is the planform area of the wing. Thus, for a given wing section (constant C_L and constant chord c), longer wings (larger aspect ratios) have smaller induced-drag coefficients. The induced drag is a significant portion of the total drag of an airplane at low velocities and must be given careful consideration in airplane design. Aircraft (such as gliders) and even birds (such as the albatross and gull) that are required to be airborne for long periods of time with minimum energy expenditure are noted for their long, slender wings. Such a wing is more efficient because the induced drag is small. To illustrate the effect of finite span, look at Fig. 11.24, which shows C_L and C_D versus α for wings with several aspect ratios.

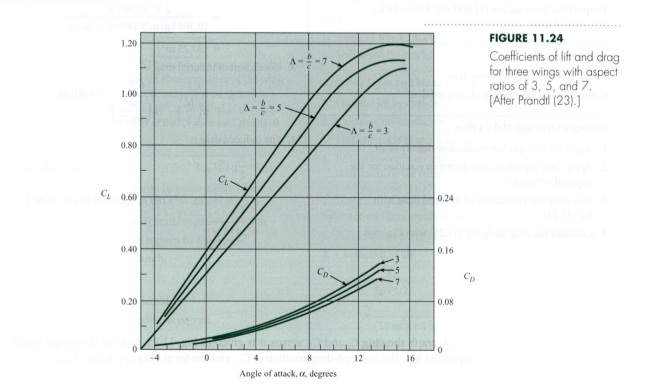

Angle of attack, α, degrees

FIGURE 11.24

Coefficients of lift and drag for three wings with aspect ratios of 3, 5, and 7. [After Prandtl (23).]

- (C_D) for a sphere can be found from charts and equations:
  - ▶ Stokes flow (Reynolds numbers < 0.5)

$$C_D = \frac{24}{\text{Re}}$$

  - ▶ Clift and Gauvin correlation (Re < 3×10^5)

$$C_D = \frac{24}{\text{Re}}(1 + 0.15\,\text{Re}^{0.687}) + \frac{0.42}{1 + 4.25 \times 10^4\,\text{Re}^{-1.16}}$$

- Drag of bluff bodies and streamlined bodies differs.
  - ▶ A *bluff body* is a body with flow separation when the Reynolds number is high enough. When flow separation occurs, the drag is mostly form drag.
  - ▶ A *streamlined body* does not have separated flow. Consequently, the drag force is mostly friction drag.
- (C_D) for cylinders and spheres drops dramatically at Reynolds numbers near 10^5 because the boundary layer changes from laminar to turbulent, moving the separation point downstream, reducing the wake region, and decreasing the form drag. This effect is called the *drag crisis*.

Rolling Resistance and Power

- To calculate the power to move a body such as a car or an airplane at a steady speed through a fluid, the usual approach is
  - ▶ Step 1. Draw a free body diagram.
  - ▶ Step 2. Apply the power equation in the form $P = FV$, where F, the force in the direction of motion, is evaluated from the free body diagram.
- The rolling resistance is the frictional force that occurs when an object such as a ball or tire rolls. The rolling resistance is calculated using

$$F_{\text{Rolling resistance}} = F_r = C_r N \tag{11.26}$$

where C_r is the coefficient of rolling resistance and N is the normal force.

Finding Terminal Velocity

- *Terminal velocity* is the steady-state speed of a body that is falling through a fluid.
- When a body has reached terminal velocity, the forces are balanced. These forces typically are weight, drag, and buoyancy.
- To find terminal velocity, sum the forces in the direction of motion and solve the resulting equation. The solution process often needs to be done using iteration (traditional method) or using a computer program (modern method).

Vortex Shedding, Streamlining, Compressible Flow

- Vortex shedding can cause beneficial effects (better mixing, better heat transfer) and detrimental effects (unwanted structural vibrations, noise, etc.).
  - ▶ *Vortex shedding* is when cylinders and bluff bodies in a cross-flow produce vortices that are released alternately from each side of the body.
  - ▶ The frequency of vortex shedding depends on a π-group called the *Strouhal number*.

- *Streamlining* involves designing a body to minimize the drag force. Usually, streamlining involves designing to reduce or minimize flow separation for a bluff body.
- In high-speed air flows, compressibility effects increase the drag.

The Lift Force

- The lift force on a body depends on four factors: shape, size, density of the flowing fluid, and speed squared. The working equation is.

$$F_L = C_L A \left(\frac{\rho V_0^2}{2} \right)$$

- The *coefficient of lift* (C_L) is a π-group defined by

$$C_L \equiv \frac{F_L}{A_{\text{Ref}}(\rho V_0^2/2)} = \frac{(\text{drag force})}{(\text{reference area})(\text{kinetic pressure})}$$

- *Circulation Theory of Lift.* The lift on an airfoil is due to the circulation produced by the airfoil on the surrounding fluid. This circulatory motion causes a change in the momentum of the fluid and a lift on the airfoil.
- The lift coefficient for a symmetric two-dimensional wing (no tip effect) is

$$C_L = 2\pi\alpha$$

where α is the angle of attack (expressed in radians) and the reference area is the product of the chord and a unit length of wing.

- As the angle of attack increases, the flow separates, the airfoil stalls, and the lift coefficient decreases.
- A wing of finite span produces trailing vortices that reduce the angle of attack and produce an induced drag.
- The drag coefficient corresponding to the minimum induced drag is

$$C_{Di} = \frac{C_L^2}{\pi(b^2/S)} = \frac{C_L^2}{\pi\Lambda}$$

where b is the wing span and S is the planform area of the wing.

REFERENCES

1. Bullivant, W. K. "Tests of the NACA 0025 and 0035 Airfoils in the Full Scale Wind Tunnel." *NACA Rept.,* 708 (1941).

2. DeFoe, G. L. "Resistance of Streamline Wires." *NACA Tech. Note,* 279 (March 1928).

3. Goett, H. J., and W. K. Bullivant. "Tests of NACA 0009, 0012, and 0018 Airfoils in the Full Scale Tunnel." *NACA Rept.,* 647 (1938).

4. Jacobs, E. N. "The Drag of Streamline Wires." *NACA Tech. Note,* 480 (December 1933).

5. Jones, G. W., Jr. "Unsteady Lift Forces Generated by Vortex Shedding about a Large, Stationary, and Oscillating Cylinder at High Reynolds Numbers." *Symp. Unsteady Flow, ASME* (1968).

6. Lindsey, W. F. "Drag of Cylinders of Simple Shapes." *NACA Rept.,* 619 (1938).

7. Miller, B. L., J. F. Mayberry, and I. J. Salter. "The Drag of Roughened Cylinders at High Reynolds Numbers." *NPL Rept. MAR Sci.,* R132 (April 1975).

8. Roshko, A. "Turbulent Wakes from Vortex Streets." *NACA Rept.,* 1191 (1954).

9. Abbott, I. H. "The Drag of Two Streamline Bodies as Affected by Protuberances and Appendages." *NACA Rept.,* 451 (1932).

10. Brevoort, M. J., and U. T. Joyner. "Experimental Investigation of the Robinson-Type Cup Anemometer." *NACA Rept.,* 513 (1935).

11.17 What drag is produced when a disk 0.75 m in diameter is submerged in water at 10°C and towed behind a boat at a speed of 4 m/s? Assume orientation of the disk so that maximum drag is produced.

11.18 PLUS A circular billboard having a diameter of 7 m is mounted so as to be freely exposed to the wind. Estimate the total force exerted on the structure by a wind that has a direction normal to the structure and a speed of 50 m/s. Assume $T = 10°C$ and $p = 101$ kPa absolute.

11.19 Consider a large rock situated at the bottom of a river and acted on by a strong current. Estimate a typical speed of the current that will cause the rock to move downstream along the bottom of the river. List and justify all your major assumptions.

11.20 Compute the overturning moment exerted by a 35 m/s wind on a smokestack that has a diameter of 2.5 m and a height of 75 m. Assume that the air temperature is 20°C and that p_a is 99 kPa absolute.

11.21 PLUS What is the moment at the bottom of a flagpole 20 m high and 8 cm in diameter in a 37.5 m/s wind? The atmospheric pressure is 100 kPa, and the temperature is 20°C.

11.22 PLUS A cylindrical anchor (vertical axis) made of concrete ($\gamma = 15$ kN/m³) is reeled in at a rate of 1.0 m/s by a man in a boat. If the anchor is 30 cm in diameter and 30 cm long, what tension must be applied to the rope to pull it up at this rate? Neglect the weight of the rope.

11.23 GO A Ping-Pong ball of mass 2.6 g and diameter 38 mm is supported by an air jet. The air is at a temperature of 18°C and a pressure of 91.4 kPa. What is the minimum speed of the air jet?

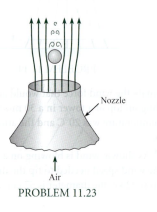

PROBLEM 11.23

11.24 Estimate the moment at ground level on a signpost supporting a sign measuring 3 m by 2 m if the wind is normal to the surface and has a speed of 35 m/s and the center of the sign is 4 m above the ground. Neglect the wind load on the post itself. Assume $T = 10°C$ and $p = 1$ atm.

11.25 PLUS Windstorms sometimes blow empty boxcars off their tracks. The dimensions of one type of boxcar are shown. What minimum wind velocity normal to the side of the car would be required to blow the car over?

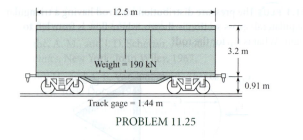

PROBLEM 11.25

11.26 A semiautomatic popcorn popper is shown. After the unpopped corn is placed in screen S, the fan F blows air past the heating coils C and then past the popcorn. When the corn pops, its projected area increases; thus it is blown up and into a container. Unpopped corn has a mass of about 0.15 g per kernel and an average diameter of approximately 6 mm. When the corn pops, its average diameter is about 18 mm. Within what range of airspeeds in the chamber will the device operate properly?

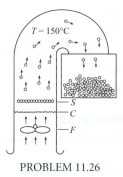

PROBLEM 11.26

11.27 Hoerner (15) presents data that show that fluttering flags of moderate-weight fabric have a drag coefficient (based on the flag area) of about 0.14. Thus the total drag is about 14 times the skin friction drag alone. Design a flagpole that is 30 m high and is to fly a flag 2 m high. Make your own assumptions regarding other required data.

Power, Energy, and Rolling Resistance (§11.2)

11.28 GO How much power is required to move a spherical-shaped submarine of diameter 1.5 m through seawater at a speed of 18.5 km/h? Assume the submarine is fully submerged.

11.29 A blimp flies at 9 m/s at an altitude where the specific weight of the air is 11 N/m³ and the kinematic viscosity is 1.2×10^{-5} m/s. The blimp has a length-to-diameter ratio of 5 and has a drag coefficient corresponding to the streamlined body in Fig. 11.9 (on p. 413 in §11.3). The diameter of the blimp is 24 m. What is the power required to propel the blimp at this speed?

11.30 PLUS Estimate the energy in joules and kcal (food calories) that a runner supplies to overcome aerodynamic drag during a 10 km race. The runner runs a 4:30 pace (i.e., each kilometer takes 4 minutes and 30 seconds). The product of frontal area and coefficient of drag is $C_D A = 0.74$ m². (One "food calorie" is equivalent to 4186 J.) Assume an air density of 1.22 kg/m³.

11.31 PLUS A cylindrical rod of diameter d and length L is rotated in still air about its midpoint in a horizontal plane. Assume the drag force at each section of the rod can be calculated assuming a two-dimensional flow with an oncoming velocity equal to the relative velocity component normal to the rod. Assume C_D is constant along the rod.

a. Derive an expression for the average power needed to rotate the rod.

b. Calculate the power for $\omega = 50$ rad/s, $d = 2$ cm, $L = 1.5$ m, $\rho = 1.2$ kg/m^3, and $C_D = 1.2$.

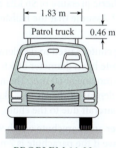

PROBLEM 11.31

11.32 PLUS Estimate the additional power (in kW) required for the truck when it is carrying the rectangular sign at a speed of 30 m/s over that required when it is traveling at the same speed but is not carrying the sign.

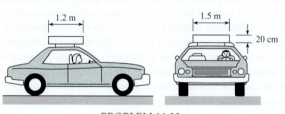

PROBLEM 11.32

11.33 Estimate the added power (in kW) required for the car when the cartop carrier is used and the car is driven at 100 km/h in a 25 km/h headwind over that required when the carrier is not used in the same conditions.

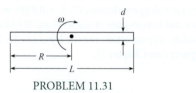

PROBLEM 11.33

11.34 PLUS The resistance to motion of an automobile consists of rolling resistance and aerodynamic drag. The weight of an automobile is 13,350 N, and it has a frontal area of 1.86 m^2. The drag coefficient is 0.30, and the coefficient of rolling friction is 0.02. Determine the percentage savings in gas mileage that one

achieves when one drives at 90 km/h instead of 105 km/h on a level road. Assume an air temperature of 15°C.

11.35 PLUS A car coasts down a very long hill. The weight of the car is 8900 N, and the slope of the grade is 6%. The rolling friction coefficient is 0.01. The frontal area of the car is 1.67 m^2, and the drag coefficient is 0.29. The density of the air is 1.03 kg/m^3. Find the maximum coasting speed of the car in km/h.

11.36 PLUS An automobile with a mass of 1000 kg is driven up a hill where the slope is 3° (5.2% grade). The automobile is moving at 30 m/s. The coefficient of rolling friction is 0.02, the drag coefficient is 0.4, and the cross-sectional area is 4 m^2. Find the power (in kW) needed for this condition. The air density is 1.2 kg/m^3.

11.37 PLUS A bicyclist is coasting down a hill with a slope of 4° into a headwind (measured with respect to the ground) of 7 m/s. The mass of the cyclist and bicycle is 80 kg, and the coefficient of rolling friction is 0.02. The drag coefficient is 0.5, and the projected area is 0.5 m^2. The air density is 1.2 kg/m^3. Find the speed of the bicycle in meters per second.

11.38 PLUS A bicyclist is capable of delivering 275 W of power to the wheels. How fast can the bicyclist travel in a 3 m/s headwind if his or her projected area is 0.5 m^2, the drag coefficient is 0.3, and the air density is 1.2 kg/m^3? Assume the rolling resistance is negligible.

11.39 PLUS Assume that the kilowatt power of the engine in the original 1932 Fiat Balillo (see Table 11.2 on p. 433 of §11.10) was 30 brake kW and that the maximum speed at sea level was 100 km/h. Also assume that the projected area of the automobile is 2.8 m^2. Assume that the automobile is now fitted with a modern 164 brake kW motor with a weight equal to the weight of the original motor; thus the rolling resistance is unchanged. What is the maximum speed of the "souped-up" Balillo at sea level?

11.40 One way to reduce the drag of a blunt object is to install vanes to suppress the amount of separation. Such a procedure was used on model trucks in a wind-tunnel study by Kirsch and Bettes. For tests on a van-type truck, they noted that without vanes the C_D was 0.78. However, when vanes were installed around the top and side leading edges of the truck body (see the figure), a 25% reduction in C_D was achieved. For a truck with a projected area of 8.36 m^2, what reduction in drag force will be effected by installation of the vanes when the truck travels at 100 km/h? Assume standard atmospheric pressure and a temperature of 20°C.

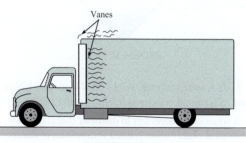

PROBLEM 11.40

11.41 For the truck of Prob. 11.40, assume that the total resistance is given by $R = F_D + C$, where F_D is the air drag and C is the resistance due to bearing friction. If C is constant at 350 N for the given truck, what fuel-savings percentage will be effected by the installation of the vanes when the truck travels at 100 km/h?

Terminal Velocity (§11.4)

11.42 Suppose you are designing an object to fall through seawater with a terminal velocity of exactly 1 m/s. What variables will have the most influence on the terminal velocity? List these variables and justify your decisions.

11.43 *PLUS* As shown, a 35-cm-diameter emergency medicine parachute supporting a mass of 20 g is falling through air (20°C). Assume a coefficient of drag of $C_D = 2.2$, and estimate the terminal velocity V_0. Use a projected area of $(\pi D^2)/4$.

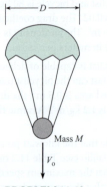

PROBLEM 11.43

11.44 Consider a small air bubble (approximately 4 mm diameter) rising in a very tall column of liquid. Will the bubble accelerate or decelerate as it moves upward in the liquid? Will the drag of the bubble be largely skin friction or form drag? Explain.

11.45 *GO* Determine the terminal velocity in water ($T = 10°C$) of a 8-cm ball that weighs 15 N in air.

11.46 *PLUS* This cube is weighted so that it will fall with one edge down as shown. The cube weighs 22.2 N in air. What will be its terminal velocity in water?

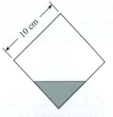

PROBLEM 11.46

11.47 *PLUS* A spherical rock weighs 30 N in air and 5 N in water. Estimate its terminal velocity as it falls in water (20°C).

11.48 A spherical balloon 2 m in diameter that is used for meteorological observations is filled with helium at standard conditions. The empty weight of the balloon is 3 N. What velocity of ascent will it attain under standard atmospheric conditions?

11.49 A sphere 2 cm in diameter rises in oil at a velocity of 1.5 cm/s. What is the specific weight of the sphere if the oil density is 900 kg/m³ and the dynamic viscosity is 0.096 N · s/m²?

11.50 *PLUS* Estimate the terminal velocity of a 1.5-mm plastic sphere in oil. The oil has a specific gravity of 0.95 and a kinematic viscosity of 10^{-4} m²/s. The plastic has a specific gravity of 1.07. The volume of a sphere is given by $\pi D^3/6$.

11.51 *GO* A 534 N skydiver is free-falling at an altitude of 1980 m. Estimate the terminal velocity in mph for minimum and maximum drag conditions. At maximum drag conditions, the product of frontal area and coefficient of drag is $C_D A = 0.743$ m². At minimum drag conditions, $C_D A = 0.0929$ m². Assume the pressure and temperature at sea level are 101 kPa abs and 15°C. To calculate air properties, use the lapse rate for the U.S. standard atmosphere (see Chapter 3).

PROBLEM 11.51

11.52 What is the terminal velocity of a 0.5-cm hailstone in air that has an atmospheric pressure of 96 kPa absolute and a temperature of 0°C? Assume that the hailstone has a specific weight of 6 kN/m³.

11.53 *PLUS* A drag chute is used to decelerate an airplane after touchdown. The chute has a diameter of 3.6 m and is deployed when the aircraft is moving at 60 m/s. The mass of the aircraft is 9000 kg, and the density of the air is 1.2 kg/m³. Find the initial deceleration of the aircraft due to the chute.

11.54 A paratrooper and parachute weigh 900 N. What rate of descent will they have if the parachute is 7 m in diameter and the air has a density of 1.20 kg/m³?

11.55 If a balloon weighs 0.10 N (empty) and is inflated with helium to a diameter of 60 cm, what will be its terminal velocity in air (standard atmospheric conditions)? The helium is at standard conditions.

11.56 A 2-cm plastic ball with a specific gravity of 1.2 is released from rest in water at 20°C. Find the time and distance needed to achieve 99% of the terminal velocity. Write out the equation of motion by equating the mass times acceleration to the buoyant force, weight, and drag force and solve by developing a computer program or using available software. Use Eq. (11.9) on p. 414 in §11.3, for the drag coefficient. [*Hint*: The equation of motion can be expressed in the form

$$\frac{dv}{dt} = -\left(\frac{C_D \, \text{Re}}{24}\right) \frac{18\mu}{\rho_b d^2} v + \frac{\rho_b - \rho_w}{\rho_b} g$$

where ρ_b is the density of the ball and ρ_w is the density of the water. This form avoids the problem of the drag coefficient approaching infinity when the velocity approaches zero because $C_D \text{Re}/24$ approaches unity as the Reynolds number approaches zero. An "if-statement" is needed to avoid a singularity in Eq. (11.9) when the Reynolds number is zero.]

Theory of Lift (§11.8)

11.57 From the following list, select one topic that is interesting to you. Then, use references such as the Internet to research your topic and prepare one page of written documentation that you could use to present your topic to your peers.

    a. Explain how an airplane works.

    b. Describe the aerodynamics of a flying bird.

    c. Explain how a propeller produces thrust.

    d. Explain how a kite flies.

11.58 Apply the grid method to each situation that follows.

    a. Use Eq. (11.17), on p. 424 in §11.8, to predict the lift force in newtons for a spinning baseball. Use a coefficient of lift of $C_L = 1.2$. The speed of the baseball is 145 km/h. Calculate area using $A = \pi r^2$, where the radius of a baseball is $r = 37$ mm. Assume a hot summer day.

    b. Use Eq. (11.17), on p. 424 in §11.8, to predict the size of wing in mm^2 needed for a model aircraft that has a mass of 570 g. Wing size is specified by giving the wing area (A) as viewed by an observer looking down on the wing. Assume the airplane is traveling at 130 km/h on a hot summer day. Use a coefficient of lift of $C_L = 1.2$. Assume straight and level flight so lift force balances weight.

11.59 Using §11.8 and other resources, answer the following questions. Strive for depth, clarity, and accuracy. Also, use effective sketches, words, and equations.

    a. What is circulation? Why is it important?

    b. What is lift force?

    c. What variables influence the magnitude of the lift force?

11.60 PLUS The baseball is thrown from west to east with a spin about its vertical axis as shown. Under these conditions it will "break" toward the (a) north, (b) south, or (c) neither.

Plan view

PROBLEM 11.60

11.61 Analyses of pitched baseballs indicate that C_L of a rotating baseball is approximately three times that shown in Fig. 11.18 (on p. 425 in §11.8). This greater C_L is due to the added circulation caused by the seams of the ball. What is the lift of a ball pitched at a speed of 137 km/h and with a spin rate of 35 rps? Also, how much will the ball be deflected from its original path by the time it gets to the plate as a result of the lift force? *Note:* The mound-to-plate distance is 18 m, the weight of the baseball is 150 g, and the circumference is 22.5 cm. Assume standard atmospheric conditions, and assume that the axis of rotation is vertical.

Lift and Drag on Airfoils (§11.9)

11.62 As shown, a glider traveling at a constant velocity will move along a straight glide path that has an angle θ with respect to the horizontal. The angle θ, also called the glide ratio, is given by $\theta = (C_D/C_L)$. Use basic principles to prove the preceding statement.

PROBLEM 11.62

11.63 PLUS A sphere of diameter 100 mm, rotating at a rate of 286 rpm, is situated in a stream of water (15°C) that has a velocity of 1.5 m/s. Determine the lift force (in newtons) on the rotating sphere.

11.64 An airplane wing having the characteristics shown in Fig. 11.24 (on p. 429 in §11.9) is to be designed to lift 8000 N when the airplane is cruising at 60 m/s with an angle of attack of 3°. If the chord length is to be 1 m, what span of wing is required? Assume $\rho = 1.24$ kg/m^3.

11.65 A boat of the hydrofoil type has a lifting vane with an aspect ratio of 4 that has the characteristics shown in Fig. 11.24 (on p. 429 in §11.9). If the angle of attack is 4° and the weight of the boat is 9535 kg, what foil dimensions are needed to support the boat at a velocity of 18 m/s?

11.66 One wing (wing A) is identical (same cross section) to another wing (wing B) except that wing B is twice as long as wing A. Then for a given wind speed past both wings and with the same angle of attack, one would expect the total lift of wing B to be (a) the same as that of wing A, (b) less than that of wing A, (c) double that of wing A, or (d) more than double that of wing A.

11.67 What happens to the value of the induced drag coefficient for an aircraft that increases speed in level flight? (a) it increases, (b) it decreases, (c) it does not change.

11.68 PLUS The total drag coefficient for an airplane wing is $C_D = C_{D0} + C_L^2/\pi\Lambda$, where C_{D0} is the form drag coefficient, C_L is the lift coefficient and Λ is the aspect ratio of the wing. The power is given by $P = F_D V = 1/2\, C_D \rho V^3 S$. For level flight the lift is equal to the weight, so $W/S = 1/2\rho C_L V^2$, where W/S is called the "wing loading." Find an expression for V for which the power is a minimum in terms of $V_{\text{MinPower}} = f(\rho, \Lambda, W/S, C_{D0})$, and find the V for minimum power when $\rho = 1$ kg/m^3, $\Lambda = 10$, $W/S = 600$ N/m^2, and $C_D = 0.02$.

11.69 The airstream affected by the wing of an airplane can be considered to be a cylinder (stream tube) with a diameter equal to the wingspan, b. Far downstream from the wing, the tube is deflected through an angle θ from the original direction. Apply the momentum equation to the stream tube between sections 1 and 2 and find the lift of the wing as a function of b, ρ, V, and θ. Relating the lift to the lift coefficient, find θ as a function of b, C_L, and wing area, S. Using the relation for induced drag, $F_{Di} = F_L\theta/2$, show that $C_{Di} = C_L^2/\pi\Lambda$, where Λ is the wing aspect ratio.

End view   Side view
PROBLEM 11.69

11.70 The landing speed of an airplane is 8 m/s faster than its stalling speed. The lift coefficient at landing speed is 1.2, and the maximum lift coefficient (stall condition) is 1.4. Calculate both the landing speed and the stalling speed.

11.71 An airplane has a rectangular-planform wing that has an elliptical spanwise lift distribution. The airplane has a mass of 1000 kg, a wing area of 16 m², and a wingspan of 10 m, and it is flying at 50 m/s at 3000 m altitude in a standard atmosphere. If the form drag coefficient is 0.01, calculate the total drag on the wing and the power ($P = F_D V$) necessary to overcome the drag.

11.72 The figure shows the relative pressure distribution for a Göttingen 387-FB lifting vane (19) when the angle of attack is 8°. If such a vane with a 20-cm chord were used as a hydrofoil at a depth of 70 cm, at what speed in 10°C freshwater would

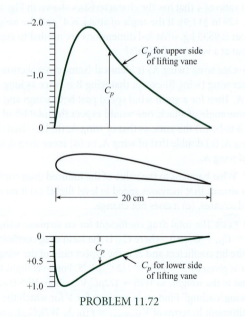

PROBLEM 11.72

cavitation begin? Also, estimate the lift per unit of length of foil at this speed.

11.73 Consider the distribution of C_p as given for the wing section in Prob. 11.72. For this distribution of C_p, the lift coefficient C_L will fall within which range of values: (a) $0 < C_L < 1.0$; (b) $1.01 < C_L < 2.0$; (c) $2.01 < C_L < 3.0$; or (d) $3.0 < C_L$?

11.74 The total drag coefficient for a wing with an elliptical lift distribution is $C_D = C_{D_0} + C_L^2/\pi\Lambda$, where Λ is the aspect ratio. Derive an expression for C_L that corresponds to minimum C_D/C_L (maximum C_L/C_D) and the corresponding C_L/C_D.

11.75 PLUS A glider at 800 m altitude has a mass of 180 kg and a wing area of 20 m². The glide angle is 1.7°, and the air density is 1.2 kg/m³. If the lift coefficient of the glider is 0.83, how many minutes will it take to reach sea level on a calm day?

11.76 The wing loading on an airplane is defined as the aircraft weight divided by the wing area. An airplane with a wing loading of 2000 N/m² has the aerodynamic characteristics given by Fig. 11.25 (on p. 431 in §11.9). Under cruise conditions the lift coefficient is 0.3. If the wing area is 10 m², find the drag force.

11.77 An ultralight airplane has a wing with an aspect ratio of 5 and with lift and drag coefficients corresponding to Fig. 11.24 (on p. 429 in §11.9). The planform area of the wing is 18.6 m². The weight of the airplane and pilot is 1800 N. The airplane flies at 15 m/s in air with a density of 1.03 kg/m³. Find the angle of attack and the drag force on the wing.

11.78 Your objective is to design a human-powered aircraft using the characteristics of the wing in Fig. 11.24 (on p. 429 in §11.9). The pilot weighs 600 N and is capable of outputting 373 W of continuous power. The aircraft without the wing has a weight of 180 N, and the wing can be designed with a weight of 5.7 N/m² of wing area. The drag consists of the drag of the structure plus the drag of the wing. The drag coefficient of the structure, C_{D0} is 0.05, so that the total drag on the craft will be

$$F_D = (C_{D_0} + C_D)\frac{1}{2}\rho V_0^2 S$$

where C_D is the drag coefficient from Fig. 11.24 (on p. 429 in §11.9). The power required is equal to $F_D V_0$. The air density is 1.23 kg/m³. Assess whether the airfoil is adequate, and if it is, find the optimum design (wing area and aspect ratio).

COMPRESSIBLE FLOW 12

FIGURE 12.1

The de Laval nozzle is used to accelerate a gas to supersonic speeds. This nozzle is used in turbines, rocket engines, and supersonic jet engines.

This particular nozzle was designed by Andrew Donelick under the guidance of Dr. John Crepeau, Professor of Mechanical Engineering at the University of Idaho. The nozzle was built by Russ Porter, also at the University of Idaho. (Photo by Donald Elger.)

Chapter Road Map

The compressibility effects in gas flows become significant when the Mach number exceeds 0.3. The performance of high-speed aircraft, the flow in rocket nozzles, and the reentry mechanics of spacecraft require inclusion of compressible flow effects. This chapter introduces topics in compressible flow.

Learning Objectives

STUDENTS WILL BE ABLE TO

- Describe the propagation of a sound wave. (§12.1)
- Explain the significance of the Mach number. (§12.1)
- Calculate the speed of sound and Mach number. (§12.1)
- Describe how pressure and temperature vary for flow along a streamline in compressible flow. (§12.2)
- Describe a normal shock wave. (§12.3)
- Calculate property change across a normal shock wave. (§12.3)
- In a de Laval nozzle, describe how flow properties vary. Also, calculate the mass flow rate and Mach number. (§12.4).

12.1 Wave Propagation in Compressible Fluids

Wave propagation in a fluid is the mechanism through which the presence of boundaries is communicated to the flowing fluid. In a liquid the propagation speed of the pressure wave is much higher than the flow velocities, so the flow has adequate time to adjust to a change in boundary shape. Gas flows, on the other hand, can achieve speeds that are comparable to and even exceed the speed at which pressure disturbances are propagated. In this situation, with compressible fluids, the propagation speed is an important parameter and must be incorporated into the flow analysis. In this section it will be shown how the speed of an

Example 12.1 illustrates the calculation of sound speed for a given temperature.

EXAMPLE 12.1

Speed of Sound Calculation

Define the Situation

Air is at 15°C.

Assume: Air is an ideal gas.

Air: Table A.2: $R = 287$ J/kg K, and $k = 1.4$

State the Goal

Calculate the speed of sound.

Generate Ideas and Make a Plan

Apply the speed of sound equation, Eq. (12.11), with $T = 288$ K.

Take Action (Execute the Plan)

$$c = \sqrt{kRT}$$

$$c = [(1.4)(287 \text{ J/kg K})(288 \text{ K})]^{1/2} = \boxed{340 \text{ m/s}}$$

Review the Solution and the Process

Knowledge. The absolute temperature must always be used in speed of sound equation.

It is possible to demonstrate, in a very simple way, the significance of sound in a compressible flow. Consider the airfoil traveling at speed V in Fig. 12.4. As this airfoil travels through the fluid, the pressure disturbance generated by the airfoil's motion propagates as a wave at sonic speed ahead of the airfoil. These pressure disturbances travel a considerable distance ahead of the airfoil before being attenuated by the viscosity of the fluid, and they "warn" the upstream fluid that the airfoil is coming. In turn, the fluid particles begin to move apart in such a way that there is a smooth flow over the airfoil by the time it arrives. If a pressure disturbance created by the airfoil is essentially attenuated in time Δt, then the fluid at a distance $\Delta t(c - V)$ ahead is alerted to prepare for the airfoil's impending arrival.

FIGURE 12.4

Propagation of a sound wave by an airfoil.

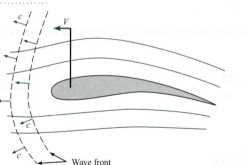

Wave front

What happens as the speed of the airfoil is increased? Obviously, the relative velocity $c - V$ is reduced, and the upstream fluid has less time to prepare for the airfoil's arrival. The flow field is modified by smaller streamline curvatures, and the form drag on the airfoil is increased. If the airfoil speed increases to the speed of sound or greater, the fluid has no warning whatsoever that the airfoil is coming and cannot prepare for its arrival. Nature, at this point, resolves the problem by creating a shock wave that stands off the leading edge, as shown in Fig. 12.5. As the fluid passes through the shock wave near the leading edge, it is decelerated to a speed less than sonic speed and therefore has time to divide and flow around the airfoil. Shock waves will be treated in more detail in Section 12.3.

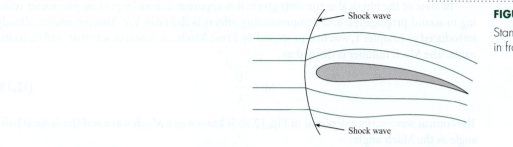

FIGURE 12.5
Standing shock wave
in front of an airfoil.

Another approach to appreciating the significance of sound propagation in a compressible fluid is to consider a point source of sound moving in a quiescent fluid, as shown in Fig. 12.6. The sound source is moving at a speed less than the local sound speed in Fig. 12.6a and faster than the local sound speed in Fig. 12.6b. At time $t = 0$ a sound pulse is generated and propagates radially outward at the local speed of sound. At time t_1 the sound source has moved a distance Vt_1, and the circle representing the sound wave emitted at $t = 0$ has a radius of ct_1. The sound source emits a new sound wave at t_1 that propagates radially outward. At time t_2 the sound source has moved to Vt_2, and the sound waves have moved outward as shown.

When the sound source moves at a speed less than the speed of sound, the sound waves form a family of nonintersecting eccentric circles, as shown in Fig. 12.6a. For an observer stationed at A the frequency of the sound pulses would appear higher than the emitted frequency because the sound source is moving toward the observer. In fact, the observer at A will detect a frequency of

$$f = f_0/(1 - V/c)$$

where f_0 is the emitting frequency of the moving sound source. This change in frequency is known as the **Doppler effect**.

When the sound source moves faster than the local sound speed, the sound waves intersect and form the locus of a cone with a half-angle of

$$\theta = \sin^{-1}(c/V)$$

The observer at A will not detect the sound source until it has passed. In fact, only an observer within the cone is aware of the moving sound source.

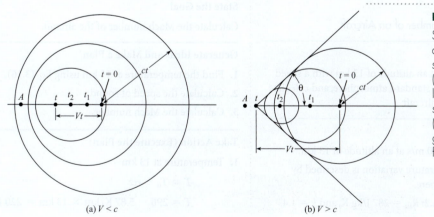

(a) $V < c$ (b) $V > c$

FIGURE 12.6
Sound field generated by
a moving point source of
sound.
(a) Speed of sound source
less than local sound
source.
(b) Speed of sound source
greater than local sound
force.

In view of the physical arguments given, it is apparent that an important parameter relating to sound propagation and compressibility effects is the ratio V/c. This parameter, already introduced in Chapter 1, was first proposed by Ernst Mach, an Austrian scientist, and bears his name. The Mach number is defined as

$$M = \frac{V}{c}$$

(12.12)

The conical wave surface depicted in Fig. 12.6b is known as a **Mach wave** and the conical half-angle as the **Mach angle**.

✔ **CHECKPOINT PROBLEM 12.1**

Consider two airplanes, A and B, flying at identical speeds. Airplane B is at a higher altitude where the atmospheric pressure and temperature are lower. Which statement is correct?

a. (Mach number of plane A) > (Mach number of plane B)

b. (Mach number of plane A) < (Mach number of plane B)

c. (Mach number of plane A) = (Mach number of plane B)

Besides the qualitative argument presented for the Mach number, it is also recalled from Chapter 8 that the Mach number is the ratio of the inertial to elastic forces acting on the fluid. If the Mach number is small, the inertial forces are ineffective in compressing the fluid, and the fluid can be regarded as incompressible.

Compressible flows are characterized by their Mach number regimes as follows:

$$M < 1 \qquad \text{subsonic flow}$$
$$M \approx 1 \qquad \text{transonic flow}$$
$$M > 1 \qquad \text{supersonic flow}$$

Flows with Mach numbers exceeding 5 are sometimes referred to as **hypersonic**. Airplanes designed to travel near sonic speeds and faster are equipped with Mach meters because of the significance of the Mach number with respect to aircraft performance.

Evaluation of the Mach number of an airplane flying at altitude is demonstrated in Example 12.2.

EXAMPLE 12.2

Calculating the Mach Number of an Aircraft

Problem Statement

An F-16 fighter is flying at an altitude of 13 km with a speed of 470 m/s. Assume a U.S. standard atmosphere, and calculate the Mach number of the aircraft.

Define the Situation

A fighter jet is flying at 470 m/s at an altitude of 13 km.

Assumptions: The temperature variation is described by the U.S. standard atmosphere.

Properties: From Table A.2: $R_{air} = 287$ J/kg K, and $k = 1.4$.

State the Goal

Calculate the Mach number of the aircraft.

Generate Ideas and Make a Plan

1. Find the temperature at 13 km using Eq. (3.16).
2. Calculate the speed of sound.
3. Calculate the Mach number.

Take Action (Execute the Plan)

1. Temperature at 13 km

$$T = T_0 - \alpha z$$

$$T = 296 - 5.87 \text{ K/km} \times 13 \text{ km} = 220 \text{ K}$$

2. Speed of sound

$$c = \sqrt{kRT} = \sqrt{1.4 \times 287 \times 220} = 297 \text{ m/s}$$

3. Mach number

$$M = \frac{V}{c} = \frac{470 \text{ m/s}}{297 \text{ m/s}} = \boxed{1.58}$$

Review the Solution and the Process

Discussion. The aircraft is flying at supersonic speed.

12.2 Mach Number Relationships

In this section it will be shown how fluid properties vary the Mach number in compressible flows. Consider a control volume bounded by two streamlines in a steady compressible flow, as shown in Fig. 12.7. Applying the energy equation, to this control volume, gives

$$-\dot{m}_1\left(h_1 + \frac{V_1^2}{2} + gz_1\right) + \dot{m}_2\left(h_2 + \frac{V_2^2}{2} + gz_2\right) = \dot{Q} \qquad (12.13)$$

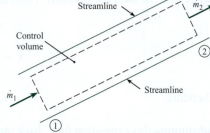

FIGURE 12.7

Control volume enclosed by streamlines.

The elevation terms (z_1 and z_2) can usually be neglected for gaseous flows. If the flow is adiabatic ($\dot{Q} = 0$), the energy equation reduces to

$$\dot{m}_1\left(h_1 + \frac{V_1^2}{2}\right) = \dot{m}_2\left(h_2 + \frac{V_2^2}{2}\right) \qquad (12.14)$$

From the principle of continuity, the mass flow rate is constant, $\dot{m}_1 = \dot{m}_2$, so

$$h_1 + \frac{V_1^2}{2} = h_2 + \frac{V_2^2}{2} \qquad (12.15)$$

Because positions 1 and 2 are arbitrary points on the same streamline, one can say that

$$h + \frac{V^2}{2} = \text{constant along a streamline in an adiabatic flow} \qquad (12.16)$$

The constant in this expression is called the **total enthalpy**, h_t. It is the enthalpy that would arise if the flow velocity were brought to zero in an adiabatic process. Thus the energy equation along a streamline under adiabatic conditions is

$$h + \frac{V^2}{2} = h_t \qquad (12.17)$$

If h_t is the same for all streamlines, the flow is **homenergic**.

It is instructive at this point to compare Eq. (12.17) with the Bernoulli equation. Expressing the specific enthalpy as the sum of the specific internal energy and p/ρ, Eq. (12.17) becomes

$$u + \frac{p}{\rho} + \frac{V^2}{2} = \text{constant}$$

If the fluid is incompressible and there is no heat transfer, the specific internal energy is constant and the equation reduces to the Bernoulli equation (excluding the pressure change due to elevation change).

Temperature

The enthalpy of an ideal gas can be written as

$$h = c_p T \tag{12.18}$$

where c_p is the specific heat at constant pressure. Substituting this relation into Eq. (12.17) and dividing by $c_p T$, results in

$$1 + \frac{V^2}{2c_p T} = \frac{T_t}{T} \tag{12.19}$$

where T_t is the **total temperature**. From thermodynamics (1) it is known for an ideal gas that

$$c_p - c_v = R \tag{12.20}$$

or

$$k - 1 = \frac{R}{c_v} = \frac{kR}{c_p}$$

Therefore

$$c_p = \frac{kR}{k-1} \tag{12.21}$$

Substituting this expression for c_p back into Eq. (12.19) and realizing that kRT is the speed of sound squared results in the *total temperature* equation

$$T_t = T\left(1 + \frac{k-1}{2}M^2\right) \tag{12.22}$$

The temperature T is called the **static temperature**—the temperature that would be registered by a thermometer moving with the flowing fluid. **Total temperature** is analogous to total enthalpy in that it is the temperature that would arise if the velocity were brought to zero adiabatically. If the flow is adiabatic, the total temperature is constant along a streamline. If not, the total temperature varies according to the amount of thermal energy transferred.

Example 12.3 illustrates the evaluation of the total temperature on an aircraft's surface.

EXAMPLE 12.3

Total Temperature Calculation

Problem Statement

An aircraft is flying at M = 1.6 at an altitude where the atmospheric temperature is −50°C. The temperature on the aircraft's surface is approximately the total

temperature. Estimate the surface temperature, taking k = 1.4.

Define the Situation

An aircraft is flying at M = 1.6. The static temperature is 50°C.

State the Goal

Calculate the total temperature.

Generate Ideas and Make a Plan

This problem can be visualized as the aircraft being stationary and an airstream with a static temperature of $-50°C$ flowing past the aircraft at a Mach number of 1.6.

1. Convert the local static temperature to degrees K.

2. Use total temperature equation, Eq. (12.22).

Take Action (Execute the Plan)

1. Static temperature in absolute temperature units

$$T = 273 - 50 = 223 \text{ K}$$

2. Total temperature

$$T_t = 223[1 + 0.2(1.6)^2] = \boxed{337 \text{ K or } 64°C}$$

If the flow is isentropic, thermodynamics shows that the following relationship for pressure and temperature of an ideal gas between two points on a streamline is valid (1):

$$\frac{p_1}{p_2} = \left(\frac{T_1}{T_2}\right)^{k/(k-1)} \tag{12.23}$$

Isentropic flow means that there is no heat transfer, so the total temperature is constant along the streamline. Therefore

$$T_t = T_1\left(1 + \frac{k-1}{2}M_1^2\right) = T_2\left(1 + \frac{k-1}{2}M_2^2\right) \tag{12.24}$$

Solving for the ratio T_1/T_2 and substituting into Eq. (12.23) shows that the pressure variation with the Mach number is given by

$$\frac{p_1}{p_2} = \left\{\frac{1 + [(k-1)/2]M_2^2}{1 + [(k-1)/2]M_1^2}\right\}^{k/(k-1)} \tag{12.25}$$

In the ideal gas law used to derive Eq. (12.23), absolute pressures must always be used in calculations with these equations.

The **total pressure** in a compressible flow is given by

$$p_t = p\left(1 + \frac{k-1}{2}M^2\right)^{k/(k-1)} \tag{12.26}$$

which is the pressure that would result if the flow were decelerated to zero speed reversibly and adiabatically. Unlike total temperature, total pressure may not be constant along streamlines in adiabatic flows. For example, it will be shown that flow through a shock wave, although adiabatic, is not reversible and, therefore, not isentropic. The total pressure variation along a streamline in an adiabatic flow can be obtained by substituting Eqs. (12.26) and (12.24) into Eq. (12.25) to give

$$\frac{p_{t_1}}{p_{t_2}} = \frac{p_1}{p_2}\left\{\frac{1 + [(k-1)/2]M_1^2}{1 + [(k-1)/2]M_2^2}\right\}^{k/(k-1)} = \frac{p_1}{p_2}\left(\frac{T_2}{T_1}\right)^{k/(k-1)} \tag{12.27}$$

Unless the flow is also reversible and Eq. (12.23) is applicable, the total pressures at points 1 and 2 will not be equal. However, if the flow is isentropic, total pressure is constant along streamlines.

Density

Analogous to the total pressure, the **total density** in a compressible flow is given by

$$\rho_t = \rho\left(1 + \frac{k-1}{2}M^2\right)^{1/(k-1)} \tag{12.28}$$

where ρ is the local or static density. If the flow is isentropic, then ρ_t is a constant along streamlines, and Eq. (12.28) can be used to determine the variation of gas density with the Mach number.

In literature dealing with compressible flows, one often finds reference to "stagnation" conditions—that is, "**stagnation temperature**" and "**stagnation pressure**." By definition, *stagnation* refers to the conditions that exist at a point in the flow where the velocity is zero, regardless of whether or not the zero velocity has been achieved by an adiabatic, or reversible, process. For example, if one were to insert a Pitot-static tube into a compressible flow, strictly speaking one would measure stagnation pressure, not total pressure, because the deceleration of the flow would not be reversible. In practice, however, the difference between stagnation and total pressure is insignificant.

Kinetic Pressure

The kinetic pressure, $q = \rho V^2/2$, is often used, as seen in Chapter 11, to calculate aerodynamic forces with the use of appropriate coefficients. It can also be related to the Mach number. Using the ideal gas law to replace ρ gives

$$q = \frac{1}{2}\frac{pV^2}{RT} \tag{12.29}$$

Then using the equation for the speed of sound, Eq. (12.11), results in

$$q = \frac{k}{2}pM^2 \tag{12.30}$$

where p must always be an absolute pressure because it derives from the ideal gas law.

The use of the equation for kinetic pressure to evaluate the drag force is shown in Example 12.4.

EXAMPLE 12.4

Calculating the Drag Force on a Sphere

Problem Statement

The drag coefficient for a sphere at a Mach number of 0.7 is 0.95. Determine the drag force on a sphere 10 mm in diameter in air if $p = 101$ kPa.

Define the Situation

A sphere is moving at a Mach number of 0.7 in air.

Properties: From Table A.2, $k_{air} = 1.4$.

State the Goal

Find the drag force (in newtons) on the sphere.

Generate Ideas and Make a Plan

The drag force on a sphere is $F_D = qC_D A$.

1. Calculate the kinetic pressure q from Eq. (12.30).
2. Calculate the drag force.

Take Action (Execute the Plan)

1. Kinetic pressure

$$q = \frac{k}{2}pM^2 = \frac{1.4}{2}(101 \text{ kPa})(0.7)^2 = 34.6 \text{ kPa}$$

2. Drag force:

$$F_D = C_D q\left(\frac{\pi}{4}\right)D^2 = 0.95\left(34.6 \times 10^3 \frac{\text{N}}{\text{m}^2}\right)\left(\frac{\pi}{4}\right)(0.01 \text{ m})^2$$

$$= \boxed{2.58 \text{ N}}$$

The Bernoulli equation is not valid for compressible flows. Consider what would happen if one decided to measure the Mach number of a high-speed air flow with a Pitot-static tube, assuming that the Bernoulli equation was valid. Assume a total pressure of 180 kPa and a static pressure of 100 kPa were measured. By the Bernoulli equation, the kinetic pressure is equal to the difference between the total and static pressures, so

$$\frac{1}{2}\rho V^2 = p_t - p \quad \text{or} \quad \frac{k}{2}pM^2 = p_t - p$$

Solving for the Mach number,

$$M = \sqrt{\frac{2}{k}\left(\frac{p_t}{p} - 1\right)}$$

and substituting in the measured values, one obtains

$$M = 1.07$$

The correct approach is to relate the total and static pressures in a compressible flow using Eq. (12.26). Solving that equation for the Mach number gives

$$M = \left\{ \frac{2}{k-1} \left[\left(\frac{p_t}{p} \right)^{(k-1)/k} - 1 \right] \right\}^{1/2} \tag{12.31}$$

and substituting in the measured values yields

$$M = 0.96$$

Thus applying the Bernoulli equation would have led one to say that the flow was supersonic, whereas the flow was actually subsonic. In the limit of low velocities ($p_t/p \to 1$), Eq. (12.31) reduces to the expression derived using the Bernoulli equation, which is indeed valid for very low ($M \ll 1$) Mach numbers.

It is instructive to see how the pressure coefficient at the stagnation (total pressure) condition varies with Mach number. The pressure coefficient is defined by

$$C_p = \frac{p_t - p}{\frac{1}{2}\rho V^2}$$

Using Eq. (12.30) for the kinetic pressure enables one to express C_p as a function of the Mach number and the ratio of specific heats.

$$C_p = \frac{2}{kM^2} \left[\left(1 + \frac{k-1}{2} M^2 \right)^{k/(k-1)} - 1 \right]$$

The variation of C_p with Mach number is shown in Fig. 12.8. At a Mach number of zero, the pressure coefficient is unity, which corresponds to incompressible flow. The pressure coefficient begins to depart significantly from unity at a number of about 0.3. From this observation it is inferred that compressibility effects in the flow field are unimportant for Mach numbers less than 0.3.

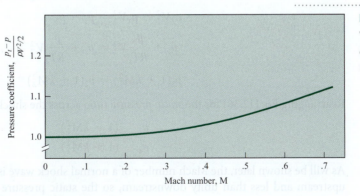

FIGURE 12.8

Variation of the pressure coefficient with Mach number.

12.3 Normal Shock Waves

Normal shock waves are wave fronts normal to the flow across which a supersonic flow is decelerated to a subsonic flow with an attendant increase in static temperature, pressure, and density. The purpose of this section is to develop relations for property changes across normal shock waves.

Change in Flow Properties across a Normal Shock Wave

The most straightforward way to analyze a normal shock wave is to draw a control surface around the wave, as shown in Fig. 12.9, and write down the continuity, momentum, and energy equations.

FIGURE 12.9

Control volume enclosing a normal shock wave.

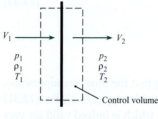

The net mass flux into the control volume is zero because the flow is steady. Therefore

$$-\rho_1 V_1 A + \rho_2 V_2 A = 0 \tag{12.32}$$

where A is the cross-sectional area of the control volume. Equating the net pressure forces acting on the control surface to the net efflux of momentum from the control volume gives

$$\rho_1 V_1 A(-V_1 + V_2) = (p_1 - p_2)A \tag{12.33}$$

The energy equation can be expressed simply as

$$T_{t_1} = T_{t_2} \tag{12.34}$$

because the temperature gradients on the control surface are assumed negligible, and thus heat transfer is neglected (adiabatic).

Using the equation for the speed of sound, Eq. (12.11), and the ideal gas law, the continuity equation can be rewritten to include the Mach number as follows:

$$\frac{p_1}{RT_1} M_1 \sqrt{kRT_1} = \frac{p_2}{RT_2} M_2 \sqrt{kRT_2} \tag{12.35}$$

The Mach number can be introduced into the momentum equation in the following way:

$$\rho_2 V_2^2 - \rho_1 V_1^2 = p_1 - p_2$$

$$p_1 + \frac{p_1}{RT_1} V_1^2 = p_2 + \frac{p_2}{RT_2} V_2^2 \tag{12.36}$$

$$p_1(1 + kM_1^2) = p_2(1 + kM_2^2)$$

Rearranging Eq. (12.36) for the *static-pressure ratio* across the shock wave results in

$$\frac{p_2}{p_1} = \frac{(1 + kM_1^2)}{(1 + kM_2^2)} \tag{12.37}$$

As will be shown later, the Mach number of a normal shock wave is always greater than unity upstream and less than unity downstream, so the static pressure always increases across a shock wave.

Rewriting the energy equation in terms of the temperature and Mach number, as done in Eq. (12.22), by utilizing the fact that $T_{t_2}/T_{t_1} = 1$, yields the static *temperature ratio* across the shock wave.

$$\frac{T_2}{T_1} = \frac{\{1 + [(k-1)/2]M_1^2\}}{\{1 + [(k-1)/2]M_2^2\}} \tag{12.38}$$

Substituting Eqs. (12.37) and (12.38) into Eq. (12.35) gives the following relationship for the Mach numbers upstream and downstream of a normal shock wave:

$$\frac{M_1}{1 + kM_1^2}\left(1 + \frac{k-1}{2}M_1^2\right)^{1/2} = \frac{M_2}{1 + kM_2^2}\left(1 + \frac{k-1}{2}M_2^2\right)^{1/2} \tag{12.39}$$

Solving this equation for M_2 as a function of M_1, results in two solutions. One solution is trivial, $M_1 = M_2$ which corresponds to no shock wave in the control volume. The other solution gives the Mach number downstream of the shock wave:

$$M_2^2 = \frac{(k-1)M_1^2 + 2}{2kM_1^2 - (k-1)} \tag{12.40}$$

Note: Because of the symmetry of Eq. (12.39), one can also use Eq. (12.40) to solve for M_1 given M_2 by simply interchanging the subscripts on the Mach numbers.

Setting $M_1 = 1$ in Eq. (12.40) results in M_2 also being equal to unity. Equations (12.38) and (12.39) also show that there would be no pressure or temperature increase across such a wave. In fact, the wave corresponding to $M_1 = 1$ is the sound wave across which, by definition, pressure and temperature changes are infinitesimal. Thus the sound wave represents a degenerate normal shock wave.

Example 12.5 demonstrates how to calculate properties downstream of a normal shock wave given the upstream Mach number.

EXAMPLE 12.5

Property Changes across a Normal Shock Wave

Problem Statement

A normal shock wave occurs in air flowing at a Mach number of 1.6. The static pressure and temperature of the air upstream of the shock wave are 100 kPa absolute and 15°C. Determine the Mach number, pressure, and temperature downstream of the shock wave.

Define the Situation

The Mach number upstream of a normal shock wave in air is 1.6.

$M_1 = 1.6$ $\longrightarrow$ $\longrightarrow$ M_2

$p_1 = 100$ kPa abs $\qquad p_2$
$T_1 = 15°C$ $\qquad T_2$

Properties: From Table A.2, $k = 1.4$.

State the Goal

Calculate the downstream Mach number, pressure, and temperature.

Generate Ideas and Make a Plan

1. Use Eq. (12.40) to calculate M_2.
2. Use Eq. (12.37) to calculate p_2.

Convert upstream temperature to degrees Kelvin and use Eq. (12.38) to find T_2.

Take Action (Execute the Plan)

1. Downstream Mach number

$$M_2^2 = \frac{(k-1)M_1^2 + 2}{2kM_1^2 - (k-1)} = \frac{(0.4)(1.6)^2 + 2}{(2.8)(1.6)^2 - 0.4} = 0.447$$

$$M_2 = \boxed{0.668}$$

2. Downstream pressure

$$p_2 = p_1\left(\frac{1 + kM_1^2}{1 + kM_2^2}\right)$$

$$= (100 \text{ kPa})\left[\frac{1 + (1.4)(1.6)^2}{1 + (1.4)(0.668)^2}\right] = \boxed{282 \text{ kPa, absolute}}$$

3. Downstream temperature

$$T_2 = T_1\left\{\frac{1 + [(k-1)/2]M_1^2}{1 + [(k-1)/2]M_2^2}\right\}$$

$$= (288 \text{ K})\left[\frac{1 + (0.2)(2.56)}{1 + (0.2)(0.447)}\right] = \boxed{400 \text{ K or } 127°C}$$

Review the Solution and the Process

Knowledge. Note that absolute values for the pressure and temperature have to be used in the equations for property changes across shock waves.

The changes in flow properties across a shock wave are presented in Table A.1 for a gas, such as air, for which $k = 1.4$.

A shock wave is an adiabatic process in which no shaft work is done. Thus for ideal gases the total temperature (and total enthalpy) is unchanged across the wave. The total pressure, however, does change across a shock wave. The total pressure upstream of the wave in Example 12.5 is

$$p_{t_1} = p_1\left(1 + \frac{k-1}{2}M_1^2\right)^{k/(k-1)}$$

$$= 100 \text{ kPa}[1 + (0.2)(1.6^2)]^{3.5} = 425 \text{ kPa}$$

The total pressure downstream of the same wave is

$$p_{t_2} = p_2\left(1 + \frac{k-1}{2}M_2^2\right)^{k/(k-1)}$$

$$= 282 \text{ kPa}[1 + (0.2)(0.668^2)]^{3.5} = 380 \text{ kPa}$$

Thus the total pressure decreases through the wave, which occurs because the flow through the shock wave is not an isentropic process. Total pressure remains constant along streamlines only in isentropic flow. Values for the ratio of total pressure across a normal shock wave are also provided in Table A.1.

Existence of Shock Waves Only in Supersonic Flows

Refer back to Eq. (12.40), which gives the Mach number downstream of a normal shock wave. If one were to substitute a value for M_1 less than unity, it is easy to see that a value for M_2 would be larger than unity. For example, if $M_1 = 0.5$ in air, then

$$M_2^2 = \frac{(0.4)(0.5)^2 + 2}{(2.8)(0.5)^2 - 0.4}$$

$$M_2 = 2.65$$

Is it possible to have a shock wave in a subsonic flow across which the Mach number becomes supersonic? In this case the total pressure would also increase across the wave; that is,

$$\frac{p_{t_2}}{p_{t_1}} > 1$$

The only way to determine whether such a solution is possible is to invoke the second law of thermodynamics, which states that for any process the entropy of the universe must remain unchanged or increase.

$$\Delta s_{\text{univ}} \geq 0 \qquad\qquad\qquad\qquad \textbf{(12.41)}$$

Because the shock wave is an adiabatic process, there is no change in the entropy of the surroundings; thus the entropy of the system must remain unchanged or increase.

$$\Delta s_{\text{sys}} \geq 0 \qquad\qquad\qquad\qquad \textbf{(12.42)}$$

The entropy change of an ideal gas between pressures p_1 and p_2 and temperatures T_1 and T_2 is given by (1)

$$\Delta s_{1 \to 2} = c_p \ln \frac{T_2}{T_1} - R \ln \frac{p_2}{p_1} \qquad\qquad \textbf{(12.43)}$$

Using the relationship between c_p and R, Eq. (12.21), one can express the entropy change as

$$\Delta s_{1 \to 2} = R \ln\left[\frac{p_1}{p_2}\left(\frac{T_2}{T_1}\right)^{k/(k-1)}\right] \qquad\qquad \textbf{(12.44)}$$

Note that the quantity in the square brackets is simply the total pressure ratio as given by Eq. (12.27). Therefore the entropy change across a shock wave can be rewritten as

$$\Delta s = R \ln \frac{p_{t_1}}{p_{t_2}} \qquad \textbf{(12.45)}$$

A shock wave across which the Mach number changes from subsonic to supersonic would give rise to a total pressure ratio less than unity and a corresponding decrease in entropy,

$$\Delta s_{sys} < 0$$

which violates the second law of thermodynamics. Therefore shock waves can exist only in supersonic flow.

The total pressure ratio approaches unity for $M_1 \rightarrow 1$, which conforms with the definition that sound waves are isentropic ($\ln 1 = 0$). Example 12.6 demonstrates the increase in entropy across a normal shock wave.

EXAMPLE 12.6

Entropy Increase across Shock Wave

Problem Statement

A normal shock wave occurs in air flowing at a Mach number of 1.5. Find the change in entropy across the wave.

Define the Situation

A normal shock wave in air with upstream Mach number of 1.5.

Properties: From Table A.2, $R_{air} = 287$ J/kg K, and $k = 1.4$.

State the Goal

Find the change in entropy (in J/kg K) across the wave.

Generate Ideas and Make a Plan

1. Calculate downstream Mach number using Eq. (12.40).
2. Calculate pressure ratio across wave using Eq. (12.37).
3. Calculate temperature across the wave using Eq. (12.38).
4. Calculate entropy change using Eq. (12.44).

Take Action (Execute the Plan)

1. Downstream Mach number

$$M_2^2 = \frac{(k-1)M_1^2 + 2}{2kM_1^2 - (k-1)} = \frac{(0.4)(1.5)^2 + 2}{(2.8)(1.5)^2 - 0.4} = 0.492$$

$$M_2 = 0.701$$

2. Pressure ratio

$$\frac{p_2}{p_1} = \left(\frac{1 + kM_1^2}{1 + kM_2^2} \right) = \left[\frac{1 + (1.4)(1.5)^2}{1 + (1.4)(0.701)^2} \right] = 2.46$$

3. Temperature ratio

$$\frac{T_2}{T_1} = \left\{ \frac{1 + [(k-1)/2]M_1^2}{1 + [(k-1)/2]M_2^2} \right\}$$

$$= \left[\frac{1 + (0.2)(2.25)}{1 + (0.2)(0.492)} \right] = 1.32$$

4. Entropy change

$$\Delta s = R \ln \left[\left(\frac{p_1}{p_2} \right) \left(\frac{T_2}{T_1} \right)^{k/(k-1)} \right]$$

$$= 287 \,(\text{J/kg K}) \ln \left[\left(\frac{1}{2.46} \right) (1.32)^{3.5} \right]$$

$$= \boxed{20.5 \text{ J/kg K}}$$

More examples of shock waves will be given in the next section. This section is concluded by qualitatively discussing other features of shock waves.

Besides the normal shock waves studied here, there are oblique shock waves that are inclined with respect to the flow direction. Look once again at the shock wave structure in front of a blunt body, as depicted qualitatively in Fig. 12.10. The portion of the shock wave immediately in front of the body behaves like a normal shock wave. As the shock wave bends in the free-stream direction, oblique shock waves result. The same relationships

FIGURE 12.10

Shock wave structure
in front of a blunt body.

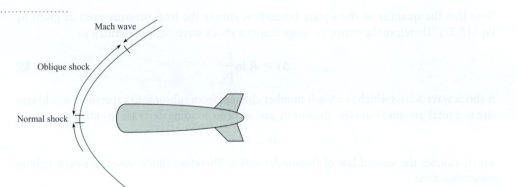

FIGURE 12.10

Shock wave structure
in front of a blunt body.

derived earlier for the normal shock waves are valid for the velocity components normal to oblique waves. The oblique shock waves continue to bend in the downstream direction until the Mach number of the velocity component normal to the wave is unity. Then the oblique shock has degenerated into a so-called Mach wave across which changes in flow properties are infinitesimal.

The familiar sonic booms are the result of weak oblique shock waves that reach ground level. One can appreciate the damage that would ensue from stronger oblique shock waves if aircraft were permitted to travel at supersonic speeds near ground level.

12.4 Isentropic Compressible Flow Through a Duct with Varying Area

With the flow of incompressible fluids through a venturi configuration, as the flow approaches the throat (smallest area), the velocity increases and the pressure decreases; then as the area again increases, the velocity decreases. The same velocity-area relationship is not always found for compressible flows. The purpose of this section is to show the dependence of flow properties on changes in cross-sectional area with compressible flow in variable area ducts.

Dependence of the Mach Number on Area Variation

Consider the duct of varying area shown in Fig. 12.11. It is assumed that the flow is isentropic and that the flow properties at each section are uniform. This type of analysis, in which the flow properties are assumed to be uniform at each section yet in which the cross-sectional area is allowed to vary (nonuniform), is classified as "quasi one-dimensional."

FIGURE 12.11

Duct with variable area.

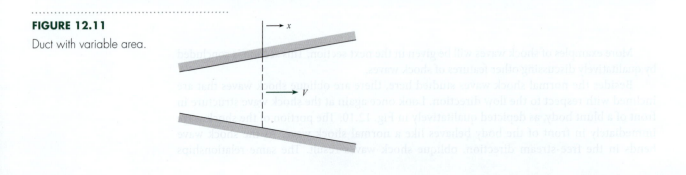

The mass flow through the duct is given by

$$\dot{m} = \rho A V \qquad \text{(12.46)}$$

where A is the duct's cross-sectional area. Because the mass flow is constant along the duct,

$$\frac{d\dot{m}}{dx} = \frac{d(\rho A V)}{dx} = 0 \qquad \text{(12.47)}$$

which can be written as*

$$\frac{1}{\rho}\frac{d\rho}{dx} + \frac{1}{A}\frac{dA}{dx} + \frac{1}{V}\frac{dV}{dx} = 0 \qquad \text{(12.48)}$$

The flow is assumed to be inviscid, so Euler's equation is valid. For steady flow

$$\rho V \frac{dV}{dx} + \frac{dp}{dx} = 0$$

Making use of Eq. (12.7), which relates $dp/d\rho$ to the speed of sound in an isentropic flow, gives

$$\frac{-V}{c^2}\frac{dV}{dx} = \frac{1}{\rho}\frac{d\rho}{dx} \qquad \text{(12.49)}$$

Using this relationship to eliminate ρ in Eq. (12.48) results in

$$\frac{1}{V}\frac{dV}{dx} = \frac{1}{M^2 - 1}\frac{1}{A}\frac{dA}{dx} \qquad \text{(12.50a)}$$

which can be written in an alternate form as

$$\frac{dV}{dA} = \frac{V}{A}\frac{1}{M^2 - 1} \qquad \text{(12.50b)}$$

This equation, although simple, leads to the following important, far-reaching conclusions.

Subsonic Flow

For subsonic flow, $M^2 - 1$ is negative so $dV/dA < 0$, which means that a decreasing area leads to an increasing velocity, and correspondingly, an increasing area leads to a decreasing velocity. This velocity area relationship parallels the trend for incompressible flows.

Supersonic Flow

For supersonic flow, $M^2 - 1$ is positive so $dV/dA > 0$, which means that a decreasing area leads to a decreasing velocity, and an increasing area leads to an increasing velocity. Thus the velocity at the minimum area of a duct with supersonic compressible flow is a minimum. This is the principle underlying the operation of diffusers on jet engines for supersonic aircraft, as shown in Fig. 12.12. The purpose of the diffuser is to decelerate the flow so that there is sufficient time for combustion in the chamber. Then the diverging nozzle accelerates the flow again to achieve a larger kinetic energy of the exhaust gases and an increased engine thrust.

*This step can easily be seen by first taking the logarithm of Eq. (12.46):

$$\ln(\rho A V) = \ln\rho + \ln A + \ln V$$

and then taking the derivative of each term:

$$\frac{d}{dx}[\ln(\rho A V)] = 0 = \frac{1}{\rho}\frac{d\rho}{dx} + \frac{1}{A}\frac{dA}{dx} + \frac{1}{V}\frac{dV}{dx}$$

FIGURE 12.12

Engine for supersonic
aircraft.

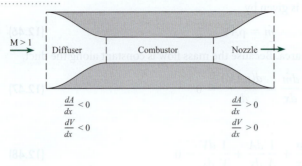

$$\frac{dA}{dx} < 0 \qquad\qquad \frac{dA}{dx} > 0$$

$$\frac{dV}{dx} < 0 \qquad\qquad \frac{dV}{dx} > 0$$

FIGURE 12.12

Engine for supersonic
aircraft.

Transonic Flow (M ≈ 1)

Stations along a duct corresponding to $dA/dx = 0$ represent either a local minimum or a local maximum in the duct's cross-sectional area, as illustrated in Fig. 12.13. If at these stations the flow were either subsonic (M < 1) or supersonic (M > 1), then by Eq. (12.50a) $dV/dx = 0$, so the flow velocity would have either a maximum or a minimum value. In particular, if the flow were supersonic through the duct of Fig. 12.l3a, then the velocity would be a minimum at the throat; if subsonic, a maximum.

Now, what happens if the Mach number is unity? Equation (12.50a) states that if the Mach number is unity and dA/dx is not equal to zero, the velocity gradient dV/dx is infinite—a physically impossible situation. Therefore, dA/dx must be zero where the Mach number is unity for a finite, physically reasonable velocity gradient to exist.*

FIGURE 12.13

Duct contours for which
dA/dx is zero.

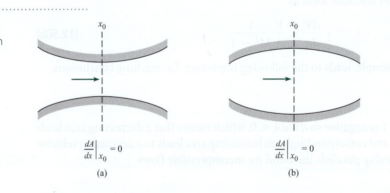

$$\left.\frac{dA}{dx}\right|_{x_0} = 0 \qquad\qquad \left.\frac{dA}{dx}\right|_{x_0} = 0$$

(a) (b)

The argument can be taken one step further here to show that sonic flow can occur only at a minimum area. Consider Fig. 12.13a. If the flow is initially subsonic, the converging duct accelerates the flow toward a sonic velocity. If the flow is initially supersonic, the converging duct decelerates the flow toward a sonic velocity. Using this same reasoning, one can prove that sonic flow is impossible in the duct depicted in Fig. 12.13b. If the flow is initially supersonic, the diverging duct increases the Mach number even more. If the flow is initially subsonic, the diverging duct decreases the Mach number; thus sonic flow cannot be achieved at a maximum area. Hence the Mach number in a duct of varying cross-sectional area can be unity only at a local area minimum (throat). This does not imply, however, that the Mach number must always be unity at a local area minimum.

*Actually, the velocity gradient is indeterminate because the numerator and denominator are both zero. It can be shown by application of L'Hôpital's rule, however, that the velocity gradient is finite.

de Laval Nozzle

The de Laval nozzle is a duct of varying area that produces supersonic flow. The nozzle is named after its inventor, de Laval (1845–1913), a Swedish engineer. According to the foregoing discussion, the nozzle must consist of a converging section to accelerate the subsonic flow, a throat section for transonic flow, and a diverging section to further accelerate the supersonic flow. Thus the shape of the de Laval nozzle is as shown in Fig. 12.14.

One very important application of the de Laval nozzle is the supersonic wind tunnel, which has been an indispensable tool in the development of supersonic aircraft. Basically, the supersonic wind tunnel, as illustrated in Fig. 12.15, consists of a high-pressure source of gas, a de Laval nozzle to produce supersonic flow, and a test section. The high-pressure source may be from a large pressure tank, which is connected to the de Laval nozzle through a regulator valve to maintain a constant upstream pressure, or from a pumping system that provides a continuous high-pressure supply of gas.

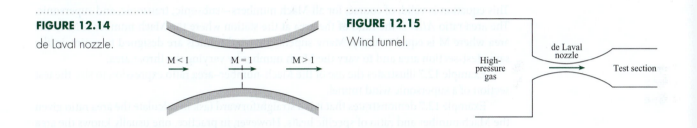

FIGURE 12.14
de Laval nozzle.

FIGURE 12.15
Wind tunnel.

The equations relating to the compressible flow through a de Laval nozzle have already been developed. Because the mass flow rate is the same at every cross section,

$$\rho V A = \text{constant}$$

and the constant is usually evaluated corresponding to those conditions that exist when the Mach number is unity. Thus

$$\rho V A = \rho_* V_* A_* \tag{12.51}$$

where the asterisk signifies conditions wherein the Mach number is equal to unity. Rearranging Eq. (12.51) gives

$$\frac{A}{A_*} = \frac{\rho_* V_*}{\rho V}$$

However, the velocity is the product of the Mach number and the local speed of sound. Therefore

$$\frac{A}{A_*} = \frac{\rho_*}{\rho} \frac{M_* \sqrt{kRT_*}}{M \sqrt{kRT}} \tag{12.52}$$

By definition $M_* = 1$, so

$$\frac{A}{A_*} = \frac{\rho_*}{\rho} \left(\frac{T_*}{T}\right)^{1/2} \frac{1}{M} \tag{12.53}$$

Because the flow in a de Laval nozzle is assumed to be isentropic, the total temperature and total pressure (and total density) are constant throughout the nozzle. From Eq. (12.28),

$$\frac{\rho_*}{\rho} = \left\{ \frac{1 + [(k - 1)/2]\,M^2}{(k + 1)/2} \right\}^{1/(k-1)}$$

and from Eq. (12.24)

$$\frac{T_*}{T} = \frac{1 + [(k - 1)/2]\,M^2}{(k + 1)/2}$$

Substituting these expressions into Eq. (12.53) yields the following relationship for *area ratio* as a function of Mach number in a variable area duct:

$$\frac{A}{A_*} = \frac{1}{M} \left\{ \frac{1 + [(k - 1)/2]\,M^2}{(k + 1)/2} \right\}^{(k+1)/2(k-1)} \tag{12.54}$$

This equation is valid, of course, for all Mach numbers—subsonic, transonic, and supersonic. The area ratio A/A_* is the ratio of the area at the station where the Mach number is M to the area where M is equal to unity. Many supersonic wind tunnels are designed to maintain the same test-section area and to vary the Mach number by varying the throat area.

Example 12.7 illustrates the use of the Mach-number–area ratio expression to size the test section of a supersonic wind tunnel.

Example 12.7 demonstrates that it is a straightforward task to calculate the area ratio given the Mach number and ratio of specific heats. However, in practice, one usually knows the area ratio and wishes to determine the Mach number. It is not possible to solve Eq. (12.54) for the Mach number as an explicit function of the area ratio. For this reason, compressible-flow tables have been developed that allow one to obtain the Mach number easily given the area ratio (as shown in Table A.1).

EXAMPLE 12.7

Finding the Test Section Size in a Supersonic Wind Tunnel

Problem Statement

Suppose a supersonic wind tunnel is being designed to operate with air at a Mach number of 3. If the throat area is 10 cm², what must the cross-sectional area of the test section be?

Define the Situation

Design of supersonic wind tunnel with air for Mach number 3 in test section.

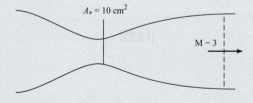

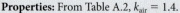

Properties: From Table A.2, $k_{air} = 1.4$.

State the Goal

Find the cross-sectional area (in cm²) of the test section.

Generate Ideas and Make a Plan

1. Use Eq. (12.54), which gives area ratio with respect to the throat section.
2. Calculate area of test section.

Take Action (Execute the Plan)

1. Area ratio

$$\frac{A}{A_*} = \frac{1}{M} \left\{ \frac{1 + [(k - 1)/2]\,M^2}{(k + 1)/2} \right\}^{(k+1)/2(k-1)}$$

$$= \frac{1}{3} \left[\frac{1 + (0.2)3^2}{1.2} \right]^3 = 4.23$$

2. Cross-sectional area of test section

$$A = 4.23 \times 10 \text{ cm}^2 = \boxed{42.3 \text{ cm}^2}$$

Consider again Table A.1. This table has been developed for a gas, such as air, for which $k = 1.4$. The symbols that head each column are defined at the beginning of the table. Tables for both subsonic and supersonic flow are provided. Example 12.8 shows how to use the tables to find flow properties at a given area ratio.

EXAMPLE 12.8

Flow Properties in a Supersonic Wind Tunnel

Problem Statement

The test section of a supersonic wind tunnel using air has an area ratio of 10. The absolute total pressure and temperature are 4 MPa and 350 K. Find the Mach number, pressure, temperature and velocity at the test section.

Define the Situation

Situation. A supersonic wind tunnel has an area ratio of 10 at the test section.

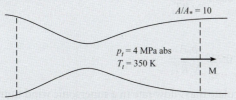

$A/A_* = 10$

$p_t = 4$ MPa abs
$T_t = 350$ K

M

Properties: From Table A.2, $k_{air} = 1.4$, $R_{air} = 287$ J/kg K.

State the Goal

Find the Mach number, pressure, temperature, and velocity at the test section.

Generate Ideas and Make a Plan

1. Use Table A.1 and interpolate to find the Mach number at test section.

2. Use Table A.1 to find the pressure and temperature ratios at test section.

3. Evaluate the pressure and temperature in test section.

4. Calculate the speed of sound using Eq. (12.11).

5. Find the velocity using $V = MC$.

Take Action (Execute the Plan)

1. From Table A.1

M	A/A_*
3.5	6.79
4.0	10.72

Interpolating between the two points gives $\boxed{M = 3.91}$ at $A/A_* = 10.0$.

2. Interpolation using Table A.1 to find the pressure and temperature ratios:

$$\frac{p}{p_t} = 0.00743 \qquad \text{and} \qquad \frac{T}{T_t} = 0.246$$

3. In the test section

$$p = 0.00743 \times 4 \text{ MPa} = \boxed{29.7 \text{ kPa}}$$
$$T = 0.246 \times 350 \text{ K} = \boxed{86 \text{ K}}$$

4. Speed of sound

$$c = \sqrt{kRT} = \sqrt{1.4 \times 287 \times 86} = 186 \text{ m/s}$$

5. Velocity

$$V = 3.91 \times 186 \text{ m/s} = \boxed{727 \text{ m/s}}$$

Review the Solution and the Process

Knowledge. Low temperatures can cause problems. Notice that the temperature of air in the test section is only 86 K, or $-187°$C. At this temperature, the water vapor in the air can condense out, creating fog in the tunnel and compromising tunnel utility.

Mass Flow Rate through a de Laval Nozzle

An important consideration in the design of a supersonic wind tunnel is size. A large wind tunnel requires a large mass flow rate, which, in turn, requires a large pumping system for a continuous-flow tunnel or a large tank for sufficient run time in an intermittent tunnel. The purpose of this section is to develop an equation for the mass flow rate.

The easiest station at which to calculate the mass flow rate is the throat because there the Mach number is unity.

$$\dot{m} = \rho_* A_* V_* = \rho_* A_* \sqrt{kRT_*}$$

It is more convenient, however, to express the mass flow in terms of total conditions. The local density and static temperature at sonic velocity are related to the total density and temperature by

$$\frac{T_*}{T_t} = \left(\frac{2}{k+1}\right)$$

$$\frac{\rho_*}{\rho_t} = \left(\frac{2}{k+1}\right)^{1/(k-1)}$$

which, when substituted into the foregoing equation, give

$$\dot{m} = \rho_t \sqrt{kRT_t}\, A_* \left(\frac{2}{k+1}\right)^{(k+1)/2(k-1)} \tag{12.55}$$

Usually, the total pressure and temperature are known. Using the ideal gas law to eliminate ρ_t, yields the expression for *critical mass flow rate*

$$\dot{m} = \frac{p_t A_*}{\sqrt{RT_t}} k^{1/2}\left(\frac{2}{k+1}\right)^{(k+1)/2(k-1)} \tag{12.56}$$

For gases with a ratio of specific heats of 1.4,

$$\dot{m} = 0.685 \frac{p_t A_*}{\sqrt{RT_t}} \tag{12.57}$$

For gases with $k = 1.67$,

$$\dot{m} = 0.727 \frac{p_t A_*}{\sqrt{RT_t}} \tag{12.58}$$

Example 12.9 illustrates how to calculate mass flow rate in a supersonic wind tunnel given the conditions in the test section.

EXAMPLE 12.9

Mass Flow Rate in Supersonic Wind Tunnel

Problem Statement

A supersonic wind tunnel with a square test section 15 cm by 15 cm is being designed to operate at a Mach number of 3 using air. The static temperature and pressure in the test section are $-20°C$ and 50 kPa abs, respectively. Calculate the mass flow rate.

Define the Situation

A Mach-3 supersonic wind tunnel has a 15 cm by 15 cm test section.

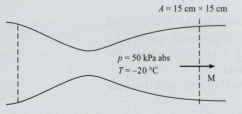

$A = 15\text{ cm} \times 15\text{ cm}$

$p = 50$ kPa abs
$T = -20\,°C$

M

Properties: From Table A.2, $k_{air} = 1.4$ and $R_{air} = 287$ J/kg K.

State the Goal

Calculate the mass flow rate (kg/s) in tunnel.

Generate Ideas and Make a Plan

1. Use Eq. (12.54) to find area ratio and calculate throat area.

2. Use Eq. (12.22) to find total temperature.

3. Use Eq. (12.26) to find total pressure.

4. Use Eq. (12.56) to find the mass flow rate.

Take Action (Execute the Plan)

1. Area ratio

$$\frac{A}{A_*} = \frac{1}{M}\left\{\frac{1+[(k-1)/2]\,M^2}{(k+1)/2}\right\}^{(k+1)/2(k-1)}$$

$$= \frac{1}{3}\left[\frac{1+0.2\times 3^2}{1.2}\right]^3 = 4.23$$

Throat area

$$A_* = \frac{225\text{ cm}^2}{4.23} = 53.2\text{ cm}^2 = 0.00532\text{ m}^2$$

2. Total temperature

$$T_t = T\left(1 + \frac{k-1}{2}M^2\right) = 253 \text{ K} (2.8) = 708 \text{ K}$$

3. Total pressure

$$p_t = p\left(1 + \frac{k-1}{2}M^2\right)^{k/(k-1)} = (50 \text{ kPa})(36.7)$$

$$= 1840 \text{ kPa} = 1.84 \text{ MPa}$$

4. Mass flow rate

$$\dot{m} = 0.685 \frac{p_t A_*}{\sqrt{RT_t}} = \frac{(0.685)[1.840(10^6 \text{ N/m}^2)](0.00532 \text{ m}^2)}{[(287 \text{ J/kg K})(708 \text{ K})]^{1/2}}$$

$$= \boxed{14.9 \text{ kg/s}}$$

Review the Solution and the Process

1. *Discussion.* An alternate way to solve this problem is to calculate the density in the test section using the ideal gas law, calculate the speed of sound with the speed of sound equation, find the air speed using the Mach number, and finally determine the mass flow rate with $\dot{m} = \rho VA$.

2. *Discussion.* A pump capable of moving air at this rate against a 1.8 MPa pressure would require over 6000 kW of power input. Such a system would be large and costly to build and to operate.

Classification of Nozzle Flow by Exit Conditions

Nozzles are classified by the conditions at the nozzle exit. Consider the de Laval nozzle depicted in Fig. 12.16 with the corresponding pressure and Mach-number distributions plotted beneath it. The pressure at the nozzle entrance is very near the total pressure because the Mach number is small. As the area decreases toward the throat, the Mach number increases and the pressure decreases. The static-to-total-pressure ratio at the throat, where conditions are sonic, is called the **critical pressure ratio**. It has a value of

$$\frac{p_*}{p_t} = \left(\frac{2}{k+1}\right)^{k/(k-1)}$$

which for air with $k = 1.4$ is

$$\frac{p_*}{p_t} = 0.528$$

It is called a critical pressure ratio because to achieve sonic flow with air in a nozzle, it is necessary that the exit pressure be equal to or less than 0.528 times the total pressure. The pressure continues to decrease until it reaches the exit pressure corresponding to the nozzle-exit area ratio. Similarly, the Mach number monotonically increases with distance down the nozzle.

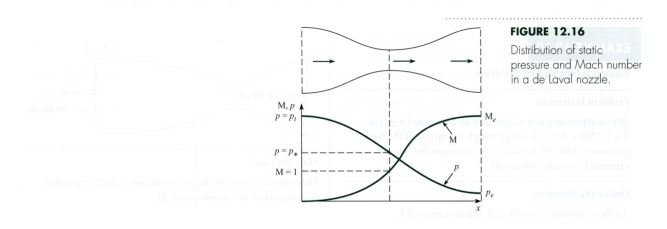

FIGURE 12.16

Distribution of static pressure and Mach number in a de Laval nozzle.

Mach-Number Relationships (§12.2)

12.17 Early passenger aircraft used to fly at a cruising speed of 400 km/h at 4500 m altitude. Did the designers of these aircraft have to be concerned about compressible flow effects? Explain.

12.18 A total heat tube inserted in the flow of a compressible fluid measures the stagnation pressure. Explain the difference between the total and stagnation pressure.

12.19 An object is immersed in an airflow with a static pressure of 200 kPa absolute, a static temperature of 20°C, and a velocity of 250 m/s. What are the pressure and temperature at the stagnation point?

12.20 An airflow at M = 0.85 passes through a conduit with a cross-sectional area of 60 cm². The total absolute pressure is 360 kPa, and the total temperature is 10°C. Calculate the mass flow rate through the conduit.

12.21 PLUS Oxygen flows from a reservoir in which the temperature is 200°C and the pressure is 300 kPa absolute. Assuming isentropic flow, calculate the velocity, pressure, and temperature when the Mach number is 0.9.

12.22 PLUS One problem in creating high-Mach-number flows is condensation of the oxygen component in the air when the temperature reaches 50 K. If the temperature of the reservoir is 300 K and the flow is isentropic, at what Mach number will condensation of oxygen occur?

12.23 GO Hydrogen flows from a reservoir where the temperature is 20°C and the pressure is 500 kPa absolute to a duct 2 cm in diameter where the velocity is 250 m/s. Assuming isentropic flow, calculate the temperature, pressure, Mach number, and mass flow rate at the 2 cm section.

12.24 PLUS The total pressure in a Mach-2.5 wind tunnel operating with air is 547 kPa absolute. A sphere 3 cm in diameter, positioned in the wind tunnel, has a drag coefficient of 0.95. Calculate the drag of the sphere.

12.25 Using Eq. (12.26) on p. 453 in §12.2, develop an expression for the pressure coefficient at stagnation conditions—that is, $C_p = (p_t - p)/[(1/2)\rho V^2]$—in terms of Mach number and ratio of specific heats, $C_p = f(k, M)$. Evaluate C_p at M = 0, 2, and 4 for k = 1.4. What would its value be for incompressible flow?

12.26 For low velocities, the total pressure is only slightly larger than the static pressure. Thus one can write $p_t/p = 1 + \epsilon$, where ϵ is a small positive number ($\epsilon \ll 1$). Using this approximation, show that as $\epsilon \to 0$ (M → 0), Eq. (12.31) on p. 455 in §12.2 reduces to

$$M = \left[\frac{2(p_t/p - 1)}{k}\right]^{1/2}$$

Normal Shock Waves (§12.3)

12.27 PLUS Which of the following statements are true?

   a. Shock waves only occur in supersonic flows.

   b. The static pressure increases across a normal shock wave.

   c. The Mach number downstream of a normal shock wave can be supersonic.

12.28 Can normal shock waves occur in subsonic flows? Explain your answer.

12.29 PLUS A normal shock wave exists in a 500 m/s stream of nitrogen having a static temperature of −50°C and a static pressure of 70 kPa. Calculate the Mach number, pressure, and temperature downstream of the wave and the entropy increase across the wave.

12.30 GO A normal shock wave exists in a Mach 3 stream of air having a static temperature and pressure of 2°C and 207 kPa. Calculate the Mach number, pressure, and temperature downstream of the shock wave.

12.31 A Pitot-static tube is used to measure the Mach number on a supersonic aircraft. The tube, because of its bluntness, creates a normal shock wave as shown. The absolute total pressure downstream of the shock wave (p_{t_2}) is 150 kPa. The static pressure of the free stream ahead of the shock wave (p_1) is 40 kPa and is sensed by the static pressure tap on the probe. Determine the Mach number (M_1) graphically.

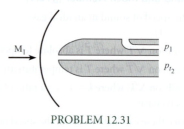

PROBLEM 12.31

12.32 PLUS A shock wave occurs in a methane stream in which the Mach number is 3, the static pressure is 89 kPa absolute, and the static temperature is 20°C. Determine the downstream Mach number, static pressure, static temperature, and density.

12.33 The Mach number downstream of a shock wave in helium is 0.85, and the static temperature is 110°C. Calculate the velocity upstream of the wave.

12.34 Show that the lowest Mach number possible downstream of a normal shock wave is

$$M_2 = \sqrt{\frac{k - 1}{2k}}$$

and that the largest density ratio possible is

$$\frac{\rho_2}{\rho_1} = \frac{k + 1}{k - 1}$$

What are the limiting values of M_2 and ρ_2/ρ_1 for air?

12.35 Show that the Mach number downstream of a weak wave (M ≈ 1) is approximated by

$$M_2^2 = 2 - M_1^2$$

[*Hint:* Let $M_1^2 = 1 + \epsilon$, where $\epsilon \ll 1$, and expand Eq. (12.40) on p. 457 in §12.3 in terms of ϵ.] Compare values for M_2 obtained using this equation with values for M_2 from Table A.1 for $M_1 = 1$, 1.05, 1.1, and 1.2.

Mass Flow in Truncated Nozzle (§12.4)

12.36 What is meant by "back pressure"?

12.37 Develop a computer program for calculating the mass flow in a truncated nozzle. The input to the program would be total pressure, total temperature, back pressure, ratio of specific heats, gas constant, and nozzle diameter. Run the program and compare the results with Example 12.12 in §12.4. Run the program for back pressures of 80, 90, 110, 120, and 130 kPa and make a table for the variation of mass flow rate with back pressure. What trends do you observe?

This program will be useful for Probs. 12.38, 12.39, 12.41, and 12.42.

12.38 PLUS The truncated nozzle shown in the figure is used to meter the mass flow of air in a pipe. The area of the nozzle is 3 cm². The total pressure and total temperature measured upstream of the nozzle in the pipe are 300 kPa absolute and 20°C. The pressure downstream of the nozzle (back pressure) is 90 kPa absolute. Calculate the mass flow rate.

12.39 GO The truncated nozzle shown in the figure is used to monitor the mass flow rate of methane. The area of the nozzle is 3 cm², and the area of the pipe is 12 cm². The upstream total pressure and total temperature are 150 kPa absolute and 30°C. The back pressure is 100 kPa.

   a. Calculate the mass flow rate of methane.

   b. Calculate the mass flow rate assuming the Bernoulli equation is valid, with the density being the density of the gas at the nozzle exit.

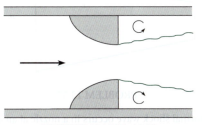

PROBLEMS 12.38, 12.39

12.40 A truncated nozzle with an exit area of 8 cm² is used to measure a mass flow of air of 0.40 kg/s. The static temperature of the air at the exit is 0°C, and the back pressure is 100 kPa. Determine the total pressure.

12.41 PLUS A truncated nozzle with a 10 cm² exit area is supplied from a helium reservoir in which the absolute pressure is first 130 kPa and then 350 kPa. The temperature in the reservoir is 28°C, and the back pressure is 100 kPa. Calculate the mass flow rate of helium for the two reservoir pressures.

12.42 GO A sampling probe is used to draw gas samples from a gas stream for analysis. In sampling, it is important that the velocity entering the probe equal the velocity of the gas stream (isokinetic condition). Consider the sampling probe shown, which has a truncated nozzle inside it to control the mass flow rate. The probe has an inlet diameter of 4 mm and a truncated

nozzle diameter of 2 mm. The probe is in a hot-air stream with a static temperature of 600°C, a static pressure of 100 kPa absolute, and a velocity of 60 m/s. Calculate the pressure required in the probe (back pressure) to maintain the isokinetic sampling condition.

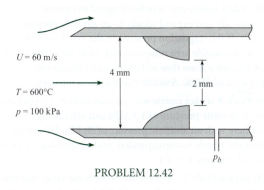

PROBLEM 12.42

12.43 Truncated nozzles are often used for flow-metering devices. Assume that you have to design a truncated nozzle, or a series of truncated nozzles, to measure the performance of an air compressor. The compressor is rated at 0.05 m³/s (Standard conditions) at 830 kPa. A performance curve for the compressor would be a plot of flow rate versus supply pressure. Explain how you would carry out the test program.

Flow in de Laval Nozzles (§12.4)

12.44 Sketch how the Mach number and velocity vary through a de Laval nozzle from the entrance to the exit. How is the velocity variation different from a venturi configuration?

12.45 When a de Laval nozzle has expansion ratio of 4, what does that mean?

12.46 Develop a computer program that requires the Mach number and ratio of specific heats as input and prints out the area ratio, the ratio of static to total pressure, the ratio of static to total temperature, the ratio of density to total density, and the ratio of pressure before and after a shock wave. Run the program for a Mach number of 2 and a ratio of specific heats of 1.4, and compare with results in Table A.1. Then run the program for the same Mach number with ratios of specific heats of 1.3 (carbon dioxide) and 1.67 (helium).

This program will be useful for Probs. 12.48, 12.49, 12.52, 12.56, 12.57, 12.59, and 12.60.

12.47 Develop a computer program that, given the area ratio, ratio of specific heats, and flow condition (subsonic or supersonic) as input, provides the Mach number. This will require some numerical root-finding scheme. Run the program for an area ratio of 5 and ratio of specific heats of 1.4. Compare the results with those in Table A.1. Then run the program for the same area ratio but with the ratios of specific heats of 1.67 (helium) and 1.31 (methane).

This program will be useful for Probs. 12.50, 12.51, and 12.54–12.60.

FIGURE 13.4

Static tube.

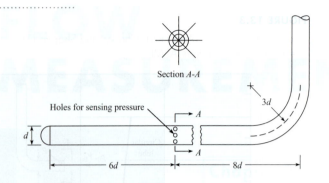

Pitot-Static Tube

The Pitot-static tube, Fig. 13.2c, measures velocity by using concentric tubes to measure static pressure and dynamic pressure. Application of the Pitot-static tube is presented in Chapter 4.

Yaw Meters

A **yaw meter**, Fig. 13.5, is an instrument for measuring velocity by using multiple pressure ports to determine the magnitude and direction of fluid velocity. The first two yaw meters in Fig. 13.5 can be used for two-dimensional flow, where flow direction in only one plane needs to be found. The third yaw meter in Fig. 13.5 is used for determining flow direction in three dimensions. In all these devices, the tube is turned until the pressure on symmetrically opposite

FIGURE 13.5

Various types of yaw meters.
(a) Cylindrical-tube yaw meter.
(b) Two-tube yaw meter.
(c) Three-dimensional yaw meter.

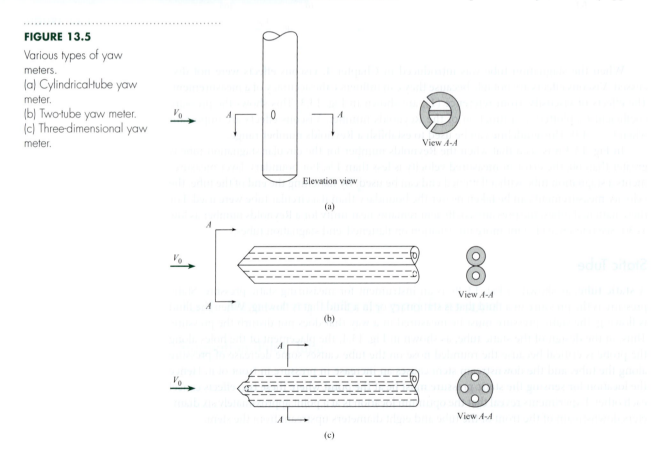

openings is equal. This pressure is sensed by a differential pressure gage or manometer connected to the openings in the yaw meter. The flow direction is sensed when a null reading is indicated on the differential gage. The velocity magnitude is found by using equations that depend on the type of yaw meter that is used.

The Vane or Propeller Anemometer

The term **anemometer** originally meant an instrument that was used to measure the velocity of the wind. However, anemometer now means an instrument that is used to measure fluid velocity because anemometers are used in water, air, nitrogen, blood, and many other fluids. See (18) for an overview of the many types of anemometers.

The **vane anemometer** (Fig. 13.6a) and the **propeller anemometer** (Fig. 13.6b) measure velocity by using vanes typical of a fan or propeller, respectively. These blades rotate with a speed of rotation that depends on the wind speed. Typically, an electronic circuit converts the rotational speed into a velocity reading. On some older instruments the rotor drives a low-friction gear train that, in turn, drives a pointer that indicates meter on a dial. Thus if the anemometer is held in an airstream for 1 min and the pointer indicates a 100-m change on the scale, the average airspeed is 100 m/min.

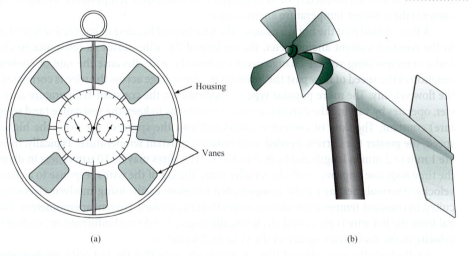

Housing

Vanes

(a)   (b)

Cup Anemometer

Instead of using vanes, the cup anemometer, in Fig. 13.7, is a device that uses the drag on cup-shaped objects to spin a rotor around a central axis. Because the rotational speed of the rotor is related to drag force, the frequency of rotation is related to the fluid velocity by appropriate calibration data. A typical rotor comprises three to five hemispherical or conical cups. In addition to applications in air, engineers use a cup anemometer to measure the velocity in streams and rivers.

Hot-Wire and Hot-Film Anemometers

The **hot-wire anemometer** (HWA), Fig. 13.8, is an instrument for measuring velocity by sensing the heat transfer from a heated wire. As velocity increases, more energy is needed to keep the wire hot, and the corresponding changes in electrical characteristics can be used to determine the velocity of the fluid that is passing by the wire.

FIGURE 13.7

Cup anemometer.

In water flows, adding a few drops of milk is common. In gaseous flows, it is common to "seed" the flow with small particles. Smoke is often used for this seeding.

Laser-Doppler anemometers that provide two or three velocity components of a particle traveling through the measuring volume are now available. This is accomplished by using laser-beam pairs of different colors (wavelengths). The measuring volumes for each color are positioned at the same physical location but oriented differently to measure a different component. The signal-processing system can discriminate the signals from each color and thereby provide component velocities.

Another recent technological advance in laser-Doppler anemometry is the use of fiber optics. The fiber optics transmit the laser beams from the laser to a probe that contains optical elements to cross the beams and generate a measuring volume. Thus measurements at different locations can be made by moving the probe and without moving the laser. For more applications of the laser-Doppler technique see Durst (4).

Marker Methods

The marker method for determining velocity involves particles that are placed in the stream. By analyzing the motion of these particles, one can deduce the velocity of the flow itself. Of course, this requires that the markers follow virtually the same path as the surrounding fluid elements. It means, then, that the marker must have nearly the same density as the fluid or that it must be so small that its motion relative to the fluid is negligible. Thus for water flow it is common to use colored droplets from a liquid mixture that has nearly the same density as the water. For example, Macagno (6) used a mixture of *n*-butyl phthalate and xylene with a bit of white paint to yield a mixture that had the same density as water and could be photographed effectively. Solid particles, such as plastic beads, that have densities near that of the liquid being studied can also be used as markers.

Hydrogen bubbles have also been used for markers in water flow. Here an electrode placed in flowing water causes small bubbles to be formed and swept downstream, thus revealing the motion of the fluid. The wire must be very small so that the resulting bubbles do not have a significant rise velocity with respect to the water. By pulsing the current through the electrode, it is possible to add a time frame to the visualization technique, thus making it a useful tool for velocity measurements. Figure 13.10 shows patches of tiny hydrogen bubbles that were released with a pulsing action from noninsulated segments of a wire located to the left of the picture. Flow is from left to right, and the necked-down section of the flow passage has higher water velocity. Therefore, the patches are longer in that region. Next to the walls the patches of

FIGURE 13.10

Combined time-streak markers (hydrogen bubbles); flow is from left to right. [After Kline (5) Courtesy of Education Development Center, Inc., Waltham, MA.]

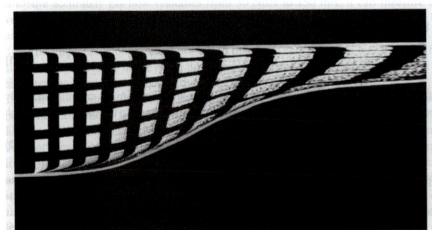

bubbles are shorter, indicating less distance traveled per unit of time. Other details concerning the marker methods of flow visualization are described by Macagno (6).

A relatively new marker method is particle image velocimetry (PIV), which provides a measurement of the velocity field. In PIV, the marker or seeding particles may be minuscule spheres of aluminum, glass, or polystyrene. Or they may be oil droplets, oxygen bubbles (liquids only), or smoke particles (gases only). The seeding particles are illuminated to produce a photographic record of their motion. In particular, a sheet of light passing through a cross section of the flow is pulsed on twice, and the scattered light from the particles is recorded by a camera. The first pulse of light records the position of each particle at time t, and the second pulse of light records the position at time $t + \Delta t$. Thus, the displacement $\Delta \mathbf{r}$ of each particle is recorded on the photograph. Dividing $\Delta \mathbf{r}$ by Δt yields the velocity of each particle. Because PIV uses a sheet of light, the method provides a simultaneous measurement of velocity at locations throughout a cross section of the flow. Hence, PIV is identified as a whole-field technique. Other velocity measurements, the LDA method, for example, are limited to measurements at one location.

PIV measurement of the velocity field for flow over a backward-facing step is shown in Fig. 13.11. This experiment was carried out in water using 15-μm-diameter, silver-coated hollow spheres as seeding particles. Notice that the PIV method provided data over the cross section of the flow. Although the data shown in Fig. 13.11 are qualitative, numerical values of the velocity at each location are also available.

The PIV method is typically performed using digital hardware and computers. For example, images may be recorded with a digital camera. Each resulting digital image is evaluated with software that calculates the velocity at points throughout the image. This evaluation proceeds by dividing the image into small subareas called "interrogation areas." Within a given interrogation area, the displacement vector ($\Delta \mathbf{r}$) of each particle is found by using statistical techniques (auto- and cross-correlation). After processing, the PIV data are typically available on a computer screen. Additional information on PIV systems is provided by Raffel et al. (7).

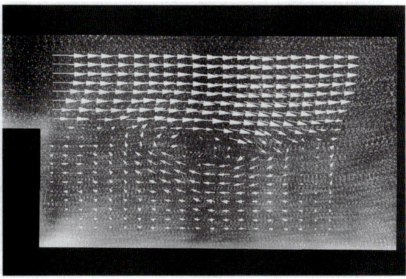

FIGURE 13.11

Velocity vectors from PIV measurements. (Courtesy of TSI Incorporated and Florida State University)

Smoke is often used as a marker in flow measurement. One technique is to suspend a wire vertically across the flow field and allow oil to flow down the wire. The oil tends to accumulate in droplets along the wire. Applying a voltage to the wire vaporizes the oil, creating streaks from the droplets. Figure 13.12 is an example of a flow pattern revealed by such a method. Smoke generators that provide smoke by heating oils are also commercially

FIGURE 13.14

Flow through a sharp-
edged pipe orifice.

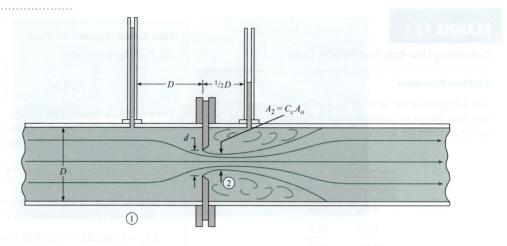

natural gas in pipelines. Because large quantities of natural gas are measured and the associated costs are high, accuracy is very important. This section describes the main ideas associated with orifice meters. Details about using orifice meters are presented in standards such as
reference (10).

Flow through a sharp-edged orifice is shown in Fig. 13.14. Note that the streamlines continue to converge a short distance downstream of the plane of the orifice. Hence the minimum-
flow area is actually smaller than the area of the orifice. To relate the minimum-flow area, often
called the contracted area of the jet, or **vena contracta**, to the area of the orifice A_o, one uses
the contraction coefficient, which is defined as

$$A_j = C_c A_o$$

$$C_c = \frac{A_j}{A_o}$$

Then, for a circular orifice,

$$C_c = \frac{(\pi/4)d_j^2}{(\pi/4)d^2} = \left(\frac{d_j}{d}\right)^2$$

Because d_j and d_2 are identical, $C_c = (d_2/d)^2$. At low values of the Reynolds number, C_c is a
function of the Reynolds number. However, at high values of the Reynolds number, C_c is only
a function of the geometry of the orifice. For d/D ratios less than 0.3, C_c has a value of
approximately 0.62. However, as d/D is increased to 0.8, C_c increases to a value of 0.72.

To derive the orifice equation, consider the situation shown in Fig. 13.14. Apply the
Bernoulli equation between section 1 and section 2:

$$\frac{p_1}{\gamma} + \frac{V_1^2}{2g} + z_1 = \frac{p_2}{\gamma} + \frac{V_2^2}{2g} + z_2$$

V_1 is eliminated by means of the continuity equation $V_1 A_1 = V_2 A_2$. Then solving for V_2 gives

$$V_2 = \left\{ \frac{2g\left[(p_1/\gamma + z_1) - (p_2/\gamma + z_2)\right]}{1 - (A_2/A_1)^2} \right\}^{1/2} \qquad (13.4a)$$

However, $A_2 = C_c A_o$ and $h = p/\gamma + z$, so Eq. (13.4a) reduces to

$$V_2 = \sqrt{\frac{2g(h_1 - h_2)}{1 - C_c^2 A_o^2/A_1^2}} \qquad (13.4b)$$

Our primary objective is to obtain an expression for discharge in terms of h_1, h_2, and the geometric characteristics of the orifice. The discharge is given by $V_2 A_2$. Hence, multiply both sides of Eq. (13.4b) by $A_2 = C_c A_o$, to give the desired result:

$$Q = \frac{C_c A_o}{\sqrt{1 - C_c^2 A_o^2 / A_1^2}} \sqrt{2g(h_1 - h_2)} \qquad \textbf{(13.5)}$$

Equation (13.5) is the discharge equation for the flow of an incompressible inviscid fluid through an orifice. However, it is valid only at relatively high Reynolds numbers. For low and moderate values of the Reynolds number, viscous effects are significant, and an additional coefficient called the *coefficient of velocity*, C_v, must be applied to the discharge equation to relate the ideal to the actual flow.* Thus for viscous flow through an orifice, we have the following discharge equation:

$$Q = \frac{C_v C_c A_o}{\sqrt{1 - C_c^2 A_o^2 / A_1^2}} \sqrt{2g(h_1 - h_2)}$$

The product $C_v C_c$ is called the **discharge coefficient**, C_d, and the combination $C_v C_c / (1 - C_c^2 A_o^2 / A_1^2)^{1/2}$ is called the **flow coefficient**, K. Thus, $Q = KA_o \sqrt{2g(h_1 - h_2)}$,where

$$K = \frac{C_d}{\sqrt{1 - C_c^2 A_o^2 / A_1^2}} \qquad \textbf{(13.6)}$$

If Δh is defined as $h_1 - h_2$, then the final form of the orifice equation reduces to

$$Q = KA_o \sqrt{2g\Delta h} \qquad \textbf{(13.7a)}$$

If a differential pressure transducer is connected across the orifice, it will sense a piezometric pressure change that is equivalent to $\gamma \Delta h$, so the orifice equation becomes

$$Q = KA_o \sqrt{2 \frac{\Delta p_z}{\rho}} \qquad \textbf{(13.7b)}$$

Experimentally determined values of K as a function of d/D and Reynolds number based on orifice size are given in Fig. 13.15. If Q is given, Re_d is equal to $4Q/\pi \, dv$. Then K is obtained from Fig. 13.15 (using the vertical lines and the bottom scale), and Δh is computed from Eq. (13.7a), or Δp_z can be computed from Eq. (13.7b). However, one is often confronted with the problem of determining the discharge Q when a certain value of Δh or a certain value of Δp_z is given. When Q is to be determined, there is no direct way to obtain K by entering Fig. 13.15 with Re, because Re is a function of the flow rate, which is still unknown. Hence another scale, which does not involve Q, is constructed on the graph of Fig. 13.15. The variables for this scale are obtained in the following manner: Because $\text{Re}_d = 4Q/\pi \, dv$ and $Q = K(\pi d^2/4) \sqrt{2g\Delta h}$, write Re_d in terms of Δh:

$$\text{Re}_d = K \frac{d}{v} \sqrt{2g\Delta h}$$

or

$$\frac{\text{Re}_d}{K} = \frac{d}{v} \sqrt{2g\Delta h} = \frac{d}{v} \sqrt{\frac{2\Delta p_z}{\rho}}$$

---

*At low Reynolds numbers the coefficient of velocity may be quite small; however, at Reynolds numbers above 10^5, C_v typically has a value close to 0.98. See Lienhard (8) for C_v analyses.

FIGURE 13.15

Flow coefficient K and Re_d/K versus the Reynolds number for orifices, nozzles, and venturi meters. [After Tuve and Sprenkle (9) and ASME (10). Permission to use Tuve granted by *Instrumentation & Control Systems* magazine, formerly *Instruments* magazine.]

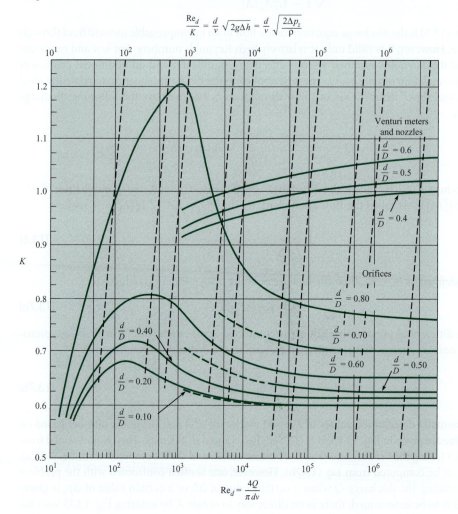

$$\frac{\text{Re}_d}{K} = \frac{d}{\nu} \sqrt{2g\Delta h} = \frac{d}{\nu} \sqrt{\frac{2\Delta p_z}{\rho}}$$

$$\text{Re}_d = \frac{4Q}{\pi\, d\nu}$$

Thus the slanted dashed lines and the top scale are used in Fig. 13.15 when Δh is known and the flow rate is to be determined. If a certain value of Δp is given, one can apply Fig. 13.15 by using $\Delta p_z/\rho$ in place of $g\Delta h$ in the parameter at the top of Fig. 13.15.

The literature on orifice flow contains numerous discussions concerning the optimum placement of pressure taps on both the upstream side and the downstream side of an orifice. The data given in Fig. 13.15 are for "corner taps." That is, on the upstream side the pressure readings were taken immediately upstream of the orifice plate (at the corner of the orifice plate and the pipe wall), and the downstream tap was at a similar downstream location. However, pressure data from flange taps (1 in. upstream and 1 in. downstream) and from the taps shown in Fig. 13.14 all yield virtually the same values for K—the differences are no greater than the deviations involved in reading Fig. 13.15. For more precise values of K with specific types of taps, see the ASME report on fluid meters (10).

Head Loss for Orifices

Some head loss occurs between the upstream side of the orifice and the vena contracta. However, this head loss is very small compared with the head loss that occurs downstream of the vena contracta. This downstream portion of the head loss is like that for an abrupt expansion. Neglecting all head loss except that due to the expansion of the flow, gives

$$h_L = \frac{(V_2 - V_1)^2}{2g} \tag{13.8}$$

where V_2 is the velocity at the vena contracta and V_1 is the velocity in the pipe. It can be shown that the ratio of this expansion loss, h_L, to the change in head across the orifice, Δh, is given as

$$\frac{h_L}{\Delta h} = \frac{\dfrac{V_2}{V_1} - 1}{\dfrac{V_2}{V_1} + 1} \tag{13.9}$$

Table 13.1 shows how the ratio increases with increasing values of V_2/V_1. It is obvious that an orifice is very inefficient from the standpoint of energy conservation. Examples 13.2 and 13.3 illustrate how to make calculations when orifice meters are used.

TABLE 13.1 Relative Head Loss for Orifices

$V_2/V_1 \rightarrow$	1	2	4	6	8	10
$h_L/\Delta h \rightarrow$	0	0.33	0.60	0.71	0.78	0.82

EXAMPLE 13.3

Applying an Orifice Meter to Measure the Flow Rate of Water

Problem Statement

A 15-cm orifice is located in a horizontal 24-cm water pipe, and a water-mercury manometer is connected to either side of the orifice. When the deflection on the manometer is 25 cm, what is the discharge in the system, and what head loss is produced by the orifice? Assume the water temperature is 20°C.

Define the Situation

Water flows through an orifice ($d = 0.15$ m) in a pipe ($D = 0.24$ m). A mercury-water manometer is used to measure pressure drop.

Properties:

- Water (20°C), Table A.5: $\nu = 1 \times 10^{-6}$ m²/s.
- Mercury (20°C), Table A.4: $S = 13.6$.

State the Goal

- Calculate discharge (in m³/s) in pipe.
- Calculate head loss (in meters) produced by the orifice.

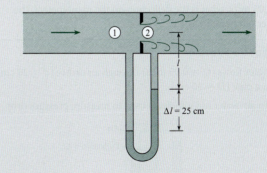

$\Delta l = 25$ cm

Generate Ideas and Make a Plan

1. Calculate $\Delta h = h_1 - h_2$ using the manometer equation.
2. Find the flow coefficient K using Fig. 13.15.

3. Find discharge Q using Eq. (13.7a).

4. Calculate the coefficient of contraction C_c using Eq. (13.6).

5. Solve for the velocity V_2 at the vena contracta.

6. Calculate head loss using Eq. (13.8).

Take Action (Execute the Plan)

1. Change in piezometric head

   • Apply manometer equation from 1 to 2.

   $$p_1 + \gamma_w(l + \Delta l) - \gamma_{Hg}\Delta l - \gamma_w l = p_2$$

   • Solve for Δh.

   $$\Delta h = \frac{p_1 - p_2}{\gamma_w} = \Delta l \frac{\gamma_{Hg} - \gamma_w}{\gamma_w} = \Delta l\left(\frac{\gamma_{Hg}}{\gamma_w} - 1\right)$$

   $$\Delta h = (0.25 \text{ m})(13.6 - 1) = 3.15 \text{ m of water}$$

2. Flow coefficient

   • Calculate (Re_d/K).

   $$\frac{\text{Re}_d}{K} = \frac{d\sqrt{2g\Delta h}}{\nu} = \frac{0.15 \text{ m}\sqrt{2(9.81 \text{ m/s}^2)(3.15 \text{ m})}}{1.0 \times 10^{-6} \text{ m}^2/\text{s}}$$

   $$= 1.2 \times 10^6$$

   • From Fig. 13.15 with $d/D = 0.625$, $K = 0.66$ (interpolated).

3. Discharge

   $$Q = 0.66A_o\sqrt{2g\Delta h}$$

   $$= 0.66\frac{\pi}{4}d^2\sqrt{2(9.81 \text{ m/s}^2)(3.15 \text{ m})}$$

   $$= 0.66(0.785)(0.15^2 \text{ m}^2)(7.86 \text{ m/s}) = \boxed{0.092 \text{ m}^3/\text{s}}$$

4. Coefficient of contraction C_c

   $$K = \frac{C_d}{\sqrt{1 - C_c^2 A_o^2/A_1^2}}$$

   Let $K = 0.66$. The ratio $(A_o/A_1)^2 = (0.625)^4 = 0.1526$ and $C_d = C_v C_c$. Assuming $C_v = 0.98$ (see the footnote on page 489) and solving for C_c, gives $C_c = 0.633$.

5. Velocity at the vena contracta

   $$V_2 = Q/(C_c A_o)$$

   $$(0.092 \text{ m}^3/\text{s})/[(0.633)(\pi/4)(0.15^2 \text{ m}^2)] = 8.23 \text{ m/s}$$

   $$V_1 = Q/A_{\text{pipe}}$$

   $$(0.092 \text{ m}^3/\text{s})/[(\pi/4)(0.24^2 \text{ m}^2)] = 2.03 \text{ m/s}$$

6. Head loss

   $$h_L = (V_2 - V_1)^2/2g = (8.23 - 2.03)^2/(2 \times 9.81)$$

   $$= \boxed{1.96 \text{ m}}$$

EXAMPLE 13.4

Applying an Orifice Meter

Problem Statement

An air-water manometer is connected to either side of a 20-cm orifice in a 30-cm water pipe. If the maximum flow rate is 0.05 m³/s, what is the deflection on the manometer? The water temperature is 15°C.

Define the Situation

• Water flows ($Q = 0.05$ m³/s) through an orifice ($d = 20$ cm) in a pipe ($D = 30$ cm)

• An air-water manometer is used to measure pressure drop.

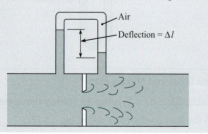

Properties: Water (15°C), Table A.5: $\nu = 1.14 \times 10^{-6}$ m²/s.

State the Goal

Calculate the deflection of water in the manometer.

Generate Ideas and Make a Plan

1. Calculate Reynolds number.

2. Find the flow coefficient K from Fig. 13.15.

3. Solve for Δh by using Eq. (13.7a).

4. Solve for Δl by using the manometer equation.

Take Action (Execute the Plan)

1. Reynolds number.

   $$\text{Re} = \frac{4Q}{\pi d\nu} = \frac{4(0.05 \text{ m}^3/\text{s})}{\pi(0.2 \text{ m})(1.14 \times 10^{-6} \text{ m}^2/\text{s})} = \boxed{2.8 \times 10^5}$$

2. Flow coefficient.

   • Use Fig. 13.15. Interpolate for $d/D = 8/12 = 0.667$ to find $K \approx 0.68$.

3. Change in piezometric head

- From $Q = KA_o\sqrt{2g\Delta h}$, solve for Δh:

$$\Delta h = \frac{Q^2}{2gK^2A_o^2} = \frac{(0.05 \text{ m}^3/\text{s})^2}{2(9.81 \text{ m/s}^2)(0.68)^2((3.14/4)(0.2)^2)^2} = 0.28 \text{ m}$$

4. Manometer deflection

- The deflection is related to Δh by

$$\Delta h = \Delta l\left(\frac{\gamma_w - \gamma_{\text{air}}}{\gamma_w}\right)$$

- Because $\gamma_w \gg \gamma_{\text{air}}$, $\Delta l = \Delta h = 0.28$ m. $\boxed{\Delta l = 0.28 \text{ m}}$

The sharp-edged orifice can also be used to measure the mass flow rate of gases. The discharge equation [Eq. (13.7b)] is multiplied by the upstream gas density and an empirical factor to account for compressibility effects (10). The resulting equation is

$$\dot{m} = YA_oK\sqrt{2\rho_1(p_1 - p_2)} \tag{13.10}$$

where K, the flow coefficient, is found using Fig. 13.15 and Y is the compressibility factor given by the empirical equation

$$Y = 1 - \left\{\frac{1}{k}\left(1 - \frac{p_2}{p_1}\right)\left[0.41 + 0.35\left(\frac{A_o}{A_1}\right)^2\right]\right\} \tag{13.11}$$

In this case both the pressure difference across the orifice and the absolute pressure of the gas are needed. One must remember when using the equation for the compressibility factor that the absolute pressure must be used.

EXAMPLE 13.5

Applying an Orifice Meter to Measure the Flow Rate of Natural Gas

Problem Statement

The mass flow rate of natural gas is to be measured using a sharp-edged orifice. The upstream pressure of the gas is 101 kPa absolute, and the pressure difference across the orifice is 10 kPa. The upstream temperature of the methane is 15°C. The pipe diameter is 10 cm, and the orifice diameter is 7 cm. What is the mass flow rate?

Define the Situation

- Natural gas (methane) is flowing through a sharp-edged orifice.
- Pipe diameter is $D = 0.1$ m. Orifice diameter is $d = 0.07$ m.
- Pressure difference across orifice is 10 kPa.

Properties: Natural gas (15°C, 1 atm), Table A.2:

$\rho = 0.678 \text{ kg/m}^3$, $v = 1.59 \times 10^{-5} \text{ m}^2/\text{s}$, $K = 1.31$.

State the Goal

Find the mass flow rate (in kg/s).

Generate Ideas (Make a Plan)

1. Find the flow coefficient K from Fig. 13.15.
2. Calculate the compressibility factor Y using Eq. (13.11).
3. Calculate the mass flow rate using Eq. (13.10).

Take Action (Execute the Plan)

1. Flow coefficient

   - Calculate (Re_d/K):

$$\frac{\text{Re}_d}{K} = \frac{d}{v}\sqrt{2\frac{\Delta p}{\rho_1}} = \frac{0.07}{1.59(10^{-5})}\sqrt{2\frac{10^4}{0.678}} = 7.56 \times 10^5$$

   - Using Fig. 13.15, $K = 0.7$.

2. Compressibility factor

$$Y = 1 - \left\{\frac{1}{1.31}\left(1 - \frac{91}{101}\right)(0.41 + 0.35 \times 0.7^4)\right\} = 0.962$$

3. Mass flow rate of methane

$$\dot{m} = YA_oK\sqrt{2\rho_1(p_1 - p_2)}$$

$$= 0.962\left(\frac{\pi}{4}0.07^2\right)(0.7)\sqrt{2(0.678)(10^4)}$$

$$= \boxed{0.302 \text{ kg/s}}$$

The foregoing examples involved the determination of either Q or Δh for a given size of orifice. Another type of problem is determination of the diameter of the orifice for a given Q and Δh. For this type of problem a trial-and-error procedure is required. Because one knows the approximate value of K, that is guessed first. Then the diameter is solved for, after which a better value of K can be determined, and so on.

Venturi Meter

The **venturi meter**, Fig. 13.16, is an instrument for measuring flow rate by using measurements of pressure across a converging-diverging flow passage. The main advantage of the venturi meter as compared to the orifice meter is that the head loss for a venturi meter is much smaller. The lower head loss results from streamlining the flow passage, as shown in Fig. 13.16. Such streamlining eliminates any jet contraction beyond the smallest flow section. Consequently, the coefficient of contraction has a value of unity, and the venturi equation is

$$Q = \frac{A_t C_d}{\sqrt{1 - (A_t/A_p)^2}}\sqrt{2g(h_p - h_t)} \tag{13.12}$$

$$Q = KA_t\sqrt{2g\Delta h} \tag{13.13}$$

where A_t is the throat area and Δh is the difference in piezometric head between the venturi entrance (pipe) and the throat. Note that the venturi equation is the same as the orifice equation. However, K for the venturi meter approaches unity at high values of the Reynolds number and small d/D ratios. This trend can be seen in Fig. 13.15, where values of K for the venturi meter are plotted along with similar data for the orifice.

FIGURE 13.16

Typical venturi meter.

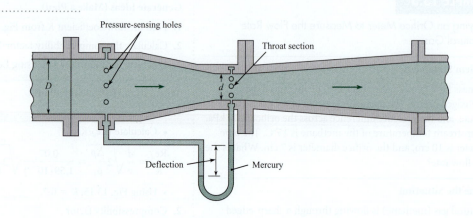

Flow Nozzles

The **flow nozzle**, Fig. 13.17, is an instrument for measuring flow rate by using the pressure drop across a nozzle that is typically placed inside a conduit. Similar to an orifice meter, design and application of the flow nozzle is described by engineering standards (10). As compared to an orifice meter, the flow nozzle is better in flows that cause wear (e.g., particle-laden flow). The reason is that erosion of an orifice will produce more change in the pressure-drop versus flow-rate relationship. Both the flow nozzle and orifice meter will produce about the same overall head loss.

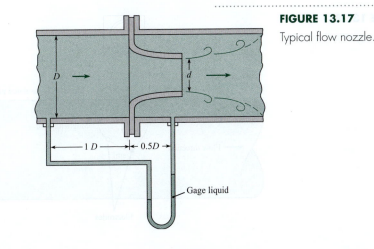

FIGURE 13.17

Typical flow nozzle.

Gage liquid

$1\,D$ — $0.5D$

EXAMPLE 13.6

Applying a Venturi Meter to Measure the Flow Rate of Water

Problem Statement

The pressure difference between the taps of a horizontal venturi meter carrying water is 35 kPa. If $d = 20$ cm and $D = 40$ cm, what is the discharge of water at 10°C?

Define the Situation

- Water flows through a horizontal venturi meter.
- Pipe diameter is $D = 0.40$ m. Venturi throat diameter is $d = 0.2$ m.

Properties: Water (10°C), Table A.5: $\nu = 1.31 \times 10^{-6}$ m²/s, and $\gamma = 9810$ N/m³.

State the Goal

Find the discharge (m³/s).

Generate Ideas and Make a Plan

1. Compute $\Delta h = h_1 - h_2$.
2. Find the flow coefficient K from Fig. 13.15.
3. Find discharge Q using Eq. (13.7a).

Take Action (Execute the Plan)

1. Change in piezometric head

$$\Delta h = \frac{\Delta p}{\gamma} + \Delta z = \frac{\Delta p}{\gamma} + 0 = \frac{35{,}000 \text{ N/m}^2}{9810 \text{ N/m}^3} = 3.57 \text{ m of water}$$

2. Flow coefficient

   - Calculate (Re_d/K):

$$\frac{\mathrm{Re}_d}{K} = \frac{d\sqrt{2g\Delta h}}{\nu} = \frac{0.20\sqrt{2(9.81)(3.57)}}{1.31(10^{-6})} = 1.28 \times 10^6$$

   - From Fig. 13.15, find that $K = 1.02$.

3. Discharge

$$Q = 1.02 A_2 \sqrt{2g\Delta h}$$

$$= 1.02(0.785)(0.20^2)\sqrt{2(9.81)(3.57)} = \boxed{0.268 \text{ m}^3/\text{s}}$$

Electromagnetic Flowmeter

All the flowmeters described so far require that some sort of obstruction be placed in the flow. The obstruction may be the rotor of a vane anemometer or the reduced cross-section of an orifice or venturi meter. A meter that neither obstructs the flow nor requires pressure taps, which are subject to clogging, is the **electromagnetic flowmeter**. Its basic principle is that a conductor that moves in a magnetic field produces an electromotive force. Hence liquids having a degree of conductivity generate a voltage between the electrodes, as in Fig. 13.18, and this voltage is proportional to the velocity of flow in the conduit. It is interesting to note that the basic principle of the electromagnetic flowmeter was investigated by Faraday in 1832. However,

FIGURE 13.18

Electromagnetic flowmeter.

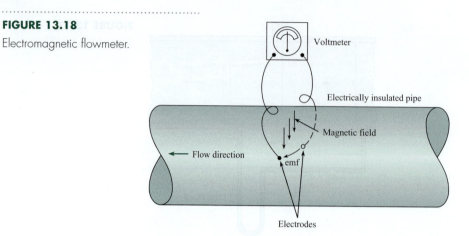

practical application of the principle was not made until approximately a century later, when it was used to measure blood flow. Recently, with the need for a meter to measure the flow of liquid metal in nuclear reactors and with the advent of sophisticated electronic signal detection, this type of meter has found extensive commercial use.

The main advantages of the electromagnetic flowmeter are that the output signal varies linearly with the flow rate and that the meter causes no resistance to the flow. The major disadvantages are its high cost and its unsuitability for measuring gas flow.

For a summary of the theory and application of the electromagnetic flowmeter, the reader is referred to Shercliff (11). This reference also includes a comprehensive bibliography on the subject.

Ultrasonic Flowmeter

Another form of nonintrusive flowmeter that is used in diverse applications ranging from blood flow measurement to open-channel flow is the **ultrasonic flowmeter**. Basically, there are two different modes of operation for ultrasonic flowmeters. One mode involves measuring the difference in travel time for a sound wave traveling upstream and downstream between two measuring stations. The difference in travel time is proportional to flow velocity. The second mode of operation is based on the Doppler effect. When an ultrasonic beam is projected into an inhomogeneous fluid, some acoustic energy is scattered back to the transmitter at a different frequency (Doppler shift). The measured frequency difference is related directly to the flow velocity.

Turbine Flowmeter

The **turbine flowmeter** consists of a wheel with a set of curved vanes (blades) mounted inside a duct. The volume rate of flow through the meter is related to the rotational speed of the wheel. This rotational rate is generally measured by a blade passing an electromagnetic pickup mounted in the casing. The meter must be calibrated for the given flow conditions. The turbine meter is versatile in that it can be used for either liquids or gases. It has an accuracy of better than 1% over a wide range of flow rates, and it operates with small head loss. The turbine flowmeter is used extensively in monitoring flow rates in fuel-supply systems.

Vortex Flowmeter

The **vortex flowmeter**, shown in Fig. 13.19, measures flow rate by relating vortex shedding frequency to flow rate. The vortices are shed from a sensor tube that is situated in the center

FIGURE 13.19

Vortex flowmeter.

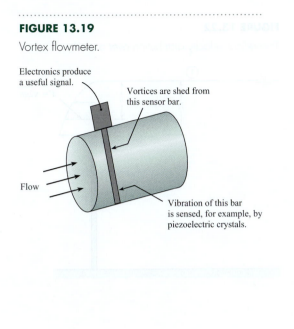

Electronics produce a useful signal.

Vortices are shed from this sensor bar.

Flow

Vibration of this bar is sensed, for example, by piezoelectric crystals.

FIGURE 13.20

Rotameter.

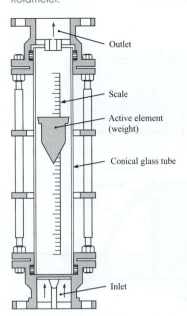

Outlet

Scale

Active element (weight)

Conical glass tube

Inlet

of a pipe. These vortices cause vibrations, which are sensed by piezoelectric crystals that are located inside the sensor tube, and are converted to an electronic signal that is directly proportional to flow rate. This vortex meter gives accurate and repeatable measurements with no moving parts. However, the corresponding head loss is comparable to that from other obstruction-type meters.

Rotameter

The **rotameter**, Fig. 13.20, is an instrument for measuring flow rate by sensing the position of an active element (weight) that is situated in a tapered tube. The equilibrium position of the weight is related to the flow rate. Because the velocity is lower at the top of the tube (greater flow section there) than at the bottom, the rotor seeks a neutral position where the drag on it just balances its weight. Thus the rotor "rides" higher or lower in the tube depending on the rate of flow. The weight is designed so that it spins, thus it stays in the center of the tube. A calibrated scale on the side of the tube indicates the rate of flow. Although venturi and orifice meters have better accuracy (approximately 1% of full scale) than the rotameter (approximately 5% of full scale), the rotameter offers other advantages, such as ease of use and low cost.

Rectangular Weir

A **weir**, shown in Fig. 13.21, is an instrument for determining flow rate in liquids by measuring the height of water relative to an obstruction in an open channel. The discharge over the weir is a function of the weir geometry and of the head on the weir. Consider flow over the weir in a rectangular channel, shown in Fig. 13.21. The head H on the weir is defined as the vertical distance between the weir crest and the liquid surface taken far enough upstream of the weir to avoid local free-surface curvature (see Fig. 13.21).

FIGURE 13.21

Definition sketch for sharp-crested weir.
(a) Plan view.
(b) Elevation view.

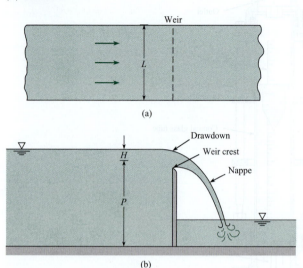

FIGURE 13.22

Theoretical velocity distribution over a weir.

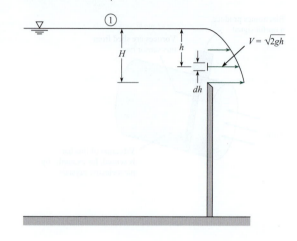

The discharge equation for the weir is derived by integrating $V\,dA = VL\,dh$ over the total head on the weir. Here L is the length of the weir and V is the velocity at any given distance h below the free surface. Neglecting streamline curvature and assuming negligible velocity of approach upstream of the weir, one obtains an expression for V by writing the Bernoulli equation between a point upstream of the weir and a point in the plane of the weir (see Fig. 13.22). Assuming the pressure in the plane of the weir is atmospheric, this equation is

$$\frac{p_1}{\gamma} + H = (H - h) + \frac{V^2}{2g} \tag{13.14}$$

Here the reference elevation is the elevation of the crest of the weir, and the reference pressure is atmospheric pressure. Therefore $p_1 = 0$, and Eq. (13.14) reduces to

$$V = \sqrt{2gh}$$

Then $dQ = \sqrt{2gh}\,Ldh$, and the discharge equation becomes

$$Q = \int_0^H \sqrt{2gh}\,Ldh \tag{13.15}$$

$$= \frac{2}{3}L\sqrt{2g}\,H^{3/2}$$

In the case of actual flow over a weir, the streamlines converge downstream of the plane of the weir, and viscous effects are not entirely absent. Consequently, a discharge coefficient C_d must be applied to the basic expression on the right-hand side of Eq. (13.15) to bring the theory in line with the actual flow rate. Thus the rectangular weir equation is

$$Q = \frac{2}{3}C_d\sqrt{2g}\,LH^{3/2} \tag{13.16}$$

$$= K\sqrt{2g}\,LH^{3/2}$$

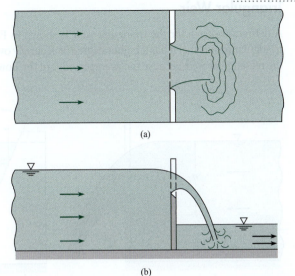

FIGURE 13.23
Rectangular weir with
end contractions.
(a) Plan view.
(b) Elevation view.

(a)

(b)

For low-viscosity liquids, the flow coefficient K is primarily a function of the relative head on the weir, H/P. An empirically determined equation for K adapted from Kindsvater and Carter (12) is

$$K = 0.40 + 0.05\frac{H}{P}$$ (13.17)

This is valid up to an H/P value of 10 as long as the weir is well ventilated so that atmospheric pressure prevails on both the top and the bottom of the weir nappe.

When the rectangular weir does not extend the entire distance across the channel, as in Fig. 13.23, additional end contractions occur. Therefore, K will be smaller than for the weir without end contractions. The reader is referred to King (13) for additional information on flow coefficients for weirs.

EXAMPLE 13.7

Applying a Rectangular Weir to Measure
the Flow Rate of Water

Problem Statement

The head on a rectangular weir that is 60 cm high in a rectangular channel that is 1.3 m wide is measured to be 21 cm. What is the discharge of water over the weir?

Define the Situation

- Water flows over a rectangular weir.
- The weir has a height of $P = 0.6$ m and a width of $L = 1.3$ m.
- Head on the weir is $H = 0.21$ m.

State the Goal

Find the discharge (m³/s).

Generate Ideas and Make a Plan

1. Calculate the flow coefficient K using Eq. (13.17).
2. Calculate flow rate using the rectangular weir equation (13.16).

Take Action (Execute the Plan)

1. Flow coefficient

$$K = 0.40 + 0.05\frac{H}{P} = 0.40 + 0.05\left(\frac{21}{60}\right) = 0.417$$

2. Discharge

$$Q = K\sqrt{2g}\,LH^{3/2} = 0.417\sqrt{2(9.81)}\,(1.3)(0.21^{3/2})$$

$$= \boxed{0.23 \text{ m}^3/\text{s}}$$

Triangular Weir

A definition sketch for the triangular weir is shown in Fig. 13.24. The primary advantage of the triangular weir is that it has a higher degree of accuracy over a much wider range of flow than does the rectangular weir because the average width of the flow section increases as the head increases.

FIGURE 13.24

Definition sketch for the triangular weir.

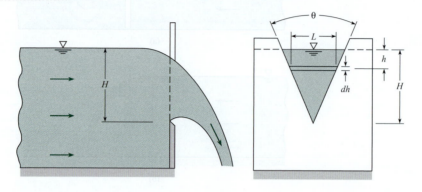

The discharge equation for the triangular weir is derived in the same manner as that for the rectangular weir. The differential discharge $dQ = V\,dA = VL\,dh$ is integrated over the total head on the weir to give

$$Q = \int_0^H \sqrt{2gh}\,(H - h)\,2\tan\left(\frac{\theta}{2}\right)dh$$

which integrates to

$$Q = \frac{8}{15}\sqrt{2g}\tan\left(\frac{\theta}{2}\right)H^{5/2}$$

However, a coefficient of discharge must still be used with the basic equation. Hence

$$Q = \frac{8}{15}C_d\sqrt{2g}\tan\left(\frac{\theta}{2}\right)H^{5/2} \tag{13.18}$$

Experimental results with water flow over weirs with $\theta = 60°$ and $H > 2$ cm indicate that C_d has a value of 0.58. Hence the triangular weir equation with these limitations is

$$Q = 0.179\sqrt{2g}\,H^{5/2} \tag{13.19}$$

EXAMPLE 13.8

Flow Rate for a Triangular Weir

Problem Statement

The head on a 60° triangular weir is measured to be 43 cm. What is the flow of water over the weir?

Define the Situation

- Water flows over a 60° triangular weir.
- Head on the weir is $H = 0.43$ m.

State the Goal

Calculate the discharge (m³/s).

Generate Ideas and Make a Plan

Apply the triangular weir equation (13.19).

Take Action (Execute the Plan)

$$Q = 0.179\sqrt{2g}\,H^{5/2} = 0.179 \times \sqrt{2 \times 9.81} \times (0.43)^{5/2}$$

$$= \boxed{0.096 \text{ m}^3/\text{s}}$$

More details about flow-measuring devices for incompressible flow can be found in references (14) and (15).

13.3 Measurement in Compressible Flow

This section describes how to measure velocity, pressure, and flow rate in compressible flows. Because fluid density is changing in these flows, the Bernoulli equation is invalid. Thus, compressible flow theory from Chapter 12 will be applied to develop valid measurement techniques.

FIGURE 13.25

Stagnation tube in supersonic flow.

Pressure Measurements

Static-pressure measurements can be made using the conventional static-pressure taps of a probe. However, if the boundary layer is disturbed by the presence of a shock wave in the vicinity of the pressure tap, the reading may not give the correct static pressure. The effect of the shock wave on the boundary layer is smaller if the boundary layer is turbulent. Therefore an effort is sometimes made to trip the boundary layer and ensure a turbulent boundary layer in the region of the pressure tap.

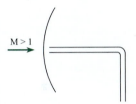

The stagnation pressure can be measured with a stagnation tube aligned with the local velocity vector. If the flow is supersonic, however, a shock wave forms around the tip of the probe, as shown in Fig. 13.25, and the stagnation pressure measured is that downstream of the shock wave and not that of the free stream. The stagnation pressure in the free stream can be calculated using the normal shock relationships, provided the free-stream Mach number is known. See Chapter 12 for more details about normal shock waves.

Mach Number and Velocity Measurements

A Pitot-static tube can be used to measure Mach numbers in compressible flows. Taking the measured stagnation pressure as the total pressure, one can calculate the Mach number in subsonic flows from the total-to-static-pressure ratio according to Eq. (12.31):

$$M = \left\{ \frac{2}{k-1} \left[\left(\frac{p_t}{p} \right)^{(k-1)/k} - 1 \right] \right\}^{1/2}$$

It is interesting to note here that one must measure the stagnation and static pressures separately to determine the pressure ratio, whereas one needs only the pressure difference to calculate the velocity of a flow.

If the flow is supersonic, then the indicated stagnation pressure is the pressure behind the shock wave standing off the tip of the tube. By taking this pressure as the total pressure downstream of a normal shock wave and the measured static pressure as the static pressure upstream of the shock wave, one can determine the Mach number of the free stream (M_1) from the static-to-total-pressure ratio (p_1/p_{t_2}) according to the expression

$$\frac{p_1}{p_{t_2}} = \frac{\{[2k/(k+1)]M_1^2 - [(k-1)/(k+1)]\}^{1/(k-1)}}{\{[(k+1)/2]M_1^2\}^{k/(k-1)}} \tag{13.20}$$

which is called the Rayleigh supersonic Pitot formula. Note, however, that M_1 is an implicit function of the pressure ratio and must be determined graphically or by some numerical procedure. Many normal-shock tables, such as those in reference (16), have p_1/p_{t_2} tabulated versus M_1, which enables one to find M_1 quite easily by interpolation.

Once the Mach number is determined, more information is needed to evaluate the velocity—namely, the local speed of sound. This can be done by inserting a probe into the flow to measure total temperature and then calculating the static temperature using Eq. (12.22):

$$T = \frac{T_t}{1 + [(k-1)/2]\,M_1^2}$$

The local speed of sound is then determined by Eq. (12.11):

$$c = \sqrt{kRT}$$

and the velocity is calculated from

$$V = M_1 c$$

The hot-wire anemometer can also be used to measure velocity in compressible flows, provided it is calibrated to account for Mach-number effects.

Mass Flow Measurement

Measuring the flow rate of a compressible fluid using a truncated nozzle was discussed in some detail in Chapter 12. Basically, the flow nozzle is a truncated nozzle located in a pipe, so the equations developed in Chapter 12 can be used to determine the flow rate through the flow nozzle. Strictly speaking, the flow rate so calculated should be multiplied by the discharge coefficient. For the high Reynolds numbers characteristic of compressible flows, however, the discharge coefficient can be taken as unity. If the flow at the throat of the flow nozzle is sonic (i.e., Mach number at the throat is 1.0), it is conceivable that the complex flow field existing downstream of the nozzle will make the reading from the downstream pressure tap difficult to interpret. That is, there can be no assurance that the measured pressure is the true back pressure. In such a case, it is advisable to use a venturi meter because the pressure is measured directly at the throat.

The mass flow rate of a compressible fluid through a venturi meter can easily be analyzed using the equations developed in Chapter 12. Consider the venturi meter shown in Fig. 13.26. Writing the energy equation, Eq. (12.15), for the flow of an ideal gas between stations 1 and 2 gives

$$\frac{V_1^2}{2} + \frac{kRT_1}{k-1} = \frac{V_2^2}{2} + \frac{kRT_2}{k-1} \tag{13.21}$$

By conservation of mass, the velocity V_1 can be expressed as

$$V_1 = \frac{\rho_2 A_2 V_2}{\rho_1 A_1}$$

FIGURE 13.26

Venturi meter.

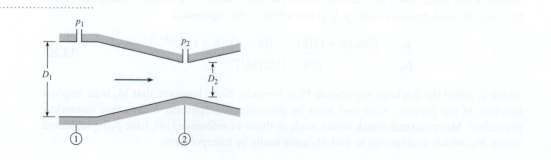

Substituting this result into Eq. (13.21), using the ideal-gas law to eliminate temperature, and solving for V_2 gives

$$V_2 = \left\{ \frac{[2k/(k-1)][(p_1/\rho_1) - (p_2/\rho_2)]}{1 - (\rho_2 A_2/\rho_1 A_1)^2} \right\}^{1/2} \tag{13.22}$$

Assuming that the flow is isentropic,

$$\frac{p_1}{p_2} = \left(\frac{\rho_1}{\rho_2}\right)^k$$

the equation for the velocity at the throat can be rewritten as

$$V_2 = \left\{ \frac{[2k/(k-1)](p_1/\rho_1)[1 - (p_2/p_1)^{(k-1)/k}]}{1 - (p_2/p_1)^{2/k}(D_2/D_1)^4} \right\}^{1/2} \tag{13.23}$$

The mass flow is obtained by multiplying V_2 by $\rho_2 A_2$. This analysis, however, has been based on a one-dimensional flow, and two-dimensional effects can be accounted for by the discharge coefficient C_d. The final result is

$$\dot{m} = C_d \rho_2 A_2 V_2 = C_d A_2 \left(\frac{p_2}{p_1}\right)^{1/k} \left\{ \frac{[2k/(k-1)]p_1\rho_1[1 - (p_2/p_1)^{(k-1)/k}]}{1 - (p_2/p_1)^{2/k}(D_2/D_1)^4} \right\}^{1/2} \tag{13.24}$$

This equation is valid for all flow conditions, subsonic or supersonic, provided no shock waves occur between station 1 and station 2. It is good design practice to avoid supersonic flows in the venturi meter to prevent the formation of shock waves and the attendant total pressure losses. Also, the discharge coefficient can generally be taken as unity if no shock waves occur between 1 and 2.

EXAMPLE 13.9

Applying the Venturi Meter to Find the Flow Rate of a Compressible Flow

Problem Statement

Calculate the mass flow rate of air (inlet static temperature = 27°C) flowing through a venturi meter. The venturi throat is 1 cm in diameter (D_2), and the pipe is 3 cm in diameter (D_1). Upstream static pressure is 150 kPa, and throat pressure is 100 kPa.

Define the Situation

- Air flows through a venturi meter. (see sketch on next page)
- Pipe diameter is $D = 0.03$ m. Venturi throat diameter is $d = 0.01$ m.
- Upstream conditions: Static temperature is 27°C; static pressure is 150 kPa.
- Pressure in throat = 100 kPa.

Properties: Air (27°C), Table A.2: $k = 1.4$, and $R = 287$ J/kg · K.

State the Goal

Calculate the mass flow rate (in kg/s).

Generate Ideas and Make a Plan

1. Calculate density of air in the pipe (upstream) using the ideal gas law.
2. Calculate mass flow rate using Eq. (13.24).

Take Action (Execute the Plan)

1. Ideal gas law

$$\rho_1 = \frac{p_1}{RT_1} = \frac{150 \times 10^3 \, \text{N/m}^2}{(287 \, \text{J/kg K})(300 \, \text{K})} = 1.74 \, \text{kg/m}^3$$

2. Mass flow rate

$$\dot{m} = 1 \times 0.785 \times 10^{-4} \, \text{m}^2 \left(\frac{1}{1.5}\right)^{0.714}$$

$$\times \left\{ \frac{7 \times 150 \times 10^3 \, \text{N/m}^2 \times 1.74 \, \text{kg/m}^3 [1 - (1/1.5)^{0.286}]}{[1 - (1/1.5)^{1.43}(1/3)^4]} \right\}^{1/2}$$

$$= \boxed{0.0264 \, \text{kg/s}}$$

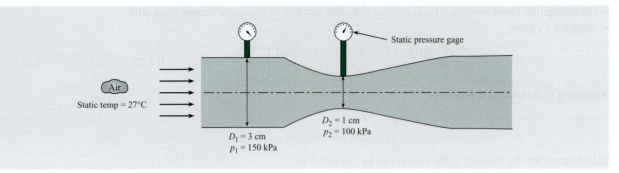

Shock Wave Visualization

When studying supersonic flow in a wind tunnel, it is important to be able to locate and identify the shock wave pattern. Unfortunately, shock waves cannot be seen with the naked eye, so the application of some type of optical technique is necessary. There are three techniques by which shock waves can be seen: the shadowgraph, the interferometer, and the schlieren system. Each technique has its special application related to the type of information on density variation that is desired. The schlieren technique, however, finds frequent use in shock wave visualization.

An illustration of the essential features of the schlieren system is given in Fig. 13.27. Light from the source s is collimated by lens L_1 to produce a parallel light beam. The light then passes through a second lens L_2 and produces an image of the source at plane f. A third lens L_3 focuses the image on the display screen. A sharp edge, usually called the knife edge, is positioned at plane f so as to block out a portion of the light.

If a shock wave occurs in the test section, the light is refracted by the density change across the wave. As illustrated by the dashed line in Fig. 13.27, the refracted ray escapes the blocking effect of the knife edge, and the shock wave appears as a lighter region on the screen. Of course, if the beam is refracted in the other direction, the knife edge blocks out more light, and the shock wave appears as a darker region. The contrast can be increased by intercepting more light with the knife edge.

FIGURE 13.27

Schlieren system.

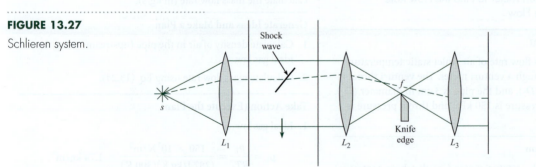

Interferometry

The interferometer allows one to map contours of constant density and to measure the density changes in the flow field. The underlying principle is the phase shift of a light beam on passing through media of different densities. The system now employed almost universally is the Mach-Zender interferometer, shown in Fig. 13.28. Light from a common source is split into two beams as it passes through the first half-silvered mirror. One beam passes through the test

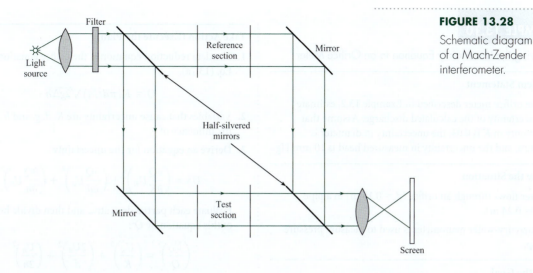

FIGURE 13.28

Schematic diagram of a Mach-Zender interferometer.

section, the other through the reference section. The two beams are then recombined and projected onto a screen or photographic plate. If the density in the test section and that in the reference section are the same, there is no phase shift between the two beams, and the screen is uniformly bright. However, a change of density in the test section changes the light speed of the test section beam, and a phase shift is generated between the two beams. Upon recombination of the beams, this phase shift gives rise to a series of dark and light bands on the screen. Each band represents a uniform-density contour, and the change in density across each band can be determined for a given system.

13.4 Accuracy of Measurements

When a parameter is measured, it is important to assess the accuracy of the measurement. The resulting analysis, called an **uncertainty analysis**, provides an estimate of the upper and lower bounds of the parameter. For example, if Q is a measured value of discharge, uncertainty analysis provides an estimate of the uncertainty U_Q in this measurement. The measurement would then be reported as $Q \pm U_Q$.

Commonly, a parameter of interest is not directly measured but is calculated from other variables. For example, discharge for an orifice meter is calculated using Eq. (13.7a). Such an equation is called a data reduction equation. Consider a data reduction equation of the form

$$x = f(y_1, y_2, \ldots, y_n)$$

where x is the parameter of interest and y_1 through y_n are the independent variables. Then, the uncertainty in x, which is written as U_x, is given by

$$U_x = \left[\left(\frac{\partial x}{\partial y_1} U_{y_1} \right)^2 + \left(\frac{\partial x}{\partial y_2} U_{y_2} \right)^2 + \cdots + \left(\frac{\partial x}{\partial y_n} U_{y_n} \right)^2 \right]^{0.5} \tag{13.25}$$

where U_{y_i} is the uncertainty in variable y_i. Equation (13.25), known as the uncertainty equation, is very useful for quantifying the accuracy of an experimental measurement, and for planning experiments. Additional information about uncertainty analysis is provided by Coleman and Steele (17).

EXAMPLE 13.10

Applying the Uncertainty Equation to an Orifice Meter

Problem Statement

For the orifice meter described in Example 13.2, estimate the uncertainty of the calculated discharge. Assume that uncertainty in K is 0.03, the uncertainty in diameter is 0.15 mm, and the uncertainty in measured head is 10 mm Hg.

Define the Situation

• Water flows through an orifice ($d = 0.15$ m) in a pipe ($D = 0.24$ m).

• A mercury-water manometer is used to measure pressure drop.

State the Goal

Find the uncertainty (in m³/s) for the calculated discharge Q.

Generate Ideas and Make a Plan

1. Identify the data reduction equation (DRE).

2. Within the DRE, identify each variable that contributes to uncertainty.

3. Develop an equation for uncertainty by applying Eq. (13.25).

4. Calculate uncertainty by using the equation developed in step 3.

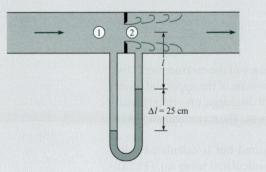

$\Delta l = 25$ cm

Take Action (Execute the Plan)

1. The data reduction equation is the orifice equation, Eq. (13.7a).

$$Q = K(\pi d^2/4)\sqrt{2g\Delta h}$$

2. Variables that cause uncertainty are K, d, g, and h. Neglect the influence of g.

3. Derive an equation for the uncertainty

$$U_Q^2 = \left(\frac{\partial Q}{\partial K}U_K\right)^2 + \left(\frac{\partial Q}{\partial d}U_d\right)^2 + \left(\frac{\partial Q}{\partial h}U_h\right)^2$$

Evaluate each partial derivative and then divide both sides of this equation by Q^2:

$$\left(\frac{U_Q}{Q}\right)^2 = \left(\frac{U_K}{K}\right)^2 + \left(\frac{2U_d}{d}\right)^2 + \left(\frac{U_h}{2h}\right)^2$$

4. Substitute values from Example 13.2:

$$\left(\frac{U_Q}{Q}\right)^2 = \left(\frac{0.03}{0.66}\right)^2 + \left(\frac{2 \times 0.15}{150}\right)^2 + \left(\frac{10}{2 \times 250}\right)^2$$

$$\left(\frac{U_Q}{Q}\right)^2 = (20.7 \times 10^{-4}) + (0.04 \times 10^{-4}) + (4 \times 10^{-4})$$

$$U_Q = 0.0497\, Q = 0.0497 \times (0.092 \text{ m}^3/\text{s})$$
$$= 0.0046 \text{ m}^3/\text{s}$$

Thus

$$\boxed{Q = (0.092 \pm 0.046) \text{ m}^3/\text{s}}$$

Review the Solution and the Process

The primary source of uncertainty in the discharge is due to U_K. The term U_h has a small effect, and U_d has a negligible effect.

13.5 Summarizing Key Knowledge

Measuring Velocity and Pressure

• Instruments for *velocity measurement* include the stagnation tube, Pitot-static tube, yaw meter, vane and cup anemometers, hot-wire and hot-film anemometers, laser-Doppler anemometer, and particle image velocimeter.

• Instruments for *pressure measurement* include the static tube, piezometer, differential manometer, Bourdon-tube gage, and several types of pressure transducers.

Measuring Flow Rate (Discharge)

- To measure flow rate, one can use several direct methods
  - ▸ Measure volume (or weight) and divide by time.
  - ▸ Measure velocities at points on a cross section and integrate using $Q = \int V dA$.
- Common instruments for *flow measurement* include the orifice meter, flow nozzle, venturi meter, electromagnetic flow meter, ultrasonic flow meter, turbine flow meter, vortex flow meter, rotameter, and weir.
- Flow rate or discharge for a flowmeter that uses a restricted opening (i.e., an orifice, flow nozzle, or venturi) is calculated using

$$Q = KA_o\sqrt{2g\Delta h} = KA_o\sqrt{2\Delta p_z/\rho}$$

where K is a flow coefficient that depends on Reynolds number and the type of flowmeter, A_o is the area of the opening, Δh is the change in piezometric head across the flowmeter, and Δp_z is drop in piezometric pressure across the flowmeter.
- Discharge for a rectangular weir of length L is given by

$$Q = K\sqrt{2g}\,LH^{3/2}$$

where K is the flow coefficient that depends on H/P. The term H is the height of the water above the crest of the weir, as measured upstream of the weir, and P is the height of the weir.
- Discharge for a 60° triangular weir with $H > 2$ cm is given by

$$Q = 0.179\sqrt{2g}\,H^{5/2}$$

Measurements in Compressible Flow

- When flow is compressible, instruments such as the stagnation tube, hot-wire anemometer, Pitot tube, and flow nozzle may be used. However, equations correlating velocity and discharge need to be altered to account for the effects of compressibility.
- To observe shock waves in compressible flow, a schlieren technique or an interferometer may be used.

Uncertainty Analysis

- Uncertainty analysis provides a way to quantify the accuracy of a measurement. When a parameter of interest x is evaluated using an equation of the form $x = f(y_1, y_2, \ldots, y_n)$, where y_1 through y_n are the independent variables, the uncertainty in x is given by

$$U_x = \left[\left(\frac{\partial x}{\partial y_1}U_{y_1}\right)^2 + \left(\frac{\partial x}{\partial y_2}U_{y_2}\right)^2 + \cdots + \left(\frac{\partial x}{\partial y_n}U_{y_n}\right)^2\right]^{0.5}$$

where U_{y_i} is the uncertainty in variable y_i. This equation, known as the uncertainty equation, is very useful for estimating uncertainty and for planning experiments.

REFERENCES

1. Hurd, C. W., K. P. Chesky, and A. H. Shapiro. "Influence of Viscous Effects on Impact Tubes." *Trans. ASME J. Applied Mechanics,* vol. 75 (June 1953).

2. King, H. W., and E. F. Brater. *Handbook of Hydraulics.* New York: McGraw-Hill, 1963.

3. Lomas, Charles C. *Fundamentals of Hot Wire Anemometry*. New York: Cambridge University Press, 1986.

4. Durst, Franz. *Principles and Practice of Laser-Doppler Anemometry*. New York: Academic Press, 1981.

5. Kline, J. J. "Flow Visualization." In *Illustrated Experiments in Fluid Mechanics: The NCFMF Book of Film Notes*. Cambridge, MA: MIT Press, 1972.

6. Macagno, Enzo O. "Flow Visualization in Liquids." *Iowa Inst. Hydraulic Res. Rept.*, 114 (1969).

7. Raffel, M., C. Wilbert, and J. Kompenhans. *Particle Image Velocimetry*. New York: Springer, 1998.

8. Lienhard, J. H., V, and J. H. Lienhard, IV. "Velocity Coefficients for Free Jets from Sharp-Edged Orifices." *Trans. ASME J. Fluids Engineering*, 106, (March 1984).

9. Tuve, G. L., and R. E. Sprenkle. "Orifice Discharge Coefficients for Viscous Liquids." *Instruments*, vol. 8 (1935).

10. ASME. *Fluid Meters*, 6th ed. New York: ASME, 1971.

11. Shercliff, J. A. *Electromagnetic Flow-Measurement*. New York: Cambridge University Press, 1962.

12. Kindsvater, Carl E., and R. W. Carter. "Discharge Characteristics of Rectangular Thin-Plate Weirs." *Trans. Am. Soc. Civil Eng.*, 124 (1959), 772–822.

13. King, L. V. *Phil. Trans. Roy. Soc. London, Ser. A*, 14 (1914), 214.

14. Miller, R. W. *Flow Measurement Engineering Handbook*. New York: McGraw-Hill, 1983.

15. Scott, R. W. W., ed. *Developments in Flow Measurement–1*. Englewood Cliffs, NJ: Applied Science, 1982.

16. NACA. "Equations, Tables, and Charts for Compressible Flow." TR 1135 (1953).

17. Coleman, Hugh W., and W. Glenn Steele, *Experimentation and Uncertainty Analysis for Engineers*. New York: Wiley, 1989.

18. Wikipedia contributors, "Anemometer," Wikipedia, The Free Encyclopedia, http://en.wikipedia.org/w/index.php?title=Anemometer&oldid= 156121777 (accessed September 8, 2007).

PROBLEMS

PLUS Problem available in *WileyPLUS* at instructor's discretion.

GO Guided Online (GO) Problem, available in *WileyPLUS* at instructor's discretion.

Velocity and Pressure Measurements (§13.1)

13.1 List five different instruments or approaches that engineers use to measure fluid velocity, and five more that are used to measure pressure. For each instrument or approach, list two advantages and two disadvantages, using this text or sources on the internet.

Flow Velocity: Stagnation Tubes (§13.1)

13.2 Consider measuring the speed of an automobile by building a stagnation tube from a drinking straw and then using this device with a water-filled *U*-tube manometer.

   a. Make a sketch that illustrates how you would propose making this measurement.

   b. Determine the lowest velocity that could be measured. Assume that the lower limit is based on the resolution of the manometer.

13.3 Without exceeding an error of 2.5%, what is the minimum air velocity that can be obtained using a 1 mm circular stagnation tube if the formula

$$V = \sqrt{2\Delta p_{stag}/\rho} = \sqrt{2gh_{stag}}$$

is used for computing the velocity? Assume standard atmospheric conditions.

13.4 Without exceeding an error of 1%, what is the minimum water velocity that can be obtained using a 1.5-mm circular stagnation tube if the formula

$$V = \sqrt{2\Delta p_{stag}/\rho} = \sqrt{2gh_{stag}}$$

is used for computing the velocity? Assume the water temperature is 20°C.

13.5 PLUS A stagnation tube 2 mm in diameter is used to measure the velocity in a stream of air as shown. What is the air velocity if the deflection on the air-water manometer is 1.0 mm? Air temperature = 10°C, and p = 1 atm.

13.6 PLUS If the velocity in an airstream (p_a = 98 kPa; T = 10°C) is 24 m/s, what deflection will be produced in an air-water manometer if the stagnation tube is 2 mm in diameter?

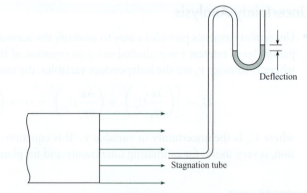

Deflection

Stagnation tube

PROBLEMS 13.5, 13.6

13.7 What would be the error in velocity determination if one used a C_p value of 1.00 for a circular stagnation tube instead of the true value? Assume the measurement is made with a stagnation tube 2 mm in diameter that is measuring air (T = 25°C, p = 1 atm) velocity for which the stagnation pressure reading is 5.00 Pa.

13.8 (WILEY GO) A velocity-measuring probe used frequently for measuring smokestack gas velocities is shown. The probe consists of two tubes bent away from and toward the flow direction and cut off on a plane normal to the flow direction, as shown. Assume the pressure coefficient is 1.0 at A and -0.4 at B. The probe is inserted in a stack where the temperature is 300°C and the pressure is 100 kPa absolute. The gas constant of the stack gases is 410 J/kg K. The probe is connected to a water manometer, and a 1.0 cm deflection is measured. Calculate the stack gas velocity.

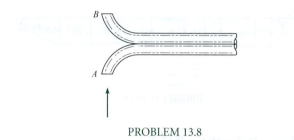

PROBLEM 13.8

Flow Velocity: Laser-Doppler Anemometers (§13.1)

13.9 On the Internet, locate technically sound resources relevant to the LDA. Skim these resources, and then

   a. Write down five findings that are relevant to engineering practice and interesting to you.

   b. Write down two questions about LDAs that are interesting and insightful.

13.10 (WILEY PLUS) A laser-Doppler anemometer (LDA) system is being used to measure the velocity of air in a tube. The laser is an argon-ion laser with a wavelength of 4880 angstroms. The angle between the laser beams is 20°. The time interval is determined by measuring the time between five spikes, as shown, on the signal from the photodetector. The time interval between the five spikes is 12 microseconds. Find the velocity.

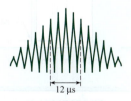

12 µs

PROBLEM 13.10

Volume Flow Rate or Discharge (§13.2)

13.11 (WILEY PLUS) Classify the following devices as to whether they are used to measure velocity (V), pressure (P), or discharge (Q).

   a. hot-wire anemometer

   b. venturi meter

   c. differential manometer

   d. orifice meter

   e. stagnation tube

   f. rotameter

   g. ultrasonic flow meter

   h. Bourdon-tube gage

   i. weir

   j. laser-Doppler anemometer

13.12 List five different instruments or approaches that engineers use to measure flow rate (discharge). For each instrument or approach, list two advantages and two disadvantages.

13.13 (WILEY PLUS) Water from a pipe is diverted into a tank for 3 min. If the weight of diverted water is measured to be 8 kN, what is the discharge in cubic meters per second? Assume the water temperature is 20°C.

13.14 (WILEY PLUS) Water from a test apparatus is diverted into a calibrated volumetric tank for 6 min. If the volume of diverted water is measured to be 67 m³, what is the discharge in cubic meters per second?

13.15 A velocity traverse in a 24-cm oil pipe yields the data in the table. What are the discharge, mean velocity, and ratio of maximum to mean velocity? Does the flow appear to be laminar or turbulent?

r (cm)	V (m/s)	r (cm)	V (m/s)
0	8.7	7	5.8
1	8.6	8	4.9
2	8.4	9	3.8
3	8.2	10	2.5
4	7.7	10.5	1.9
5	7.2	11.0	1.4
6	6.5	11.5	0.7

13.16 A velocity traverse inside a 40-cm-circular air duct yields the data in the table. What is the rate of flow in m³/s and m³/min? What is the ratio of V_{max} to V_{mean}? Does it appear that the flow is laminar or turbulent? If $p = 98.6$ kPa abs and $T = 21$°C, what is the mass flow rate?

y^*	V (m/s)	y^*	V (m/s)
0.0	0	5.0	33.5
0.25	22	7.5	35.7
0.5	24	10.0	37.2
1.0	27	12.5	38.4
1.5	28.3	15.0	39.3
2.5	30.5	17.5	40.2
3.75	32.5	20	41.1

*Distance from pipe wall, cm.

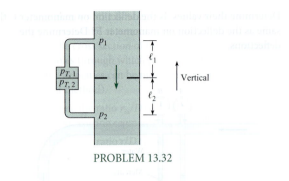

PROBLEM 13.32

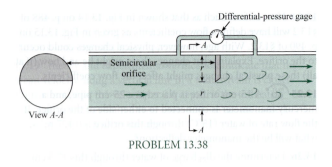

View A-A

PROBLEM 13.38

13.33 Water ($T = 10°C$) is pumped at a rate of 0.57 m³/s through the system shown in the figure. What differential pressure will occur across the orifice? What power must the pump supply to the flow for the given conditions? Also, draw the HGL and the EGL for the system. Assume $f = 0.015$ for the pipe.

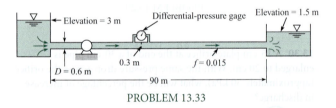

PROBLEM 13.33

13.34 $\overset{\text{WILEY}}{\text{GO}}$ Determine the size of orifice required in a 15-cm pipe to measure 0.03 m³/s of water with a deflection of 1 m on a mercury-water manometer.

13.35 What is the discharge of gasoline ($S = 0.68$) in a 12-cm horizontal pipe if the differential pressure across a 6-cm orifice in the pipe is 50 kPa?

13.36 What size orifice is required to produce a change in head of 6 m for a discharge of 2 m³/s of water in a pipe 1 m in diameter?

13.37 An orifice is to be designed to have a change in pressure of 48 kPa across it (measured with a differential-pressure transducer) for a discharge of 4.0 m³/s of water in a pipe 1.2 m in diameter. What diameter should the orifice have to yield the desired results?

13.38 Semicircular orifices such as the one shown are sometimes used to measure the flow rate of liquids that also transport sediments. The opening at the bottom of the pipe allows free passage of the sediment. Derive a formula for Q as a function of Δp, D, and other relevant variables associated with the problem. Then, using that formula and guessing any unknown data, estimate the water discharge through such an orifice when Δp is read as 80 kPa and flow is in a 30-cm pipe.

Discharge: Venturi Meters (§13.2)

13.39 What is the main advantage of a venturi meter versus an orifice meter? The main disadvantage?

13.40 Water flows through a venturi meter that has a 40-cm throat. The venturi meter is in a 70-cm pipe. What deflection will occur on a mercury-water manometer connected between the upstream and throat sections if the discharge is 0.75 m³/s? Assume $T = 20°C$.

13.41 $\overset{\text{WILEY}}{\text{PLUS}}$ What is the throat diameter required for a venturi meter in a 61-cm horizontal pipe carrying water with a discharge of 0.76 m³/s if the differential pressure between the throat and the upstream section is to be limited to 200 kPa at this discharge? For a first iteration, assume $K = 1.02$.

13.42 Estimate the rate of flow of water through the venturi meter shown.

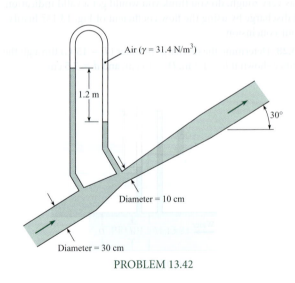

PROBLEM 13.42

13.43 $\overset{\text{WILEY}}{\text{PLUS}}$ When no flow occurs through the venturi meter, the indicator on the differential-pressure gage is straight up and indicates a Δp of zero. When 0.142 m³/s of water flows to the right, the differential-pressure gage indicates $\Delta p = +70$ kPa. If the flow is now reversed and 0.142 m³/s flow to the left through the venturi meter, in which range would Δp fall? (a) $\Delta p < -70$ kPa, (b) -70 kPa $< \Delta p < 0$, (c) $0 < \Delta p < 70$ kPa, or (d) $\Delta p = 70$ kPa?

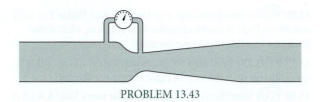

PROBLEM 13.43

13.44 [WILEY PLUS] The pressure differential across this venturi meter is 92 kPa. What is the discharge of water ($T = 20°C$) through it? [*Hint*: The value of flow coefficient you calculate should be $K = 1.02$]

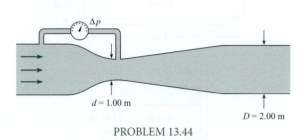

$d = 1.00$ m

$D = 2.00$ m

PROBLEM 13.44

13.45 Engineers are calibrating a poorly designed venturi meter for the flow of petroleum by relating the pressure difference between taps 1 and 2 to the discharge. By applying the Bernoulli equation and assuming a quasi-one-dimensional flow (velocity uniform across every cross section), the engineers find that

$$Q_0 = A_2[2(p_1 - p_2)/\rho]^{0.5}[1 - (d/D)^4]^{-0.5}$$

where D and d are the duct diameters at stations 1 and 2. However, they realize that the flow is not quasi-one-dimensional and that the pressure at tap 2 is not equal to the average pressure in the throat because of streamline curvature. Thus the engineers introduce a correction factor K into the foregoing equation to yield

$$Q = KQ_0$$

Use your knowledge of pressure variation across curved streamlines to decide whether K is larger or smaller than unity, and support your conclusion by presenting a rational argument.

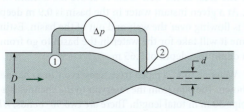

PROBLEM 13.45

13.46 The differential-pressure gage on the venturi meter shown reads 37.2 kPa, $h = 63.5$ cm, $d = 18$ cm, and $D = 30$ cm. What is the discharge of water in the system? Assume $T = 10°C$.

13.47 The differential-pressure gage on the venturi meter reads 40 kPa, $d = 20$ cm, $D = 40$ cm, and $h = 75$ cm. What is the discharge of gasoline ($S = 0.69$; $\mu = 3 \times 10^{-4}$ N · s/m²) in the system?

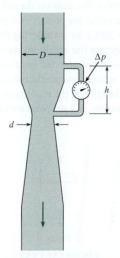

PROBLEMS 13.46, 13.47

13.48 A flow nozzle has a throat diameter of 2 cm and a beta ratio (d/D) of 0.5. Water flows through the nozzle, creating a pressure difference across the nozzle of 8 kPa. The viscosity of the water is 10^{-6} m²/s, and the density is 1000 kg/m³. Find the discharge.

13.49 Water flows through an annular venturi consisting of a body of revolution mounted inside a pipe. The pressure is measured at the minimum area and upstream of the body. The pipe is 5 cm in diameter, and the body of revolution is 2.5 cm in diameter. A head difference of 1 m is measured across the pressure taps. Find the discharge in cubic meters per second.

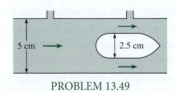

PROBLEM 13.49

Other Discharge Measurement Techniques (§13.2)

13.50 What is the head loss in terms of $V_0^2/2g$ for the flow nozzle shown?

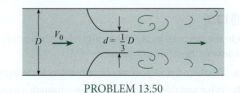

PROBLEM 13.50

13.51 ⓦ A vortex flowmeter is used to measure the discharge in a duct 5 cm in diameter. The diameter of the shedding element is 1 cm. The Strouhal number based on the shedding frequency from one side of the element is 0.2. A signal frequency of 50 Hz is measured by a pressure transducer mounted downstream of the element. What is the discharge in the duct?

13.52 A rotameter operates by aerodynamic suspension of a weight in a tapered tube. The scale on the side of the rotameter is calibrated in cubic meter per minute at standard conditions ($p = 1$ atm and $T = 20°C$). By considering the balance of weight and aerodynamic force on the weight inside the tube, determine how the readings would be corrected for nonstandard conditions. In other words, how would the actual cubic meter per minute be calculated from the reading on the scale, given the pressure, temperature, and gas constant of the gas entering the rotameter?

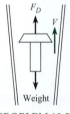

PROBLEM 13.52

13.53 ⓦ A rotameter is used to measure the flow rate of a gas with a density of 1.0 kg/m³. The scale on the rotameter indicates 5 liters/s. However, the rotameter is calibrated for a gas with a density of 1.2 kg/m³. What is the actual flow rate of the gas (in liters per second)?

13.54 Ultrasonic flowmeters are used to measure velocity in systems where it is important to not disrupt the flow, such as for blood velocity. One mode of operation of ultrasonic flowmeters is to measure the travel times between two stations for a sound wave traveling upstream and then downstream with the flow. The downstream propagation speed with respect to the measuring stations is $c + V$, where c is the sound speed and V is the flow velocity. Correspondingly, the upstream propagation speed is $c - V$.

   a. Derive an expression for the flow velocity in terms of the distance between the two stations, L; the difference in travel times, Δt; and the sound speed.

   b. The sound speed is typically much larger than V ($c \gg V$). With this approximation, express V in terms of L, c, and Δt.

   c. A 10-ms time difference is measured for waves traveling 20 m in a gas where the speed of sound is 300 m/s. Calculate the flow velocity.

Weirs (§13.2)

13.55 On the Internet, locate technically reliable resources about weirs to answer the following questions.

   a. What are five important considerations for using weirs?

   b. What variables influence flow rate through a rectangular weir?

13.56 ⓦ Water flows over a rectangular weir that is 2 m wide and 30 cm high. If the head on the weir is 10 cm, what is the discharge in cubic meters per second?

13.57 ⓦ The head on a 60° triangular weir is 25 cm. What is the discharge over the weir in cubic meters per second?

13.58 ⓦ Water flows over two rectangular weirs. Weir A is 1.5 m long in a channel 3 m wide; weir B is 1.5 m long in a channel 1.5 m wide. Both weirs are 0.6 m high. If the head on both weirs is 0.3 m, then one can conclude that (a) $Q_A = Q_B$, (b) $Q_A > Q_B$, or (c) $Q_A < Q_B$.

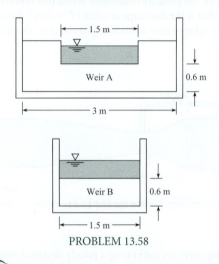

PROBLEM 13.58

13.59 ⓦ A 30-cm-high rectangular weir (weir 1) is installed in a 60-cm-wide rectangular channel, and the head on the weir is observed for a discharge of 0.283 m³/s. Then the 30-cm weir is replaced by a 60-cm-high rectangular weir (weir 2), and the head on the weir is observed for a discharge of 0.283 m³/s. The ratio H_1/H_2 should be (a) equal to 1.00, (b) less than 1.00, or (c) greater than 1.00.

13.60 ⓦ A 3-m-long rectangular weir is to be constructed in a 3-m-wide rectangular channel, as shown (a). The maximum flow in the channel will be 4 m³/s. What should be the height P of the weir to yield a depth of water of 2 m in the channel upstream of the weir?

13.61 Consider the rectangular weir described in Prob. 13.60. When the head is doubled, the discharge is (a) doubled, (b) less than doubled, or (c) more than doubled.

13.62 A basin is 15 m long, 0.6 m wide, and 1.2 m deep. A sharp-crested rectangular weir is located at one end of the basin, and it spans the width of the basin (the weir is 0.6 m long). The crest of the weir is 0.6 m above the bottom of the basin. At a given instant water in the basin is 0.9 m deep; thus water is flowing over the weir and out of the basin. Estimate the time it will take for the water in the basin to go from the 0.9-m depth to a depth of 0.65 m.

13.63 Water at 10°C is piped from a reservoir to a channel like that shown. The pipe from the reservoir to the channel is a 10-cm steel pipe 30 m in total length. There are two 90° bends, $r/D = 1$, in the line, and the entrance and exit are sharp edged. The weir is 0.6 m long. The elevation of the water surface in the reservoir is 30 m, and the elevation of the bottom of the channel is 21 m. The crest of the weir is 0.9 m above the bottom of the channel. For steady flow conditions determine the water surface elevation in the channel and the discharge in the system.

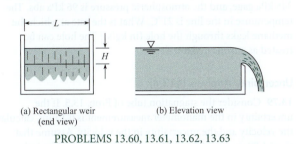

(a) Rectangular weir
(end view)

(b) Elevation view

PROBLEMS 13.60, 13.61, 13.62, 13.63

13.64 At one end of a rectangular tank 1 m wide is a sharp-crested rectangular weir 1 m high. In the bottom of the tank is a 10-cm sharp-edged orifice. If 0.10 m^3/s of water flows into the tank and leaves the tank both through the orifice and over the weir, what depth will the water in the tank attain?

13.65 What is the water discharge over a rectangular weir 0.9 m high and 3 m long in a rectangular channel 3 m wide if the head on the weir is 0.5 m?

13.66 `WILEY GO` A reservoir is supplied with water at 15°C by a pipe with a venturi meter as shown. The water leaves the reservoir through a triangular weir with an included angle of 60°. The flow coefficient of the venturi is unity, the area of the venturi throat is 0.09 m^2, and the measured Δp is 70 kPa. Find the head, H, of the triangular weir.

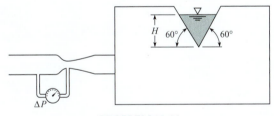

PROBLEM 13.66

13.67 At a particular instant water flows into the tank shown through pipes A and B, and it flows out of the tank over the rectangular weir at C. The tank width and weir length (dimensions normal to page) are 0.6 m. Then, for the given conditions, is the water level in the tank rising or falling?

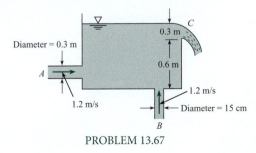

PROBLEM 13.67

13.68 Water flows from the first reservoir to the second over a rectangular weir with a width-to-head ratio of 3. The height P of the weir is twice the head. The water from the second reservoir flows over a 60° triangular weir to a third reservoir. The discharge across both weirs is the same. Find the ratio of the head on the rectangular weir to the head on the triangular weir.

13.69 Given the initial conditions of Prob. 13.68, tell, qualitatively and quantitatively, what will happen if the flow entering the first reservoir is increased 50%.

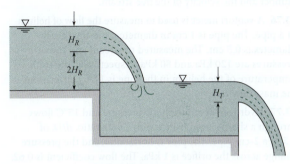

PROBLEMS 13.68, 13.69

13.70 The head on a 60° triangular weir is 55 cm. What is the discharge of water over the weir?

13.71 An engineer is designing a triangular weir for measuring the flow rate of a stream of water that has a discharge of 0.0028 m^3/s. The weir has an included angle of 45° and a coefficient of discharge of 0.6. Find the head on the weir.

13.72 A pump is used to deliver water at 10°C from a well to a tank. The bottom of the tank is 2 m above the water surface in the well. The pipe is commercial steel 2.5 m long with a diameter of 5 cm. The pump develops a head of 20 m. A triangular weir with an included angle of 60° is located in a wall of the tank with the bottom of the weir 1 m above the tank floor. Find the level of the water in the tank above the floor of the tank.

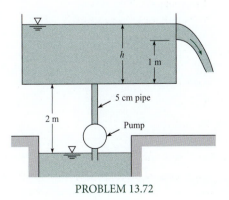

PROBLEM 13.72

Measurements in Compressible Flow (§13.3)

13.73 A Pitot-static tube is used to measure the Mach number in a compressible subsonic flow of air. The stagnation pressure is 140 kPa, and the static pressure is 100 kPa. The total temperature of the flow is 300 K. Determine the Mach number and the flow velocity.

13.74 Use the normal shock wave relationships developed in Chapter 12 to derive the Rayleigh supersonic Pitot formula.

13.75 The static and stagnation pressures measured by a Pitot-static tube in a supersonic air flow are 54 kPa and 200 kPa, respectively. The total temperature is 350 K. Determine the Mach number and the velocity of the free stream.

13.76 A venturi meter is used to measure the flow of helium in a pipe. The pipe is 1 cm in diameter, and the throat diameter is 0.5 cm. The measured upstream and throat pressures are 120 kPa and 80 kPa, respectively. The static temperature of the helium in the pipe is 17°C. Determine the mass flow rate.

13.77 Hydrogen at atmospheric pressure and 15°C flows through a sharp-edged orifice with a beta ratio, d/D, of 0.5 in a 2-cm pipe. The pipe is horizontal, and the pressure change across the orifice is 1 kPa. The flow coefficient is 0.62.

Find the mass flow (in kilograms per second) through the orifice.

13.78 A hole 5 mm in diameter is accidentally punctured in a line carrying natural gas (methane). The pressure in the pipe is 345 kPa gage, and the atmospheric pressure is 96 kPa abs. The temperature in the line is 21°C. What is the rate at which the methane leaks through the hole (in kg/s)? The hole can be treated as a truncated nozzle.

Uncertainty Analysis (§13.4)

13.79 Consider the stagnation tube of Prob. 13.5. If the uncertainty in the manometer measurement is 0.1 mm, calculate the velocity and the uncertainty in the velocity. Assume that $C_p = 1.00$, $\rho_{air} = 1.25$ kg/m³, and the only uncertainty is due to the manometer measurement.

13.80 Consider the orifice meter in Prob. 13.26. Calculate the flow rate and the uncertainty in the flow rate. Assume the following values of uncertainty: 0.03 in flow coefficient, 0.125 cm in orifice diameter, and 1.25 cm in height of mercury.

13.81 Consider the weir in Prob. 13.65. Calculate the discharge and the uncertainty in the discharge. Assume the uncertainty in K is 5%, in H is 7.5 cm, and in L is 2.5 cm.

TURBOMACHINERY 14

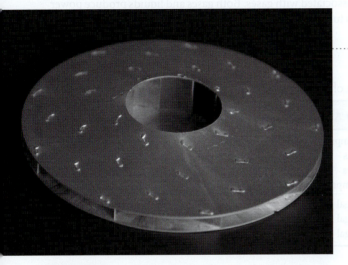

FIGURE 14.1

This figure shows the impeller from the blower inside a vacuum cleaner. This impeller rotates inside a housing. This rotational motion creates a suction pressure that draws air into the center hole. The air is flung outward by the spinning blades of the impeller.

This impeller was "liberated" from the vacuum cleaner by Jason Stirpe, while he was an engineering student. Jason used this impeller with a DC motor and a homemade housing to fabricate a blower for a design that he was creating. Being resourceful is at the heart of technology innovation. (Photo by Donald Elger.)

Chapter Road Map

Machines to move fluids or to extract power from moving fluids have been designed since the beginning of recorded history. Fluid machines are everywhere. They are the essential components of the automobiles we drive, the supply systems for the water we drink, the power generation plants for the electricity we use, and the air-conditioning and heating systems that provide the comfort we enjoy. Thus, this chapter introduces the concepts underlying various types of machines.

Learning Objectives

STUDENTS WILL BE ABLE TO

- Describe the factors that influence the thrust and efficiency of a propeller. (§14.1)
- Calculate the thrust and efficiency of a propeller. (§14.1)
- Describe axial flow and radial flow pumps. (§14.2, 14.3)
- Define the head coefficient and the discharge coefficient. (§14.2)
- Sketch a pump performance curve and describe the relevant π-groups that appear. (§14.2, 14.3)
- Explain how specific speed is used to select an appropriate type of pump for an application. (§14.4)
- Describe an impulse turbine and a reaction turbine. (§14.8)
- Describe the maximum power that can be produced by a wind turbine. (§14.8)

Fluid machines are separated into two broad categories: positive-displacement machines and turbomachines. **Positive-displacement machines** operate by forcing fluid into or out of a chamber. Examples include the bicycle tire pump, the gear pump, the peristaltic pump, and the human heart. **Turbomachines** involve the flow of fluid through rotating blades or rotors that remove or add energy to the fluid. Examples include propellers, fans, water pumps, windmills, and compressors.

Axial-flow turbomachines operate with the flow entering and leaving the machine in the direction that is parallel to the axis of rotation of blades. A radial-flow machine can have the flow either entering or leaving the machine in the radial direction that is normal to the axis of rotation.

Table 14.1 provides a classification for turbomachinery. Power-absorbing machines require power to increase head (or pressure). A power-producing machine provides shaft power at the expense of head (or pressure) loss. Pumps are associated with liquids, whereas fans (blowers) and compressors are associated with gases. Both gases and liquids produce power through turbines. Oftentimes the gas turbine refers to an engine that has both a compressor and turbine and produces power.

TABLE 14.1 Categories of Turbomachinery

	Power Absorbing	**Power Producing**
Axial machines	Axial pumps Axial fans Propellers Axial compressors	Axial turbine (Kaplan) Wind turbine Gas turbine
Radial machines	Centrifugal pump Centrifugal fan Centrifugal compressor	Impulse turbine (Pelton wheel) Reaction turbine (Francis turbine)

14.1 Propellers

A propeller is a fan that converts rotational motion into thrust. The design of a propeller is based on the fundamental principles of airfoil theory. For example, consider a section of the propeller in Fig. 14.2, and notice the analogy between the lifting vane and the propeller. This propeller is rotating at an angular speed ω, and the speed of advance of the airplane and propeller is V_0. Focusing on an elemental section of the propeller, Fig. 14.2c, it is noted that the given section has a velocity with components V_0 and V_t. Here V_t is tangential velocity, $V_t = r\omega$, resulting from the rotation of the propeller. Reversing and adding the velocity vectors V_0 and V_t yield the velocity of the air relative to the particular propeller section (Fig. 14.2d).

The angle θ is given by

$$\theta = \arctan\left(\frac{V_0}{r\omega}\right) \tag{14.1}$$

For a given forward speed and rotational rate, this angle is a minimum at the propeller tip $(r = r_0)$ and increases toward the hub as the radius decreases. The angle β is known as the **pitch angle**. The local angle of attack of the elemental section is

$$\alpha = \beta - \theta \tag{14.2}$$

The propeller can be analyzed as a series of elemental sections (of width dr) producing lift and drag, which provide the propeller thrust and create resistive torque. This torque multiplied by the rotational speed is the power input to the propeller.

The propeller is designed to produce thrust, and because the greatest contribution to thrust comes from the lift force F_L, the goal is to maximize lift and minimize drag, F_D. For

FIGURE 14.2

Propeller motion.
(a) Airplane motion.
(b) View A-A.
(c) View B-B.
(d) Velocity relative to blade element.

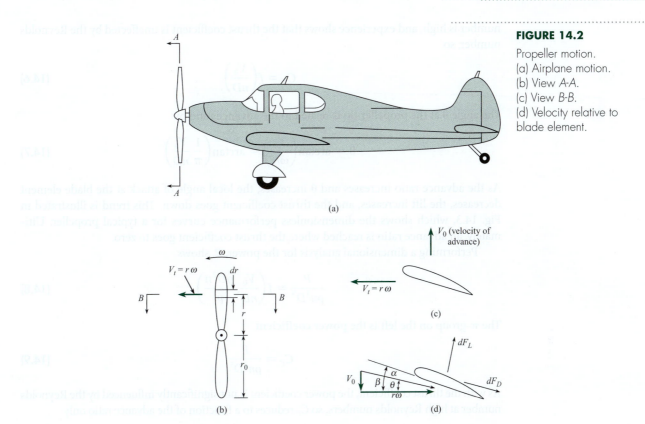

a given shape of propeller section, the optimum angle of attack can be determined from data such as are given in Fig. 11.24. Because the angle θ increases with decreasing radius, the local pitch angle has to change to achieve the optimum angle of attack. This is done by twisting the blade.

A dimensional analysis can be performed to determine the π-groups that characterize the performance of a propeller. For a given propeller shape and pitch distribution, the thrust of a propeller T, will depend on the propeller diameter D, the rotational speed n, the forward speed V_0, the fluid density ρ, and the fluid viscosity μ.

$$T = f(D, \omega, V_0, \rho, \mu) \tag{14.3}$$

Performing a dimensional analysis results in

$$\frac{T}{\rho n^2 D^4} = f\left(\frac{V_0}{nD}, \frac{\rho D^2 n}{\mu}\right) \tag{14.4}$$

It is conventional practice to express the rotational rate, n, as revolutions per second (rps). The π-group on the left is called the **thrust coefficient**,

$$C_T = \frac{T}{\rho n^2 D^4} \tag{14.5}$$

The first π-group on the right is the **advance ratio**. The second group is a Reynolds number based on the tip speed and diameter of the propeller. For most applications, the Reynolds number is high, and experience shows that the thrust coefficient is unaffected by the Reynolds number, so

$$C_T = f\left(\frac{V_0}{nD}\right)$$ (14.6)

The angle θ at the propeller tip is related to the advance ratio by

$$\theta = \arctan\left(\frac{V_0}{\omega r_0}\right) = \arctan\left(\frac{1}{\pi}\frac{V_0}{nD}\right)$$ (14.7)

As the advance ratio increases and θ increases, the local angle of attack at the blade element decreases, the lift increases, and the thrust coefficient goes down. This trend is illustrated in Fig. 14.3, which shows the dimensionless performance curves for a typical propeller. Ultimately, an advance ratio is reached where the thrust coefficient goes to zero.

Performing a dimensional analysis for the power, P, shows

$$\frac{P}{\rho n^3 D^5} = f\left(\frac{V_0}{nD}, \frac{\rho D^2 n}{\mu}\right)$$ (14.8)

The π-group on the left is the **power coefficient,**

$$C_P = \frac{P}{\rho n^3 D^5}$$ (14.9)

As with the thrust coefficient, the power coefficient is not significantly influenced by the Reynolds number at high Reynolds numbers, so C_P reduces to a function of the advance ratio only

$$C_P = f\left(\frac{V_0}{nD}\right)$$ (14.10)

FIGURE 14.3

Dimensionless performance curves for a typical propeller; D = 2.90 m, n = 1400 rpm. [After Weick (1).]

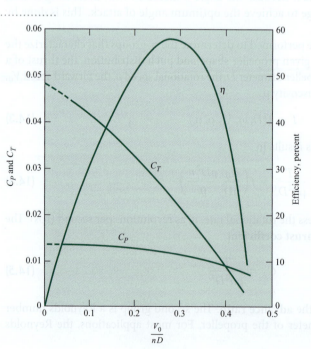

The functional relationship between C_P and V_0/nD for an actual propeller is also shown in Fig. 14.3. Even though the thrust coefficient approaches zero for a given advance ratio, the power coefficient shows little decrease because it still takes power to overcome the torque on the propeller blade.

The curves for C_T and C_P are evaluated from performance characteristics of a given propeller operating at different values of V_0 as shown in Fig. 14.4. Although the data for the curves are obtained for a given propeller, the values for C_T and C_P, as a function of advance ratio, can be applied to geometrically similar propellers of different sizes and angular speeds.[*] Example 14.1 illustrates such an application.

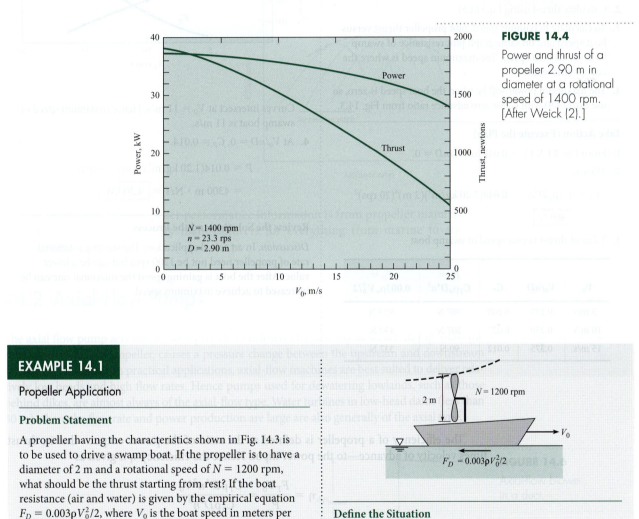

FIGURE 14.4

Power and thrust of a propeller 2.90 m in diameter at a rotational speed of 1400 rpm. [After Weick (2).]

EXAMPLE 14.1

Propeller Application

Problem Statement

A propeller having the characteristics shown in Fig. 14.3 is to be used to drive a swamp boat. If the propeller is to have a diameter of 2 m and a rotational speed of $N = 1200$ rpm, what should be the thrust starting from rest? If the boat resistance (air and water) is given by the empirical equation $F_D = 0.003\rho V_0^2/2$, where V_0 is the boat speed in meters per second, F_D is the drag, and ρ is the mass density of the water, what will be the maximum speed of the boat and what power will be required to drive the propeller? Assume $\rho_{air} = 1.20$ kg/m³ and $\rho_{water} = 1000$ kg/m³.

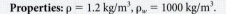

Define the Situation

A propeller is being used to drive a swamp boat.

Properties: $\rho = 1.2$ kg/m³, $\rho_w = 1000$ kg/m³.

---

[*]The speed of sound was not included in the dimensional analysis. However, the propeller performance is reduced because the Mach number based on the propeller tip speed leads to shock waves and other compressible-flow effects.

Head and Discharge Coefficients for Pumps

The thrust coefficient is defined as $F_T/\rho D^4 n^2$ for use with propellers, and if the same variables are applied to flow in an axial pump, the thrust can be expressed as $F_T = \Delta pA = \gamma\Delta HA$ or

$$C_T = \frac{\gamma\Delta HA}{\rho D^4 n^2} = \frac{\pi}{4}\frac{\gamma\Delta HD^2}{\rho D^4 n^2} = \frac{\pi}{4}\frac{g\Delta H}{D^2 n^2} \tag{14.12}$$

A new parameter, called the **head coefficient** C_H, is defined using the variables of Eq. (14.12), as

$$C_H = \frac{4}{\pi}C_T = \frac{\Delta H}{D^2 n^2/g} \tag{14.13}$$

which is a π-group that relates head delivered to fan diameter and rotational speed.

The independent π-group relating to propeller operation is V_0/nD; however, multiplying the numerator and denominator by the diameter squared gives $V_0 D^2/nD^3$, and $V_0 D^2$ is proportional to the discharge, Q. Thus the π-group for pump similarity studies is Q/nD^3 and is identified as the **discharge coefficient** C_Q. The power coefficient used for pumps is the same as the power coefficient used for propellers. Summarizing, the π-groups used in the similarity analyses of pumps are

$$C_H = \frac{\Delta H}{D^2 n^2/g} \tag{14.14}$$

$$C_P = \frac{P}{\rho D^5 n^3} \tag{14.15}$$

$$C_Q = \frac{Q}{nD^3} \tag{14.16}$$

where C_H and C_P are functions of C_Q for a given type of pump.

Figure 14.7 is a set of curves of C_H and C_P versus C_Q for a typical axial-flow pump. Also plotted on this graph is the efficiency of the pump as a function of C_Q. The dimensional curves

FIGURE 14.7

Dimensionless performance curves for a typical axial-flow pump. [After Stepanoff (3).]

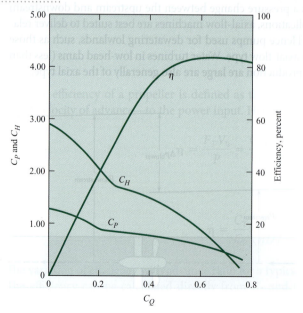

FIGURE 14.8

Performance curves for a typical axial-flow pump. [After Stepanoff (3).]

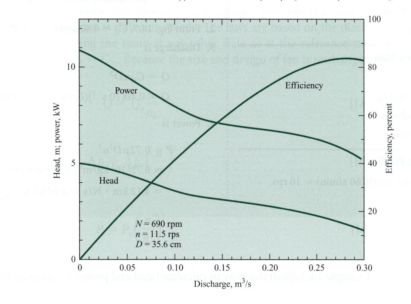

(head and power versus Q for a constant speed of rotation) from which Fig. 14.7 was developed are shown in Fig. 14.8. Because curves like those shown in Fig. 14.7 or Fig. 14.8 characterize pump performance, they are often called **characteristic curves** or **performance curves**. These curves are obtained by experiment.

There can be a problem with overload when operating axial-flow pumps. As seen in Fig. 14.7, when the pump flow is throttled below maximum-efficiency conditions, the required power increases with decreasing flow, thus leading to the possibility of overloading at low-flow conditions. For very large installations, special operating procedures are followed to avoid such overloading. For instance, the valve in the bypass from the pump discharge back to the pump inlet can be adjusted to maintain a constant flow through the pump. However, for small-scale applications, it is often desirable to have complete flexibility in flow control without the complexity of special operating procedures.

Performance curves are used to predict prototype operation from model tests or the effect of changing the speed of the pump. Example 14.2 shows how to use pump curves to calculate discharge and power.

EXAMPLE 14.2

Discharge and Power for Axial-Flow Pump

Define the Situation

For the pump represented by Figs. 14.7 and 14.8, what discharge of water in cubic meters per second will occur when the pump is operating against a 2-m head and at a speed of 600 rpm? What power in kilowatts is required for these conditions?

Define the Situation

This problem involves an axial flow pump with water.

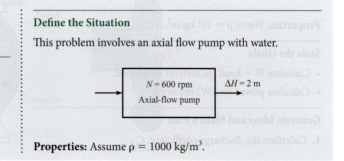

Properties: Assume $\rho = 1000 \text{ kg/m}^3$.

FIGURE 14.10

Performance curves for a typical centrifugal pump; $D = 37.1$ cm. [After Daugherty and Franzini (4). Used with the permission of the McGraw-Hill Companies.]

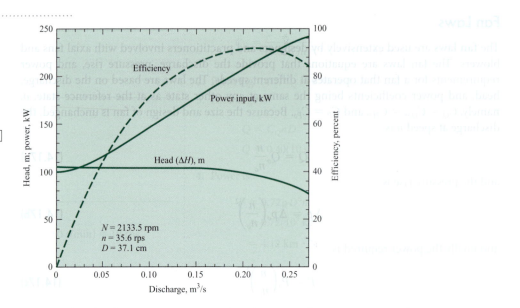

pump in that the change in pressure results in large part from rotary action (pressure increases outward like that in the rotating tank in Section 4.4 produced by the rotating impeller). Additional pressure increase is produced in the radial-flow pump when the high velocity of the flow leaving the impeller is reduced in the expanding section of the casing.

Although the basic designs are different for radial- and axial-flow pumps, it can be shown that the same similarity parameters (C_Q, C_P, and C_H) apply for both types. Thus the methods that have already been discussed for relating size, speed, and discharge in axial-flow machines also apply to radial-flow machines.

The major practical difference between axial- and radial-flow pumps so far as the user is concerned is the difference in the performance characteristics of the two designs. The dimensional performance curves for a typical radial-flow pump operating at a constant speed of rotation are shown in Fig. 14.10. The corresponding dimensionless performance curves for the same pump are shown in Fig. 14.11. Note that the power required at shutoff flow is less than

FIGURE 14.11

Dimensionless performance curves for a typical centrifugal pump, from data given in Fig. 14.9. [After Daugherty and Franzini (4).]

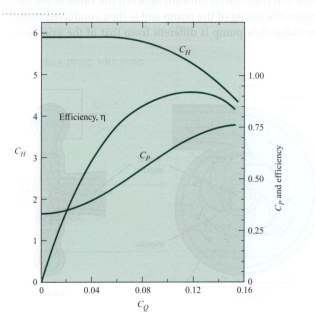

that required for flow at maximum efficiency. Normally, the motor used to drive the pump is chosen for conditions corresponding to maximum pump efficiency. Hence the flow can be throttled between the limits of shutoff condition and normal operating conditions with no chance of overloading the pump motor. In this latter case, a radial-flow pump offers a distinct advantage over axial-flow pumps.

Radial-flow pumps are manufactured with power levels ranging from 0.5 kW to thousands of kilowatts and heads from 15 m to several hundred meters. Figure 14.12 shows a cutaway view of a single-suction, single-stage, horizontal-shaft radial pump. Fluid enters in the direction of the rotating shaft and is accelerated outward by the rotating impeller. There are many other configurations designed for specific applications.

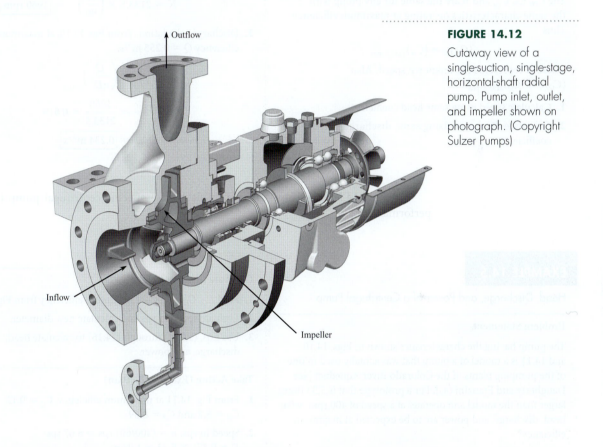

Outflow

Inflow

Impeller

FIGURE 14.12

Cutaway view of a single-suction, single-stage, horizontal-shaft radial pump. Pump inlet, outlet, and impeller shown on photograph. (Copyright Sulzer Pumps)

Example 14.4 shows how to find the speed and discharge for a centrifugal pump needed to provide a given head.

EXAMPLE 14.4

Speed and Discharge of Centrifugal Pump

Problem Statement

A pump that has the characteristics given in Fig. 14.10 when operated at 2133.5 rpm is to be used to pump water

at maximum efficiency under a head of 76 m. At what speed should the pump be operated, and what will the discharge be for these conditions?

Define the Situation

A centrifugal pump operated at 2133.5 rpm pumps water to head of 76 m at maximum efficiency.

Assumptions: Assume pump is the same size as that corresponding to Fig. 14.10 and water properties are the same.

State the Goal

1. Find the operational speed of pump (rpm).
2. Calculate discharge (m³/s).

Generate Ideas and Make a Plan

The C_H, C_P, C_Q, and η are the same for any pump with the same characteristics operating at maximum efficiency. Thus

$$(C_H)_N = (C_H)_{2133.5 \text{ rpm}}$$

where N represents the unknown speed. Also $(C_Q)_N = (C_Q)_{2133.5 \text{ rpm}}$.

1. Calculate speed using same head coefficient.
2. Calculate discharge using same discharge coefficient.

Take Action (Execute the Plan)

1. Speed calculation: From Fig. 14.10, at maximum efficiency $\Delta H = 90$ m.

$$\left(\frac{g\Delta H}{n^2 D^2}\right)_N = \left(\frac{g\Delta H}{n^2 D^2}\right)_{2133.5}$$

$$\frac{76 \text{ m}}{N^2} = \frac{90 \text{ m}}{2133.5^2 \text{ rpm}^2}$$

$$N = 2133.5 \times \left(\frac{76}{90}\right)^{1/2} = \boxed{1960 \text{ rpm}}$$

2. Discharge calculation: From Fig. 14.10, at maximum efficiency $Q = 0.255$ m³/s.

$$\left(\frac{Q}{nD^3}\right)_N = \left(\frac{Q}{nD^3}\right)_{2133.5}$$

$$\frac{Q_{1960}}{Q_{2133.5}} = \frac{1960}{2133.5} = 0.919$$

$$Q_{1960} = \boxed{0.234 \text{ m}^3/\text{s}}$$

Example 14.5 shows how to scale up data for a specific centrifugal pump to predict performance.

EXAMPLE 14.5

Head, Discharge, and Power of a Centrifugal Pump

Problem Statement

The pump having the characteristics shown in Figs. 14.10 and 14.11 is a model of a pump that was actually used in one of the pumping plants of the Colorado River Aqueduct [see Daugherty and Franzini (4)]. For a prototype that is 5.33 times larger than the model and operates at a speed of 400 rpm, what head, discharge, and power are to be expected at maximum efficiency?

Define the Situation

A prototype pump is 5.33 times larger than the corresponding model. The prototype operates at 400 rpm.

Assumptions: Pumping water with $\rho = 10^3$ kg/m³.

State the Goal

Find (at maximum efficiency)

1. Head (in meters)
2. Discharge (in m³/s)
3. Power (in kW)

Generate Ideas and Make a Plan

1. Find C_Q, C_H, and C_P at maximum efficiency from Fig. 14.11.
2. Evaluate speed in rps and calculate new diameter.
3. Use Eqs. (14.14) through (14.16) to calculate head, discharge, and power.

Take Action (Execute the Plan)

1. From Fig. 14.11 at maximum efficiency, $C_Q = 0.12$, $C_H = 5.2$ and $C_P = 0.69$.

2. Speed in rps: $n = (400/60)$ rps $= 6.67$ rps
   $D = 0.371 \times 5.33 = 1.98$ m.

3. Pump performance
   • Head

$$\Delta H = \frac{C_H D^2 n^2}{g} = \frac{5.2(1.98 \text{ m})^2 (6.67 \text{ s}^{-1})^2}{(9.81 \text{ m/s}^2)} = \boxed{92.4 \text{ m}}$$

   • Discharge

$$Q = C_Q nD^3 = 0.12(6.67 \text{ s}^{-1})(1.98 \text{ m})^3 = \boxed{6.21 \text{ m}^3/\text{s}}$$

   • Power

$$P = C_P \rho D^5 n^3 = 0.69((10^3 \text{ kg})/\text{m}^3)(1.98 \text{ m})^5 (6.67 \text{ s}^{-1})^3$$

$$= \boxed{6230 \text{ kW}}$$

14.4 Specific Speed

From the discussion in the preceding sections it was pointed out that axial-flow pumps are best suited for high discharge and low head, whereas radial machines perform better for low discharge and high head. A tool for selecting the best pump is the value of a π-group called the specific speed, n_s. The **specific speed** is obtained by combining both C_H and C_Q in such a manner that the diameter D is eliminated:

$$n_s = \frac{C_Q^{1/2}}{C_H^{3/4}} = \frac{(Q/nD^3)^{1/2}}{[\Delta H/(D^2 n^2/g)]^{3/4}} = \frac{n Q^{1/2}}{g^{3/4} \Delta H^{3/4}}$$

Thus specific speed relates different types of pumps without reference to their sizes.

As shown in Fig. 14.13, when efficiencies of different types of pumps are plotted against n_s, it is seen that certain types of pumps have higher efficiencies for certain ranges of n_s. For low specific speeds, the radial-flow pump is more efficient, whereas high specific speeds favor axial-flow machines. In the range between the completely axial-flow machine and the completely radial-flow machine, there is a gradual change in impeller shape to accommodate the particular flow conditions with maximum efficiency. The boundaries between axial, mixed, and radial machines are somewhat vague, but the value of the specific speed provides some guidance on which machine would be most suitable. The final choice would depend on which pumps were commercially available as well as their purchase price, operating cost, and reliability.

FIGURE 14.13

Optimum efficiency and impeller design versus specific speed.

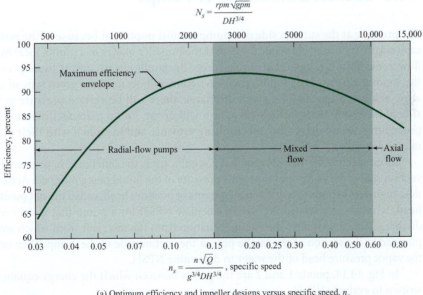

(a) Optimum efficiency and impeller designs versus specific speed, n_s.

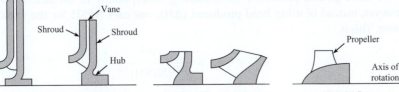

(b) Radial-flow impellers.   (c) Mixed-flow impellers.   (d) Axial flow.

Example 14.6 illustrates the use of specific speed to select a pump type.

EXAMPLE 14.6

Using Specific Speed to Select a Pump

Problem Statement

What type of pump should be used to pump water at the rate of 0.3 m³/s and under a head of 180 m? Assume $N = 1100$ rpm.

Define the Situation

A pump will be pumping water at 0.3 m³/s for a head of 180 m.

State the Goal

Find the best type of pump for this application.

Generate Ideas and Make a Plan

1. Calculate specific speed.
2. Use Fig. 14.13 to select pump type.

Take Action (Execute the Plan)

1. Rotational rate in rps

$$n = \frac{1100}{60} = 18.33 \text{ rps}$$

Specific speed

$$n_s = \frac{n\sqrt{Q}}{(g\Delta H)^{3/4}}$$

$$= \frac{18.33 \text{ rps} \times (0.3 \text{ m}^3/\text{s})^{1/2}}{(9.81 \text{ m/s}^2 \times 180 \text{ m})^{3/4}} = 0.037$$

2. From Fig. 14.13, a radial-flow pump is the best choice.

14.5 Suction Limitations of Pumps

The pressure at the suction side of a pump is most important because of the possibility that cavitation may occur. As water flows past the impeller blades of a pump, local high-velocity flow zones produce low relative pressures (Bernoulli effect), and if these pressures reach the vapor pressure of the liquid, then cavitation will occur. For a given type of pump operating at a given speed and a given discharge, there will be certain pressure at the suction side of the pump below which cavitation will occur. Pump manufacturers in their testing procedures always determine this limiting pressure and include it with their pump performance curves.

More specifically, the pressure that is significant is the difference in pressure between the suction side of the pump and the vapor pressure of the liquid being pumped. Actually, in practice, engineers express this difference in terms of pressure head, called the **net positive suction head**, which is abbreviated NPSH. To calculate NPSH for a pump that is delivering a given discharge, one first applies the energy equation from the reservoir from which water is being pumped to the section of the intake pipe at the suction side of the pump. Then one subtracts the vapor pressure head of the water to determine NPSH.

In Fig. 14.14, points 1 and 2 are the points between which the energy equation would be written to evaluate NPSH.

A more general parameter for indicating susceptibility to cavitation is specific speed. However, instead of using head produced (ΔH), one uses NPSH for the variable to the 3/4 power. This is

$$n_{ss} = \frac{nQ^{1/2}}{g^{3/4}(\text{NPSH})^{3/4}}$$

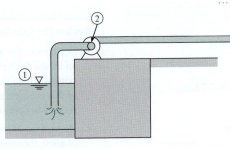

FIGURE 14.14

Locations used to evaluate NPSH for a pump.

Here n_{ss} is called the suction specific speed. The more traditional suction specific speed used in the United States is $N_{ss} = NQ^{1/2}/(NPSH)^{3/4}$, where N is in rpm, Q is in gallons per minute (gpm), and NPSH is in feet. Analyses of data from pump tests show that the value of the suction specific speed is a good indicator of whether cavitation may be expected. For example, the Hydraulic Institute (5) indicates that the critical value of N_{ss} is 8500. The reader is directed to manufacturer's data or the Hydraulic Institute for more details about critical NPSH or N_{ss}.

An analysis to find NPSH for a pump system is illustrated in Example 14.7.

EXAMPLE 14.7

Calculating Net Positive Suction Head

Problem Statement

In Fig. 14.14 the pump delivers 0.06 m³/s flow of 25°C water, and the intake pipe diameter is 20 cm. The pump intake is located 2 m above the water surface level in the reservoir. The pump operates at 1750 rpm. What are the net positive suction head and the traditional suction specific speed for these conditions?

Define the Situation

A pump delivers 0.06 m³/s flow of 25°C water.

Assumptions:

1. Entrance loss coefficient = 0.10.

2. Bend loss coefficient = 0.20.

Properties: Table A.5, (Water at 25°C) $\gamma = 9781$ N/m³, and $p_{vap} = 3500$ Pa.

State the Goal

- Calculate the positive suction head (NPSH).
- Calculate the traditional suction specific speed (N_{ss}).

Generate Ideas and Make a Plan

The net positive suction head is the difference between pressure at pump inlet and the vapor pressure.

1. Determine the atmospheric pressure in head of water for reservoir surface.

2. Determine velocity in 20 cm pipe.

3. Apply the energy equation between the reservoir and pump entrance.

4. Calculate NPSH.

5. Calculate N_{ss} with $N_{ss} = (NQ^{1/2})/(NPSH)^{3/4}$.

Take Action (Execute the Plan)

1. Pressure head at reservoir

$$\frac{p_1}{\gamma} = \frac{101.3 \times 10^3 \text{ Pa}}{9781 \text{ N/m}^3} = 10.36 \text{ m}$$

2. Velocity in pipe

$$V_2 = \frac{Q}{A} = \frac{0.06 \text{ m}^3/\text{s}}{\pi \times (0.1 \text{ m})^2} = 1.9 \text{ m/s}$$

3. Energy equation between points 1 and 2:

$$\frac{p_1}{\gamma} + \frac{V_1^2}{2g} + z_1 = \frac{p_2}{\gamma} + \frac{V_2^2}{2g} + z_2 + \sum h_L$$

- Input values

$$V_1 = 0, \quad z_1 = 0, \quad z_2 = 2$$

- Head loss

$$\sum h_L = (0.1 + 0.2)\frac{V_2^2}{2g}$$

- Head at pump entrance

$$\frac{p_2}{\gamma} = \frac{p_1}{\gamma} - z_2 - \frac{V_2^2}{2g}(1 + 0.3)$$

$$= 10.36 - 2 - 1.3 \times \frac{1.9^2}{2 \times 9.81} = 8.12 \text{ m}$$

4. Vapor pressure in meters of head

$$3500/9781 = 0.358 \text{ m}$$

Net positive suction head

$$\text{NPSH} = 8.12 - 0.358 = 7.76 \text{ m}$$

5. Traditional suction specific speed

$$Q = 0.06 \text{ m}^3/\text{s} = 3600 \text{ lpm}$$

$$N_{ss} = (1750)(3600)^{1/2}/(7.76)^{3/4} = \boxed{22,600}$$

Review the Solution and the Process

1. *Discussion.* For a typical single-stage centrifugal pump with an intake diameter of 20 cm and pumping 3600 lpm, the critical NPSH is normally about 3 m; therefore, the pump of this example is operating well within the safe range with respect to cavitation susceptibility.

2. *Discussion.* This value of N_{ss} is much below the critical limit of 8500; therefore, it is in a safe operating range so far as cavitation is concerned.

A typical pump performance curve for a centrifugal pump that would be supplied by a pump manufacturer is shown in Fig. 14.15. The solid lines labeled from 12.7 cm to 17.78 cm represent different impeller sizes that can be accommodated by the pump housing. These curves give the head delivered as a function of discharge. The dashed lines represent the power required by the pump for a given head and discharge. Lines of constant efficiency are also shown. Obviously, when selecting an impeller, one would like to have the operating point as close as possible to the point of maximum efficiency. The NPSH value gives the minimum head (absolute head) at the pump intake for which the pump will operate without cavitation.

FIGURE 14.15

Centrifugal pump performance curve. [After McQuiston and Parker (6). Used with permission of John Wiley and Sons.]

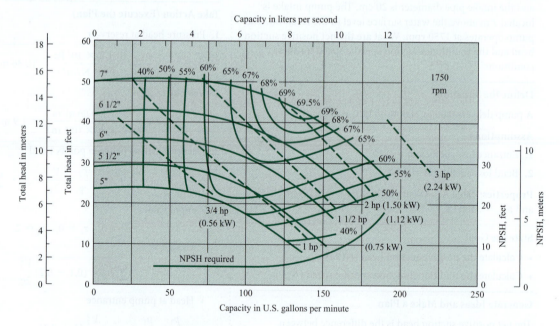

14.6 Viscous Effects

In the foregoing sections, similarity parameters were developed to predict prototype results from model tests, neglecting viscous effects. The latter assumption is not necessarily valid, especially if the model is quite small. To minimize the viscous effects in modeling pumps, the

Hydraulic Institute standards (5) recommend that the size of the model be such that the model impeller is not less than 30 cm in diameter. These same standards state that "the model should have complete geometric similarity with the prototype, not only in the pump proper, but also in the intake and discharge conduits."

Even with complete geometric similarity, one can expect the model to be less efficient than the prototype. An empirical formula proposed by Moody (7) is used for estimating prototype efficiencies of radial- and mixed-flow pumps and turbines from model efficiencies. That formula is

$$\frac{1 - e_1}{1 - e} = \left(\frac{D}{D_1}\right)^{1/5} \tag{14.18}$$

Here e_1 is the efficiency of the model and e is the efficiency of the prototype.

Example 14.8 shows how to estimate the efficiency due to viscous effects.

EXAMPLE 14.8

Calculating Viscous Effects on Pump Efficiency

Problem Statement

A model having an impeller diameter of 45 cm is tested and found to have an efficiency of 85%. If a geometrically similar prototype has an impeller diameter of 1.80 m, estimate its efficiency when it is operating under conditions that are dynamically similar to those in the model test ($C_{Q, \text{model}} = C_{Q, \text{prototype}}$).

Define the Situation

A pump with a 45-cm diameter impeller has 85% efficiency.

Assumptions: The efficiency differences are due to viscous effects.

State the Goal

Find the efficiency of a pump with a 1.6-m impeller.

Generate Ideas and Make a Plan

Use Eq. (14.18) to determine viscous effects.

Take Action (Execute the Plan)

Efficiency

$$e = 1 - \frac{1 - e_1}{(D/D_1)^{1/5}} = 1 - \frac{0.15}{1.32} = 1 - 0.11 = 0.89$$

or

$$\boxed{e = 89\%}$$

14.7 Centrifugal Compressors

Centrifugal compressors are similar in design to centrifugal pumps. Because the density of the air or gases used is much less than the density of a liquid, the compressor must rotate at much higher speeds than the pump to effect a sizable pressure increase. If the compression process were isentropic and the gases ideal, the power necessary to compress the gas from p_1 to p_2 would be

$$P_{\text{theo}} = \frac{k}{k - 1} Q_1 p_1 \left[\left(\frac{p_2}{p_1}\right)^{(k-1)/k} - 1\right] \tag{14.19}$$

where Q_1 is the volume flow rate into the compressor and k is the ratio of specific heats. The power calculated using Eq. (14.19) is referred to as the **theoretical adiabatic power**. The efficiency of a compressor with no water cooling is defined as the ratio of the theoretical adiabatic power to the actual power required at the shaft. Ordinarily the efficiency improves with higher inlet-volume flow rates, increasing from a typical value of 0.60 at 0.6 m³/s to 0.74 at 40 m³/s. Higher efficiencies are obtainable with more expensive design refinements.

Example 14.9 shows how to calculate shaft power required to operate a compressor.

EXAMPLE 14.9

Calculating Shaft Power for a Centrifugal Compressor

Problem Statement

Determine the shaft power required to operate a compressor that compresses air at the rate of 1 m³/s from 100 kPa to 200 kPa. The efficiency of the compressor is 65%.

Define the Situation

The inlet flow rate to a compressor is 1.0 m³/s. The pressure change is from 100 kPa to 200 kPa.

$Q = 1$ m³/s
100 kPa
200 kPa

From Table A.2, $k = 1.4$.

State the Goal

$P_{shaft}(\text{kW}) \blacktriangleleft$ Required shaft power (in kW).

Generate Ideas and Make a Plan

1. Use Eq. (14.19) to calculate theoretical power.
2. Divide theoretical power by efficiency to find shaft (required) power.

Take Action (Execute the Plan)

1. Theoretical power

$$P_{theo} = \frac{k}{k-1} Q_1 p_1 \left[\left(\frac{p_2}{p_1} \right)^{(k-1)/k} - 1 \right]$$

$$= (3.5)(1 \text{ m}^3/\text{s})(10^5 \text{ N/m}^2)[(2)^{0.286} - 1]$$

$$= 0.767 \times 10^5 \text{ N} \cdot \text{m/s} = 76.7 \text{ kW}$$

2. Shaft power

$$P_{shaft} = \frac{76.7}{0.65} \text{kW} = \boxed{118 \text{ kW}}$$

Cooling is necessary for high-pressure compressors because of the high gas temperatures resulting from the compression process. Cooling can be achieved through the use of water jackets or intercoolers that cool the gases between stages. The efficiency of water-cooled compressors is based on the power required to compress ideal gases isothermally, or

$$P_{theo} = p_1 Q_1 \ln \frac{p_2}{p_1} \tag{14.20}$$

which is usually called the **theoretical isothermal power**. The efficiencies of water-cooled compressors are generally lower than those of noncooled compressors. If a compressor is cooled by water jackets, its efficiency characteristically ranges between 55% and 60%. The use of intercoolers results in efficiencies from 60% to 65%.

Application to Fluid Systems

The selection of a pump, fan, or compressor for a specific application depends on the desired flow rate. This process requires the acquisition or generation of a system curve for the flow system of interest and a performance curve for the fluid machine. The intersection of these two curves provides the operating point as discussed in Chapter 10.

For example, consider using the centrifugal pump with the characteristics shown in Fig. 14.15 to pump water at 15°C from a wall into a tank as shown in Fig. 14.16. A pumping capacity of at least 300 lpm is required. Sixty meters of standard schedule-40 50-mm galvanized iron pipe are to be used. There is a check valve in the system as well as an open gate valve. There is a 6-m

FIGURE 14.16

System for pumping water from a well into a tank.

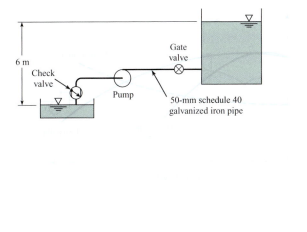

FIGURE 14.17

System and pump performance curves for pumping application.

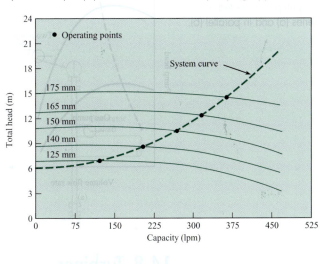

elevation between the well and the top of the fluid in the tank. Applying the energy equation, the head required by the pump is

$$h_p = \Delta z + \frac{V^2}{2g}\left(\frac{fL}{D} + \sum K_L\right)$$

where K_L represents the head loss coefficients for the entrance, check valve, gate valve, and sudden-expansion loss entering the tank. Using representative values for the loss coefficients and evaluating the friction factor from the Moody diagram in Chapter 10 leads to

$$h_p = 20 + 0.00305\, Q^2$$

where Q is the flow rate in gpm. This is the system curve.

The result of plotting the system curve on the pump performance curves is shown in Fig. 14.17. The locations where the lines cross are the operating points. One notes that a discharge of just over 300 lpm is achieved with the 165-mm impeller. Also, referring back to Fig. 14.15, the efficiency at this point is about 62%. To ensure that the design requirements are satisfied, the engineer may select the larger impeller, which has an operating point of 360 lpm. If the pump is to be used in continuous operation and the efficiency is important to operating costs, the engineer may choose to consider another pump that would have a higher efficiency at the operation point. An engineer experienced in the design of pump systems would be very familiar with the trade-offs for economy and performance and could make a design decision relatively quickly.

In some systems it may be advantageous to use two pumps in series or in parallel. If two pumps are used in series, the performance curve is the sum of the pump heads of the two machines at the same flow rate, as shown in Fig. 14.18a. This configuration would be desirable for a flow system with a steep system curve, as shown in the figure. If two pumps are connected in parallel, the performance curve is obtained by adding the flow rates of the two pumps at the same pump heads, as shown in Fig. 14.18b. This configuration would be advisable for flow systems with shallow system curves, as shown in the figure. The concepts presented here for pumps also apply to fans and compressors.

FIGURE 14.21

Control-volume approach for the impulse turbine using the angular-momentum principle.

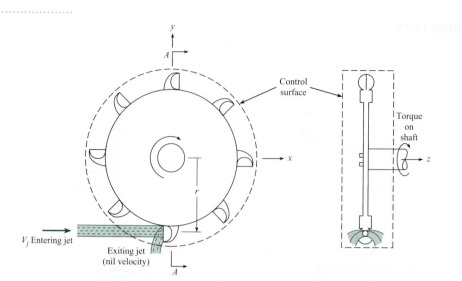

where Q is the discharge of the incoming jet, γ is the specific weight of jet fluid, and $h_t = V_j^2/2g$, or the velocity head of the jet. Thus Eq. (14.21) reduces to

$$P = \rho Q \frac{V_j^2}{2} \qquad (14.22)$$

To obtain the torque on the turbine shaft, the angular-momentum equation is applied to a control volume, as shown in Fig. 14.21. For steady flow

$$\sum M = \sum_{cs} \mathbf{r}_o \times (\dot{m}_o \mathbf{v}_o) - \sum_{cs} \mathbf{r}_i \times (\dot{m}_i \mathbf{v}_i)$$

Generally it is assumed that the exiting fluid has negligible angular momentum. The moment acting on the system is the torque T acting on the shaft. Thus the angular-momentum equation reduces to

$$T = -\dot{m} r V_j \qquad (14.23)$$

The mass flow rate across the control surface is ρQ, so the torque is

$$T = -\rho Q V_j r$$

The minus sign indicates that the torque applied to the system (to keep it rotating at constant angular velocity) is in the clockwise direction. However, the torque applied by the system to the shaft is in the counterclockwise direction, which is the direction of wheel rotation, so

$$T = \rho Q V_j r \qquad (14.24)$$

The power developed by the turbine is $T\omega$, or

$$P = \rho Q V_j r \omega \qquad (14.25)$$

Furthermore, if the velocity of the turbine vanes is $(1/2)V_j$ for maximum power, as noted earlier, then $P = \rho Q V_j^2/2$, which is the same as Eq. (14.22).

The calculation of torque for an impulse turbine is illustrated in Example 14.10.

EXAMPLE 14.10

Analyzing an Impulse Turbine

Problem Statement

What power in kilowatts can be developed by the impulse turbine shown if the turbine efficiency is 85%? Assume that the resistance coefficient f of the penstock is 0.015 and the head loss in the nozzle itself is negligible. What will be the angular speed of the wheel, assuming ideal conditions ($V_j = 2V_{bucket}$), and what torque will be exerted on the turbine shaft?

Define the Situation

This problem involves an impulse turbine with an efficiency of 85%.

Assumptions:

1. There is no entrance loss.

2. Head loss in nozzle is negligible.

3. Water density is 1000 kg/m³.

Elevation = 1670 m

Penstock

$D = 1$ m

6 km

$d_{jet} = 18$ cm

3 m

Elevation = 1000 m

State the Goal

Find:

- Power (kW) developed by turbine
- Angular speed (rpm) of wheel for maximum efficiency
- Torque (kN · m) on turbine shaft

Generate Ideas and Make a Plan

1. Apply energy equation, to find nozzle velocity.

2. Use Eq. (14.22) for power.

3. For maximum efficiency, $\omega r = (V_j/2)$.

4. Calculate torque from $P = T\omega$.

Take Action (Execute the Plan)

1. Energy equation

$$\frac{p_1}{\gamma} + \frac{V_1^2}{2g} + z_1 = \frac{p_j}{\gamma} + \frac{V_j^2}{2g} + z_j + h_L$$

- Values in energy equation

$$p_1 = 0, z_1 = 1670 \text{ m}, V_1 = 0, p_j = 0, z_j = 1000 \text{ m}$$

- Penstock-supply pipe velocity ratio

$$V_{penstock} = \frac{V_j A_j}{A_{penstock}} = V_j \left(\frac{0.18 \text{ m}}{1 \text{ m}}\right)^2 = 0.0324 \, V_j$$

- Head loss

$$h_L = f \frac{L}{D} \frac{1}{2g} V_{penstock}^2$$

$$= \frac{0.015 \times 6000}{1} (0.0324)^2 \frac{V_j^2}{2g} = 0.094 \frac{V_j^2}{2g}$$

- Jet velocity

$$z_1 - z_2 = 1.094 \frac{V_j^2}{2g}$$

$$V_j = \left(\frac{2 \times 9.81 \text{ m/s}^2 \times 670 \text{ m}}{1.094}\right)^{1/2} = 109.6 \text{ m/s}$$

2. Gross power

$$P = Q\gamma \frac{V_j^2}{2g} = \frac{\gamma A_j V_j^3}{2g}$$

$$= \frac{9810(\pi/4)(0.18)^2(109.6)^3}{2 \times 9.81} = 16,750 \text{ kW}$$

Power delivered

$$P = 16,750 \times \text{efficiency} = \boxed{14,240 \text{ kW}}$$

3. Angular speed of wheel

$$V_{bucket} = \frac{1}{2}(109.6 \text{ m/s}) = 54.8 \text{ m/s}$$

$$r\omega = 54.8 \text{ m/s}$$

$$\omega = \frac{54.8 \text{ m/s}}{1.5 \text{ m}} = 36.5 \text{ rad/s}$$

Wheel speed

$$N = (36.5 \text{ rad/s}) \frac{1 \text{ rev}}{2\pi \text{ rad}} (60 \text{ s/min}) = \boxed{349 \text{ rpm}}$$

4. Torque

$$T = \frac{\text{power}}{\omega} = \frac{14,240 \text{ kW}}{36.5 \text{ rad/s}} = \boxed{390 \text{ kN} \cdot \text{m}}$$

Vane Angles

It should be apparent that the head loss in a turbine will be less if the flow enters the runner with a direction tangent to the runner vanes than if the flow approaches the vane with an angle of attack. In the latter case, separation will occur with consequent head loss. Thus vanes of an impeller designed for a given speed and discharge and with fixed guide vanes will have a particular optimum blade angle β_1. However, if the discharge is changed from the condition of the original design, the guide vanes and impeller vane angles will not "match" the new flow condition. Most turbines for hydroelectric installations are made with movable guide vanes on the inlet side to effect a better match at all flows. Thus α_1 is increased or decreased automatically through governor action to accommodate fluctuating power demands on the turbine.

To relate the incoming-flow angle α_1 and the vane angle β_1, first assume that the flow entering the impeller is tangent to the blades at the periphery of the impeller. Likewise, the flow leaving the stationary guide vane is assumed to be tangent to the guide vane. To develop the desired equations, consider both the radial and the tangential components of velocity at the outer periphery of the wheel ($r = r_1$). It is easy to compute the radial velocity, given Q and the geometry of the wheel, by the continuity equation:

$$V_{r_1} = \frac{Q}{2\pi r_1 B} \tag{14.28}$$

where B is the height of the turbine blades. The tangential (tangent to the outer surface of the runner) velocity of the incoming flow is

$$V_{t_1} = V_{r_1} \cot \alpha_1 \tag{14.29}$$

However, this tangential velocity is equal to the tangential component of the relative velocity in the runner, $V_{r_1} \cot \beta_1$, plus the velocity of the runner itself, ωr_1. Thus the tangential velocity, when viewed with respect to the runner motion, is

$$V_{t_1} = r_1 \omega + V_{r_1} \cot \beta_1 \tag{14.30}$$

Now, eliminating V_{t_1} between Eqs. (14.29) and (14.30) results in

$$V_{r_1} \cot \alpha_1 = r_1 \omega + V_{r_1} \cot \beta_1 \tag{14.31}$$

Equation (14.31) can be rearranged to yield

$$\alpha_1 = \text{arccot}\left(\frac{r_1 \omega}{V_{r_1}} + \cot \beta_1 \right) \tag{14.32}$$

Example 14.11 illustrates how to calculate the inlet blade angle to avoid separation.

EXAMPLE 14.11

Analyzing a Francis Turbine

Problem Statement

A Francis turbine is to be operated at a speed of 600 rpm and with a discharge of 4.0 m³/s. If $r_1 = 0.60$ m, $\beta_1 = 110°$, and the blade height B is 10 cm, what should be the guide vane angle α_1 for a nonseparating flow condition at the runner entrance?

Define the Situation

A Francis turbine is operating with an angular speed of 600 rpm and a discharge of 4.0 m³/s.

State the Goal

Find the inlet guide vane angle, α_1.

Generate Ideas and Make a Plan

Use Eq. (14.32) for inlet guide angle.

Take Action (Execute the Plan) Radial velocity at inlet	Inlet guide vane angle

$$\alpha_1 = \text{arccot}\left(\frac{r_1\omega}{V_{r_1}} + \cot\beta_1\right)$$

$$r_1\omega = 0.6 \times 600 \text{ rpm} \times 2\pi \text{ rad/rev} \times 1/60 \text{ min/s}$$

$$= 37.7 \text{ m/s}$$

$$V_{r_1} = \frac{Q}{2\pi r_1 B} = \frac{4.00 \text{ m}^3/\text{s}}{2\pi \times 0.6 \text{ m} \times 0.10 \text{ m}} = 10.61 \text{ m/s}$$

$$\cot\beta_1 = \cot(110°) = -0.364$$

$$\alpha_1 = \text{arccot}\left(\frac{37.7}{10.61} - 0.364\right) = \boxed{17.4°}$$

Specific Speed for Turbines

Because of the attention focused on the production of power by turbines, the specific speed for turbines is defined in terms of power:

$$n_s = \frac{nP^{1/2}}{g^{3/4}\gamma^{1/2}h_t^{5/4}}$$

It should also be noted that large water turbines are innately more efficient than pumps. The reason for this is that as the fluid leaves the impeller of a pump, it decelerates appreciably over a relatively short distance. Also, because guide vanes are generally not used in the flow passages with pumps, large local velocity gradients develop, which in turn cause intense mixing and turbulence, thereby producing large head losses. In most turbine installations, the flow that exits the turbine runner is gradually reduced in velocity through a gradually expanding **draft tube**, thus producing a much smoother flow situation and less head loss than for the pump. For additional details of hydropower turbines, see Daugherty and Franzini (4).

Gas Turbines

The conventional gas turbine consists of a compressor that pressurizes the air entering the turbine and delivers it to a combustion chamber. The high-temperature, high-pressure gases resulting from combustion in the combustion chamber expand through a turbine, which both drives the compressor and delivers power. The theoretical efficiency (power delivered/rate of energy input) of a gas turbine depends on the pressure ratio between the combustion chamber and the intake; the higher the pressure ratio, the higher the efficiency. The reader is directed to Cohen et al. (8) for more detail.

Wind Turbines

Wind energy is discussed frequently as an alternative energy source. The application of wind turbines* as potential sources for power becomes more attractive as utility power rates increase and the concern over greenhouse gases grows. In many European countries, especially northern Europe, the wind turbine is playing an ever-increasing role in power generation.

In essence, the wind turbine is just a reverse application of the process of introducing energy into an airstream to derive a propulsive force. The wind turbine extracts energy from

---

*The phrase "wind turbine" is used to convey the idea of conversion of wind to electrical energy. A windmill converts wind energy to mechanical energy.

- The efficiency of a propeller is the ratio of the power delivered by the propeller to the power provided to the propeller.

$$\eta = \frac{F_T V_0}{P}$$

Pumps

- Pumps can be *axial flow* or *radial flow*
  ▶ An axial-flow pump consists of an impeller, much like a propeller, mounted in a housing.
  ▶ In a radial-flow pump, fluid enters near the eye of the impeller, passes through the vanes, and exits at the edge of the vanes.
- The head provided by a pump is quantified by the *head coefficient*, C_H, defined as

$$C_H = \frac{g\Delta H}{n^2 D^2}$$

  where ΔH is the head across the pump.
- The head coefficient is a function of the *discharge coefficient*, which is

$$C_Q = \frac{Q}{nD^3}$$

  where Q is the discharge.
- Pump performance curves show head delivered, power required, and efficiency as a function of discharge.
- The specific speed of a pump can be used to select an appropriate type of pump for a given application.
  ▶ Axial-flow pumps are best suited for high-discharge, low-head applications.
  ▶ Radial-flow pumps are best suited for low-discharge, high-head applications.

Water Turbines

- Turbines convert the energy associated with a moving fluid to shaft work.
- Turbines are classified into two categories.
  ▶ The *impulse turbine* consists of a liquid jet impinging on vanes of a turbine wheel or runner.
  ▶ A *reaction turbine* consists of a series of rotating vanes immersed in a flowing fluid. The pressure on the vanes provides the torque for the power.

Wind Turbines

- Wind turbines are classified based on the axis of the rotor
  ▶ The rotor of a turbine can revolve around a *horizontal axis*. Most commercial wind turbines use this design.
  ▶ The rotor of a turbine can revolve around a *vertical axis*. Two types of turbine in this category are the Darrieus turbine and the Savonius turbine.

- The maximum power derivable from a wind turbine is

$$P_{\max} = \frac{16}{27}\left(\frac{1}{2}\rho V_0^3 A\right)$$

where A is the capture area of the wind turbine (projected area from direction of wind) and V_0 is the wind speed.

REFERENCES

1. Weick, F. E. *Aircraft Propeller Design.* New York: McGraw-Hill, 1930.

2. Weick, Fred E. "Full Scale Tests on a Thin Metal Propeller at Various Pit Speeds." *NACA Report*, 302 (January 1929).

3. Stepanoff, A. J. *Centrifugal and Axial Flow Pumps*, 2nd ed. New York: John Wiley, 1957.

4. Daugherty, Robert L., and Joseph B. Franzini. *Fluid Mechanics with Engineering Applications.* New York: McGraw-Hill, 1957.

5. Hydraulic Institute. *Centrifugal Pumps.* Parsippany, NJ: Hydraulic Institute, 1994.

6. McQuiston, F. C., and J. D. Parker. *Heating, Ventilating and Air Conditioning.* New York: John Wiley, 1994.

7. Moody, L. F. "Hydraulic Machinery." In *Handbook of Applied Hydraulics*, ed. C. V. Davis. New York: McGraw-Hill, 1942.

8. Cohen, H., G. F. C. Rogers, and H. I. H. Saravanamuttoo. *Gas Turbine Theory.* New York: John Wiley, 1972.

9. Glauert, H. "Airplane Propellers." *Aerodynamic Theory*, vol. IV, ed. W. F. Durand. New York: Dover, 1963.

10. Manwell, J. F., J. G. McGowan, and A. L. Rogers. *Wind Energy Explained: Theory, Design and Application.* Chichester, UK: John Wiley, 2002.

PROBLEMS

PLUS Problem available in *WileyPLUS* at instructor's discretion.

GO Guided Online (GO) Problem, available in *WileyPLUS* at instructor's discretion.

Propellers (§14.1)

14.1 Explain why the thrust of a fixed-pitch propeller decreases with increasing forward speed.

14.2 What limits the rotational speed of a propeller?

14.3 PLUS What thrust is obtained from a propeller 3 m in diameter that has the characteristics given in Fig. 14.3 on p. 520 of §14.1 when the propeller is operated at an angular speed of 1100 rpm and an advance velocity of zero? Assume $\rho = 1.05$ kg/m³.

14.4 PLUS What thrust is obtained from a propeller 3 m in diameter that has the characteristics given in Fig. 14.3 on p. 520 of §14.1 when the propeller is operated at an angular speed of 1400 rpm and an advance velocity of 80 km/h? What power is required to operate the propeller under these conditions? Assume $\rho = 1.05$ kg/m³.

14.5 A propeller 2.4 m in diameter has the characteristics shown in Fig. 14.3 on p. 520 of §14.1. What thrust is produced by the propeller when it is operating at an angular speed of 1200 rpm and a forward speed of 48 km/h? What power input is required under these operating conditions? If the forward speed is reduced to zero, what is the thrust? Assume $\rho = 1.24$ kg/m³.

14.6 PLUS A propeller 2.4 m in diameter, like the one for which characteristics are given in Fig. 14.3, on p. 520 of §14.1, is to be used on a swamp boat and is to operate at maximum efficiency

when cruising. If the cruising speed is to be 48 km/h, what should the angular speed of the propeller be? Assume $\rho = 1.24$ kg/m³.

14.7 For the propeller and conditions described in Prob. 14.6, determine the thrust and the power input.

14.8 GO A propeller is being selected for an airplane that will cruise at 2000 m altitude, where the pressure is 60 kPa absolute and the temperature is 10°C. The mass of the airplane is 1200 kg, and the planform area of the wing is 10 m². The lift-to-drag ratio is 30:1. The lift coefficient is 0.4. The engine speed at cruise conditions is 3000 rpm. The propeller is to operate at maximum efficiency, which corresponds to a thrust coefficient of 0.025. Calculate the diameter of the propeller and the speed of the aircraft.

14.9 PLUS If the tip speed of a propeller is to be kept below $0.8c$, where c is the speed of sound, what is the maximum allowable angular speed of propellers having diameters of 2 m, 3 m, and 4 m? Take the speed of sound as 335 m/s.

14.10 A propeller 2 m in diameter, like the one for which characteristics are given in Fig. 14.3, on p. 520 of §14.1, is to be used on a swamp boat and is to operate at maximum efficiency when cruising. If the cruising speed is to be 40 km/h, what should the angular speed of the propeller be?

14.11 For the propeller and conditions described in Prob. 14.10, determine the thrust and the power input. Assume $\rho = 1.2$ kg/m³.

14.12 PLUS A propeller 2 m in diameter and like the one for which characteristics are given in Fig. 14.3 on p. 520 of §14.1 is used on a swamp boat. If the angular speed is 1000 rpm and if the boat and passengers have a combined mass of 300 kg, estimate the initial acceleration of the boat when starting from rest. Assume $\rho = 1.1$ kg/m³.

Axial Flow Pumps and Fans (§14.2)

14.13 Answer the following questions about axial-flow pumps.

    a. Axial-flow pumps are best suited for what conditions of head produced and discharge?

    b. For an axial-flow pump, how does the head produced by the pump and the power required to operate a pump vary with flow rate through the pump?

14.14 PLUS If a pump having the characteristics shown in Fig. 14.7 on p. 524 of §14.2 has a diameter of 40 cm and is operated at a speed of 1000 rpm, what will be the discharge when the head is 3 m?

14.15 The pump used in the system shown has the characteristics given in Fig. 14.8 on p. 525 of §14.2. What discharge will occur under the conditions shown, and what power is required?

14.16 If the conditions are the same as in Prob. 14.15 except that the speed is increased to 900 rpm, what discharge will occur, and what power is required for the operation?

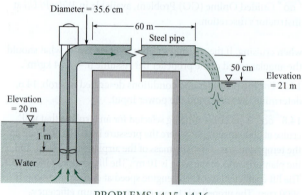

PROBLEMS 14.15, 14.16

14.17 For a pump having the characteristics given in Fig. 14.7 or 14.8 of §14.2, what water discharge and head will be produced at maximum efficiency if the pump diameter is 50 cm and the angular speed is 1100 rpm? What power is required under these conditions?

14.18 GO A pump has the characteristics given by Fig. 14.7 on p. 524 of §14.2. What discharge and head will be produced at maximum efficiency if the pump size is 50 cm and the angular speed is 45 rps? What power is required when pumping water at 10°C under these conditions?

14.19 For a pump having the characteristics of Fig. 14.7 on p. 524 of §14.2, plot the head-discharge curve if the pump is 35 cm in diameter and is operated at a speed of 1000 rpm.

14.20 For a pump having the characteristics of Fig. 14.7 on p. 524 of §14.2, plot the head-discharge curve if the pump diameter is 60 cm and the speed is 690 rpm.

14.21 An axial-flow blower is used for a wind tunnel that has a test section measuring 60 cm by 60 cm and is capable of airspeeds up to 40 m/s. If the blower is to operate at maximum efficiency at the highest speed and if the rotational speed of the blower is 2000 rpm at this condition, what are the diameter of the blower and the power required? Assume that the blower has the characteristics shown in Fig. 14.7 on p. 524 of §14.2. Assume $\rho = 1.2$ kg/m³.

14.22 PLUS An axial-flow blower is used to air-condition an office building that has a volume of 10^5 m³. It is decided that the air at 15°C in the building must be completely changed every 15 min. Assume that the blower operates at 600 rpm at maximum efficiency and has the characteristics shown in Fig. 14.7 on p. 524 of §14.2. Calculate the diameter and power requirements for two blowers operating in parallel.

14.23 An axial fan 2 m in diameter is used in a wind tunnel as shown (test section 1.2 m in diameter; test section velocity of 60 m/s). The rotational speed of the fan is 1800 rpm. Assume the density of the air is constant at 1.2 kg/m³. There are negligible losses in the tunnel. The performance curve of the fan is identical to that shown in Fig. 14.7 on p. 524 of §14.2. Calculate the power needed to operate the fan.

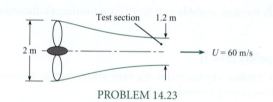

PROBLEM 14.23

Radial Flow Pumps (§14.3)

14.24 The radial flow pump is best suited for what conditions of head produced and discharge?

14.25 A pump is used to pump water out of a reservoir. What limits the depth for which the pump can draw water?

14.26 If a pump having the characteristics given in Fig. 14.10 on p. 528 of §14.3 is doubled in size but halved in speed, what will be the head and discharge at maximum efficiency?

14.27 A pump having the characteristics given in Fig. 14.10 on p. 528 of §14.3 pumps water at 20°C from a reservoir at an elevation of 366 m to a reservoir at an elevation of 450 m through a 36-cm steel pipe. If the pipe is 610 m long, what will be the discharge through the pipe?

14.28 GO If a pump having the characteristics given in Fig. 14.10 or 14.11 (both in §14.3) is operated at a speed of 1600 rpm, what will be the discharge when the head is 41 m?

14.29 If a pump having the performance curve shown is operated at a speed of 1600 rpm, what will be the maximum possible head developed?

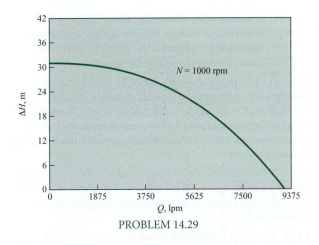

PROBLEM 14.29

14.30 If a pump having the characteristics given in Fig. 14.10 on p. 528 of §14.3 is operated at a speed of 30 rps, what will be the shutoff head?

14.31 If a pump having the characteristics given in Fig. 14.11 on p. 528 of §14.3 is 40 cm in diameter and is operated at a speed of 25 rps, what will be the discharge when the head is 50 m?

14.32 PLUS A centrifugal pump 20 cm in diameter is used to pump kerosene at a speed of 5000 rpm. Assume that the pump has the characteristics shown in Fig. 14.11 on p. 528 of §14.3. Calculate the flow rate, the pressure rise across the pump, and the power required if the pump operates at maximum efficiency.

Pump Selection (§14.4)

14.33 Answer the following questions regarding pump sizing and selection.

    a. What is the difference between a system curve and a pump curve? Explain.

    b. The operating point for a pump system is established by what condition?

14.34 The value of the specific speed suggests the type of pump to be used for a given application. A high specific speed suggests the use of what kind of pump?

14.35 PLUS The pump curve for a given pump is represented by

$$h_{p,pump} = 6\left[1 - \left(\frac{Q}{400}\right)^2\right]$$

where $h_{p,pump}$ is the head provided by the pump in meters and Q is the discharge in liters per minute (lpm). The system curve for a pumping application is

$$h_{p,sys} = 2 + 1.25 \times 10^{-4}Q^2$$

where $h_{p,sys}$ is the head in meters required to operate the system and Q is the discharge in L/min. Find the operating point (Q) for (a) one pump, (b) two identical pumps connected in series, and (c) two identical pumps connected in parallel.

14.36 What is the suction specific speed for the pump that is operating under the conditions given in Prob. 14.15? Is this a safe operation with respect to susceptibility to cavitation?

14.37 What type of pump should be used to pump water at a rate of 0.3 m³/s and under a head of 9 m? Assume $N = 1500$ rpm.

14.38 For the most efficient operation, what type of pump should be used to pump water at a rate of 0.10 m³/s and under a head of 30 m? Assume $n = 25$ rps.

14.39 What type of pump should be used to pump water at a rate of 0.40 m³/s and under a head of 70 m? Assume $N = 1100$ rpm.

14.40 An axial-flow pump is to be used to lift water against a head (friction and static) of 4.5 m. If the discharge is to be 15000 lpm, what maximum speed in revolutions per minute is allowed if the suction head is 1.5 m?

14.41 A pump is needed to pump water at a rate of 0.2 m³/s from the lower to the upper reservoir shown in the figure. What type of pump would be best for this operation if the impeller speed is to be 600 rpm? Assume $f = 0.02$ and $K_e = 0.5$.

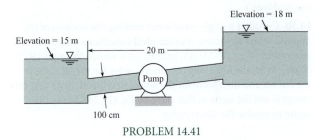

PROBLEM 14.41

14.42 Plot the five performance curves from Fig. 14.15 on p. 534 of §14.5 for the different impeller diameters in terms of the head and discharge coefficients. Use impeller diameter for D.

Compressors (§14.7)

14.43 PLUS The pressure rise associated with gases in a compressor causes the gas temperature to increase as well. The ratio of final temperature to initial temperature is less than the ratio of final pressure to initial pressure. Will the final density be (a) less or (b) greater than the initial density?

14.44 PLUS Methane flowing at the rate of 1 kg/s is to be compressed by a noncooled centrifugal compressor from 100 kPa to 165 kPa. The temperature of the methane entering the compressor is 27°C. The efficiency of the compressor is 70%. Calculate the shaft power necessary to run the compressor.

14.45 A 36 kW (shaft output) motor is available to run a noncooled compressor for carbon dioxide. The pressure is to be increased from 100 kPa to 150 kPa. If the compressor is 60% efficient, calculate the volume flow rate into the compressor.

14.46 PLUS A water-cooled centrifugal compressor is used to compress air from 100 kPa to 600 kPa at the rate of 2 kg/s. The temperature of the inlet air is 15°C. The efficiency of the compressor is 50%. Calculate the necessary shaft power.

Impulse Turbines (§14.8)

14.47 An impulse turbine will produce no power if the velocity of the jet striking the bucket is the same as the bucket velocity. Explain.

14.48 A penstock 1 m in diameter and 10 km long carries water at 10°C from a reservoir to an impulse turbine. If the turbine is 85% efficient, what power can be produced by the system if the upstream reservoir elevation is 650 m above the turbine jet and the jet diameter is 16.0 cm? Assume that $f = 0.016$ and neglect head losses in the nozzle. What should the diameter of the turbine wheel be if it is to have an angular speed of 360 rpm? Assume ideal conditions for the bucket design $[V_{bucket} = (1/2)V_j]$.

14.49 Consider an idealized bucket on an impulse turbine that turns the water through 180°. Prove that the bucket speed should be one-half the incoming jet speed for a maximum power production. (*Hint:* Set up the momentum equation to solve for the force on the bucket in terms of V_j and V_{bucket}; then the power will be given by this force times V_{bucket}. You can use your mathematical talent to complete the problem.)

14.50 Consider a single jet of water striking the buckets of the impulse wheel as shown. Assume ideal conditions for power generation $[V_{bucket} = (1/2)V_j$ and the jet is turned through 180° of arc]. With the foregoing conditions, solve for the jet force on the bucket and then solve for the power developed. Note that this power is not the same as that given by Eq. (14.24)! Study the figure to resolve the discrepancy.

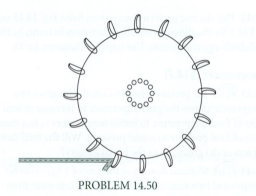

PROBLEM 14.50

Reaction Turbines (§14.8)

14.51 Answer the following questions about reaction turbines.

    a. How does a reaction turbine differ from a centrifugal pump?

    b. What is meant by the "runner" in a reaction turbine?

14.52 For a given Francis turbine, $\beta_1 = 60°$, $\beta_2 = 90°$, $r_1 = 5$ m, $r_2 = 3$ m, and $B = 1$ m. The discharge is 126 m³/s, and the rotational speed is 60 rpm. Assume $T = 10°C$.

    a. What should α_1 be for a nonseparating flow condition at the entrance to the runner?

    b. What is the maximum attainable power with the conditions noted?

c. If you were to redesign the turbine blades of the runner, what changes would you suggest to increase the power production if the discharge and overall dimensions are to be kept the same?

14.53 To produce a discharge of 3.3 m³/s, a Francis turbine will be operated at a speed of 60 rpm, $r_1 = 1.5$ m, $r_2 = 1.20$ m, $B = 33$ cm, $\beta_1 = 85°$, and $\beta_2 = 165°$. What should (a) α_1 be for nonseparating flow to occur through the runner? What (b) power and (c) torque should result with this operation? Assume $T = 10°C$.

14.54 A Francis turbine is to be operated at a speed of 120 rpm and with a discharge of 200 m³/s. If $r_1 = 3$ m, $B = 0.90$ m, and $\beta_1 = 45°$, what should α_1 be for nonseparating flow at the runner inlet?

14.55 Shown is a preliminary layout for a proposed small hydroelectric project. The initial design calls for a discharge of 0.23 m³/s through the penstock and turbine. Assume 80% turbine efficiency. For this setup, what power output could be expected from the power plant? Draw the HGL and EGL for the system.

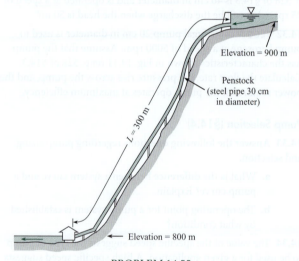

Elevation = 900 m

Penstock (steel pipe 30 cm in diameter)

$L = 300$ m

Elevation = 800 m

PROBLEM 14.55

Wind Turbines (§14.8)

14.56 What determines the minimum and maximum wind speeds at which a wind turbine can operate?

14.57 Using the Internet and other resources, identify at least four types of wind turbines. For each type, describe its distinguishing characteristics and its relative advantages and disadvantages.

14.58 Calculate the minimum capture area necessary for a wind turbine that will be required to power the 2 kW demands of an energy-efficient home. Assume a wind velocity of 16 km/h and an air density of 1.2 kg/m³.

14.59 Calculate the maximum power derivable from a conventional horizontal-axis wind turbine with a propeller 2.3 m in diameter in a 47 km/h wind with density 1.2 kg/m³.

14.60 A wind farm consists of 20 Darrieus turbines, each 15 m high. The total output from the turbines is to be 2 MW in a wind of 20 m/s and an air density of 1.2 kg/m³. The Darrieus turbine shown has the shape of an arc of a circle. Find the minimum width, W, of the turbine needed to provide this power output.

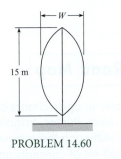

PROBLEM 14.60

14.61 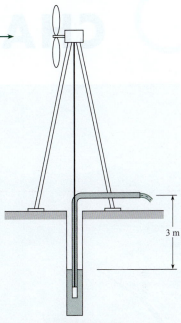 A windmill is connected directly to a mechanical pump that is to pump water from a well 3 m deep as shown. The windmill is a conventional horizontal-axis type with a fan diameter of 3 m. The efficiency of the mechanical pump is 80%. The density of the air is 1.12 kg/m³. Assume the windmill delivers the maximum power available. There is 6 m of 5 cm galvanized pipe in the system. What would the discharge of the pump be (in liters per minute) for a 48 km/h wind?

PROBLEM 14.61

15 FLOW IN OPEN CHANNELS

FIGURE 15.1

Aerial view of the California Aqueduct at the southwest end of the Tehachapi Mountains. (Macduff Everton/The Image Bank/Getty Images).

Chapter Road Map

The flow of water in open channels can be observed in aqueducts, rivers, flumes, irrigation ditches, and other contexts. Although these contexts are quite different, a small set of concepts and a few equations generalize to most applications of open channel flow. These ideas are introduced in this chapter.

Learning Objectives

STUDENTS WILL BE ABLE TO

- Define an open channel. Define uniform flow and nonuniform flow. (§15.1)

- Define the Froude number, the hydraulic radius, and the Reynolds number. List the criteria for laminar and turbulent flow. (§15.1)

- For steady flow, explain the physics of the energy equation and also explain the corresponding HGL and EGL. (§15.2)

- For uniform flow, calculate flow rate with the (a) Darcy-Weisbach approach, (b) Chezy equation, and (c) Manning equation. (§15.3)

- Define and explain the best hydraulic section. (§15.3)

- Describe and compare rapidly varied flow and gradually varied flow. (§15.4)

- Describe critical depth, specific energy, supercritical flow, and subcritical flow. (§15.5)

- Describe a hydraulic jump. Perform calculations. (§15.6)

- Describe the factors used to classify surface profiles that occur in gradually varied flow. (§15.7)

An **open channel** is one in which a liquid flows with a free surface. A **free surface** means that the liquid surface is exposed to the atmosphere. Examples of open channels are natural creeks and rivers, artificial channels such as irrigation ditches and canals, and pipelines or sewers flowing less than full. In most cases, water or wastewater is the flowing liquid.

15.1 Description of Open-Channel Flow

Flow in an open channel is described as uniform or nonuniform, as distinguished in Fig. 15.2. As defined in Chapter 4, **uniform flow** means that the velocity is constant along a streamline, which in open-channel flow means that depth and cross section are constant along the length of a channel. The depth for uniform-flow conditions is called **normal depth** and is designated by y_n. For **nonuniform flow**, the velocity changes from section to section along the channel, thus one observes changes in depth. The velocity change may be due to a change in channel configuration, such as a bend, change in cross-sectional shape, or change in channel slope. For example, Fig. 15.2 shows steady flow over a spillway of constant width, where the water must flow progressively faster as it goes over the brink of the spillway (from A to B), caused by the suddenly steeper slope. The faster velocity requires a smaller depth, in accordance with conservation of mass (continuity). From reach B to C, the flow is uniform because the velocity, and thus depth, are constant. After reach C the abrupt flattening of channel slope requires the velocity to suddenly, and turbulently, slow down. Thus there is a deeper depth downstream of C than in reach B to C.

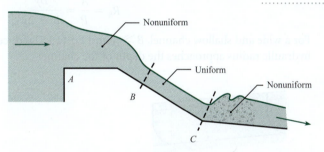

FIGURE 15.2

Distinguishing uniform and nonuniform flow. This example shows steady flow over a spillway, such as the emergency overflow channel of a dam.

The most complicated open-channel flow is unsteady nonuniform flow. An example of this is a breaking wave on a sloping beach. Theory and analysis of unsteady nonuniform flow are reserved for more advanced courses.

Dimensional Analysis in Open-Channel Flow

Open-channel flow results from gravity moving water from higher to lower elevations and is impeded by friction forces caused by the roughness of the channel. Thus the functional equation $Q = f(\mu, \rho, V, L)$ and dimensional analysis leads to two important π-groups: the Froude number and the Reynolds number. The Froude number squared is the ratio of kinetic force to gravity force:

$$\text{Fr}^2 = \frac{\text{kinetic force}}{\text{gravity force}} = \frac{\rho L^2 V^2}{\gamma L^3} = \frac{V^2}{L\gamma/\rho} \qquad (15.1)$$

$$\text{Fr} = \frac{V}{\sqrt{gL}} \qquad (15.2)$$

The Froude number is important if the gravitational force influences the direction of flow, such as in flow over a spillway, or the formation of surface waves. However, it is unimportant when gravity causes only a hydrostatic pressure distribution, such as in a closed conduit.

The use of Reynolds number for determining whether the flow in open channels will be laminar or turbulent depends on the **hydraulic radius**, given by

$$R_h = \frac{A}{P} \tag{15.3}$$

where A is the cross-sectional area of flow and P is the wetted perimeter. The characteristic length R_h is analogous to diameter D in pipe flow. Recall that for pipe flow (Chapter 10), if the Reynolds number ($VD\rho/\mu = VD/\nu$) is less than 2000, the flow will be laminar, and if it is greater than about 3000, one can expect the flow to be turbulent. The Reynolds number criterion for open-channel flow would be 2000 if one replaced D in the Reynolds number by $4R_h$, where R_h is the hydraulic radius. For this definition of Reynolds number, laminar flow would occur in open channels if $V(4R_h)/\nu < 2000$.

However, the standard convention in open-channel flow analysis is to define the Reynolds number as

$$Re = \frac{VR_h}{\nu} \tag{15.4}$$

Therefore, in open channels, if the Reynolds number is less than 500, the flow is laminar, and if Re is greater than about 750, one can expect to have turbulent flow. A brief analysis of this turbulent criterion (see Example 15.1) will show that water flow in channels will usually be turbulent unless the velocity and/or the depth is very small.

It should be noted that for rectangular channels (see Fig. 15.3), the hydraulic radius is

$$R_h = \frac{A}{P} = \frac{By}{B + 2y} \tag{15.5}$$

For a wide and shallow channel, $B \gg y$ and Eq. (15.5) reduces to $R_h \approx y$ which means that the hydraulic radius approaches the depth of the channel.

FIGURE 15.3

Open-channel relations.

Side view

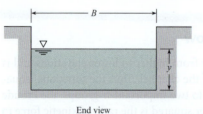

End view

✔CHECKPOINT PROBLEM 15.1

What is the hydraulic radius for this channel?

a. $\pi R/(4 + 2\pi)$

b. $\pi R/(2 + \pi)$

c. $R/4$

d. $R/2$

e. R

Most open-channel flow problems involve turbulent flow. If one calculates the conditions needed to maintain laminar flow, as in Example 15.1, one sees that laminar flow is uncommon.

EXAMPLE 15.1

Calculating Reynolds Number and Classifying Flow for a Rectangular Open Channel

---

Problem Statement

Water (15°C) flows in a 3 m-wide rectangular channel at a depth of 2 m. What is the Reynolds number if the mean velocity is 0.03 m/s? With this velocity, at what maximum depth can one be assured of having laminar flow?

Define the Situation

Water flows in a rectangular channel.

$$B = 3 \text{ m}, y = 2 \text{ m}, V = 0.03 \text{ m/s}.$$

Properties:

Water (15°C, 1 atm, Table A.5): $\nu = 1.14 \times 10^{-6} \text{ m}^2/\text{s}$.

State the Goal

1. Re ← Reynolds number
2. y_m(m) ← Maximum depth for laminar flow

Generate Ideas and Make a Plan

To find Re, apply Eq. (15.4). To find y_m, apply the criteria that laminar flow occurs for Re < 500. The plan is:

1. Calculate hydraulic radius using Eq. (15.5).
2. Calculate Reynolds number using Eq. (15.4).
3. Let Re = 500, solve for R_h, and then solve for y_m.

Take Action (Execute the Plan)

1. Hydraulic radius

$$R_h = \frac{By}{B + 2y} = \frac{(3 \text{ m})(2 \text{ m})}{(3 \text{ m}) + 2(2 \text{ m})} = 0.86 \text{ m}$$

2. Reynolds number

$$\text{Re} = \frac{VR_h}{\nu} = \frac{(0.03 \text{ m/s})(0.86 \text{ m})}{(1.14 \times 10^{-6} \text{ m}^2/\text{s})} = \boxed{22632}$$

3. Laminar Flow Criteria (Re < 500).

$$\text{Re} = VR_h/\nu = (0.03 \text{ m/s})R_h/(1.14 \times 10^{-6} \text{ m}^2/\text{s}) = 500$$

$$R_h = (500)(1.14 \times 10^{-6} \text{ m}^2/\text{s})/(0.03 \text{ m/s}) = 0.019 \text{ m}$$

For a rectangular channel,

$$R_h = (By)/(B + 2y)$$

$$(By)/(B + 2y) = (3y)/(3 + 2y) = 0.019 \text{ m}$$

$$y_m = \boxed{0.019 \text{ m}}$$

Review the Solution and the Process

1. *Knowledge.* Velocity or depth must be very small to yield laminar flow of water in an open channel.

2. *Knowledge.* Depth and hydraulic radius are virtually the same when depth is very small relative to width.

15.2 Energy Equation for Steady Open-Channel Flow

To derive the energy equation for flow in an open channel, begin with Eq. (7.29) and let the pump head and turbine head equal zero: $h_p = h_t = 0$. Equation (7.29) becomes

$$\frac{p_1}{\gamma} + \alpha_1 \frac{V_1^2}{2g} + z_1 = \frac{p_2}{\gamma} + \alpha_2 \frac{V_2^2}{2g} + z_2 + h_L \qquad \textbf{(15.6)}$$

Use Fig. 15.4 to show that

$$\frac{p_1}{\gamma} + z_1 = y_1 + S_0\Delta x \qquad \text{and} \qquad \frac{p_2}{\gamma} + z_2 = y_2$$

where S_0 is the slope of the channel bottom, and y is the depth of flow. Assume the flow in the channel is turbulent, so $\alpha_1 = \alpha_2 \approx 1.0$. Equation (15.6) becomes

$$y_1 + \frac{V_1^2}{2g} + S_0\Delta x = y_2 + \frac{V_2^2}{2g} + h_L \qquad \textbf{(15.7)}$$

FIGURE 15.4

Definition sketch for flow in open channels.

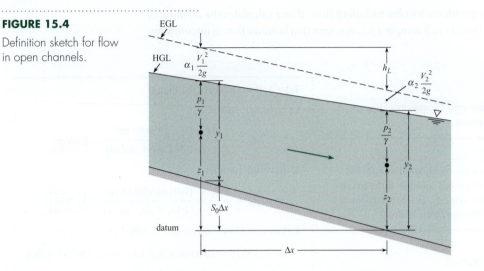

In addition to the foregoing assumptions, Eq. (15.7) also requires that the channel have a uniform cross section, and the flow be steady.

15.3 Steady Uniform Flow

Uniform flow requires that velocity be constant in the flow direction, so the shape of the channel and the depth of fluid is the same from section to section. Consideration of the foregoing slope equations shows that for uniform flow, the slope of the HGL will be the same as the channel slope because the velocity and depth are the same in both sections. The HGL, and thus the slope of the water surface, is controlled by head loss. If one restates the Darcy-Weisbach equation introduced in Chapter 10 with D replaced by $4R_h$, the head loss is

$$h_f = \frac{fL}{4R_h} \frac{V^2}{2g} \quad \text{or} \quad \frac{h_f}{L} = \frac{f}{4R_h} \frac{V^2}{2g} \tag{15.8}$$

From Fig. 15.4, $S_0 = $ [slope of EGL], which is a function of the head loss, so $S_0 = (h_f/L)$, yielding the following equation for velocity:

$$V = \sqrt{\frac{8g}{f} R_h S_0} \tag{15.9}$$

To solve Eq. (15.9) for velocity, the friction factor f can be found from the Moody diagram (Fig. 10.14) and can then be used to solve iteratively for the velocity for a given uniform-flow condition. This is demonstrated in Example 15.2.

EXAMPLE 15.2

Applying the Darcy-Weisbach Equation to Find the Flow Rate in a Rectangular Open Channel

Problem Statement

Estimate the discharge of water that a concrete channel 3 m wide can carry if the depth of flow is 2 m and the slope of the channel is 0.0016.

Define the Situation

- Water flows in a rectangular channel.
- $B = 3$ m, $y = 2$ m, $S_0 = 0.0016$.

Assumptions. Uniform flow

Properties.

- Water (15°C, 1 atm, Table A.5): $\nu = 1.14 \times 10^{-6}$ m²/s
- Concrete (Table 10.4): $k_s \approx 0.0015$ m

State the Goal

$Q \longleftarrow$ Discharge in the channel

Generate Ideas and Make a Plan

Because the goal is Q, apply the flow rate equation

$$Q = VA \qquad \text{(a)}$$

To find V in Eq. (a), apply Eq. (15.9):

$$V = \sqrt{\frac{8g}{f} R_h S_0} \qquad \text{(b)}$$

To find R_h in Eq. (b), apply Eq. (15.5).

$$R_h = \frac{By}{B + 2y} = \frac{(3 \text{ m})(2 \text{ m})}{(3 \text{ m}) + 2(2 \text{ m})} = 0.86 \text{ m} \qquad \text{(c)}$$

To find f in Eq. (b), use an iterative approach with the Moody diagram. This is a Case 2 problem from Chapter 10. The plan is:

1. Calculate relative roughness. Then, guess a value of f.

2. Calculate V using Eq. (b).

3. Calculate Reynolds number, then look up f on the Moody diagram and compare to the guess in step 1. If needed, go back to step 2.

4. Calculate Q using Eq. (a).

Take Action (Execute the Plan)

1. Calculate relative roughness.

$$\frac{k_s}{4R_h} = \frac{0.0015 \text{ m}}{4(6 \text{ m}^2/7 \text{ m})} = 0.00044$$

Use value of $k_s/4R_h = 0.00044$ as a guide to estimate $f = 0.016$.

2. Calculate V based on guess of f.

$$V = \sqrt{\frac{8(9.81 \text{ m/s}^2)(0.86 \text{ m})(0.0016)}{0.016}}$$

$$= 2.59 \text{ m/s}$$

3. Calculate a new value of f based on V from step 2.

$$\text{Re} = V\frac{4R_h}{\nu} = \frac{(2.59 \text{ m/s})(3.44 \text{ m})}{(1.14 \times 10^{-6} \text{ m}^2/\text{s})} = 7.8 \times 10^6$$

Using this new value of Re and $k_s/4R_h = 0.00044$, read f as 0.016. This value of f is the same as the previous estimate. Thus, we conclude that

$$V = 2.59 \text{ m/s}$$

4. Flow rate equation

$$Q = VA = (2.59 \text{ m/s})(6 \text{ m}^2) = \boxed{15.54 \text{ m}^3/\text{s}}$$

Review the Solution and the Process

1. *Notice.* The approach to solving this problem parallels the approach presented in Chapter 10 for solving problems that involve flow in conduits.

2. *Knowledge.* Hydraulic diameter is four times the hydraulic radius. This is why the relative roughness formula in step 1 is $k_s/(4R_h)$.

Rock-Bedded Channels

For rock-bedded channels such as those in some natural streams or unlined canals, the larger rocks produce most of the resistance to flow, and essentially none of this resistance is due to viscous effects. Thus, the friction factor is independent of the Reynolds number. This is analogous to the fully rough region of the Moody diagram for pipe flow. For a rock-bedded channel, Limerinos (1) has shown that the resistance coefficient f can be given in terms of the size of rock in the stream bed as

$$f = \frac{1}{\left[1.2 + 2.03 \log\left(\dfrac{R_h}{d_{84}}\right) \right]^2} \qquad \textbf{(15.10)}$$

where d_{84} is a measure of the rock size.*

---

*Most river-worn rocks are somewhat elliptical in shape. Limerinos (1) showed that the intermediate dimension d_{84} correlates best with f. The d_{84} refers to the size of rock (intermediate dimension) for which 84% of the rocks in the random sample are smaller than the d_{84} size. Details for choosing the sample are given by Wolman (3).

EXAMPLE 15.3

Resistance Coefficient for Boulders

Problem Statement

Determine the value of the resistance coefficient, f, for a natural rock-bedded channel that is 30 m wide and has an average depth of 1.3 m. The d_{84} size of boulders in the stream bed is 0.22 m.

Define the Situation

A natural channel is lined with boulders.

State the Goal

Find the friction factor, f.

Generate Ideas and Make a Plan

1. Since the channel is wide, approximate R_h as the depth of the channel.

2. Use Eq. (15.10) to find f on the basis of the d_{84} boulder size.

Take Action (Execute the Plan)

1. R_h is 1.2 m.

2. Evaluate f.

$$f = \frac{1}{\left[1.2 + 2.03 \log\left(\dfrac{1.3}{0.22}\right)\right]^2} = \boxed{0.130}$$

The Chezy Equation

Leaders in open-channel research have recommended the use of the methods already presented (involving the Reynolds number and relative roughness k_s) for channel design (2). However, many engineers continue to use two traditional methods, the Chezy equation and the Manning equation.

As noted earlier, the depth in uniform flow, called normal depth, y_n, is constant. Consequently, h_f/L is the slope S_0 of the channel, and Eq. (15.8) can be written as

$$R_h S_0 = \frac{f}{8g} V^2$$

or

$$V = C\sqrt{R_h S_0} \tag{15.11}$$

where

$$C = \sqrt{8g/f} \tag{15.12}$$

Because $Q = VA$, the discharge in a channel is given by

$$Q = CA\sqrt{R_h S_0} \tag{15.13}$$

This equation is known as the **Chezy equation** after a French engineer of that name. For practical application, the coefficient C must be determined. One way to determine C is by knowing an acceptable value of the friction factor f and using Eq. (15.12).

The Manning Equation

The second, and more common, way to determine C in the SI system of units is given as:

$$C = \frac{R_h^{1/6}}{n} \tag{15.14}$$

where n is a resistance coefficient called **Manning's** n, which has different values for different types of boundary roughness. When this expression for C is inserted into Eq. (15.13), the result is a common form of the discharge equation for uniform flow in open channels for SI units, referred to as the Manning equation:

$$Q = \frac{1.0}{n} A R_h^{2/3} S_0^{1/2} \tag{15.15}$$

Table 15.1 gives values of n for various types of boundary surfaces. The major limitation of this approach is that the viscous or relative-roughness effects are not present in the design formula. Hence, application outside the range of normal-sized channels carrying water is not recommended.

Manning Equation—Traditional System of Units

The form of the Manning equation depends on the system of units because Manning's equation is not dimensionally homogeneous. In Eq. (15.15), notice that the primary dimensions on the left side of the equation are L^3/T and the primary dimensions on the right side are $L^{8/3}$.

To convert the Manning equation from SI to traditional units, one must apply a factor equal to 1.49 if the same value of n is used in the two systems. Thus in the traditional system the discharge equation using Manning's n is

$$Q = \frac{1.49}{n} A R_h^{2/3} S_0^{1/2} \tag{15.16}$$

TABLE 15.1 Typical Values of Roughness Coefficient, Manning's n

Lined Canals	n
Cement plaster	0.011
Untreated gunite	0.016
Wood, planed	0.012
Wood, unplaned	0.013
Concrete, troweled	0.012
Concrete, wood forms, unfinished	0.015
Rubble in cement	0.020
Asphalt, smooth	0.013
Asphalt, rough	0.016
Corrugated metal	0.024
Unlined Canals	
Earth, straight and uniform	0.023
Earth, winding and weedy banks	0.035
Cut in rock, straight and uniform	0.030
Cut in rock, jagged and irregular	0.045
Natural Channels	
Gravel beds, straight	0.025
Gravel beds plus large boulders	0.040
Earth, straight, with some grass	0.026
Earth, winding, no vegetation	0.030
Earth, winding, weedy banks	0.050
Earth, very weedy and overgrown	0.080

In Example 15.4, a value for Manning's n is calculated from known information about a channel and compared to tabulated values for n in Table 15.1.

EXAMPLE 15.4

Apply the Chezy Equation to find Manning's Value of n for Flow in a Channel

Problem Statement

If a channel with boulders has a slope of 0.0030, is 30 m wide, has an average depth of 1.3 m, and is known to have a friction factor of 0.130, what is the discharge in the channel, and what is the numerical value of Manning's n for this channel?

Define the Situation

Water flows in an channel with boulders

$$S_0 = 0.003, B = 30 \text{ m}, y = 1.3 \text{ m}, f = 0.13$$

Assumptions. $R_h \approx y = 1.3$ m (because the channel is wide).

State the Goal

1. $Q(\text{m}^3/\text{s})$ ◀ Discharge in the channel
2. n ◀ Manning's n

Generate Ideas and Make a Plan

To find Q, apply the flow rate equation

$$Q = VA \qquad \text{(a)}$$

To find V in Eq. (a), apply Eq. (15.9):

$$V = \sqrt{\frac{8g}{f} R_h S_0} \qquad \text{(b)}$$

To find n, apply Eq. (15.15):

$$Q = \frac{1}{n} A R_h^{2/3} S_0^{1/2} \qquad \text{(c)}$$

Because Eqs. (a) to (c) form a set of three equations with three unknowns, they can be solved. The plan is:

1. Calculate V using Eq. (b).
2. Calculate Q using Eq. (a).
3. Calculate n using Eq. (c).

Take Action (Execute the Plan)

1. Velocity

$$V = \left[\sqrt{\frac{(8)(9.81 \text{ m/s}^2)}{0.130}} \right] \left[\sqrt{(1.3 \text{ m})(0.0030)} \right] = 1.53 \text{ m/s}$$

2. Flow Rate Equation

$$Q = VA = (1.53 \text{ m/s})(30 \text{ m} \times 1.3 \text{ m}) = \boxed{60 \text{ m}^3/\text{s}}$$

3. Manning's n (SI units).

$$n = \frac{1}{Q} A R_h^{2/3} S_0^{1/2}$$

$$n = \left(\frac{1}{60 \text{ m}^3/\text{s}} \right)(30 \text{ m} \times 1.3 \text{ m})(1.3 \text{ m})^{2/3}(0.003)^{1/2}$$

$$n = \boxed{0.0424}$$

Review the Solution and the Process

1. *Validation.* This calculated value of n is within the range of typical values in Table 15.1 under the category of "Unlined Canals, Cut in rock."

2. *Notice.* For uniform flow, f in the Darcy-Weisbach equation can be related to Manning's n (as shown by this example).

In Example 15.5 the Chezy equation for traditional units is used to compute discharge.

EXAMPLE 15.5

Discharge Using Chezy Equation

Problem Statement

Using the Chezy equation with Manning's n, compute the discharge in a concrete channel 3 m wide if the depth of flow is 2 m and the slope of the channel is 0.0016.

Define the Situation

Water flows in a concrete channel. Width = 3 m. Depth = 2 m. Slope = 0.0016.

Properties: $n = 0.015$ for concrete channel (Table 15.1).

State the Goal

Find the discharge, Q.

Generate Ideas and Make a Plan

Use the Chezy equation for traditional units, Eq. (15.15).

Take Action (Execute the Plan)

$$Q = \frac{1}{n} A R_h^{2/3} S_0^{1/2}$$

$$R_h = \frac{6 \text{ m}^2}{7 \text{ m}} = 0.86 \text{ m} \quad \text{and} \quad R_h^{2/3} = 0.90$$

$$S_0^{1/2} = 0.04 \quad \text{and} \quad A = 6 \text{ m}^2$$

$$Q = \frac{1}{0.015}(6)(0.90)(0.04) = \boxed{14.4 \text{ m}^3/\text{s}}$$

The two results (Examples 15.5 and 15.5) are within expected engineering accuracy for this type of problem. For a more complete discussion of the historical development of Manning's equation and the choice of n values for use in design or analysis, refer to Yen (4) and Chow (5).

Best Hydraulic Section for Uniform Flow

The **best hydraulic section** is the channel geometry that gives the maximum discharge for a given cross sectional area. Maximum discharge occurs when a geometry has the minimum wetted perimeter. Therefore, it yields the least viscous energy loss for a given area. Consider the quantity $A R_h^{2/3}$ in Manning's equation given in Eqs. (15.15 and 15.16), which is referred to as the section factor. Because $R_h = A/P$, the section factor relating to uniform flow is given by $A(A/P)^{2/3}$. Thus, for a channel of given resistance and slope, the discharge will increase with increasing cross-sectional area but decrease with increasing wetted perimeter P. For a given area, A, and a given shape of channel—for example, rectangular cross section—there will be a certain ratio of depth to width (y/B) for which the section factor will be maximum. This ratio is the best hydraulic section.

Example 15.6 shows that the best hydraulic section for a rectangular channel occurs when $y = \frac{1}{2}B$. It can be shown that the best hydraulic section for a trapezoidal channel is half a hexagon as shown; for the circular section, it is the half circle with depth equal to radius; and for the triangular section, it is a triangle with a vertex of 90° (Fig. 15.5). Of all the various shapes, the half circle has the best hydraulic section because it has the smallest perimeter for a given area.

The best hydraulic section can be relevant to the cost of the channel. For example, if a trapezoidal channel were to be excavated and if the water surface were to be at adjacent ground level, the minimum amount of excavation (and excavation cost) would result if the channel of best hydraulic section were used.

FIGURE 15.5

Best hydraulic sections for different geometries.

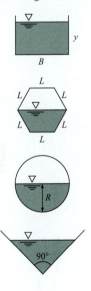

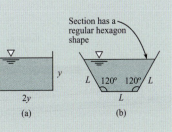

EXAMPLE 15.6

Finding the Best Hydraulic Section for a Rectangular Channel

Problem Statement

Determine the best hydraulic section for a rectangular channel with depth y and width B.

Define the Situation

Water flows in a rectangular channel. Depth = y. Width = B.

State the Goal

Find the best hydraulic section (relate B and y).

Generate Ideas and Make a Plan

1. Set $A = By$ and $P = B + 2y$ so that both are a function of y.
2. Let A be constant, and minimize P.
   - Differentiate P with respect to y and set the derivative equal to zero.
   - Express the result of minimizing P as a relation between y and B.

Take Action (Execute the Plan)

1. Relate A and P in terms of y.

$$P = \frac{A}{y} + 2y$$

2a. Minimize P.

$$\frac{dP}{dy} = \frac{-A}{y^2} + 2 = 0$$

$$\frac{A}{y^2} = 2$$

2b. Express result in terms of y and B.

$A = By$, so

$$\frac{By}{y^2} = 2 \quad \text{or} \quad \boxed{y = \frac{1}{2}B}$$

Review the Solution and the Process

Knowledge. The best hydraulic section for a rectangular channel occurs when the depth is one-half the width of the channel, see Fig. 15.5.

Uniform Flow in Culverts and Sewers

Sewers are conduits that carry sewage (liquid domestic, commercial, or industrial waste) from households, businesses, and factories to sewage disposal sites. These conduits are often circular in cross section, but elliptical and rectangular conduits are also used. The volume rate of sewage varies throughout the day and season, but of course sewers are designed to carry the maximum design discharge flowing full or nearly full. At discharges less than the maximum, the sewers will operate as open channels.

Sewage usually consists of about 99% water and 1% solid waste. Because most sewage is so dilute, it is assumed that it has the same physical properties as water for purposes of discharge computations. However, if the velocity in the sewer is too small, the solid particles may settle out and cause blockage of the flow. Therefore, sewers are usually designed to have a minimum velocity of about 0.60 m/s at times when the sewer is flowing full. This condition is met by choosing a slope on the sewer line to achieve the desired velocity.

A culvert is a conduit placed under a fill such as a highway embankment. It is used to convey stream-flow from the uphill side of the fill to the downhill side. Figure 15.6 shows the

FIGURE 15.6

Culvert under a highway embankment.

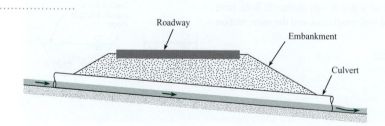

Roadway

Embankment

Culvert

essential features of a culvert. A culvert should be able to convey runoff from a **design storm** without overtopping the fill and without erosion of the fill at either the upstream or downstream end of the culvert. The design storm, for example, might be the maximum storm that could be expected to occur once in 50 years at the particular site.

The flow in a culvert is a function of many variables, including cross-sectional shape (circular or rectangular), slope, length, roughness, entrance design, and exit design. Flow in a culvert may occur as an open channel throughout its length, it may occur as a completely full pipe, or it may occur as a combination of both. The complete design and analysis of culverts are beyond the scope of this text; therefore, only simple examples are included here (Examples 15.7 and 15.8). For more extensive treatment of culverts, please refer to Chow (5), Henderson (6), and American Concrete Pipe Assoc. (7).

EXAMPLE 15.7

Sizing a Round Concrete Sewer Line

Problem Statement

A sewer line is to be constructed of concrete pipe to be laid on a slope of 0.006. If $n = 0.013$ and if the design discharge is 3 m³/s, what size pipe (commercially available) should be selected for a full-flow condition? What will be the mean velocity in the sewer pipe for these conditions? (It should be noted that concrete pipe is readily available in commercial sizes of 20 cm, 25 cm, and 30 cm diameter and then in 7.5 cm increments up to 90 cm diameter. From 90 cm diameter up to 365 cm the sizes are available in 15 cm increments.)

Define the Situation

Sewer line, $S_0 = 0.006$, Q (design) = 3 m³/s.

Assumptions: Can only use a standard pipe size.

State the Goal

Find: Pipe diameter large enough to carry design discharge and that allows $V \geq 0.66$ m/s at full-flow condition.

Generate Ideas and Make a Plan

1. Use Chezy equation for SI units, Eq. (15.15).
2. Solve for $AR^{2/3}$.
3. For pipe flowing full, relate A and P to diameter through R_h.
4. Solve for diameter, and use the next commercial size larger.
5. Check that velocity for full flow is greater than 0.66 m/s.

Take Action (Execute the Plan)

1. Chezy equation for SI units is

$$Q = \frac{1}{n} AR^{2/3} S_0^{1/2}$$

$$Q = 3.0 \text{ m}^3/\text{s}$$

$$n = 0.013$$

$$S_0 = 0.006 \text{ (assume atmospheric pressure in the pipe)}$$

2. Solve for $AR^{2/3}$. Note that units of $AR^{2/3}$ are m$^{8/3}$ because A is in m² and R_h is in m$^{2/3}$.

$$AR^{2/3} = \frac{(3.0 \text{ m}^3/\text{s})(0.0130)}{(1)(0.006)^{1/2}} = 0.50 \text{ m}^{8/3}$$

3. Relate A and P to diameter by relating to R_h.

$$R_h = \frac{A}{P} \quad \text{and} \quad R_h^{2/3} = \left(\frac{A}{P}\right)^{2/3}$$

$$AR_h^{2/3} = \frac{A^{5/3}}{P^{2/3}} = 0.50 \text{ m}^{8/3}$$

For a pipe flowing full, $A = \pi D^2/4$ and $P = \pi D$, or

$$\frac{(\pi D^2/4)^{5/3}}{(\pi D)^{2/3}} = 0.50 \text{ m}^{8/3}$$

4. Solving for diameter yields $D = 119$ cm. Use the next commercial size larger, which is $\boxed{D = 120 \text{ cm}.}$

$$A = \frac{\pi D^2}{4} = 4.52 \text{ m}^2 \text{ (for pipe flowing full)}$$

5. Verify that velocity of full flow is greater than 0.6 m/s.

$$V = \frac{Q}{A} = \frac{(3 \text{ m}^3/\text{s})}{(4.52 \text{ m}^2)} = \boxed{0.66 \text{ m/s}}$$

Example 15.8 demonstrates the calculation of necessary slope given all sources of head loss and a required discharge.

EXAMPLE 15.8

Culvert Design

Problem Statement

A 140 cm-diameter culvert laid under a highway embankment has a length of 60 m and a slope of 0.01. This was designed to pass a 50-year flood flow of 6.4 m³/s under full-flow conditions (see figure). For these conditions, what head H is required? When the discharge is only 1.4 m³/s, what will be the uniform flow depth in the culvert? Assume $n = 0.012$.

Define the Situation

Situation: Culvert has been designed to carry 6.4 m³/s with given dimensions.

Assumptions: Uniform flow, so that pipe head loss h_f can be related to S_0.

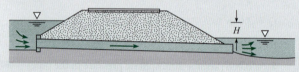

State the Goal

Find:

1. The height H required between the two free surfaces when flowing full.
2. The uniform flow depth in the culvert when $Q = 1.4$ m³/s.

Generate Ideas and Make a Plan

1. Use energy equation between the two end sections, accounting for head loss.
2. Document all sources of head loss.
3. Find pipe head loss h_f using Eq. (15.17) and the fact that

$$S_0 = \frac{h_f}{L}.$$

4. Use continuity equation to find V, the uniform flow velocity, needed to calculate head loss.
5. Solve for H.
6. Solve for depth of flow, for $Q = 1.4$ m³/s, using Eq. (15.15) and pipe geometry relations for pipe flowing partly full.

Take Action (Execute the Plan)

1. Energy equation

$$\frac{p_1}{g} + \frac{V_1^2}{2g} + z_1 = \frac{p_2}{g} + \frac{V_2^2}{2g} + z_2 + \sum h_L$$

Let points 1 and 2 be at the upstream and downstream water surfaces, respectively.

Thus, $(p_1 = p_2 = 0 \text{ gage} \quad \text{and} \quad V_1 = V_2 = 0)$

Also, $(z_1 - z_2 = H)$

Therefore, $(H = \sum h_L)$

2. Head losses occur at culvert entrance and exit, as well as over the length of pipe.

$H = $ pipe head loss + entrance head loss + exit head loss

$$H = \frac{V^2}{2g}(K_e + K_E) + \text{pipe head loss}$$

$$K_e = 0.50 \text{ (from Table 10.5)}$$

$$K_E = 1.00 \text{ (from Table 10.5)}$$

3. Pipe head loss is

$$Q = \frac{1}{n} A R_h^{2/3} S_0^{1/2}$$

$$Q = 6.4 \text{ m}^3/\text{s}$$

$$A = \frac{\pi D^2}{4} = 1.54 \text{ m}^2$$

$$R_h = \frac{A}{P} = \frac{\pi D^2/4}{\pi D} = \frac{D}{4} = 0.35 \text{ m}$$

$$R_h^{2/3} = (0.35 \text{ m})^{2/3} = 0.497 \text{ m}^{2/3}$$

$$S_0 = \frac{h_f}{L}$$

$$225 = \frac{1}{0.012}(1.54 \text{ m}^2)(0.497 \text{ m}^{2/3})\left(\frac{h_f}{60}\right)^{1/2}$$

$$h_f = 0.60 \text{ m}$$

4. Continuity equation yields

$$V = \frac{Q}{A} = \frac{6.4 \text{ m}^3/\text{s}}{1.54 \text{ m}^2} = 4.16 \text{ m/s}$$

5. Solve for H.

$$H = \frac{4.16^2}{19.62}(0.50 + 1.0) + 0.6$$

$$H = 1.32 \text{ m} + 0.6 \text{ m} = \boxed{1.92 \text{ m}}$$

6. Depth of flow for $Q = 1.4$ m³/s is

$$50 = \frac{1}{0.012} A R_h^{2/3}(0.01)^{1/2}$$

Values of A and R_h will depend on geometry of partly full pipe, as shown:

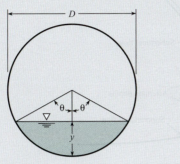

Area A if angle θ is given in degrees

$$A = \left[\left(\frac{\pi D^2}{4}\right)\left(\frac{2\theta}{360°}\right)\right] - \left(\frac{D}{2}\right)^2 (\sin\theta \cos\theta)$$

Wetted perimeter will be $P = \pi D(\pi/180°)$, so

$$R_h = \frac{A}{P} = \left(\frac{D}{4}\right)\left[1 - \left(\frac{\sin\theta \cos\theta}{(\pi\theta/180°)}\right)\right]$$

Substituting these relations for A and R_h into the discharge equation and solving for θ yields $\theta = 70°$. Therefore, y is

$$y = \frac{D}{2} - \frac{D}{2}\cos\theta = \left(\frac{140\text{ cm}}{2}\right)(1 - 0.342) = \boxed{46.06\text{ cm}}$$

15.4 Steady Nonuniform Flow

As stated in the beginning of this chapter, and shown in Fig. 15.2, all open-channel flows are classified as either uniform or nonuniform. Recall that uniform flow has constant velocity along a streamline and thus has constant depth for a constant cross section. In steady nonuniform flow, the depth and velocity change over distance (but not with time). For all such cases, the energy equation as generally introduced in Section 15.2 is invoked to compare two cross sections. However, for analysis of nonuniform flow, it is useful to distinguish whether the depth and velocity change occurs over a short distance, referred to as **rapidly varied flow**, or over a long reach of the channel, referred to as **gradually varied flow** (Fig. 15.7). The head loss term is different for these two cases. For rapidly varied flow, one can neglect the resistance of the channel walls and bottom because it occurs over a short distance. For gradually varied flow, because of the long distances involved, the surface resistance is a significant variable in the energy balance.

FIGURE 15.7

Classifying nonuniform flow.

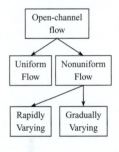

15.5 Rapidly Varied Flow

Rapidly varied flow is analyzed with the energy equation presented previously for open-channel flow, Eq. (15.7), with the additional assumptions that the channel bottom is horizontal ($S_0 = 0$) and the head loss is zero ($h_L = 0$). Therefore, Eq. (15.7) becomes

$$y_1 + \frac{V_1^2}{2g} = y_2 + \frac{V_2^2}{2g} \qquad (15.17)$$

Specific Energy

The sum of the depth of flow and the velocity head is defined as the **specific energy**:

$$E = y + \frac{V^2}{2g} \qquad (15.18)$$

Note that specific energy has dimensions [L]; that is, it is an energy head. Equation (15.17) states that the specific energy at section 1 is equal to the specific energy at section 2, or $E_1 = E_2$. The continuity equation between sections 1 and 2 is

$$A_1 V_1 = A_2 V_2 = Q \qquad (15.19)$$

FIGURE 15.8

Relation between depth and specific energy.

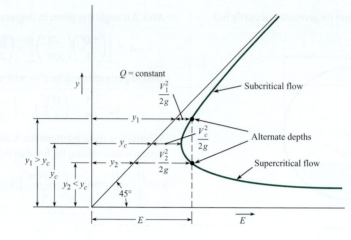

Therefore, Eq. (15.17) can be expressed as

$$y_1 + \frac{Q^2}{2gA_1^2} = y_2 + \frac{Q^2}{2gA_2^2} \tag{15.20}$$

Because A_1 and A_2 are functions of the depths y_1 and y_2, respectively, the magnitude of the specific energy at section 1 or 2 is solely a function of the depth at each section. If, for a given channel and given discharge, one plots depth versus specific energy, a relationship such as that shown in Fig. 15.8 is obtained. By studying Fig. 15.8 for a given value of specific energy, one can see that the depth may be either large or small. This means that for the small depth, the bulk of the energy of flow is in the form of kinetic energy—that is, $Q^2/(2gA^2) \gg y$—whereas for a larger depth, most of the energy is in the form of potential energy. Flow under a **sluice gate** (Fig. 15.9) is an example of flow in which two depths occur for a given value of specific energy. The large depth and low kinetic energy occur upstream of the gate; the low depth and large kinetic energy occur downstream. The depths as used here are called **alternate depths**. That is, for a given value of E, the large depth is alternate to the low depth, or vice versa. Returning to the flow under the sluice gate, one finds that if the same rate of flow is maintained, but the gate is set with a larger opening, as in Fig. 15.9b, the upstream depth will drop, and the downstream depth will rise. This results in different alternate depths and a smaller value of specific energy than before. This is consistent with the diagram in Fig. 15.8.

FIGURE 15.9

Flow under a sluice gate. (a) Smaller gate opening. (b) Larger gate opening.

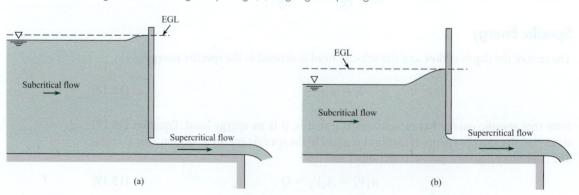

Finally, it can be seen in Fig. 15.8 that a point will be reached where the specific energy is minimum and only a single depth occurs. At this point, the flow is termed critical. Thus one definition of **critical flow** is the flow that occurs when the specific energy is minimum for a given discharge. The flow for which the depth is less than critical (velocity is greater than critical) is termed **supercritical flow**, and the flow for which the depth is greater than critical (velocity is less than critical) is termed **subcritical flow**. Therefore, subcritical flow occurs upstream and supercritical flow occurs downstream of the sluice gate in Fig. 15.9. Subcritical flows corresponds to a Froude number less than one (Fr < 1), and supercritical flow corresponds to (Fr > 1). Some engineers refer to subcritical and supercritical flow as **tranquil** and **rapid** flow, respectively. Other aspects of critical flow are shown in the next section.

Characteristics of Critical Flow

Critical flow occurs when the specific energy is minimum for a given discharge. The depth for this condition may be determined by solving for dE/dy from $E = y + Q^2/2gA^2$ and setting dE/dy equal to zero:

$$\frac{dE}{dy} = 1 - \frac{Q^2}{gA^3} \cdot \frac{dA}{dy} \tag{15.21}$$

However, $dA = T\,dy$, where T is the width of the channel at the water surface, as shown in Fig. 15.10. Then Eq. (15.21), with $dE/dy = 0$, will reduce to

$$\frac{Q^2 T_c}{gA_c^3} = 1 \tag{15.22}$$

or

$$\frac{A_c}{T_c} = \frac{Q^2}{gA_c^2} \tag{15.23}$$

If the **hydraulic depth**, D, is defined as

$$D = \frac{A}{T} \tag{15.24}$$

then Eq. (15.23) will yield a critical hydraulic depth D_c, given by

$$D_c = \frac{Q^2}{gA_c^2} = \frac{V^2}{g} \tag{15.25}$$

Dividing Eq. (15.25) by D_c and taking the square root yields

$$1 = \frac{V}{\sqrt{gD_c}} \tag{15.26}$$

Note: $V/\sqrt{gD_c}$ is the Froude number. Therefore, it has been shown that the Froude number is equal to unity when critical flow prevails.

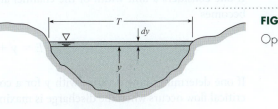

FIGURE 15.10

Open-channel relations.

If a channel is of rectangular cross section, then A/T is the actual depth, and $Q^2/A^2 = q^2/y^2$, so the condition for **critical depth** (Eq. 15.23) for a rectangular channel becomes

$$y_c = \left(\frac{q^2}{g}\right)^{1/3}$$

(15.27)

where q is the discharge per unit width of channel.

EXAMPLE 15.9

Calculating Critical Depth in a Channel

Problem Statement

Determine the critical depth in this trapezoidal channel for a discharge of 14 m³/s. The width of the channel bottom is $B = 6$ m, and the sides slope upward at an angle of 45°.

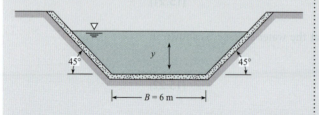

Define the Situation

Water flows in a trapezoidal channel with known geometry.

State the Goal

Calculate the critical depth.

Generate Ideas and Make a Plan

1. For critical flow, Eq. (15.22) must apply.
2. Relate this channel geometry to width T and area A in Eq. (15.22).
3. By iteration (choose y and compute A^3/T), find y that will yield A^3/T equal to 20 m². This y will be critical depth y_c.

Take Action (Execute the Plan)

1. Apply Eq. (15.22) or Eq. (15.23).

$$\frac{Q^2 T_c}{g A_c^3} = 1 \text{ or } \frac{Q^2}{g} = \frac{A_c^3}{T_c}$$

2. For $Q = 14$ m³/s,

$$\frac{A_c^3}{T_c} = \frac{14^2}{9.81} = 20 \text{ m}^2$$

For this channel, $A = y(B + y)$ and $T = B + 2y$.
3. Iterate to find y_c.

$$y_c = \boxed{0.7 \text{ m}}$$

Critical flow may also be examined in terms of how the discharge in a channel varies with depth for a given specific energy. For example, consider flow in a rectangular channel where

$$E = y + \frac{Q^2}{2gA^2}$$

or

$$E = y + \frac{Q^2}{2gy^2B^2}$$

If one considers a unit width of the channel and lets $q = Q/B$, then the foregoing equation becomes

$$E = y + \frac{q^2}{2gy^2}$$

If one determines how q varies with y for a constant value of specific energy, one sees that critical flow occurs when the discharge is maximum (see Fig. 15.11).

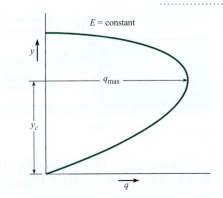

FIGURE 15.11

Variation of q and y with constant specific energy.

Originally, the term **critical flow** probably related to the unstable character of the flow for this condition. Referring to Fig. 15.8, one can see that only a slight change in specific energy will cause the depth to increase or decrease a significant amount; this is a very unstable condition. In fact, observations of critical flow in open channels show that the water surface consists of a series of standing waves. Because of the unstable nature of the depth in critical flow, designing canals so that normal depth is either well above or well below critical depth is usually best. The flow in canals and rivers is usually subcritical; however, the flow in steep chutes or over spillways is supercritical.

In this section, various characteristics of critical flow have been explored. The main ones can be summarized as follows:

1. Critical flow occurs when specific energy is minimum for a given discharge (Fig. 15.8).

2. Critical flow occurs when the discharge is maximum for a given specific energy.

3. Critical flow occurs when

$$\frac{A^3}{T} = \frac{Q^2}{g}$$

4. Critical flow occurs when Fr $= 1$. Subcritical flow occurs when Fr < 1. Supercritical flow occurs when Fr > 1.

5. For rectangular channels, critical depth is given as $y_c = (q^2/g)^{1/3}$.

Common Occurrence of Critical Flow

Critical flow occurs when a liquid passes over a broad-crested weir (Fig. 15.12). The principle of the broad-crested weir is illustrated by first considering a closed sluice gate that prevents water from being discharged from the reservoir, as shown in Fig. 15.12a. If the gate is opened a small amount (gate position a'-a'), the flow upstream of the gate will be subcritical, and the flow downstream will be supercritical (as in the condition shown in Fig. 15.9). As the gate is opened further, a point is finally reached where the depths immediately upstream and downstream of the gate are the same. This is the critical condition. At this gate opening and beyond, the gate has no influence on the flow; this is the condition shown in Fig. 15.12b, the broad-crested weir. If the depth of flow over the weir is measured, the rate of flow can easily be computed from Eq. (15.27):

$$q = \sqrt{gy_c^3}$$

or

$$Q = L\sqrt{gy_c^3} \tag{15.28}$$

where L is the length of the weir crest normal to the flow direction.

FIGURE 15.12

Flow over a broad-crested weir.
(a) Depth of flow controlled by sluice gate.
(b) Depth of flow is controlled by weir, and is y_c.

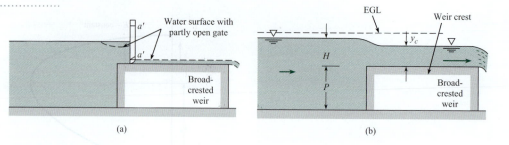

(a)

(b)

Because $y_c/2 = (V_c^2/2g)$, from Eq. (15.25), it can be shown that $y_c = (2/3E)$, where E is the total head above the crest $(H + V_{approach}^2/2g)$; hence Eq. (15.28) can be rewritten as

$$Q = L\sqrt{g}\left(\frac{2}{3}\right)^{3/2} E^{3/2}$$

or

$$Q = 0.385 L \sqrt{2g} E_c^{3/2} \tag{15.29}$$

For high weirs, the upstream velocity of approach is almost zero. Hence Eq. (15.29) can be expressed as

$$Q_{theor} = 0.385 L \sqrt{2g} H^{3/2} \tag{15.30}$$

If the height P of the broad-crested weir is relatively small, then the velocity of approach may be significant, and the discharge produced will be greater than that given by Eq. (15.30). Also, head loss will have some effect. To account for these effects, a discharge coefficient C is defined as

$$C = Q/Q_{theor} \tag{15.31}$$

Then

$$Q = 0.385 CL \sqrt{2g} H^{3/2} \tag{15.32}$$

where Q is the actual discharge over the weir. An analysis of experimental data by Raju (15) shows that C varies with $H/(H + P)$ as shown in Fig. 15.13. The curve in Fig. 15.13 is for a weir with a vertical upstream face and a sharp corner at the intersection of the upstream face and the weir crest. If the upstream face is sloping at a 45° angle, the discharge coefficient should be increased 10% over that given in Fig. 15.13. Rounding of the upstream corner will also produce a coefficient of discharge as much as 3% greater.

Equation (15.32) reveals a definite relationship for Q as a function of the head, H. This type of discharge-measuring device is in the broad class of discharge meters called

FIGURE 15.13

Discharge coefficient for a broad-crested weir for $0.1 < H/L < 0.8$.

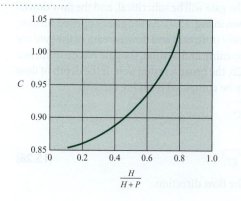

critical-flow flumes. Another very common critical-flow flume is the **venturi flume**, which was developed and calibrated by Parshall (8). Figure 15.14 shows the essential features of the venturi flume. The discharge equation for the venturi flume is in the same form as Eq. (15.32), the only difference being that the experimentally determined coefficient C will have a different value from the C for the broad-crested weir. For more details on the venturi flume, you may refer to Roberson et al. (9), Parshall (8), and Chow (5). The venturi flume is especially useful for discharge measurement in irrigation systems because little head loss is required for its use, and sediment is easily flushed through if the water happens to be silty.

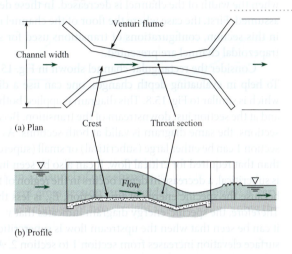

FIGURE 15.14

Flow through a venturi flume.

The depth also passes through a critical stage in channel flow where the slope changes from a mild one to a steep one. A **mild slope** is defined as a slope for which the normal depth y_n is greater than y_c. Likewise, a **steep slope** is one for which $y_n < y_c$. This condition is shown in Fig. 15.15. Note that y_c is the same for both slopes in the figure because y_c is a function of the discharge only. However, normal depth (uniform-flow depth) for the mild upstream channel is greater than critical, whereas the normal depth for the steep downstream channel is less than critical; hence it is obvious that the depth must pass through a critical stage. Experiments show that critical depth occurs a very short distance upstream of the intersection of the two channels.

Another place where critical depth occurs is upstream of a free overfall at the end of a channel with a mild slope Fig. 15.16. Critical depth will occur at a distance of $3y_c$ to $4y_c$ upstream of

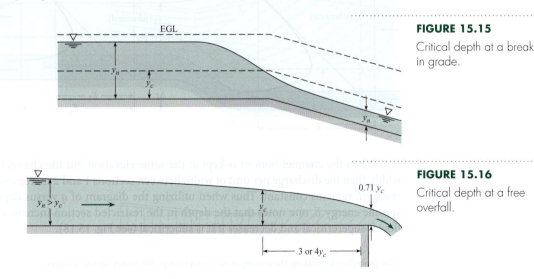

FIGURE 15.15

Critical depth at a break in grade.

FIGURE 15.16

Critical depth at a free overfall.

elevation through the transition. Analyses of transitions utilizing the one-dimensional form of the energy equation are applicable only if the flow is subcritical. If the flow is supercritical, then a much more involved analysis is required. For more details on the design and analysis of transitions, you are referred to Hinds (10), Chow (5), U.S. Bureau of Reclamation (11), and Rouse (12).

Wave Celerity

Wave celerity is the velocity at which an infinitesimally small wave travels relative to the velocity of the fluid. It can be used to characterize the velocity of waves in the ocean or propagation of a flood wave following a dam failure. A derivation of wave celerity, c, follows.

Consider a small solitary wave moving with velocity c in a calm body of liquid of small depth (Fig. 15.22a). Because the velocity in the liquid changes with time, this is a condition of unsteady flow. However, if one referred all velocities to a reference frame moving with the wave, the shape of the wave would be fixed, and the flow would be steady. Then the flow is amenable to analysis with the Bernoulli equation. The steady-flow condition is shown in Fig. 15.22b. When the Bernoulli equation is written between a point on the surface of the undisturbed fluid and a point at the wave crest, the following equation results:

$$\frac{c^2}{2g} + y = \frac{V^2}{2g} + y + \Delta y \tag{15.33}$$

FIGURE 15.22

Solitary wave (exaggerated vertical scale).
(a) Unsteady flow.
(b) Steady flow.

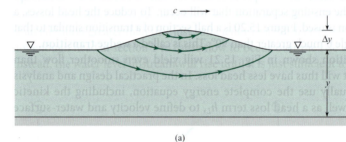

(a)

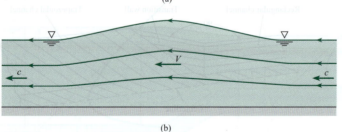

(b)

In Eq. (15.33), V is the velocity of the liquid in the section where the crest of the wave is located. From the continuity equation, $cy = V(y + \Delta y)$. Hence

$$V = \frac{cy}{y + \Delta y}$$

and

$$V^2 = \frac{c^2 y^2}{(y + \Delta y)^2} \tag{15.34}$$

When Eq. (15.34) is substituted into Eq. (15.33), it yields

$$\frac{c^2}{2g} + y = \frac{c^2 y^2}{2g[y^2 + 2y\Delta y + (\Delta y)^2]} + y + \Delta y \tag{15.35}$$

Solving Eq. (15.35) for c after discarding terms with $(\Delta y)^2$, assuming an infinitesimally small wave, yields the **wave celerity equation**

$$c = \sqrt{gy} \tag{15.36}$$

It has thus been shown that the speed of a small solitary wave is equal to the square root of the product of the depth and g.

15.6 Hydraulic Jump

Occurrence of the Hydraulic Jump

An interesting and important case of rapidly varied flow is the hydraulic jump. A **hydraulic jump** occurs when the flow is supercritical in an upstream section of a channel and is then forced to become subcritical in a downstream section (the change in depth can be forced by a sill in the downstream part of the channel or just by the prevailing depth in the stream further downstream), resulting in an abrupt increase in depth, and considerable energy loss. Hydraulic jumps (Fig. 15.23) are often considered in the design of open channels and spillways of dams. If a channel is designed to carry water at supercritical velocities, the designer must be certain that the flow will not become subcritical prematurely. If it did, overtopping of the channel walls would undoubtedly occur, with consequent failure of the structure. Because the energy loss in the hydraulic jump is initially not known, the energy equation is not a suitable tool for analysis of the velocity-depth relationships. Because there is a significant difference in hydrostatic head on both sides of the equation causing opposing pressure forces, the momentum equation can be applied to the problem, as developed in the following sections.

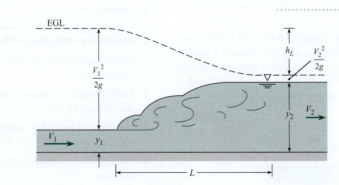

FIGURE 15.23

Definition sketch for the hydraulic jump.

Derivation of Depth Relationships in Hydraulic Jumps

Consider flow as shown in Fig. 15.23. Here it is assumed that uniform flow occurs both upstream and downstream of the jump and that the resistance of the channel bottom over the relatively short distance L is negligible. The derivation is for a horizontal channel, but experiments show that the results of the derivation will apply to all channels of moderate slope

FIGURE 15.24

Control-volume analysis for the hydraulic jump.

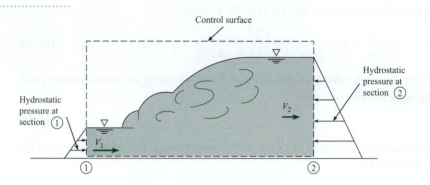

$(S_0 < 0.02)$. The derivation is started by applying the momentum equation in the x-direction to the control volume shown in Fig. 15.24:

$$\sum F_x = \dot{m}_2 V_2 - \dot{m}_1 V_1$$

The forces are the hydrostatic forces on each end of the system; thus the following is obtained:

$$\bar{p}_1 A_1 - \bar{p}_2 A_2 = \rho V_2 A_2 V_2 - \rho V_1 A_1 V_1$$

or

$$\bar{p}_1 A_1 + \rho Q V_1 = \bar{p}_2 A_2 + \rho Q V_2 \tag{15.37}$$

In Eq. (15.37), $\bar{p}_1$ and $\bar{p}_2$ are the pressures at the centroids of the respective areas A_1 and A_2.

A representative problem (e.g., Example 15.10) is to determine the downstream depth y_2 given the discharge and upstream depth. The left-hand side of Eq. (15.37) would be known because V, A, and p are all functions of y and Q, and the right-hand side is a function of y_2; therefore, y_2 can be determined.

EXAMPLE 15.10

Calculating Downstream Depth for a Hydraulic Jump

Problem Statement

Water flows in a trapezoidal channel at a rate of 8.5 m³/s. The channel has a bottom width of 3 m and side slopes of 1 vertical to 1 horizontal. If a hydraulic jump is forced to occur where the upstream depth is 0.3 m, what will be the downstream depth and velocity? What are the values of Fr_1 and Fr_2?

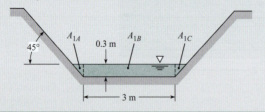

Define the Situation

A hydraulic jump is forced in a trapezoidal channel.

Properties: Water (10°C), Table A.5:

$\gamma = 9810$ N/m³, and $\rho = 1000$ kg/m³.

State the Goal

1. Downstream depth and velocity
2. Values of Fr_1 and Fr_2

Generate Ideas and Make a Plan

1. Find cross section, velocity, and hydraulic depth in the upstream section.

2. Find pressure in the upstream section to use for left-hand side of Eq. (15.37).

3. Use channel geometry information to solve for y_2 in right-hand side of Eq. (15.37).

4. Use Eq. (15.2) to solve for the Froude number at both sections.

Take Action (Execute the Plan)

1. By inspection, for the upstream section, the cross-sectional flow area is 1 m².

   Therefore, the mean velocity is $V_1 = Q/A_1 = 8.5$ m/s.

   The hydraulic depth is $D_1 = A_1/T_1 = 1$ m²/3.6 m = 0.28 m.

2. The location of the centroid ($\bar{y}$) of the area A_1 can be obtained by taking moments of the subareas about the water surface (see example sketch).

$A_1 \bar{y}_1 = A_{1A} \times 0.0999 \text{ m} + A_{1B} \times 0.045 \text{ m} + A_{1C} \times 0.0999 \text{ m}$

$(1 \text{ m}^2)\bar{y}_1 = (0.0999 \text{ m})(0.045 \text{ m}^2 \times 2) + (0.15 \text{ m})(0.9 \text{ m}^2)$

$\bar{y} = 0.144 \text{ m}$

Pressure $p_1 = 9810 \text{ N/m}^3 \times 0.144 \text{ m} = 1412.64 \text{ N/m}^2$

Therefore,

$1412.64 \times 1 + 1000 \times 8.5 \times 8.5 = \bar{p}_2 A_2 + \rho Q V_2$

3. Using right-hand side of Eq. (15.37), solve for y_2.

$$\bar{p}_2 A_2 + \rho Q V_2 = 73{,}663 \text{ N}$$

$$\gamma \bar{y}_2 A_2 + \frac{\rho Q^2}{A_2} = 73{,}663$$

$$\bar{y}_2 = \frac{\sum A_i y_i}{A_2} = \frac{B y_2^2/2 + y_2^3/3}{A_2}$$

Using $B = 3$ m, $Q = 8.5$ m^2/s, and material properties assumed earlier,

$$y_2 = \boxed{1.75 \text{ m}}$$

4. Froude numbers at both sections are

$$\text{Fr}_1 = \frac{V_1}{\sqrt{gD_1}} = \frac{8.5 \text{ m/s}}{(9.81 \text{ m/s}^2 \times 0.28 \text{ m})^{1/2}} = \boxed{5.12}$$

$$V_2 = \frac{Q}{A_2} = \frac{8.5}{5.34 + 1.75^2} = 1.01 \text{ m/s}$$

$$D_2 = \frac{A_2}{T_2} = \frac{8.41 \text{ m}^2}{6.5} = 1.29 \text{ m}$$

$$\text{Fr}_2 = \frac{V}{\sqrt{gD}} = \frac{1.01 \text{ m/s}}{(9.81 \times 1.29)^{1/2}} = \boxed{0.284}$$

Hydraulic Jump in Rectangular Channels

If one writes Eq. (15.37) for a unit width of a rectangular channel where $\bar{p}_1 = \gamma y_1/2$, $\bar{p}_2 = \gamma y_2/2$, $Q = q$, $A_1 = y_1$, and $A_2 = y_2$, this will yield

$$\gamma \frac{y_1^2}{2} + \rho q V_1 = \gamma \frac{y_2^2}{2} + \rho q V_2 \qquad \textbf{(15.38a)}$$

but $q = Vy$, so Eq. (15.38a) can be rewritten as

$$\frac{\gamma}{2}(y_1^2 - y_2^2) = \frac{\gamma}{g}(V_2^2 y_2 - V_1^2 y_1) \qquad \textbf{(15.38b)}$$

The preceding equation can be further manipulated to yield

$$\frac{2V_1^2}{gy_1} = \left(\frac{y_2}{y_1}\right)^2 + \frac{y_2}{y_1} \qquad \textbf{(15.39)}$$

The term on the left-hand side of Eq. (15.39) will be recognized as twice Fr_1^2. Hence Eq. (15.39) is written as

$$\left(\frac{y_2}{y_1}\right)^2 + \frac{y_2}{y_1} - 2\text{Fr}_1^2 = 0 \qquad \textbf{(15.40)}$$

By use of the quadratic formula, it is easy to solve for y_2/y_1 in terms of the upstream Froude number. This yields an equation for **depth ratio** across a hydraulic jump:

$$\frac{y_2}{y_1} = \frac{1}{2}(\sqrt{1 + 8\text{Fr}_1^2} - 1) \qquad \textbf{(15.41)}$$

or

$$y_2 = \frac{y_1}{2}(\sqrt{1 + 8\text{Fr}_1^2} - 1) \qquad \textbf{(15.42)}$$

The other solution of Eq. (15.40) gives a negative downstream depth, which is not physically possible. Hence the downstream depth is expressed in terms of the upstream depth and the upstream Froude number. In Eqs. (15.41) and (15.42), the depths y_1 and y_2 are said to be **conjugate** or **sequent** (both terms are in common use) to each other, in contrast to the alternate depths obtained from the energy equation. Numerous experiments show that the relation represented by Eqs. (15.41) and (15.42) is valid over a wide range of Froude numbers.

Although no theory has been developed to predict the length of a hydraulic jump, experiments [see Chow (5)] show that the relative length of the jump, L/y_2, is approximately 6 for Fr_1 ranging from 4 to 18.

Head Loss in a Hydraulic Jump

In addition to determining the geometric characteristics of the hydraulic jump, it is often desirable to determine the head loss produced by it. This is obtained by comparing the specific energy before the jump to that after the jump, the head loss being the difference between the two specific energies. It can be shown that the head loss for a jump in a rectangular channel is

$$h_L = \frac{(y_2 - y_1)^3}{4 y_1 y_2} \tag{15.43}$$

For more information on the hydraulic jump, see Chow (5). The following example shows that Eq. (15.43) yields a magnitude that equals the difference between the specific energies at the two ends of the hydraulic jump.

EXAMPLE 15.11

Calculating Head Loss in a Hydraulic Jump

Problem Statement

Water flows in a rectangular channel at a depth of 30 cm with a velocity of 16 m/s, as shown in the sketch that follows. If a downstream sill (not shown) forces a hydraulic jump, what will be the depth and velocity downstream of the jump? What head loss is produced by the jump?

Define the Situation

A hydraulic jump is occurring in a rectangular channel.

State the Goal

- Calculate downstream depth and velocity.
- Calculate head loss produced by the jump.

Generate Ideas and Make a Plan

1. To calculate h_L using Eq. (15.43), one calculates y_2 from the depth ratio equation (Eq. 15.42). This requires Fr_1.

2. Check validity of head loss by comparing to $E_1 - E_2$.

Take Action (Execute the Plan)

1. Calculate Fr_1, y_2, V_2, and h_L from Eqs. (Eq. 15.42) and (15.43).

$$Fr_1 = \frac{V}{\sqrt{gy_1}} = \frac{16}{\sqrt{9.81\,(0.30)}} = 9.33$$

$$y_2 = \frac{0.30}{2}[\sqrt{1 + 8(9.33)^2} - 1] = \boxed{3.81 \text{ m}}$$

$$V_2 = \frac{q}{y_2} = \frac{(16 \text{ m/s})(0.30 \text{ m})}{3.81 \text{ m}} = \boxed{1.26 \text{ m/s}}$$

$$h_L = \frac{(3.81 - 0.30)^3}{4(0.30)(3.81)} = \boxed{9.46 \text{ m}}$$

2. Compare the head loss to $E_1 - E_2$.

$$h_L = \left(0.30 + \frac{16^2}{2 \times 9.81}\right) - \left(3.81 + \frac{1.26^2}{2 \times 9.81}\right) = 9.46 \text{ m}$$

The value is the same, so $\boxed{\text{validity of } h_L \text{ equation is verified.}}$

Use of Hydraulic Jump on Downstream End of Dam Spillway

Previously it was shown that the transition from supercritical to subcritical flow produces a hydraulic jump, and that the relative height of the jump (y_2/y_1) is a function of Fr_1. Because flow over the spillway of a dam invariably results in supercritical flow at the lower end of the spillway, and because flow in the channel downstream of a spillway is usually subcritical, it is obvious that a hydraulic jump must form near the base of the spillway (see Fig. 15.25). The downstream portion of the spillway, called the spillway **apron**, must be designed so that the hydraulic jump always forms on the concrete structure itself. If the hydraulic jump were allowed to form beyond the concrete structure, as in Fig. 15.26, severe erosion of the foundation material as a result of the high-velocity supercritical flow could undermine the dam and cause its complete failure. One way to solve this problem might be to incorporate a long, sloping apron into the design of the spillway, as shown in Fig. 15.27. A design like this would work

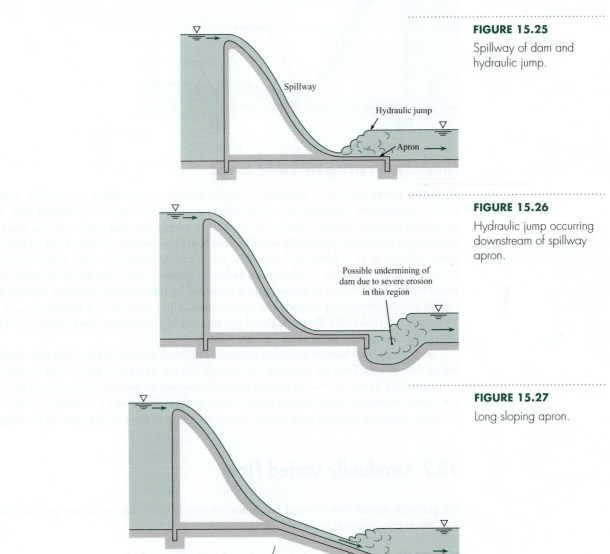

FIGURE 15.25
Spillway of dam and hydraulic jump.

FIGURE 15.26
Hydraulic jump occurring downstream of spillway apron.

FIGURE 15.27
Long sloping apron.

very satisfactorily from the hydraulics point of view. For all combinations of Fr_1 and water-surface elevation in the downstream channel, the jump would always form on the sloping apron. However, its main drawback is cost of construction. Construction costs will be reduced as the length, L, of the stilling basin is reduced. Much research has been devoted to the design of stilling basins that will operate properly for all upstream and downstream conditions and yet be relatively short to reduce construction cost. Research by the U.S. Bureau of Reclamation (13) has resulted in sets of standard designs that can be used. These designs include sills, baffle piers, and chute blocks, as shown in Fig. 15.28.

FIGURE 15.28

Spillway with stilling basin Type III as recommended by the USBR (13).

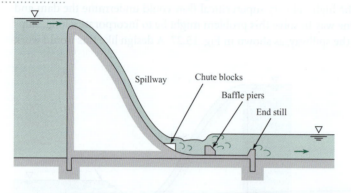

Naturally Occurring Hydraulic Jumps

Hydraulic jumps can occur naturally in creeks and rivers, providing spectacular standing waves, called rollers. Kayakers and white-water rafters must exercise considerable skill when navigating hydraulic jumps because the significant energy loss that occurs over a short distance can be dangerous, potentially engulfing the boat in turbulence. A special case of hydraulic jump, referred to as a submerged hydraulic jump, can be deadly to white-water enthusiasts because it is not easy to see. A **submerged hydraulic jump** occurs when the downstream depth predicted by conservation of momentum is exceeded by the tailwater elevation, and the jump cannot move upstream in response to this disequilibrium because of a buried obstacle [see Valle and Pasternak (14)]. Thus, the visual markers of a hydraulic jump, particularly the rolling waves depicted in Figs. 15.23 and 15.24, are hidden.

A **surge**, or **tidal bore**, is a moving hydraulic jump that may occur for a high tide entering a bay or river mouth. Tides are generally low enough that the waves they produce are smooth and nondestructive. However, in some parts of the world the tides are so high that their entry into shallow bays or mouths of rivers causes a surge to form. Surges may be very hazardous to small boats. The same analytical methods used for the jump can be used to solve for the speed of the surge.

15.7 Gradually Varied Flow

For gradually varied flow, channel resistance is a significant factor in the flow process. Therefore, the energy equation is invoked by comparing S_0 and S_f.

Basic Differential Equation for Gradually Varied Flow

There are a number of cases of open-channel flow in which the change in water-surface profile is so gradual that it is possible to integrate the relevant differential equation from one section to another to obtain the desired change in depth. This may be either an analytical integration or, more commonly, a numerical integration. In Section 15.2, the energy equation was written between two sections of a channel Δx distance apart. Because the only head loss here is the channel resistance, the h_L is given by Δh_f, and Eq. (15.7) becomes

$$y_1 + \frac{V_1^2}{2g} + S_0 \Delta x = y_2 + \frac{V_2^2}{2g} + \Delta h_f \tag{15.44}$$

The friction slope S_f is defined as the slope of the EGL, or $\Delta h_f / \Delta x$. Thus $\Delta h_f = S_f \Delta x$, and defining $\Delta y = y_2 - y_1$, then

$$\frac{V_2^2}{2g} - \frac{V_1^2}{2g} = \frac{d}{dx}\left(\frac{V^2}{2g}\right)\Delta x \tag{15.45}$$

Therefore, Eq. (15.44) becomes

$$\Delta y = S_0 \Delta x - S_f \Delta x - \frac{d}{dx}\left(\frac{V^2}{2g}\right)\Delta x$$

Dividing through by Δx and taking the limit as Δx approaches zero gives us

$$\frac{dy}{dx} + \frac{d}{dx}\left(\frac{V^2}{2g}\right) = S_0 - S_f \tag{15.46}$$

The second term is rewritten as $[d(V^2/2g)/dy]\, dy/dx$, so that Eq. (15.46) simplifies to

$$\frac{dy}{dx} = \frac{S_0 - S_f}{1 + d(V^2/2g)/dy} \tag{15.47}$$

To put Eq. (15.47) in a more usable form, the denominator is expressed in terms of the Froude number. This is accomplished by observing that

$$\frac{d}{dy}\left(\frac{V^2}{2g}\right) = \frac{d}{dy}\left(\frac{Q^2}{2gA^2}\right) \tag{15.48}$$

After differentiating the right side of Eq. (15.48), the equation becomes

$$\frac{d}{dy}\left(\frac{V^2}{2g}\right) = \frac{-2Q^2}{2gA^3}\cdot\frac{dA}{dy}$$

But $dA/dy = T$ (top width), and $A/T = D$ (hydraulic depth); therefore,

$$\frac{d}{dy}\left(\frac{V^2}{2g}\right) = \frac{-Q^2}{gA^2 D}$$

or

$$\frac{d}{dy}\left(\frac{V^2}{2g}\right) = -\text{Fr}^2$$

Hence, when the expression for $d(V^2/2g)/dy$ is substituted into Eq. (15.47), the result is

$$\frac{dy}{dx} = \frac{S_0 - S_f}{1 - \text{Fr}^2} \tag{15.49}$$

This is the general differential equation for gradually varied flow. It is used to describe the various types of water-surface profiles that occur in open channels. Note that, in the derivation of the equation, S_0 and S_f were taken as positive when the channel and energy grade lines, respectively, were sloping downward in the direction of flow. Also note that y is measured from the bottom of the channel. Therefore, $dy/dx = 0$ if the slope of the water surface is equal to the slope of the channel bottom, and dy/dx is positive if the slope of the water surface is less than the channel slope.

Introduction to Water-Surface Profiles

In the design of projects involving the flow in channels (rivers or irrigation canals, for example), the engineer must often estimate the **water-surface profile** (elevation of the water surface along the channel) for a given discharge. For example, when a dam is being designed for a river project, the water-surface profile in the river upstream must be defined so that the project planners will know how much land to acquire to accommodate the upstream pool. The first step in defining a water-surface profile is to locate a point or points along the channel where the depth can be computed for a given discharge. For example, at a change in slope from mild to steep, critical depth will occur just upstream of the break in grade (see Fig. 15.32). At that point one can solve for y_c with Eq. (15.25) or (15.27). Also, for flow over the spillway of a dam, there will be a discharge equation for the spillway from which one can calculate the water-surface elevation in the reservoir at the face of the dam. Such points where there is a unique relationship between discharge and water-surface elevation are called **controls**. Once the water-surface elevations at these controls are determined, then the water-surface profile can be extended upstream or downstream from the control points to define the water-surface profile for the entire channel. The completion of the profile is done by numerical integration. However, before this integration is performed, it is usually helpful for the engineer to sketch in the profiles. To assist in the process of sketching the possible profiles, the engineer can refer to different categories of profiles (water-surface profiles have unique characteristics depending on the relationship between normal depth, critical depth, and the actual depth of flow in the channel). This initial sketching of the profiles helps the engineer to scope the problem and to obtain a solution, or solutions, in a minimum amount of time. The next section describes the various types of water-surface profiles.

Types of Water-Surface Profiles

There are 12 different types of water-surface profiles for gradually varied flow in channels, and these are shown schematically in Fig. 15.29. Each profile is identified by a letter and number designator. For example, the first water-surface profile in Fig. 15.29 is identified as an M1 profile. The letter indicates the type of slope of the channel—that is, whether the slope is mild (M), critical (C), steep (S), horizontal (H), or adverse (A). The slope is defined as mild if the uniform flow depth, y_n, is greater than the critical flow depth, y_c. Conversely, if y_n is less than y_c, the channel would be termed steep. Or if $y_n = y_c$, this would be a channel with critical slope. The designation M, S, or C is determined by computing y_n and y_c for the given channel for a given discharge. Equations (15.11) through (15.15) are used to compute y_n, and Eq. (15.27) is used to compute y_c. Figure 15.30 shows the relationship between y_n and y_c for the H, M, S, C, and A designations. As the name implies, a horizontal slope is one where the channel actually has a zero slope, and an adverse slope is one where the slope of the channel is upward in the direction of flow. Normal depth does not exist for these two cases (for example, water cannot flow at uniform depth in either a horizontal channel or one with adverse slope); therefore, they are given the special designations H and A, respectively.

FIGURE 15.29

Classification of water-surface profiles of gradually varied flow.

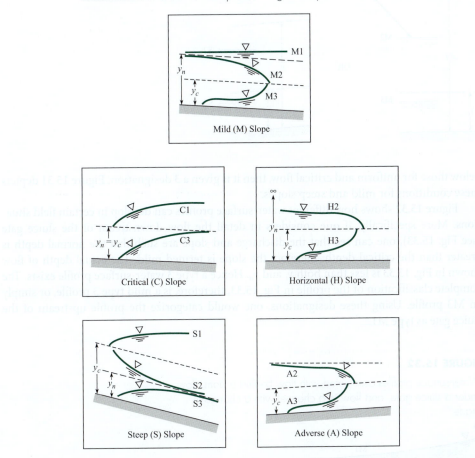

The number designator for the type of profile relates to the position of the *actual* water surface in relation to the position of the water surface for uniform and critical flow in the channel. If the actual water surface is above that for uniform and critical flow ($y > y_n$; $y > y_c$), then that condition is given a 1 designation; if the actual water surface is between those for uniform and critical flow, then it is given a 2 designation; and if the actual water surface lies

FIGURE 15.30

Letter designators as a function of the relationship between y_n and y_c.

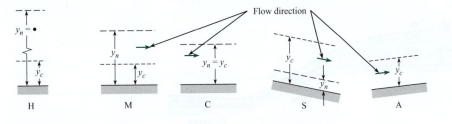

FIGURE 15.31

Number designator as a function of the location of the actual water surface in relation to y_n and y_c.

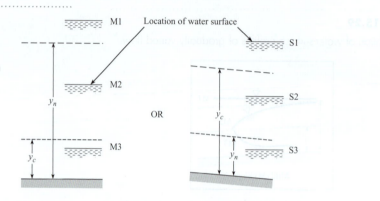

below those for uniform and critical flow, then it is given a 3 designation. Figure 15.31 depicts these conditions for mild and steep slopes.

Figure 15.32 shows how different water-surface profiles can develop in certain field situations. More specifically, if one considers in detail the flow downstream of the sluice gate (see Fig. 15.33), one can see that the discharge and slope are such that the normal depth is greater than the critical depth; therefore the slope is termed mild. The actual depth of flow shown in Fig. 15.33 is less than both y_c and y_n. Hence a type 3 water-surface profile exists. The complete classification of the profile in Fig. 15.33, therefore, is a mild type 3 profile, or simply an M3 profile. Using these designations, one would categorize the profile upstream of the sluice gate as type M1.

FIGURE 15.32

Water-surface profiles associated with flow behind a dam, flow under a sluice gate, and flow in a channel with a change in grade.

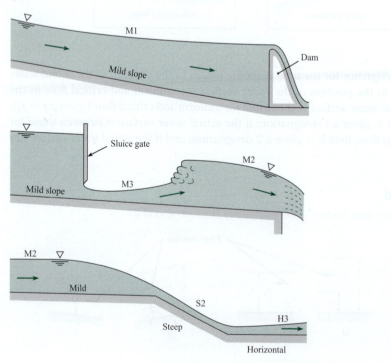

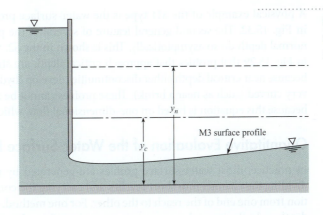

FIGURE 15.33

Water-surface profile, M3 type.

M3 surface profile

EXAMPLE 15.12

Classification of Water-Surface Profiles

Problem Statement

Classify the water-surface profile for the flow downstream of the sluice gate in Fig. 15.9 if the slope is horizontal, and that for the flow immediately downstream of the break in grade in Fig. 15.15.

Define the Situation

Nonuniform flow is occurring in a channel.

State the Goal

Find the water-surface profile classification for the two different flow situations.

Generate Ideas and Make a Plan

1. Select a number designator based on the location of the actual water surface relative to y_n and y_c (see Fig. 15.31).

2. Select a letter designator to describe the steepness of the slopes, which can also be characterized by the relative size of y_n and y_c (see Fig. 15.30).

Take Action (Execute the Plan)

For Fig. 15.9

1. The actual depth is less than critical; thus the profile is type 3.

2. The channel is horizontal; hence the profile is designated type H3.

For Fig. 15.15

1. The actual depth is greater than normal but less than critical, so the profile is type 2.

2. The uniform-flow depth (normal depth y_n) is less than the critical depth; hence the slope is steep. Therefore the water-surface profile is designated type S2.

With the previous introduction to the classification of water-surface profiles, one can refer to Eq. (15.49) to describe the shapes of the profiles. Again, for example, if one considers the M3 profile, it is known that Fr > 1 because the flow is supercritical ($y < y_c$), and that $S_f > S_0$ because the velocity is greater than normal velocity. Hence a head loss greater than that for normal flow must exist. Inserting these relative values into Eq. (15.49) reveals that both the numerator and the denominator are negative. Thus dy/dx must be positive (the depth increases in the direction of flow), and as critical depth is approached, the Froude number approaches unity. Hence the denominator of Eq. (15.49) approaches zero. Therefore, as the depth approaches critical depth, $dy/dx \rightarrow \infty$. What actually occurs in cases where the critical depth is approached in supercritical flow is that a hydraulic jump forms and a discontinuity in profile is thereby produced.

Certain general features of profiles, as shown in Fig. 15.29, are evident. First, as the depth becomes very great, the velocity of flow approaches zero. Hence Fr $\rightarrow 0$ and $S_f \rightarrow 0$ and dy/dx approaches S_0 because $dy/dx = (S_0 - S_f)(1 - \text{Fr}^2)$. In other words, the depth increases at the same rate at which the channel bottom drops away from the horizontal. Thus the water surface approaches the horizontal. The profiles that show this tendency are types M1, S1, and C1.

A physical example of the M1 type is the water-surface profile upstream of a dam, as shown in Fig. 15.32. The second general feature of several of the profiles is that those that approach normal depth do so asymptotically. This is shown in the S2, S3, M1, and M2 profiles. Also note in Fig. 15.29 that profiles that approach critical depth are shown by dashed lines. This is done because near critical depth either discontinuities develop (hydraulic jump), or the streamlines are very curved (such as near a brink). These profiles cannot be accurately predicted by Eq. (15.49) because this equation is based on one-dimensional flow, which, in these regions, is invalid.

Quantitative Evaluation of the Water-Surface Profile

In practice, most water-surface profiles are generated by numerical integration, that is, by dividing the channel into short reaches and carrying the computation for water-surface elevation from one end of the reach to the other. For one method, called the **direct step method**, the depth and velocity are known at a given section of the channel (one end of the reach), and one arbitrarily chooses the depth at the other end of the reach. Then the length of the reach is solved for. The applicable equation for quantitative evaluation of the water-surface profile is the energy equation written for a finite reach of channel, Δx:

$$y_1 + \frac{V_1^2}{2g} + S_0\,\Delta x = y_2 + \frac{V_2^2}{2g} + S_f\Delta x$$

or

$$\Delta x(S_f - S_0) = \left(y_1 + \frac{V_1^2}{2g}\right) - \left(y_2 + \frac{V_2^2}{2g}\right)$$

or

$$\Delta x = \frac{(y_1 + V_1^2/2g) - (y_2 + V_2^2/2g)}{S_f - S_0} = \frac{(y_1 - y_2) + (V_1^2 - V_2^2)/2g}{S_f - S_0} \tag{15.50}$$

The procedure for evaluation of a profile starts by ascertaining which type applies to the given reach of channel (using the methods of the preceding subsection). Then, starting from a known depth, one computes a finite value of Δx for an arbitrarily chosen change in depth. The process of computing Δx, step by step, up (negative Δx) or down (positive Δx) the channel is repeated until the full reach of channel has been covered. Usually small changes of y are taken, so that the friction slope is approximated by the following equation:

$$S_f = \frac{h_f}{\Delta x} = \frac{fV^2}{8gR_h} \tag{15.51}$$

Here V is the mean velocity in the reach, and R_h is the mean hydraulic radius. That is, $V = (V_1 + V_2)/2$, and $R_h = (R_{h1} + R_{h2})/2$. It is obvious that a numerical approach of this type is ideally suited for solution by computer.

EXAMPLE 15.13

Classification and Numerical Analysis of a Water-Surface Profile

---

Problem Statement

Water discharges from under a sluice gate into a horizontal rectangular channel at a rate of 1 m³/s per meter of width, as

shown in the following sketch. What is the classification of the water-surface profile? Quantitatively evaluate the profile downstream of the gate and determine whether it will extend all the way to the abrupt drop 80 m downstream. Make the simplifying assumptions that the resistance factor f is equal to 0.02 and that the hydraulic radius R_h is equal to the depth y.

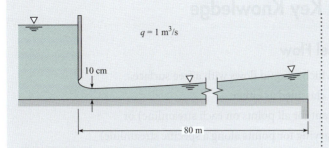

$q = 1 \text{ m}^3/\text{s}$

10 cm

80 m

Define the Situation

Water discharges underneath a sluice gate.

Assumptions:

1. Resistance factor f is equal to 0.02.
2. Hydraulic radius R_h is equal to the depth y.

State the Goal

- Classify of the downstream profile.
- Determine if increasing slope will prevail all the way to a point of interest 80 m downstream.

Generate Ideas and Make a Plan

1. Determine the letter designation of channel using Fig. 15.30.
2. For flow leaving sluice gate, determine critical depth y_c, and compare to actual depth of flow. Use this information to refine the classification.
3. Solve for depth versus distance using Eqs. (15.50) and (15.51).

Take Action (Execute the Plan)

1. Channel is horizontal, so letter designation is H.
2. Determine critical depth y_c using Eq. (15.27).

$$y_c = (q^2/g)^{1/3} = [(1^2 \text{ m}^4/\text{s}^2)/(9.81 \text{ m/s}^2)]^{1/3}$$

$$= \boxed{0.467 \text{ m}}$$

Thus, the depth of flow from sluice gate is less than the critical depth. Therefore the water-surface profile is classified as

$$\boxed{\text{type H3.}}$$

3. To determine depth versus distance along the channel, apply Eqs. (15.50) and (15.51) using the numerical approach given in Table 15.2. Then, plot the results as shown. From the plot, conclude that the

$$\boxed{\text{profile extends to the abrupt drop.}}$$

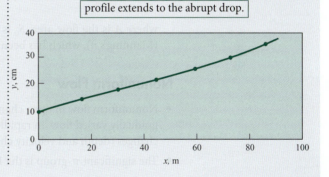

TABLE 15.2 Solution To Example 15.13

Section Number Downstream of Gate	Depth y, m	Velocity at Section V, m/s	Mean Velocity in Reach, $(V_1 + V_2)/2$	V^2	Mean Hydraulic Radius, $R_m = (y_1 + y_2)/2$	$S_f = \dfrac{fV^2_{\text{mean}}}{8gR_m}$	$\Delta x = \dfrac{(y_1 - y_2) + \dfrac{(V_1^2 - V_2^2)}{2g}}{(S_f - S_0)}$	Distance from Gate x, m
1 (at gate)	0.1	10	...	100	...	...	...	0
	...	...	8.57	73.4	0.12	0.156	15.7	
2	0.14	7.14	...	51.0	...	...	...	15.7
	...	...	6.35	40.3	0.16	0.064	15.3	
3	0.18	5.56	...	30.9	...	...	...	31.0
	...	...	5.05	25.5	0.20	0.032	15.1	
4	0.22	4.54	...	20.6	...	...	...	46.1
	...	...	4.19	17.6	0.24	0.019	13.4	
5	0.26	3.85	...	14.8	...	...	...	59.5
	...	...	3.59	12.9	0.28	0.012	12.4	
6	0.30	3.33	...	11.1	...	...	...	71.9
	...	...	3.13	9.8	0.32	0.008	10.9	
7	0.34	2.94	...	8.6	...	...	...	82.8

15.8 Summarizing Key Knowledge

Describing Open Channel Flow

- An open channel is one in which a liquid flows with a free surface.
- Steady open-channel flow is classified as either
  - ▸ *Uniform* (velocity is constant for all points on each streamline) or
  - ▸ *Nonuniform* (velocity is varying for points along a specific streamline)

Steady and Uniform Flow

- The head loss corresponds to the potential energy change associated with the slope of the channel.
- The discharge is given by the Manning equation:

$$Q = \frac{1}{n}AR_h^{2/3}S_0^{1/2}$$

where A is the flow area, S_0 is the slope of the channel, and n is the resistance coefficient (Manning's n), which has been tabulated for different surfaces.

Nonuniform Flow

- Nonuniform flow in open channels is characterized as either rapidly varied flow or gradually varied flow. In rapidly varied flow the channel resistance is negligible, and flow changes (depth and velocity changes) occur over relatively short distances.
- The significant π-group is the Froude number

$$\mathrm{Fr} = \frac{V}{\sqrt{gD_c}}$$

where D_c is the hydraulic depth, A/T. When the Froude number is equal to unity, the flow is critical.
- Subcritical flow occurs when the Froude number is less than unity, and supercritical when the Froude number is greater than unity.

Hydraulic Jump

- A hydraulic jump usually occurs when the flow along the channel changes from supercritical to subcritical.
- The governing equation for hydraulic jump in a horizontal, rectangular channel is

$$y_2 = \frac{y_1}{2}(\sqrt{1 + 8\,\mathrm{Fr}_1^2} - 1)$$

- The corresponding head loss in the hydraulic jump is

$$h_L = \frac{(y_2 - y_1)^3}{4y_1 y_2}$$

- When the flow along the channel changes from subcritical to supercritical flow, the head loss is assumed to be negligible, and the depth and velocity relationship is governed by the change in elevation of the channel bottom and the specific energy, $y + V^2/2g$. Typical cases of this type of flow are

1. Flow under a sluice gate
2. An upstep in the channel bottom
3. Reduction in width of the channel

Gradually Varied Flow

• For gradually varied flow the governing differential equation is

$$\frac{dy}{dx} = \frac{S_0 - S_f}{1 - \text{Fr}^2}$$

When this equation is integrated along the length of the channel, the depth y is determined as a function of distance x along the channel. This yields the water surface profile for the reach of the channel.

REFERENCES

1. Limerinos, J. T. "Determination of the Manning Coefficient from Measured Bed Roughness in Natural Channels." Water Supply Paper 1898-B, U.S. Geological Survey, Washington, D.C., 1970.

2. Committee on Hydromechanics of the Hydraulics Division of American Society of Civil Engineers. "Friction Factors in Open Channels." *J. Hydraulics Div., Am. Soc. Civil Eng.* (March 1963).

3. Wolman, M. G. "The Natural Channel of Brandywine Creek, Pennsylvania." Prof. Paper 271, U.S. Geological Survey, Washington D.C., 1954.

4. Yen, B. C. (ed.) *Channel Flow Resistance: Centennial of Manning's Formula.* Littleton, CO: Water Resources Publications, 1992.

5. Chow, Ven Te. *Open Channel Hydraulics.* New York: McGraw-Hill, 1959.

6. Henderson, F. M. *Open Channel Flow.* New York: Macmillan, 1966.

7. American Concrete Pipe Assoc. *Concrete Pipe Design Manual.* Vienna, VA: American Concrete Pipe Assoc., 1980.

8. Parshall, R. L. "The Improved Venturi Flume." *Trans. ASCE,* 89 (1926), 841–851.

9. Roberson, J. A., J. J. Cassidy, and M. H. Chaudhry. *Hydraulic Engineering.* New York: John Wiley, 1988.

10. Hinds, J. "The Hydraulic Design of Flume and Siphon Transitions." *Trans. ASCE,* 92 (1928), pp. 1423–1459.

11. U.S. Bureau of Reclamation. *Design of Small Canal Structures.* U.S. Dept. of Interior, Washington, DC: U.S. Govt. Printing Office, 1978.

12. Rouse, H. (ed.). *Engineering Hydraulics.* New York: John Wiley, 1950.

13. U.S. Bureau of Reclamation. *Hydraulic Design of Stilling Basin and Bucket Energy Dissipators.* Engr. Monograph no. 25, U.S. Supt. of Doc., 1958.

14. Valle, B. L., and G. B. Pasternak. "Submerged and Unsubmerged Natural Hydraulic Jumps in a Bedrock Step-Pool Mountain Channel." *Geomorphology,* 82 (2006), pp. 146-159.

15. Raju, K. G. R. *Flow Through Open Channels.* New Delhi: Tata McGraw-Hill, 1981.

PROBLEMS

PLUS Problem available in *WileyPLUS* at instructor's discretion.

GO Guided Online (GO) Problem, available in *WileyPLUS* at instructor's discretion.

Describing Open-Channel Flow (§15.1)

15.1 Why is the Reynolds number for onset of turbulence given by Re > 2000 in fully flowing pipes and Re > 500 in partly flowing pipes and other open channels?

15.2 A rectangular open channel has a base of length $2b$, and the water is flowing with a depth of b.

  a. Sketch this channel.
  b. What is the hydraulic radius of this channel?

15.3 PLUS Two channels have the same cross-sectional area, but different geometry, as shown.

   a. Which channel has the largest wetted perimeter?

   b. Which channel has more contact between water and channel wall?

   c. Which channel will have more energy loss to friction?

Uniform Open-Channel Flow (§15.3)

15.4 PLUS Consider uniform flow of water in the two channels shown. They both have the same slope, the same wall roughness, and the same cross-sectional area. Then one can conclude that (a) $Q_A = Q_B$, (b) $Q_A < Q_B$, or (c) $Q_A > Q_B$.

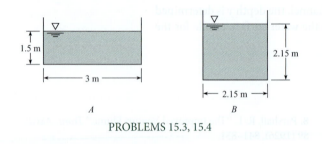

PROBLEMS 15.3, 15.4

15.5 PLUS This wood flume has a slope of 0.0019. What will be the discharge of water in it for a depth of 1 m? The wood is planed.

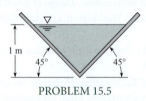

PROBLEM 15.5

15.6 Estimate the discharge in a rock-bedded stream ($d_{84} = 30$ cm) that has an average depth of 1.8 m, a slope of 0.0037, and a width of 52 m. Assume $k_s = d_{84}$.

15.7 Estimate the discharge of water ($T = 10°C$) that flows 1.5 m deep in a long rectangular concrete channel that is 3 m wide and is on a slope of 0.001.

15.8 A rectangular concrete channel is 4 m wide and has uniform water flow. If the channel drops 1.5 m in a length of 2440 m, what is the discharge? Assume $T = 15°C$. The depth of flow is 1 m.

15.9 Consider channels of rectangular cross section carrying 3 m³/s of water flow. The channels have a slope of 0.001. Determine the cross-sectional areas required for widths of 0.6 m, 1.2 m, 1.8 m, 2.4 m, and 4.5 m. Plot A versus y/b, and see how the results compare with the accepted result for the best hydraulic section.

15.10 A concrete sewer pipe 1 m in diameter is laid so it has a drop in elevation of 30 cm per 300 m of length. If sewage (assume the properties are the same as those of water) flows at a depth of 50 cm in the pipe, what will be the discharge?

15.11 Determine the discharge in a 1.5 m-diameter concrete sewer pipe on a slope of 0.001 that is carrying water at a depth of 1.2 m.

15.12 Water flows at a depth of 2 m in the trapezoidal, concrete-lined channel shown. If the channel slope is 30 cm in 600 m, what is the average velocity, and what is the discharge?

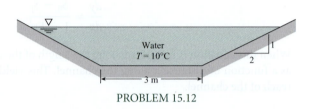

PROBLEM 15.12

15.13 What will be the depth of flow in a trapezoidal concrete-lined channel that has a water discharge of 30 m³/s? The channel has a slope of 30 cm in 150 m. The bottom width of the channel is 3 m, and the side slopes are 1 vertical to 1 horizontal.

15.14 PLUS What discharge of water will occur in a trapezoidal channel that has a bottom width of 6 m and side slopes of 1 vertical to 1 horizontal if the slope of the channel is 0.8 m/km and the depth is 1.5 m? The channel is lined with troweled concrete.

15.15 A rectangular concrete channel 4 m wide on a slope of 0.004 is designed to carry a water ($T = 10°C$) discharge of 25 m³/s. Estimate the uniform flow depth for these conditions. The channel has a rectangular cross section.

15.16 A rectangular troweled concrete channel 2.4 m wide with a slope of 3 m in 1000 m is designed for a discharge of 11.3 m³/s For a water temperature of 4°C, estimate the depth of flow.

15.17 A concrete-lined trapezoidal channel having a bottom width of 3 m and side slopes of 1 vertical to 2 horizontal is designed to carry a flow of 85 m³/s. If the slope of the channel is 0.001, what will be the depth of flow in the channel?

15.18 GO Design a canal having a trapezoidal cross section to carry a design discharge of irrigation water of 25 m³/s. The slope of the canal is to be 0.002. The canal is to be lined with concrete, and it is to have the best hydraulic section for the design flow.

Nonuniform Open-Channel Flow (§15.5)

15.19 How are head loss and slope related for nonuniform flow, as compared to uniform flow?

15.20 Is critical flow a desirable or undesirable flow condition? Why?

15.21 PLUS Critical flow _____. (Select all of the following that are correct.)

    a. occurs when specific energy is a minimum for a given discharge.

    b. occurs when the discharge is maximum for a given specific energy.

    c. occurs when Fr < 1.

    d. occurs when Fr = 1.

15.22 PLUS Water flows at a depth of 20 cm with a velocity of 10 m/s in a rectangular channel. (a) Is the flow subcritical or supercritical? (b) What is the alternate depth?

15.23 PLUS The water discharge in a rectangular channel 5 m wide is 25 m^3/s. If the depth of water is 1 m, is the flow subcritical or supercritical?

15.24 PLUS The discharge in a rectangular channel 6 m wide is 12 m^3/s. If the water velocity is 3 m/s, is the flow subcritical or supercritical?

15.25 GO Water flows at a rate of 8 m^3/s in a rectangular channel 2 m wide. Determine the Froude number and the type of flow (subcritical, critical, or supercritical) for depths of 30 cm, 1.0 m, and 2.0 m. What is the critical depth?

15.26 For a rectangular channel 3 m wide and discharge of 12 m^3, what is the alternate depth to the 30 cm depth? What is the specific energy for these conditions?

15.27 Water flows at the critical depth with a velocity of 10 m/s. What is the depth of flow?

15.28 PLUS Water flows uniformly at a rate of 9 m^3/s in a rectangular channel that is 4 m wide and has a bottom slope of 0.005. If n is 0.014, is the flow subcritical or supercritical?

15.29 GO The discharge in a trapezoidal channel is 10 m^3/s. The bottom width of the channel is 3.0 m, and the side slopes are 1 vertical to 1 horizontal. If the depth of flow is 1.0 m, is the flow supercritical or subcritical?

15.30 For the channel of Prob. 15.29, determine the critical depth for a discharge of 20 m^3/s.

15.31 A rectangular channel is 6 m wide, and the discharge of water in it is 18 m^3/s. Plot depth versus specific energy for these conditions. Let specific energy range from E_{min} to $E = 7$ m. What are the alternate and sequent depths to the 30-cm depth?

15.32 GO A long rectangular channel that is 8 m wide and has a mild slope ends in a free outfall. If the water depth at the brink is 0.55 m, what is the discharge in the channel?

15.33 A rectangular channel that is 4.5 m wide and has a mild slope ends in a free outfall. If the water depth at the brink is 36 cm, what is the discharge in the channel?

15.34 A horizontal rectangular channel 4 m wide carries a discharge of water of 14.2 m^3/s. If the channel ends with a free outfall, what is the depth at the brink?

15.35 What discharge of water will occur over a 1 m-high, broad-crested weir that is 3 m long if the head on the weir is 0.5 m?

15.36 What discharge of water will occur over a 2-m-high, broad-crested weir that is 5 m long if the head on the weir is 60 cm?

15.37 The crest of a high, broad-crested weir has an elevation of 100 m. If the weir is 10 m long and the discharge of water over the weir is 25 m^3/s, what is the water-surface elevation in the reservoir upstream?

15.38 The crest of a high, broad-crested weir has an elevation of 90 m. If the weir is 12 m long and the discharge of water over the weir is 34 m^3/s, what is the water-surface elevation in the reservoir upstream?

15.39 Water flows with a velocity of 3 m/s and at a depth of 3 m in a rectangular channel. What is the change in depth and in water-surface elevation produced by a gradual upward change in bottom elevation (upstep) of 30 cm? What would be the depth and elevation changes if there were a gradual downstep of 30 cm? What is the maximum size of upstep that could exist before upstream depth changes would result?

15.40 Water flows with a velocity of 2 m/s and at a depth of 3 m in a rectangular channel. What is the change in depth and in water-surface elevation produced by a gradual upward change in bottom elevation (upstep) of 60 cm? What would be the depth and elevation changes if there were a gradual downstep of 15 cm? What is the maximum size of upstep that could exist before upstream depth changes would result?

15.41 Assuming no energy loss, what is the maximum value of Δz that will permit the unit flow rate of 6 m^2/s to pass over the hump without increasing the upstream depth? Sketch carefully the water-surface shape from section 1 to section 2. On the sketch give values for Δz, the depth, and the amount of rise or fall in the water surface from section 1 to section 2.

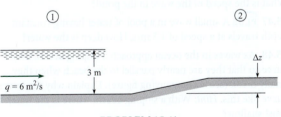

PROBLEM 15.41

15.42 Water flows with a velocity of 3 m/s in a rectangular channel 3 m wide at a depth of 3 m. What is the change in depth and in water-surface elevation produced when a gradual contraction in the channel to a width of 2.6 m takes place? Determine the greatest contraction allowable without altering the specified upstream conditions.

15.43 Because of the increased size of ships, the phenomenon called "ship squat" has produced serious problems in harbors where the draft of vessels approaches the depth of the ship channel. When a ship steams up a channel, the resulting flow situation is analogous to open-channel flow in which a constricting flow section exists (the ship reduces the cross-sectional area of the channel). The problem may be analyzed by referencing the water velocity to the ship and applying the energy equation. Thus, at the section of the channel where the ship is located, the relative water velocity in the channel will be greatest, and the water level in the channel will be reduced as dictated by the energy equation. Consequently, the ship itself will be at a lower elevation than if it were stationary; this lowering is referred to as "ship squat." Estimate the squat of the fully loaded supertanker *Bellamya* when it is steaming at 5 kt (1 kt = 0.515 m/s) in a channel that is 35 m deep and 200 m wide. The draft of the *Bellamya* when fully loaded is 29 m. Its width and length are 63 m and 414 m, respectively.

15.44 A rectangular channel that is 3 m wide is very smooth except for a small reach that is roughened with angle irons attached to the bottom. Water flows in the channel at a rate of 5.7 m³/s and at a depth of 30 cm upstream of the rough section. Assume frictionless flow except over the roughened part, where the total drag of all roughness (all of the angle irons) is assumed to be 8900 N. Determine the depth downstream of the roughness for the assumed conditions.

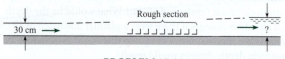

PROBLEM 15.44

15.45 Water flows from a reservoir into a steep rectangular channel that is 4 m wide. The reservoir water surface is 3 m above the channel bottom at the channel entrance. What discharge will occur in the channel?

15.46 A small wave is produced in a pond that is 15 cm deep. What is the speed of the wave in the pond?

15.47 PLUS A small wave in a pool of water having constant depth travels at a speed of 1.5 m/s. How deep is the water?

15.48 As waves in the ocean approach a sloping beach, they curve so that they are nearly parallel to the beach when they finally break (see accompanying figure). Explain why the waves curve like this. *Hint*: With a sloping beach, where is the water most shallow?

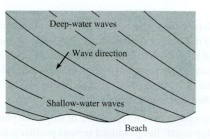

PROBLEM 15.48

Hydraulic Jumps (§15.6)

15.49 PLUS For a hydraulic jump, _____. (Select all of the following that are correct.)

  a. the flow changes from subcritical to supercritical.

  b. the flow changes from supercritical to subcritical.

  c. significant energy is lost.

  d. the height of the water abruptly increases from the upstream to the downstream cross-section.

  e. the downstream and upstream depth are related quantitatively in terms of the upstream Fr.

  f. the energy equation is a better tool for analysis than the momentum equation.

15.50 The baffled ramp shown is used as an energy dissipator in a two-dimensional open channel. For a discharge of 2 m³/s per meter of width, calculate the head lost, the power dissipated, and the horizontal component of force exerted by the ramp on the water.

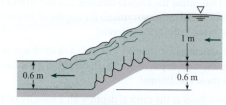

PROBLEM 15.50

15.51 PLUS The spillway shown has a discharge of 2.9 m³/s per meter of width occurring over it. What depth y_2 will exist downstream of the hydraulic jump? Assume negligible energy loss over the spillway.

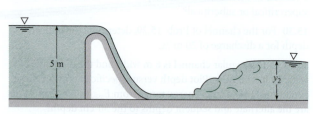

PROBLEM 15.51

15.52 GO The flow of water downstream from a sluice gate in a horizontal channel has a depth of 32 cm and a flow rate of 5.2 m³/s per meter of width. The sluice gate is 2 m wide.

  a. Could a hydraulic jump be caused to form downstream of this section?

  b. If so, what would be the depth downstream of the jump?

15.53 It is known that the discharge per unit width is 6 m³/s per meter and that the height (H) of the hydraulic jump is 4 m. What is the depth y_1?

PROBLEM 15.53

15.54 Water flows in a channel at a depth of 40 cm and with a velocity of 8 m/s. An obstruction causes a hydraulic jump to be formed. What is the depth of flow downstream of the jump?

15.55 Water flows in a trapezoidal channel at a depth of 40 cm and with a velocity of 10 m/s. An obstruction causes a hydraulic jump to be formed. What is the depth of flow downstream of the jump? The bottom width of the channel is 5 m, and the side slopes are 1 vertical to 1 horizontal.

15.56 PLUS A hydraulic jump occurs in a wide rectangular channel. If the depths upstream and downstream are 0.15 m and 3 m, respectively, what is the discharge per meter of width of channel?

15.57 The 6 m-wide rectangular channel shown has three different reaches. $S_{0,1} = 0.01$; $S_{0,2} = 0.0004$; $S_{0,3} = 0.00317$; $Q = 14$ m³/s; $n_1 = 0.015$; normal depth for reach 2 is 1.6 m and that for reach 3 is 0.82 m. Determine the critical depth and normal depth for reach 1 (use Manning's equation from §15.3). Then classify the flow in each reach (supercritical, subcritical, critical), and determine whether a hydraulic jump could occur. In which reach(es) might it occur if it does occur?

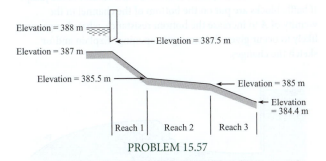

PROBLEM 15.57

15.58 Water flows from under the sluice gate as shown and continues on to a free overfall (also shown). Upstream from the overfall the flow soon reaches a normal depth of 1.1 m. The profile immediately downstream of the sluice gate is as it would be if there were no influence from the part nearer the overfall. Will a hydraulic jump form for these conditions? If so, locate its position. If not, sketch the full profile and label each part. Draw the energy grade line for the system.

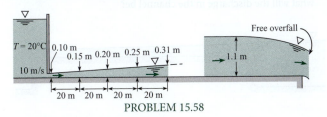

PROBLEM 15.58

15.59 PLUS Water is flowing as shown under the sluice gate in a horizontal rectangular channel that is 1.5 m wide. The depths of y_0 and y_1 are 20 m and 0.3 m, respectively. What will be the power in kilowatts in the hydraulic jump?

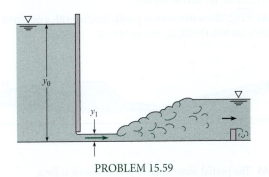

PROBLEM 15.59

15.60 Water flows uniformly at a depth $y_1 = 32$ cm in the concrete channel shown, which is 8 m wide. Estimate the height of the hydraulic jump that will form when a sill is installed to force it to form. Assume Manning's n value is $n = 0.012$.

15.61 For the derivation of Eq. (15.28) on p. 571 of §15.5 it is assumed that the bottom shearing force is negligible. For water flowing uniformly at a depth $y_1 = 40$ cm in the concrete channel shown, which is 10 m wide, a sill is installed to force a hydraulic jump to form. Estimate the magnitude of the shearing force F_s associated with the hydraulic jump and then determine F_s/F_H, where F_H is the net hydrostatic force on the hydraulic jump. Assume Manning's n value is $n = 0.012$.

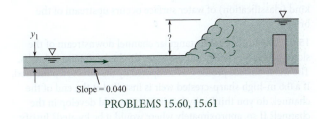

PROBLEMS 15.60, 15.61

15.62 The normal depth in the channel downstream of the sluice gate shown is 1 m. What type of water-surface profile occurs downstream of the sluice gate? Also, estimate the shear stress on the smooth bottom at a distance 0.5 m downstream of the sluice gate.

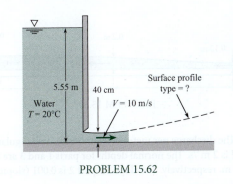

PROBLEM 15.62

15.63 PLUS Water flows at a rate of 3 m³/s in a rectangular channel 3 m wide. The normal depth in that channel is 0.6 m. The actual depth of flow in the channel is 1.2 m. The water-surface profile in the channel for these conditions would be classified as (a) S1, (b) S2, (c) M1, or (d) M2.

15.64 PLUS The water-surface profile labeled with a question mark is (a) M2, (b) S2, (c) H2, or (d) A2.

PROBLEM 15.64

15.65 The partial water-surface profile shown is for a rectangular channel that is 3 m wide and has water flowing in it at a rate of 5 m³/s. Sketch in the missing part of the water-surface profile and identify the type(s).

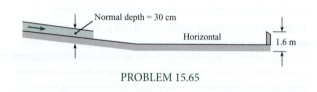

PROBLEM 15.65

15.66 A very long 3 m-wide concrete rectangular channel with a slope of 0.0001 ends with a free overfall. The discharge in the channel is 4 m³/s. One mile upstream the flow is uniform. What kind (classification) of water surface occurs upstream of the brink?

15.67 The horizontal rectangular channel downstream of the sluice gate is 3 m wide, and the water discharge therein is 3 m³/s. The water-surface profile was computed by the direct step method. If a 0.6 m-high sharp-crested weir is installed at the end of the channel, do you think a hydraulic jump would develop in the channel? If so, approximately where would it be located? Justify your answers by appropriate calculations. Label any water-surface profiles that can be classified.

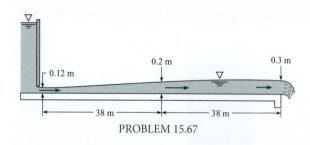

PROBLEM 15.67

15.68 The discharge per meter of width in this rectangular channel is 2 m³/s. The normal depths for parts 1 and 3 are 0.15 m and 0.3 m, respectively. The slope for part 2 is 0.001 (sloping

upward in the direction of flow). Sketch all possible water-surface profiles for flow in this channel, and label each part with its classification.

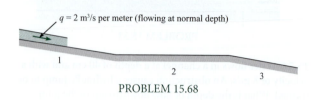

PROBLEM 15.68

15.69 Water flows from under a sluice gate into a horizontal rectangular channel at a rate of 3 m³/s per meter of width. The channel is concrete, and the initial depth is 20 cm. Apply Eq. (15.42) on p. 579 of §15.6 to construct the water-surface profile up to a depth of 60 cm. In your solution, compute reaches for adjacent pairs of depths given in the following sequence: $d = 20$ cm, 30 cm, 40 cm, 50 cm, and 60 cm. Assume that f is constant with a value of 0.02. Plot your results.

15.70 A horizontal rectangular concrete channel terminates in a free outfall. The channel is 4 m wide and carries a discharge of water of 12 m³/s. What is the water depth 300 m upstream from the outfall?

15.71 Consider the hydraulic jump shown for the long horizontal rectangular channel. What kind of water-surface profile (classification) is located upstream of the jump? What kind of water-surface profile is located downstream of the jump? If baffle blocks are put on the bottom of the channel in the vicinity of A to increase the bottom resistance, what changes are likely to occur given the same gate opening? Explain and/or sketch the changes.

PROBLEM 15.71

15.72 The steep rectangular concrete spillway shown is 4 m wide and 500 m long. It conveys water from a reservoir and delivers it to a free outfall. The channel entrance is rounded and smooth (negligible head loss at the entrance). If the water-surface elevation in the reservoir is 2 m above the channel bottom, what will the discharge in the channel be?

PROBLEM 15.72

15.73 The concrete rectangular channel shown is 3.5 m wide and has a bottom slope of 0.001. The channel entrance is rounded and smooth (negligible head loss at the entrance), and the reservoir water surface is 2.5 m above the bed of the channel at the entrance.

  a. Estimate the discharge in it if the channel is 3000 m long.

  b. Tell how you would solve for the discharge in it if the channel were only 100 m long.

PROBLEM 15.73

15.74 A dam 50 m high backs up water in a river valley as shown. During flood flow, the discharge per meter of width, q, is equal to 10 m³/s. Making the simplifying assumptions that $R = y$ and $f = 0.030$, determine the water-surface profile upstream from the dam to a depth of 6 m. In your numerical calculation, let the first increment of depth change be y_c; use increments of

depth change of 10 m until a depth of 10 m is reached; and then use 2 m increments until the desired limit is reached.

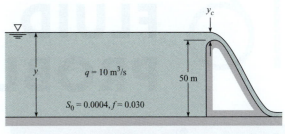

PROBLEM 15.74

15.75 Water flows at a steady rate of 1 m³/s per meter of width in the wide rectangular concrete channel shown. Determine the water-surface profile from section 1 to section 2.

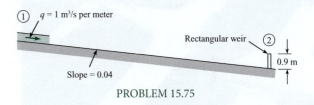

PROBLEM 15.75

16 MODELING OF FLUID DYNAMICS PROBLEMS

FIGURE 16.1

The Eagle X-TS and the workers at the assembly plant where the plane was built. The Eagle X-TS was designed by John Roncz using CFD. Roncz, a world-class designer, is responsible for some portion of 50 aircraft designs. Two of Roncz's designs are on display at the National Air and Space Museum in the United States. (Photo courtesy of John Roncz.)

Roncz describes how he learned fluid mechanics:

> *"The main advantage I have is that I've never taken a single course in aeronautical engineering. . . . As a result, I've had to figure it all out myself. You understand things better that way."* (1).

Chapter Road Map

This chapter describes modeling and introduces two methods that are useful for modeling:

- **Partial Differential Equations (PDEs).** This method involves formulating the governing scientific laws as partial differential equations.
- **Computational Fluid Dynamics (CFD).** This method involves approximating the partial differential equations with algebraic equations and then using a computer algorithm to solve these equations.

Learning Objectives

STUDENTS WILL BE ABLE TO

- Describe how engineers build models. (§16.1)
- Explain how engineers apply PDEs. Explain these topics: velocity field, Taylor series, invariant notation, mathematical operators, the material derivative, and the acceleration field. (§16.2)
- List the steps to derive the continuity equation. (§16.3)
- List and describe the various forms of the continuity equation. (§16.3)
- List the steps to derive the Navier-Stokes equation. (§16.4)
- Describe CFD. Describe how engineers select a CFD code. (§16.5)
- Describe how CFD codes work. Explain these topics: grid, time step, boundary condition, validation, verification, and turbulence models. (§16.5)

Paths cannot be taught, they can only be taken.—Traditional Zen saying

In the first chapter of this book, we invited students to take the first steps of their journey in learning fluid mechanics. In this last chapter, we present new material and suggest a path for moving forward.

16.1 Models in Fluid Mechanics

Engineers create models of systems because this process saves money and results in better designs. Modeling involve analyses, experiments, and computer simulations. These topics are introduced in this section.

The Concept of a Model

In engineering, there is something real, for example, a dam and associated power plant, and there is an idealization (i.e., a model) of this real thing. A model, according to Wang (2) is a *tool to represent a simplified version of reality*. Ford (3) suggests that the model is *a substitute for a real system*. Some examples of models include

- A road map is a model because a map represents a complex array of roads.
- Architects' drawings are models because they represent buildings that will be built.
- A table of contents is a model because the table of contents represents the subject matter of a book.

Some examples of models relevant to fluid mechanics:

- The ideal gas law is a model because it is an idealized (simplified) description of how the variables of density, pressure, and temperature are related.
- A collection of equations can be a model. For example, the energy equation together with the Darcy-Weisbach equation and suitable minor loss coefficients can be used to predict the flow rate for water through a siphon. Using the equations is a substitute for building a system and then correlating experimental data.
- A small-scale car that is used in a wind tunnel to estimate drag acting on a full-scale car is a model.

    To advance the discussion of modeling, we next describe an engineering project.

Example of an Engineering Project. The slow sand filter (Fig. 16.2) is a widely used technology for producing clean drinking water. Water enters the filter at the top, and naturally occurring organisms that live in the topmost layer of the filter remove the biological contaminants. This topmost layer, called the *schmutzdecke*, is found in the top few millimeters of the sand layer. The sand and gravel below the schmutzdecke collects dirt and clay particles.

    Several years ago students from the University of Idaho designed a slow sand filter for applications in Kenya. Because slow sand filters do not require chemicals or electricity, this technology is especially suitable to applications in the developing world.

    The team choose to develop various models of the slow sand filter. The model-building process is described in the next subsections.

Summary. A **model** is an idealization or simplified version of reality. Models are valuable when they help engineers and other professionals reach goals in an economical way.

FIGURE 16.2

The slow sand filter.

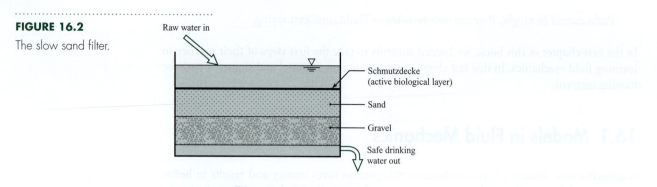

How to Build a Model of a System

The reason for building a model is to solve a problem (Fig. 16.3). The process of model building, according to Montgomery et al. (4), involves identifying relevant variables, determining the relationships between these variables, and then testing the model to ensure that it is accurate (i.e., does the model faithfully capture what happens in reality?). As shown, the process of model building is iterative.

FIGURE 16.3

The model in the context of engineering problem solving.

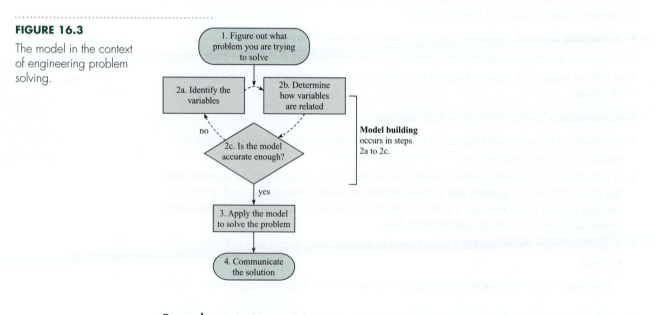

Example. To build a model of a slow sand filter (Fig. 16.2), the modeling process involves the following steps.

• **Step 2a. Identify the variables** Determine which variables characterize performance. Then, classify the variables into two groups.

  ▸ **Performance variables** characterize how well the product performs. Examples of these variables include the flow rate through the filter, the clarity of the water that leaves the filter, and the time period between maintenance for filter cleaning. Performance variables are dependent variables, meaning that they depend on the values of the design variables.

  ▸ **Design variables** are the factors that engineers can select. Examples of these variables are depth of water on top of the filter, the thickness of the sand layer, and the distribution of sand and gravel sizes.

- **Step 2b. Determine how the variables are related**. The purpose of this step is to identify cause and effect. For example, if one changes the size of the sand particles, does this make the filter perform better or worse? Why? There are two approaches for determining how the variables are related (4).

  ▸ **Mechanistic**. Mechanistic models are based on scientific knowledge of the phenomena. For example, Darcy's Law describes flow of fluids through a porous medium such as sand and gravel, and the equation itself tells us the relationship among the variables.

  ▸ **Empirical**. Empirical models involve relating the variables by using curve fits of experimental data. For example, experiments and correlation could be used to determine the time it takes for the schmutzdecke to develop.

- **Step 2c. Test the model for accuracy**. The result of step 2a is an ability to predict the relationship between design variables (e.g., dimensions, particle sizes) and performance variables (e.g., water quality or flow rate). The purpose of step 2c is to check to see how accurate the predictions are. Much of this time, this step is done by comparing experimental data with predictions.

- **Iterate back to step 2a**. In practice, modeling building is iterative. Iteration involves repeating a process with the aim of reaching a desired goal. Each repetition of the process is called an iteration, and the results of one iteration are used as the starting point for the next iteration. Iterations are ended when the model has enough accuracy for the purposes of the engineers.

Example of Iteration (Slow Sand Filter). To build a model of a slow sand filter, one might start out with a model comprised of a few equations and a simple, bench-top experiment. The model would be highly simplified, and the purpose of the first iteration would be to gain experience with modeling and measuring the flow of water through sand. In subsequent iterations, the analytical and experimental models would be developed and continually compared. After analytical models had been developed, the team might create a CFD model to perform parametric studies on the design.

After the model has been validated through the iterative process, the next steps are to apply the model to solve the problem (step 3 of Fig. 16.3) and to communicate the solution (step 4).

Summary. Models are built in an iterative process that involves identifying the variables, classifying these variables into performance variables and design variables, and determining how the variables are related. Last the model is validated to see if model predictions are accurate enough for the needs of the problem. The most important aspect of model building is to start simple and then use sequential iterations to improve accuracy. Model building was introduced in Chapter 1. When models are based on scientific laws and equations, then the Wales-Woods approaches describes how experts build math models.

Three Methods for Model Building

Model building involves three methods.

Analytical Fluid Dynamics (AFD) involves knowledge and equations that are commonly found in engineering textbooks and references.

Experimental Fluid Dynamics (EFD) involves experiments to gather information about variables. EFD is often used to validate calculations, to validate computer solutions, and to determine performance characteristics of systems that are not easily modeled using calculations or computers.

Computational Fluid Dynamics (CFD) involves computer solutions of the governing partial differential equations. That is, engineers run a computer program to understand how the variables interact.

engineers can calculate nearly anything of engineering interest such as drag force, head loss, and power requirements.

Thus, solving the PDEs is the ultimate solution technique. But there is a catch! In practice, the PDEs have nonlinear terms that prevent direct mathematical solutions, except in a limited number of special cases. These special cases were solved many years ago, and today's engineers do not solve problems by directly solving the PDEs. Nevertheless, there are two benefits to learning PDEs.

Understanding Existing Solutions (Benefit #1). The literature has many existing solutions of the PDEs. These solutions classify into two categories:

- **Exact Solutions**. An exact solution involves a physical situation in which the equations of motion reduce to equations that can be solved. There are about 100 such solutions in existence. Examples include Poiseuille flow and Couette flow.

- **Idealized Solutions**. An idealized solution involves a physical situation where assumptions are made that allow the governing equations to be simplified and solved mathematically. Two examples of idealized solution are

  ▸ **Potential Flow**. When an external flow around a body is assumed to be inviscid (i.e., frictionless) and irrotational (i.e., the fluid particles are not rotating), the equations reduce to equations that can be solved analytically. This situation is called potential flow.

  ▸ **Laminar Boundary Layer Flow**. When laminar viscous flow near a wall is simplified by making the boundary layer assumptions, the equations can be solved. The resulting solution, called the Blasius solution, describes flow in the laminar boundary layer.

Engineers use existing solutions to gain understanding of more complex problems. For example, a bicycle rider was severely injured in a collision caused by a bus that passed too close to him. When a large vehicle passes closely by a cyclist, this causes side forces on the cyclist. To gain insight into these side forces, an engineer used the solution for potential flow around an elliptical body to predict the magnitude and direction of the side force.

A second example involves modeling blood flow in the human abdominal aorta. Sometimes the aorta loses its structural integrity and bulges out to form an aneurysm. If an aneurysm ruptures, death is common. Thus, the researchers wanted to understand the forces exerted by the flow on the aneurysm walls. Two existing solutions were used to gain insight into this problem: the Poiseuille solution for steady laminar flow in round tube and Womersley solution for oscillatory laminar flow in a round tube.

Understanding and Validating CFD (Benefit #2). Because CFD codes solve PDEs, the first step in learning CFD is to learn about the PDEs.

Existing solutions are used to validate CFD codes. For example, when a CFD model of blood flow in an aneurysm was developed, the code was validated in part by modeling an existing analytical solution (i.e., the Womersley solution) and then checking to make sure that the CFD solution matched the analytical solution.

Summary. Three reasons for learning PDEs are to (a) be able to understand and apply existing solutions that are found in the literature, (b) to understand the equations that are being solved by CFD codes, and (c) to validate CFD codes by ensuring that the CFD code can correctly predict the results given by a known classical solution.

The remainder of this section introduces mathematics that are useful in the development of PDEs.

Velocity Field: Cartesian Coordinates

FIGURE 16.5

Cartesian coordinates.

The solution of the equations of motion are fields such as the pressure field, the density field, the temperature field, and the velocity field. Thus, understanding fields is important. This section introduces the velocity field.

In the Cartesian coordinate system, a point in space is identified by specifying coordinates $(x, y, z$; Fig. 16.5). The associated unit vectors are $\mathbf{i}$ in x-direction, $\mathbf{j}$ in y-direction, and $\mathbf{k}$ in z-direction. Notice that the coordinate system is *right-handed,* which means that the cross product of $\mathbf{i}$ and $\mathbf{j}$ is the $\mathbf{k}$ unit vector:

$$\mathbf{i} \times \mathbf{j} = \mathbf{k}$$

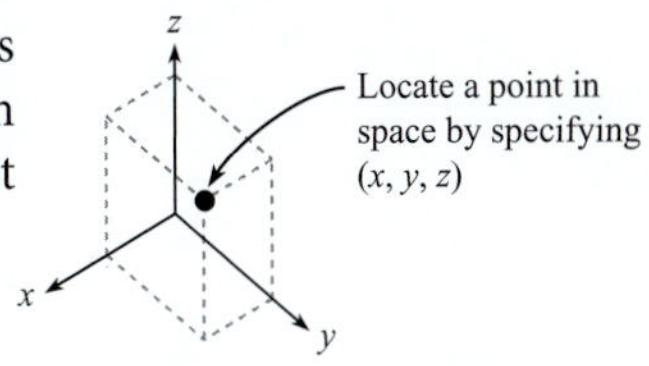

The velocity field is given by

$$\mathbf{V} = u(x, y, z, t)\mathbf{i} + v(x, y, z, t)\mathbf{j} + w(x, y, z, t)\mathbf{k} \qquad (16.1)$$

where $u = u(x, y, z, t)$ is the x-direction component of the velocity vector, and v and w have similar meanings. The independent variables are position (x, y, z) and time (t).

The next two examples show how to reduce the general form of the velocity field so that it applies to a specific situation. Notice the process steps.

EXAMPLE. Consider steady flow in a plane (Fig. 16.6). Reduce the general equation for the velocity field so that is applies to this situation.

Ideas/Action.

1. Write the general equation for the velocity field.

$$\mathbf{V} = u(x, y, z, t)\mathbf{i} + v(x, y, z, t)\mathbf{j} + w(x, y, z, t)\mathbf{k}$$

2. Analyze the dependent variables. Because the flow is planar, $w = 0$. Thus, the dependent variables reduce to u and v.

3. Analyze the independent variables. Because the flow is planar, z is not parameter. Because the flow is steady, time is not a parameter. Thus, the independent variables are (x, y), and the velocity field reduces to $\mathbf{V} = u(x, y)\mathbf{i} + v(x, y)\mathbf{j}$.

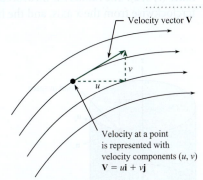

Velocity vector **V**

Velocity at a point is represented with velocity components (u, v)
$\mathbf{V} = u\mathbf{i} + v\mathbf{j}$

FIGURE 16.6

Example of velocity components for planar flow.

EXAMPLE. Consider steady flow entering a channel (Fig. 16.7) formed by plates that extend to $\pm\infty$ in the z-direction. Such plates are called *infinite plates.* Reduce the general equation for the velocity field so that it applies to this situation.

Ideas/Action

1. Write the general equation for the velocity field.

$$\mathbf{V} = u(x, y, z, t)\mathbf{i} + v(x, y, z, t)\mathbf{j} + w(x, y, z, t)\mathbf{k}$$

FIGURE 16.7

Flow between infinite plates.

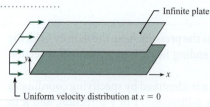
Infinite plate
Uniform velocity distribution at $x = 0$

2. Analyze the dependent variables. Let $w = 0$ because there is no flow in the z-direction.

3. Analyze the independent variables. Because the flow is planar, the velocity does not vary with z. Because the flow is steady, the velocity does not vary this time. Thus, the reduced equation for the velocity field is

$$\mathbf{V} = u(x, y)\mathbf{i} + v(x, y)\mathbf{j}$$

This equation means that both u and v will be nonzero and both u and v will vary with x and y. The reason is that flow in the entrance to the channel is developing (see Chapter 10). Once the flow is fully developed, then the velocity field will reduce to the form.

$$\mathbf{V} = u(y)\mathbf{i}$$

Summary: The general form of the velocity field in Cartesian coordinates is given by Eq. (16.1). To reduce this equation so it applies to a given situation, analyze the independent and the dependent variables and eliminate terms that are not relevant or are zero.

Velocity Field: Cylindrical Coordinates

Because cylindrical coordinates are widely used in application, this system is introduced next.

In cylindrical coordinates (Fig. 16.8), a point in space is described by specifying coordinates (r, θ, z). The radius r is measured from the origin, the azimuth angle θ is measured counterclockwise from the x axis, and the height z is measured from the x-y plane.

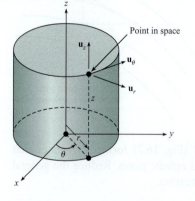
Point in space

The velocity field in general form is

$$\mathbf{V} = v_r(r, \theta, z, t)\mathbf{u}_r + v_\theta(r, \theta, z, t)\mathbf{u}_\theta + v_z(r, \theta, z, t)\mathbf{u}_z \qquad \text{(16.2)}$$

EXAMPLE. Fig. 16.9 shows ideal flow over a circular cylinder. Reduce the general form of the velocity field so that it applies to this situation.

Ideas/Action

Represent the velocity vector at a point of interest, (see Fig. 16.9).

- Step 1. Sketch an x- and y-coordinate axis.
- Step 2. Sketch a radius vector of length r.
- Step 3. Sketch unit vectors u_r and u_θ.
- Step 4. Represent the velocity vector with components v_r and v_θ.

Next, do a term-by-analysis of Eq. (16.2). Eliminate z and v_z because the flow is planar. Eliminate t because the flow is steady. Eq. (16.2) reduces to:

$$\mathbf{V} = v_r(r, \theta)\mathbf{u}_r + v_\theta(r, \theta)$$

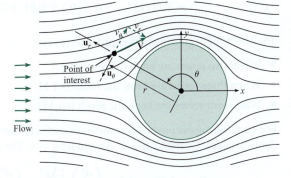

FIGURE 16.9

Using polar coordinates to represent the velocity vector at a point. The pictured flow is ideal flow (i.e., inviscid and irrotational) over a circular cylinder.

For flow in a plane (e.g., Fig. 16.9), the z-direction is not needed, and one uses only the r and θ coordinates. This two-dimensional coordinate system is called **polar coordinates**.

Summary: When cylindrical coordinates are used, the general form of the velocity field is given by Eq. (16.2). For the flow in a plane, the coordinates can be simplified to a 2-D flow, and the coordinates are called polar coordinates.

Taylor Series

Engineers learn Taylor series because Taylor series are used to

- Derive both ordinary and partial differential equations
- Convert differential equations into algebraic equations that can be solved using a computer algorithm by a CFD program

A **Taylor series** is a series expansion of a function about a point. The general formula for the function $f(x)$ expanded around the point $x = a$ is

$$f(x) = f(a) + \left(\frac{df}{dx}\right)_a \frac{(x-a)}{1!} + \left(\frac{d^2f}{dx^2}\right)_a \frac{(x-a)^2}{2!} + \dots \qquad (16.3)$$

For example, when the function is $f(x) = e^x$, Eq. (16.3) becomes:

$$e^x = e^a\left[1 + (x-a) + \frac{(x-a)^2}{2} + \frac{(x-a)^3}{6} + \dots\right] \qquad (16.4)$$

Kojima et al. (5) suggest that a Taylor series is an *imitation* of an equation, just as equations are imitations of the physical world (See Fig. 16.10).

FIGURE 16.10

Data (observations) are
idealized with equations.
Then, equations are
idealized with a Taylor
series.

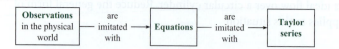

Taylor series are commonly truncated. This means that higher order terms are neglected. For
example, consider the following Taylor series:

$$\underbrace{f(x) = \frac{1}{1-x}}_{\text{Function}} = \overbrace{1 + x + x^2 + x^3 + x^4 + \ldots (\text{H.O.T.})}^{\substack{\text{Taylor series approximation} \\ \text{(The acronym H.O.T. stands for = Higher Order Terms)}}} \tag{16.5}$$

When $x = 0.1$, Eq. (16.5) gives

$$\underbrace{f(x) = \frac{1}{1-0.1} = 1.11111\ldots}_{\text{Function}} = \overbrace{1 + 0.1 + 0.01 + 0.001 + 0.0001\ldots + \ldots (\text{H.O.T.})}^{\text{Taylor series approximation}} \tag{16.6}$$

The effects of neglecting higher order terms are:

• When two terms are kept, the result is 1.1.
• When three terms are kept, the result is 1.11.
• When four terms are kept, the result is 1.111.

For engineering purposes it is sometimes useful to modify Eq. (16.3) by changing the independent variables. Change x to $x + \Delta x$ and let $a = x$. The result is

$$f(x + \Delta x) = f(x) + \left(\frac{df}{dx}\right)_x \frac{(\Delta x)}{1!} + \left(\frac{d^2 f}{dx^2}\right)_x \frac{(\Delta x)^2}{2!} + \ldots (\text{H.O.T.}) \tag{16.7}$$

In fluid mechanics, one uses the Taylor series for a function of several variables. The
general form of Taylor series for a function of two variables $f(x, y)$ expanded about point
(a, b) is

$$f(x, y) = f(a, b) + \left(\frac{\partial f}{\partial x}\right)_{a,b} \frac{(x - a)}{1!} + \left(\frac{\partial f}{\partial y}\right)_{a,b} \frac{(y - b)}{1!}$$
$$+ \left(\frac{\partial^2 f}{\partial x^2}\right)_{a,b} \frac{(x - a)^2}{2!} + \left(\frac{\partial^2 f}{\partial x \partial y}\right)_{a,b} \frac{2(x - a)(y - b)}{2!} + \left(\frac{\partial^2 f}{\partial y^2}\right)_{a,b} \frac{(y - b)^2}{2!} + \ldots \tag{16.8}$$

To modify Eq. (16.8) so it is more useful for fluid mechanics, let $x = x + \Delta x$, $y = y$, $a = x$, and
$b = y$.

$$f(x + \Delta x, y) = f(x, y) + \left(\frac{\partial f}{\partial x}\right)_{x,y} \frac{\Delta x}{1!} + \left(\frac{\partial^2 f}{\partial x^2}\right)_{x,y} \frac{(\Delta x)^2}{2!} + \left(\frac{\partial^3 f}{\partial x^3}\right)_{x,y} \frac{(\Delta x)^3}{3!} + \ldots \tag{16.9}$$

Next, introduce the variables used in fluid mechanics:

$$f(x + \Delta x, y, z, t) = f(x, y, z, t) + \left(\frac{\partial f}{\partial x}\right)_{x,y,z,t} \frac{\Delta x}{1!} + \left(\frac{\partial^2 f}{\partial x^2}\right)_{x,y,z,t} \frac{(\Delta x)^2}{2!} + \ldots \text{H.O.T.} \tag{16.10}$$

Summary. In fluid mechanics, engineers commonly expand functions into a Taylor series,
which is a series expansion of about a point. Often higher order terms (H.O.T.) are neglected.
A useful form of the Taylor series for fluid mechanics is Eq. (16.10).

Mathematical Notation (Invariant Notation and Operators)

In addition to Cartesian, cylindrical, and polar coordinates, engineers use other systems such as spherical coordinates, toroidal coordinates, and generalized curvilinear coordinates. Because there is a large amount of detail, engineers sometimes write equations in ways that apply to any coordinate system. **Invariant notation** is a mathematical notation that applies (i.e., generalizes) to any coordinate system.

To introduce invariant notation, consider the gradient of the pressure field (Table 16.1). As shown, the gradient can be written several ways. Also, the mathematical notation can be classified into two categories.

- **Coordinate specific notation**. Terms in equations are written so that they apply to a specific coordinate system. For example, Table 16.1 shows Cartesian and cylindrical coordinates.

- **Invariant notation**. Terms are written so that they apply to any coordinate system; that is, they generalize.

TABLE 16.1 Alternative Ways to Write the Gradient of the Pressure Field

Category	Description	Mathematical Form
Coordinate-specific notation	Cartesian coordinates	$\frac{\partial p}{\partial x}\mathbf{i} + \frac{\partial p}{\partial y}\mathbf{j} + \frac{\partial p}{\partial z}\mathbf{k}$
	Cylindrical coordinates	$\frac{\partial p}{\partial r}\mathbf{u}_r + \frac{1}{r}\frac{\partial p}{\partial \theta}\mathbf{u}_\theta + \frac{\partial p}{\partial z}\mathbf{u}_z$
Invariant notation	Del notation	∇p
	Gibbs notation	$\mathrm{grad}(p)$
	Indicial notation (Einstein summation convention)	$\frac{\partial p}{\partial x_i}$ or $\partial_i p$

Table 16.1 shows three types of invariant notation.

- *Del notation* is represented by the nabla symbol ∇. Del notation is the most common approach used in engineering.
- *Gibbs notation* uses words to represent operators, for example *grad* to represent the gradient. Gibbs notation is common in mathematics.
- *Indicial notation* is a shorthand approach that is common in both engineering and physics.

Mathematical Operators

In mathematics, collections of terms called **operators** are given names because they appear commonly in the equations of mathematical physics. In the equations of fluid mechanics, some of the common operators are

- *Gradient*: for example, the gradient of the pressure field or the gradient of the velocity field
- *Divergence*: for example, the divergence of the velocity field
- *Curl*: for example, the curl of the velocity field
- *LaPlacian*: for example, the LaPlacian of the velocity field
- *Material derivative*: for example, the time derivative of the temperature field

The Acceleration Field

The acceleration field describes the acceleration of each fluid particle:

$$\begin{pmatrix} \text{acceleration at a} \\ \text{point in a field} \end{pmatrix} = \begin{pmatrix} \text{acceleration of} \\ \text{the fluid particle} \\ \text{at this location} \end{pmatrix} = \begin{pmatrix} \text{material derivative} \\ \text{of the} \\ \text{velocity field} \end{pmatrix} \qquad (16.20)$$

Therefore, introduce the material derivative to describe the acceleration field.

$$\mathbf{a} = \frac{d\mathbf{V}}{dt} \qquad (16.21)$$

To represent Eq. (16.21) in Cartesian coordinates, insert the velocity field from Eq. (16.1).

$$\mathbf{a} = \frac{d}{dt}(u\mathbf{i} + v\mathbf{j} + w\mathbf{k}) = \frac{du}{dt}\mathbf{i} + \frac{dv}{dt}\mathbf{j} + \frac{dw}{dt}\mathbf{k} \qquad (16.22)$$

Because du/dt is the material derivative of a scalar field, this term can be evaluated using Eq. (16.16). When this is done for each term on the right side of Eq. (16.22), the acceleration in Cartesian coordinates is given by

$$\begin{pmatrix} \text{acceleration} \\ \text{of a fluid particle} \end{pmatrix} = \mathbf{a} = \frac{d\mathbf{V}}{dt} = \begin{bmatrix} \left\{\left(\frac{\partial u}{\partial t}\right) + u\left(\frac{\partial u}{\partial x}\right) + v\left(\frac{\partial u}{\partial y}\right) + w\left(\frac{\partial u}{\partial z}\right)\right\}\mathbf{i} \\ \left\{\left(\frac{\partial v}{\partial t}\right) + u\left(\frac{\partial v}{\partial x}\right) + v\left(\frac{\partial v}{\partial y}\right) + w\left(\frac{\partial v}{\partial z}\right)\right\}\mathbf{j} \\ \left\{\left(\frac{\partial w}{\partial t}\right) + u\left(\frac{\partial w}{\partial x}\right) + v\left(\frac{\partial w}{\partial y}\right) + w\left(\frac{\partial w}{\partial z}\right)\right\}\mathbf{k} \end{bmatrix} \qquad (16.23)$$

When acceleration is derived in cylindrical coordinates, the result is

$$\mathbf{a} = \begin{bmatrix} a_r \\ a_\theta \\ a_z \end{bmatrix} = \begin{bmatrix} \dfrac{dv_r}{dt} - \dfrac{v_\theta^2}{r} \\ \dfrac{dv_\theta}{dt} + \dfrac{v_r v_\theta}{r} \\ \dfrac{dv_z}{dt} \end{bmatrix} = \begin{bmatrix} \dfrac{\partial v_r}{\partial t} + v_r\dfrac{\partial v_r}{\partial r} + \dfrac{v_\theta}{r}\dfrac{\partial v_r}{\partial \theta} + v_z\dfrac{\partial v_r}{\partial z} - \dfrac{v_\theta^2}{r} \\ \dfrac{\partial v_\theta}{\partial t} + v_r\dfrac{\partial v_\theta}{\partial r} + \dfrac{v_\theta}{r}\dfrac{\partial v_\theta}{\partial \theta} + v_z\dfrac{\partial v_\theta}{\partial z} + \dfrac{v_r v_\theta}{r} \\ \dfrac{\partial v_z}{\partial t} + v_r\dfrac{\partial v_z}{\partial r} + \dfrac{v_\theta}{r}\dfrac{\partial v_z}{\partial \theta} + v_z\dfrac{\partial v_z}{\partial z} \end{bmatrix} \qquad (16.24)$$

Summary. The acceleration field is given by the material derivative of the velocity field. This can be written in Cartesian coordinates (16.23) or cylindrical coordinates (16.24).

16.3 The Continuity Equation

The continuity equation, according to Frank White (7), is one of five partial differential equations that are needed to model a flowing fluid. The set of five equations is

- **The continuity equation.** This is the law of conservation of mass applied to a fluid and expressed as a partial differential equation.
- **The momentum equation.** This is Newton's second law of motion applied to fluid. This equation is mostly commonly developed for a Newtonian fluid, and the equation is called the Navier-Stokes equation.

- **The energy equation.** This is the law of conservation of energy applied to a fluid.
- **Equations of state (two equations).** An equation of state describes how thermodynamic variables are related. For example, an equation of state for density describes how density varies with temperature and pressure.

The continuity equation is described in this section, the Navier-Stokes equation is described in the next section. The other three equations are described in the books by White (7, 8).

In practice, there are multiple ways to write the continuity equation as a partial differential equation. This can be quite confusing when learning. Thus, the main purpose of this section is to

- Introduce various forms of the continuity equation
- Introduce the language and ideas for understanding how and why engineers use these different forms.

Derivation Using a Control Volume (Conservation Form)

This section introduces one of the ways to derive the continuity equation.

Step 1. Select a Control Volume (CV). Select a CV (Fig. 16.13) centered around the point (x, y, z). Assume that the CV is stationary and nondeforming.

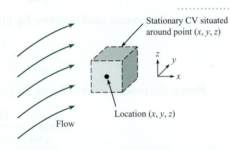

Stationary CV situated around point (x, y, z)

Location (x, y, z)

Flow

FIGURE 16.13

A stationary, nondeforming, infinitesimal CV that is situated at point (x, y, z) in a moving fluid.

Let the CV have dimensions $(\Delta x, \Delta y, \Delta z)$, where each dimension is infinitesimal in size. *Infinitesimal* means that dimensions are approaching zero in the sense of the limit in calculus (e.g., limit $\Delta x \rightarrow 0$).

Step 2. Apply Conservation of Mass. Apply conservation of mass to the CV. The physics are

$$\text{(rate of accumulation of mass)} + \text{(net outflow of mass)} = \text{(zero)} \quad \textbf{(16.25)}$$

These physics can represented by this equation:

$$\frac{dm_{cv}}{dt} + \dot{m}_{net} = 0 \quad \textbf{(16.26)}$$

Step 3: Analyze the Accumulation. The accumulation term is

$$\frac{dm_{cv}}{dt} = \frac{\partial(\text{mass in cv})}{\partial t} = \frac{\partial(\rho \forall)}{\partial t} = \left(\frac{\partial \rho}{\partial t}\right)\forall = \left(\frac{\partial \rho}{\partial t}\right)\Delta x \Delta y \Delta z \quad \textbf{(16.27)}$$

Eq. (16.27) uses a partial derivative because the control volume is fixed in place, which means that the variables x, y and z have fixed values. The volume term was pulled out of the derivative because the volume of the CV is constant with time.

Step 4. Analyze the Outflow To analyze $\dot{m}_{net}$, consider flow through the x-faces (Fig. 16.14) of the CV. An x-face is defined to as a face of the cube that is perpendicular to the x-axis. As shown, there is outflow through the positive x-face and inflow through the negative x-face.

FIGURE 16.14
Inflow and outflow of mass through the x-faces of the control volume.

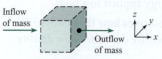

The mass flow rates through the x-faces are

$$\begin{aligned}\dot{m}_{\substack{\text{positive}\\ \text{x-face}}} &= (\rho A u)_{x+\Delta x/2} = (\rho u)_{x+\Delta x/2}(\Delta y \Delta z)\\ \dot{m}_{\substack{\text{negative}\\ \text{x-face}}} &= (\rho A u)_{x-\Delta x/2} = (\rho u)_{x-\Delta x/2}(\Delta y \Delta z)\end{aligned} \tag{16.28}$$

The net flow rate through the x-faces is

$$\begin{aligned}\dot{m}_{\text{net}} &= \dot{m}_{\substack{\text{positive}\\ \text{x-face}}} - \dot{m}_{\substack{\text{negative}\\ \text{x-face}}}\\ &= ((\rho u)_{x+\Delta x/2} - ((\rho u)_{x-\Delta x/2}))(\Delta y \Delta z)\end{aligned} \tag{16.29}$$

Simplify Eq. (16.29) by expanding the derivatives in a Taylor series to give

$$\dot{m}_{\substack{\text{net}\\ \text{x-face}}} = \frac{\partial(\rho u)}{\partial x}(\Delta x \Delta y \Delta z) \tag{16.30}$$

Repeat the process used to derive Eq. (16.30) for the y-face to give

$$\dot{m}_{\substack{\text{net}\\ y\text{-face}}} = \frac{\partial(\rho v)}{\partial y}(\Delta x \Delta y \Delta z) \tag{16.31}$$

Repeat the process used to derive Eq. (16.30) for the z-face to give

$$\dot{m}_{\substack{\text{net}\\ z\text{-face}}} = \frac{\partial(\rho w)}{\partial z}(\Delta x \Delta y \Delta z) \tag{16.32}$$

To sum the mass flow rates through all faces, add up the terms in Eqs. (16.30) to (16.32).

$$\dot{m}_{\text{net}} = \left(\frac{\partial(\rho u)}{\partial x} + \frac{\partial(\rho v)}{\partial y} + \frac{\partial(\rho w)}{\partial z}\right)(\Delta x \Delta y \Delta z) \tag{16.33}$$

Step 5. Combine Results. Insert terms from Eqs. (16.27) and (16.33) into Eq. (16.26):

$$\left(\frac{\partial \rho}{\partial t}\right)(\Delta x \Delta y \Delta z) + \left(\frac{\partial(\rho u)}{\partial x} + \frac{\partial(\rho v)}{\partial y} + \frac{\partial(\rho w)}{\partial z}\right)(\Delta x \Delta y \Delta z) = 0 \tag{16.34}$$

Divide through by the volume of the CV to give the final result:

$$\frac{\partial \rho}{\partial t} + \frac{\partial(\rho u)}{\partial x} + \frac{\partial(\rho v)}{\partial y} + \frac{\partial(\rho w)}{\partial z} = 0 \tag{16.35}$$

Step 6. Interpret the Physics The meaning of Eq. (16.35) is

$$\underbrace{\frac{\partial \rho}{\partial t}}_{\substack{\text{rate of accumulation of mass}\\ \text{in a differential CV divided by}\\ \text{the volume of the CV}\\ (\text{kg/s per m}^3)}} + \underbrace{\frac{\partial(\rho u)}{\partial x} + \frac{\partial(\partial v)}{\partial y} + \frac{\partial(\rho w)}{\partial z}}_{\substack{\text{net rate of mass flow}\\ \text{out of the CV divided by the}\\ \text{volume of the CV}\\ (\text{kg/s per m}^3)}} = 0 \tag{16.36}$$

Note the dimensions and units of the terms that appear in the continuity equation:

$$\frac{\text{(mass/time)}}{\text{(volume)}} = \frac{\text{kg/s}}{\text{m}^3} \tag{16.37}$$

Derivation Using a Fluid Particle (The Nonconservation Form)

The literature uses two forms of the continuity equation:

- **Conservation Form.** The conservation form, developed in the last subsection, is derived by starting with a differential control volume and applying conservation of mass to this CV. This is an Eulerian approach.
- **Nonconservation Form.** The nonconservation form, developed in this subsection, is derived by starting with a differential fluid particle and applying conservation of mass to this particle. This is a Lagrangian approach.

Next, we derive the *nonconservation form* of the continuity equation.

Step 1. Select a Fluid Particle. Select a fluid particle (Fig. 16.15) centered around a point (x, y, z) in space. Because a particle moves with a flowing fluid, this particle is at this location only at a specific instant in time.

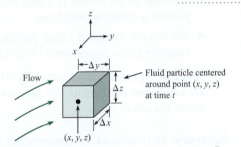

FIGURE 16.15

Fluid particle (infinitesimal in size) located at point (x, y, z) in a flowing fluid.

Step 2. Apply Conservation of Mass. By definition, the *mass of the particle must stay constant with time.* To say this mathematically:

$$\frac{d(\text{mass})}{dt} = \frac{d[(\text{density})(\text{volume})]}{dt} = \frac{d(\rho \mathcal{V})}{dt} = 0 \tag{16.38}$$

Step 3. Apply the Product Rule. Eq. (16.38) becomes

$$\rho \frac{d\mathcal{V}}{dt} + \mathcal{V} \frac{d\rho}{dt} = 0 \tag{16.39}$$

Step 4. Analyze the Change in Volume Term. The change in volume term describes how the volume of the fluid particle changes with time. To analyze this term, apply the definition of the derivative:

$$\frac{d\mathcal{V}}{dt} = \lim_{\Delta t \to 0} \frac{\mathcal{V}(t + \Delta t) - \mathcal{V}(t)}{\Delta t} \tag{16.40}$$

In Eq. (16.40), the volume at time t is

$$\mathcal{V}(t) = \Delta x \Delta y \Delta z \tag{16.41}$$

and the volume at time $t + \Delta t$ is

$$\mathcal{V}(t + \Delta t) = (\Delta x + \Delta x')(\Delta y + \Delta y')(\Delta z + \Delta z') \tag{16.42}$$

where each term of the form $\Delta x'$ represents a change in the length of the side of the particle. Next, multiply out the terms on the right side of Eq. (16.42) and neglect higher order terms (H.O.T.). The equation becomes

$$\mathcal{V}(t + \Delta t) \approx \Delta x \Delta y \Delta z + \Delta x' \Delta y \Delta z + \Delta x \Delta y' \Delta z + \Delta x \Delta y \Delta z' \quad \text{(16.43)}$$

Next, combine Eqs. (16.41) and (16.43) and apply Taylor's series:

$$\mathcal{V}(t + \Delta t) - \mathcal{V}(t) \approx \left(\frac{\partial u}{\partial x}\Delta x\right)\Delta y \Delta z \Delta t + \Delta x \left(\frac{\partial v}{\partial y}\Delta y\right)\Delta z \Delta t + \Delta x \Delta y \left(\frac{\partial w}{\partial z}\Delta z\right)\Delta t \quad \text{(16.44)}$$

Then, substitute Eq. (16.44) into (16.40) to give

$$\frac{d\mathcal{V}}{dt} = \left(\frac{\partial u}{\partial x} + \frac{\partial v}{\partial y} + \frac{\partial w}{\partial z}\right)\Delta x \Delta y \Delta z = \left(\frac{\partial u}{\partial x} + \frac{\partial v}{\partial y} + \frac{\partial w}{\partial z}\right)\mathcal{V} \quad \text{(16.45)}$$

Step 5. Combine Results. Substitute Eq. (16.45) into Eq. (16.39), divide each term by the volume of the particle, and rearrange to give

$$\frac{d\rho}{dt} + \rho\left(\frac{\partial u}{\partial x} + \frac{\partial v}{\partial y} + \frac{\partial w}{\partial z}\right) = 0 \quad \text{(16.46)}$$

Step 6. Interpret the Physics. The derivation of Eq. (16.46), reveals two main ideas:

- A change in the density of a fluid particle occurs if, and only if, the volume of the fluid particle is changing with time.

- The volume change of a fluid particle is represented mathematically by the bracketed variables in the second term of Eq. (16.46).

Note that the conservation form (Eq. (16.35)) and the nonconservation form (Eq. (16.46)) are equivalent mathematically because one can start with one of these equations and derive the other.

Summary. Derivation of the conservation and the nonconservation forms of the continuity equation gives equations that are equivalent mathematically. However, these equations have different physical interpretations.

Cylindrical Coordinates

The continuity equation can also be derived in cylindrical coordinates; see Pritchard (9). The result (conservation form) is

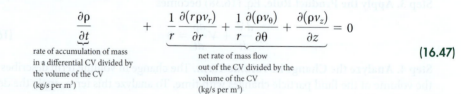

$$\underbrace{\frac{\partial\rho}{\partial t}}_{\substack{\text{rate of accumulation of mass} \\ \text{in a differential CV divided by} \\ \text{the volume of the CV} \\ \text{(kg/s per m}^3\text{)}}} + \underbrace{\frac{1}{r}\frac{\partial(r\rho v_r)}{\partial r} + \frac{1}{r}\frac{\partial(\rho v_\theta)}{\partial\theta} + \frac{\partial(\rho v_z)}{\partial z}}_{\substack{\text{net rate of mass flow} \\ \text{out of the CV divided by the} \\ \text{volume of the CV} \\ \text{(kg/s per m}^3\text{)}}} = 0 \quad \text{(16.47)}$$

One can also derive the continuity equation in spherical coordinates and in other coordinate systems.

Invariant Notation

This subsection shows how to modify the continuity equation to an invariant form. The "del" operator is defined as:

$$\nabla \equiv \mathbf{i}\frac{\partial}{\partial x} + \mathbf{j}\frac{\partial}{\partial y} + \mathbf{k}\frac{\partial}{\partial z} \quad \text{(16.48)}$$

Thus, start with the continuity equation in Cartesian components and introduce the del operator using the dot product.

$$\frac{\partial \rho}{\partial t} + \frac{\partial(\rho u)}{\partial x} + \frac{\partial(\rho v)}{\partial y} + \frac{\partial(\rho w)}{\partial z} = 0$$

$$\frac{\partial \rho}{\partial t} + \left(\mathbf{i}\frac{\partial}{\partial x} + \mathbf{j}\frac{\partial}{\partial y} + \mathbf{k}\frac{\partial}{\partial z} \right) \cdot ((\rho u)\mathbf{i} + (\rho v)\mathbf{j} + (\rho w)\mathbf{k}) = 0 \qquad \text{(16.49)}$$

$$\frac{\partial \rho}{\partial t} + \nabla \cdot (\rho \mathbf{V}) = 0$$

The last line in Eq. (16.49) is the continuity equation in an invariant form. The physics are

$$\underbrace{\frac{\partial \rho}{\partial t}}_{\text{Accumlation}} + \underbrace{\nabla \cdot (\rho \mathbf{V})}_{\text{Net Outflow of Mass}} = 0 \qquad \text{(16.50)}$$

The term $\nabla \cdot (\rho \mathbf{V})$ is the divergence. Eq. (16.50) can also be written with the Gibbs notation.

$$\underbrace{\frac{\partial \rho}{\partial t}}_{\text{Accumlation}} + \underbrace{\text{div}(\rho \mathbf{V})}_{\text{Net Outflow of Mass}} = 0 \qquad \text{(16.51)}$$

A useful aspect of invariant notion is that it provides a way to describe the physics of a math operator. Example: the physics of the divergence operator can be established from Eq. (16.50):

$$\text{div}(\rho \mathbf{V}) = \nabla \cdot (\rho \mathbf{V}) = \frac{\left(\begin{array}{c} \text{net rate of outflow of mass} \\ \text{from a differential CV centered} \\ \text{about point } (x, y, z) \end{array} \right)}{(\text{volume of the CV})} \qquad \text{(16.52)}$$

The physics of the divergence operator can also be found another way. Step 1 is to write Eq. (16.46) in this form:

$$\frac{d\rho}{dt} + \rho(\nabla \cdot \mathbf{V}) = 0$$

$$\frac{d\rho}{dt} + \rho\,\text{div}(\mathbf{V}) = 0 \qquad \text{(16.53)}$$

Step 2 is to go back through the derivation of Eq. (16.46). This will reveal that

$$\nabla \cdot \mathbf{V} = \text{div}(\mathbf{V}) = \frac{(\text{time rate of change of the volume of a fluid particle})}{(\text{volume of the fluid particle})} \qquad \text{(16.54)}$$

Summary. The continuity equation can be written in an invariant form. This approach provides a method for developing a physical interpretation of the divergence operator. As shown, the divergence operator has two different physical interpretations.

Continuity for Incompressible (Constant Density) Flow

Because it is common to assume a constant value of density, the continuity equation is often written for the case of constant density. This is usually called incompressible flow.

When density is constant, the continuity equation written in Cartesian coordinates (Eq. (16.36) or Eq. (16.46)) reduces to:

$$\frac{\partial u}{\partial x} + \frac{\partial v}{\partial y} + \frac{\partial w}{\partial z} = 0 \qquad (16.55)$$

Similarly, the continuity equation for cylindrical coordinates (Eq. (16.47)) reduces to:

$$\frac{1}{r}\frac{\partial(r v_r)}{\partial r} + \frac{1}{r}\frac{\partial v_\theta}{\partial \theta} + \frac{\partial v_z}{\partial z} = 0 \qquad (16.56)$$

When density is constant, Eq. (16.51) reduces to:

$$\nabla \cdot \mathbf{V} = \text{div}(\mathbf{V}) = 0 \qquad (16.57)$$

Summary. When flow is modeled as incompressible, the continuity equation reduces to $\nabla \cdot \mathbf{V} = \text{div}(\mathbf{V}) = 0$, which means that the divergence of the velocity field is zero. This equation can also be written in Cartesian coordinates (Eq. 16.55) and cylindrical coordinates (Eq. 16.56).

Summary of the Mathematical Forms of the Continuity Equation

Table 16.2 lists some of the ways to write the continuity equation as a PDE. Recognize that the math simply reflects alternative ways to describe the physics.

TABLE 16.2 Alternative Ways to Write the Continuity Equation as a PDE

	Description	Equation
General equation	Cartesian coordinates (conservation form)	$\dfrac{\partial \rho}{\partial t} + \dfrac{\partial(\rho u)}{\partial x} + \dfrac{\partial(\rho v)}{\partial y} + \dfrac{\partial(\rho w)}{\partial z} = 0$
	Cartesian coordinates (nonconservation form)	$\dfrac{d\rho}{dt} + \rho\left(\dfrac{\partial u}{\partial x} + \dfrac{\partial v}{\partial y} + \dfrac{\partial w}{\partial z}\right) = 0$ $\dfrac{\partial \rho}{\partial t} + \left(u\dfrac{\partial \rho}{\partial x} + v\dfrac{\partial \rho}{\partial y} + w\dfrac{\partial \rho}{\partial z}\right) + \rho\left(\dfrac{\partial u}{\partial x} + \dfrac{\partial v}{\partial y} + \dfrac{\partial w}{\partial z}\right) = 0$
	Cylindrical coordinates (conservation form)	$\dfrac{\partial \rho}{\partial t} + \dfrac{1}{r}\dfrac{\partial(r\rho v_r)}{\partial r} + \dfrac{1}{r}\dfrac{\partial(\rho v_\theta)}{\partial \theta} + \dfrac{\partial(\rho v_z)}{\partial z} = 0$
	Invariant (conservation form)	$\dfrac{\partial \rho}{\partial t} + \nabla \cdot (\rho \mathbf{V}) = 0$
	Invariant (nonconservation form)	$\dfrac{d\rho}{dt} + \rho(\nabla \cdot \mathbf{V}) = 0$
Incompressible flow equation	Invariant form	$\nabla \cdot \mathbf{V} = \text{div}(\mathbf{V}) = 0$
	Cartesian coordinates	$\dfrac{\partial u}{\partial x} + \dfrac{\partial v}{\partial y} + \dfrac{\partial w}{\partial z} = 0$
	Cylindrical coordinates	$\dfrac{1}{r}\dfrac{\partial(r v_r)}{\partial r} + \dfrac{1}{r}\dfrac{\partial v_\theta}{\partial \theta} + \dfrac{\partial v_z}{\partial z} = 0$

As shown in the next example, the continuity equation can be applied in two steps.

- **Step 1. Selection.** From Table 16.2, select an applicable form of the continuity equation.
- **Step 2. Reduction.** Eliminate the variables in the continuity equation that are equal to zero or are negligible.

EXAMPLE. Consider developing laminar flow in a round pipe (Fig. 16.16). At the entrance to the pipe, the velocity profile is uniform. As the flow proceeds down the pipe, the velocity profile becomes fully developed. Assume the flow is steady and constant density. Reduce the general equation for the continuity equation so that it applies to this situation.

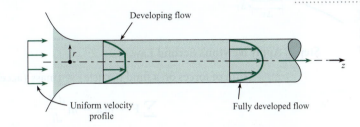

FIGURE 16.16

Developing laminar flow in a round pipe.

Action

Step 1. Selection. Because the flow is constant density and the geometry is a round pipe, select the incompressible flow form of continuity in cylindrical coordinates (Eq. 16.56).

$$\frac{1}{r}\frac{\partial(rv_r)}{\partial r} + \frac{1}{r}\frac{\partial v_\theta}{\partial \theta} + \frac{\partial v_z}{\partial z} = 0$$

Step 2. Reduction. Assume that the flow is symmetric about the z axis. Thus, let $v_\theta = 0$. The continuity equation reduces to

$$\frac{1}{r}\frac{\partial(rv_r)}{\partial r} + \frac{\partial v_z}{\partial z} = 0$$

Review. One could solve this equation to give the velocity field for developing flow in a round pipe. Because this equation has two unknown variables (v_r and v_z), one would also need to solve the Navier-Stokes equation.

16.4 The Navier-Stokes Equation

The Navier-Stokes equation is widely used in both theory and in application. Thus, this section introduces this equation.

The Navier-Stokes equation represents Newton's second law of motion as applied to viscous flow of a Newtonian fluid. In this text, we assume incompressible flow and constant viscosity. In the literature, one can find more general derivations.

Derivation

Similar to the continuity equations, there are multiple ways to derive the Navier-Stokes equation. This section shows how to derive the equation by starting with a fluid particle and applying Newton's second law. Thus, the result will be the nonconservation form of the equation. Because the derivation is complex, we omit some of the technical details; to access these details, we recommend the text *Viscous Fluid Flow* (8).

Step 1. Select a Fluid Particle. Select a fluid particle in a flowing fluid (Fig. 16.17). Let the particle have the shape of a cube. Assume the dimensions are infinitesimal and that the particle is at the position (x, y, z) at the instant shown.

FIGURE 16.17

A fluid particle situated in a flowing fluid.

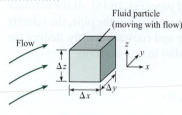

Step 2. Apply Newton's Second Law.

(sum of forces on a fluid particle) = (mass)(acceleration)

$$\sum \mathbf{F} = m\mathbf{a} = m\frac{d\mathbf{V}}{dt} \tag{16.58}$$

Regarding the forces, the two categories are body forces and surfaces forces. The only possible surface forces are the pressure force and the shear force. Assume that the only body force is the weight **W**. Eq. (16.58) becomes

(weight) + (pressure force) + (shear force) = (density)(volume)(acceleration)

$$\mathbf{W} + \mathbf{F}_p + \mathbf{F}_s = \rho \mathcal{V}\frac{d\mathbf{V}}{dt} \tag{16.59}$$

The weight is given by

$$\mathbf{W} = (\text{mass})(\text{gravity vector}) = \rho \mathcal{V}\mathbf{g} \tag{16.60}$$

Insert Eq. (16.60) into Eq. (16.59) to give

$$\rho \mathcal{V}\mathbf{g} + \mathbf{F}_p + \mathbf{F}_s = \rho \mathcal{V}\frac{d\mathbf{V}}{dt} \tag{16.61}$$

Step 3. Analyze the Pressure Force. To begin, consider the forces on the x-faces of the particle (Fig. 16.18).

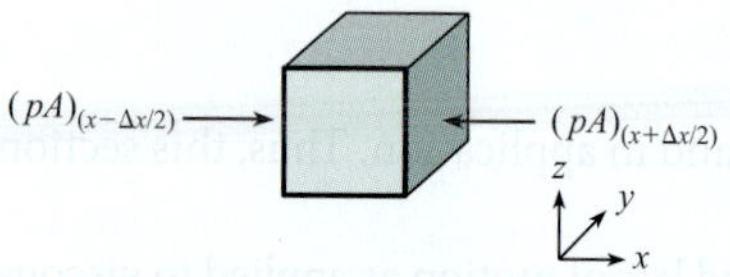

FIGURE 16.18

The pressure forces on the x-faces of a fluid particle.

$(pA)_{(x-\Delta x/2)} \longrightarrow$ ◻ $\longleftarrow (pA)_{(x+\Delta x/2)}$

The net force due to pressure on the x-faces is:

$$\mathbf{F}_{\text{pressure} \atop x\text{-faces}} = -((pA)_{x+\Delta x/2} - (pA)_{x-\Delta x/2})\mathbf{i} = -(p_{x+\Delta x/2} - p_{x-\Delta x/2})(\Delta y \Delta z)\mathbf{i} \tag{16.62}$$

Simplify Eq. (16.62) by applying a Taylor series expansion (twice) and neglecting higher order terms to give

$$\mathbf{F}_{\text{pressure} \atop x\text{-faces}} = \frac{\partial p}{\partial x}(\Delta x \Delta y \Delta z)\mathbf{i} \tag{16.63}$$

Repeat this process for the y- and z-faces, and combine results to give

$$\mathbf{F}_{\text{pressure} \atop \text{all faces}} = -\left(\frac{\partial p}{\partial x}(\Delta x \Delta y \Delta z)\mathbf{i} + \frac{\partial p}{\partial y}(\Delta x \Delta y \Delta z)\mathbf{j} + \frac{\partial p}{\partial z}(\Delta x \Delta y \Delta z)\mathbf{k}\right) \tag{16.64}$$

Simplify Eq. (16.64) and then introduce vector notation to give

$$\mathbf{F}_{\text{pressure}} = -\left(\frac{\partial p}{\partial x}\mathbf{i} + \frac{\partial p}{\partial y}\mathbf{j} + \frac{\partial p}{\partial z}\mathbf{k}\right)(\Delta x \Delta y \Delta z) = -\nabla p(\Delta x \Delta y \Delta z) \qquad \textbf{(16.65)}$$

Physics of the Gradient. Eq. (16.65) reveals a physical interpretation of the gradient:

$$\begin{pmatrix} \text{gradient of the} \\ \text{pressure field} \\ \text{at a point} \end{pmatrix} = \frac{\begin{pmatrix} \text{net pressure force} \\ \text{on a fluid particle} \end{pmatrix}}{(\text{volume of the particle})} \qquad \textbf{(16.66)}$$

Step 4. Analyze the Shear Force. The shear force is the net force on the fluid particle due to shear stresses. Shear stress is caused by viscous effects and is represented mathematically as shown in Fig. 16.19. This figure shows that each face of the fluid particle has three stress components. For example the positive x-face has three stress components, which are τ_{xx}, τ_{xy}, and τ_{xz}. The double subscript notation describes the direction of the stress component and the face on which the component acts. For example,

- τ_{xx} is the shear stress on the x-face in the x-direction.
- τ_{xy} is the shear stress on the x-face in the y-direction.
- τ_{xz} is the shear stress on the x-face in the z-direction.

Shear stress is a type of mathematical entity called a second order tensor. A *tensor* is analogous to but more general than a vector. Examples: A zeroth order tensor is a scalar, a first order tensor is a vector. A second order tensor has magnitude, direction, and orientation (where orientation describes which face the stress acts on).

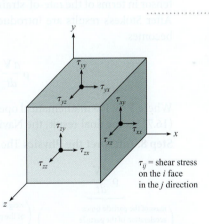

FIGURE 16.19

Shear stresses that act on a fluid particle.

τ_{ij} = shear stress on the i face in the j direction

To find the net shear force on the particle, each stress component is be multiplied by area, and the forces are added. Then, a Taylor series expansion is applied and the result is that

$$\mathbf{F}_{\text{shear}} = \begin{bmatrix} F_{x,\text{shear}} \\ F_{y,\text{shear}} \\ F_{z,\text{shear}} \end{bmatrix} = \begin{bmatrix} \left(\dfrac{\partial \tau_{xx}}{\partial x} + \dfrac{\partial \tau_{xy}}{\partial x} + \dfrac{\partial \tau_{xz}}{\partial x}\right) \\ \left(\dfrac{\partial \tau_{yx}}{\partial y} + \dfrac{\partial \tau_{yy}}{\partial y} + \dfrac{\partial \tau_{yz}}{\partial y}\right) \\ \left(\dfrac{\partial \tau_{zx}}{\partial z} + \dfrac{\partial \tau_{zy}}{\partial z} + \dfrac{\partial \tau_{zz}}{\partial z}\right) \end{bmatrix}(\Delta x \Delta y \Delta z) \qquad \textbf{(16.67)}$$

Eq. (16.67) can be written in invariant notation as

$$\mathbf{F}_{\text{shear}} = (\nabla \cdot \tau)\text{\textcrv} = (\text{div}(\tau))\text{\textcrv} \tag{16.68}$$

where the terms on the right side represent the divergence of the stress tensor times the volume of the fluid particle.

Eq. (16.68) reveals the physics of the divergence when it operates on the stress tensor. Note that this is the third physical interpretation of the divergence operator in this chapter. This is because the physics of a mathematical operator depend on the context in which the operator is used.

$$\begin{pmatrix} \text{divergence of} \\ \text{the stress tensor} \end{pmatrix} = \frac{\begin{pmatrix} \text{net shear force} \\ \text{on a fluid particle} \end{pmatrix}}{(\text{volume of the particle})} \tag{16.69}$$

Step 6. Combine Terms. Substitute the shear force Eq. (16.68) and the pressure force Eq. (16.65), into Newton's second law of motion Eq. (16.61). Then, divide by the volume of the fluid particle to give

$$\rho \frac{d\mathbf{V}}{dt} = \rho\mathbf{g} - \nabla p + \nabla \cdot \tau_{ij} \tag{16.70}$$

Eq. (16.70) is the differential form of the linear momentum equation without any assumption about the nature of the fluid. The next step involves modifying this equation so that it applies to a Newtonian fluid.

Step 7. Assume a Newtonian Fluid. Stokes in 1845 figured out a way to write the stress tensor in terms of the rate-of-strain tensor of the flowing fluid. The details are omitted here. After Stokes's results are introduced, assume constant density and viscosity. Eq. (16.70) becomes

$$\rho \frac{d\mathbf{V}}{dt} = \rho\mathbf{g} - \nabla p + \mu\nabla^2\mathbf{V} \tag{16.71}$$

Where $\nabla^2\mathbf{V}$ is a mathematical operator that is called the Laplacian of the velocity field. Eq. (16.71) is the final result, the Navier-Stokes equation.

Step 8. Interpret the Physics The physics of the Navier-Stokes equation are

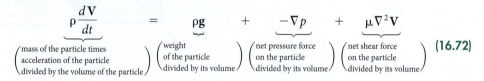

$$\underbrace{\rho \frac{d\mathbf{V}}{dt}}_{\substack{\text{mass of the particle times} \\ \text{acceleration of the particle} \\ \text{divided by the volume of the particle}}} = \underbrace{\rho\mathbf{g}}_{\substack{\text{weight} \\ \text{of the particle} \\ \text{divided by its volume}}} + \underbrace{-\nabla p}_{\substack{\text{net pressure force} \\ \text{on the particle} \\ \text{divided by its volume}}} + \underbrace{\mu\nabla^2\mathbf{V}}_{\substack{\text{net shear force} \\ \text{on the particle} \\ \text{divided by its volume}}} \tag{16.72}$$

Note the dimensions and units:

$$\text{dimensions} = \frac{\text{force}}{\text{volume}} \sim \frac{\text{N}}{\text{m}^3} = \frac{\text{kg}}{\text{m}^2 \cdot \text{s}^2} \tag{16.73}$$

Cartesian and Cylindrical Coordinates

To write Eq. (16.72) in Cartesian coordinates, find a suitable reference (e.g., the Internet, an advanced fluids text, an engineering handbook) and look up the material derivative ($d\mathbf{V}/dt$), the gradient, and the Laplacian operator in Cartesian coordinates. After substitution, the Navier-Stokes equation (constant properties) for Cartesian coordinates is

$$\rho\left(\frac{\partial u}{\partial t} + u\frac{\partial u}{\partial x} + v\frac{\partial u}{\partial y} + w\frac{\partial u}{\partial z}\right) = \rho g_x - \frac{\partial p}{\partial x} + \mu\left(\frac{\partial^2 u}{\partial x^2} + \frac{\partial^2 u}{\partial y^2} + \frac{\partial^2 u}{\partial z^2}\right)$$

$$\rho\left(\frac{\partial v}{\partial t} + u\frac{\partial v}{\partial x} + v\frac{\partial v}{\partial y} + w\frac{\partial v}{\partial z}\right) = \rho g_y - \frac{\partial p}{\partial y} + \mu\left(\frac{\partial^2 v}{\partial x^2} + \frac{\partial^2 v}{\partial y^2} + \frac{\partial^2 v}{\partial z^2}\right) \qquad (16.74)$$

$$\rho\left(\frac{\partial w}{\partial t} + u\frac{\partial w}{\partial x} + v\frac{\partial w}{\partial y} + w\frac{\partial w}{\partial z}\right) = \rho g_z - \frac{\partial p}{\partial z} + \mu\left(\frac{\partial^2 w}{\partial x^2} + \frac{\partial^2 w}{\partial y^2} + \frac{\partial^2 w}{\partial z^2}\right)$$

The Navier-Stokes equation cannot be solved in general because of the nonlinear terms. An example of a nonlinear term is

$$u\frac{\partial u}{\partial x}$$

This term is nonlinear because a dependent variable (u) is multiplied by its first derivative ($\partial u/\partial x$). In general, nonlinear terms in differential equations involve functions of the dependent variables.

The Navier-Stokes equation (constant properties) for cylindrical coordinates is

$$r: \rho\left(\frac{\partial v_r}{\partial t} + v_r\frac{\partial v_r}{\partial r} + \frac{v_\theta}{r}\frac{\partial v_r}{\partial \theta} + v_z\frac{\partial v_r}{\partial z} - \frac{v_\theta^2}{r}\right) = \rho g_r - \frac{\partial p}{\partial r} + \mu\left(\frac{1}{r}\frac{\partial}{\partial r}\left(r\frac{\partial v_r}{\partial r}\right) + \frac{1}{r^2}\frac{\partial^2 v_r}{\partial \theta^2} + \frac{\partial^2 v_r}{\partial z^2} - \frac{v_r}{r^2} - \frac{2}{r^2}\frac{\partial v_\theta}{\partial \theta}\right)$$

$$\theta: \rho\left(\frac{\partial v_\theta}{\partial t} + v_r\frac{\partial v_\theta}{\partial r} + \frac{v_\theta}{r}\frac{\partial v_\theta}{\partial \theta} + v_z\frac{\partial v_\theta}{\partial z} + \frac{v_r v_\theta}{r}\right) = \rho g_\theta - \frac{1}{r}\frac{\partial p}{\partial \theta} + \mu\left(\frac{1}{r}\frac{\partial}{\partial r}\left(r\frac{\partial v_\theta}{\partial r}\right) + \frac{1}{r^2}\frac{\partial^2 v_\theta}{\partial \theta^2} + \frac{\partial^2 v_\theta}{\partial z^2} - \frac{v_\theta}{r^2} + \frac{2}{r^2}\frac{\partial v_\theta}{\partial \theta}\right) \qquad (16.75)$$

$$z: \rho\left(\frac{\partial v_z}{\partial t} + v_r\frac{\partial v_z}{\partial r} + \frac{v_\theta}{r}\frac{\partial v_z}{\partial \theta} + v_z\frac{\partial v_z}{\partial z}\right) = \rho g_z - \frac{\partial p}{\partial z} + \mu\left(\frac{1}{r}\frac{\partial}{\partial r}\left(r\frac{\partial v_z}{\partial r}\right) + \frac{1}{r^2}\frac{\partial^2 v_z}{\partial \theta^2} + \frac{\partial^2 v_z}{\partial z^2}\right)$$

Summary. The Navier-Stokes equation represents Newton's second law of motion applied to the viscous flow of a Newtonian fluid. The Navier-Stokes equation has nonlinear terms that prevent an exact mathematical solution for most problems.

✔ **CHECKPOINT PROBLEMS**

Regarding the Navier-Stokes equation (Eq. 16.74), which statements are true?

   a. The terms on the right side are linear.

   b. The equation is invariant.

   c. The equation applies to all fluids (all liquids and all gases).

16.5 Computational Fluid Dynamics (CFD)

Computational fluid dynamics (CFD) is a method for obtaining approximate solutions to problems in fluid mechanics and heat transfer by using numerical solutions of the governing PDEs. This section describes

- Why CFD is valuable

- How CFD is used in practice

- What CDF programs are and how they work

Why CFD Is Valuable

CFD gives engineers a modeling tool that greatly extends their abilities. For example, there is not a straightforward way to develop and solve equations that will predict the pressure field and streamline patterns for flow around a building. One might use an experimental approach, but this has issues such as matching the Reynolds number and the difficulty in doing parametric studies.

Thus, CFD provides a way to simulate physical phenomena that are impossible for analysis and difficult for experiments. Some examples where CFD is a useful modeling tool include

- Complex systems (e.g., the ink-jet printer, the human heart, mixing tanks)
- Full-scale simulations (e.g., ships, airplanes, dams)
- Environmental effects (e.g., hurricanes, weather, pollution dispersion)
- Hazards (e.g., explosions, radiation dispersion)
- Physics (e.g., planetary boundary layer, stellar evolution)

CFD is also useful for studying the effects of design perturbations. For example, to design a propeller, one could systemically vary design variables such as the blade profile, blade pitch, and rotation speed and see the effect on performance variables such as efficiency, thrust, and power.

CFD is used in many industries and fields of study: aerospace, automotive, biomedical, chemical processing, HVAC, hydraulics, hydrology, marine, oil and gas, and power generation.

Summary. CFD is valuable to the engineer because:

- CFD provides a method for modeling complex problems that cannot effectively be modeled with analytical or experimental fluid mechanics.
- CFD provides a way to consider design perturbations on complex problems such as propeller design and the design of spillways.
- CFD is widely used in industry.

CFD Codes in Professional Practice

A *code* is engineering lingo for a computer program. In professional practice and most research projects, engineers can

- *Option 1.* Write their own code. This is rarely done.
- *Option 2.* Apply a code that has been developed by others. This is the most common practice because code development requires years of effort.

This subsection describes three commonly used codes and provides suggestions about selecting a code.

Ground Water Modeling. MODFLOW (10) is a computer program for analyzing groundwater flow. This code has been under development since the early 1980s. MODFLOW is considered the de facto standard for simulation of ground flows. The program is well validated and is considered as legally defensible in U.S. courts.

MODFLOW is available in noncommerical (i.e., free) versions. However, the licensing is limited to government and academic entities. For commercial use, implementations of MODFLOW cost from $1,000 to $7,000 USD (10).

Internal Combustion Engine Modeling. The KIVA codes (11, 12) were originally developed in 1985 by Los Alamos National Laboratory to simulate the processes taking place inside an internal combustion engine. The KIVA programs have become the most widely used CFD (computational fluid dynamics) programs for multidimensional combustion modeling. KIVA can be applied to understand combustion chemistry processes, such as autoignition of fuels, and to optimize diesel engines for high efficiency and low emissions. Hence, KIVA has been used by engine manufacturers to improve the performance of engines.

Modeling of Flows with Free Surfaces. In 1963, Tony Hirt of the Los Alamos National Laboratory pioneered a computational method called the volume of fluid (VOF) approach that is useful for tracking and locating a free surface or a fluid–fluid interface. Thus, the VOF method is useful for modeling flows such as flow from a reservoir or the flow of metal into a mold. Dr. Hirt left Los Alamos and founded a company called Flow Science that now markets a code called FLOW-3D.

Some examples of the capabilities of FLOW-3D, according to the company's Web site (13), include

- Modeling of a coffer dam and spillway of a hydroelectric power plant
- Design of a canoe chute for passage around a low head dam
- Modeling of the molding of foamed polyurethane resin, which can expand in volume by more than 30 times during molding

The examples of Flow-3D, KIVA, and MODFlOW suggest some common ideas:

CFD programs can be very useful for applications. The three codes just described provide technologies for modeling (a) groundwater flow, (b) internal combustion engines and (c) open channel flows. There are other codes available that allow one to model other applications. Thus, CFD is a powerful technology for modeling problems that involve fluids.

Select a CFD code to match your problem. CFD codes are developed to solve specific types of problems. FLOW-3D is for open channel flow, whereas KIVA is for internal-combustion engines and MODFLOW is for modeling groundwater flow. Thus, make sure that the CFD code is well suited for the type of problem you want to solve.

Use an Existing Code. In the three examples, the codes have been under development since the 1980s or earlier. Many years of work have gone into these codes. Thus, it is cost effective to take advantage of this legacy instead of writing a code from scratch.

Features of CFD Programs

This subsection describes the vocabulary and ideas that used by most CFD programs.

Approximation of PDEs. CFD codes apply mathematical methods to develop approximate solutions to the governing PDEs. Approximate solutions (estimates) can be close to reality or far away from reality, depending on the details of how the estimate is made. The accuracy of the estimate is determined in part by how the code was developed. However, most of the accuracy is based on decisions made by the user of the code.

There are many ways to develop approximate solutions of partial differential equations. Three common approaches are called the *finite difference method*, the *finite element method*, and the *finite volume method*.

When a partial differential equation is approximated, the result is a set of *many* algebraic equations that are solved at points in space. These points in space are defined using a grid.

Grid Generation. A **grid** (Fig. 16.20) is a set of points in space at which a code solves for values of velocity and other variables of interest. The grid is set up by the user. There are two trade-offs:

- **Accuracy.** If the grid is closely spaced, which is called a fine grid, then the solution is generally more accurate. In the grid shown in Fig. 16.20, notice how the user set a fine grid near the wall of the cylinder.

- **Computational Time.** If the grid is coarse (grid lines are widely spaced), then the amount of time for the code to run deceases. Decreasing the computer run time is important because CFD codes can take a long time (i.e., days) to run one simulation.

FIGURE 16.20

A grid used to model subsonic flow past a circular cylinder at a Reynolds number of 10,000. From NASA (14)

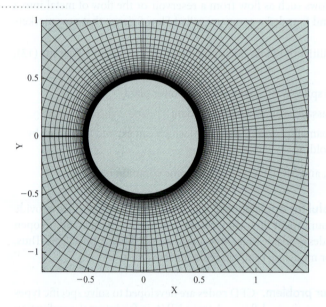

Grid generation capability is set up by the code developers, and the grid generation itself is done by the user. Wyman (15) describes three approaches available for grid generation.

- **Structured Grid Methods.** With this method, the grid is laid out in a regular repeating pattern called a block (Fig. 16.20). Details (fine grid, coarse grid, etc.) are specified by the user. The advantage of a structured grid is that the user can set up the grid to maximize accuracy while achieving acceptable run time. A drawback of a structured grid is that it can take significant time for the user to input the parameters needed to create the grid. Also, a structured grid requires user expertise for proper layout.

- **Unstructured Grid Methods.** An unstructured grid is based on a computer algorithm that selects an arbitrary collection of elements to fill the solution domain. Because the elements lack a pattern, the grid is called unstructured. An unstructured grid method is well suited for novices because the grid can be set up easily and quickly and does not require much user expertise. The drawbacks are that the grid may not be good as a structured grid in terms of accuracy and solution time.

- **Hybrid Grid Methods.** Hybrid grid methods are designed to take advantage of the positive aspects of both structured and unstructured grids. Hybrid grids use a structured grid in local regions while using an unstructured grid in the bulk of the domain.

Time Steps. Because PDEs are being solved by CFD codes, the approximation methods involve solving for variables at specific instances in time. The interval between each solution time is called a **time step**.

Accuracy versus Solution Time. In general, if one selects a fine grid and small time steps, the CFD solution is more accurate. However, fine resolution of space and time drive up the required solution time for the computer. This might seem like a nonissue with today's fast computers, but CFD programs can require days or weeks of solution time. Thus, there is a trade-off between accuracy of a solution versus the time that the computer needs for calculations.

Boundary Conditions and Initial Conditions. Solving PDEs, which includes using CFD programs to develop approximate solutions to PDEs, involves the specification of boundary conditions and initial conditions.

- Specifying a **boundary condition** involves *giving numerical values* for the *dependent variables* on the *physical boundaries* that describe that spatial region in which the differential equations are to be solved. Examples:
  - ▸ When flow enters a boundary, the user might specify a known value of velocity at each point. This is known as a *velocity boundary condition*.
  - ▸ When flow enters a boundary, the user might specify a known value of pressure at each point. This is known as a *pressure boundary condition*.
- Specifying an **initial condition** involves giving numerical values for the *dependent variables* at all spatial points at the starting time of the solutions. Nearly always, this starting time is time equals zero.

Turbulence (Direct Numerical Simulation). Because most flows of engineering interest involve turbulent flow, CFD codes have methods for analyzing turbulent flow. The most accurate approach, which is called **direct numerical simulation (DNS),** involves setting the grid and time steps fine enough to resolve the features of the turbulent flow. As a result Hussan (16) asserts that a DNS solution is very accurate but is also "*unrealistic for 99.9% of CFD problems because it is computationally unrealistic.*" This is because the required computer time is too large for today's computers. Thus, DNS is not used for most problems.

Turbulence Modeling. **Turbulence modeling** involves the prediction of effects of turbulence by applying simplified equations. These equations are simpler that the full time-dependent Navier-Stokes equations. CFD Online (www.cfd-online.com) describes twenty-seven turbulence models and is a good source for details.

One of the most widely used turbulence models is called the **k-epsilon model or k-ε model.** This model uses one equation for the turbulent kinetic energy (k) and another equation for the rate of dissipation of the turbulent energy ($ε$). These equations are used together with the **Reynolds-averaged Navier-Stokes (RANS) equations**. The RANS equations are developed by starting with the equations of motion and then taking the time average. According to Hussan (16), the k-$ε$ method *can be very accurate, but it is not suitable for transient flows because the averaging process wipes out most of the important characteristics of a time-dependent solution.* The main advantage of the k-$ε$ model is that it is computationally efficient.

Another widely used turbulence model is called **large eddy simulation** (LES). Large eddy simulation is a compromise between DNS and k-$ε$. LES uses enough detail to resolve the large-scale structures of the turbulence but uses the k-$ε$ equations to resolve the small scales. The LES method allows one to solve problems that are not well modeled with the k-$ε$ model by using an approach that is more computationally efficient than DNS.

Solver. A solver is the computer algorithm that solves the algebraic equations used by the CFD code. The outputs of the solver are the values of the velocity, pressure, and other relevant fields.

Post Processing. After the solver has generated a solution, the code uses this solution to calculate other parameters of interest. This process is called post processing, and the software that does this work is called the **post processor**. Some common functions of a post processor:

- Calculate derivative variables such as vorticity or shear stress
- Calculate integral variables such as pressure force, shear force, lift, drag, coefficient of lift, and coefficient of drag
- Calculate turbulent quantities such as Reynolds stresses and energy spectra
- Develop plots and other visual representations of data; for example
  - ▸ Plots showing time history; for example, time history of forces or wave heights
  - ▸ 2-D contour plots of variables such as pressure, velocity, or vorticity
  - ▸ 2-D velocity vector plots
  - ▸ 3-D iso-surface plots of parameters such as pressure or vorticity
  - ▸ Plots showing streamlines, pathlines, or streaklines
  - ▸ Animations of the flow

Verification and Validation

Engineers are very interested in assessing the trustworthiness of solutions. To this end, the CFD community has adopted methods for assessing correctness.

Validation examines the degree to which CFD predictions agree with real-world observations. A common validation strategy is to systematically compare CFD predictions to experimental data or to solutions to well-known problems, called *benchmark solutions*.

Verification examines the degree to which the numerical methods used by the code result in accurate answers. Verification can involve varying the spacing in the grid and ensuring that the predicted results are not dependent on the grid spacing. Similarly, verification can involve varying the time step to ensure that results are time step independent.

16.6 Examples of CFD

This section presents three examples showing how professionals apply and think about CFD.

Flow through a Spillway

Problem Definition. This study by Li et al. (18) involved the Canton Dam (see Fig. 16.21), which is located on the North Canadian River in Oklahoma. When the dam was built in 1948, the design was based on maximum flowrate (during a flood) of about 10,000 m³/s. Since this time, improved hydrology data have suggested that the dam should be able to pass a peak flood discharge of 17,700 m³/s. Thus, a new auxiliary spillway was proposed, and the study presents an analysis of the proposed spillway.

Methods. A commercial CFD code, Fluent, was used to solve the time-dependent Reynolds-averaged Navier-Stokes (RANS) equations. The turbulence model was a k-ε model with wall functions. The CFD code was used to develop a tentative design. This design was then built in a 1:54 scale physical model, and the experimental data was used to validate the CFD code.

FIGURE 16.21

The Canton Dam showing the proposed new auxiliary spillway.

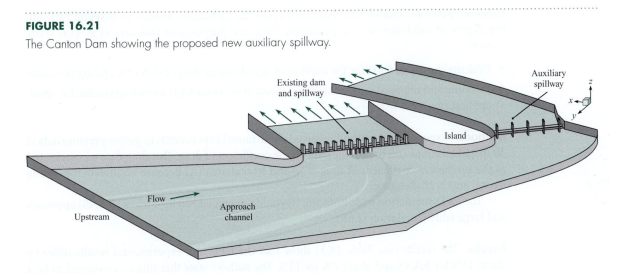

Results. Li et al. stated,

> *"The physical model results were compared to the CFD model results, and found to be in good agreement. The CFD model was thus validated, which in turn validated the methodology used and the results of all the CFD runs, not just the one configuration constructed and tested in the physical model. Averaged water surface elevation at the control plane for the favorable geometry design is still 1.08 m higher than the allowed maximum pool elevation. Since the CFD model runs gave insight into how each geometry modification affected flow patterns and water surface elevations, further modifications to lower the reservoir water surface can be undertaken with confidence directly in the physical model. This study shows a successful application of a CFD model in the design process of an auxiliary spillway. The encouraging result gives hydraulic engineers and CFD modelers more information on integrating numerical model and physical model to the process of design of hydraulic structures."*

In this quote, several useful ideas to notice:

- The engineers concluded that the CFD model was trustworthy.
- The engineers suggest that integrating CFD and experimental modeling is a viable approach for the design of hydraulic structures.

Drag on a Cyclist

Problem Definition. Bicycle racers and coaches want to understand how to reduce aerodynamic drag (see Fig. 16.22) because 90% or more of the resistive forces on the cyclist is due to this drag.

FIGURE 16.22

Cyclist Positions: (a) upright position, (b) dropped position, and (c) time-trail position.

(a) (b) (c)

However, past CFD studies have issues with how the turbulence models were set up and with the degree of validation with experiment. Thus, the purposes of Defraeye et al. (14) study were to

- Evaluate the use of CFD for the analysis of aerodynamic drag of different cycling positions
- Examine and improve some of the limitations of previous CFD modeling studies for sport applications

Methods. Experimental method involved wind-tunnel experiments to gather pressure data at 30 spatial locations and to provide data on the coefficient of drag. This drag data was measured as the product of coefficient of drag (C_D) and frontal area (A) because accurately measuring frontal area is challenging.

The CFD simulation used both the Reynolds-averaged Navier-Stokes (RANS) approach and large eddy simulation (LES).

Results. The results (see Table 16.3) show that the CFD and experimental results differ by about 11% for RANS and about 7% for LES. The authors state that this is considered to be a close agreement in CFD studies. The authors report a fair agreement for the predicted surface pressures, especially with LES. Despite the higher accuracy of LES, the authors suggest that the higher computational cost make RANS more attractive for practical use.

The authors conclude that CFD is a valuable tool for evaluating the drag corresponding to different cyclist positions and for investigating the influence of small adjustments in the cyclist's position. A strong advantage of CFD is that detailed flow field information is obtained, which cannot easily be obtained from wind-tunnel tests. These details provides insights about the drag force and guidance for position improvements.

TABLE 16.3 Predicted Drag for Cyclists from Defraeye et al. (14)

Cyclist Position (see Fig. 16.22)	Turbulence Model	AC_D	Comparison with experiment (%)[a]
Upright	RANS	0.219	13
	LES	0.219	13
Dropped	RANS	0.179	7
	LES	0.172	3
Time Trial	RANS	0.150	12
	LES	0.142	6

[a]The comparison with experiment is calculated using this formula:

$$\frac{(AC_D \text{ predicted from CFD}) - (AC_D \text{ measured from experiment})}{(AC_D \text{ measured from experiment})}$$

Predicting Wind Loads on a Telescope Structure

Problem. Because the next generation of optical telescopes are large, wind loading on the structure becomes more significant. Thus, Mamou et al. (19) conducted a study to investigate the wind-loading on the prototype Canadian/United States Very Large Optical Telescope (VLOT) structure. The study was done during the first phase of design to assess wind loads, vortex shedding, and cavity resonances caused by wind blowing over the opening of the telescope. The structure (Fig. 16.23) is 51 m in diameter, with a 24-meter-diameter opening through which the telescope views the sky. The purpose of the study was to assess the capability of a CFD model.

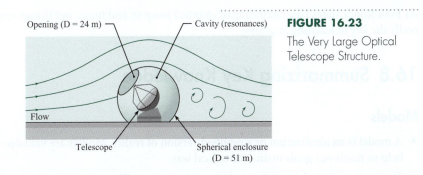

FIGURE 16.23

The Very Large Optical Telescope Structure.

Methods. The code was a fully unsteady Lattice-Boltzmann CFD program. Wind tunnel data were used to validate the code.

Results. The authors noted that cavity resonance due to flow over the opening and vortex shedding from the spherical structure were observed in the wind tunnel experiments and the CFD computations. The CFD code predicted three simultaneously excited cavity modes that were identical to those measured.

16.7 A Path for Moving Forward

Because some students want to learn more fluid mechanics, this section gives ideas for moving forward.

Study at the Graduate Level

Some useful graduate courses include partial differential equations, continuum mechanics, numerical methods, fluid mechanics, and computational fluid mechanics. While taking classes, some useful ways to expand one's horizons are to

- Read the research literature
- Read technical books
- Read on the Internet (for example, see the online CFD community at http://www.cfd-online.com/)

Learn via Application (Jump into the Swimming Pool)

Some ideas for application include

- Find a CFD code and learn to run this code.
- Do projects for companies.
- Get involved in research. Take the lead role in writing a research paper.

For students who become involved in research, consider going to conferences and presenting your work. Submit your papers for publication. Sometimes work will get criticized, but peer reviews are an opportunity for learning.

At research meetings, get to know the members of the community. Most people who attend research meeting have passion for their technical work, and many enjoy helping new people who are becoming engaged in the discipline.

Follow John Roncz's Advice

As John Roncz states (see beginning of chapter) jump in and figure out things yourself. This is really the key to learning anything.

16.8 Summarizing Key Knowledge

Models

- A model is an idealization or simplified version of reality. Models are valuable when they help us reach our goals in an economical way.
- The process of model building is an iterative process. The steps are
  - ▶ Identify the variables.
  - ▶ Classify the variables into performance variables (dependent variables) and design variables (independent variables).
  - ▶ Determine how to relate the variables. When variables can be related by applying engineering equations, apply the Wales-Woods model. When variables can be related by correlating experimental data, apply regression analysis and other methods from statistics.
  - ▶ Validate to determine if model predictions are accurate enough.
- In fluid mechanics, there are three approaches to model building: analytical fluid mechanics, experimental fluid mechanics, and computational fluid mechanics. Most models involve two or three of these approaches working synergistically.
- Model building is best done by starting with simple models and then evolving these models through an iterative process. Multiple trade-offs in model building involve resources, benefits, solution accuracy, and solution detail.

Foundations for Learning Partial Differential Equations (PDEs)

- The PDEs that govern flowing fluids can be solved for only a few special cases because nonlinear terms preclude a general solution. Problems that can be solved are called *exact solutions*. These exact solutions were discovered many years ago.
- Two reasons for learning PDEs are
  - ▶ To understand and apply existing solutions (found in the literature)
  - ▶ To understand the equations that are being solved by CFD codes
- The solution of the PDEs are fields. The general form of a field is exemplified by the velocity field. The velocity field is

Cartesian	$\mathbf{V} = u(x, y, z, t)\mathbf{i} + v(x, y, z, t)\mathbf{j} + w(x, y, z, t)\mathbf{k}$
Cylindrical	$\mathbf{V} = v_r(r, \theta, z, t)\mathbf{u}_r + v_\theta(r, \theta, z, t)\mathbf{u}_\theta + v_z(r, \theta, z, t)\mathbf{u}_z$

- Notice that the velocity field involves
  - ▶ *Independent variables.* The independent variables are the three position variables and time.
  - ▶ *Dependent variables.* The dependent variables are the three velocity components.

- Taylor series are commonly applied in fluid mechanics for developing derivations and for developing CFD programs. A useful form of the Taylor series is

$$f(x + \Delta x, y, z, t) = f(x, y, z, t) + \left(\frac{\partial f}{\partial x}\right)_{x,y,z,t} \frac{\Delta x}{1!} + \left(\frac{\partial^2 f}{\partial x^2}\right)_{x,y,z,t} \frac{(\Delta x)^2}{2!} + \dots \text{H.O.T.}$$

  where H.O.T. stands for "Higher Order Terms." For a small change (i.e., Δx is small), higher order terms are often neglected.

- PDEs are written in two ways.
  - ▶ *Coordinate specific form.* Terms apply to a specific coordinate system. This approach is useful for specific applications.
  - ▶ *Invariant form.* Terms apply to any coordinate system; that is, they generalize. This approach is useful for writing (e.g., thesis, research paper) and presentations because the equations are compact, and they illustrate the physics.

- *Invariant notation* is a mathematical notation that applies (i.e., generalizes) to multiple coordinate systems. Three common forms of invariant notation are
  - ▶ *Del notation* uses the nabla symbol ∇.
  - ▶ *Gibbs notation* uses words (e.g., grad, div, curl) to represent operators.
  - ▶ *Indicial notation* uses subscripted letters to represent vector components and summations.

- An *operator* is a named collection of mathematical terms. Common operators in fluid mechanics equations are
  - ▶ *Gradient*: for example the gradient of the pressure field
  - ▶ *Divergence*: for example, the divergence of the velocity field
  - ▶ *Curl*: for example, the curl of the velocity field
  - ▶ *LaPlacian*: for example, the LaPlacian of the velocity field
  - ▶ *Material Derivative*: for example, the time derivative of the temperature field

- Each operator has one or more physical interpretations. These interpretations can be developed by working through the derivations of the PDEs.

- The *material derivative*
  - ▶ Has multiple names in the literature (e.g., substantial derivative, Lagrangian derivative, and derivative following the particle)
  - ▶ Represents the time rate of change of a property of a fluid particle
  - ▶ In symbols is

$$\underbrace{\frac{dJ}{dt}}_{\substack{\text{time derivative} \\ \text{of property } J \text{ of} \\ \text{a fluid particle}}} = \underbrace{\left(\frac{\partial J}{\partial t}\right) + \mathbf{V} \cdot \nabla J = \left(\frac{\partial J}{\partial t}\right) + \mathbf{V} \cdot \text{grad}\,(J)}_{\substack{\text{mathematics needed to do the derivative when} \\ \text{a field (i.e., an Eulerian approach) is being used}}}$$

- *Acceleration*, defined at a point in space, means the acceleration of the fluid particle at this point at the given instant in time. Acceleration in Cartesian coordinates is

$$\binom{\text{acceleration}}{\text{of a fluid particle}} = \mathbf{a} = \frac{d\mathbf{V}}{dt} = \begin{bmatrix} \left\{\left(\frac{\partial u}{\partial t}\right) + u\left(\frac{\partial u}{\partial x}\right) + v\left(\frac{\partial u}{\partial y}\right) + w\left(\frac{\partial u}{\partial z}\right)\right\}\mathbf{i} \\ \left\{\left(\frac{\partial v}{\partial t}\right) + u\left(\frac{\partial v}{\partial x}\right) + v\left(\frac{\partial v}{\partial y}\right) + w\left(\frac{\partial v}{\partial z}\right)\right\}\mathbf{j} \\ \left\{\left(\frac{\partial w}{\partial t}\right) + u\left(\frac{\partial w}{\partial x}\right) + v\left(\frac{\partial w}{\partial y}\right) + w\left(\frac{\partial w}{\partial z}\right)\right\}\mathbf{k} \end{bmatrix}$$

The Continuity Equation

- Any problem involving a flowing fluid can, in principle, be solved by solving a coupled set of five partial differential equations comprised of the continuity equation, the momentum equation, the energy equation, and two equations of state.

- The *conservation form* of the continuity equation is derived by applying the law of conservation of mass to a differential control volume. The resulting equation, in Cartesian coordinates, is

$$\underbrace{\frac{\partial \rho}{\partial t}}_{\substack{\text{rate of accumulation of mass} \\ \text{in a differential CV divided by} \\ \text{the volume of the CV} \\ (\text{kg/s per m}^3)}} + \underbrace{\frac{\partial(\rho u)}{\partial x} + \frac{\partial(\rho v)}{\partial y} + \frac{\partial(\rho w)}{\partial z}}_{\substack{\text{net rate of mass flow} \\ \text{out of the CV divided by the} \\ \text{volume of the CV} \\ (\text{kg/s per m}^3)}} = 0$$

- The continuity equation can be expressed using two forms.

  ▸ The *conservation form* is derived by starting with a differential control volume and applying conservation of mass to this CV.

  ▸ The *nonconservation form* is derived by starting with a differential fluid particle and applying conservation of mass to this particle.

  ▸ The conservation and nonconservation forms are mathematically equivalent because one can start with one form of the equation and derive the other form.

- The nonconservation form of the continuity equation in Cartesian coordinates is

$$\frac{d\rho}{dt} + \rho\left(\frac{\partial u}{\partial x} + \frac{\partial v}{\partial y} + \frac{\partial w}{\partial z}\right) = 0$$

- Derivation of the continuity equation provides two interpretations of the divergence operator.

$$\text{div}(\rho\mathbf{V}) = \nabla \cdot (\rho\mathbf{V}) = \frac{\left(\begin{array}{c}\text{net rate of outflow of mass} \\ \text{from a differential CV centered} \\ \text{about point } (x, y, z)\end{array}\right)}{(\text{volume of the CV})}$$

$$\nabla \cdot \mathbf{V} = \text{div}(\mathbf{V}) = \frac{(\text{time rate of change of the volume of a fluid particle})}{(\text{volume of the fluid particle})}$$

- When density is constant, the flow is called incompressible, and the continuity equation can be written as:

Invariant form	$\nabla \cdot \mathbf{V} = \text{div}(\mathbf{V}) = 0$
Cartesian coordinates	$\dfrac{\partial u}{\partial x} + \dfrac{\partial v}{\partial y} + \dfrac{\partial w}{\partial z} = 0$

The Navier-Stokes Equation

- The Navier-Stokes equation is derived by applying Newton's second law of motion to a viscous flow while also assuming that the fluid is Newtonian.

- In invariant form, the Navier-Stokes equation for an incompressible flow with constant density and viscosity is.

$$\underbrace{\rho \frac{d\mathbf{V}}{dt}}_{\begin{pmatrix}\text{mass of the particle times}\\\text{acceleration of the particle}\\\text{divided by the volume of the particle}\end{pmatrix}} = \underbrace{\rho\mathbf{g}}_{\begin{pmatrix}\text{weight}\\\text{of the particle}\\\text{divided by its volume}\end{pmatrix}} + \underbrace{-\nabla p}_{\begin{pmatrix}\text{net pressure force}\\\text{on the particle}\\\text{divided by its volume}\end{pmatrix}} + \underbrace{\mu\nabla^2\mathbf{V}}_{\begin{pmatrix}\text{net shear force}\\\text{on the particle}\\\text{divided by its volume}\end{pmatrix}}$$

- Derivation of the Navier-Stokes equation reveals the physics of operators.
  - ▸ The gradient of the pressure field describes the net pressure force on a fluid particle divided by the volume of the particle.
  - ▸ The divergence of the shear stress tensor describes the net viscous force on a fluid particle divided by the volume of the particle.
- Nonlinear terms (see Fig. 16.24) appear in the acceleration term of the Navier-Stokes equation.

FIGURE 16.24

Nonlinear terms in the Navier-Stokes equation contain the product of velocity and its derivative.

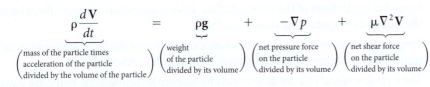

velocity ⟶ derivative of velocity

Computational Fluid Dynamics (CFD)

- *Computational fluid dynamics* (CFD) is a method for solving fluid mechanics problems by developing approximate solutions to the governing PDEs. Benefits of learning CFD include
  - ▸ CFD can be applied to model complex problems that cannot be modeled effectively with experiment or analysis
  - ▸ CFD provides a way to vary design parameters and learn what happens to the performance of the system under study.
  - ▸ CFD is widely used in industry.
- Regarding CFD codes
  - ▸ Engineers typically apply an existing code rather than writing their own code because many excellent codes are available, and the process of developing a code takes years of effort.
  - ▸ Engineers select codes that fit the type of problem that they are trying to solve (e.g., for modeling groundwater, engineers might select MODFLOW; for modeling an internal combustion engine, engineers might select KIVA).
- CFD codes have an associated language.
  - ▸ A *grid* is a set of points in space at which a code solves for values of velocity and other variables of interest.
  - ▸ A *time step* is the interval between each solution time.
  - ▸ *Boundary conditions* are specified values of the dependent variables (e.g., pressure, velocity) on the physical boundaries of the problem.
  - ▸ Specifying an *initial condition* involves giving numerical values for the *dependent variables* at all spatial points at the starting time of the solution.
  - ▸ A *solver* is a label for the computer algorithm that solves the algebraic equations that approximate the PDEs that are being solved by the CFD code.
  - ▸ A *post processor* is a computer algorithm that uses the solution from the solver to generate plots and calculate parameters such as drag force and shear stress.
  - ▸ *Validation* assesses the degree to which CFD predictions agree with experimental data.
  - ▸ *Verification* examines the degree to which the numerical methods used by the code result in accurate answers.

- Three common approaches to modeling turbulent flow are:

  ▸ *Direct numerical simulation* (DNS) involves setting the grid and time steps fine enough to resolve the features of the turbulent flow. DNS is unrealistic for most flows because the required computational time is too large.

  ▸ *Large eddy simulation* (LES) involves direct simulation of the large-scale eddies in the turbulence and approximate simulation of the smaller eddies.

  ▸ The *k-epsilon model* (k-ε model) models turbulence by introducing two extra equations. As compared to DNS and LES, the k-ε model is computationally efficient.

REFERENCES

1. Noland, David, Wing Man, *Air & Space: Smithsonian*, 5(5), December 1990/January 1991, p. 34–40.

2. Wang, Herbert, and Mary P. Anderson. *Introduction to Groundwater Modeling : Finite Difference and Finite Element Methods*. San Diego: Academic Press, 1982.

3. Ford, Andrew. *Modeling the Environment*. Washington, DC: Island Press, 2010.

4. Montgomery, Douglas C., George C. Runger, and Norma Faris Hubele. *Engineering Statistics*. Hoboken, NJ: John Wiley, 2011.

5. Kojima, H., S. Togami, and B. Co. *The Manga Guide to Calculus*. San Francisco: No Starch Press, 2009.

6. Schey, H. M. *Div, Grad, Curl, and All That: An Informal Text on Vector Calculus*. New York: W.W. Norton, 2005.

7. White, Frank M. *Fluid Mechanics*. 4e, Boston; London: McGraw-Hill, 2011.

8. White, Frank M. *Viscous Fluid Flow*. New York: McGraw-Hill, 2006.

9. Pritchard, P. J., *Fox and McDonald's Introduction to Fluid Mechanics*, 8e. Hoboken, NJ: John Wiley, 2011.

10. "MODFLOW—Wikipedia, the free encyclopedia." Downloaded on 1/3/12 from http://en.wikipedia.org/wiki/Modflow

11. "KIVA." Downloaded on 1/3/12 from http://www.lanl.gov/orgs/t/t3/codes/kiva.shtml

12. "KIVA (software)—Wikipedia, the free encyclopedia." Downloaded on 1/3/12 from http://en.wikipedia.org/wiki/KIVA_(software)

13. "Computational Fluid Dynamics Software | FLOW-3D from Flow Science, CFD." Downloaded on 1/3/12 from http://www.flow3d.com/

14. "Test Cases." Downloaded on 1/4/12 from http://cfl3d.larc.nasa.gov/Cfl3dv6/cfl3dv6_testcases.html#cylinder

15. Wyman, Nick, "CFD Review | State of the Art in Grid Generation." Downloaded on 1/14/12 from http://www.cfdreview.com/article.pl?sid=01/04/28/2131215

16. http://piv.tamu.edu/CFD/les.htm, downloaded on 2/14/12.

17. Defraeye, Thijs, Bert Blocken, Erwin Koninckx, Peter Hespel, and Jan Carmeliet. "Aerodynamic Study of Different Cyclist Positions: CFD Analysis and Full-Scale Wind-Tunnel Tests." *Journal of Biomechanics* 43, no. 7 (2010).

18. Li, S., S. Cain, M. Wosnik, C. Miller, H. Kocahan, and P. E. Russell Wyckoff. "Numerical Modeling of Probable Maximum Flood Flowing Through a System of Spillways." *Journal of Hydraulic Engineering* 137 (2011).

19. Mamou, M., K. Cooper, A. Benmeddour, M. Khalid, J. Fitzsimmons, and R. Sengupta. "Correlation of CFD Predictions and Wind Tunnel Measurements of Mean and Unsteady Wind Loads on a Large Optical Telescope." *Journal of Wind Engineering and Industrial Aerodynamics* 96, no. 6–7 (2008).

PROBLEMS

PLUS Problem available in *WileyPLUS* at instructor's discretion.

GO Guided Online (GO) Problem, available in *WileyPLUS* at instructor's discretion.

Models in Fluid Mechanics (§16.1)

16.1 Which of the following could be considered a model? Why? (select all that apply).

- **a.** The ideal-gas law
- **b.** A set of instructions for using a pitot-static tube to measure velocity
- **c.** An airplane built from a kit
- **d.** A computer program to predict the force on a pipe bend

16.2 Apply the modeling building process to the *Balloon Payload* problem described here.

- **a.** What are the relevant variables?
- **b.** How are the variables related? What are the relevant equations? How can you apply these equations to develop a single algebraic equation to solve for your goal?
- **c.** What is a simple and low-cost way to test your math model using experimental data?

THE BALLOON PAYLOAD PROBLEM

Your team is designing a helium-filled balloon that will travel to at least 24,000 meter elevation in the atmosphere. The balloon will transport a payload comprised of a camera and a data acquisition system. Right now, you choose to solve a simpler problem, which is to develop a model that predicts the weight on the earth's surface (at your location) such that a helium-filled balloon is neutrally buoyant. This simpler problem can be easily tested with an experiment in your classroom.

16.3 Apply the modeling building process to the *Rocket Problem* described here.

    a. What are the relevant variables?

    b. How are the variables related? What are the relevant equations?

    c. What is a simple and low-cost way to test your math model using experimental data?

THE ROCKET PROBLEM

Your team is designing a two-stage, solid-fuel rocket that is intended to travel to 4,500 meter and take photos from this elevation. Right now, you choose to solve a simpler problem, which is to develop a model that predicts the height that a small, low-cost rocket will fly because a small rocket can be purchased from manufacturers such as Estes® or Pitsco®, and it is relatively easy to measure elevation for such a rocket.

Foundations for Learning PDEs (§16.2)

16.4 Why do you think that engineers make the effort to learn partial differential equations? What are the benefits to them?

16.5 Consider the function $f(x) = \dfrac{1}{1-x}$. Show how to find the Taylor series expansion for the function $f(x)$ about the point $x = 0$. Evaluate the numerical value of the Taylor series for $x = 0.1$ using 5 terms.

16.6 Consider the function $f(x) = \ln(x)$. Show how to find the Taylor series expansion for the function $f(x)$ about the point $x = a$. Then, find the numerical value for $x = 1.5$ and six terms of the Taylor series expansion.

16.7 Consider a flat horizontal plate that is infinite in size in both dimensions. Above the plate is a fluid of viscosity μ. The plate is at rest. Then, at time equals zero seconds, the plate is set in motion to the right with a constant velocity V acting to the

right. Consider the velocity field in the fluid above the plate and simplify the general form of the velocity field by answering the following questions.

    a. Which velocity components (u, v, w) are zero? Which are nonzero? Why?

    b. Which spatial variables (x, y, z) are parameters? Which can be ignored? Why?

    c. Is time a parameter? Or, can time be ignored? Why?

    d. What is the reduced equation that represents the velocity field?

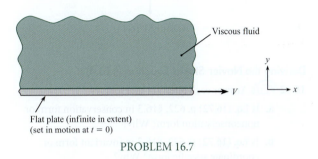

Flat plate (infinite in extent)
(set in motion at $t = 0$)

PROBLEM 16.7

The Continuity Equation (§16.3)

16.8 Compare and contrast the integral-form of the continuity equation (Eq. 5.28), p. 183, §5.3 with the PDE form of the continuity equation (Eq. 16.36) p. 614, §16.3. Address the following questions.

    a. Are the units and dimensions the same? Or different?

    b. How do the physics compare? What is the same? What is different?

    c. How do the derivations compare? What is the same? What is different?

    d. When would you want to apply the integral-form of the continuity equation (Chapter 5)? When would you want to apply the PDE-form of the continuity equation (Chapter 16)?

16.9 Start with the conservation form of the continuity equation in Cartesian coordinates and derive the nonconservation form.

16.10 Start with the nonconservation form of the continuity equation in Cartesian coordinates and derive the conservation form.

16.11 Consider water draining out of round hole in the bottom of a round tank. Assume constant density and also assume that the water does not swirl. Then,

    a. Select the general form of the continuity equation that best applies to this problem.

    b. Show how to simplify the general equation from part (a) to develop the reduced form.

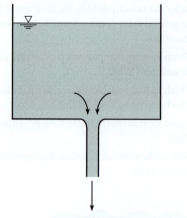

PROBLEM 16.11

Deriving the Navier-Stokes Equation (§16.4)

16.12 PLUS Answer each question that follows.

    a. Is Eq. (16.72) p. 622, §16.3 in conservation form or nonconservation form? Why?

    b. Is Eq. (16.72) p. 622, §16.3 in invariant form or coordinate specific form? Why?

16.13 PLUS What are the physics of the gradient of the pressure field? What are the units? What are the dimensions?

16.14 What are the physics of the divergence of the shear stress tensor? What are the units? What are the dimensions.?

16.15 Compare the Navier-Stokes equation to Euler's equation.

    a. What are the two important similarities?

    b. What are two important differences?

16.16 Stress, as introduced in the derivation of the Navier-Stokes equation, is a second-order tensor. Using the Internet, find some articles on tensors and answer the following questions:

    a. Why do people use tensors? What are the benefits?

    b. What does tensor mean? How is a tensor defined?

    c. What are five examples of tensors as they are applied in engineering and physics?

Computational Fluid Dynamics (§16.5)

16.17 If someone asked you, *Why are CFD codes useful for engineers?* how would you answer? List your top three reasons in priority order.

16.18 Would you prefer to write your own CFD programs, or would you prefer to use codes that have been written by others? Discuss the advantages and disadvantages of each approach.

16.19 Using the Internet, find one example of a publicly available CFD program (either a commercial or noncommercial code) and describe the code so that others can understand the code. Address the following questions in your response.

    a. What is the history of the code? When was the code developed? By whom?

    b. What is the main purpose of the code? What type of flow is the code well suited for?

    c. How much does the code cost?

    d. What training and resources are available to help you learn the code?

16.20 Briefly explain the meaning of each of the following ideas.

    a. Grid

    b. Time Step

    c. Solution time for a CFD program versus the accuracy

    d. Boundary condition

    e. Initial condition

16.21 Briefly explain the meaning of each of the following ideas.

    a. DNS

    b. *k*-epsilon method

    c. LES

16.22 Briefly explain the meaning of each of the following ideas.

    a. Post processor

    b. Verification

    c. Validation

APPENDIX

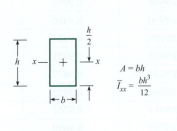

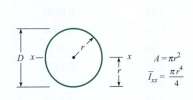

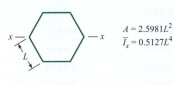

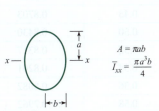

Volume and Area Formulas:

$$A_{circle} = \pi r^2 = \pi D^2/4$$

$$A_{sphere\ surface} = \pi D^2$$

$$V_{sphere} = \frac{1}{6}\pi D^3 = \frac{4}{3}\pi r^3$$

$$V_{cone} = \frac{1}{12}\pi D^2 h = \frac{1}{3}\pi r^3 h$$

TABLE A.1 Compressible Flow Tables for an Ideal Gas with $k = 1.4$

M or M_1 = local number or Mach number upstream of a normal shock wave; p/p_t = ratio of static pressure to total pressure; ρ/ρ_t = ratio of static density to total density; T/T_t = ratio of static temperature to total temperature; A/A_* = ratio of local cross-sectional area of an isentropic stream tube to cross-sectional area at the point where M = 1; M_2 = Mach number downstream of a normal shock wave; p_2/p_1 = static pressure ratio across a normal shock wave; T_2/T_1 = static pressure ratio across a normal shock wave; p_{t_2}/p_{t_1} = total pressure ratio across normal shock wave.

Subsonic Flow				
M	p/p_t	ρ/ρ_t	T/T_t	A/A_*
0.00	1.0000	1.0000	1.0000	∞
0.05	0.9983	0.9988	0.9995	11.5914
0.10	0.9930	0.9950	0.9980	5.8218
0.15	0.9844	0.9888	0.9955	3.9103
0.20	0.9725	0.9803	0.9921	2.9630
0.25	0.9575	0.9694	0.9877	2.4027
0.30	0.9395	0.9564	0.9823	2.0351
0.35	0.9188	0.9413	0.9761	1.7780
0.40	0.8956	0.9243	0.9690	1.5901
0.45	0.8703	0.9055	0.9611	1.4487
0.50	0.8430	0.8852	0.9524	1.3398
0.52	0.8317	0.8766	0.9487	1.3034
0.54	0.8201	0.8679	0.9449	1.2703
0.56	0.8082	0.8589	0.9410	1.2403
0.58	0.7962	0.8498	0.9370	1.2130
0.60	0.7840	0.8405	0.9328	1.1882
0.62	0.7716	0.8310	0.9286	1.1657
0.64	0.7591	0.8213	0.9243	1.1452
0.66	0.7465	0.8115	0.9199	1.1265
0.68	0.7338	0.8016	0.9153	1.1097
0.70	0.7209	0.7916	0.9107	1.0944
0.72	0.7080	0.7814	0.9061	1.0806
0.74	0.6951	0.7712	0.9013	1.0681
0.76	0.6821	0.7609	0.8964	1.0570
0.78	0.6691	0.7505	0.8915	1.0471
0.80	0.6560	0.7400	0.8865	1.0382
0.82	0.6430	0.7295	0.8815	1.0305
0.84	0.6300	0.7189	0.8763	1.0237
0.86	0.6170	0.7083	0.8711	1.0179
0.88	0.6041	0.6977	0.8659	1.0129
0.90	0.5913	0.6870	0.8606	1.0089
0.92	0.5785	0.6764	0.8552	1.0056
0.94	0.5658	0.6658	0.8498	1.0031
0.96	0.5532	0.6551	0.8444	1.0014
0.98	0.5407	0.6445	0.8389	1.0003
1.00	0.5283	0.6339	0.8333	1.0000

(Continued)

TABLE A.1 Compressible Flow Tables for an Ideal Gas with $k = 1.4$ (Continued)

	Supersonic Flow				Normal Shock Wave			
M_1	p/p_t	ρ/ρ_t	T/T_t	A/A_*	M_2	p_2/p_1	T_2/T_1	p_{t_2}/p_{t_1}
1.00	0.5283	0.6339	0.8333	1.000	1.0000	1.000	1.000	1.0000
1.01	0.5221	0.6287	0.8306	1.000	0.9901	1.023	1.007	0.9999
1.02	0.5160	0.6234	0.8278	1.000	0.9805	1.047	1.013	0.9999
1.03	0.5099	0.6181	0.8250	1.001	0.9712	1.071	1.020	0.9999
1.04	0.5039	0.6129	0.8222	1.001	0.9620	1.095	1.026	0.9999
1.05	0.4979	0.6077	0.8193	1.002	0.9531	1.120	1.033	0.9998
1.06	0.4919	0.6024	0.8165	1.003	0.9444	1.144	1.039	0.9997
1.07	0.4860	0.5972	0.8137	1.004	0.9360	1.169	1.046	0.9996
1.08	0.4800	0.5920	0.8108	1.005	0.9277	1.194	1.052	0.9994
1.09	0.4742	0.5869	0.8080	1.006	0.9196	1.219	1.059	0.9992
1.10	0.4684	0.5817	0.8052	1.008	0.9118	1.245	1.065	0.9989
1.11	0.4626	0.5766	0.8023	1.010	0.9041	1.271	1.071	0.9986
1.12	0.4568	0.5714	0.7994	1.011	0.8966	1.297	1.078	0.9982
1.13	0.4511	0.5663	0.7966	1.013	0.8892	1.323	1.084	0.9978
1.14	0.4455	0.5612	0.7937	1.015	0.8820	1.350	1.090	0.9973
1.15	0.4398	0.5562	0.7908	1.017	0.8750	1.376	1.097	0.9967
1.16	0.4343	0.5511	0.7879	1.020	0.8682	1.403	1.103	0.9961
1.17	0.4287	0.5461	0.7851	1.022	0.8615	1.430	1.109	0.9953
1.18	0.4232	0.5411	0.7822	1.025	0.8549	1.458	1.115	0.9946
1.19	0.4178	0.5361	0.7793	1.026	0.8485	1.485	1.122	0.9937
1.20	0.4124	0.5311	0.7764	1.030	0.8422	1.513	1.128	0.9928
1.21	0.4070	0.5262	0.7735	1.033	0.8360	1.541	1.134	0.9918
1.22	0.4017	0.5213	0.7706	1.037	0.8300	1.570	1.141	0.9907
1.23	0.3964	0.5164	0.7677	1.040	0.8241	1.598	1.147	0.9896
1.24	0.3912	0.5115	0.7648	1.043	0.8183	1.627	1.153	0.9884
1.25	0.3861	0.5067	0.7619	1.047	0.8126	1.656	1.159	0.9871
1.30	0.3609	0.4829	0.7474	1.066	0.7860	1.805	1.191	0.9794
1.35	0.3370	0.4598	0.7329	1.089	0.7618	1.960	1.223	0.9697
1.40	0.3142	0.4374	0.7184	1.115	0.7397	2.120	1.255	0.9582
1.45	0.2927	0.4158	0.7040	1.144	0.7196	2.286	1.287	0.9448
1.50	0.2724	0.3950	0.6897	1.176	0.7011	2.458	1.320	0.9278
1.55	0.2533	0.3750	0.6754	1.212	0.6841	2.636	1.354	0.9132
1.60	0.2353	0.3557	0.6614	1.250	0.6684	2.820	1.388	0.8952
1.65	0.2184	0.3373	0.6475	1.292	0.6540	3.010	1.423	0.8760
1.70	0.2026	0.3197	0.6337	1.338	0.6405	3.205	1.458	0.8557
1.75	0.1878	0.3029	0.6202	1.386	0.6281	3.406	1.495	0.8346
1.80	0.1740	0.2868	0.6068	1.439	0.6165	3.613	1.532	0.8127
1.85	0.1612	0.2715	0.5936	1.495	0.6057	3.826	1.569	0.7902

(Continued)

TABLE A.1 Compressible Flow Tables for an Ideal Gas with $k = 1.4$ (Continued)

	Supersonic Flow				Normal Shock Wave			
M_1	p/p_t	ρ/ρ_t	T/T_t	A/A_*	M_2	p_2/p_1	T_2/T_1	p_{t_2}/p_{t_1}
1.90	0.1492	0.2570	0.5807	1.555	0.5956	4.045	1.608	0.7674
1.95	0.1381	0.2432	0.5680	1.619	0.5862	4.270	1.647	0.7442
2.00	0.1278	0.2300	0.5556	1.688	0.5774	4.500	1.688	0.7209
2.10	0.1094	0.2058	0.5313	1.837	0.5613	4.978	1.770	0.6742
2.20	0.9352^{-1}†	0.1841	0.5081	2.005	0.5471	5.480	1.857	0.6281
2.30	0.7997^{-1}	0.1646	0.4859	2.193	0.5344	6.005	1.947	0.5833
2.50	0.5853^{-1}	0.1317	0.4444	2.637	0.5130	7.125	2.138	0.4990
2.60	0.5012^{-1}	0.1179	0.4252	2.896	0.5039	7.720	2.238	0.4601
2.70	0.4295^{-1}	0.1056	0.4068	3.183	0.4956	8.338	2.343	0.4236
2.80	0.3685^{-1}	0.9463^{-1}	0.3894	3.500	0.4882	8.980	2.451	0.3895
2.90	0.3165^{-1}	0.8489^{-1}	0.3729	3.850	0.4814	9.645	2.563	0.3577
3.00	0.2722^{-1}	0.7623^{-1}	0.3571	4.235	0.4752	10.330	2.679	0.3283
3.50	0.1311^{-1}	0.4523^{-1}	0.2899	6.790	0.4512	14.130	3.315	0.2129
4.00	0.6586^{-2}	0.2766^{-1}	0.2381	10.72	0.4350	18.500	4.047	0.1388
4.50	0.3455^{-2}	0.1745^{-1}	0.1980	16.56	0.4236	23.460	4.875	0.9170^{-1}
5.00	0.1890^{-2}	0.1134^{-1}	0.1667	25.00	0.4152	29.000	5.800	0.6172^{-1}
5.50	0.1075^{-2}	0.7578^{-2}	0.1418	36.87	0.4090	35.130	6.822	0.4236^{-1}
6.00	0.6334^{-2}	0.5194^{-2}	0.1220	53.18	0.4042	41.830	7.941	0.2965^{-1}
6.50	0.3855^{-2}	0.3643^{-2}	0.1058	75.13	0.4004	49.130	9.156	0.2115^{-1}
7.00	0.2416^{-3}	0.2609^{-2}	0.9259^{-1}	104.1	0.3974	57.000	10.47	0.1535^{-1}
7.50	0.1554^{-3}	0.1904^{-2}	0.8163^{-1}	141.8	0.3949	65.460	11.88	0.1133^{-1}
8.00	0.1024^{-3}	0.1414^{-2}	0.7246^{-1}	190.1	0.3929	74.500	13.39	0.8488^{-2}
8.50	0.6898^{-4}	0.1066^{-2}	0.6472^{-1}	251.1	0.3912	84.130	14.99	0.6449^{-2}
9.00	0.4739^{-4}	0.8150^{-3}	0.5814^{-1}	327.2	0.3898	94.330	16.69	0.4964^{-2}
9.50	0.3314^{-4}	0.6313^{-3}	0.5249^{-1}	421.1	0.3886	105.100	18.49	0.3866^{-2}
10.00	0.2356^{-4}	0.4948^{-3}	0.4762^{-1}	535.9	0.3876	116.500	20.39	0.3045^{-2}

†x^{-n} means $x \cdot 10^{-n}$.

Source: Abridged with permission from R. E. Bolz and G. L. Tuve, *The Handbook of Tables for Applied Engineering Sciences,* CRC Press, Inc., Cleveland, 1973. Copyright © 1973 by The Chemical Rubber Co., CRC Press, Inc.

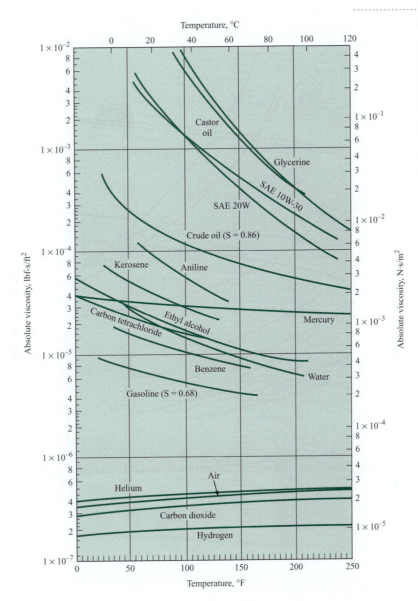

FIGURE A.2

Absolute viscosities of certain gases and liquids [Adapted from Fluid Mechanics, 5th ed., by V. L. Streeter. Copyright © 1971, McGraw-Hill Book Company, New York. Used with permission of the McGraw-Hill Book Company.]

FIGURE A.3

Kinematic viscosities of certain gases and liquids. The gases are at standard pressure. [Adapted from Fluid Mechanics, 5th ed., by V. L. Streeter. Copyright © 1971, McGraw-Hill Book Company, New York. Used with permission of the McGraw-Hill Book Company.]

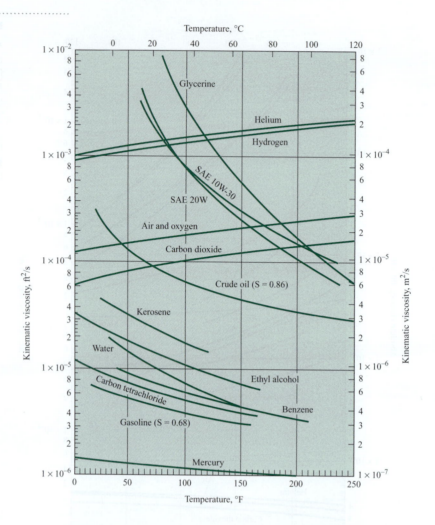

TABLE A.2 Physical Properties of Gases [$T = 15°C$ (59°F), $p = 1$ atm]

Gas	Density kg/m³ (slugs/ft³)	Kinematic Viscosity m²/s (ft²/s)	R Gas Constant J/kg K (ft-lbf/slug-°R)	c_p $\frac{J}{kg\ K}$ $\left(\frac{Btu}{lbm\text{-}°R}\right)$	$k = \dfrac{c_p}{c_v}$	S Sutherland's Constant K(°R)
Air	1.22 (0.00237)	1.46×10^{-5} (1.58×10^{-4})	287 (1716)	1004 (0.240)	1.40	111 (199)
Carbon dioxide	1.85 (0.0036)	7.84×10^{-6} (8.48×10^{-5})	189 (1130)	841 (0.201)	1.30	222 (400)
Helium	0.169 (0.00033)	1.14×10^{-4} (1.22×10^{-3})	2077 (12,419)	5187 (1.24)	1.66	79.4 (143)
Hydrogen	0.0851 (0.00017)	1.01×10^{-4} (1.09×10^{-3})	4127 (24,677)	14,223 (3.40)	1.41	96.7 (174)
Methane (natural gas)	0.678 (0.0013)	1.59×10^{-5} (1.72×10^{-4})	518 (3098)	2208 (0.528)	1.31	198 (356)
Nitrogen	1.18 (0.0023)	1.45×10^{-5} (1.56×10^{-4})	297 (1776)	1041 (0.249)	1.40	107 (192)
Oxygen	1.35 (0.0026)	1.50×10^{-5} (1.61×10^{-4})	260 (1555)	916 (0.219)	1.40	

Source: V. L. Streeter (ed.), *Handbook of Fluid Dynamics,* McGraw-Hill Book Company, New York, 1961; also R. E. Bolz and G. L. Tuve, *Handbook of Tables for Applied Engineering Science,* CRC Press, Inc. Cleveland, 1973; and *Handbook of Chemistry and Physics,* Chemical Rubber Company, 1951.

TABLE A.3 Mechanical Properties of Air at Standard Atmospheric Pressure

Temperature	Density	Specific Weight	Dynamic Viscosity	Kinematic Viscosity
	kg/m^3	N/m^3	N · s/m^2	m^2/s
−20°C	1.40	13.70	1.61×10^{-5}	1.16×10^{-5}
−10°C	1.34	13.20	1.67×10^{-5}	1.24×10^{-5}
0°C	1.29	12.70	1.72×10^{-5}	1.33×10^{-5}
10°C	1.25	12.20	1.76×10^{-5}	1.41×10^{-5}
20°C	1.20	11.80	1.81×10^{-5}	1.51×10^{-5}
30°C	1.17	11.40	1.86×10^{-5}	1.60×10^{-5}
40°C	1.13	11.10	1.91×10^{-5}	1.69×10^{-5}
50°C	1.09	10.70	1.95×10^{-5}	1.79×10^{-5}
60°C	1.06	10.40	2.00×10^{-5}	1.89×10^{-5}
70°C	1.03	10.10	2.04×10^{-5}	1.99×10^{-5}
80°C	1.00	9.81	2.09×10^{-5}	2.09×10^{-5}
90°C	0.97	9.54	2.13×10^{-5}	2.19×10^{-5}
100°C	0.95	9.28	2.17×10^{-5}	2.29×10^{-5}
120°C	0.90	8.82	2.26×10^{-5}	2.51×10^{-5}
140°C	0.85	8.38	2.34×10^{-5}	2.74×10^{-5}
160°C	0.81	7.99	2.42×10^{-5}	2.97×10^{-5}
180°C	0.78	7.65	2.50×10^{-5}	3.20×10^{-5}
200°C	0.75	7.32	2.57×10^{-5}	3.44×10^{-5}
	slugs/ft^3	lbf/ft^3	lbf-s/ft^2	ft^2/s
0°F	0.00269	0.0866	3.39×10^{-7}	1.26×10^{-4}
20°F	0.00257	0.0828	3.51×10^{-7}	1.37×10^{-4}
40°F	0.00247	0.0794	3.63×10^{-7}	1.47×10^{-4}
60°F	0.00237	0.0764	3.74×10^{-7}	1.58×10^{-4}
80°F	0.00228	0.0735	3.85×10^{-7}	1.69×10^{-4}
100°F	0.00220	0.0709	3.96×10^{-7}	1.80×10^{-4}
120°F	0.00213	0.0685	4.07×10^{-7}	1.91×10^{-4}
150°F	0.00202	0.0651	4.23×10^{-7}	2.09×10^{-4}
200°F	0.00187	0.0601	4.48×10^{-7}	2.40×10^{-4}
300°F	0.00162	0.0522	4.96×10^{-7}	3.05×10^{-4}
400°F	0.00143	0.0462	5.40×10^{-7}	3.77×10^{-4}

Source: Reprinted with permission from R. E. Bolz and G. L. Tuve, *Handbook of Tables for Applied Engineering Science*, CRC Press, Inc., Cleveland, 1973. Copyright © 1973 by The Chemical Rubber Co., CRC Press, Inc.

TABLE A.4 Approximate Physical Properties of Common Liquids at Atmospheric Pressure

Liquid and Temperature	Density kg/m³ (slugs/ft³)	Specific Gravity	Specific Weight N/m³ (lbf/ft³)	Dynamic Viscosity N · s/m² (lbf-s/ft²)	Kinematic Viscosity m²/s (ft²/s)	Surface Tension N/m* (lbf/ft)
Ethyl alcohol[1][3] 20°C (68°F)	799 (1.55)	0.79	7,850 (50.0)	1.2×10^{-3} (2.5×10^{-5})	1.5×10^{-6} (1.6×10^{-5})	2.2×10^{-2} (1.5×10^{-3})
Carbon tetrachloride[3] 20°C (68°F)	1,590 (3.09)	1.59	15,600 (99.5)	9.6×10^{-4} (2.0×10^{-5})	6.0×10^{-7} (6.5×10^{-6})	2.6×10^{-2} (1.8×10^{-3})
Glycerine[3] 20°C (68°F)	1,260 (2.45)	1.26	12,300 (78.5)	1.41 (2.95×10^{-2})	1.12×10^{-3} (1.22×10^{-2})	6.3×10^{-2} (4.3×10^{-3})
Kerosene[1][2] 20°C (68°F)	814 (1.58)	0.81	8,010 (51)	1.9×10^{-3} (4.0×10^{-5})	2.37×10^{-6} (2.55×10^{-5})	2.9×10^{-2} (2.0×10^{-3})
Mercury[1][3] 20°C (68°F)	13,550 (26.3)	13.55	133,000 (847)	1.5×10^{-3} (3.1×10^{-5})	1.2×10^{-7} (1.3×10^{-6})	4.8×10^{-1} (3.3×10^{-2})
Sea water 10°C at 3.3% salinity	1,026 (1.99)	1.03	10,070 (64.1)	1.4×10^{-3} (2.9×10^{-5})	1.4×10^{-6} (1.5×10^{-5})	
Oils—38°C (100°F) SAE 10W[4]	870 (1.69)	0.87	8,530 (54.4)	3.6×10^{-2} (7.5×10^{-4})	4.1×10^{-5} (4.4×10^{-4})	
SAE 10W-30[4]	880 (1.71)	0.88	8,630 (55.1)	6.7×10^{-2} (1.4×10^{-3})	7.6×10^{-5} (8.2×10^{-4})	
SAE 30[4]	880 (1.71)	0.88	8,630 (55.1)	1.0×10^{-1} (2.1×10^{-3})	1.1×10^{-4} (1.2×10^{-3})	

*Liquid–air surface tension values.

Source: (1) V. L. Streeter, *Handbook of Fluid Dynamics*, McGraw-Hill, New York, 1961; (2) V. L. Streeter, *Fluid Mechanics*, 4th ed., McGraw-Hill, New York, 1966; (3) A. A. Newman, *Glycerol*, CRC Press, Cleveland, 1968; (4) R. E. Bolz and G. L. Tuve, *Handbook of Tables for Applied Engineering Sciences*, CRC Press, Cleveland, 1973.

TABLE A.5 Approximate Physical Properties of Water* at Atmospheric Pressure

Temperature	Density	Specific Weight	Dynamic Viscosity	Kinematic Viscosity	Vapor Pressure
	kg/m³	N/m³	N · s/m²	m²/s	N/m² abs
0°C	1000	9810	1.79×10^{-3}	1.79×10^{-6}	611
5°C	1000	9810	1.51×10^{-3}	1.51×10^{-6}	872
10°C	1000	9810	1.31×10^{-3}	1.31×10^{-6}	1,230
15°C	999	9800	1.14×10^{-3}	1.14×10^{-6}	1,700
20°C	998	9790	1.00×10^{-3}	1.00×10^{-6}	2,340
25°C	997	9781	8.91×10^{-4}	8.94×10^{-7}	3,170
30°C	996	9771	7.97×10^{-4}	8.00×10^{-7}	4,250
35°C	994	9751	7.20×10^{-4}	7.24×10^{-7}	5,630
40°C	992	9732	6.53×10^{-4}	6.58×10^{-7}	7,380
50°C	988	9693	5.47×10^{-4}	5.53×10^{-7}	12,300
60°C	983	9643	4.66×10^{-4}	4.74×10^{-7}	20,000
70°C	978	9594	4.04×10^{-4}	4.13×10^{-7}	31,200
80°C	972	9535	3.54×10^{-4}	3.64×10^{-7}	47,400
90°C	965	9467	3.15×10^{-4}	3.26×10^{-7}	70,100
100°C	958	9398	2.82×10^{-4}	2.94×10^{-7}	101,300
	slugs/ft³	lbf/ft³	lbf-s/ft²	ft²/s	psia
40°F	1.94	62.43	3.23×10^{-5}	1.66×10^{-5}	0.122
50°F	1.94	62.40	2.73×10^{-5}	1.41×10^{-5}	0.178
60°F	1.94	62.37	2.36×10^{-5}	1.22×10^{-5}	0.256
70°F	1.94	62.30	2.05×10^{-5}	1.06×10^{-5}	0.363
80°F	1.93	62.22	1.80×10^{-5}	0.930×10^{-5}	0.506
100°F	1.93	62.00	1.42×10^{-5}	0.739×10^{-5}	0.949
120°F	1.92	61.72	1.17×10^{-5}	0.609×10^{-5}	1.69
140°F	1.91	61.38	0.981×10^{-5}	0.514×10^{-5}	2.89
160°F	1.90	61.00	0.838×10^{-5}	0.442×10^{-5}	4.74
180°F	1.88	60.58	0.726×10^{-5}	0.385×10^{-5}	7.51
200°F	1.87	60.12	0.637×10^{-5}	0.341×10^{-5}	11.53
212°F	1.86	59.83	0.593×10^{-5}	0.319×10^{-5}	14.70

*Notes: Bulk modulus E_v of water is approximately 2.2 GPa (3.2×10^5 psi).

Source: Reprinted with permission from R. E. Bolz and G. L. Tuve, *Handbook of Tables for Applied Engineering Science*, CRC Press, Inc., Cleveland, 1973. Copyright © 1973 by The Chemical Rubber Co., CRC Press, Inc.

TABLE A.6 Nomenclature

Symbol	Dimensions	Description	Symbol	Dimensions	Description
A	L^2	Area	F_D	ML/T^2	Drag force
A_j	L^2	Jet area	F_L	ML/T^2	Lift force
A_0	L^2	Orifice area	F_S	ML/T^2	Surface resistance
A_*	L^2	Nozzle area at M = 1	f	...	Friction factor
$\mathbf{a}, a$	L/T^2	Acceleration	G	...	Giga, multiple = 10^9
b	...	Intensive property	g	L/T^2	Acceleration due to gravity
B	L	Linear measure	H	L	Head
B	...	Extensive property	h	L	Height
b	L	Linear measure; wing span	h	L	Piezometric head
C_c	...	Coefficient of contraction	h	$L^2/T^2\theta$	Specific enthalpy
C_D	...	Coefficient of drag	h_f	L	Head loss in pipe
C_d	...	Coefficient of discharge	h_L	L	Head loss
C_f	...	Average shear stress coefficient	h_p	L	Head supplied by pump
C_F	...	Force coefficient	h_t	L	Head given up to turbine
C_H	...	Head coefficient	$\bar{I}$	L^4	Area moment of inertia, centroidal
C_L	...	Coefficient of lift	$\mathbf{i}$	...	Unit vector in x direction
C_P	...	Power coefficient	$\mathbf{j}$	...	Unit vector in y direction
C_p	...	Pressure coefficient	$\mathbf{k}$	...	Unit vector in z direction
C_Q	...	Discharge coefficient	K	...	Minor loss coefficient
C_T	...	Thrust coefficient	k	...	Ratio of specific heats
C_v	...	Coefficient of velocity	k_s	L	Equivalent sand roughness
c	L/T	Speed of sound	L	L	Linear measure
c_f	...	Local shear stress coefficient	l	L	Linear measure
c_p	$L^2/T^2\theta$	Specific heat at constant pressure	ℓ	L	Linear measure
c_v	$L^2/T^2\theta$	Specific heat at constant volume	M	...	Mach number
CP	...	Center of pressure	M	ML^2/T^2	Moment
cs	...	Control surface	$\mathcal{M}$	M/mol	Molar mass
cv	...	Control volume	m	M	Mass
D	L	Diameter	$\dot{m}$	M/T	Mass flow rate
D	L	Hydraulic depth	N	T^{-1}	Rotational speed
D_h	L	Hydraulic diameter	N_s	$L^{3/4}/T^{3/2}$	Specific speed
d	L	Diameter	N_{ss}	$L^{3/4}/T^{3/2}$	Suction specific speed
d	L	Depth	n	...	Manning's roughness coefficient
E	ML^2/T^2	Energy	n	T^{-1}	Rotational speed
E	L	Specific energy	n_s	...	Specific speed
E_v	M/LT^2	Elasticity, bulk	n_{ss}	...	Suction specific speed
e	L^2/T^2	Energy per unit mass	p	M/LT^2	Pressure
Fr	...	Froude number	Δp	M/LT^2	Change in pressure
$\mathbf{F}, F$	ML/T^2	Force	P	ML^2/T^3	Power

(Continued)

TABLE A.6 Nomenclature (*Continued*)

Symbol	Dimensions	Description	Symbol	Dimensions	Description
p_*	M/LT^2	Pressure at M = 1	v'	L/T	Velocity fluctuation in y direction
p_t	M/LT^2	Total pressure	W	ML^2/T^2	Work
p_v	M/LT^2	Vapor pressure	W	ML/T^2	Weight
p_z	M/LT^2	Piezometric pressure	We	$\ldots$	Weber number
Q	L^3/T	Discharge, volumetric flow rate	w	L/T	Velocity component, z direction
Q	ML^2/T^2	Heat transferred	x	L	Linear measure
q	L^2/T	Discharge per unit width	y	L	Linear measure
q	M/LT^2	Kinetic pressure	y_c	L	Critical depth
R_h	L	Hydraulic radius	y_n	L	Normal depth
R	ML/T^2	Reaction or resultant force	z	L	Elevation
R	$L^2/\theta T^2$	Gas constant	Δz	L	Change in elevation
Re	$\ldots$	Reynolds number			
r	L	Linear measure in radial direction	**Greek Letters**		
S	L^2	Planform area	α	$\ldots$	Angular measure
St	$\ldots$	Strouhal number	α	$\ldots$	Lapse rate
S_0	$\ldots$	Channel slope	α	$\ldots$	Kinetic energy correction factor
s	$L^2/T^2\theta$	Specific entropy	α	$\ldots$	Angle of attack
S, S.G.	$\ldots$	Specific gravity	β	$\ldots$	Angular measure
s	L	Linear measure	Γ	L^2/T	Circulation
T	ML^2/T^2	Torque	γ	M/L^2T^2	Specific weight
T	θ	Temperature	Δ	$\ldots$	Increment
T_t	θ	Total temperature	δ	L	Boundary layer thickness
T_*	θ	Temperature at M = 1	δ'	L	Laminar sublayer thickness
t	T	Time	δ'_N	L	Nom. laminar sublayer thickness
U_0	L/T	Free-stream velocity	η	$\ldots$	Efficiency
u	L/T	Velocity component, x direction	θ	$\ldots$	Angular measure
u	L^2/T^2	Internal energy per unit of mass	κ	$\ldots$	Turbulence constant
u_*	L/T	Shear velocity	Λ	$\ldots$	Aspect ratio of a wing
u'	L/T	Velocity fluctuation in x direction	μ	M/LT	Dynamic viscosity
$\mathbf{u}_n$	$\ldots$	Unit vector, normal direction	τ	M/LT^2	Shear stress
$\mathbf{u}_t$	$\ldots$	Unit vector, tangential direction	ν	L^2/T	Kinematic viscosity
$\mathbf{u}_r$	$\ldots$	Unit vector, radial direction	π	$\ldots$	Dimensionless group
$\mathbf{u}_\theta$	$\ldots$	Unit vector, azmuthal direction	ρ	M/L^3	Mass density
$\mathbf{u}_z$	$\ldots$	Unit vector, axial direction	ρ_*	M/L^3	Density at M = 1
V	L/T	Velocity	ρ_t	M/L^3	Total density
V_0	L/T	Free-stream velocity	Ω	T^{-1}	Rate of rotation
$\forall$	L^3	Volume	ω	T^{-1}	Angular speed
$\overline{V}$	L/T	Area-averaged velocity	ω	T^{-1}	Vorticity
v	L/T	Velocity component, y direction	σ	M/T^2	Surface tension

ANSWERS

Solutions to Checkpoint Problems

➡ To review or download detailed solutions, go to www.wiley.com/college/elger.

Chapter 1

1.1   The Point^{TM} is a unit.

Chapter 2

2.1   Glycerin has a higher viscosity at both temperatures.

2.2 $v = 2.07 \times 10^{-6}\,\text{m}^2/\text{s}$

2.3   (c)

Chapter 3

3.1   (d)

3.2   (d)

3.3   (c)

Chapter 4

4.1   (a)

4.2   (b)

4.3   The water level in the sight glass will go down.

Chapter 5

5.1   (c)

5.2   (b)

5.3   (b)

Chapter 6

6.1   (h)

6.2   (e)

6.3   (a) and (d)

Chapter 7

7.1   (a), (c) and (f)

7.2   (b), (f) and (g)

7.3   (b)

Chapter 8

8.1   (b)

Chapter 9

9.1   Dimensions. Secondary dimensions = (force)/(volume). Primary dimensions = $M/(L^2T^2)$.

Meaning of Terms and Physics. Picture an infinitesimal CV located at point (x, y, z). The sum of forces acting on the matter inside this CV is zero. That is, the forces balance like this:

$$\frac{(\text{Resultant force due to Viscosity})}{(\text{Volume of the CV})}$$
$$= \frac{(\text{Resultant force due to Pressure} + \text{Weight})}{(\text{Volume of the CV})}$$

9.2   (d)

Chapter 10

10.1   (a)

10.2   Re = 79,000, $k_s/D = 1.73 \times 10^{-2}$, $f = 0.047$, $h_f = 5.75$ m

10.3   Equation for V

$$V = \sqrt{\frac{2D p_{\text{supply}}}{f\rho L}}$$

If the length is doubled, the velocity will decrease to 70.7% of its initial value.

Chapter 11

11.1   (b) and (f)

11.2   Idealize a football as a streamlined body. Rotation of the football lowers the drag.

11.3   (d)

Chapter 12

12.1   (b)

Chapter 13

13.1   No solution provided.

Chapter 15

15.1   (a)   15.2   (b)

Chapter 16

16.1   True, False, False.

Solutions to Selected Even-Numbered Problems

Chapter 1

1.8   (b), (c), and (d).

1.14 $\rho = 3.926$ kg/m^3

1.16   (a) $F = 100$ N

1.18 $E = 1.09 \times 10^5$ kWh, $C = \$10,900$

1.20   (a) $N = 3.30 \times 10^{22}$ molecules

       (b) $N = 2.48 \times 10^{19}$ molecules

1.22 $R_{N_2} = 297 \frac{J}{kg \cdot K}$

1.24 $\rho = 1.09$ kg/m^3 (local conditions), $\rho = 1.22$ kg/m^3 (table value)

1.26 $\rho_{Methane} = 1.74$ kg/m^3, $\gamma_{Methane} = 17.1$ N/m^3

1.28 $\frac{M_2}{M_1} = 1.5$

1.30 $W_{total} = 460.14$ N

1.32   (a) $\rho_{air} = 6.47$ kg/m^3, (b) $\gamma_{air} = 63.5$ N/m^3

1.36 $W = 13.1$ N

Chapter 2

2.4   a and b

2.6 $\rho_l = 1700$ kg/m^3; $\gamma = 16,660$ N/m^3

2.8 $\rho = 1.53 \frac{kg}{m^3} = 3.20 \times 10^{-3} \frac{slug}{ft^3}$

2.10   c

2.12 $V_{final} = 1998$ cm^3

2.14   volume % change = 2.8%; water level rise = 0.002 m

2.16   a. Seawater;   b. $\rho = 1030$ kg/m^3;
       c. p = standard atmospheric pressure

2.20   b

2.22   d

2.24   in water: $\Delta\mu = -9.06 \times 10^{-4}$ N · s/m^2; $\Delta\rho = -22$ kg/m^3;
       in air, $\Delta\mu = 2.8 \times 10^{-6}$ N · s/m^2; $\Delta\rho = -0.22$ kg/m^3

2.28   more viscous than water

2.30 $\tau_{y=1\,mm} = 1.49$ Pa

2.32   a. $\tau_{max} = 1.0$ N/m^2;   b. midway between the two walls

2.34 $\tau = 60$ N/m^2

2.36 τ_{max} at $y = H$; zero stress at $y = \dfrac{H}{2} - \dfrac{\mu u_t}{H dp/ds}$;
       $u_t = (1/2\mu)\dfrac{dp}{ds}H^2$

2.38 $\dfrac{\tau_2}{\tau_3} = \dfrac{2}{3}$; $V = 0.060$ m/s; $\tau = 0.30$ N/m^2

2.42   a and b

2.44   b

2.46 $\dfrac{v}{v_o} = \dfrac{p_o}{p}\left(\dfrac{T}{T_o}\right)^{5/2}\dfrac{T_o + S}{T + S}$

2.48 $v = 1.99 \times 10^{-5}$ m^2/s

2.50 $v = 0.69 \times 10^{(-4)}$ m^2/s

2.52 $\mu = 0.032$ N · s/m^2

2.54 $\mu_{air} = 1.91 \times 10^{-5} \frac{N \cdot s}{m^2}$; $v_{air} = 10.1 \times 10^{-6}$ m^2/s; $\mu_{water} = 6.53 \times 10^{-5}$ N·s/m^2; $v_{water} = 6.58 \times 10^{-7}$ m^2/s

2.56   a, b, d, e, and f

2.58 $\Delta p = \frac{4\sigma}{R}$; $\Delta p_{4mm\ rad.} = 73.0$ N/m^2

2.60 $d_{6.35} = 4.69$ mm; $d_{3.2} = 9.31$ mm; $d_{0.8} = 37.3$ mm

2.62 $\Delta p = 292$ N/m^2

2.64   a. $\Delta h = 2.98$ m;   b. higher in clay;   c. pores in roots need to be smaller than in soil

2.66 $D = 9.46$ mm

2.68   b

2.70 $P = 2340$ Pa abs

2.72 $T_{boiling,\ 3000m} = 89.7\,°C$

Chapter 3

3.2 $P_{abs} = 101.3 - 30 = 71.3$ kPa, $P_{gage} = -30$ kPa pr 30 kPa vacuum

3.6   (a) $W_2/W_1 = (D_2/D_1)^2$, (b) $D_1 = 70$ mm, $D_2 = 38.3$ m

3.14   The height has decreased by 2.55 m.

3.16   a

3.18   a, b, d, e and f

3.20 $p_{btm} = 10.3$ kPa gage

3.24 $h_2 = 4w/(S)(\gamma_{water})(\pi D_{21})$

3.26 $p_{max} = 127.5$ kPa, maximum pressure will be at the bottom of the liquid that has a specific gravity of $S = 3$.

3.28 $F_{bolts} = 288$ kN (acting downward)

3.30 5.46×10^{-4} m^3

3.32 $d = 2.80$ m

3.34 $T_{boiling,\ 2000\ m} \approx 93.2\,°C$, $T_{boiling,\ 4000\ m} \approx 86.7\,°C$

3.38 $T = 287$ K; $p_a = 86.0$ kPa; $\rho = 1.04$ kg/m^3

3.40 $p(z = 8\ km) = 3.31$ mbar, $p(z = 30\ km) = 0.383$ mbar

3.46 $p_A = 2734$ Pa

3.48 $p_B = -1.00$ kPa gage

3.50 $p_{container} = 1441$ Pa

3.52 $\gamma_{liq.} = 4995$ N/m^3

3.54 $p_A = 89.2$ kPa gage

3.56 $h_{new} = 4.0h$

3.58 $\Delta h_b - \Delta h_a = 1.02$ m

3.62   Part 1, b; Part 2, c

3.64   a. Tank 1;   b. Tank 2

3.66   b and c

3.68   a. $\bar{z} = 2\,\text{m}$
        b. $F = 78.4\,\text{kN}$
        c. $y_{cp} - \bar{y} = 0.167\,\text{m}$

3.70 $F_{\text{gate}} = 105\,\text{kN (acts to the right)}$

3.72 $F = 11{,}380\,\text{N to left, or } F = -11{,}380\,\text{N}$

3.74 $P = 300\,\text{kN}$

3.76 $T = 33{,}685\,\text{N-m}$

3.78   gate will fall

3.80 $M = 18\,\text{kN·m per meter of form}$

3.82 $F_{\text{pull}} = 2670\,\text{N}$

3.84   a. $R_A = 0.510\gamma W\ell^2$;   b. H-component same;
        V-component less

3.86   a. $F_V = 17{,}515\,\text{N}$
        b. $F_H = 14{,}715\,\text{N}$
        c. $F_R = 22{,}876\,\text{N at } \theta = 40°2'$

3.88 $F_h = 2465\,\text{N}; F_v = 321\,\text{N}$

3.90 $\mathbf{F}_{\text{result}} = 22{,}072\mathbf{i} + 34{,}654\mathbf{j}\,\text{N}$

3.94   c

3.96 $W_{\text{scrap}} = 3420\,\text{N}$

3.98 $\forall = 14.7\,\text{L}$

3.100 $\forall = 14.7\,\text{L}; \gamma_{\text{block}} = 17.2\,\text{kN/m}^3$

3.102 $\Delta h = 0.37\,\text{cm}$

3.104 $d = 2.105L$

3.106 $\Delta h = 8\,\text{mm; no subsequent change}$

3.108 $\ell = 8.59\,\text{m}$

3.110 $z = 22.8\,\text{km}$

3.112 $W = 1.50 \times 10^{-2}\,\text{N}$

3.114 $1.27 \leq S \leq 1.59$

3.116 $\dfrac{\ell}{w} = 0.211; S = 0.211$

3.118   is not stable

Chapter 4

4.2   streakline

4.4   c

4.8   c

4.12   a

4.14   Point A: Unsteady, uniform
        Point B: Non-uniform, unsteady

4.16   Non-uniform; could be steady or unsteady

4.18   a. 2D;   b. 1D;   c. 1D;   d. 2D;   e. 3D;   f. 3D;   g. 2D

4.20   centripetal

4.22   a

4.24   a. Steady;   b. 2D;   c. No;   d. Yes, at streamline
        curvature

4.26 $a_x = -\left(3\,U_0^2\,\dfrac{r_0^3}{x^4}\right)\left(1 - \dfrac{r_0^3}{x^3}\right)$

4.28 $a_c = 1.5\,\text{m/s}^2$

4.30 $a_l = \dfrac{8q_o}{3Bt_o}$

4.32 $a_\ell = 3.56\,\text{m/s}^2; a_c = 63\,\text{m/s}^2$

4.36 $\dfrac{\partial p}{\partial z} = -75.8\,\text{lbf/ft}^3$

4.38 $p_2 = 4120\,\text{Pa}$

4.40 $\dfrac{\partial p}{\partial s} = -8{,}000\,\text{N/m}^3$

4.42 $a_z = -43.6\,\text{m/s}^2$

4.44 $\dfrac{\partial p}{\partial x} = -825\,\text{kPa/m}$

4.46   a. $p_B - p_A = 12.7\,\text{kPa}$;
        b. $p_C - p_A = 44.6\,\text{kPa}$

4.48   a and c

4.50 $V_2 = 5.46\,\text{m/s}$

4.52 $V_1 = 7.13\,\text{m/s}$

4.54 $V = 74\,\text{m/s}$

4.56   c

4.58 $V_2 = 60.7\,\text{m/s}$

4.60 $V_0 = 0.88\,\text{m/s}$

4.62   b

4.64 $V = 57.7\,\text{m/s}$

4.66 $V = 111\,\text{m/s}$

4.68 $p_B - p_C = 18\,\text{kPa}$

4.70 $V_0 = 13.0\,\text{m/s}$

4.72 $C_p = -0.687; p_{\text{gage}} = -4.12\,\text{kPa gage}$

4.74 $V = 0.098\,\text{km/s}$

4.76   b

4.78   No

4.80   rate of rotation $= \omega_z = \omega$; vorticity $= 2\omega$

4.82   irrotational

4.84 $\frac{\Delta\theta}{2\pi}\,(\text{rad}) = r^2\,\dfrac{\exp(-r^2)}{1 - \exp(-r^2)}$

4.86 $p_A = 13{,}440\,\text{Pa}$

4.88 $p_h - p_\ell = 3.84\,\text{kPa}$

4.90 $p_A - p_B = 2817\,\text{Pa}$

4.92   No

4.94   c

4.96   No

4.98 $p = -33{,}143\,\text{Pa}$

4.100	$a_r = 88,800$ m/s^2; $RCF = 9060$
4.102	$\omega = 6.26$ rad/s
4.104	$S = 2.5$
4.106	$p_A = 149$ Pa
4.108	$F = 15.7$ N
4.110	$z_2 = 12.6$ m
4.112	$\Delta p_{max} = \dfrac{\rho\omega^2 r_0^2}{2} + \gamma r_0 + \dfrac{\gamma g}{2\omega^2} = 34,319$ pa; $z_{min} = 0.063$ above axis

Chapter 5

5.2	$d = 15.7$ m
5.4	a
5.6	b
5.8	$Q = 0.1993$ m^3/s; $Q = 11,958$ L/min
5.10	$\dot{m} = 0.115$ kg/s
5.12	$D = 1.15$ m
5.14	$\dfrac{\bar{V}}{V_o} = \dfrac{1}{3}$
5.16	$Q = 4.41$ m^3/s; $Q = 264,600$ L/min
5.18	a. $Q = 5.0$ m^3/s
	b. $V = 5.0$ m/s
	c. $\dot{m} = 6.0$ kg/s
5.20	$Q = 0.743$ m^3/s
5.22	$Q = 1.70 \times 10^{-3}$ m^3/s
5.24	$q = 3.09$ m^2/s; $V = 2.57$ m/s
5.26	$V = \left(\dfrac{1}{n+1}\right) V_c$
5.28	$V = 0.073$ m/s
5.30	$Q = 0.0024$ m^3/s $= 144$ L/min
5.32	$Q = 0.0276$ m^3/s
5.36	a. extensive; b. extensive; c. intensive; d. extensive; e. intensive
5.42	a. True; b. True; c. True; d. True; e. True,
5.44	a
5.46	No
5.50	Rising
5.52	$V_{exit} = 4.6$ m/s; $a_{exit} = 5.2$ m/s^2
5.54	$a_{2s} = -5,060$ m/s^2; $a_{3s} = -11,400$ m/s^2
5.56	$t = 14$ s
5.58	$V_R = 0.2$ m/s
5.60	$V_1 = 12$ m/s; $V_2 = 36$ m/s
5.62	$V_{15} = 5.66$ m/s; $V_{20} = 6.37$ m/s
5.64	$V_B = 5.00$ m/s
5.66	$Q_B = 0.095$ m^3/min; flow is leaving via pipe B
5.68	Rising; $\dfrac{dh}{dt} = 0.04$ m/s

5.70	$Q_p = 0.21$ m^3/s
5.72	$\dot{m} = 104.55$ kg/s; $V_C = 6.4$ m/s; $S = 0.925$
5.74	$V = 10.8$ m/s
5.76	$Q = 0.658 A_o \sqrt{\dfrac{2(p_1 - p_2)}{\rho}}$; $Q = 5.20 \times 10^{-4}$ m^3/s
5.78	$A = 8.59 \times 10^{-4}$ mm^2
5.80	$t = 185$ s
5.82	$\Delta t = 329$ s; $\Delta t = 10.6$ min
5.84	$\Delta t = 1491$ s
5.88	$\rho_e = 0.0676$ kg/m^3
5.90	$\dfrac{d\rho}{dt} = 250$ kg/m$^3 \cdot$ s
5.92	b
5.94	$p_B = 124$ kPa
5.96	$Q = 0.2$ m^3/s
5.98	$\Delta h = 0.046$ m
5.100	a. $V_{e\,max} = 3.62$ m/s
	b. $Q_{max} = 0.00362$ m^3/s
	c. Lift$_{max} = 8,850$ N
5.102	$V_2 = 7.2$ m/s; $F = 191$ N
5.104	cavitation; pump no longer effective
5.106	vapor pressures
5.108	$V_0 = 7.3$ m/s
5.110	$V_0 = 15.9$ m/s
5.112	$V_0 = 15.9$ m/s
5.114	$V_0 = 9.54$ m/s

Chapter 6

6.4	a. in MD; b. in FD; c. in FD if borne by a reaction force; d. in FD (if significant); e. in FD
6.6	$F = 0.122$ N
6.8	a. True b. False
6.10	$\mu = 1.095$
6.12	$v_1 = 18$ m/s
6.14	F_1 is larger because fluid is launched to the right; $F_1 = 182$ N; $F_2 = 169$ N;
6.16	$\dot{m} = 200$ kg/s; $d = 7.14$ cm
6.18	$p_{air} = 8.25$ atm
6.20	$T = 3422$ N
6.24	$F_x = -1442$ N (acts to the left); $F_y = -374$ N (acts downward)
6.26	$\mathbf{F} = (19.3$ N$)\,\mathbf{i} + (72$ N$)\,\mathbf{j}$
6.28	$v = 7.79$ m/s
6.30	$F_{on\,wall} = \rho t v^2 \sin 45°$ (acting to the right)
6.32	$\mathbf{F} = (816\mathbf{i} - 338\mathbf{j})$ N
6.34	$F_{rolling} = -280$ N (acting to the left)

6.36	$P = 83{,}331$ N-m/s/m $= 83{,}331$ W/m
6.38	$P = 29{,}724$ W
6.40	$T = 22{,}780$ N; $T = 16{,}004$ N
6.42	$V_e = 910$ m/s
6.44	$p_1 = 312$ kPa; Force on nozzle $= 2.26$ kN to the left
6.46	$F_x = -40.7$ kN (acts to the left)
6.48	$F_x = -75.4$ kN
6.50	Force per bolt $= 1410$ N
6.52	$p_{\text{gage}} = 13.3$ kPa; $F_x = -1.38$ kN/m
6.54	$\mathbf{F} = (-2158.8\mathbf{i} - 61.2\mathbf{j})$ N
6.56	$\mathbf{F} = (-36.8\mathbf{i} + 119\mathbf{j})$ N
6.58	d
6.60	$\mathbf{F} = (-14.1\mathbf{i} + 0\mathbf{j} + 1.38\mathbf{k})$ kN
6.62	$\mathbf{F} = (-3610\mathbf{i} - 1632\mathbf{j} + 1426\mathbf{k})$ N
6.64	$F_{a,v} = 14{,}595$ N
6.66	$F_y = -59{,}640$ N (acting downward)
6.68	$F_x = -5360$ N (acting to the left, opposite of inlet flow)
6.70	$F_x = 1140$ N
6.72	$F_x = -805$ N (acting to the left)
6.74	$F_x = -9.96$ kN (acts to the left); $F_y = -1.8$ kN (acts downward)
6.76	$F_x = -2.86$ kN (acts to the left)
6.78	$F_\tau = \frac{\pi D^2}{4}[p_1 - p_2 - (1/3)\rho U^2]$
6.80	$T = 688$ N (acting to the right)
6.84	$\dot{m} = 30.9$ kg/s; $v_{\max} = 54$ m/s; $F_D = 466$ N
6.86	$T = 124$ k N
6.88	$a_r = 0.034$ m/s^2
6.90	$F_x = -73.97$ N (acting to the left)
6.92	$F_r = -100$ N (acting to the left)
6.94	$\Delta t = 2.22$ s
6.96	$v_{\max} = 15.2$ m/s
6.98	$P = 1.11$ kW
6.100	$\mathbf{F} = (1750\mathbf{i} - 1950\mathbf{j})$ N; $\mathbf{M} = (-5850\mathbf{k})$ N-m
6.102	$\mathbf{F} = -4128\mathbf{i}$ N; $\mathbf{M} = (-2480\mathbf{j} + 398\mathbf{k})$ N-m

Chapter 7

7.4	a and d
7.6	0.25 (estimate)
7.8	41.0 W
7.10	a and b
7.14	$\alpha = 2.7$
7.16	$\alpha = \dfrac{1}{4}\left[\dfrac{[(n+2)(n+1)]^3}{(3n+2)(3n+1)}\right],\ \alpha = 1.06$
7.18	1.19

7.24	$Q = 0.311$ m^3/s, $p_B = 86.4$ kPa
7.26	$p_A = -39.2$ kPa, $V_2 = 14.0$ m/s
7.28	20.75 Mpa
7.30	132 kPa
7.32	82 kPa
7.36	1116 kPa
7.38	$Q = 0.266$ m^3/s, $p_B = -45.3$ kPa, gage
7.40	$p_A = 59$ kPa; $p_B = 17.9$ kPa
7.42	$Q = 0.258$ m^3/s, $H = 0.919$ m
7.44	$t = 9260$s $= 2.57$ h
7.48	1.76 MW
7.50	19. kW
7.52	$P = 61.6$ kW
7.54	1155 kW
7.56	$P_{\text{electrical}} = 1.49$ W
7.58	10.1 m
7.60	204 kW
7.64	$h_L = 0.125$ m
7.66	$Q = 0.0149$ m^3/s
7.68	17.1 N acting to the left
7.70	850 N acting upward
7.72	$p_{80} = 1211$ kPa, $F_x = -910$ kN
7.76	Possible if the fluid is being accelerated to the left.
7.78	a. A to E; b. D; c. constant; d. no; f. nothing else
7.84	0.23 m^3/s
7.86	58 kW
7.88	12.9 m^3/s; 44,609 N/m^2
7.90	21.5 MW

Chapter 8

8.2	3 dimensionless variables (or 3 π-groups)
8.4	a. homogeneous; b. not homogeneous; c. homogeneous; d. homogeneous
8.6	$\dfrac{\Delta h}{d} = f\left(\dfrac{D}{d}, \dfrac{\gamma t^2}{\rho d}, \dfrac{h_1}{d}\right)$
8.8	$\dfrac{F_D}{\mu V d} = C$
8.10	$\dfrac{F}{\rho V^2 D^2} = f\left(\dfrac{\rho V D}{\mu}, \dfrac{k_s}{D}, \dfrac{\omega D}{V}\right)$
8.12	$\dfrac{\Delta p}{n^2 \rho D^2} = f\left(\dfrac{Q}{nD^3}\right)$
8.14	$\dfrac{h}{d} = f\left(\dfrac{\sigma t^2}{\rho d^3}, \dfrac{\gamma t^2}{\rho d}, \dfrac{\mu t}{\rho d^2}\right)$
8.16	$\dfrac{\Delta p d^4}{\rho Q^2} = f\left(\dfrac{\mu d}{\rho Q}, \dfrac{D}{d}\right)$
8.18	$\dfrac{\dot{m}}{\sqrt{\rho \Delta p}\, D^2} = f\left(\dfrac{\mu D}{\dot{m}}\right)$

8.20	$\dfrac{F_D}{\rho V^2 S} = f\left(\dfrac{\omega^2 S}{V^2}\right)$	9.26	$\tau = 0.951\dfrac{\mu_o U}{L}$
8.22	$\dfrac{V_h}{\sqrt{gD}} = f\left(\dfrac{\mu}{\rho_f g^{1/2} D^{3/2}}, \dfrac{\rho_f - \rho_b}{\rho_f}\right)$	9.28	$\dfrac{dp}{ds} = -1.80 \times 10^4$ Pa/m; $P = 1.8 \times 10^{-4}$ W
8.24	$\dfrac{F_D}{\rho V^2 B^2} = f\left(\dfrac{\mu}{\rho VB}, \dfrac{u_{rms}}{V}, \dfrac{L_x}{B}\right)$	9.34	$\dfrac{\delta}{x} = 0.0071$
8.26	a. $\alpha_{aorta} = 19.9$ b. $\alpha_{capillary} = 0.035$ c. yes, aorta; yes, capillary	9.36	a
		9.38	$F_x = 5.15$ N
8.34	$V_w = 0.05$ m/s	9.40	$u = U_0 = 0.85$ m/s
8.36	$\dfrac{Q_m}{Q_p} = \dfrac{1}{10}; \Delta p_p = 4.0$ kPa	9.42	$F_s = 0.0943$ N; $\delta = 15.8$ mm
8.38	$F_p = 33$ N	9.44	b. T; d. L; f. T; h. T; j. T
8.40	$\rho_m = 10.5$ kg/m^3	9.46	$F_s = 1.83 \times 10^{-4}$ N
8.42	$\dfrac{F_p}{F_m} = 935$	9.48	a. $F_{s,\,wing} = 230$ N; b. $P = 12.8$ kW; c. $x_{cr} = 14.4$ cm; d. $\dfrac{F_{tripped\ B.L.}}{F_{normal}} = 16.2\%$ increase
8.44	$V_m = 0.37$ m/s		
8.46	$\nu_e = 1.12 \times 10^{-5}$ m^2/s		
8.48	$V_p = 1.78$ m/s; $\Delta p_p = 3.19$ kPa	9.50	$F_s = 2.82$ N/m; $\dfrac{du}{dy} = 5.97 \times 10^4$ s^{-1}
8.50	$V_{tunnel} = 25$ m/s; $F_{tunnel} = F_{prot.} = 2400$ N	9.52	$\delta^* = \int_0^\delta \left(1 - \dfrac{\rho u}{\rho_\infty U_\infty}\right) dy$
8.52	$Q_m = 0.0455$ m^3/s; $C_p = 1.07$	9.54	$\dfrac{\delta^*}{\delta} = \dfrac{1}{8}$
8.54	$F_p = 25$ kN		
8.56	$V_m = 178$ m/s; Yes	9.56	$P = 10.4$ kW
8.58	$d = 0.406$ mm	9.58	$U_0 = 0.805$ m/s
8.60	$L_p = 1.60$ m; $t_p = 8.94$ s	9.60	$U_0 = 103$ m/s.
8.62	$V_m = 13.3$ m/s $= 47.8$ km/hr	9.62	$F_s = 59.1$ N
8.64	$V_m = 0.56$ m/s; $d_m = 0.094$ m	9.64	$P = 73,550$ W
8.66	$V_p = 3.12$ m/s; $L_p = 12 \times 2.5$ cm $= 30$ cm	9.66	$F_{s_{100}} = 534$ N; $F_{s_{200}} = 3020$ N
8.68	$V_m = 0.147$ m/s; $t_m = 30.6$ min	9.68	$\dfrac{\delta}{W}\bigg|_{case\ 1} = 0.0406; \dfrac{\delta}{W}\bigg|_{case\ 2} = 0.0362$
8.70	$Q_m = 0.014$ m^3/s; $F_p = 26,100$ kN		
8.72	$V_p = 7.5$ m/s	9.70	$F_s = 7390$ N
8.74	d	9.72	Wave Drag on Ship $= 2.34 \times 10^5$ N
8.76	$\dfrac{F}{\frac{1}{2}\rho V^2 L^2} = f\left(\dfrac{VL}{\nu}\right)$; completed figure	9.74	$\tau_{0_{min}} = 106$ N/m^2
		9.76	$F_D = 287$ kN; $\delta = 0.678$ m

Chapter 9

9.4	$V = 1.38$ m/s
9.6	$\mu = 5.43 \times 10^{-2}$ N $\cdot$ s/m^2
9.8	Pressure flow in x-direction and upper plate moves to left; τ_{min} at $du/dy = 0$
9.10	a. $u = 150y$ (m/s); b. rotational; c. yes; d. $F_s = 180$ N
9.12	$V_{lower} = \dfrac{V\mu_1/t_1}{\mu_2/t_2 + \mu_1/t_1}$
9.14	a. sketch; b. $F = 1130$ N
9.18	$\mu = 2.40$ N $\cdot$ s/m^2
9.20	downward
9.22	$u_{max} = +0.035$ m/s
9.24	$Q = 6.62 \times 10^{-8}$ m^3/s

Chapter 10

10.2	Turbulent; $L_e = 7.5$ m
10.6	$p_{tank} = 1.75$ kPa gage
10.8	$V = 2.19$ m/s; $Q = 0.110$ L/s; sketch
10.12	a. $\dot{m} = 0.0141$ kg/s; b. $f = 0.064$; c. $\dfrac{h_f}{L} = 0.00108$ m per m of pipe length; d. $\dfrac{\Delta p}{L} = 10.6$ Pa per m of pipe length
10.14	$V_2 = 0.215$ m/s
10.16	$h_f = 22.7$ m per 30 m of pipe
10.18	$V = 0.22$ m/s; $Q = 6.2 \times 10^{-6}$ m^3/s
10.20	$P = 1340$ W
10.22	d
10.24	$V_2 = 0.0409$ m/s

10.26	$\nu = 8.91 \times 10^{-5}$ m^2/s
10.28	downward (from right to left); $f = 0.076$; laminar; $\mu = 0.071$ N $\cdot$ s/m^2
10.30	$\Delta p = 570$ Pa
10.32	$f = 0.0139$
10.34	$f = 0.016$
10.36	a. $V_{max} = 0.632$ m/s; b. $f = 0.041$; c. $u_* = 0.0358$ m/s; d. $\tau_{25\,mm} = 0.513$ N/m^2; e. No, will nearly quadruple
10.38	$f = 0.038$
10.40	$p_A = 768$ kPa
10.42	quadrupled
10.44	a. $\Delta p = 29.1$ kPa; b. $h_f = 2.97$ m; c. $P = 177$ W
10.46	From B to A; $q = 0.25$ m^3/s
10.48	$p_2 = 458$ kPa
10.50	a. $\Delta p = 2270$ N/m^2; b. $P = 240{,}563$ W; c. student sketch
10.52	a. Case 1; b. Case 3; c. Case 3
10.54	$D = 0.022$ m
10.56	$V = 3.15$ m/s
10.58	$Q = 6.59 \times 10^{-3}$ m^3/s
10.60	a. $D = 22$ cm (next largest available diameter); b. $P = 45.6$ kW for each kilometer of pipe length
10.62	$K = 12.4$
10.64	$t = 22$ s
10.66	a. $V_2 = 1.81$ m/s; b. $h = 16.7$ cm
10.68	$P = 29{,}336$ W
10.72	$Q = 0.18$ m^3/s
10.74	$P = 5.61$ W
10.76	$Q_{2m} = 0.000356$ m^3/s; $Q_{10m} = 0.000629$ m^3/s; $Q_{20m} = 0.000755$ m^3/s
10.78	$K_v = 18.5$
10.80	$Q = 0.0129$ m^3/s; $p_A = -92.0$ kPa
10.82	$z_1 - z_2 = 44.5$ m
10.84	$p_A = 274$ kPa
10.86	$\Delta p_f = 77.97$ N/m^2
10.88	a. $\Delta p = 0.6$ in-H$_2$O; b. $P = 38.9$ W
10.90	$P = 188$ kW
10.92	$Q \approx 11{,}166$ L/min
10.94	$\dfrac{Q_B}{Q_A} = 0.283$
10.96	$V_2 = 3.346$ m/s; $V_1 = 4.472$ m/s
10.98	a
10.100	$Q_{0.3} = 0.15$ m^3/s; $h_{L_{A-B}} = 8.25$ m
10.102	$Q_1 = 0.27$ m^3/s; $Q_2 = 0.33$ m^3/s; $h_\ell = 5.05$ m
10.104	$Q_v = 0.456$ m^3/s; $Q_p = 0.656$ m^3/s

Chapter 11

11.2	d
11.4	$C_D = 2.0$
11.6	a, b and d
11.10	$F_D = 27{,}148$ N
11.12	$C_D \approx 1.18 \approx 1.2$
11.14	$F_D = 1.26 \times 10^7$ N
11.16	$V = 19.7$ m/s
11.18	$F_D = 70.4$ kN
11.20	$M_o = 3.12$ MN $\cdot$ m
11.22	$T = 142$ N
11.24	$M = 21.7$ kN $\cdot$ m
11.26	$(5.9$ m/s$) \le V \le (17.7$ m/s$)$
11.28	$P = 55.5$ kW
11.30	Energy $= 61.9$ kJ $= 14.8$ Food Calories
11.32	Additional power $= 16.4$ kW
11.34	Energy savings are 13.95%
11.36	$P = 47.2$ kW
11.38	$V_c = 12.6$ m/s
11.40	$F_{D_{reduction}} = 756$ N
11.42	Specific gravity S_{obj}, the ratio $\forall/A$, and C_D
11.44	Bubble will accelerate; Drag is form drag
11.46	$V_o = 1.47$ m/s
11.48	$V_o = 6.70$ m/s
11.50	$V_o = 1.55$ mm/s
11.52	$V_o = 9.13$ m/s
11.54	$V_o = 5.70$ m/s
11.56	$t = 0.54$ seconds and $d = 14.2$ cm
11.58	$A = 6.1 \times 10^3$ mm^2
11.60	a
11.64	$b = 6.48$ m
11.66	d
11.68	$V = \left[\frac{4}{3}(W/S)^2(1/(\pi\Lambda\rho^2 C_{D_0}))\right]^{1/4}$; $V = 29.6$ m/s
11.70	$V_L = 107.8$ m/s; $V_s = 99.8$ m/s
11.72	$V_o = 10.5$ m/s; $F_{L/\text{length}} = 16{,}000$ N/m
11.74	$C_L = \sqrt{\pi\Lambda C_{D_0}}$; $C_L/C_D = (1/2)\sqrt{\pi\Lambda/C_{D_0}}$
11.76	$F_D = 4000$ N

Chapter 12

12.2	a. $V = 761$ mph; b. $s = 1.36$ km
12.6	$c = 427$ m/s
12.8	$c_{H_2} = 1267$ m/s
12.10	$c = \sqrt{RT}$

12.12 $T_t = 185\,°C$

12.14   a. $V = 1{,}970$ km/hr; $T_t = 104\,°C$
       b. Speed for $M = 1$ is $V = 1090$ km/hr

12.16 $W = 948$ Pa

12.20 $\dot{m} = 5.08$ kg/s

12.22 $M = 5$

12.24 $F_D = 94.1$ N

12.28   No; would violate 2nd law of thermodynamics

12.30 $M_2 = 0.475$; $T_2 = 737$ K $= 464°C$; $p_2 = 2138$ kPa

12.32 $M_2 = 0.454$; $p_2 = 680$ kPa, abs;

       $T_2 = 680$ K $= 407\,°C$; $\rho_2 = 2.55$ kg/m^3

12.34 $M_2 \rightarrow \sqrt{(k-1)/2k}$; $\rho_2/\rho_1 \rightarrow (k+1)/(k-1)$;

       M_2 (air) $= 0.378$; ρ_2/ρ_1 (air) $= 6.0$

12.38 $\dot{m} = 0.213$ kg/s

12.40 $p_t = 224$ kPa

12.42 $p_b = 87.2$ kPa

12.48 $A/A_* = 4.23$; $T_t = 700$ K $= 427°C$; $p_t = 367.3$ kPa

12.50   Underexpanded

12.52   a. Proof;   b. $p = 413.4$ kPa and $T = -31\,°C$;
       c. overexpanded;   d. $p_t = 173.7$ kPa

12.54   a. $M_e = 2.62$, $p_e = 52.1 \times 10^3$ Pa, and $\rho_e = 0.0671$ kg/m^3
       b. $\dot{m} = 6.78$ kg/s
       c. $T = 18.0$ kN
       d. $p_t = 691$ kPa; $T = 9.86$ kN

12.56 $A/A_* = 2.46$

12.58 $\Delta s = 5230$ J/kgK

12.60 $p_2 = 321$ kPa

Chapter 13

13.4 $V \geq 0.06$ m/s

13.6 $\Delta h = 35.4$ mm

13.8 $V_0 = 18.1$ m/s

13.10 $V = 0.468$ m/s

13.14 $Q = 0.186$ m^3/s

13.16   a. $Q = 4.3$ m^3/s $= 258$ m^3/min
       b. $V_{max}/V_{mean} = 1.2$;   c. turbulent;
       d. $\dot{m} = 5.03$ kg/s

13.18   a. $r_m/D = 0.189$;   b. $r_c/D = 0.351$;
       c. $\dot{m} = 16.4$ kg/s

13.22 $C_c = 0.856$

13.26 $Q = 0.13$ m^3/s

13.28 $Q = 0.0135$ m^3/s

13.30   Percent increase in discharge $= 96\%$

13.32 $p_{T,1} - p_{T,2} = (p_1 + \gamma z_1) - (p_2 + \gamma z_2)$

13.34 $d = 6.26$ cm

13.36 $d = 0.601$ m

13.38 $Q = KA_0\sqrt{2\Delta p/\rho}$; $Q = 0.290$ m^3/s

13.40 $h = 0.13$ m

13.42 $Q = 0.04$ m^3/s

13.44 $Q = 10.9$ m^3/s

13.46 $Q = 0.22$ m^3/s

13.48 $Q = 0.00124$ m^3/s

13.50 $h_L = 64V_0^2/2g$

13.54   a. $V = (L/\Delta t)[-1 + \sqrt{1 + (c\Delta t/L)^2}]$
       b. $V = \frac{c^2\Delta t}{2L}$
       c. $V = 22.5$ m/s

13.56 $Q = 0.117$ m^3/s

13.58   c

13.60   Weir height $P = 1.22$ m

13.62 $\Delta t = 43.4$ s

13.64   Depth $= 1.124$ m

13.66 $H = 42$ cm

13.68 $H_R/H_T = 0.456$

13.70 $Q = 0.18$ m^3/s

13.72 $h = 1.24$ m

13.76 $\dot{m} = 0.0021$ kg/s

13.78 $\dot{m} = 0.015$ kg/s

13.80 $Q = 0.095$ m^3/s; $U_Q = 0.00523$ m^3/s

Chapter 14

14.4 $F_T = 926$ N; $P = 35.7$ kW

14.6 $N = 1160$ rpm

14.8 $D = 1.71$ m; $V_0 = 89.4$ m/s

14.10 $N = 1170$ rpm

14.12 $a = 0.783$ m/s^2

14.14 $Q = 0.667$ m^3/s

14.16 $Q = 0.32$ m^3/s; $P = 13.5$ kW

14.18 $Q = 3.60$ m^3/s; $\Delta H = 38.7$ m; $P = 1710$ kW

14.22 $D = 2.07$ m; $P = 27.8$ kW

14.26 $\Delta H = 91.3$ m; $Q = 0.878$ m^3/s

14.28 $Q = 0.18$ m^3/s

14.30 $H_{30} = 73.8$ m

14.32 $Q = 0.0833$ m^3/s; $\Delta h = 146$ m; $P = 104$ kW

14.36 $N_{ss} = 2{,}760$, which is much below 8,500; therefore, safe.

14.38   Radial flow pump

14.40 $N = 2070$ rpm

14.44 $P_{ref} = 118.0$ kW

14.46 $P_{ref} = 592.4$ kW

14.48 $P = 10.6$ MW; $D = 2.85$ m

14.52   a. $\alpha_1 = 6.78°$;   b. $P = 88.9$ MW;   c. increase β_2

14.54 $\alpha_1 = 13.6°$

14.58 $A = 64.3$ m^2

14.60 $W = 3.46$ m

Chapter 15

15.2   a. Student sketch;   b. $R_h = \dfrac{b}{2}$

15.4   c

15.6 $Q = 188.1$ m^3/s

15.8 $Q_{DarcyWeisbach} = 9.3$ m^3/s; $Q_{Manning} = 8.1$ m^3/s

15.10 $Q = 0.26$ m^3/s

15.12 $V = 2.3$ m/s; $Q = 27.6$ m^3/s

15.14 $Q = 30.2$ m^3/s

15.16 $d = 1.3$ m

15.18 $b = 2.99$ m

15.20   Critical flow is undesirable because it is unstable.

15.22   a. Supercritical;   b. $y_2 = 5.5$ m

15.24   Subcritical

15.26 $y_2 = 9.35$ m; $E = 9.36$ m

15.28   Supercritical

15.30 $y_c = 1.40$ m

15.32 $Q = 17.1$ m^3/s

15.34 $d_{brink} = 0.78$ m

15.36 $Q = 3.43$ m^3/s

15.38   Elev. $= 91.6$ m

15.40   For upstep of 60 cm: $\Delta y = -0.76$ m and drop in elev'n is 0.16 m; For downstep of 15 cm: $\Delta y = 0.17$ m and increase in elev'n is 0.02 m; For max. upstep without upstream depth changes: $y_2 = y_c$, $\Delta z_{step} = 0.89$ m

15.42 $y_2 = 2.71$ m; $\Delta z_{water\ surface} = 0.29$ m; width for max. contraction $= 2.46$ m

15.44 $y_2 = 0.5$ m

15.46 $V = 1.21$ m/s

15.50 $h_L = 0.65$ m; $P = 12{,}753$ W; $F_r = 542$ N opposite direction of flow

15.52   a. Yes;   b. $y_2 = 3.99$ m

15.54 $y_2 = 2.09$ m

15.56 $q = 2.9$ m^2/s

15.60 Δ Elev $= 1.42$ m (increase)

15.62   S3; $\tau_0 = 143$ N/m^2

15.64   d

15.66   M2

15.72 $Q = 19.2$ m^3/s

Chapter 16

16.12   True; False; False

INDEX

A

Abrupt/sudden expansion, 270–271
Absolute pressure, 51, 62
Absolute viscosity, 35
Acceleration
  calculating, when velocity field is
    specified, 127
  centripetal, 126
  convective and location, 126–127
  defined, 123
  field, 612
  mathematical description of, 124–126
  moving objects and, 231–233
  physical interpretation of, 124
Accumulation,
  mass, 183
  momentum, 215–216
Adhesion, 47–50
Adiabatic process, 447, 451
Advance ratio, 519–520
Adverse pressure gradient, 148–149
Airfoils
  drag and lift on, 426–432
  sound propagation, 448–449
Airplanes
  drag and lift on, 426–432
  Mach number for, calculating, 450–451
  total temperature calculation, 452–453
  wind tunnel applications, 306–307
Alternate depths, 568
Analytical fluid dynamics (AFD), 601
Anemometers
  cup, 481
  hot-wire or hot-film, 481–482
  laser-Doppler, 483–484
  vane or propeller, 481
Angular momentum equation, 233–236
Apparent shear stresses, 337
Apron, 581
Archimedes' principle, 87
Area-averaged velocity, 171
Atmospheric variations, pressure, 69–71
Attached flow, 122, 123
Automobiles
  drag and lift on, 312, 432–435
  model tests for drag force on, 312
Avogadro's law, 13
Axial-flow pumps, 523–527
Axisymmetric bodies, drag and, 413–417

B

Barometers, 72
Bernoulli equation (math)
  inviscid flow, 135
  inviscid and irrotational flow, 147
Bernoulli equation (physics)
  compared with the energy equation, 270
  continuity equation and, 191
  derivation of, 132–133, 270
  examples, 136–138
  head form, 133
  irrotational form, 146–147
  physical interpretation-energy is conserved, 133–134
  physical interpretation-velocity and pressure vary, 134–135
  pressure form, 133
  summary of, 135
Best hydraulic section, 563–564
Blasius, H., 331–332
Body force, 209–210
Boundary, 28, 29
Boundary conditions, 627
Boundary layer
  defined, 330
  development and growth of, 330–331
  pressure gradient effects, 347–349
  separation, 348–349
  transition, 335–336
  tripped, 336
Boundary layer, laminar
  equations, 331–333
  resistance calculation for, on a flat plate, 335
  shear force, 333–334
  shear stress, 333
  shear-stress coefficients, 334–335
  summary of equations, 346
  thickness, 332, 333–334
Boundary layer, turbulent
  applications, 343
  logarithmic velocity distribution, 337–341
  mixing-length theory, 338–339
  power-law equation, 342
  shear-stress coefficients, 344–346
  thickness, 344–346
  summary of equations, 346
  summary of zones, 342
  tripped, 336, 347
  velocity defect law/region, 341–342
  velocity distribution, 336–343
  viscous sublayer, 337

Bourdon-tube gages, 72–73
Boyle's law, 13
Buckingham π theorem, 294
Bulk modulus of elasticity, 32–33, 54
Buoyancy, center of, 89
Buoyant forces
    calculating, 85–88
    equations, 86–87

C

Calculations. *See* Wales-Woods Model (WWM)
Capillary action, 48–50
Capture area, 546, 547
Cartesian components, 117
Cartesian coordinates, 605–606, 622–623
Cavitation
    benefits of, 192
    defined, 192
    degradation from, 192
    sites, identifying, 193–193
    Tunnel, William P. Morgan, 193
Center of buoyancy, 89
Center of pressure (CP), 78, 80–81
Centrifugal compressors, 535–538
Centrifugal pumps, 386–387, 527–530
Centripetal acceleration, 126
Characteristic curves, 525
Charles's law, 13
Chemical energy, 253
Chezy equation, 560, 562–563
Circulation
    combined with uniform flow, 423–424
    defined, 422–423
Closed system (control mass)
    conservation of energy and, 256–257
    description of, 176–177, 178
Coefficients
    discharge, 489, 524
    of drag, 409, 410
    flow, 489
    force, 300, 301
    head, 524
    of lift, 424–425
    minor loss, 380–382
    power, 520
    pressure, 147, 300, 301, 308–309
    roughness, 561
    shear stress, 300, 301
    shear stress, laminar boundary layer, 334–335
    shear stress, turbulent boundary layer, 344–346
    thrust, 519, 520
    of velocity, 489
Cohesive force, 46
Colebrook-White formula, 373
Combined head loss, 364, 379–383
Combined head loss equation, 382

Component head loss, 364, 379–383
Compressible flows/fluids
    density, 453–454
    drag and, 421–422
    isentropic, through a duct with varying area, 460–471
    kinetic, 454–455
    Mach number relationships, 451–455
    measuring, 501–505
    normal shock waves, 455–460
    speed of sound, 446–450
    temperature, 452–453
    wave propagation, 445–451
Compressors, centrifugal, 535–538
Computational fluid dynamics (CFD), 311, 601, 623
    computer programs (codes), 624–625
    examples of, 628–631
    features of, 625–628
    importance of, 624
    validation and verification, 628
Conduit, defined, 360
Conduits, flow in
    centrifugal pumps, 386–387
    classifying, 362
    developing versus fully developed, 361–362
    entry or entrance length, 362
    flow problems, strategies for solving, 375–379
    laminar flow in round tubes, 367–371
    laminar versus turbulent, 360–361
    nonround, 384–385
    parallel pipes, 387–388
    pipe head loss, 363–366
    pipe networks, 388–391
    pipe sizes, specifying, 363
    stress distributions in pipe flow, 366–367
    turbulent flow, 371–375
Conjugate depth, 580
Conservation of energy, 255–257
Conservation of mass. *See* Continuity equation
Consistent units, 8–9
Constant
    density, 32–34, 617–618
    primary dimensions of a, 22
    universal gas, 13
    velocity, 229–231
    volume versus pressure, 51
Continuity equation (math)
    differential equation form, 618
    algebraic form, 184
Continuity equation (physics)
    applications, 184–191
    Bernoulli equation and, 191
    constant density, 617–618
    cylindrical coordinates, 616
    derivation, 182–183, 613–616
    description of, 612–619
    invariant notation, 616–617

physical interpretation of, 183–184
pipe flow form, 189–191
summary of, 184, 618–619
units, 184
Continuum assumption, 5–6
Control mass (closed system), 176–177, 178
Controls, 584
Control surface (CS)
defined, 177
transport across, 179–180
Control volume approach
closed system (control mass), 176–177, 178
open system (control volume), 177–178
properties, intensive and extensive, 178–179
Reynolds transport theorem, 180–182
transport across control surface, 179–180
Control volume (CV) (open system)
conservation of energy and, 257
description of, 177–178
linear momentum equation for stationary,
218–228
Convective acceleration, 126
Conversion ratios, 10
Couette, M., 327
Couette flow, 40–42, 326–327
Critical depth, 570
Critical flow, 569–574
Critical-flow flumes, 573
Critical mass flow rate, 466
Critical pressure ratio, 467
Cross, H., 389
Culverts, uniform flow in, 564–567
Cup anemometers, 481
Curved surfaces, calculating forces on, 83–85
Curves, characteristic or performance, 525
Cyclist, drag on a, 629–630
Cylinders, drag on, 411–412
Cylindrical coordinates, 606–607, 616, 623

D

Dam spillways, hydraulic jumps on, 581–582
Darcy-Weisbach equation (DWE), 364–366, 558–559
Deforming CV, 177
de Laval nozzles
flow in, 463–464
mass flow rate, 465–467, 470–471, 502–503
nozzle flow classification by exit conditions, 467–470
shock waves in, 469–470
truncated, 470–471
Density
of common liquids, 31
constant versus variable, 32–34
description of fluid, 30–31, 53
total, 453–454
units for, 30

Depth, conjugate versus sequent, 580
Depth ratio, 579
Derivatives, 21
Design storm, 565
Developing flow, 361–362
Dimensional analysis
approximate, at high Reynolds numbers,
309–312
Buckingham π theorem, 294
common π groups, 299–301
defined, 295
exponent method, 297–298
flows without free-surface effects, 305–308
free-surface model studies, 312–315
model-prototype performance, 308–309
need for, 292–294
open-channel flow, 555–556
similitude, 302–305
step-by-step method, 295–297
variables, selection of significant, 298–299
Dimensional homogeneity (DH), 19–22
Dimensionality, flow, 121
Dimensionless groups, 20
Dimensions
defined, 7
organizing, 9
primary, 7–8, 21–22
relationship between units and, 8
secondary, 7
Direct numerical simulation (DNS), 627
Direct step method, 588
Discharge coefficient, 489, 524
Discharge (volume) flow rate, 170–173, 174, 369
measuring, 486–500
in a pipe network, 390–391
Distributed force, 35
Dividing streamline, 112–113
Doppler effect, 449
Draft tube, 545
Drag curve, standard, 414
Drag force
airfoil, 426–432
automobile, 312, 432–435
axisymmetric bodies and, 413–417
calculating, 408–412
coefficient of, 409, 410
compressible flows and, 421–422
on a cyclist, 629–630
cylinder, 411–412
defined, 406–407
equation, 409
induced, 428
on a sphere, 416
on a sphere, calculating, 454
on a sphere using exponent method, 298
on a sphere using step-by-step method, 296–297

Drag force (*continued*)
　streamlining to reduce, 420–421
　stress distribution and, 407–408
　surface roughness, 412
　three-dimensional bodies and, 413–417
　two-dimensional bodies and, 410
Dynamic similitude, 303–305
Dynamic viscosity, 35

E

Efficiency
　equation, 267
　mechanical, 267–269
　thermal, 267
Efflux, 183
Elasticity, bulk modulus of, 32–33, 54
Electrical energy, 254
Electromagnetic flowmeters, 495–496
Elevation changes, calculating pressure changes associated with, 65–71
Empirical equations, 13
Energy
　categories of, 253–254
　concepts, basic, 253–255
　conservation of, 255–257
　defined, 253
　specific, 567–569
　units, 253
Energy equation (math), 263
Energy equation (physics)
　applications, 264–269
　compared with the Bernoulli equation, 270
　derivation of, 257–258, 261–262
　flow and shaft work, 258–259
　kinetic energy correction factor, 259–261
　physical interpretation of, 262–263
　for steady open-channel flow, 557–558
　summary of, 263–264
Energy grade line (EGL), 273–277
Engineering fluid mechanics, defined, 2–3
Engineers, role of, 2
Enthalpy, 51
　total, 451
Entry or entrance length, 362
Equations
　See also name of
　Bernoulli, 132–138, 270
　buoyant force, 86–87
　continuity, 182–191
　empirical, 13
　energy, 257–269
　Euler's, 127–132
　flow rate, 170–171, 173, 174
　hydrostatic (differential), 65–69
　hydrostatic (algebraic), 66–67

ideal gas law, 14–15
　manometer, 76
　momentum, 208–236
　panel, 81
　power, 265–267
　of state, 613
　Sutherland's, 44
　vector, 211–212
　viscosity, 35, 39–42
Equivalent sand roughness, 373, 375
Ericsson, A., 15
Eulerian approach, 116–117
Euler's equation
　calculation, 130–132
　derivation of, 127–129
　physical interpretation of, 129–130
　pressure variation, due to changing speed of particle, 129
　pressure variation, normal to curved streamlines, 130
　pressure variation, normal to rectilinear streamlines, 129–130
Expansion, abrupt/sudden, 270–271
Experimental fluid dynamics (EFD), 601
Exponent method, 297–298
Extensive properties, 178–179

F

Fan laws, 527
Fanning friction factor, 365
Favorable pressure gradient, 148–149, 348
Feynman, R., 3
Field, 115–116
Fixed CV, 177
Fixed identity, 6
Floating bodies, 89–92
Flow
　See also Conduits, flow in; Open channels, flow in
　coefficient, 489
　critical, 569–574
　critical mass, 466
　gradually varied, 567, 582–589
　nonuniform, 118, 555, 567
　nozzles, 494–495
　rapid, 569
　rapidly varied, 567–577
　subcritical, 569
　supercritical, 569
　tranquil, 569
　uniform, 118, 555
　uniform laminar, 325–329
　without free-surface effects, model studies for, 305–308
　work, 258–259
Flowing fluids
　acceleration, 123–127
　circular cylinders and, 147–149

dimensionality, 121
describing, 117–123, 153–154
Eulerian and Lagrangian approaches, 116–117
how engineers describe, 43–44, 153–154
inviscid, 122
laminar, 119–121
pathlines, streaklines, and streamlines, 112–114
regions, 122
rotational motion, 142–146
separation, 122–123
steady and unsteady, 119
turbulent, 119–121
uniform and nonuniform, 118
velocity, 114–117
viscous, 122
Flowing gases, thermal energy in, 51–52
Flow measurements. *See* Measuring; Measuring devices
Flowmeters
electromagnetic, 495–496
turbine, 496
ultrasonic, 496
vortex, 496–497
Flow rate
critical mass, 466
differential areas for determining, 175–176
equations, 170–171, 173, 174
example problems, 174–176
mass, 173, 174, 465–467
measuring, 486–500
units, 170, 173
volume (discharge), 170–173, 174
Flow Science, 625
FLOW-3D, 625
Fluid in a solid body rotation, 150
Fluid jets, 218–221
Fluid mechanics, defined, 3
Fluid particle, 6
Fluid properties
bulk modulus of elasticity, 32–33, 54
density, 30–31, 53
finding, 34
kinematic viscosity, 38, 54
specific gravity, 32, 53
specific weight, 31, 53
summary of, 53–54
surface tension, 45–50, 54
vapor pressure, 50–51, 54
viscosity, 35–45, 54
Fluids
See also Flowing fluids
constant versus variable density, 32–34
nature of, 3–4
Newtonian versus non-Newtonian, 44–45
Force (forces)
body, 209–210
buoyant, calculating, 85–88

coefficient, 300, 301
cohesive, 46
on curved surfaces, calculating, 83–85
defined, 209
diagram, 217–218
distributed, 35
on plane surfaces (panels), calculating, 77–83
shear, 35–36, 324–350
summary of, 210
surface, 209
transitions and, 272–273
Force equilibrium
to calculate pressure rise inside a water droplet, 47
to calculate sewing needle size supported by surface tension, 49–50
hydraulic jack and, 64–65
Form drag, 408
Francis turbines, 235–236, 538, 544–545
Free surface, defined, 555
Free-surface effects, model studies for flows without, 305–308
Free-surface model studies, 312–315, 625
Friction drag, 408
Friction factor, 365
laminar flow, 369–370
Moody diagram, 374
turbulent flow, 372–373, 374
Friction velocity, 337
Froude number, 300, 301, 555
Fully developed flow, 361–362

G

Gage pressure, 62
Gases
attributes of, 4
defined, 4
Gas turbines, 545
Geometric similitude, 303
Gradually varied flow, 567, 582–589
Grid generation, 626
Grid method, 9–10, 11, 12
Ground-effect pod, 434
Ground water modeling, 624

H

Hagen-Poiseuille flow, 368
Hardy Cross method, 389
Head
coefficient, 524
defined, 134
total, 273
Head loss
component/combined, 364, 379–383
defined, 262
flow problems, strategies for solving, 375–379
in hydraulic jumps, 580

Head loss (*continued*)
    laminar flow and, 369–371
    in a nozzle in reverse flow, 311–312
    for orifices, 491, 493
    pipe, 363–366
Heat transfer, 256
Hele-Shaw flow, 327–329
Hirt, T., 625
Homenergic flows, 451
Hot-wire or hot-film anemometers, 481–482
HVAC duct, pressure drop in, 385
Hydraulic depth, 569
Hydraulic diameter, 384
Hydraulic grade line (HGL), 273–277
Hydraulic jumps
    dam spillways and, 581–582
    depth relationships, 577–579
    head loss in, 580
    naturally occurring, 582
    occurrences of, 577
    in rectangular channels, 579–580
Hydraulic machines, 63–65
Hydraulic radius, 384, 556
Hydrometers, 87
Hydrophillic, 48
Hydrophobic, 48
Hydrostatic equation, 66–67
Hydrostatic differential equation, 65–69
Hydrostatic equilibrium
    See also Pressure
    defined, 61
Hydrostatic force
    on curved surface, 84
    due to concrete, 82
    to open an elliptical gate, 82–83
Hydrostatic force (panel) equations, 80–81
Hydrostatic pressure distribution, 78–79
Hypersonic flows, 450

I

Ideal fluid, 148
Ideal gas law (IGL)
    applying, to predict weight, 16–17
    development of, 13–14
    dimensional homogeneity application, 20
    equations, 14–15
    validity of, 14
Idealized models, 5–6
Ideally expanded nozzles, 468
Immersed bodies, 89
Impulse turbines, 538–541
Incompressible, 33
Induced drag, 428
Induction, 13
Inertial reference frames, 228–229

Infinitesimal particle, 6
Inflow, 183
Initial conditions, 627
Integrals, 21
Intensive properties, 178–179
Interferometers, 504–505
Internal combustion engine modeling, 624–625
International System of Units (SI), 8, 9
Invariant notation, 609, 616–617
Inviscid flow, 122
Irrotational flow, 143, 146–147
Isentropic process
    compressible fluids through a duct with varying
        area, 460–471
    defined, 447

J

Joule, J. P., 255–257

K

Kaplan turbines, 538
Karman's constant, 340
k-epsilon model, 627
Kinematic viscosity, 38, 54
Kinetic energy correction factor, 259–261
Kinetic pressure, 139, 454–455
KIVA, 624–625
Kutta condition, 426–427

L

Lagrangian approach, 116–117
Laminar boundary layer. *See* Boundary layer, laminar
Laminar flow
    Couette, 326–327
    defined, 367–368
    description of, 119–121, 360–361
    discharge and mean velocity, 369
    head loss and friction factor, 369–371
    Hele-Shaw, 327–329
    kinetic energy correction factor, 260–261
    in round tubes, 367–371
    uniform, 325–329
    velocity profile, 368–369
Large eddy simulation (LES), 627
Laser-Doppler anemometers (LDAs), 483–484
Law of the wall, 341
Learning, defined, 3
Length scale, 360
Lift force
    airfoil, 426–432
    automobile, 432–435
    circulation, 422–423
    circulation combined with uniform flow, 423–424
    coefficient, 424–425
    defined, 406–407
    equation, 424

on rotating sphere, 425
stress distribution and, 407–408
Limit concept, 5
Linear momentum equation
applications, 216–218
applications for moving objects, 228–233
for stationary control volume, 218–228
summary of, 216
theory, 213–216
Liquids
attributes of, 4
defined, 4
Location acceleration, 126
Logarithmic velocity distribution, 337–341
Los Alamos National Laboratory, 624–625

M

Mach, E., 450
Mach angle, 450
Mach number, 300, 301
for airplanes, calculating, 450–451
area variation and, 460–461
critical, 421–422
defined, 450
relationships and compressible flows, 451–455
velocity measurements and, 501–502
Mach wave, 450
Mach-Zender interferometers, 504–505
Macroscopic viewpoint, 5
Magnus effect, 424
Manning equation, 560–563
Manning's *n*, 560–562
Manometer equations, 76
Manometers, 73–75
Marker methods, 484–486
Mass balance equation, 184
Mass density. *See* Density
Mass flow rate, 173, 174
critical, 466
de Laval nozzles and, 465–467, 470–471, 502–503
equation, 174
Material derivative, 610–611
Math, defined, 2
Matter, viewpoints of, 5–6
Mean velocity, 171–172, 369
Measuring
accuracy of, 505–506
in compressible flow, 501–505
flow rate, 486–500
marker methods, 484–486
pressure, 72–77, 139–142, 478–486
shock wave visualization, 504
velocity, 139–142, 478–486, 501–502
Measuring devices
anemometers, cup, 481

anemometers, hot-wire or hot-film, 481–482
anemometers, laser-Doppler, 483–484
anemometers, vane or propeller, 481
flowmeters, electromagnetic, 495–496
flowmeters, turbine, 496
flowmeters, ultrasonic, 496
flowmeters, vortex, 496–497
flow nozzles, 494–495
interferometers, 504–505
orifice meters, 487–490, 491–493
Pitot-static tube, 140–142, 480
rotameters, 497
stagnation (Pitot) tube, 478–479
static tube, 479–481
venturi meters, 494, 495, 503–504
weirs, rectangular, 497–499
weirs, triangular, 500–501
yaw meters, 480–481
Mechanical advantage, 63
Mechanical efficiency, 267–269
Mechanical energy, 253
Mechanics, defined, 3
Metacenter, 90
Metacentric height, 90
Microscopic viewpoint, 5
Mild slope, 573
Minor loss coefficient, 380–382
Mixing-length theory, 338–339
Models (modeling)
assessing the value of, 602–603
computational fluid dynamics, 601, 623–631
continuity equation, 612–619
defined, 599
methods for building, 601–602
Navier-Stokes equation, 612, 619–623
partial differential equations, 603–612
process, 599–603
Model testing
applications, 302–303
approximate similitude at high Reynolds
numbers, 309–312
dynamic similitude, 303–305
for flows without free-surface effects,
305–308
free-surface, 312–315
geometric similitude, 303
model-prototype performance, 308–309
ship, 302, 314–315
spillway, 312–314
MODFLOW, 624, 625
Mole, 13
Moment-of-momentum equation, 233–236
Momentum
accumulation, 215–216
diagram, 217, 218
flow, 213–215

Momentum equation (math)
    linear momentum, 216
    angular momentum, 233
Momentum equation (physics)
    angular, 233–236
    linear, applications, 216–218, 228–233
    linear, for stationary control volume, 218–228
    linear, theory, 213–216
    Newton's law of motion, 209–212
    nozzles and, 224–225
    visual solution method, 211–212, 217–218
Moody diagram, 373, 375
Moving objects
    accelerating, 231–233
    constant velocity, 229–231
    linear momentum equations and, 228–233
    reference frames, 228–229
Multidimensional flow, 121

N

Nardi, A., 15
National Research Council, 2
Navier-Stokes equation, 325, 326, 612
    Cartesian and cylindrical coordinates, 622–623
    derivation, 619–622
Net positive suction head (NPSH), 532–534
Newtonian versus non-Newtonian fluids, 44–45
Newton's Law of Motion, 209–212, 325
Newton's Law of Viscosity, 35
Nominal Pipe Size (NPS), 363
Noninertial reference frames, 228–229
Non-Newtonian fluids, 45
Nonround conduits, 384–385
Nonuniform flow, 118, 555, 567
Normal depth, 555, 560
Normal shock waves
    defined, 455
    in de Laval nozzles, 469–470
    property changes across, 456–458
    in supersonic flows, 458–460
No-slip condition, 37–38, 330
Nozzles
    See also de Laval nozzles
    applications, 223
    flow classification by exit conditions, 467–470
    ideally expanded, 468
    momentum equation and, 224–225
    overexpanded, 468
    underexpanded, 468
Nuclear energy, 254

O

One-dimensional flow, 121
Open channels, defined, 555

Open channels, flow in
    best hydraulic section, 563–564
    Chezy equation, 560, 562–563
    critical flow, 569–574
    description of, 555–557
    dimensional analysis, 555–556
    energy equation for steady, 557–558
    gradually varied, 567, 582–589
    hydraulic jump, 577–582
    Manning equation, 560–563
    rapidly varied, 567–577
    Reynolds number, 556, 557
    rock-bedded channels, 559–560
    steady nonuniform, 567
    steady uniform, 558–567
    transitions, 574–576
    wave celerity, 576–577
Open system (control volume), 177–178
Orifice meters, 487–490, 491–493
Orifices, head loss for, 491, 493
Outflow, 183
Overexpanded nozzles, 468

P

Panel (hydrostatic force) equations, 81
Panels, calculating forces on, 77–83
Parallel pipes, 387–388
Parallel plates
    Couette flow, 326–327
    Hele-Shaw flow, 327–329
    pressure gradient between, 329
Partial differential equations (PDEs)
    acceleration field, 612
    approximation of, 625
    Cartesian coordinates, 605–606
    cylindrical coordinates, 606–607
    material derivative, 610–611
    notation, invariant, 609
    operators, 609–610
    reasons for learning, 603–604
    Taylor series, 607–608
Particle image velocimetry (PIV), 485
Pascal's principle, 63–64
Pathlines, 112–114
Pelton wheel, 538
Performance curves, 525
Phase diagrams, 50
π-groups, 20
    common, 299–301
    exponent method, 297–298
    step-by-step method, 295–297
    use of term, 294
π theorem, Buckingham, 294
Piezometer, 73
Piezometric head, 66, 67

Piezometric pressure, 66, 67
Pipes
  bends, 225–226
  expansion, abrupt/sudden, 270–271
  flow form, continuity equation and, 189–191
  forces on, 272–273
  head loss, 363–366, 375–379
  networks, 388–391
  parallel, 387–388
  sizes, specifying, 363
  stress distributions in pipe flow, 366–367
Pitch angle, 518
Pitot (stagnation tube), 139–140, 478–479
Pitot-static tube, 140–142, 480
Plane surfaces, calculating forces on, 77–83
Poiseuille flow, 39–40, 368
Post processor, 628
Pounds-mass, 11–12
Power
  coefficient, 520
  defined, 254–255
  equation, 265–267
  resistance, 416–417
Power-law equation, 342, 371, 372
Prandtl, L., 331–332, 338–339, 372, 428
Pressure
  absolute, 51, 62
  atmospheric variations, 69–71
  center of, 78, 80–81
  changes associated with elevation changes,
      calculating, 65–71
  coefficient, 147, 300, 301, 308–309
  defined, 61–65
  distribution, 77, 148–149
  forces, calculating buoyant, 85–88
  forces on curved surfaces, calculating, 83–85
  forces on plane surfaces (panels), calculating, 77–83
  gage, 62
  hydraulic machines, 63–65
  hydrostatic differential equation, 65–69
  hydrostatic pressure distribution, 78–79
  kinetic, 139, 454–455
  measuring, 72–77, 139–142, 478–486
  piezometric, 66, 67
  static, 139
  stratosphere, 70, 71
  tap, 139
  total, 453
  transducers, 76–77
  troposphere, 69, 70
  uniform pressure distribution, 78
  units, 61–62
  vacuum, 62
Pressure field
  circular cylinders and, 147–149
  rotating flow and, 149–152

Pressure gradient
  between parallel plates, 329
  effects on boundary layers, 347–349
  favorable and adverse, 148–149, 348
Pressure ratio, critical, 467
Pressure variation. *See* Euler's equation
Primary dimensions, 7–8, 21–22
Process
  defined, 3
  system, 29
Projected area, 409
Propeller anemometers, 481
Propellers, 518–523
Properties
  See also Fluid properties
  defined, 29
  intensive and extensive, 178–179
Pump curve, 386
Pump head, 262
Pumps
  axial-flow, 523–527
  centrifugal, 386–387, 527–530
  curves, characteristic or performance, 525
  defined and types of, 255
  discharge coefficient, 524
  head coefficient, 524
  mechanical efficiency, 267–269
  power equation, 265–267
  suction limitations of, 532–534

R

Radial-flow machines, 527–530
Rapid flow, 569
Rapidly varied flow, 567–577
Rate of shear strain, 36–37
Reaction turbines, 538, 542–543
Reference frames, 228–229
Relative roughness, 373
Resistance, power and rolling, 416–417
Resistance coefficient, 365, 560
Reynolds, O., 300, 360
Reynolds-averaged Navier-Stokes (RANS)
      equations, 627
Reynolds number, 300, 301
  approximate similitude at high, 309–312
  conduit flow type and, 360–361
  length scale, 360
  open-channel flow and, 556, 557
  similitude for flow over a blimp, 306
  similitude for a valve, 307–308
Reynolds stresses, 337
Reynolds transport theorem, 176, 180–182
Rock-bedded channels, 559–560
Rolling resistance, 416–417
Rotameters, 497

Rotational motion, 142–146
    pressure field for, 149–152
Roughness
    coefficient, 561
    drag and surface, 412
    equivalent sand, 373, 375
    relative, 373
    sand roughness height, 373
    type of flow and effects of wall, 373
Runner, 538

S

Sand roughness
    equivalent, 373, 375
    height, 373
Saturation pressure, 50
Saturation temperature, 50
Schlieren system, 504
Science, defined, 2
Secondary dimensions, 7
Sectional drag coefficients, 410
Sequent depth, 580
Sewers, uniform flow in, 564–567
Shaft work, 258–259
Shear force
    calculating, on a flat plate, 346
    defined, 35–36
    laminar boundary layer and, 333–334
    predicting, 324–350
    tripped boundary layer, 347
Shear strain, 36–37
Shear stress
    apparent, 337
    coefficients, 300, 301
    coefficients, laminar boundary layer, 334–335
    coefficients, turbulent boundary layer,
        344–346
    Couette flow and, 327
    defined, 35–36
    distributions in pipe flow, 366–367
    predicting, 324–350
    Reynolds, 337
    thickness, laminar boundary layer, 333–334
    thickness, turbulent boundary layer, 344–346
    viscosity equations used to calculate, 39–40
Shear velocity, 337
Ship model testing, 302, 314–315
Shock waves
    oblique, 459–460
    visualization, 504
Shock waves, normal
    defined, 455
    in de Laval nozzles, 469–470
    property changes across, 456–458
    in supersonic flows, 458–460

Similitude
    approximate, at high Reynolds numbers, 309–312
    defined, 302
    dynamic, 303–305
    flows without free-surface effects, 305–308
    free-surface model studies, 312–315
    geometric, 303
    model-prototype performance, 308–309
    scope of, 302–303
SI system, 8
Slugs, 11–12
Sluice gate, 568
Solid mechanics, defined, 3
Solids, attributes of, 4
Solver, 627
Sonic booms, 460
Sound
    Doppler effect, 449
    speed of, 446–450
Specific energy, 567–569
Specific gravity, 32, 53
Specific heat, 51
Specific heat ratio, 51
Specific speed, 531–532
Specific weight, 31, 53
Speed of sound, 446–450
Spheres, drag force and
    calculating, 416, 454
    using exponent method, 298
    using step-by-step method, 296–297
Spheres, lift on rotating, 425
Spillways
    hydraulic jumps on dam, 581–582
    models, 312–314, 628–629
Stability of immersed and floating bodies,
        calculating, 88–92
Stager, R., 15
Stagnation
    point, 113
    tube (Pitot), 139–140, 478–479
    use of term, 454
Stall, 427
State, system, 29, 30
Static pressure, 139
    ratio, 456
Static temperature, 452
    ratio, 456
Static tube, 479–481
Steady flow, 119
    energy equation for steady open-channel flow,
        557–558
    nonuniform, 567
    uniform, 558–567
Steep slope, 573
Step-by-step method, 295–297
Strain, shear, 36–37

Stratosphere
    defined, 70
    pressure variation in lower, 71
Streaklines, 112–114
Streamlines, 112–114
    to reduce drag, 420–421
Stress
    See also Shear stress
    distributions and drag and lift, 407–408
    distributions in pipe flow, 366–367
    Reynolds, 337
Strouhal number, 419–420
Subcritical flow, 569
Submerged hydraulic jumps, 582
Subsonic flow, 461
Supercritical flow, 569
Supersonic flows
    de Laval nozzles, 463–465
    diffusers, 461–462
    shock waves in, 458–460
Supersonic wind tunnels. *See* Wind tunnels, supersonic
Surface force, 209
Surface roughness, drag and, 412
Surface tension
    adhesion and capillary action, 47–50
    description of, 45–50
    examples of, 45, 47
    summary of, 54
Surge hydraulic jumps, 582
Surroundings, 28, 29
Sutherland's equation, 44
System
    curve, 386
    defined, 28
    examples of, 29
    process, 29
    properties, 29
    state, 29, 30

T

Taylor series, 607–608
Technology, defined, 2
Temperature
    effects, viscosity, 42–44
    saturation, 50
    static, 452
    total, 452–453
Terminal velocity, 418–419
Theoretical adiabatic power, 535
Theoretical isothermal power, 536
Thermal efficiency, 267
Thermal energy
    defined, 253
    in flowing gases, 51–52
Thinking operations, 17–18

Three-dimensional bodies, drag and, 413–417
Thrust coefficient, 519, 520
Tidal bore, 582
Time-averaged velocity, 171
Time steps, 626
Total density, 453–454
Total enthalpy, 451
Total head tube, 139–140
Total pressure, 453
Total temperature, 452–453
Traditional unit system, 8, 9
Tranquil flow, 569
Transition(s)
    abrupt/sudden expansions and, 270–271
    boundary layer, 335–336
    defined, 335, 574
    forces on, 272–273
    open channel, 574–576
    warped-wall, 575
    wedge, 575
Transonic flow, 462
Transport
    across control surface, 179–180
    Reynolds transport theorem, 176, 180–182
Tripped boundary layer, 336, 347
Troposphere
    defined, 69–70
    pressure variation in, 70
Truncated nozzles, 470–471
Turbine flowmeters, 496
Turbine head, 262
Turbines
    defined, 255, 538
    Francis, 235–236, 538, 544–545
    gas, 545
    impulse, 538–541
    Kaplan, 538
    mechanical efficiency, 267–269
    reaction, 538, 542–543
    specific speed for, 545
    types of, 255
    vane angles, 544–545
    wind, 545–547
Turbomachinery
    categories of, 518
    compressors, centrifugal, 535–538
    propellers, 518–523
    pumps, axial-flow, 523–527
    pumps, centrifugal, 527–530
    pumps, suction limitations of, 532–534
    radial-flow machines, 527–530
    specific speed, 531–532
    turbines, 538–547
    viscous effects, 534–535
Turbulence modeling, 627
Turbulent boundary layer. *See* Boundary layer, turbulent

Turbulent flow
  defined, 119–121, 360–361, 371
  friction factor, 372–373, 374
  Moody diagram, 373, 375
  velocity distribution, 371–372
Two-dimensional bodies, drag and, 410

U

Ultrasonic flowmeters, 496
Uncertainty analysis, 505–506
Underexpanded nozzles, 468
Understanding, defined, 3
Uniform flow
  best hydraulic section, 563–564
  in culverts and sewers, 564–567
  defined, 118, 555
  steady, 558–567
Uniform laminar flow, 325–329
Uniform pressure distribution, 78
U.S. Standard Atmosphere, 69
Units
  carrying and canceling, 9–12
  consistent, 8–9
  defined, 8
  grid method, 9–10, 11, 12
  organizing, 9
  pounds-mass and slugs, 11–12
  relationship between dimensions and, 8
  systems, 8, 9
  unity conversion ratios, 10
Unity conversion ratios, 10
Universal gas constant, 13
Universal turbulence constant, 340
Unsteady flow, 119

V

Vacuum pressure, 62
Validation, 628
Vane anemometers, 481
Vanes, 221–223
  angles, 544–545
Vapor pressure, 50–51, 54
Variable density, 32–34
Vector equation, 211–212
Velocity
  Cartesian components, 117
  coefficient of, 489
  constant, and moving bodies, 229–231
  defect law/region, 341–342
  defined, 114
  Eulerian and Lagrangian approaches, 116–117
  field, 115–116
  flowing fluids and, 114–117

friction, 337
  gradient, 36, 37
  mean, 171–172, 369
  measuring, 139–142, 478–486, 501–502
  profile, 37
  profile in laminar flow, 368–369
  shear, 337
  terminal, 418–419
  time-averaged, 120
Velocity distribution
  logarithmic, 337–341
  measuring, 486
  power-law equation, 342, 371, 372
  turbulent boundary layer, 336–343
  turbulent flow, 371–372
  variable, 226–228
Vena contracta, 480
Venturi flumes, 573
Venturi meters, 494, 495, 503–504
Venturi nozzles, 138
Verification, 628
Viscous effects, 534–535
Viscosity
  absolute versus dynamic, 35
  defined, 35
  equation, 35, 39–42
  finding values of, 38
  kinematic, 38, 54
  Newton's law of, 35
  no-slip condition, 37–38
  shear force and stress, 35–36
  shear strain, 36–37
  summary of, 54
  temperature effects, 42–44
  velocity profile, 37
Viscous flow, 122
  pressure distribution for, 149
Viscous sublayer, 337
Visual solution method (VSM), 211–212, 217–218
Volume (discharge) flow rate, 170–173, 174
Volume of fluid (VOF), 625
Volume flow rate equation, 174
Vortex flowmeters, 496–497
Vortex shedding, 122–123, 419–420
Vorticity, 145

W

Wales, C., 15
Wales-Woods Model (WWM)
  applications, 15–17, 18–19
  rationale and development of, 15
  structure of, 17–18
Warped-wall transitions, 575

Water hammer, 455
Water-surface profiles
    defined, 584
    evaluation of, 588–589
    types of, 584–588
Wave celerity, 576–577
Wave celerity equation, 577
Wave propagation, in compressible fluids,
    445–451
Weber number, 300, 301
Wedge transitions, 575
Weirs
    rectangular, 497–499
    triangular, 500–501
Wind loads on a telescope structure, predicting,
    630–631

Wind tunnels
    applications, 302–303, 306–307, 309
    momentum equation and finding drag force, 227–228
Wind tunnels, supersonic
    de Laval nozzles and, 463–465
    flow properties in, 465
    mass flow rate in, 466–467
    test section size in, 464
Wind turbines, 545–547
Woods, D., 15
Work
    defined, 254
    flow and shaft, 258–259

Y

Yaw meters, 480–481

Water hammer, 455
Water-surface profiles
    defined, 584
    evaluation of, 588–589
    types of, 584–588
Wave celerity, 576–577
Wave celerity equation, 577
Wave propagation in compressible fluids,
    415–451
Weber number, 300–301
Wedge transitions, 575
Weirs
    rectangular, 497–499
    triangular, 500–501
Wind loads on a telescope structure, predicting,
    630–631

Wind tunnels
    applications, 302–303, 306–307, 309
    momentum equation and finding drag force, 227–228
Wind tunnels, supersonic
    de Laval nozzles and, 463–465
    flow properties in, 465
    mass flow rate in, 466–467
    test section size in, 464
Wind turbines, 515–516
Woods, D, 15
Weak
    defined, 254
    flow and shaft, 255–259

Y

Yaw meters, 480–481